NONLINEAR SOLID MECHANICS

This book covers solid mechanics for nonlinear elastic and elastoplastic materials, describing the behavior of ductile materials subjected to extreme mechanical loading and their eventual failure. The book highlights constitutive features to describe the behavior of frictional materials such as geological media. On the basis of this theory, including large strain and inelastic behaviors, bifurcation and instability are developed with a special focus on the modeling of the emergence of local instabilities such as shear band formation and flutter of a continuum. The former is regarded as a precursor of fracture, whereas the latter is typical of granular materials. The treatment is complemented with qualitative experiments, illustrations from everyday life and simple examples taken from structural mechanics.

Davide Bigoni is a professor in the faculty of engineering at the University of Trento, where he has been head of the Department of Mechanical and Structural Engineering. He was honored as a Euromech Fellow of the European Mechanics Society. He is co-editor of the *Journal of Mechanics of Materials and Structures* (an international journal founded by C. R. Steele) and is associate editor of *Mechanics Research Communications*.

Nonlinear Solid Mechanics

BIFURCATION THEORY AND MATERIAL INSTABILITY

Davide Bigoni
University of Trento

CAMBRIDGE
UNIVERSITY PRESS

32 Avenue of the Americas, New York NY 10013-2473, USA

Cambridge University Press is part of the University of Cambridge.

It furthers the University's mission by disseminating knowledge in the pursuit of education, learning and research at the highest international levels of excellence.

www.cambridge.org
Information on this title: www.cambridge.org/9781107699502

First published 2012
First paperback edition 2014

A catalogue record for this publication is available from the British Library

Library of Congress Cataloguing in Publication data
Bigoni, Davide, 1959–
Nonlinear solid mechanics : bifurcation theory and material instability / Davide Bigoni.
p. cm.
Includes bibliographical references and index.
ISBN 978-1-107-02541-7
1. Nonlinear mechanics. 2. Materials–Mechanical properties. 3. Elastic analysis (Engineering) 4. Bifurcation theory. I. Title.
TA405.B4983 2012
620.1′292–dc23 2012013657

ISBN 978-1-107-02541-7 Hardback
ISBN 978-1-107-69950-2 Paperback

a Michela

NON·SONO·IO·AD·AVERE·UN·PROBLEMA·È·IL·PROBLEMA·AD·AVERE·ME

Contents

Preface

The purpose of this book is to present a research summary on solid mechanics at large strain, including the treatment of bifurcation and instability phenomena. The framework is crucial to the understanding of failure mechanisms in ductile materials, as connected to material instabilities, such as, shear banding.

I have employed Chapters 2 through 5 as a textbook for a graduate course on non-linear elasticity that I have offered at the University of Trento since 1999, whereas Chapters 8, 10, 11 and 13 have been the basis for a course held at CISM (no. 414, 'Material Instabilities in Elastic and Inelastic Solids', H. Petryk, ed.). Chapters 6, 7, 9, 12, 14 and 15 have been added to present the elasticity and the yield critera in detail, including a treatment on elastic bifurcation and instability, wave propagation and multiple shear banding. This material has been taught during seminars for graduate students at various universities. Chapter 16 is devoted to the perturbative approach to material instability, developed by me in a series of articles in cooperation with D. Capuani, M. Brun, F. Dal Corso, M. Gei, A. Piccolroaz and J. R. Willis. Finally, I have to admit that the Introduction of the book is overlong; in fact, I have used it for a 20-hour graduate course on stability and bifurcation. The hope is to attract attention to the main topics presented in the book.

During preparation of this book, I have enjoyed help from a number of friends, who have read and commented on parts of the manuscript: L. Argani, M. Bacca, K. Bertoldi, M. Brun, F. Dal Corso, A. Gajo, M. Gei, G. Mishuris, D. Misseroni, A. B. Movchan, N. V. Movchan, G. Noselli, H. Petryk, A. Piccolroaz, G. Puglisi, A. Reali, S. Roccabianca and D. Veber.

The photos presented in this book have been taken by me (using a Nikon FG–20 traditional camera or a Panasonic DMC–FZ5 digital camera) or by students at the University of Trento (using a Nikon D100 or a Nikon D200 digital camera). Most of the experiments presented have been performed at the University of Trento in the Laboratory for Physical Modeling of Structures and Photoelasticity.

Foreword

This book clearly exhibits some remarkable and unusual features. The central theme addresses one of the primary research challenges at present in solid and structural mechanics. In fact, research on nonlinearities owing to large deformations and inelastic behaviours of materials now has to be tackled for many systematic applications in mechanical and civil engineering because the evaluation of safety margins has become computationally possible, with obvious advantages when compared with "admissible stress" criteria, popular in past structural engineering practice.

The content of this book reflects the intensive and successful research work carried out by the author and his co-workers both at the University of Trento and at other institutions. The detailed introduction includes several clear illustrative descriptions of experiments and, hence, solid links with practical motivation and application for the book's content. It seems that in his writing, Davide Bigoni has been mindful of Cicero's admonition not always implemented in books on mechanics: *'Non enim paranda nobis solum, sed fruenda sapientia est'* ('The knowledge should not only be acquired; it should be utilized as well'). Isaac Newton expanded on Cicero's advice when he wrote, *'Exempla docent non minus quam praecepta'* ('Examples are not less instructive than theories'). In fact, the subsequent chapters include many examples to clarify notions of applied mathematics and theoretical continuum mechanics.

The mathematics and physics covered in this volume are not easily found in the existing engineering-oriented literature in the consistent manner presented herein. At present, attention should be paid more than in the past to the warning addressed to engineers (*'ingeniarii'*) by Leonardo da Vinci, namely, *'Quelli che si innamoran di pratica senza scienzia son come 'l norchier ... senza timone e bussola'* ('Those who like practice without science are like a steersman without rudder and without steering compass'). More explicitly, Leonardo underlined the important role of mathematics: *'Nessuna umana investigazione si può dimandare vera scienza se non passa per le matematiche dimostrazioni'* ('No human research can be true science if it does not go though mathematical demonstrations'). Probably the author paid attention to this master's wisdom in compiling this volume.

As a conclusion, I express the opinion that this book provides a remarkable and timely contribution both to scientific education at the doctoral level and to the updating of scientific approaches and analytical tools in several areas of mechanical, civil and materials technologies.

Giulio Maier

1 Introduction

The mechanical modelling of the behaviour of materials subject to large strain is a concern in a number of engineering applications. During deformation, the material may remain in the elastic range, as, for instance, when a rubber band is stretched, but usually inelasticity is involved, as, for instance, when a metal staple is bent. The achievement of severe deformations involves the possibility of the nucleation and development of non-trivial deformation modes—including localized deformations, shear bands and fractures—, emerging from nearly uniform fields. The description of the conditions in which these modes may appear, which can be analysed through bifurcation and stability theory, represents the key for the understanding of failure of materials and for the design of structural elements working under extreme conditions. Bifurcation and instability modes occur in a variety of geometrical forms (as can be shown with the example of a cylinder subject to axial compression) and may explain the so-called 'size effect', 'softening' and 'snap-back' even when fracture, damage and inelasticity are excluded. Shear banding can occur as an isolated event, leading to global failure, or as a repetitive mechanism of strain 'accumulation' (as can be shown through the examples of chains with softening elements). Features determining bifurcation loadings and modes strongly depend on the constitutive features of the materials involved (as can be shown with the example of the Shanley model for inelastic column buckling). The most general framework for inelastic behaviour, including metal plasticity as a particular case, is the so-called 'non-associative elastoplasticity', which permits the description of solids where Coulomb friction is the essential inelastic micromechanism, such as granular and rock-like materials. Within this framework, the perturbative approach to material instability is presented as a way to open new perspectives in the understanding of failure in ductile materials to explain, for instance, flutter instability, a form of dynamical instability related to the presence of dry friction (as will be demonstrated with a simple experiment). Final examples are included with the purpose of presenting the complete solution of a nonlinear problem of elastic bifurcation and instability, the Euler elastica, and to show different features of instability (occurring for tensile dead loading and leading to multiple bifurcations or being suppressed when the load is proportionally increased from an unstable situation).

1.1 Bifurcation and instability to explain pattern formation

An initially flat and uniform surface of mud (Fig. 1.1, *top*) or of painting (Fig. 1.1, *centre*) has deformed under the effect of uniform drying (environmental exposition), initially undergoing a homogeneous deformation and eventually suffering a highly localized deformation, later evolving into a regular crack pattern. A presumably uniform mass of melt meteor iron has cooled down, giving rise to highly inhomogeneous but regular separation of different iron/nickel phases (so-called Widmanstätten pattern; Fig. 1.1, *bottom*). A common feature of these examples is that an initially homogeneous deformation field evolves, eventually self-organizing into an inhomogeneous but still regular deformation pattern. This phenomenon can be explained in different ways. One is to invoke the unavoidable lack of initial homogeneity of the materials into consideration, which, at least at an appropriate scale, must contain randomly distributed defects (an approach which in itself can hardly explain the regularity of the observed crack patterns). We are especially interested in another explanation,[1] which is in terms of bifurcation and instability theory so that the initial homogeneous deformation pattern (called 'trivial') ceases at a certain load level to be unique and stable, setting a bifurcation point in the deformation path and leading to an inhomogeneous alternative deformation pattern. Bifurcated deformation modes break the initially high-rank symmetry of the mechanical fields, thus introducing a lower-rank symmetry deformation pattern. The most famous example of bifurcation is Euler beam buckling (see Fig. 1.2*a*, where straight beams—with different constraints at the edges—have deformed, remaining initially straight under compressive loads but evolving into bent configurations after buckling).

Euler beam buckling (explained in Section 1.13.1) is a simple example of *elastic bifurcation and instability*: elastic because the beams in Fig. 1.2 immediately return to their initially straight configuration when the load is removed (so that permanent or viscous deformation is not involved), bifurcation because in the load/end-rotation diagram (Fig. 1.2*c*) the trivial equilibrium path corresponding to pure axial load bifurcates when the solution looses its uniqueness, and 'instability' because the straight configuration cannot be maintained after the first bifurcation load (while among the bent configurations only that corresponding to the first mode and for $\alpha < 130.7099°$ is stable; see Section 1.13.1 for details) in the sense that an arbitrary small disturbance can induce a departure from the trivial configuration. Note that instability and bifurcation are different concepts, so a system can become unstable and still have a unique response to load.

Bifurcation also may involve irreversible deformation and thus be *plastic*, as is the case reported in Fig. 1.2*b*, where a steel specimen has been permanently deformed after loading in compression, and in Figs. 1.3 and 1.4, where rock layers subject to longitudinal compressive stresses have bifurcated and deformed in the plastic regime (see also Biot, 1965, and Price and Cosgrove, 1990).

[1] The dichotomy between the two approaches can be reconciled at least in part through the perturbative approach to material instability, which will be discussed later in this Introduction and detailed in Chapter 16, where a random distribution of dislocation-like defects is shown to induce regular deformation patterns in a metallic material pre-stressed near an instability threshold.

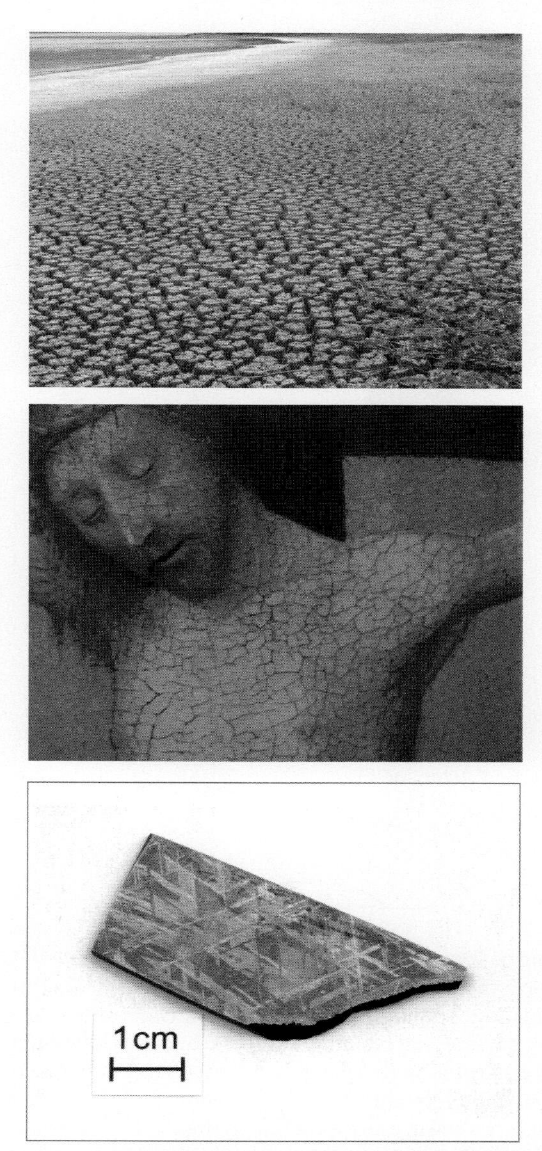

Figure 1.1. Pattern formation in initially uniformly deformed materials. *Top*: Regular crack patterns originated from drying of mud at Yaxha Lake, Petén Region, Guatemala. *Centre*: Regular crack patterns in a detail of a painting by Hieronymus Bosch (Bois-le-Duc 1450, Bois-le-Duc 1516) *La Crucifixion* (Musées Royaux des Beaux-Arts, Bruxelles). *Bottom*: A polished and etched section of an iron meteorite (a piece of the Gibeon meteorite, Namibian desert, purchased at a mineral exhibition in Trento, Italy) showing a Widmanstätten pattern. This is made up of alternate bands of kamacite (iron with 5% nickel) and taenite (iron with at least 13% nickel) developing during a cooling down process which may take thousands of years.

The fundamental difference between elastic and plastic bifurcation is that plastic deformation introduces a dissipative behaviour in the system so that the work performed to reach a certain final strain depends on the path involved (see the

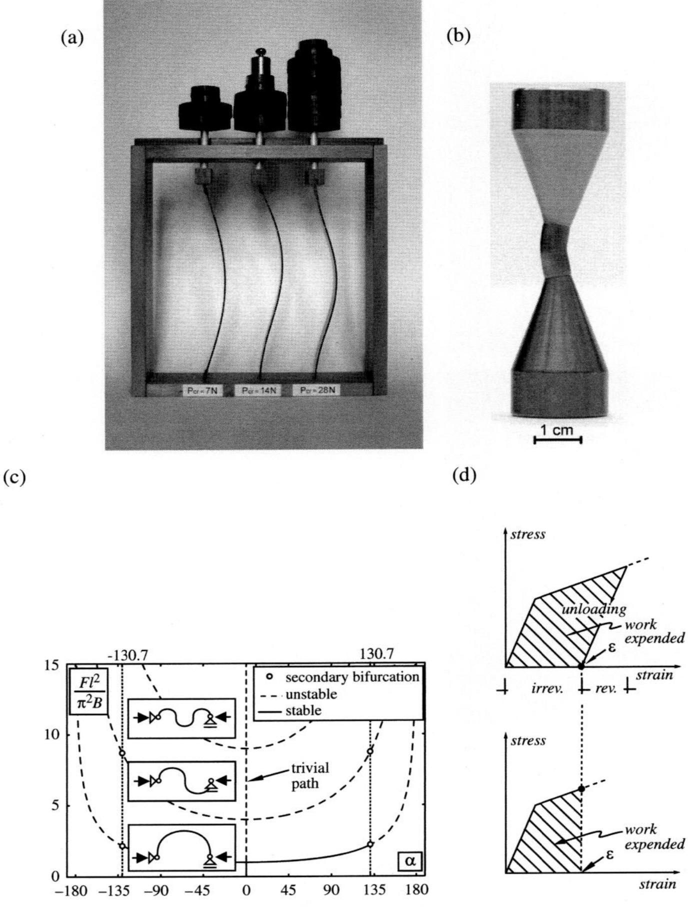

Figure 1.2. (*a*) Teaching model for buckling of beams with different end constraints (from left to right: hinged/hinged, hinged/clamped, clamped/clamped). (*b*) A metal sample subject to compressive load has buckled in the plastic range so that it is now permanently deformed. (*c*) The load (F normalized through division by π^2 times the bending stiffness B and multiplication by the square of the beam length l) versus end rotation (α in degrees) for a doubly supported inextensible Euler beam, evidencing the so-called 'pitchfork' diagram, with the 'trivial' and 'bifurcated' paths. Three bifurcation curves are shown, corresponding to the modes sketched. Note that all bifurcated paths superior to the first are unstable and that the first becomes unstable when the two supports of the beam coincide, corresponding to $\alpha = 130.7099°$ and $Fl^2/(\pi B) = 2.1834$ (details can be found in Section 1.13.1). (*d*) A uniaxial stress/strain diagram for a plastic material evidencing irreversible deformation on unloading. This material defines a dissipative system; in fact, the work performed to reach, for instance, the final deformation ϵ is path-dependent.

example reported in Fig. 1.2*d*).[2] Other examples of plastic bifurcation are the barreling of the initially cylindrical soil sample shown in Fig. 1.5 and the necking formed

[2] Dissipative terms also can be connected to the way in which the external loads are applied. For simplicity, we will always refer to conservative external load (dead loads or prescribed nominal tractions), whereas nonconservative loads are never considered in this book, with the exception of the the example on flutter instability (see the Exercise 1.13.5).

Figure 1.3. Gently folded rock layers (Silurian formation) at Constitution Hill, Aberystwyth, S. Wales, UK. The bending has been the consequence of buckling of initially straight layers.

Figure 1.4. Severe folding of metamorphic rock layers (so-called accomodation structures) initiated as buckling owing to compression stresses (Trearddur Bay, Holyhead, N. Wales, UK; the coin in the photos is a pound). (See color plates section.)

Figure 1.5. Examples of barreling of a 36.5 × 74.9-mm cylindrical silty clay (normally consolidated, namely, with an overconsolidation ratio equal to 1) sample (shown intact on the left) tested under undrained triaxial compression. Barreling has been a progressive phenomenon, initiated after a small (near 2%) homogeneous strain (the test was interrupted at an axial strain near 10%). A sample tested with a fixed top is shown in the centre, whereas a free top was used to test the sample visible on the right. No substantial difference between the behaviours of the two specimens was found in the global stress/strain response (reported in Fig. 1.18).

in the (cylindrical and dog-bone-shaped strip) metal specimens pulled in tension in Fig. 1.6.[3] While barreling and Euler buckling are examples of diffuse bifurcations, necking is an example of localized bifurcation. Note also that necking is a bifurcation occurring for tensile load.[4]

1.2 Bifurcations in elasticity: The elastic cylinder

Although the behaviour of materials is often inelastic near bifurcation stresses so that, for instance, the necking and barreling shown in Figs. 1.5 and 1.6 typically involve plastic deformation, elasticity remains a useful framework for the analysis of bifurcation.[5]

Some bifurcation thresholds and relative modes of deformation are reported in Fig. 1.7 for uniaxial compression of a cylindrical specimen made up of an elastic incompressible material, with material parameters selected to simulate the behaviour of silicon nitride at high temperature (example taken from Gei et al. 2004 and developed in detail in Chapter 12). The compression is transmitted through contact with two smooth and rigid constraints at the edges. Both the nominal and true

[3] Additional (beautiful) examples of plastic bifurcation have been given by Rittel and Roman (1989), Rittel (1990), and Rittel et al. (1991).

[4] Necking is a *bifurcation occurring for tensile axial loads*, whereas bifurcation of structural elements such as beams, columns and frames is normally believed to be associated with compressive loads. An example reported in Section 1.13.2 will be sufficient to show that buckling of structures subject to tensile dead load can occur for simple mechanical systems (see also Zaccaria et al. 2011).

[5] Chapters 10 and 11 are devoted to the general theory of bifurcation and instability in elastic and plastic solids (the former section to global and the latter to local conditions), whereas examples of bifurcations for elastic and elastoplastic solids are presented in Sections 12 and 13, respectively.

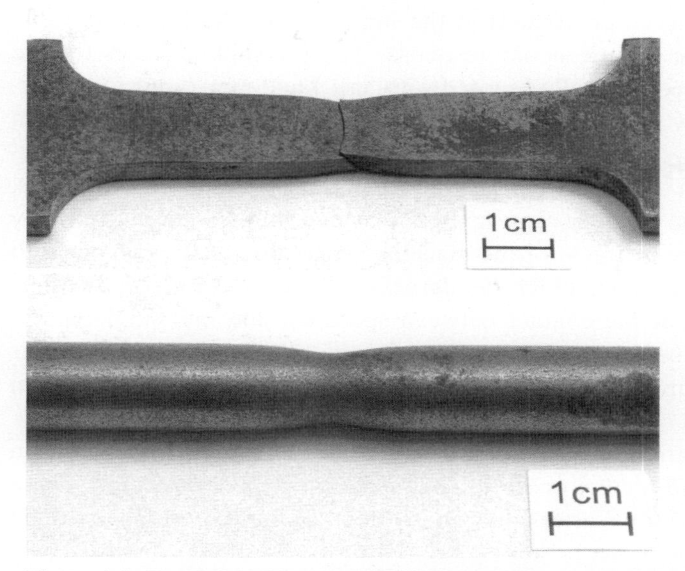

Figure 1.6. Examples of localization of deformation occurring for tensile loads in low-carbon steel samples. *Top*: Necking is the first mode occurring in the dog-bone-shaped strip shown in the figure. Later, shear banding (with a 45° inclination) develops within the necked zone, which eventually degenerates into a fracture (visible in the photo). *Bottom*: Necking in a cylindrical bar under tension.

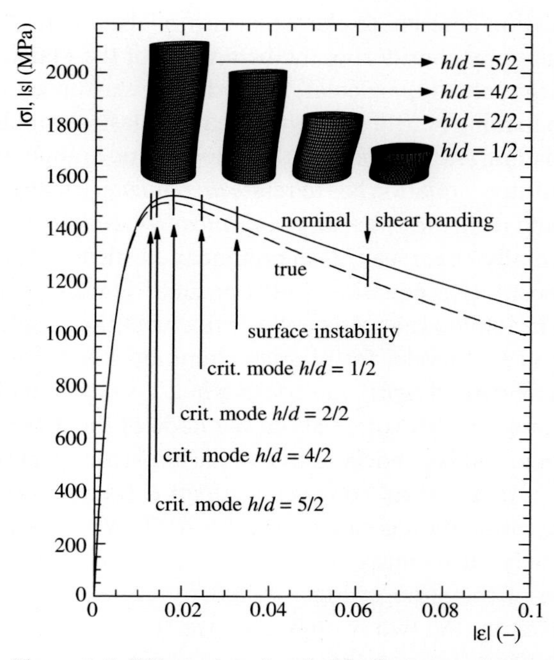

Figure 1.7. Bifurcation thresholds (in terms of true σ and nominal s stress versus axial logarithmic strain) of an elastic, incompressible cylinder uniaxially compressed through smooth and rigid constraints (adapted from Gei et al. 2004). Some modes associated with initial bifurcations for different aspect ratios ($h/d = 1/2, 1, 2, 5/2$) are sketched in the upper part. Note that surface bifurcation (or surface instability) and shear banding are extreme forms of instability, occurring 'far' in the softening regime.

stresses are reported (the latter dashed) in the figure versus the (absolute value of) logarithmic strain for homogeneous response. The threshold stresses for the first bifurcations are called 'critical' (denoted with 'crit') and may occur for different aspect ratios (namely $h/d = 1/2, 1, 2, 5/2$). These are marked on the curves in Fig. 1.7.

The different modes of bifurcation may be triggered by boundary conditions, for instance, if the elastic cylinder represents a specimen in a testing machine, by the specific test geometry, stiffness of the machine, or friction at the contact with the loading plates. For example, a high slenderness will induce a Euler-type bifurcation, whereas friction at the contact between specimen and end supports will favour development of barreling. The example of the elastic cylinder illustrates the great variety of bifurcation mechanisms often observable during mechanical tests.

Two peculiar bifurcation instabilities in Fig. 1.7 will receive a strong attention later, namely, *surface bifurcation* or *surface instability* and *shear banding*.[6] These instabilities occur at high strain, after other bifurcations, so these represent extreme forms of instabilities.

1.3 Bifurcations in elastoplasticity: The Shanley model

Bifurcation and instability theory is much simpler for elastic materials than for elastoplastic materials. The complication inherent to elastoplastic models, which are dissipative, already can be appreciated by considering the behaviour of the elastoplastic spring shown in Fig. 1.8 on the left. In particular, since during elastoplastic behaviour the response differs from loading (stiffness k_t, the so-called plastic branch of the constitutive equation) to unloading (stiffness k_e, the so-called elastic branch of the constitutive equation), the constitutive equation has to relate *incremental* (instead of finite, as in elasticity) quantities and in an *incrementally non-linear (piece-wise linear) relation* (instead of an incrementally linear way, as in nonlinear elasticity). The complication introduced by the constitutive equation yields peculiar features for problems of plastic bifurcation and instability that can be illustrated with the simple example of the celebrated Shanley (1947) model for plastic column buckling. The Shanley model is shown in Fig. 1.8 (*centre* and *right*) and consists in a T-shaped rigid rod constrained to move vertically and possibly rotate about the node of the T (see also Hutchinson 1973, 1974). The rigid rod is connected to two elastoplastic springs (behaving as in Fig. 1.8, *left*), subject to a vertical compressive load F (taken positive for compression), which before bifurcation is simply equal to $P/2$. We assume that F is (1) either at yielding, namely, at its maximum value $F_{\max}$ (in the plastic branch of the constitutive equation of the springs; Fig. 1.8, *left*) or (2) is unloading after having reached $F_{\max}$. Therefore, the two springs obey the following force

[6] Surface instability is characterized by the fact that the wavelengths of the bifurcation mode may become infinitely small. This is a tendency already visible in Fig. 12.17 of Chapter 12, where modes R, T and U correspond to increasing circumferential and longitudinal wavenumbers (see Fig. 12.18). Subsequent bifurcations involve smaller wavelengths, which approach zero at surface instability, representing an *accumulation point for bifurcation loads*.

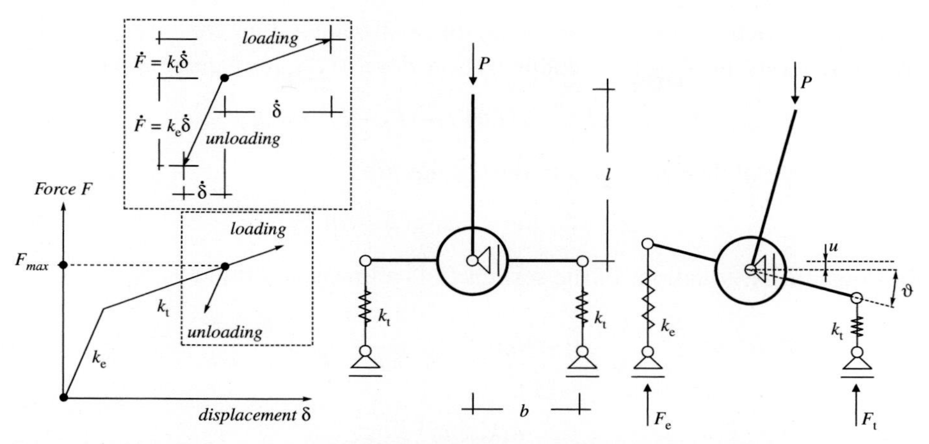

Figure 1.8. *Left*: Force-displacement relation for an elastoplastic spring. *Centre* and *right*: Shanley model for plastic column buckling. The rigid T-shaped rod has two degrees of freedom $\{u,\theta\}$ and is constrained by two elastoplastic springs, displaying the piecewise-linear force-displacement behaviour shown on the left.

F – displacement δ rate constitutive relation:

$$1.\ \text{At yielding: } F = F_{\max} \Longrightarrow \begin{cases} \dot{F} = k_t\dot{\delta}, & \text{for plastic loading } \dot{F} > 0, \\ \dot{F} = k_e\dot{\delta}, & \text{for elastic unloading } \dot{F} < 0, \end{cases} \tag{1.1}$$

$$2.\ \text{Within the elastic range: } F < F_{\max} \Longrightarrow F = k_e\delta.$$

where superposed dots denote rates.

Note the incremental non-linearity (more precisely, piecewise linearity) of the constitutive equations, which exhibit two different branches, one for plastic loading and another for elastic unloading (the so-called neutral loading corresponds to $\dot{\delta} = \dot{F} = 0$). This nonlinearity in the rate represents the main complication of elastoplasticity.

It is assumed that the system is loaded until the springs are in the plastic range, without any prior elastic instability. This loading occurs without any rotation and defines the fundamental path, where $F = P/2$. Bifurcations are sought as quasi-static (namely, involving negligible dynamical effects) departures from this configuration, in terms of a rate vertical displacement $\dot{u}$ of the centre of the T-shaped rod and a rate rotation $\dot{\theta}$. Note that we intend with 'rate' the derivative with respect to an increasing time-like parameter governing the loading of the system.

Equilibrium written for a configuration displaced a finite amount u and θ is readily obtained as

$$P = F_e + F_t \qquad Pl\sin\theta + (F_e - F_t)b\cos\theta = 0, \tag{1.2}$$

with displacements in the springs

$$u_t = u + b\sin\theta \qquad u_e = u - b\sin\theta. \tag{1.3}$$

By approximating Eqs. (1.2) and (1.3) for small θ and taking the rates, we obtain the rate equations, holding for a configuration close to the fundamental path

$$\dot{P} = \dot{F}_e + \dot{F}_t \qquad (P\theta)^{\cdot}\, l + (\dot{F}_e - \dot{F}_t)b = 0, \tag{1.4}$$

whereas the rate of displacements in the springs are

$$\dot{u}_t = \dot{u} + b\dot{\theta} \qquad \dot{u}_e = \dot{u} - b\dot{\theta}, \tag{1.5}$$

so the constitutive equations of the springs (1.1) allow us to obtain

$$(P\theta)^{\cdot} = -\frac{k_e - k_t}{k_e + k_t}\frac{b}{l}\dot{P} + P_R\dot{\theta} \qquad \dot{u} = \frac{\dot{P}}{k_e + k_t} + \frac{k_e - k_t}{k_e + k_t}b\dot{\theta}, \tag{1.6}$$

where

$$P_R = \frac{4(k_e k_t)b^2}{(k_e + k_t)l} > 0, \tag{1.7}$$

is the so-called reduced-modulus load. Equations (1.6) are valid if the appropriate loading/unloading conditions are satisfied. For positive (clockwise) $\dot{\theta}$ and using Eqs. (1.5), these conditions are

$$\dot{u} - b\dot{\theta} < 0, \;\; \dot{u} + b\dot{\theta} > 0. \tag{1.8}$$

From Eqs. (1.6) we can observe the following.

- Analysis of bifurcations emanating from the fundamental path, $\theta = 0$, is governed by the equations

$$\frac{k_e - k_t}{k_e + k_t}\frac{b}{l}\dot{P} = (P_R - P)\dot{\theta} \qquad \dot{u} = \frac{\dot{P}}{k_e + k_t} + \frac{k_e - k_t}{k_e + k_t}b\dot{\theta}. \tag{1.9}$$

  Working Eqs. (1.9) into Eqs. (1.8), we can write the conditions enforcing the satisfaction of loading/unloading constraints in the springs

$$P_T \le P \le P_E \qquad P_T = \frac{2k_t b^2}{l} \qquad P_E = \frac{2k_e b^2}{l}, \tag{1.10}$$

  where P_T is the so-called tangent modulus load, whereas P_E is the critical load corresponding to a purely elastic bifurcation.

  For a value of the load P belonging to the interval $[P_T, P_E]$, bifurcations are possible because Eqs. (1.9) always admit a non-trivial solution. There are three particular cases of interest: (1) When $P = P_T$, bifurcation occurs as if the two springs were elastic and both are characterized by a stiffness k_t; moreover, neutral loading (i.e., $\dot{u}_e = 0$) occurs for the unloading spring, which remains 'frozen' at bifurcation. (2) When $P = P_R$, bifurcation occurs at a fixed load (i.e., for $\dot{P} = 0$). (3) When $P = P_E$, a purely elastic bifurcation occurs.
- Integration of the rate Eqs. (1.6) is simple if the springs do not change their behaviour. Performing the integration from an initial state where the load is $P_{\text{bif}} \in [P_T, P_E]$ and $\theta = 0$ to a final state in which the load is $P = P_{\text{bif}} + \Delta P$ and the rotation is θ, we obtain

$$\frac{\Delta P + P_{\text{bif}}}{P_R} = \frac{P_{\text{bif}}/P_R(P_E/P_T - 1) + (P_E/P_T + 1)l/b\theta}{(P_E/P_T - 1) + (P_E/P_T + 1)l/b\theta}, \tag{1.11}$$

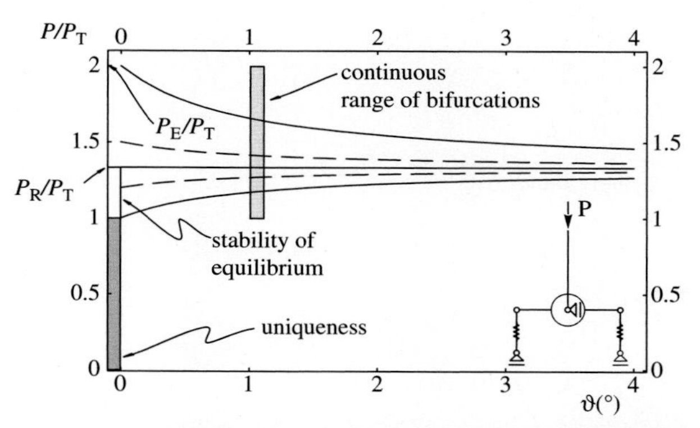

Figure 1.9. Critical and post-critical response of the Shanley model for plastic column buckling (Fig. 1.8). The vertical load is made dimensionless through division by the tangent modulus load P_T corresponding to the first possibility of bifurcation. Note that there is a continuous range of bifurcations for loads within the interval $[P_T, P_E]$, and stability of equilibrium is preserved within the interval $[P_T, P_R[$, so bifurcation does not imply loss of stability of equilibrium.

in which[7]

$$P_R = 2P_T \frac{P_E/P_T}{P_E/P_T + 1}. \tag{1.12}$$

For a given $P_{\text{bif}}/P_T \in [1, P_E/P_T]$, the dimensionless load $P/P_T = (\Delta P + P_{\text{bif}})/P_T$ can be plotted as a function of P_E/P_T and l/b. An example is reported in Fig. 1.9, for $P_E/P_T = 2$ and $l/b = 20$.

- The post-critical behaviour expressed through Eq. (1.11) and plotted in Fig. 1.9 reveals that *bifurcation occurs continuously from the bifurcation load of the 'linearized in-loading response' P_T to the elastic bifurcation load P_E.* Moreover, bifurcation occurs (1) at *increasing load*, for $P_T \leq P < P_R$, (2) at *fixed load*, when $P = P_R$, (3) at *decreasing load*, for $P_R < P \leq P_E$.

The Shanley model highlights typical features of plastic bifurcation and stability. First, we note that if we replace the real (incrementally non-linear) elastoplastic springs with linear springs corresponding to the loading branch of the constitutive equations, so-called linear comparison solid, the lowest bifurcation load is correctly calculated. Second, there is a continuous spectrum of bifurcations (while for elastic springs, there would be only one bifurcation point[8]) 'departing' from the fundamental path. Third, loss of stability of equilibrium occurs well *after* the first bifurcation load. Fourth, the fundamental deformation path becomes unstable from the first bifurcation point, a condition that should not be confused with instability of equilibrium (although the two concepts coincide in elasticity; see Section 10.2.2).

[7] Equation (1.11) coincides with Eq. (2.9) of Hutchinson (1974) by setting his $K(\theta)$ equal to zero.
[8] The fact that there are two degrees of freedom and only one critical load (with single multiplicity) is detailed in Exercise 1.13.3.

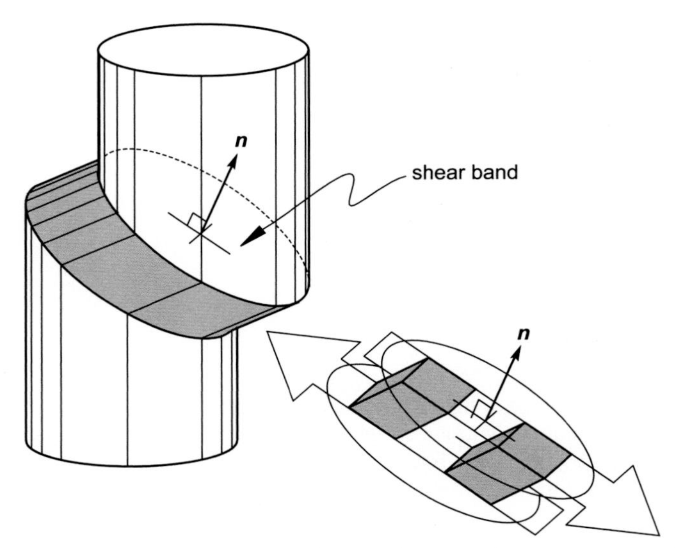

Figure 1.10. Strain localization mode in the form of a planar shear band. The unit vector ***n*** defines the normal to the shear band. Note that a material element within the shear band suffers a shear deformation (compare with Fig. 1.11, *centre*).

1.4 Shear bands and strain localization

Necking (Fig. 1.6) is a form of plastic localized bifurcation mode. An extreme form of localization occurs when one or more thin layers of severely deformed material nucleate from an essentially uniform deformation field, as shown in the sketch in Fig. 1.10. This has been the case with the cylindrical soil samples and prismatic wood samples shown in Figs. 1.11 and 1.12, in both cases subjected to compressive axial loads. In the former case, Fig. 1.11, a clear X–shaped and \–shaped layer of concentrated shear deformation, so-called shear bands, are visible, exactly as sketched in Fig. 1.10. However, thin zones of concentrated deformations are not always subject to pure shear strain as sketched in Fig. 1.10: For instance, such zones may become subject to dominant compressive deformation, which is the case for the prismatic wood samples shown in Fig. 1.12. These have been subjected to uniaxial compression and have failed with the formation of deformation bands almost perpendicular to the loading direction (so-called crushing failure, see Bodig and Jayne 1982)[9]]. Therefore, with the nomenclature 'shear bands' (or 'localization of deformation', or 'strain localization'), it is usually meant formation of thin planar zones of intense deformations. The micromechanism of localization of deformation in wood corresponds to fibre bending and subsequent collapse within the localization band. It is an example of a so-called 'kink banding', a phenomenon common in fibre-reinforced materials and rocks (Price and Cosgrove 1990).

[9] While materials usually exhibit failure tensile stress inferior to compressive stress, the reverse is often the case for wood, where local microbuckling of fibres organized within kinking bands sets failure limits in compression. Even under four-point bending, failure normally initiates at the compressive side, particularly in clear, low-density wood. A similar behaviour is exhibited by glass- or carbon-fibre reinforced materials, where the ratio between compressive and tensile failure stresses can be on the order of 0.6 so that compressive failure owing to kink band formation is the main limiting factor for design.

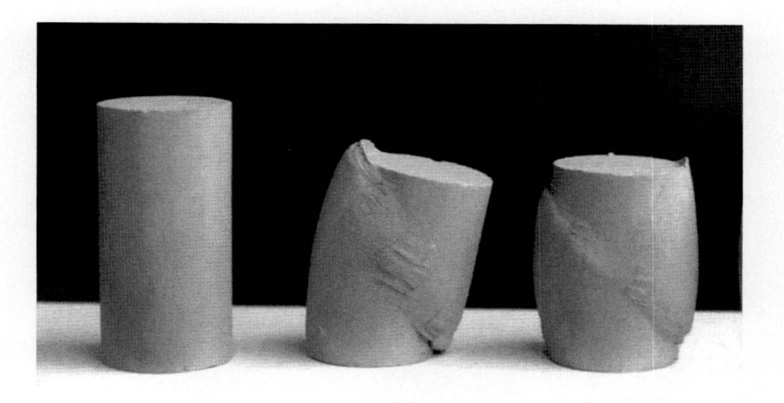

Figure 1.11. Examples of localization of deformation *occurring for compressive loads*. Shear banding in a 36.5 × 74.9-mm cylindrical silty clay sample (the same material used for the tests reported in Fig. 1.5, but now over-consolidated, with over-consolidation ratio equal to 2) subjected to an undrained triaxial compression test. The intact sample before the test is shown on the left. A sample tested with a free top is reported at the centre (in which a single shear band has developed), whereas a fixed top was used for the sample on the right (in which two twin shear bands have developed). Shear bands have emerged before the peak of the stress/strain curve was reached (near 2% axial strain, whereas the test was interrupted at an axial strain near 25%), up to this point, the results of the two tests for the two samples shown at the centre and on the right were near coincident, but the post-peak behaviour was strongly different (softening was more pronounced in the test the with free top; see Fig. 1.18). The qualitative nominal stress versus conventional strain curves corresponding to the samples shown in Figs. 1.5 and 1.11 are reported in Fig. 1.18.

Localization of deformation is an extremely common phenomenon and is considered to be the key to an understanding of failure in materials. Shear band formation is a special feature of localized deformation that (1) in metals usually preludes fracture nucleation and growth, (2) in quasi-brittle materials (such as rock and concrete) generates zones of intense damage, and (3) in granular material may continue up to extremely large strains without grain damage (Desrues et al. 1985, 1996). Localization of deformation is usually identified with the condition of loss of ellipticity of the equations governing incremental equilibrium (see Chapter 11).

The theory of shear bands viewed as a bifurcation of a homogeneously deformed material element has been pioneered by Nadai (1931, 1950), Hill (1952, 1962), Thomas (1953, 1954, 1961), Prager (1954), Mandel (1962a, 1962b), and Rice (1977); this theory is presented in detail in Section 11, whereas examples of explicit calculations are given in Chapters 12 and 13.

Shear bands are also common phenomena when dynamic loading is involved. In this case, there is a high temperature rise within the deformation bands, which are referred as 'adiabatic shear bands'. Experimental evidence of these goes back to Tresca (1878), see Bai and Dodd (1992) and Walley (2007). As an example taken from geophysics, we show in Fig. 1.13 that a very localized dynamic shearing deformation has occurred in a fault, visible at the Earthquake Fault Line Observatory near Gifu, Japan.

Figure 1.12. Examples of localisation of deformation *occurring for compressive loads*. Prismatic 94 × 97 × 170 mm (Italian) spruce wood samples (density 0.43 to 0.42 g/cm^3) compressed parallel to the fibres. Crushing bands with some splitting emerged accompanied by strong acoustic emission before the peak in the load/displacement curve, without prior visible manifestations of instability. The peak loading was reached at a 36.4 N/mm^2 (37.4 N/mm^2) nominal stress and 3.6% (1.5%) conventional strain for the sample on the left (right). The tests were interrupted at a 7.7% (2.7%) strain, when the load was decreased to 31.9 N/mm^2 (27.3 N/mm^2).

Figure 1.13. Fault active during the strongest inland earthquake in the history of Japan (October 28, 1891). Photo taken at the Neo Valley Earthquake Fault Line Observatory, Midori Neo-mura Motosu-gun (Neomidori Motosu-shi), Gifu, Japan. (See color plates section.)

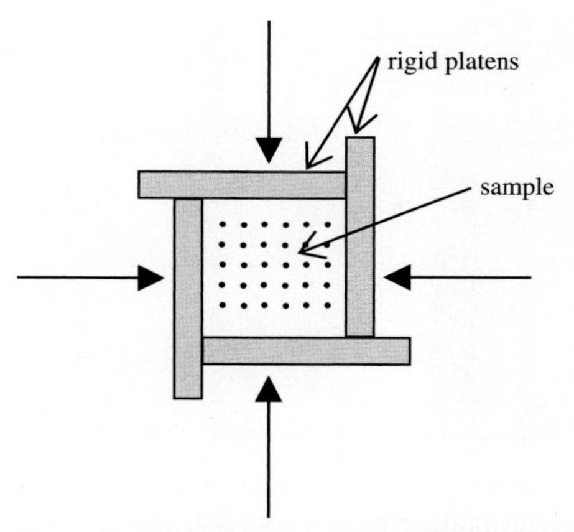

Figure 1.14. Sketch of a rigid boundary true triaxial apparatus for soils, where mechanical conditions are imposed similar to the so-called van Hove conditions. This apparatus has been designed in an attempt to investigate whether shear band can emerge, without other bifurcation modes 'disturb' the homogeneity of the sample.

The nucleation of both shear banding and surface instability is independent of the geometry of the body in which they occur, although this influences the *development* of the instabilities. The former corresponds to a deformation mode that is discontinuous across two planes, delimiting a band within which the deformation is constant (Fig. 1.10). Surface instability corresponds to a wave-like surface deformation mode that decays exponentially away from the surface. It can be concluded that

surface instability and shear banding are local modes which may 'shrink to a point', thus becoming independent of the geometry of the problem

and also, considering again Fig. 1.7, that

surface instability and, especially, strain localisation represent extreme forms of instability, usually possible only after earlier (possibly infinite) bifurcations in diffuse modes are met.

Therefore, even if localisation of deformation is a very common behaviour of material tested until failure, it is likely that this occurs as a mode superimposed on a bifurcated field.

Localisation of deformation and surface instability is an example of *material instability*, namely, instability that can start to grow within a small part, 'at a point', of in a body and grow until the global behaviour is strongly affected. A test designed to investigate strain localisation should realise mechanical conditions in which localisation may develop as an initial mode to preserve homogeneity until instability. It may be intuitive that this test should 'constrain the boundary' so that global bifurcations (such as those shown in Fig. 1.7) and surface instability remain excluded. Such a device indeed has been designed and employed to test soil specimens, with the name of 'rigid boundary true triaxial apparatus' (Fig. 1.14).

Except for the fact that the sample may slide along the rigid edges and that the contact is unilateral (for compressive loading, this unilaterality condition becomes inconsequential), the rigid boundary true triaxial apparatus shown in Fig. (1.14) and the uniaxial compression strain test shown in Fig. 1.15 tends to realise the so-called

Figure 1.15. Examples of mechanical conditions similar to the so-called van Hove. Plane uniaxial strain apparatus to test a regular packing of cylindrical hollow tubes (drinking straws). *Left*: Configuration before the test; *centre*: the test at 0.05 conventional deformation, still not showing evidence of localisation (some small distortion of the straws near the boundary owing to friction are visible); *right*: the test at 0.2 conventional deformation showing an X-shaped shear banding. Experiment performed at Trento University following Poirier et al. (1992).

van Hove conditions corresponding to homogeneous deformation of a material element with prescribed displacements along the whole boundary. From a mathematical point of view, *van Hove conditions can be proved to exclude all bifurcations when the constitutive operator is uniform in space, linear and strongly elliptic* (van Hove, 1947; Hayes, 1966, see Section 11). Moreover, a generalisation of the van Hove conditions, covering cases shown in Figs. 1.14 and 1.15 (except for the unilaterality of the contact), has been given by Ryzhak (1993, 1994), whereas Drugan (2007) has extended the results to an elastic constraint at the boundary.

In all the preceding examples of bifurcation and shear banding, inhomogeneous deformation modes are sort of 'parasitic fields' emerging within homogeneously deformed samples, when they are subject to boundary conditions compatible with continued homogeneous deformation, and soon dominating subsequent deformation; for instance, Euler buckling develops on a uniformly compressed rod. However, this is not always the case, and bifurcations can in fact emerge from inhomogeneous stress fields, such as, for instance, in the case of a column subjected to self-weight or the bending of a thin beam (Timoshenko and Gere, 1961). Another example, which will be analysed in Chapter 12, is that of the finite bending of a thick elastic plate, see the experiments reported in Figs. 12.21 and 12.24 (other experiments have been given by Gent and Cho, 1999, and Gent, 2005, see Section 12.4). The more general case of a coated layer subject to bending has been addressed experimentally and theoretically by Roccabianca et al. (2010, 2011).

Shear bands often can be observed to prelude failure in structures where the deformation is highly inhomogeneous. This is, for instance, the case of the prismatic, two-component epoxy resin sample shown in Fig. 1.16 and subject to vertical compression (see Bigoni et al., 2008, for details). This sample contains a stiff and thin aluminium inclusion, producing highly stressed/deformed near-tip fields. When the material is far from failure, photoelastic analysis reveals a surprisingly good adherence to the linear elastic (singular) solution (Dal Corso et al., 2008; Noselli et al.,

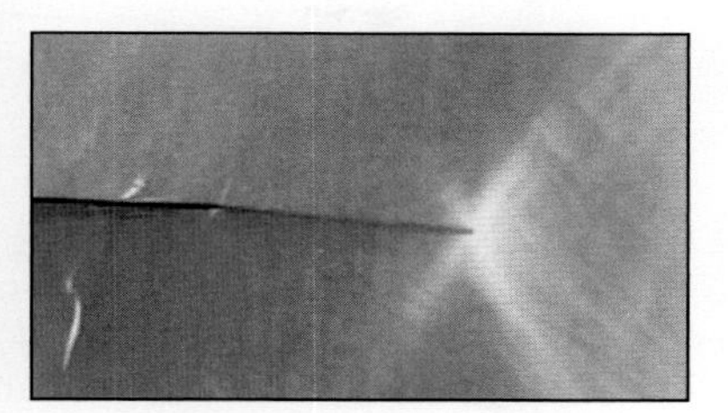

Figure 1.16. Shear bands emerging from the tip of a stiffener (a 0.3 mm thick 44×18-mm aluminium platelet) embedded in a two-component epoxy resin prismatic ($100 \times 100 \times 18$-mm) sample. The sample has been loaded uniaxially in vertical compression (photo taken in bright light at a 50 MPa compressive stress). The material exhibited a ductile behaviour and suffered an out-of plane buckling. Light reflection evidences strain localisation in the form of shear bands at the end of the platelet (clearly visible in the detail on the right). (See color plates section.)

2011, see Chapter 16, Figs. 16.18 and 16.19). However, when inelastic deformation prevails, the strain pattern evidences shear band formation at the tip of the stiffener (visible from reflected bright light in Fig. 1.16).[10] It is a remarkable feature that shear bands dominate failure deformation patterns, a circumstance motivating the fact that limit analysis solutions based on perfectly plastic behaviour usually correctly capture final deformation modes, as demonstrated in pioneering works by Nadai (1931, 1950), Hill (1952); Prager (1954), Thomas (1953, 1954, 1961). These are clear evidence that the theory of elastoplasticity represents the correct framework for capturing instabilities preluding failure in laboratory samples and structural elements.

1.5 Bifurcation, softening and size effect as the response of a structure

Loss of homogeneity of deformation of a sample in a test designed to induce continued homogeneous strain is related to the emergence of size effects. In these cases, the specimen becomes a structure, the global nominal stress versus conventional strain behaviour becomes dependent on the size of the structure, and the sample exhibits sensitivity to boundary conditions.

To illustrate this fact with a very simple example (detailed in Section 1.13.4), let us consider the structure (analyzed also by Feodosyev, 1977; Poston and Steward,

[10] A similar behaviour also has been observed at the tip of a propagating crack in a fibre-reinforced composite by Zhang and Clifton (2007).

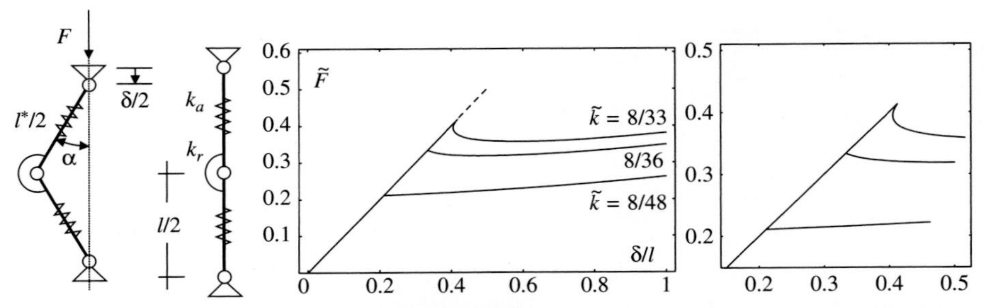

Figure 1.17. A simple elastic structure explaining how size effect inducing softening behaviour (and even snap-back) may be the consequence of bifurcation (*left*). Dimensionless load $\tilde{F} = 2F/lk_a$ versus dimensionless end displacement δ/l curve (*centre*) for different stiffness ratios $\tilde{k} = 8k_r/l^2k_a$, showing bifurcation and equilibrium branches. A detail of the curve is shown on the right, where it can be appreciated that softening may be followed by re-hardening and that there is also the possibility of a so-called snap-back instability, occurring in fact for $\tilde{k} = 8/33$.

1978, Goulbitski and Schaeffer, 1985, and Potier Ferry, 1987) sketched in Fig. 1.17 on the left, consisting of two hinged-end rods of undeformed (of deformed) length $l/2$ ($l^*/2$) which are axially elastic (with axial stiffness k_a), but flexurally rigid and connected to each other with a rotational spring of stiffness k_r.

Equilibrium in a deformed configuration of the structure sketched on the left of Fig. 1.17 corresponds to the equation

$$\tilde{F}^2 \cos\alpha - \tilde{F} + \tilde{k}\frac{\alpha}{\sin\alpha} = 0, \tag{1.13}$$

where α is the angle of inclination of the springs with the vertical direction and

$$\tilde{F} = \frac{2F}{l k_a} \qquad \text{and} \qquad \tilde{k} = \frac{8k_r}{l^2 k_a} \tag{1.14}$$

are the dimensionless force and the spring stiffness, respectively.

Equation (1.13) can be solved to provide the two non trivial equilibrium conditions

$$\tilde{F} = \frac{1 \pm \sqrt{1 - 4\tilde{k}\alpha \cot\alpha}}{2\cos\alpha}, \tag{1.15}$$

and in the limit $\alpha \longrightarrow 0$ yields the bifurcation loads

$$\tilde{F}_{cr} = \frac{1 \pm \sqrt{1 - 4\tilde{k}}}{2}. \tag{1.16}$$

The dimensionless displacement of the two edges of the structure $\tilde{\delta} = \delta/l$ is given by

$$\tilde{\delta} = 1 - \cos\alpha(1 - \tilde{F}\cos\alpha), \tag{1.17}$$

so the dimensionless force/displacement ($\tilde{F}$ vs. $\tilde{\delta}$) curve emanating from the first bifurcation load of the structure, for given values of the dimensionless stiffness $\tilde{k}$, is shown in Fig. 1.17 (a detail of the graphs is reported on the right).

Note that Eq. (1.13) admits real solutions only when $\tilde{k} \leq 1/4$, so there is no bifurcation for $\tilde{k} > 1/4$. Since $\tilde{k}$, given by Eq. $(1.14)_2$, increases when k_a decreases, this simple model provides *an example of a mechanical system in which a decrease in elastic stiffness of a component* (k_a at fixed k_r) *may increase the bifurcation load*, or in other words and as will be shown in Section 1.13.4, *a decrease in elastic axial stiffness may stabilise the structure.*

The structure considered in this example, similar to a model for kink band analysis of a stack of thin layers presented by Hunt et al. (2000) and Wadee et al. (2004), has been analysed only from the point of view of bifurcation, so it is not essential whether the displacement δ or the force F is controlled, a feature strongly influencing stability (see Section 1.13.4).

Size effect related to bifurcation in the example shown in Fig. 1.17 comes from the fact that the response depends on l^2, hidden in $\tilde{k}$, Eq. $(1.14)_2$, a parameter determining the stability of the post-critical behaviour: For $\tilde{k} = 8/48$, the post-critical behaviour is stable, for $\tilde{k} = 8/36$, it evidences softening (namely, a decrease of the applied force while axial displacement is increasing) and becomes unstable if the force F is prescribed, whereas stability is preserved if the displacement of the structure is controlled, which is not the case of $\tilde{k} = 8/33$, corresponding to the so-called snap-back,[11] a 're-entering' load/displacement curve (see Section 1.13.4).

The example shows that an elastic structure made of components having positive elastic stiffness may exhibit size effects and structural softening,[12] with possible 'snap-back'. Moreover, in certain cases, *softening may prelude a subsequent re-hardening*.

Softening and size effects are well-known consequences of necking (corresponding in fact to a peak in the nominal stress versus conventional strain curve) for a cylindrical bar subject to a tensile force and have been advocated to explain experimental results on marble, rock and concrete under uniaxial compression by Read and Hegemier (1984). In addition, *softening and size effects are clearly connected to the formation of shear bands*, whereas a sort of re-hardening after a perfectly plastic behaviour can be related to the development of a barreling mode. From the experimental point of view, the fact that softening is connected with strain localisation, whereas barreling is not, can be appreciated from the nominal stress versus conventional strain curves reported in Fig. 1.18, corresponding to the tests performed on the samples shown in Figs. 1.5 and 1.11. It is important to mention that these tests were performed with both fixed and free ends, and the responses obtained in the case of barreling bifurcation were found to be almost independent of the boundary conditions and similar to a perfectly plastic behaviour, whereas strain localisation had begun just before the peak and has determined a softening stronger for free ends (where only one shear band has formed) than for the fixed top (where two 'twin' shear bands have formed).

The fact that softening is a consequence of shear banding is fully confirmed by modelling. We may in fact consider Fig. 1.19, where the simulation is reported,

[11] The analysed structure is also interesting for other aspects, for instance, a so-called cusp and smooth bifurcation appear when $\tilde{k} = 8/32$; moreover, the rectilinear configuration returns to be stable after the vertical load has exceeded a certain value; see Section 1.13.4.

[12] During softening, the *incremental stiffness becomes negative*, so viewed as a 'black box', the structure behaves as a spring with incremental negative stiffness. Structural elements in a buckled configuration have been employed to produce the so-called negative stiffness materials (Lakes, 2001).

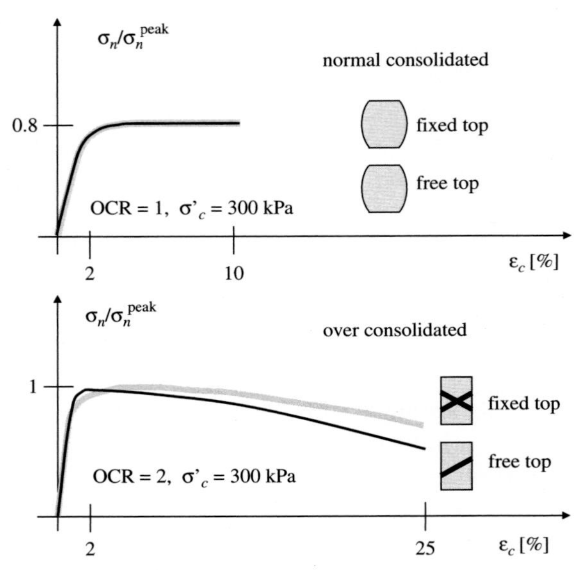

Figure 1.18. Qualitative nominal stress σ_n versus conventional strain ϵ_c curves of the cylindrical silty clay samples reported in Figs. 1.5 and 1.11. Shear band nucleation occurred just before the peak of the curve for the over-consolidated samples (OCR = 2, where OCR denotes the so-called over-consolidation ratio), so that softening is related to the growth of deformation inside the shear band. The normal consolidated material (OCR = 1) did not show softening but an ideally plastic behaviour.

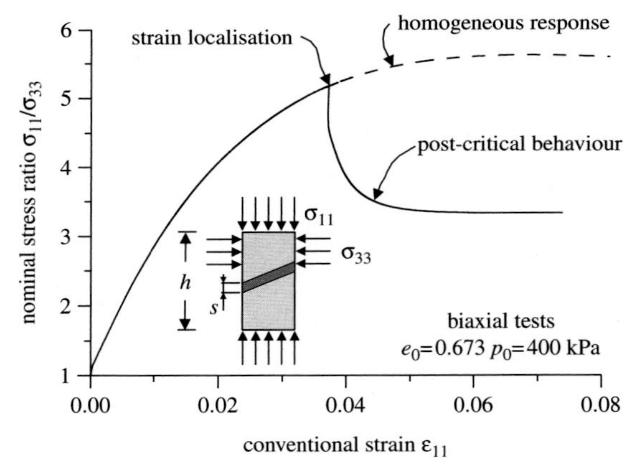

Figure 1.19. Simulation of strain localisation occurring in a biaxial (plane strain) compression test. Localisation occurs during hardening and induces a strong softening in the global nominal stress versus conventional strain response (simulation referred to a test on dry, dense sand; e_0 indicates the initial void ratio and p_0 the confining pressure; adapted from Gajo et al., 2004).

in terms of 'global' nominal-stress versus conventional strain (plotted as if the deformation were uniform) of the two-dimensional deformation of a prismatic dry, dense sand sample loaded in a so-called biaxial compression test (Drescher et al., 1990), namely, a plane strain confined compression test (vertical stress σ_{11} and

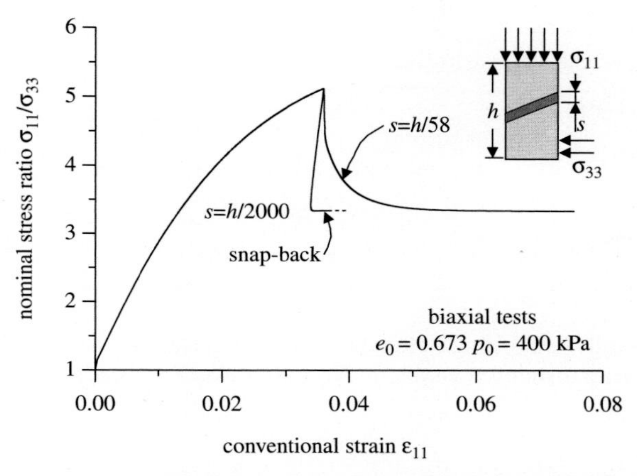

Figure 1.20. Simulation showing the effect of band thickness on post critical response. The thinner the band, the steeper is the softening: This clearly represents a size effect. Snap-back behaviour also may occur, such as, for instance, when s = h/2000 (the simulation is referred to a biaxial, plane strain compression test on dry sand; e_0 indicates the initial void ratio and p_0 the confining pressure; adapted from Gajo et al., 2004).

horizontal stress σ_{33}). Here we may see that[13] strain localisation initiates during hardening[14] and occurs with *elastic unloading outside and plastic loading inside the shear band* (see Chapter 11), precipitating the response of the material into a strong softening.

The fact that *size effect is a consequence of strain localisation* is also confirmed by modelling, so Fig. 1.20 (where the behaviour for $s/h = 1/58$ as in Fig. 1.19 is also reported) shows that softening depends on the ratio between the localisation width s and the height h of the sample. In particular, the smaller this ratio, the 'steeper' is the softening response, and if the ratio is small enough ($s/h = 1/2000$ in the figure), *snap-back* behaviour is observed.

In solids without any internal length (which will be addressed only in this book), *the width of a shear band occurring at loss of ellipticity is arbitrary*.[15] This fact makes ill-posed a problem when equations are outside the elliptic range, which, in terms of a finite elements analysis, implies that results become mesh-dependent. This peculiarity of loss of ellipticity has been the basis of an immense research effort, which, however, still appears far from being concluded (see Chapter 16 for an explanation of this point).

Many experimental evidences and theoretical considerations support the conclusion that

[13] The behaviour of the sand sample is modelled within the framework of infinitesimal elastoplastic coupling, a concept introduced in Chapter 8 and the figure is taken from Gajo et al. (2004), where the interested reader is referred for details. Note that the simulation has been possible through introduction of several simplificative assumptions explained in Chapter 15, among which is the assumption of a fixed ratio between band thickness and sample height, $s/h = 1/58$.

[14] Within a geometrically linear theory, strain localisation is possible during hardening only when a nonassociative flow rule is employed; see Section 11.

[15] The width of the shear band has been prescribed—borrowing values from experimental evidence—to obtain results plotted in Figs. 1.19 and 1.20.

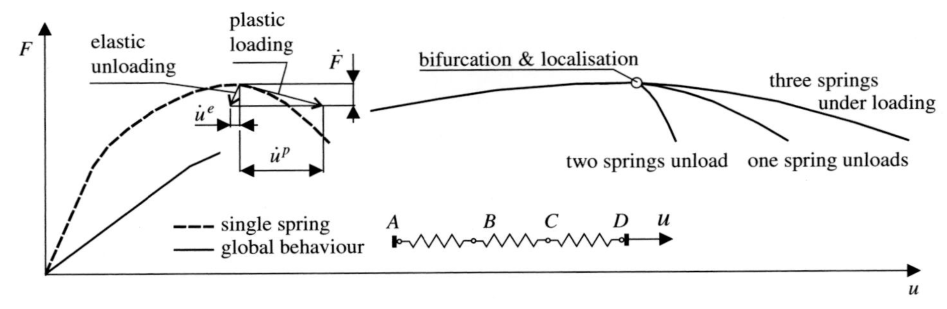

Figure 1.21. A simple three-spring model exhibiting a global behaviour substantially different from the behaviour of the single spring forming the system. Note that elastic unloading may occur at localisation.

> *size effect and softening are typically structural effects related to bifurcation and instability (possibly occurring within a microstructure), so that they do not represent the behaviour of a homogeneously deformed/stressed material element.*

1.6 Chains with softening elements

The exploitation of new materials exhibiting phase transformations has driven a strong research effort on the behaviour of elastic materials with non-monotone stress/strain curves corresponding to non-convex elastic potentials. In a pioneering paper, Ericksen (1975) has shown that when displacements are prescribed at the ends of a bar (so-called loading in a hard device) characterised by a non-monotone stress/strain curve with a spinodal region (shaped as in the upper part of Fig. 1.22), a number of configurations become possible, exhibiting jumps in strain across interfaces separating zones of uniform deformation.[16] Although our main interest is not the investigation of phase transformation processes, the analysis of concepts developed in that context[17] becomes instrumental in the understanding of size effects and multiple shear band formation, particularly because the spinodal stress/strain curve employed in these analyses involves softening and subsequent re-hardening. We focus attention on discrete systems made up of components each exhibiting softening that occurrs only when a certain stress level has been reached.

Let us begin with the simple example shown in Fig. 1.21, referred to as a three *elastoplastic* non-linear spring system (Tvergaard and Needleman, 1980; Bigoni and Zaccaria, 1993). Each spring obeys the non-linear elastoplastic constitutive law shown dashed in the figure. In particular, there is an initial linear and reversible behaviour, followed by a nonlinear elastoplastic behaviour (with different incremental response at loading and unloading), culminating at a 'peak', followed by softening so that there is no re-hardening. Now assume that imperfections are absent and that

[16] The formation of zones of different strain has been confirmed qualitatively by a number of experimental evidences, showing that martensitic phase transformation bands in NiTi shape memory alloy results from continued shear band formation (Shaw and Kyriakides, 1997).

[17] An interesting application is to dislocation models in metals, the so-called Frenkel-Kontorova formulation (Braun and Kivshar, 1998).

both the displacements at the end of the system are prescribed, one to be null and the other to be monotonically increasing.

Since the force in the three springs must remain the same by equilibrium, and employing the constitutive behaviour sketched dashed in the figure, seeking for incremental bifurcations, we conclude that the response of the system remains unique until the peak of the curve is reached. At that point, reached simultaneously by all the three springs, an incremental bifurcation may occur, so the following three behaviours become possible: (1) All the three springs continue to load, (2) one spring unloads and the other two continue to load, and (3) two springs unload and one continues to load. In all the three cases, we observe a softening in the global response of the system, but in cases 2 and 3, the softening becomes stronger (thus corresponding to a 'steeper' force/displacement curve). Cases 2 and 3 are examples of localization of deformation because the loading continues in only one or two springs. Which of the three branches will be followed by the structure cannot be predicted by a pure bifurcation analysis, but consideration of stability may provide some additional information. Assuming the requirement that the path that actually will be followed by the structure is that corresponding to the stronger softening, we find that localization with one spring subject to loading is the preferred mode. Note also that employing springs with softening stronger than that sketched in the figure, a so-called 'snap-back behaviour' also can be obtained, corresponding to a subvertical softening, in which instability occurs even under controlled end displacement, and bifurcation can occur before the peak of the curve is reached.

If in the constitutive law of the springs, after softening, a re-hardening is present, a more complicated situation may occur that can be illustrated with the simple example of a chains of *fully reversible*, in other words, *elastic,* tri-linear springs with a spinodal load/displacement curve (as in the upper part of Fig. 1.22). This constitutive behaviour, in the nomenclature of phase-transforming materials, consists of the three branches denoted as 'phases I, II and III'. The part of the graph where the force is a decreasing function of displacement (phase II) is the so-called spinodal region. Note that nonlinear springs with spinodal zones can be 'practically' realized by the structure shown in Fig. 1.17, also considered in Section 1.13.4.

Let us begin considering the case of prescribed end force on a single tri-linear spring, illustrated in Fig. 1.22 and described by the force/displacement (F–u) relation

$$F = \begin{cases} \dfrac{k}{a}u & 0 \leq u/a \leq 2 & \text{phase I} \\ -\dfrac{k\beta}{a}u + 2k(1+\beta) & 2 \leq u/a \leq 2\dfrac{1+\beta}{\beta} & \text{phase II} \\ \dfrac{k\gamma}{a}u - 2k\gamma\dfrac{1+\beta}{\beta} & 2\dfrac{1+\beta}{\beta} \leq u/a & \text{phase III} \end{cases} \tag{1.18}$$

and the (non-convex) elastic energy W

$$W = \begin{cases} \dfrac{k}{2a}u^2 & 0 \leq u/a \leq 2 & \text{phase I} \\ -\dfrac{k\beta}{2a}u^2 + 2k(1+\beta)u - 2ka(1+\beta) & 2 \leq u/a \leq 2\dfrac{1+\beta}{\beta} & \text{phase II} \\ \dfrac{k\gamma}{2a}u^2 - 2k\gamma\dfrac{1+\beta}{\beta}u + \theta ka & 2\dfrac{1+\beta}{\beta} \leq u/a & \text{phase III} \end{cases} \tag{1.19}$$

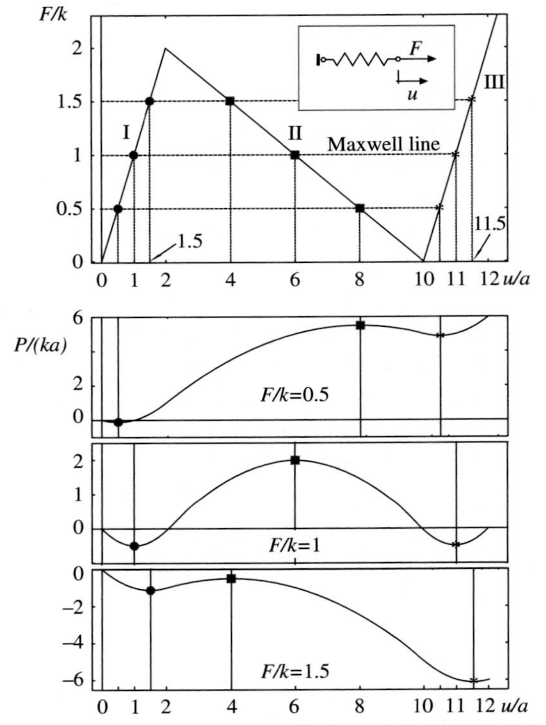

Figure 1.22. *Top*: A tri-linear elastic spring subject to a dead force F (note the three 'phases' I, II and III). *Bottom*: Potential energy P (normalised through division by ka) at different vales of applied force $F/k = \{0.5, 1, 1.5\}$. Note that the so-called Maxwell line splitting the graph into two equal area regions corresponds to $F/k = 1$. All states corresponding to phases I and III are at least locally stable, but the states below the Maxwell line on phase I and above it on phase III, such as the point $u/a = 0.5$ $(u/a = 11.5)$, are in an absolute minimum of potential energy, whereas other states in phases I and III are in a relative minimum. All states in phase II are unstable and correspond to a maximum of the potential energy.

where k, $k\alpha$ and $k\beta$ are the stiffnesses of the phases I, II and III, respectively; a is the displacement defining $F = k$; and θ can be selected to enforce continuity of the strain energy, so the example reported in Fig. 1.22 corresponds to $\beta = 1/4$, $\gamma = 1$ and $\theta = 60$.

Stability of the system at a given dead load can be decided on the basis of the potential enegry $P = W(u) - Fu$ so that a minimum (maximum) corresponds to stability (instability). A simple study of P, with W given by Eq. (1.19), shows that phase II is always unstable, whereas phases I and III may be 'locally' or 'globally' stable. 'Locally' ('globally') stable means that the equilibrium configuration corresponds to a relative (absolute) minimum of P. As sketched in Fig. 1.22, phase I (III) is globally stable only below (above) the so-called Maxwell line splitting the force/displacement graph into two equal area parts. States corresponding to absolute, or 'global', minima can be considered 'more stable' than states corresponding to relative, or 'local', minima, in the sense that for a fixed dead force, the disturbance needed to move the system from the former configuration to the latter is stronger than that needed to produce the opposite transition. States corresponding to local minima are often called 'metastable'.

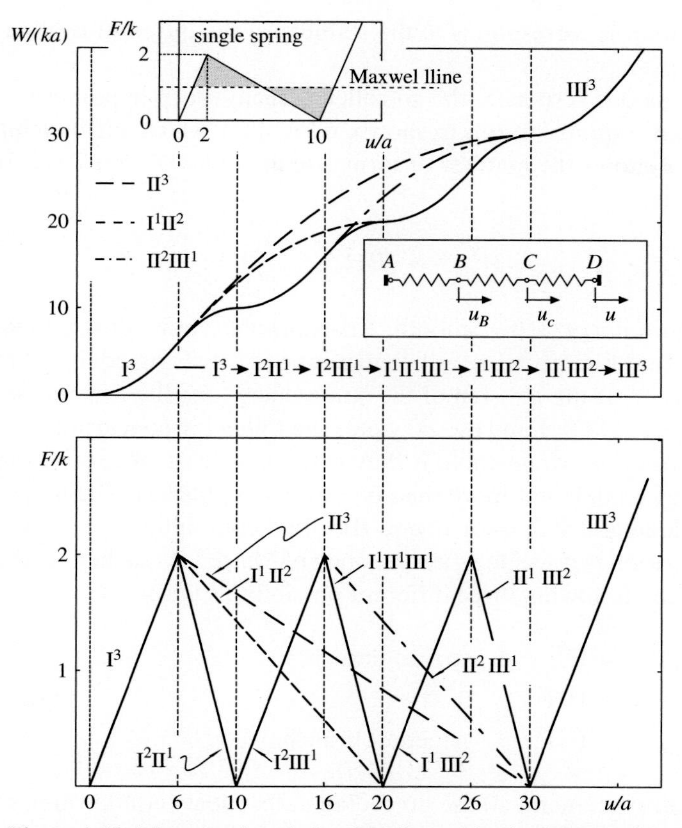

Figure 1.23. Three-spring chain of tri-linear elastic elements (shown in the inset on the upper part and identical to that shown in Fig. 1.22). Dimensionless elastic energy $W/(ka)$ and force F/k are reported for all configurations of the system satisfying equilibrium. The trivial path is $\mathrm{I}^3 \longrightarrow \mathrm{II}^3 \longrightarrow \mathrm{III}^3$, whereas that corresponding to the minimum energy is $\mathrm{I}^3 \longrightarrow \mathrm{I}^2\mathrm{II}^1 \longrightarrow \mathrm{I}^2\mathrm{III}^1 \longrightarrow \mathrm{I}^1\mathrm{II}^1\mathrm{III}^1 \longrightarrow \mathrm{I}^1\mathrm{III}^2 \longrightarrow \mathrm{II}^1\mathrm{III}^2 \longrightarrow \mathrm{III}^3$.

We consider now chains of three springs with prescribed end displacements. The first case is sketched in Fig. 1.23, where all springs obey the constitutive relation as in the case shown in Fig. 1.22.

For a chain of three springs with prescribed end displacement u, the total potential energy coincides with the elastic potential energy W, sum of the potential energies of the springs

$$W_{BA}(\Delta_{BA}) + W_{CB}(\Delta_{CB}) + W_{DC}(\Delta_{DC}),$$

functions of the elongations

$$\Delta_{BA} = u_B, \quad \Delta_{CB} = u_C - u_B, \quad \Delta_{DC} = u - u_C,$$

and the Hessian of the total potential energy results

$$\begin{bmatrix} \dfrac{\partial^2 W_{BA}}{\partial \Delta_{BA}^2} + \dfrac{\partial^2 W_{CB}}{\partial \Delta_{CB}^2} & -\dfrac{\partial^2 W_{CB}}{\partial \Delta_{CB}^2} \\ -\dfrac{\partial^2 W_{CB}}{\partial \Delta_{CB}^2} & \dfrac{\partial^2 W_{CB}}{\partial \Delta_{CB}^2} + \dfrac{\partial^2 W_{DC}}{\partial \Delta_{DC}^2} \end{bmatrix}, \tag{1.20}$$

so that its positive definiteness corresponds to the stability of the different configurations.

Assuming a homogeneous response, the so-called Cauchy-Born hypothesis, a tri-linear force-elongation response results as shown in Fig. 1.23, which, introducing the notation $\mathcal{I}^i$, where i denotes the number of springs in the phase $\mathcal{I}$, and $\mathcal{I}$ is I, II or III, can be described as

$$\mathrm{I}^3 \longrightarrow \mathrm{II}^3 \longrightarrow \mathrm{III}^3.$$

The corresponding strain energy is the smooth, 'tri-parabolic', non-convex function shown in Fig. 1.23. In the same figure, all configurations satisfying equilibrium are reported. Consideration of the Hessian of the strain energy for the constitutive equations (1.18) with $\beta = 1/4$, $\gamma = 1$ and $\theta = 60$ yields the following conclusions.

All possible configurations are unstable if they contain at least two springs in phase II and stable if they contain not more than one spring in phase II. The homogeneous response is unique and stable for I^3 and III^3. In a continuous quasi-static deformation path, bifurcation is possible when the apex of Phase I is reached, at the end of I^3. At this point, the following three different possibilities arise:

$$\begin{array}{ll} \mathrm{I}^3 \longrightarrow \mathrm{II}^3 & \text{trivial branch,} \\ \mathrm{I}^3 \longrightarrow \mathrm{I}^1\mathrm{II}^2 & \\ \mathrm{I}^3 \longrightarrow \mathrm{I}^2\mathrm{II}^1 & \text{steepest branch.} \end{array} \tag{1.21}$$

For prescribed end displacement, all the preceding paths satisfy equilibrium, so the question arises

What is the path followed by the system?

We can provide an answer by employing the definition of stability; thus, since II^3 and $\mathrm{I}^1\mathrm{II}^2$ are both unstable and $\mathrm{I}^2\mathrm{II}^1$ is stable, the steepest branch will be followed by the system, so

The system follows the path of minimum potential energy.

It is important to highlight that

the assumption of a definition of stability has consequences on the number of springs suffering increasing displacement (the volume fraction of deformation bands).

In fact, the situation $\mathrm{I}^2\mathrm{II}^1$ (involving one spring subject to elongation, the minimum band width) is privileged against the situation $\mathrm{I}^1\mathrm{II}^2$ (involving two springs subject to elongation). The preceding statement becomes more evident in the first example of Chapter 15, where a system of six springs is considered.

Increasing the deformation of the system, other bifurcation points are encoutered, but employing the notion of stability, it is not difficult to show that the path that will be followed is

$$\mathrm{I}^3 \longrightarrow \mathrm{I}^2\mathrm{II}^1 \longrightarrow \mathrm{I}^2\mathrm{III}^1 \longrightarrow \mathrm{I}^1\mathrm{II}^1\mathrm{III}^1 \longrightarrow \mathrm{I}^1\mathrm{III}^2 \longrightarrow \mathrm{II}^1\mathrm{III}^2 \longrightarrow \mathrm{III}^3,$$

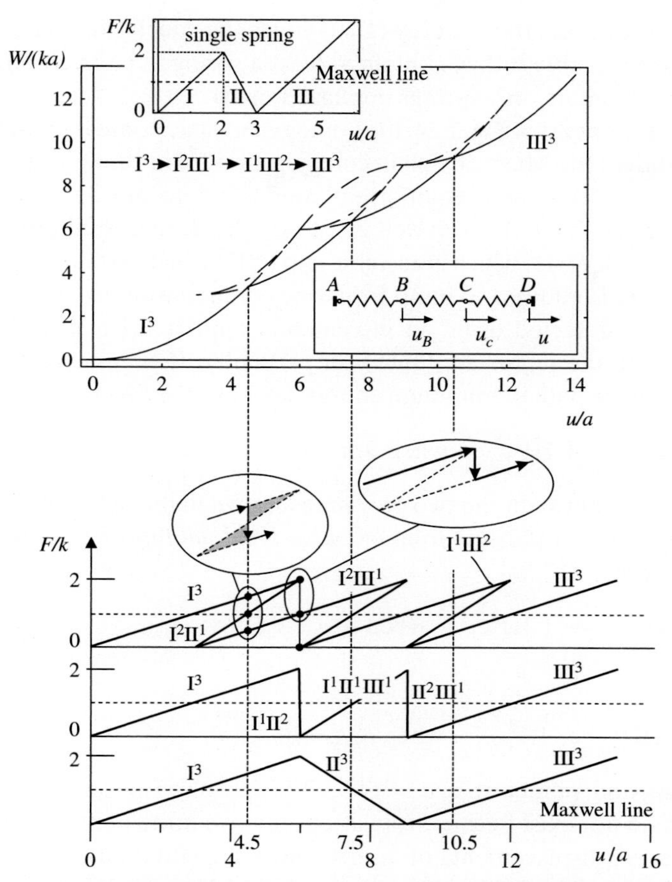

Figure 1.24. Three-spring chain of tri-linear elastic elements (shown in the inset). Dimensionless elastic energy $W/(ka)$ and force F/k are reported for all possible configurations of the system. The trivial path is $\mathrm{I}^3 \longrightarrow \mathrm{II}^3 \longrightarrow \mathrm{III}^3$, and the path corresponding to the minimum energy is $\mathrm{I}^3 \longrightarrow \mathrm{I}^2\mathrm{III}^1 \longrightarrow \mathrm{I}^1\mathrm{III}^2 \longrightarrow \mathrm{III}^3$.

always involving the attainment of absolute minima and corresponding to the steepest branches from the bifurcation points.

Note that since the response of the system has been found to always be continuous, the bifurcations in the preceding example can be detected using an incremental approach in which, at a given load and displacement, the response of the system is analysed for an incremental displacement. The fact that the response is always continuous is related to the number of springs and to the values of the constitutive parameters β and γ of the spring, so it may be instructive now to consider again a chain of three springs, but with $\beta = 2$, $\gamma = 1$ and $\theta = 15/2$, providing the trilinear spring with the behaviour shown in the inset of Fig. 1.24, where total elastic energy of the chain W (divided by ka) and the 'global' force/displacement relation $(F/k - u/a)$ are shown. The complication introduced in this new example is that the global force/displacement relation exhibits for certain configurations a re-entering branch, so-called 'snap-back', a feature which makes the incremental approach to bifurcation not always satisfactory.

Examination of the Hessian of strain energy (1.20) yields the conclusion that all possible configurations are unstable if they contain at least a spring in phase II and *at least locally stable* if they contain only springs in phase I or III.

With reference to the lower part of Fig. 1.24, the homogeneous response is unique until the load is below (above) the Maxwell line in configuration I^3 (in configuration III^3), situations in which no alternative equilibrium configurations are available. However, when the load rises above the Maxwell line in I^3, alternative equilibrium configurations become possible (I^2II^1 which is unstable and I^2III^1 which is stable) that can be reached if the system is allowed to jump from one configuration to another (a feature which cannot be detected using an incremental approach because the discontinuities involved with the jumps are finite). In particular, if we assume the rule that the system follows the path of minimum energy, we have the first transition

$$\text{I}^3 \longrightarrow \text{I}^2\text{III}^1, \qquad \text{when } u/a = 4.5,$$

corresponding to the equality between the two grey areas shown in the detail of Fig. 1.24. This transition occurs at the *transition from global to local stability of* $\textit{I}^3$. Later, we find the second transition

$$\text{I}^2\text{III}^1 \longrightarrow \text{I}^1\text{III}^2, \qquad \text{when } u/a = 7.5,$$

and the final jump

$$\text{I}^1\text{III}^2 \longrightarrow \text{III}^3, \qquad \text{when } u/a = 10.5.$$

Note that all these transition points at $u/a = \{4.5,\ 7.5,\ 10.5\}$ are points where incremental bifurcations do not exist because transitions involve finite jumps in the longitudinal force and in the displacements of internal nodes, a situation different from the example reported in Fig. 1.23, where incremental bifurcations occur at $u/a = \{6, 16\}$.

The preceding described situation changes if perturbations to the system are absent, so jumps do not occur initially because the system follows a stable (below the first transition point) and later (above the first transition point) a meta-stable I^3 path until the apex of branch I^3, occurring at $u/a = 6$. At this point, the following different possibilities arise:

$$\begin{array}{lll} \text{I}^3 \longrightarrow \text{II}^3 & \text{trivial branch,} & \\ \text{I}^3 \longrightarrow \text{I}^1\text{II}^2 & \text{vertical branch,} & \\ \text{I}^3 \longrightarrow \text{I}^2\text{II}^1 & \text{re-entering branch,} & \qquad (1.22) \\ \text{I}^3 \longrightarrow \text{I}^2\text{III}^1 & \text{1st stable branch,} & \\ \text{I}^3 \longrightarrow \text{I}^1\text{III}^2 & \text{2nd stable branch.} & \end{array}$$

For prescribed end displacement, the re-entering branch cannot be followed, and the vertical and trivial branches are unstable. If we look for incremental bifurcations, there are only unstable paths available (I^1II^2 and II^3), so if the system follows one of these, it will tend to move dynamically to another configuration, which can be reached only admitting a finite jump in the axial force and internal nodal displacements. The assumption that the response of the system remains continuous becomes unrealistic, and we have to allow the possibility of jumps to bring the system into one of the two

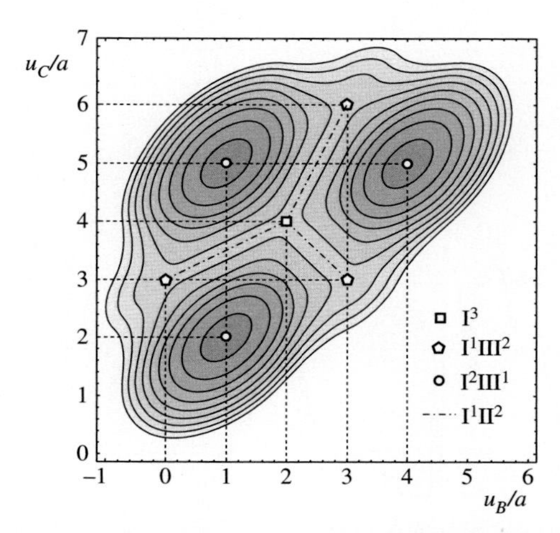

Figure 1.25. Level sets of the elastic energy for the three-spring chain of tri-linear elastic elements (shown in the inset of Fig. 1.24) at an end displacement $u/a = 6$ as a function of the dimensionless displacements of the internal nodes u_B/a and u_C/a. Note the three equivalent minima corresponding to I^2III^1 and that configurations I^3 and I^1III^2 and path I^1II^2 correspond to the same level of energy.

stable configurations, I^2III^1 or I^1III^2. Again, if we assume that the system follows the path of minimum energy, I^2III^1 is privileged, but this is not the only possibility. We have reported in Fig. 1.25 the level sets of the elastic energy of the system for an end displacement $u/a = 6$ as a function of the displacement of internal nodes u_B and u_C. We may see that there are three equivalent minima {1,2}, {1,5} and {4,5} corresponding to the three possible configurations I^2III^1 and that all configurations I^3, I^1III^2 and I^1II^2 correspond to the same level of energy, so I^3 and I^1III^2 are connected through a contour level I^1II^2, defining an unstable path. To decide what path will be followed by the system is impossible without recurring to a dynamical analysis, but this requires a definition of masses and damping, so at the present level of description, only 'reasonable conjectures' are possible. In closure of this example, we address the interested reader to the exhaustive treatment by Puglisi and Truskinovsky (2002, 2002); see also Puglisi (2006), where nonlocal interactions are also considered.

Structures behaving as chains with softening elements can be realised with simple experimental settings, for instance, by subjecting to uniaxial compression a hollow metallic cylinder (Bardi et al., 2003), or a stack of aluminium cans, or finally, subjecting to uniaxial tension a corrugated tube. We report experimental results obtained by us and designed to reproduce the behaviour of chains with softening elements, as reported in the examples of Figs. 1.21 and 1.24. In particular, we have subjected to prescribed end displacements stacks of aluminium cans and corrugated tubes,[18] providing compressive load in the former case and tensile in the latter.

[18] We have used 250-ml empty cans of Fanta, manufactured by the Coca-Cola company (externally painted to hide the company's name in the photos), and Pop Toobs, sold as toys from Slinky Toys, Inc.

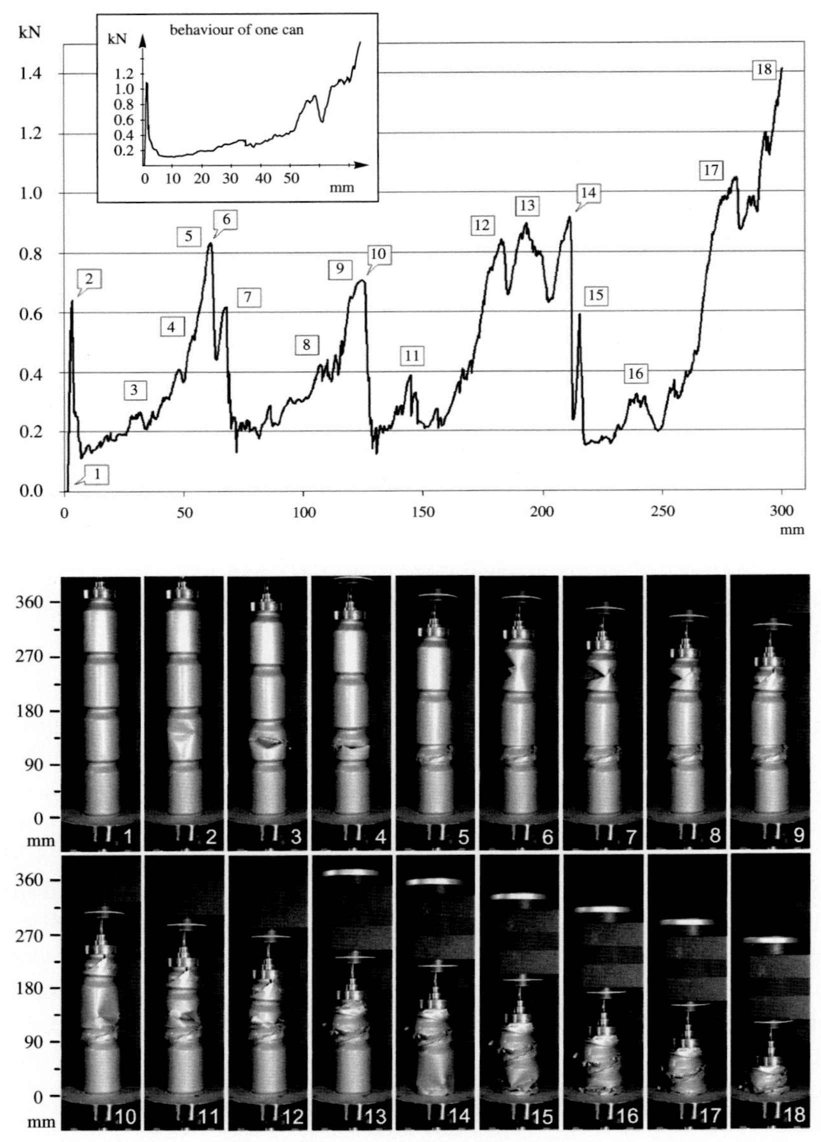

Figure 1.26. Behaviour of a stack of four cans subject to uniaxial compression. *Top*: The global force/displacement response (the behaviour of a single can is shown in the *inset*). *Bottom*: The first can to collapse is the second (photo 2) from below, followed by the fourth (photo 6), the third (photo 10) and finally, the first (photo 14). The fact that *one* can collapses while the others unload is an example of strain localisation. Since all cans collapse at different stage of global deformation, this is a form of multiple localisation, yielding an oscillation in the applied load (*top*).

Let us begin with the behaviour of the stack of cans shown in Fig. 1.26. The single can essentially behaves as an axially loaded thin-walled tube, a structure with a well-known unstable post-critical behaviour (see, e.g., Hutchinson and Koiter, 1970). Its response is shown in the inset of Fig. 1.26, where after an initially linear range, a strong softening is evidenced, followed at high deformation by re-hardening. Note that the force/displacement curve is not regular because the can 'scrunches up', giving rise to continuous stress oscillations. Stacks made up of three, four and five cans have been tested, all giving a qualitatively similar behaviour. We report in Fig. 1.26 results relative to a pile of four cans. We clearly see that the system qualitatively behaves as those previously analysed (although now the behaviour is elastoplastic), so the global load/displacement curve increases monotonically up to the first can collapses, which induces softening, and this softening continues up to a re-hardening, terminating at the collapse of the second can.

A behaviour similar to the stack of cans is that obtained during a *tensile* test of the plastic corrugated tube reported in Fig. 1.27. Here, the stress oscillations are much more regular and resemble those evidenced by the simple models analysed previously (although the corrugated tube, as the stack of cans, evidences an inelastic deformation).

1.7 Shear band saturation and multiple shear banding

Knowledge of the behaviour of chains with softening elements facilitates an understanding of band saturation and multiple shear banding phenomena. In fact, the re-hardening mechanism induces multiple localisation of deformation, as clearly evidenced by the experiments shown in Figs. 1.26 and 1.27 and the results from the simple models of Figs. 1.21 and 1.24 (see also Chapter 15). For two-dimensional deformation of an elastoplastic continuum, we refer now to results obtained by Gajo et al. (2004), summarised in Chapter 15. We consider the simulation of a biaxial compression test (namely, a plane strain compression test with confining stress) on dry, dense sand, for which 'global' nominal stress versus conventional strain (thus plotted as if the deformation were uniform) is reported in Figs. 1.19 and 1.20. In both cases, the formation of a single shear band terminates the hardening and precipitates the response in a softening regime. However, this behaviour is not the only observed. In fact, for certain values of constitutive parameters and confining pressure, a multiple shear band formation occurs, in the sense that several shear bands may form subsequently in a manner similar to the accumulation of localisation of deformation in the case of the stack of cans and of the corrugated tube. In certain cases, after a few repetitions of localisation, a differently inclined shear band occurs in the material transformed by subsequent shear banding and yields a sharp softening (a deformation mechanism explained in Chapter 15), whereas in other cases the shear band accumulation becomes 'persistent', thus yielding to the situation reported in Fig. 1.28 (referred to as a biaxial compression test on loose sand and taken from Gajo et al., 2004). The subsequent formation of shear bands induces a stress oscillation in the nominal stress versus conventional strain curve, as shown in the detail reported in Fig. 1.28. Although not completely confirmed experimentally in sand, the described behaviour is well recognised in ductile metals (Hall, 1970), in foams (Moore et al., 2006) and in shape memory alloy strips (Shaw and Kyriakides, 1997).

More complex materials than the uni-dimensional structures shown in Figs. 1.26 and 1.27, developing two-dimensional deformation, are provided by the failure mechanism of cellular materials (Shaw and Sata, 1966; Stronge and Shim, 1988; Papka and Kyriakides, 1999; Gong and Kyriakides, 2005; Moore et al., 2006) or the packaging of hollow cylinders (drinking straws) analysed by Poirier et al. (1992). The latter case

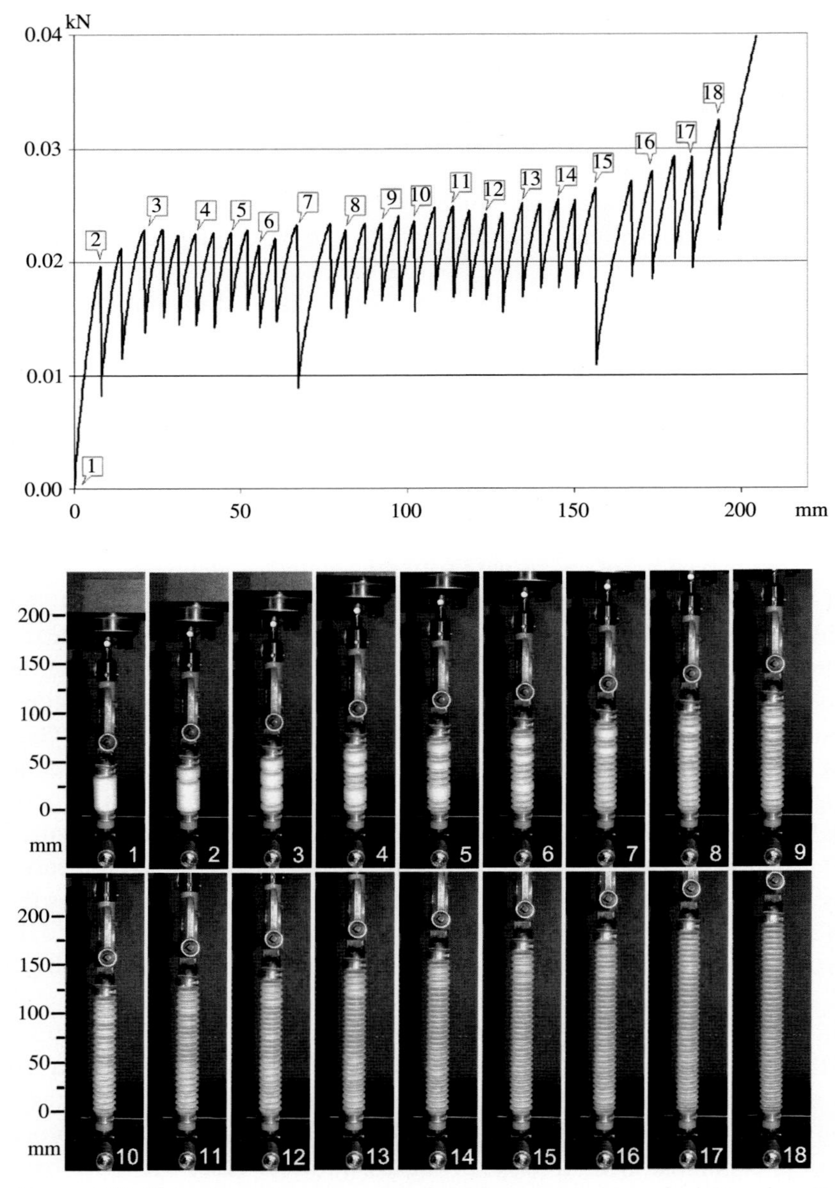

Figure 1.27. Behaviour of a corrugated plastic tube subject to uniaxial tension. *Top*: The global force/displacement response. *Bottom*: Photographs of the tube at different elongations. The fact that corrugations snap at different instants and not all together can be seen as an example of localisation of deformation. Note the mechanism of multiple localisation of deformation and the 'band accumulation' with related load oscillation.

is particularly interesting because the material behaves as a granular material (or, more precisely, as a sort of so-called Schneebeli material, Schneebeli, 1956) but with highly deformable (and strongly nonlinear) 'grains'.

Results of experiments are reported in Figs. 1.29 and 1.30 for compression of packing of drinking straws (21 cm, $\phi = 7$ mm, purchased at Leone s.r.l., art. B922.Z) in

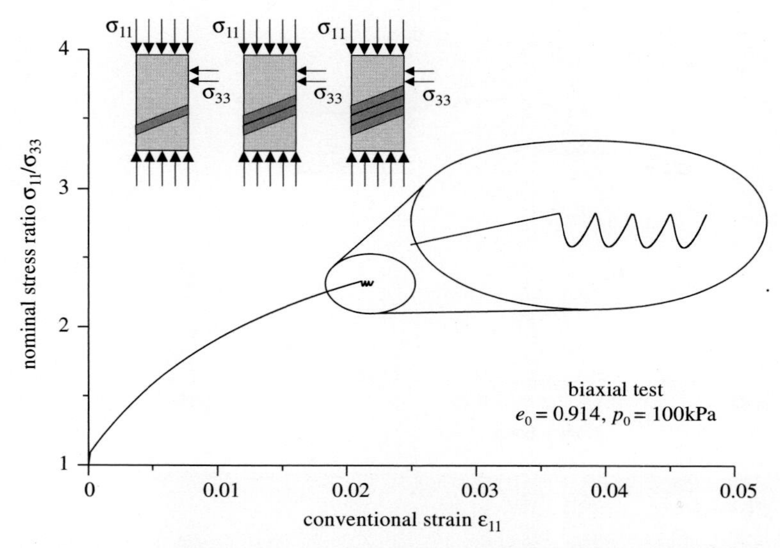

Figure 1.28. Simulation showing progressive accumulation of shear bands, giving rise to a continued stress oscillation in the global nominal stress versus conventional strain response (the simulation is referred to a biaxial, plane strain compression test on dry loose sand; e_0 indicates the initial void ratio and p_0 the confining pressure; figure adapted from Gajo et al., 2004). Compare with the simulation shown in Fig. 15.3.

a uniaxial strain device conceptually identical to that shown in Fig. 1.15 but designed to be used in a displacement-controlled electromechanical testing machine (we have used an ELE Tritest 50 from ELE International, Ltd.).

In both experiments, the nominal stress/conventional strain curves evidence a moderate softening, with strain localisation initiating in one case during softening (Fig. 1.29) and in the other case early in the hardening (Fig. 1.30). Strain localisation has been detected by visual inspection, which makes awkward the precise identification of the threshold. In the case of Fig. 1.30, an early bifurcation mode corresponding to cross-sectional ovalisation of straws[19] has been identified with strain localisation.

The nearly flat stress/strain response occurring after the peak and evidencing stress oscillation is the effect of progressive accumulation of localisation bands, an effect already encountered in the experiments shown in Figs. 1.26 and 1.27 and explained through the simulations shown in Figs. 1.24, 1.28, 15.1 and 15.3.[20]

1.8 Brittle and quasi-brittle materials

Bifurcation and instability phenomena are clearly related to failure in ductile materials, typically metals, granular materials, soils and plastics. Different from brittle materials, where failure is dominated by fracture mechanics, quasi-brittle materials

[19] A somehow similar bifurcation pattern with ovalisation of initially circular voids has been found by Mullin et al. (2007) and Michel et al. (2007).

[20] An alternative mechanism of formation of multiple bands of shear deformation in ductile metals, in which many parallel bands develop simultaneously (and not necessarily sequentially), has been revealed theoretically by Petryk and Thermann (2000, 2002).

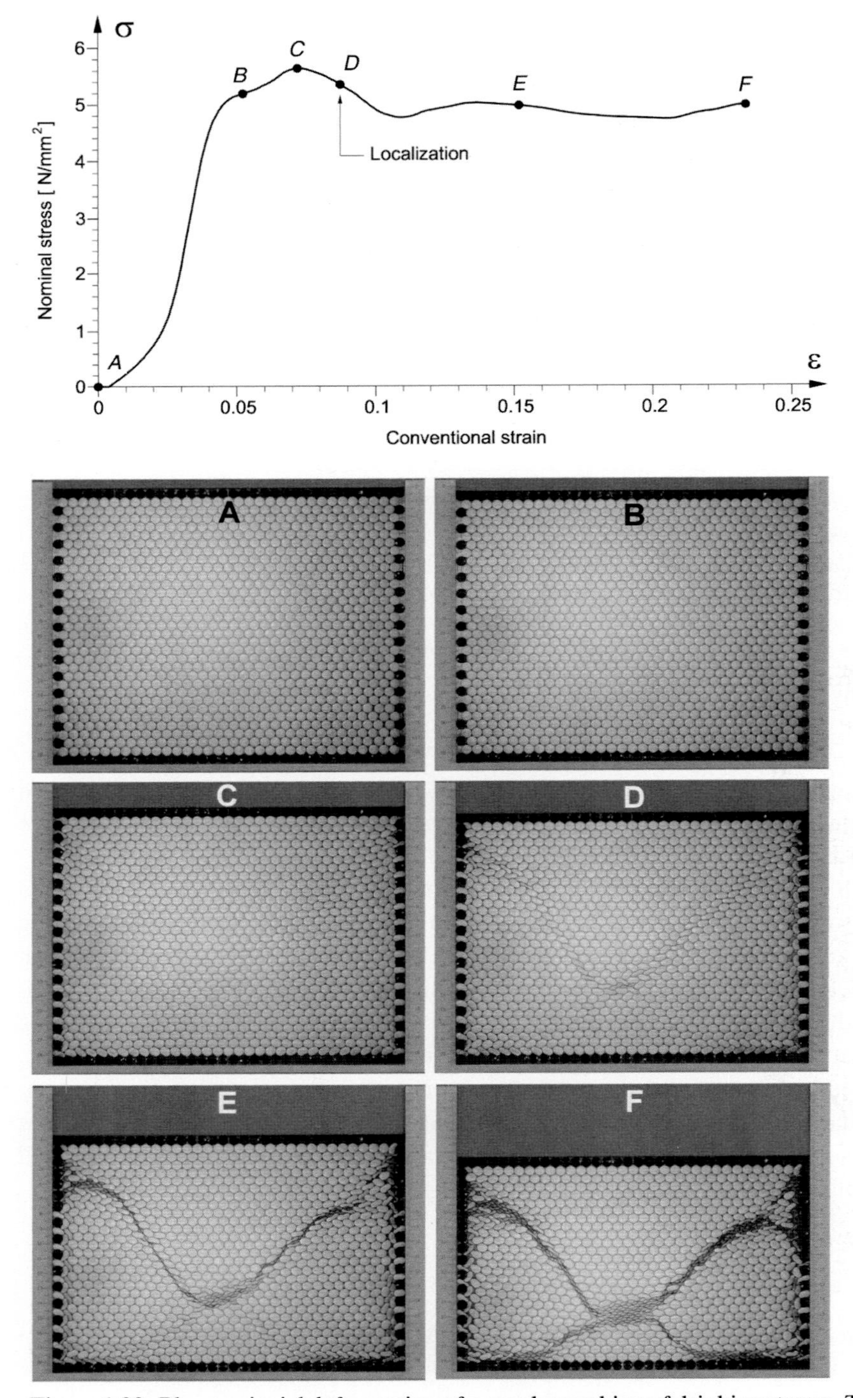

Figure 1.29. Plane uniaxial deformation of a regular packing of drinking straws. *Top*: Nominal stress versus conventional strain behaviour. *Bottom*: Starting from the unloaded state (photo A), initially the response is homogeneous (photos B and C). Strain localisation is identified when straws collapse and initiates during softening (photo D) and continues with multiple band formation and accumulation (photos E and F).

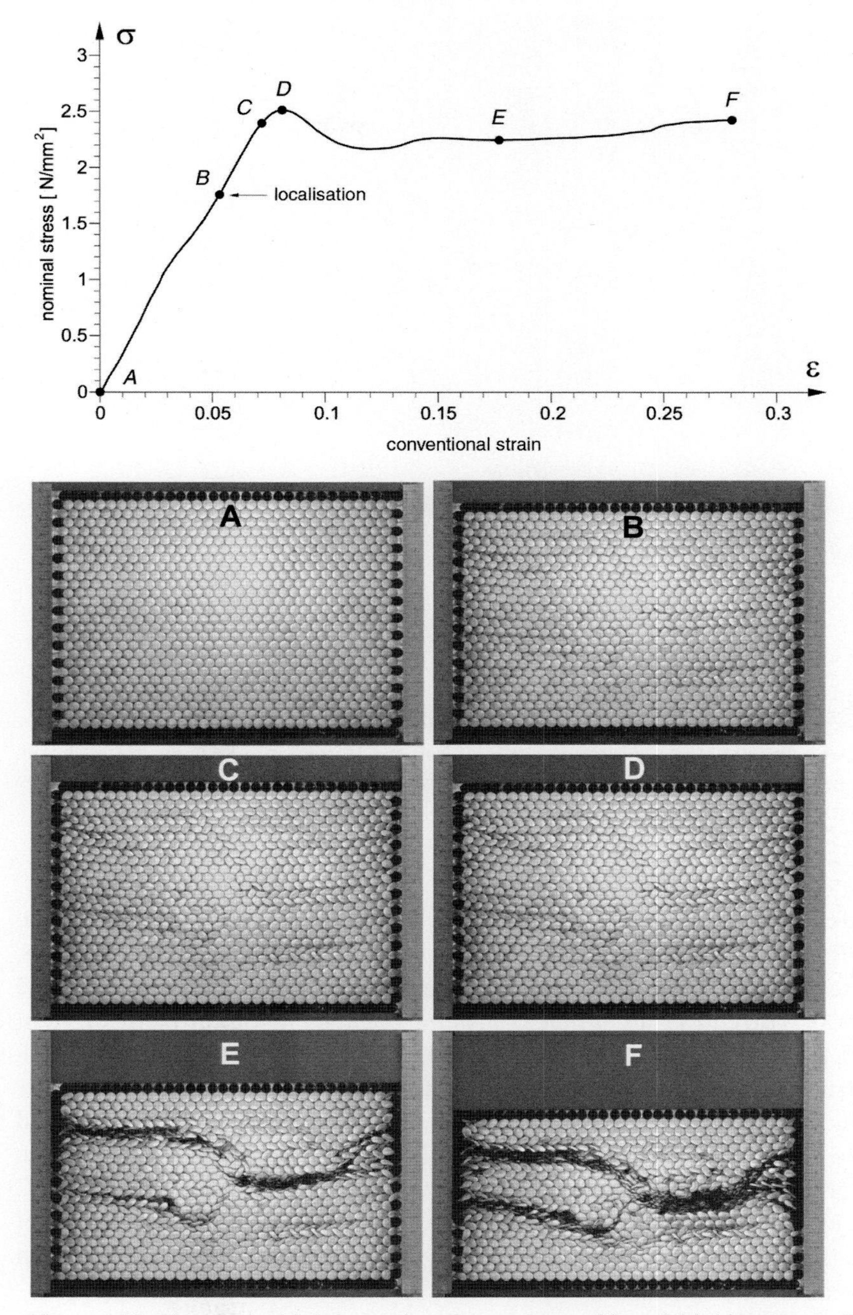

Figure 1.30. Plane uniaxial deformation of a regular packing of drinking straws. *Top*: Nominal stress versus conventional strain behaviour. *Bottom*: Photo A identifies the initial unloaded state. Strain localisation is identified with an early ovalisation of different arrays of straws (photo B), a mechanisms (still evident in photos C and D) different from straw collapse, evident at the end of the test (photos E and F, in which an accumulation of subsequent localised bands is clearly visible).

Figure 1.31. An ancient stone column ('cortile di Palazzo Bonet', Via S. Anna, 21 Palermo, Italy) evidencing a typical failure mode in compression through localisation of damage (compare with Fig. 1.10).

are a broad class of materials (including rock, concrete and ice) where *damage* localisation in different forms (microcracking, void closure and collapse, sliding on fissures, interfaces separation, aggregates segregation) is the usual failure mode (Bazant and Mazars, 1990; Torrenti et al., 1991).[21] From a phenomenological point of view, damage is a form of inelastic deformation, so that, since damage theory has a mathematical structure very similar to elastoplasticity, concepts developed for the latter can be immediately applied to the former (Benallal et al., 1988, 1993; Benallal and Bigoni, 2004). Examples of failure of quasi-brittle materials are the column shown in Fig. 1.31 and the silicon nitride cylinder subject to uniaxial compression at 1200 °C in air (Gei et al., 2004). In the former example, we do not know the exact cause of failure, but the inclined fracture plane is typical of a failure under axial compression through localisation of damage.

In the latter example, at 1200 °C, the silicon nitride is much less brittle than at room temperature, so when it is subjected to uniaxial compression, it reveals a stress/strain curve evidencing a softening (Fig. 1.32, on the left). Several experiments have been performed on this material, testing different aspect ratios of the cylinders,

[21] Several factors affect the behaviour of materials and may imply transitions between brittle, quasi-brittle and ductile deformation. The most known of these mechanisms is certainly temperature, but time of load application and mean pressure are other important factors. Moreover, sometimes a material may behave in a brittle way under tensile load, evidencing instead some ductility for compressive load.

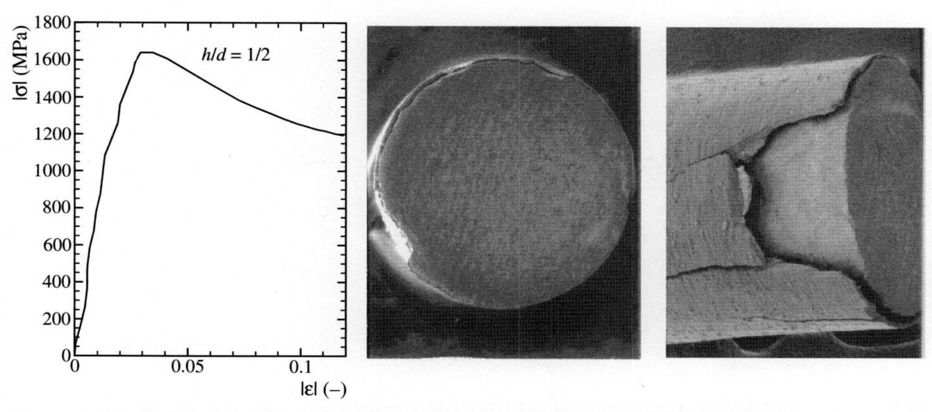

Figure 1.32. *Left*: Global nominal stress versus conventional strain curve for a silicon nitride cylinder (diameter 2 mm, height 4 mm) subject to uniaxial compression at 1200 °C in air. The behaviour is not brittle, so a softening is evident, which was found to be the effect of surface exfoliation of the sample shown in the centre and on the right; see Gei et al. (2004) for further details.

with great care in reducing the friction at the sample ends. It has been shown that the softening is the effect of a progressive surface exfoliation mechanism (the initial phase of this mechanism is visible in Fig. 1.32, where the test has been interrupted after the first exfoliated layer had detached) which can be explained as the evolution of a bifurcation pattern for which tensile stresses develop. The behaviour found for the silicon nitride cylinders is not peculiar, but it has been found in other quasi-brittle materials (in rock, by Vardoulakis and Sulem, 1995, their Fig. 1.2.6; in concrete, by Hudson et al., 1971) and represents one example of how bifurcation theory can explain failure even beyond the realm of ductile materials. This example also shows that *failure is not always directly connected to strain localisation, but other failure mechanisms may prevail.*

1.9 Coulomb friction and non-associative plasticity

Granular materials and, more generally, quasi-brittle materials have a strongly inelastic behaviour in which the inelastic deformation may be produced by several micro-mechanisms, such as sliding between grains or aggregates or at fissures, modifications in pore sizes, opening/closure of fractures and micro-structural rearrangements. Among these, the chief mechanism is certainly related to the friction occurring at micro-contacts. All these micro-mechanisms produce a collective behaviour evidencing *pressure sensitivity of yielding* (in other words, an increase with the mean pressure of shear stress producing yielding) and *dilatant* or *contractant inelastic deformation*. An appropriate consideration of these constitutive features leads to the so-called model of non-associative elastoplasticity (introduced by Mróz, 1963; 1966; Mandel, 1966a; Maier, 1970a), which is the continuous mechanics counterpart of the model of contact with Coulomb friction.

The striking analogy between the equations governing contact with Coulomb friction and the constitutive equations of non-associative elastoplasticity is revealed by the simple model illustrated below.

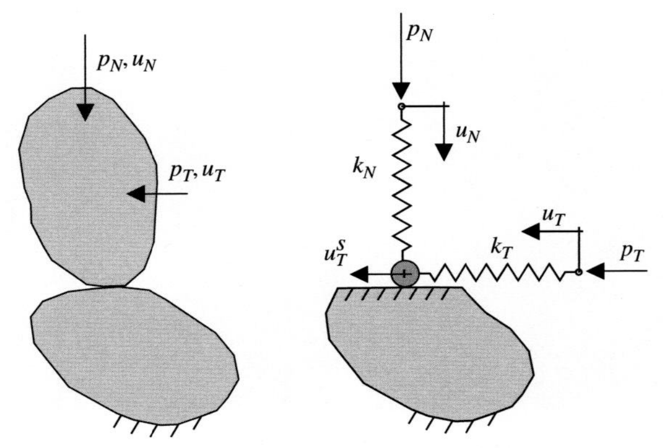

Figure 1.33. Contact between two grains (*left*); a simple model of this, involving Coulomb friction (*right*).

A typical contact between two grains is sketched on the left in Fig. 1.33, which is idealised with the simple model reported on the right in Fig. 1.33. In particular, a point mass capable of moving only in the horizontal direction is attached to two springs of stiffness k_N and k_T. The vertical movement u_N of the edge of the vertical spring (subject to the compressive force p_N) is purely reversible (and due to the spring deformation only), whereas the displacement u_T of the edge of the horizontal spring (subject to the tangential force p_T) is the sum of the reversible deformation of the spring itself plus a possible slip u_T^s of the point mass on the rigid horizontal constraint. We limit the analysis to the condition of *contact* (so that we do not consider separation and so-called grazing; see Radi et al., 1999), corresponding to a compressive (assumed positive) normal force $p_N > 0$, so the:

Coulomb friction condition (playing the role of the yield condition in plasticity) holds

$$f(p_N, p_T) = |p_T| - \mu p_N \leq 0, \tag{1.23}$$

where μ is the friction coefficient; a geometrical interpretation of criterion (1.23) is given in Fig. 1.34.

When condition (1.23) is satisfied, there are two possibilities:

- *Stick*, namely, $\dot{u}_T^s = 0$ (corresponding to elastic behaviour in plasticity), occurring when

$$|p_T| - \mu p_N < 0, \tag{1.24}$$

so that the corresponding behaviour is reversible and governed by the linear relation

$$\boldsymbol{p} = \boldsymbol{E}\boldsymbol{u} \qquad \begin{bmatrix} p_N \\ p_T \end{bmatrix} = \begin{bmatrix} k_N & 0 \\ 0 & k_T \end{bmatrix} \begin{bmatrix} u_N \\ u_T \end{bmatrix}, \tag{1.25}$$

or

- *Slip* $\left(\dot{u}_T^s \neq 0\right)$ or *stick* $\left(\dot{u}_T^s = 0\right)$ (corresponding to plastic flow or elastic unloading in plasticity), occurring when

$$|p_T| - \mu p_N = 0. \tag{1.26}$$

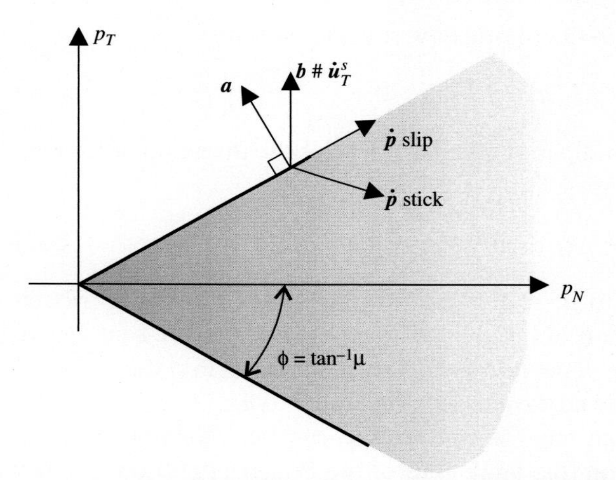

Figure 1.34. Coulomb friction condition, with normal $\boldsymbol{a}$ and direction of slip $\boldsymbol{b}$ (the symbol $\sharp$ means 'parallel'). Points $\{p_N, p_T\}$ inside the criterion (*grey zone*) correspond to stick, whereas for points at the boundary of the criterion, stick or slip may occur, depending on the direction of the rate $\dot{\boldsymbol{p}} = \{\dot{p}_N, \dot{p}_T\}$; in particular, stick (slip) corresponds to $\dot{\boldsymbol{p}} \cdot \boldsymbol{a} < 0$ $(= 0)$. Note that the fact that $\boldsymbol{a}$ is not parallel to $\boldsymbol{b}$ implies that the model does not obey the so-called normality (or associativity) rule, which is $\boldsymbol{a} \sharp \boldsymbol{b}$, so the model is 'non-associative'.

In particular, we have stick (or slip) if the rate of force is 'directed' inward (or tangential to) the friction condition (Fig. 1.34), namely,

$$\boldsymbol{a} \cdot \dot{\boldsymbol{p}} < 0, \quad \Longrightarrow \text{stick} \qquad \text{or} \qquad \boldsymbol{a} \cdot \dot{\boldsymbol{p}} = 0, \quad \Longleftarrow \text{slip}, \tag{1.27}$$

where $\boldsymbol{a}$ is the gradient of the left-hand expression in the friction condition

$$[\boldsymbol{a}] = \left[\frac{\partial f}{\partial \boldsymbol{p}} \right] = [-\mu, \mathsf{sign} p_T]^T. \tag{1.28}$$

For stick behaviour, the response is purely reversible, namely,

$$|p_T| - \mu p_N = 0 \qquad \text{and} \qquad \boldsymbol{a} \cdot \dot{\boldsymbol{p}} < 0, \quad \Longrightarrow \quad \dot{\boldsymbol{p}} = \boldsymbol{E} \dot{\boldsymbol{u}} \tag{1.29}$$

For slip behaviour, when (1.26) and $(1.27)_2$ are satisfied, the rate equations for contact can be obtained from the following three assumptions.

1. An *additive decomposition of slip into reversible and irreversible rates* (analogous to the decomposition into elastic and plastic rates in plasticity)

$$\dot{\boldsymbol{u}} = \dot{\boldsymbol{u}}^r + \dot{\boldsymbol{u}}^s, \tag{1.30}$$

   where the dot denotes derivative with respect to a time-like parameter governing the loading program.
2. The rule that *rate of force is related to the reversible rate of slip* (analogous to the rule that rate of stress is related to elastic deformation rate through the elastic fourth-order tensor in plasticity)

$$\dot{\boldsymbol{p}} = \boldsymbol{E} \dot{\boldsymbol{u}}^r. \tag{1.31}$$

3. The *slip rule* (analogous to the plastic flow rule in plasticity)

$$\dot{\boldsymbol{u}}^S = \dot{\lambda}\boldsymbol{b} \quad \dot{\lambda} \geq 0, \tag{1.32}$$

where $\dot{\lambda}$ is a non-negative slip multiplier and $\boldsymbol{b}$ rules the direction of the rate of slip, so in the case of Fig. 1.33,

$$[\boldsymbol{b}] = [0, \mathsf{sign}\, p_T]^T. \tag{1.33}$$

Note that $\boldsymbol{b}$ is not parallel to $\boldsymbol{a}$, so it is not normal to the Coulomb friction condition in the representation of Fig. 1.34. The model is therefore said 'not to follow the normality rule' or to be 'non-associative', meaning that the so-called associativity or normality rule corresponds to $\boldsymbol{b}$ parallel to $\boldsymbol{a}$.

The rate constitutive equations for contact slip can be obtained from the so-called consistency equation [the analogous of the Prager (1949) consistency condition in plasticity], expressing the fact that since the force cannot ever violate the condition of friction, slip may occur only when the rate of force is tangential to the friction condition, Eq. $(1.27)_2$

$$\boldsymbol{a} \cdot \dot{\boldsymbol{p}} = 0, \quad \Longleftrightarrow \boldsymbol{a} \cdot \left(\boldsymbol{E}\dot{\boldsymbol{u}} - \dot{\lambda}\boldsymbol{E}\boldsymbol{b}\right) = 0, \tag{1.34}$$

from which $\dot{\lambda}$ can be calculated so that

$$|p_T| - \mu p_N = 0 \qquad \text{and} \qquad \boldsymbol{a} \cdot \dot{\boldsymbol{p}} = 0, \quad \Longrightarrow \quad \dot{\lambda} = \frac{\boldsymbol{a} \cdot \boldsymbol{E}\dot{\boldsymbol{u}}}{\boldsymbol{a} \cdot \boldsymbol{E}\boldsymbol{b}}, \tag{1.35}$$

and the slip/stick condition becomes

$$\frac{\boldsymbol{a} \cdot \boldsymbol{E}\dot{\boldsymbol{u}}}{\boldsymbol{a} \cdot \boldsymbol{E}\boldsymbol{b}} \begin{cases} > 0 & \text{slip} \\ < 0 & \text{stick} \end{cases} \tag{1.36}$$

whereas the transition condition $\boldsymbol{a} \cdot \boldsymbol{E}\dot{\boldsymbol{u}} = 0$ represents so-called neutral loading.

Assuming $\boldsymbol{a} \cdot \boldsymbol{E}\boldsymbol{b} > 0$, the following rate constitutive equations for contact are finally deduced

$$\dot{\boldsymbol{p}} = \begin{cases} \boldsymbol{E}\dot{\boldsymbol{u}}, & \text{if} \quad |p_T| - \mu p_N < 0, \\ \boldsymbol{E}\dot{\boldsymbol{u}} - \dfrac{\langle \boldsymbol{a} \cdot \boldsymbol{E}\dot{\boldsymbol{u}} \rangle}{\boldsymbol{a} \cdot \boldsymbol{E}\boldsymbol{b}} \boldsymbol{E}\boldsymbol{b} & \text{if} \quad |p_T| - \mu p_N = 0, \end{cases} \tag{1.37}$$

and in the case considered in Fig. 1.33, they become the *rate equations of contact with Coulomb friction*

$$\dot{p}_N = k_N \dot{u}_N, \quad \dot{p}_T = k_T \dot{u}_T - \langle -\mu k_N \dot{u}_N + k_T \dot{u}_T \,\mathsf{sign}\, p_T \rangle \,\mathsf{sign}\, p_T. \tag{1.38}$$

The operator $\langle \cdot \rangle$ in Eqs. (1.37) and (1.38) is the so-called Macaulay bracket operator, Eq. (2.103), defined for every $\alpha \in \mathbb{R}$ as $\langle \alpha \rangle = (|\alpha| + \alpha)/2$.

The Macaulay bracket operator provides *the rate piece-wise linearity (a simple form of incremental non-linearity, distinguishing between plastic loading and elastic unloading) typical of elastoplasticity.*

Equations (1.37) are formally identical to the rate constitutive equations of non-associative ideal (i.e., with null hardening) elastoplasticity. In the equations of elastoplasticity, $\dot{\boldsymbol{p}}$ is replaced by the rate of stress, $\dot{\boldsymbol{u}}$ by the deformation rate, $\boldsymbol{a}$ by the yield function gradient, $\boldsymbol{b}$ by the plastic flow mode tensor and $\boldsymbol{E}$ by the fourth-order elastic tensor. Exactly as in the rate constitutive equations of non-associative elastoplasticity, it turns out in the problem of contact with friction that

- The contact condition (1.37) is written *in a rate form and cannot be resolved into a form involving finite quantities* (to understand this important point, it suffices to consider that the knowledge of a finite displacement u_T at a given value of vertical force p_N does not determine the tangential force p_T because the irreversible part of displacement u_T^s is not known, and this can be obtained only through integration of the rate equations).
- The rate equations are incrementally non-linear and characterised by an elastic

$$\dot{\boldsymbol{p}} = \boldsymbol{E}\dot{\boldsymbol{u}}, \qquad (1.39)$$

and a plastic

$$\dot{\boldsymbol{p}} = \boldsymbol{C}\dot{\boldsymbol{u}}, \qquad \boldsymbol{C} = \boldsymbol{E} - \frac{1}{\boldsymbol{a} \cdot \boldsymbol{E}\boldsymbol{b}} \boldsymbol{E}\boldsymbol{b} \otimes \boldsymbol{E}\boldsymbol{a}, \qquad (1.40)$$

branch [the symbol $\otimes$ is the dyadic product, defined by Eq. (2.14)].
- *The constitutive tensor **C** characterising the plastic branch is not symmetric*; therefore, the structure of problems involving friction is not self-adjoint. Note that the fact that $\boldsymbol{C}$ is not symmetric follows from the difference between $\boldsymbol{a}$ and $\boldsymbol{b}$. The former vector is *normal* to the friction criterion (see Fig. 1.34), whereas the latter is not. Therefore, the model lacks 'normality' or, in other words, is 'non-associative', in the sense that the slip rule 'associated' to the friction criterion requires $\boldsymbol{b}$ to be parallel to $\boldsymbol{a}$.

Note that while pressure sensitivity and non-associativity are both consequent to the frictional behaviour, softening is a mechanism independent of friction and therefore also of pressure sensitivity and non-associativity.

1.10 Non-associative flow rule promotes material instabilities

The pioneers of elastoplasticity theory were interested mainly in modelling the behaviour of metallic materials, which, broadly speaking, do not exhibit frictional behaviour. They therefore developed models in which pressure sensitivity is not involved and the rate of plastic flow is coaxial with the yield function gradient, the so-called normality or associativity rule.

The model of contact with Coulomb friction explains why granular materials, which are dominated by friction at the micro-scale, cannot be modelled employing pressure-insensitive yield functions and the associative flow rule; rather, pressure sensitivity and non-normality become essential ingredients. Lack of normality represents a strong generalisation of the associative law so that

> *non-associativity not only promotes instabilities which may be already qualitatively described in the framework of normality but also opens the possibility for new types of instabilities, such as, for instance, the so-called flutter.*

Some of these instabilities, and in particular, flutter instability, are not yet fully understood. The latter is believed also to occur in elastic systems with Coulomb friction (Adams, 1995; Lorang et al., 2006; Martins et al., 1990; Nguyen, 2003; Simões and Martins, 1998), and *an experimental example showing flutter as induced by friction in a simple structure has been provided recently by Bigoni and Noselli (2011)*. This example, reported in Section 1.13.5, is a clear demonstration that flutter may be induced in a system that would be stable in the absence of friction, so it should not be surprising if similar phenomena are found in materials whose behaviour is dominated by friction at the micro-scale.

To provide further examples of features connected to non-associativity, we mention that when normality holds, the linearised constitutive equations are symmetric and in a continuous quasi-static loading, failure of strong ellipticity coincides with loss of ellipticity. The latter has been shown by Rice (1977) to correspond to the possibility of strain localisation, occurring in the form of a bifurcation mode similar to that illustrated in Fig. 1.10. However, since strain localisation is usually identified with ellipticity loss, whereas under the above-mentioned van Hove conditions uniqueness is associated to strong ellipticity,

> *for non-associative flow rules, loss of strong ellipticity does not imply loss of ellipticity, so for frictional materials, the van Hove conditions (or those realized by the rigid box triaxial apparatus) are at least in principle not sufficient to prevent possible bifurcations before strain localisation.*

Moreover,

> *at the instant of ellipticity loss, strain localisation might not be a unique form of instability; rather, many others might occur simultaneously.*

Therefore, excluding from the discussion the instabilities occurring at the boundary of a body, several local criteria of stability and bifurcation need to be introduced. These are (1) positive definiteness and (2) non-singularity of the constitutive operator, (3) strong ellipticity and (4) ellipticity of the acoustic tensor and, finally, (5) *flutter instability* (complex conjugate eigenvalues of the acoustic tensor), an instability which remains excluded in the case of symmetry of the incremental constitutive operator and which *is completely unrelated to other instability criteria*, so it may occur when the constitutive operator is still positive definite or when it is strongly elliptic or elliptic (Bigoni and Loret, 1999).

Instability is a dynamic concept, so the achievement of any of the above-mentioned thresholds has strong consequences on the dynamic response of the boundary value problem under consideration. In particular,

> *flutter is a typical dynamical instability which remains undetected when quasi-static behaviour is only analysed.*

1.11 A perturbative approach to material instability

The strain localisation visible, for instance, in the experiments shown in Figs. 1.4, 1.11, 1.12, 1.13, 1.15, and 1.16 is only a 'strong and evident manifestation' that the material traverses a state prone to instability, beginning at a certain—sufficiently high—strain level, culminating with localisation and (as in the case of the drinking straw packaging) continuing with an accumulation of deformation bands (or in

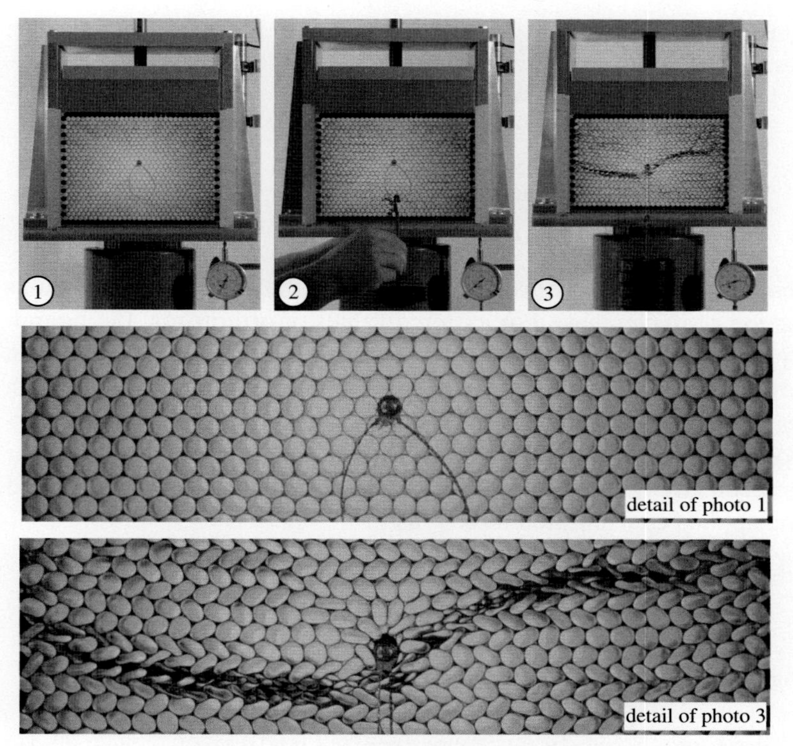

Figure 1.35. Application of a concentrated load (50 N), triggering strain localisation during uniaxial deformation of a packaging of regularly disposed drinking straws. Photo 1 was taken before the test initiation, whereas a vertical load has been applied at the instant when photo 2 was taken, provoking the shear bands visible in photo 3. The two photos reported below are details of photo 1 (initial configuration) and 3 (strain localisation already developed), where the device used to apply the concentrated load is visible.

other cases with intense deformation, damage and possibly fracture within a single deformation band).

Since the 'standard' approach to material instability is limited to determination of the threshold for the *onset* of strain localization (identified with the loss of ellipticity of the incremental governing equations), the state previously experienced by the material is usually left unexplored. However, this state can be investigated efficaciously through analysis of the material response to a *perturbation* applied at a certain level of deformation. For instance, we can perturb the sample in the experiment shown in Fig. 1.15 by applying a concentrated force when the deformation is still uniform but the peak of the stress/strain curve is approached. As a result, the deformation induced by the perturbing force becomes highly focused and localised, which would have not been the case if the perturbation were provided much before the peak of the stress/strain response (Fig. 1.35). This approach, which will be detailed in Chapter 16, has been rationalised into an analytical procedure by Bigoni and Capuani (2002), who have defined a perturbation in terms of a concentrated force acting in an infinite pre-stressed continuum.

To explain in a simple way the perturbative approach, we follow Bigoni and Noselli (2010a, 2010b) and refer to an orthotropic elastic material defined in plane strain by the constitutive equations

$$\sigma_{11} = \frac{E_1}{1-\nu_{12}\nu_{21}}(\epsilon_{11}+\nu_{21}\epsilon_{22}),$$
$$\sigma_{22} = \frac{E_2}{1-\nu_{12}\nu_{21}}(\epsilon_{22}+\nu_{12}\epsilon_{11}), \quad (1.41)$$
$$\sigma_{12} = 2\mu_{12}\epsilon_{12},$$

relating the stress σ_{ij} to the strain ϵ_{ij} through the elastic moduli E_i, the shear modulus μ_{12} and contraction coefficients ν_{ij}. For extreme values of orthotropy, the boundary of ellipticity can be approached, and this condition can be tested employing an appropriate perturbation. We can consider, in particular, an elastic half space with orthotropy axes parallel and orthogonal to the free surface and perturb this configuration by imposing a concentrated load orthogonal to the surface. The solution for this problem has been given by Lekhnitskii (1981) in a polar coordinate system r and θ centered at the applied concentrated force F as

$$\sigma_r(r,\theta) = -\frac{F(u_1+u_2)\cos\theta}{\pi\, r\,\Lambda(\theta)}\sqrt{\frac{E_2}{E_1}}, \quad (1.42)$$

where, assuming for simplicity $\nu_{ij}=0$,

$$\Lambda(\theta) = \frac{E_2}{E_1}\sin^4\theta + \frac{E_2}{\mu_{12}}\sin^2\theta\cos^2\theta + \cos^4\theta, \quad (1.43)$$

and u_1 and u_2 are two roots of the equation (to be chosen with the rule that the resultant of σ_r over a semi circle has to be F)

$$\frac{E_2}{E_1}u^4 - \frac{E_2}{\mu_{12}}u^2 + 1 = 0. \quad (1.44)$$

By definition, $\Lambda(\theta)$ is positive for all θ in the elliptic range but vanishes for some θ when the boundary of ellipticity is touched. This condition makes already evident that there is a blow up of the solution when the elliptic boundary is approached, and in fact, the interesting aspect is the behaviour of the solution *before* the threshold is attained. We report in Fig. 1.36 the response to a concentrated force of an isotropic (left, the so-called Flamant solution) and a highly orthotropic (right, for a ratio E_2/E_1 between vertical and horizontal elastic moduli equal to 300) half space. The solution pertaining to the highly orthotropic material represents the response near a material instability and is (1) highly localised and (2) strongly directional (in this case oriented parallel to the direction of the applied force).[22]

The near-vertical stress percolation shown in Fig. 1.36 on the right can be observed in reality in highly orthotropic materials, such as, for instance, masonry (Bigoni and Noselli, 2010a, 2010b). In particular, a transmission photoelastic investigation is reported in Fig. 1.37 on models made up of a PSM-9 material, purchased

[22] Everstine and Pipkin (1971) provided an asymptotic approximation to Eq. (1.42) valid for an extreme orthotropy similar to that considered here. They also noticed a 'stress channelling effect' for fibre-reinforced materials, so they envisaged a route to material instability in essence similar to the perturbative approach proposed independently by Bigoni and Capuani (2002).

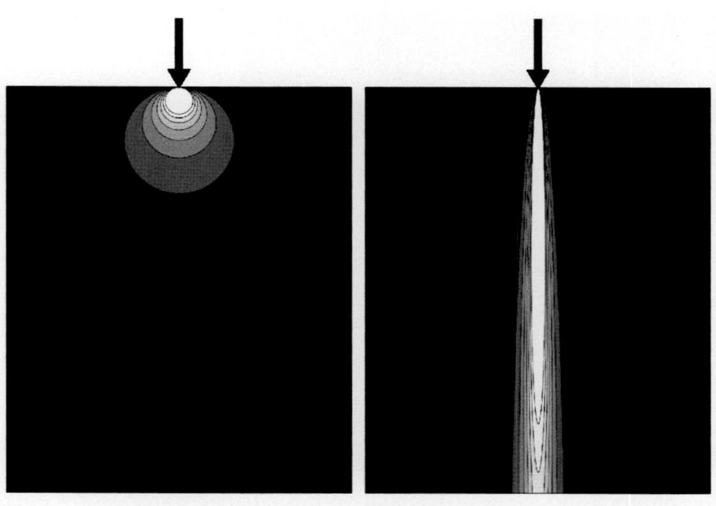

Figure 1.36. Maps of in-plane principal stress difference for two elastic materials, one isotropic (*left*) and another highly orthotropic (*right*), subject to a concentrated load (denoted by the *black arrow*). The figure has to be compared with the photoelastic experiments shown in Fig. 1.37.

from Vishay. This investigation has to be compared to Fig. 1.36, where we see that the Flamant solution is fully confirmed for a platelet of intact material (simulating a semi-infinite elastic half space), but more important to our purposes, the model of dry masonry reported on the right gives full evidence to a highly localised stress percolation, in agreement to the Lekhnitskii solution (1.42) used with high orthotropy contrast. The localised stress distribution obtained for high orthotropy contrast degenerates at the boundary of ellipticity into a set of vertical lines, transmitting the load without diffusion, as in the so-called no-tension material model introduced by Heyman (1965).

It is important to note from Fig. 1.36 on the right that

> *as a response to a perturbation, strain localisation may become visible still inside the elliptical range.*

To further substantiate this statement, we note that although loss of ellipticity is excluded (formally, it occurs at an infinite stretch) for a Mooney-Rivlin material, a horizontal shear band is clearly visible in Fig. 1.38 on the right, formed as the response to a horizontal force dipole perturbing the material subject to a uniform stretch near 3 (corresponding to a pre-stress $\sigma = 1.96\mu$, still inside the elliptical range), whereas such a shear band is not visible in the figure on the left (corresponding to a unit stretch, or null pre-stress, $\sigma = 0$). With reference to an orthotropic, incompressible pre-stressed elastic material [the orthotropy parameter μ_*/μ is taken equal to 1/4; see Section 16.1.1 or Bigoni and Capuani (2002) for details], the incremental response to a horizontal force dipole is shown in Fig. 1.39 on the left at null pre-stress and on the right at a pre-stress near loss (but still inside) of ellipticity. Localised patterns of incremental deformation in essence similar to that reported in Fig. 1.36 on the right are evident in Figs. 1.38 and 1.39, the latter inclined with respect to the orthotropy axes.

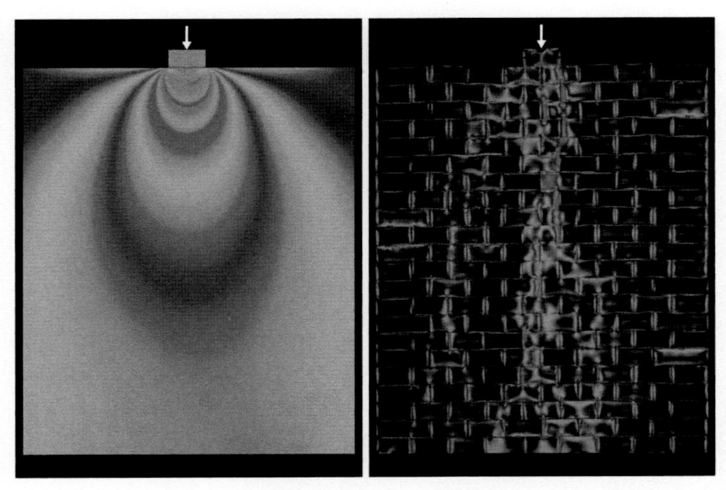

Figure 1.37. Isocromatic photoelastic fringes (detected with a circular transmission polariscope at white light) of a two-component resin platelet subject to a vertical load, denoted with a white arrow (*left*), and of a model of dry masonry (*right*). Vertical load is 500 N for the sample on the left and 125 N for the sample on the right. While the load diffuses within the material according to the Flamant solution in the case reported on the left, there is a strong vertically localised stress percolation in the case reported on the right, according with the Lekhnitskii solution (1.42) at high orthotropy contrast; see Fig. 1.36. (See color plates section.)

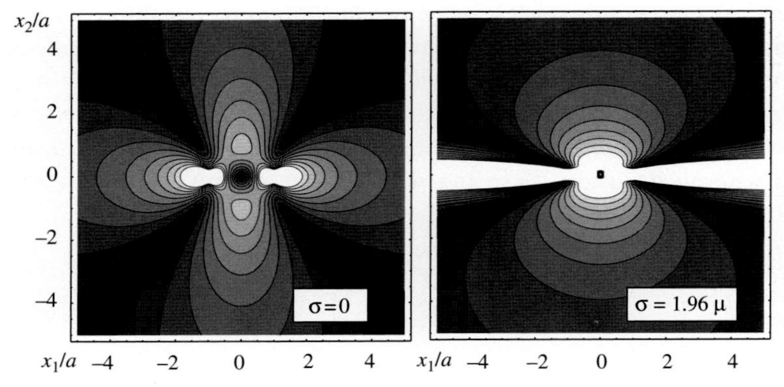

Figure 1.38. Level sets of the modulus of incremental displacement produced by a horizontal dipole (two unit forces placed at $x_1 = \pm a$) for a Mooney-Rivlin elastic incompressible material, homogeneously deformed in plane strain. Incremental displacement map at null prestress $\sigma = 0$ (*left*), far from the boundary of ellipticity loss, and at a pre-stress $\sigma = 1.96\mu$ (*right*), close to the boundary of ellipticity loss. Adapted from Bigoni and Capuani (2002).

Bigoni and Capuani (2005) have extended the perturbative approach to the dynamic range, considering the response to a time-harmonic pulsating force dipole (Fig. 1.40). This figure (where the real component of the level sets of the modulus of displacement is reported on the left and the imaginary part on the right) is relative to a circular frequency ω (multiplied by the half dipole distance a and divided by the horizontal velocity of propagation c) equal to $\pi/4$ and a pre-stress $\sigma = 1.72\mu$ for the same material considered in Fig. 1.39. It becomes clear now that localised wave

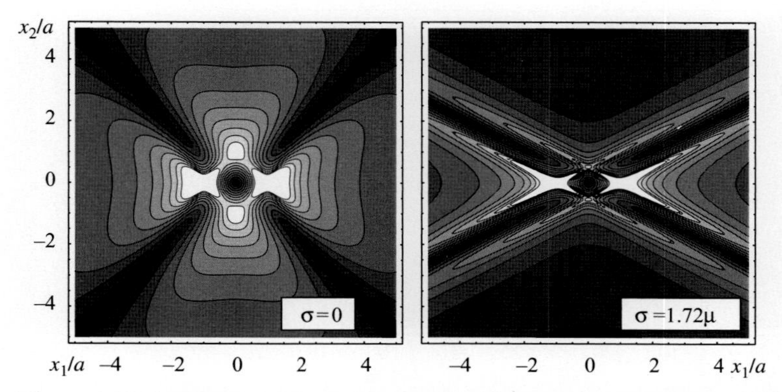

Figure 1.39. Level sets of the modulus of incremental displacement produced by a horizontal dipole (two unit forces placed at $x_1 = \pm a$) for an incompressible orthotropic elastic material with $\mu_*/\mu = 1/4$ uniformly deformed in plane strain. Incremental displacement map at null prestress, far from the boundary of ellipticity loss (*left*), and at a pre-stress $\sigma = 1.72\mu$, close to the boundary of ellipticity loss (*right*). Adapted from Bigoni and Capuani (2002).

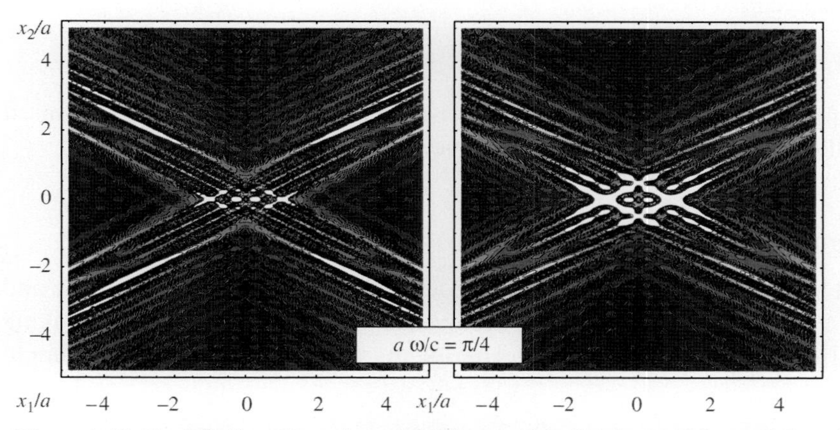

Figure 1.40. Real (*left*) and imaginary (*right*) part of the level sets of the modulus of incremental displacement produced by a horizontal time-harmonic pulsating dipole (two pulsating unit forces placed at $x_1 = \pm a$). The maps are relative to a pre-stress $\sigma = 1.72\mu$, close to the boundary of ellipticity loss, and a value of circular frequency ω multiplied by the dipole half distance a and divided by the velocity of wave propagation along the horizontal axis c equal to $\pi/4$. Adapted from Bigoni and Capuani (2005).

patterns emerge near the boundary of ellipticity loss, with focussing of signal in the direction of the shear bands.

The dynamic perturbative approach has been applied to a situation of flutter instability by Piccolroaz et al. (2006). Differently from non-linear elasticity, now the perturbative approach may become questionable because the material is expected to change response from plastic loading to elastic unloading during vibrations [see Bigoni and Petryk (2002), for a complete discussion on this point]. Nevertheless, the dynamical consequence of flutter instability is revealed in Fig. 1.41, where the map of the modulus of the real part of incremental displacement is shown for same material ‘far’ from the flutter condition (left) and ‘inside’ the flutter region (right).

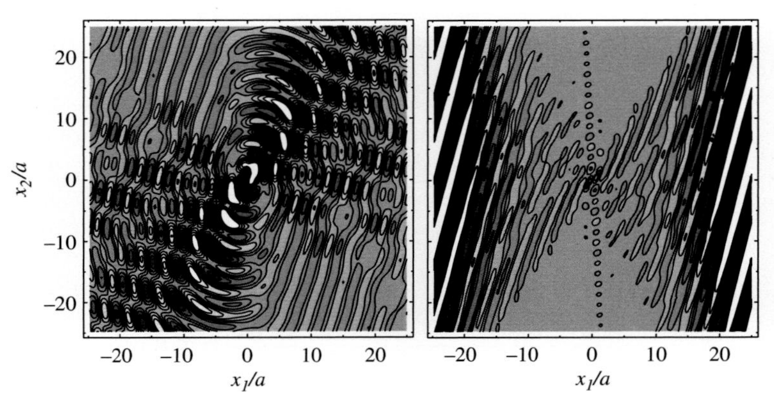

Figure 1.41. Real part of the level sets of the horizontal component u_1 of incremental displacement produced by a pulsating force dipole inclined at 45°. The maps are relative to a unit value of circular frequency ω (multiplied by the dipole half distance a and divided by the velocity of wave propagation along the horizontal axis c). Situation far from (*left*) flutter (at high hardening $g/\mu = 3$) and inside (*right*) the flutter region (at low hardening $g/\mu = 0.32$). Flutter consists of an incremental vibration organised along parallel plane waves blowing up in space. Adapted from Piccolroaz et al. (2006).

From this figure we may conclude that flutter is an incremental vibration organised along parallel plane waves blowing up in space.

1.12 A summary

Continued homogeneous deformations in 'ductile and plastic' and 'quasi-brittle and damaging' materials are limited by the occurrence of different failure modes taking place at different scales. In particular, experiments show that failure is not a simple event; rather

> *failure is a progressive phenomenon influenced by several concurrent mechanisms.*

As an example of this behaviour, we may mention the failure progression within a metallic, ductile bar pulled in tension (Fig. 1.42).

These failure mechanisms can be analysed in terms of bifurcation and instability theory for inelastic solids. In particular,

> *since mechanical tests on material elements usually consist of imposing displacements (resultant forces) on certain external surfaces of a finite volume of material and measuring resultant forces (displacements) on the same surfaces, stress and strain cannot be directly controlled, instabilities occurring in a mechanical test always should be referred to a specific boundary value problem.*

For instance, a compression test is usually performed under controlled vertical displacement imposed by a rigid, possibly smooth constraint so that several failure mechanisms can be envisaged (Fig. 1.43). Therefore, depending on the operating failure mechanism,

> *different softening and size effects are produced during near-failure testing of a sample.*

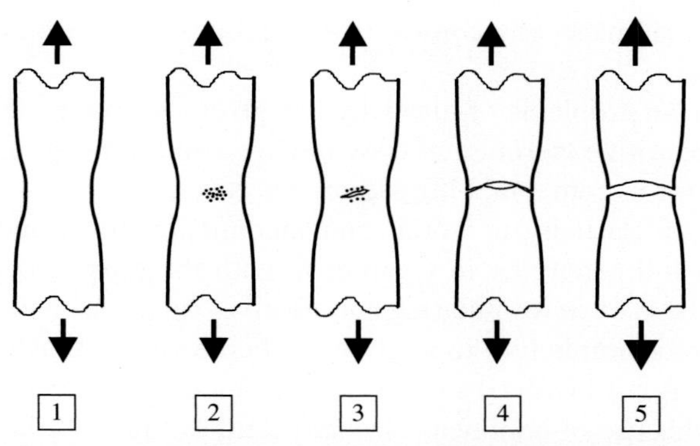

Figure 1.42. Subsequent mechanisms of failure during a tensile test on a ductile, metallic bar. (1) necking; (2) cavitation; (3) fracture; (4) shear banding joining the fracture to the external boundary; (5) separation. See Dieter (1961) for more detail.

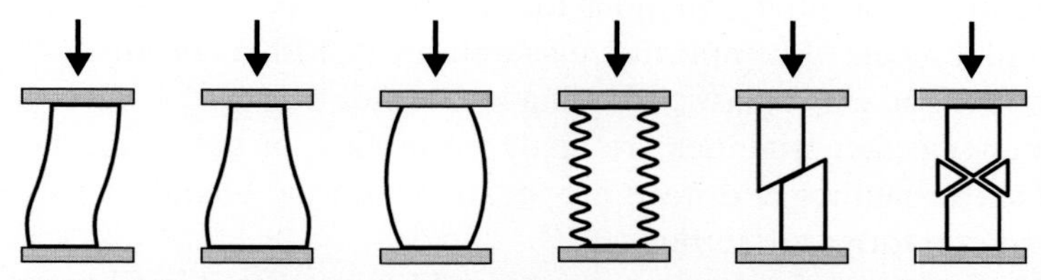

Figure 1.43. Failure in compression. *From left to right*: Euler mode, bulging, barreling, surface instability, asymmetric shear banding and symmetric shear banding; compare with Figs. 1.5 and 1.11.

Failure usually involves *large* and *inelastic* deformations, so a unified framework for the analysis of bifurcations and instabilities leading to failure is represented by large strain elastoplasticity. Consideration of large strain is in any case imperative in analysing bifurcations and instability (a small strain theory evidently cannot capture a realistic range of bifurcations, such as, for instance, Euler buckling in a sufficiently slender sample).

The class of materials considered by us includes metals and alloys, soil, granular media and—under circumstances related to temperature, mean pressure and time scale of loading application—rock, concrete, ice and plastic. Therefore, a reasonably general constitutive framework is that of elastoplasticity with smooth yield function and the non-associative flow rule. 'Non-associativity' (or 'non-normality') means that the plastic flow rate is not parallel to the yield-function gradient (which can be viewed as orthogonal to the yield surface), and since

> *for granular materials Coulomb friction between grains is the leading strength micro-mechanism and is inherently non-associative,*

elastoplasticity with non-normality plays the role of a sort of 'reasonably simple, but general, Coulomb-like constitutive equation'.

The plan of this book is as follows. After a prelude on the necessary mathematical tools (Chapter 2) and elements of continuous mechanics at large strain

(Chapter 3), we introduce in the simplest way constitutive equations for isotropic non-linear elasticity (Chapter 4).

Solutions of simple, non-linear problems of elasticity are given in Chapter 5. These are crucial for comparisons with experimental tests and are also used later as fundamental deformation paths for incremental bifurcation analyses.

The constitutive equations of elasticity in 'total' and incremental form and hypoelasticity are introduced in full generality in Chapter 6, with the purpose of highlighting general concepts to be used in formulating constitutive equations. Note that non-linear elasticity provides a simple tool to analyse the behaviour of plastic materials when elastic unloading is not involved so that they 'momentarily' behave as non-linear elastic materials (this use of nonlinear elasticity defines the so-called deformation theory of plasticity).

With the objective of analysing elastoplastic behaviour, we introduce the concept of yield function [using a generalised framework proposed by Bigoni and Piccolroaz (2004) which becomes particularly useful for pressure-sensitive materials in Chapter 7] and the constitutive equations for elastoplasticity (several models are included at different levels of complexity: elastoplasticity with associative, the non-associative flow rule and elastoplastic coupling in Chapter 8).

In Chapter 9, moving discontinuities are analyzed in view of the connection with the analysis of shear banding and wave propagation, and the boundary value problems in finite and rate forms are settled.

Global and local conditions of uniqueness and stability are treated in Chapters 10 and 11, respectively, following the theory developed by Hill (1958, 1959, 1978) and Petryk (1985a, 1991) for elastoplastic solids with the associative flow rule and the extension to non-associativity provided by Raniecki (1979), Raniecki and Bruhns (1981) and Bigoni and Zaccaria (1992a, 1992b).

The essential complication of elastoplasticity is the incremental non-linearity (piece-wise linearity) of the constitutive operator so that so-called linear comparison solids are introduced to provide bounds to bifurcation thresholds.

In elastoplasticity, we will distinguish between uniqueness, stability of equilibrium and stability of a deformation path (the two stability concepts reduce to a single notion for elastic behaviour). It will become clear that while the notion of stability of a deformation path can be extended to non-associative behaviour, the energetic criterion needed to make this notion operative is limited only to associative elastoplasticity. For the non-associative flow rule, several of sufficient or necessary local conditions for stability are given: positive definiteness and non-singularity of the constitutive operator, strong ellipticity, ellipticity, flutter and divergence instability (as proposed by Rice, 1977). These are connected to different and 'more or less known' forms of material instability. For instance,

the condition of ellipticity loss is shown to correspond to shear band formation,

following concepts proposed by Nadai (1931, 1950), Hill (1952, 1961), Thomas (1953, 1954, 1961) and Mandel (1962a, 1962b). Finally, a simple algorithm to determine the critical conditions for shear banding in elastoplastic materials under the assumption of small strain is given (Bigoni and Hueckel, 1990, 1991a, 1991b).

Examples of bifurcations and stability thresholds evaluated for finitely strained elastic solids are presented in Chapter 12, following concepts pioneered by Biot (1965). These include (1) plane strain deformation of elastic incompressible materials, with analyses of surface and interface bifurcation, bifurcations of a block, of a block on an elastic foundation and on an elastic half-space and of a multilayered structure (Bigoni et al., 1997, 2008), (2) bifurcation of an elastic incompressible cylinder loaded in axial compression (Gei et al., 2004) and (3) bifurcation of an elastic incompressible solid subject to plane strain bending (Roccabianca et al., 2010, 2011). These examples show that shear banding has to be understood as an extreme form of instability, usually occurring after other manifestations of instability.

The fact that shear banding is usually observed in many experiments is indicative that the instabilities necessarily encountered before are not 'strong enough' to completely destroy the uniformity of the stress/strain fields within the sample or to 'terminate' the possibility of the sample to sustain increasing deformation.

Examples of bifurcation and stability analyses of elastoplastic solids with the non-associative flow rule are provided in Chapter 13, where (1) different material stability thresholds are calculated in the simple case of the infinitesimal theory; (2) a complete analysis of axisymmetric bifurcations of an elastoplastic cylinder is performed, with a full account of all thresholds involving linear comparison materials and (3) a flutter and divergence instability analysis is provided for a specific material model, tailored to mimic the behaviour of granular media.

The theory of incremental plane waves is reported in Chapter 14 for non-linear elastic materials and is contrasted with the much more complicated theory pertaining to elastoplasticity (made complex by the incremental non-linearity of the constitutive operator; Bigoni and Petryk, 2002). The Chapter is concluded with analysis of acceleration waves developed by Hill (1962), Mandel (1962a, 1962b, 1969) and Raniecki (1976), presented in full generality, in view of the implications on shear band analysis.

A rather common experimental observation (e.g., in phase-transforming materials) is that

multiple shear bands may occur, giving rise to stress oscillations in the nominal stress versus global strain response.

This is analysed in Chapter 15, where simple calculations employing chains of softening elements are proposed to provide a simple description of the phenomenon. These calculations are insufficient for a correct modelling of elastoplastic materials, so a methodology is introduced for a more sophisticated treatment. Here, a material model tailored to represent the behaviour of granular materials can be used to disclose a variety of deformation mechanisms evidencing: (1) softening induced by localisation and (2) band saturation and multiple shear banding with formation of, say, 'persistent' or 'not persistent' bands (Gajo et al., 2004, 2007).

All the above-mentioned techniques for the analysis of bifurcation and stability still leave many problems open, some of which are definitely important. For instance, how do shear bands behave for dynamic loading and how do they interact with a crack or with a rigid inclusion or—simply—with another shear band? In an attempt to answer these questions, a perturbative approach has been invented by Bigoni and

Capuani (2002), which is developed in Chapter 16, assuming *different perturbative agents*, namely, a concentrated force, a pulsating dipole, a fracture, a stiff and thin lamellar inclusion, and a pre-existing shear band. The perturbative approach opens a new perspective in analyses of bifurcation and instability of ductile materials and closes our treatment.

1.13 Exercises, details and curiosities

Simple exercises are provided below to complement concepts introduced in this chapter and to stimulate interest in special features of bifurcation and stability theory. In particular, we present and solve the following exercises. (1) The problem of the elastica for a simply supported beam subject to compressive force, which represents an elegant solution to a non-linear bifurcation and post-bifurcation problem. (2) To highlight the idea that bifurcation in tension should not be considered something 'unusual', we show an example of a simple one degree of freedom elastic structure buckling under tensile, dead load. (3) As a complement to the Shanley model, we show with two simple examples that the number of bifurcation loads in a discrete structure has a relation with the number of degrees of freedom of the structure weaker than is often believed. (4) We complete the analysis of the structure sketched in Fig. 1.17, showing that at increasing loading, the vertical configuration of the structure first becomes unstable but later returns to be stable at higher load. (5) We present a structure which is stable when constraints are smooth but becomes subject to flutter and divergence instability when friction is present. This exercise shows that flutter as connected to friction is a truly real phenomenon.

1.13.1 Exercise: The Euler elastica and the double supported beam subject to compressive load

Determination of the equilibrium solutions of an elastic thin rod, straight in its unloaded configuration and subject to end thrust, is an old and fascinating problem which was posed by Jacob Bernoulli (1654–1705) and suggested by Daniel Bernoulli (1700–1782) to Leonhard Euler (1707–1783), who solved it. Elementary experimental evidence (Fig. 1.2)[23] indicates that the 'trivial' straight configuration of a compressed rod is not the unique equilibrium configuration, whereas 'non-trivial' inflexional configurations determine a special shape of the rod, the so-called elastic line or elastica.

The determination of all possible equilbrium configurations of an initially straight thin elastic rod subject to end thrust is a beautiful example of the solution of a linear and non-linear bifurcation problem with complete determination of the critical and post-critical behaviours of the structure. A brief account of the theory is given by Timoshenko and Gere (1961); we will follow the exhaustive treatments by Love (1927) and Reiss (1969). The important issue of stability has been addressed by Maddocks (1984); we will follow Kuznetsov and Levyakov (2002) (see also Levyakov and Kuznetsov, 2010, and Sachkov and Levyakov, 2010).

[23] Petris van Musschenbroek ('Introductio ad coharentiam corporum firmorum', 1729) experimentally discovered that the critical load of a compressed rod is inversely proportional to the square of its length. Todhunter and Pearson (1960) report that Euler was aware of this fact.

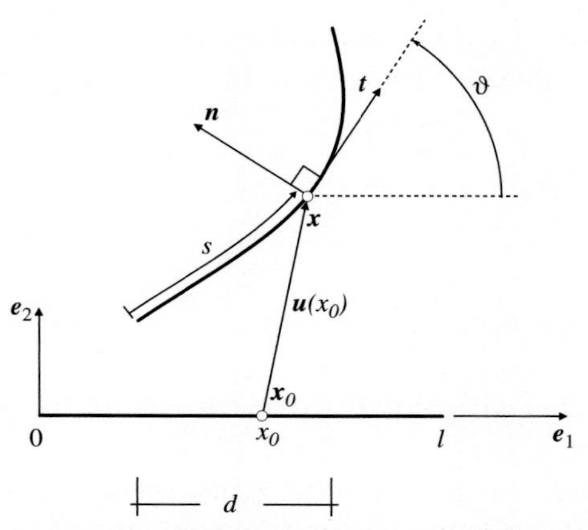

Figure 1.44. The kinematics of an elastic inextensible rod of length l, rectilinear in the reference configuration. Displacement of a point $\boldsymbol{x}_0$ of coordinate x_0 is $\boldsymbol{u}(x_0) = \boldsymbol{x} - \boldsymbol{x}_0$. Note that inextensibility implies that the curvilinear coordinate s is equal to the coordinate x_0, namely, $s = x_0$.

The kinematics of an inextensible rod in a plane We consider an inextensible rod of length l, rectilinear in a reference configuration and smoothly deformed, as shown in Fig. 1.44. In the (un-deformed) deformed configuration, the generic point can be picked up using (a coordinate $x_0 \in [0,l]$) a curvilinear coordinate $s \in [0,l]$, so inextensibility implies that $x_0 = s$ (so that $dx_0 = ds$).

The *displacement* $\boldsymbol{u}$ of the point $\boldsymbol{x}_0$ from the reference configuration is

$$\boldsymbol{u} = u_1(\boldsymbol{x}_0)\boldsymbol{e}_1 + u_2(\boldsymbol{x}_0)\boldsymbol{e}_2 = \boldsymbol{x} - \boldsymbol{x}_0, \tag{1.45}$$

which, introducing the deformation[24]

$$\boldsymbol{x} = \boldsymbol{g}(\boldsymbol{x}_0) \tag{1.46}$$

and noting that the point $\boldsymbol{x}_0$ has coordinate x_0 (so that $\boldsymbol{x}_0 = x_0\boldsymbol{e}_1$), becomes

$$\boldsymbol{u} = \boldsymbol{g}(x_0\boldsymbol{e}_1) - x_0\boldsymbol{e}_1. \tag{1.47}$$

Note that since $\boldsymbol{e}_1$ is fixed, the dependence on it of function $\boldsymbol{g}$ could be omitted, so Eq. (1.46) would become the parametric representation of the curve describing the elastica. However, we would like to proceed in a way similar to that employed in continuum mechanics (Chapter 3) so that the reader will find a useful parallel between the developments of the two theories.

Let us consider now two neighbouring points of the reference configuration at coordinates x_0 and $x_0 + \omega_0$, defining the vector $\boldsymbol{t}_0 = \omega_0\boldsymbol{e}_1$. This vector is mapped to

$$\boldsymbol{g}(\boldsymbol{x}_0 + \omega_0\boldsymbol{e}_1) - \boldsymbol{g}(\boldsymbol{x}_0), \tag{1.48}$$

[24] It is evident that the deformation has to possess certain regularity properties. In fact, a discontinuous deformation would correspond to a fracture in the elastica. It therefore will assumed that $\boldsymbol{g}$ is twice-continuously differentiable.

so assuming ω_0 to be small and performing a Taylor series expansion of the deformation around $\omega_0 = 0$ yields the transformed vector (tangent to the deformed line at $\boldsymbol{x}_0$) as

$$\boldsymbol{F}(\omega_0 \boldsymbol{e}_1), \tag{1.49}$$

where

$$\boldsymbol{F} = \frac{\partial \boldsymbol{g}}{\partial \boldsymbol{x}_0} = \left(u_1' + 1\right) \boldsymbol{e}_1 \otimes \boldsymbol{e}_1 + u_2' \boldsymbol{e}_2 \otimes \boldsymbol{e}_1 + \boldsymbol{e}_2 \otimes \boldsymbol{e}_2, \tag{1.50}$$

where the superscript $'$ denotes differentiation with respect to the coordinate $x_0 = s$ and the symbol $\otimes$ denotes the dyadic product, see Chapter 2, Eq. (2.14).

Since the elastica is assumed *inextensible*, the length of the transformed vector $\boldsymbol{F}(\omega_0 \boldsymbol{e}_1)$ must maintain the same length of the initial vector $\boldsymbol{t}_0 = \omega_0 \boldsymbol{e}_1$, therefore, from Eq. (1.49) we obtain

$$|\boldsymbol{F}\boldsymbol{e}_1| = 1, \tag{1.51}$$

which, using Eq. (1.50), yields

$$u_1' + 1 = \sqrt{1 - (u_2')^2}. \tag{1.52}$$

Taking the derivative of Eq. (1.52) finally provides the inexstensibility constraint in the form

$$u_1'' = -\frac{u_2' u_2''}{\sqrt{1 - (u_2')^2}}. \tag{1.53}$$

Since the inextensibility constraint is enforced and the tangent to the elastica at $\boldsymbol{x}$ is given by the unit vector $\boldsymbol{t}$, i.e.,

$$\boldsymbol{t} = \left(u_1' + 1\right) \boldsymbol{e}_1 + u_2' \boldsymbol{e}_2 = \sqrt{1 - (u_2')^2}\, \boldsymbol{e}_1 + u_2' \boldsymbol{e}_2, \tag{1.54}$$

the angle θ of inclination of the tangent $\boldsymbol{t}$ to the elastica at $\boldsymbol{x}$ is given by

$$\sin\theta = x_2' = u_2', \quad \cos\theta = x_1' = \sqrt{1 - (u_2')^2}, \tag{1.55}$$

and the length d of the projection of the elastica onto the $\boldsymbol{e}_1$ axis is

$$d = \int_0^l \cos\theta\, ds = \int_0^l \sqrt{1 - (u_2')^2}\, ds. \tag{1.56}$$

The unit vector $\boldsymbol{n}$ normal to the elastica at $\boldsymbol{x}$ can be obtained by differentiating (with respect to s) the scalar product $\boldsymbol{t} \cdot \boldsymbol{t}$, so $\boldsymbol{t}'$ is found normal to $\boldsymbol{t}$ [this property is detailed in Chapter 2; see Eq. (2.107) and Fig. 2.9] in the form

$$\boldsymbol{t}' = -\frac{u_2' u_2''}{\sqrt{1 - (u_2')^2}} \boldsymbol{e}_1 + u_2'' \boldsymbol{e}_2, \qquad \text{or} \qquad \boldsymbol{t}' = -\theta' \sin\theta\, \boldsymbol{e}_1 + \theta' \cos\theta\, \boldsymbol{e}_2. \tag{1.57}$$

The unit normal therefore can be obtained from Eqs. $(1.57)_1$ or $(1.57)_2$ through division by the modulus (the so-called curvature): [25]

$$|\boldsymbol{t}'| = \frac{|u_2''|}{\sqrt{1-(u_2')^2}} = |\theta'|, \tag{1.59}$$

thus obtaining

$$\boldsymbol{n} = \mathrm{sign}(u_2'')\left(-u_2'\boldsymbol{e}_1 + \sqrt{1-(u_2')^2}\,\boldsymbol{e}_2\right) \quad \text{or} \quad \boldsymbol{n} = \mathrm{sign}(\theta')\,(-\sin\theta\,\boldsymbol{e}_1 + \cos\theta\,\boldsymbol{e}_2). \tag{1.60}$$

The *signed curvature* κ is[26]

$$\kappa = \frac{u_2''}{\sqrt{1-(u_2')^2}} \quad \text{or} \quad \kappa = \theta'. \tag{1.65}$$

Loading, constitutive equation and the elastica The elastica is assumed to be loaded at its edges by forces and couples, generating a moment, a normal and a shearing force distribution along the deformed line.

The constitutive equation used for the elastica is the celebrated Jacob Bernoulli's assumption that the effects of normal and shearing forces are neglected and that the curvature of the deflection curve is proportional to the bending moment, namely,

$$\theta'(s) = \frac{M(s)}{B}, \tag{1.66}$$

in which B is the bending stiffness (equal to the product between the Young modulus and the moment of inertia of the cross-sectional area of the beam in the linear beam theory).

[25] Note that the inverse of $\boldsymbol{F}$ is

$$\boldsymbol{F}^{-1} = \frac{1}{\sqrt{1-(u_2')^2}}\boldsymbol{e}_1\otimes\boldsymbol{e}_1 + \boldsymbol{e}_2\otimes\boldsymbol{e}_2 - \frac{u_2'}{\sqrt{1-(u_2')^2}}\boldsymbol{e}_2\otimes\boldsymbol{e}_1, \tag{1.58}$$

so the unit normal (1.60) also can be obtained from Eq. (3.22).

[26] If the deformed elastica is described in a Cartesian system in which its coordinates are

$$\tilde{x} = s + u_1(s) - u_1(0) \quad \text{and} \quad \tilde{y} = f(\tilde{x}), \tag{1.61}$$

the signed curvature (1.65) can be obtained from the expression

$$\kappa = \frac{d^2f/d\tilde{x}^2}{[1+(df/d\tilde{x})^2]^{3/2}}, \tag{1.62}$$

by considering the identity

$$u_2(s) = f[s + u_1(s) - u_1(0)], \tag{1.63}$$

from which the two following equations (keeping into account the inextensibility constraint) can be derived

$$\frac{df}{d\tilde{x}} = \frac{u_2'}{\sqrt{1-(u_2')^2}} \quad \text{and} \quad \frac{d^2f}{d\tilde{x}^2} = \frac{u_2''}{[1-(u_2')^2]^2}, \tag{1.64}$$

(where a prime denotes differentiation with respect to s), which have to be substituted into Eq. (1.62) to obtain Eq. (1.65).

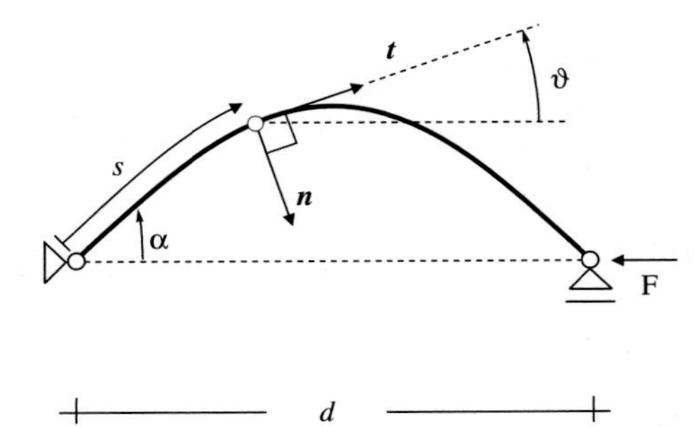

Figure 1.45. The kinematics of a simply supported elastic inextensible rod of length l, rectilinear in the reference configuration.

With reference to a simply supported beam (Fig. 1.45), *excluding for the moment the possibility that the two supports coincide, namely, that $d = 0$*,[27] the equilibrium condition for a given axial load F (assumed positive when compressive) is

$$M(s) = -Fu_2(s), \tag{1.70}$$

so, using the constitutive law (1.66), we arrive at the differential equation governing the equilibrium of the deformed configuration[28]

$$\theta'(s) + \frac{F}{B}u_2(s) = 0. \tag{1.71}$$

Taking the derivative with respect to s of Eq. (1.71) and using Eq. (1.55), we arrive at the *differential equation for the elastica*

$$\theta'' + \frac{F}{B}\sin\theta = 0, \tag{1.72}$$

an equation which is formally identical to the equation ruling the oscillation of a simple pendulum (Temme, 1996).

[27] In the special case where the two supports coincide, $d = 0$, equilibrium does not exclude the possibility of vertical reactions of the supports. If a support transmits a vertical reaction R, the equilibrium equation (1.70) is replaced by

$$M(s) = -Fu_2 + R(d - x_1), \tag{1.67}$$

so instead of Eq. (1.71), we obtain

$$\theta' + \frac{F}{B}u_2 - \frac{R}{B}(d - x_1) = 0. \tag{1.68}$$

Thus, using Eq. $(1.55)_2$, we generalise Eq. (1.72) with

$$\theta'' + \frac{F}{B}\sin\theta + \frac{R}{B}\cos\theta = 0, \tag{1.69}$$

governing the situation in which the two supports are coincident, $d = 0$.

[28] Note that since the radius of curvature R of the elastica is the inverse of the curvature, namely, $R = 1/\theta'$, Eq. (1.71) states that

$$u_2 R = -\frac{B}{F},$$

in words, 'the product of the distance of the point from axis $\boldsymbol{e}_1$ times the radius of curvature of the elastica is constant'. A formally identical law, in which R is the radius of curvature of a surface film in a fluid and $-B/F$ is replaced by the ratio between surface tension and weight of the fluid per unit volume, governs the shape of the capillary curve of a liquid (in two dimensions), so the shapes of a rod under finite flexure (Fig. 1.47) also represent possible capillary curves (Lamb, 1928).

In summary, defining

$$\lambda^2 = \frac{F}{B}, \tag{1.73}$$

the equations governing the equilibrium of the elastica for a simply supported beam are

$\theta''(s) + \lambda^2 \sin\theta(s) = 0,$	$s \in [0,l]$	governing differential equation
$\theta'(0) = \theta'(l) = 0,$		b. c.: null moment at both supports
$u_1(0) = 0$		b. c.: null horizontal displacement at the left support
$u_2(0) = u_2(l) = 0,$		b. c.: null vertical displacement at both supports
$u_1'(s) = \cos\theta(s) - 1,$	$s \in [0,l]$	differential equation for the horizontal displacement
$u_2'(s) = \sin\theta(s),$	$s \in [0,l]$	differential equation for the vertical displacement

(1.74)

Equations (1.74) define a *non-linear eigenvalue problem*, for which the *trivial solution* $\theta = 0$ is always possible, so the question arises whether non-trival solutions exist or not. Bifurcation corresponds to the situation in which the trivial solution (or possibly a bifurcated solution) of Eq. (1.74) splits into two or more as λ passes through a critical value λ_{cr} called the 'bifurcation point'.

Before embarking on the solution of the problem (1.74), let us consider its linearisation about the solution $\theta = 0$, for which the horizontal displacement is null, $u_1(s) = 0$, so the remaining equations are

$\theta''(s) + \lambda^2 \theta(s) = 0,$	$s \in [0,l]$	governing differential equation
$\theta'(0) = \theta'(l) = 0,$	$u_2(0) = u_2(l) = 0,$	boundary conditions
$u_2'(s) = \theta(s),$	$s \in [0,l]$	differential equation for the vertical displacement

(1.75)

Equations (1.75) define a *linear eigenvalue problem*, also called a 'Sturm-Liouville problem' (Broman, 1970). It admits the infinite solutions

$$\theta = A_n \cos\frac{n\pi s}{l}, \quad n = 0,1,2,... \tag{1.76}$$

and

$$\underbrace{u_2 = A_0 = 0, \quad n = 0,}_{\text{trivial solution}} \qquad \underbrace{u_2 = \frac{lA_n}{n\pi}\sin\frac{n\pi s}{l}, \quad n = 1,2,...}_{\text{bifurcation mode}} \tag{1.77}$$

where the trivial solution holds for every thrust F, whereas the non-trivial solutions hold if and only if

$$\lambda = \lambda_n = \frac{n\pi}{l}, \quad \Longleftrightarrow \quad F = F_n^{cr} = \frac{n^2\pi^2 B}{l^2}, \quad n = 1,2,... \tag{1.78}$$

which defines Euler's critical loads. As a consequence of the linearisation, the amplitudes A_n ($n = 1,2,...$) of the bifurcation modes remain undetermined; nevertheless, the critical loads correctly identify the bifurcation points on the trivial path, λ_n, as will be proven below.

Let us now solve the non-linear problem (1.74). First of all, we note that if

$$\theta, \ u_1, \ u_2,$$

represent a solution corresponding to λ^2, the fields

$$\pm\theta + 2n\pi, \quad u_1, \quad \pm u_2, \qquad n = ..., -2, -1, 0, 1, 2, ...$$

also represent other solutions (symmetric with respect to the x_1 axis) and the fields

$$\pm\theta + (2n+1)\pi, \quad -u_1 - 2s, \quad u_2, \quad n = ..., -2, -1, 0, 1, 2, ...$$

are valid for $-\lambda^2$. These solutions correspond to deformations symmetric with respect to the x_1 or the x_2 axis, which will be ignored without loss of generality. Therefore, defining

$$\alpha = \theta(0), \tag{1.79}$$

we can only consider $0 \le \alpha \le \pi$.

A multiplication of Eq. $(1.74)_1$ by θ' yields

$$\frac{d}{ds}\left[\frac{1}{2}(\theta')^2 - \lambda^2 \cos\theta\right] = 0, \tag{1.80}$$

so integration and consideration of Eqs. (1.74) and (1.79) leads to

$$\theta' = \lambda\sqrt{2(\cos\theta - \cos\alpha)}, \tag{1.81}$$

where we have selected the positive root because the two solutions differ merely in sign.

An equation formally identical to Eq. (1.81) is usually obtained in the analysis of the oscillation of a simple pendulum (Temme, 1996), so it is a standard expedient to operate the following change of variables

$$k = \sin\frac{\alpha}{2} \qquad \text{and} \qquad k\sin\phi = \sin\frac{\theta}{2}, \tag{1.82}$$

leading through trigonometric formulae to the differential problem

$$\frac{d\phi(s)}{ds} = \lambda\sqrt{1 - k^2\sin^2\phi}. \tag{1.83}$$

The boundary conditions imply that $\sin\phi(0) = 1$ and $\sin^2\phi(l) = 1$, so

$$\phi(0) = \frac{4h+1}{2}\pi \qquad \text{and} \qquad \phi(l) = \frac{2j+1}{2}\pi, \qquad h, j = 0, \pm1, \pm2, ... \tag{1.84}$$

and therefore, separating the variables and integrating Eq. (1.83) yields

$$s\lambda = \int_{\frac{4h+1}{2}\pi}^{\phi(s)} \frac{d\phi}{\sqrt{1 - k^2\sin^2\phi}}, \qquad h = 0, \pm1, \pm2, ... \tag{1.85}$$

which, for $s = l$, becomes

$$l\lambda = \int_{\frac{4h+1}{2}\pi}^{\frac{2j+1}{2}\pi} \frac{d\phi}{\sqrt{1 - k^2\sin^2\phi}}, \qquad h, j = 0, \pm1, \pm2, ... \tag{1.86}$$

Taken over one period, the integral (1.86) is equal to $2K(k)$, where

$$K(k) = \int_0^{\frac{\pi}{2}} \frac{d\phi}{\sqrt{1 - k^2\sin^2\phi}} \tag{1.87}$$

is the complete elliptic integral of the first kind or the so-called real quarter period of the elliptic function (Byrd and Friedman, 1954; Temme, 1996).

The integral in Eq. (1.86) can be rewritten as a function of an integer m as

$$l\lambda = 2mK(k), \quad \Longleftrightarrow \quad F = \frac{B}{l^2} 4m^2 \left[K\left(\sin\frac{\alpha}{2}\right)\right]^2, \tag{1.88}$$

an equation providing the relation between the applied load F and the rotation of the beam's edge on the left associated with the mth bifurcation mode [and coincident with Reiss, 1969, his eq. (3.16)].

For small α, a Taylor series expansion of Eq. (1.88) provides exactly Eq. (1.78), thus proving that

> *the critical Euler loads (1.78), calculated from the linearised theory, correctly determine the bifurcation points emanating from the trivial path.*

Let us go back now to Eq. (1.85) and note that the integral on the right-hand side always can be written as

$$\int_{\frac{4m+1}{2}\pi}^{\phi(s)} (\cdots)\, ds = -\int_0^{\frac{4m+1}{2}\pi} (\cdots)\, ds + \int_0^{\phi(s)} (\cdots)\, ds. \tag{1.89}$$

Thus, since (Byrd and Friedman, 1954; p. 12)

$$\int_0^{\frac{4m+1}{2}\pi} \frac{d\phi}{\sqrt{1-k^2\sin^2\phi}} = (4m+1)K(k), \qquad m = 0, \pm 1, \pm 2, ... \tag{1.90}$$

we obtain

$$s\lambda + (4m+1)K(k) = \int_0^{\phi(s)} \frac{d\phi}{\sqrt{1-k^2\sin^2\phi}}, \qquad m = 0, \pm 1, \pm 2, ... \tag{1.91}$$

which provides

$$\phi(s) = \mathrm{am}\,(s\lambda + (4m+1)K(k), k), \qquad m = 0, \pm 1, \pm 2, ... \tag{1.92}$$

where am denotes the Jacobi amplitude function of modulus k. Employing the property (Baker, 1890, p. 25)

$$\mathrm{am}\,(x \pm 2nK(k), k) = \mathrm{am}\,(x, k) \pm n\pi, \qquad n = 0, \pm 1, \pm 2, ...$$

Eq. (1.92) can be simplified to

$$\phi(s) = \mathrm{am}\,(s\lambda + K(k), k) + 2m\pi, \qquad m = 0, \pm 1, \pm 2, ... \tag{1.93}$$

so the definition of ϕ, Eq. $(1.82)_2$, yields

$$\sin\frac{\theta}{2} = k\,\mathrm{sn}\,(s\lambda + K(k), k), \tag{1.94}$$

where sn is the Jacobi sine amplitude function, defined as

$$\mathrm{sn}\,(x, k) = \sin\left(\mathrm{am}\,(x, k)\right).$$

A substitution of Eq. (1.94) into Eq. (1.81), where the identity $\cos\theta = 1 - 2\sin^2(\theta/2)$ is employed, yields

$$\theta' = 2\lambda k \,\mathrm{cn}\,(s\lambda + K(k), k), \tag{1.95}$$

where cn is the Jacobi cosine amplitude function, defined as

$$\mathrm{cn}\,(x,k) = \cos\,(\mathrm{am}\,(x,k))\,.$$

Note that owing to the properties

$$\mathrm{cn}\,(K(k),k) = \mathrm{cn}\,(3K(k),k) = \mathrm{cn}\,[(2m+1)K(k),k] = 0, \qquad m = 0, \pm 1, \pm 2, \pm 3$$

the boundary conditions $(1.74)_2$, namely, $\theta'(0) = \theta'(l) = 0$, are satisfied.[29]

According to Eqs. (1.47) and $(1.74)_{(5,6)}$, the differential equations determining the points $\boldsymbol{x}$ of the deformed elastica are

$$x_1' = \cos\theta \qquad \text{and} \qquad x_2' = \sin\theta, \tag{1.96}$$

which, since $\cos\theta = 1 - 2\sin^2(\theta/2)$ and $\sin\theta = 2\sin(\theta/2)\sqrt{1-\sin^2(\theta/2)}$, and using Eq. (1.94), provide the two differential equations

$$\begin{aligned} x_1' &= 1 - 2k^2\,\mathrm{sn}^2(s\lambda + K(k), k), \\ x_2' &= 2k\,\mathrm{sn}\,(s\lambda + K(k), k)\,\mathrm{dn}\,(s\lambda + K(k), k), \end{aligned} \tag{1.97}$$

where dn is the Jacobi elliptic function, defined as

$$\mathrm{dn}\,(s\lambda + K(k), k) = \sqrt{1 - k^2\,\mathrm{sn}^2(s\lambda + K(k), k)}. \tag{1.98}$$

Since the following differentiation rules are known (Byrd and Friedman, 1954, p. 284)

$$\begin{aligned} &\frac{\partial}{\partial x}\mathrm{E}(x,k) = \sqrt{1 - k^2\sin^2 x}, \\ &\frac{\partial}{\partial x}\,\mathrm{am}\,(x,k) = \mathrm{dn}\,(x,k), \\ &\frac{\partial}{\partial x}\,\mathrm{cn}\,(x,k) = -\,\mathrm{sn}\,(x,k)\,\mathrm{dn}\,(x,k), \end{aligned} \tag{1.99}$$

where $\mathrm{E}(x,k)$ is the incomplete elliptic integral of the second kind of modulus k, defined as

$$\mathrm{E}(x,k) = \int_0^x \sqrt{1 - k\sin^2 t}\,dt,$$

taking into account the boundary conditions $(1.74)_{3,4}$, we integrate Eqs. (1.97), thus arriving at the equations describing the shape of the elastica

$$\begin{aligned} x_1 &= -s + \frac{2}{\lambda}\left\{\mathrm{E}\left[\mathrm{am}\,(s\lambda + K(k), k)\,, k\right] - \mathrm{E}\left[\mathrm{am}\,(K(k), k)\,, k\right]\right\}, \\ x_2 &= -\frac{2k}{\lambda}\,\mathrm{cn}(s\lambda + K(k)), \end{aligned} \tag{1.100}$$

which are identical with those provided by Love [1927, his eqs. (12) at n. 263].

[29] Equation (1.88) has been used in the boundary condition at $\theta'(l)$.

The displacement of the point of application of the force F is negative (for $F > 0$), and its absolute value can be obtained immediately from Eqs. (1.100)$_1$ because $|u_1(l)| = l - x_1(l)$ in the form

$$|u_1(l)| = 2l - \frac{2}{\lambda}\left\{\mathrm{E}\left[\mathrm{am}\left(l\lambda + K(k), k\right), k\right] - \mathrm{E}\left[\mathrm{am}\left(K(k), k\right), k\right]\right\}, \tag{1.101}$$

so that now using Eq. (1.88), we obtain

$$\frac{|u_1(l)|}{l} = 2 - \frac{\mathrm{E}\left[\mathrm{am}\left((2m+1)K(k), k\right), k\right] - \mathrm{E}\left[\mathrm{am}\left(K(k), k\right), k\right]}{mK(k)}, \qquad m = 1, 2, ... \tag{1.102}$$

which[30] eventually can be simplified to [an equation given by Reiss (1969), his eq. (3.20)]

$$\frac{|u_1(l)|}{l} = 2 - \frac{2\mathrm{E}(k)}{K(k)}, \tag{1.103}$$

which is independent of the bifurcation mode m, so the displacement of the right pin of the rod depends only on α (through k).

The mid-span deflection of the rod is null for even values of the mode m, whereas for odd m it can be evaluated as

$$\frac{|u_2(l/2)|}{l} = \frac{k}{mK(k)}, \qquad m = 1, 3, 5, ... \tag{1.104}$$

In summary, for a given α and a given mode m, we can calculate the corresponding λ [using Eq. (1.88)] and $u_1(l)$ [using Eq. (1.103)] and plot the elastica [using Eqs. (1.100)]. The bifurcation diagram showing the load F (normalised through division by $\pi^2 B$ and multiplication by l^2) as a function of the displacement of the right pin of the rod (normalised through division by l) is shown in Fig. 1.46 on the left, whereas the load F as a function of the rotation of the end of the rod is shown on the right. In the figure, the first four critical loads and the corresponding four branches are reported. We note that the branches do not cross each other and that the load is continuously increasing during the post-critical behaviour.

The deformed elastic lines have been evaluated and plotted in Fig. 1.47 for the first four branches at fixed values of α, namely, $\{10°, 45°, 90°, 135°, 160°\}$. These values of rotation correspond to rod end displacements, respectively, equal to $\{0.008, 0.149, 0.543, 1.049, 1.340\}l$. Note in this figure that the undeformed configuration, $\alpha = 0°$, also is reported to provide the scale of the displacement.

It should be noticed that the line of thrust (joining the two forces in Fig. 1.47) intersects the elastica at points of inflexion ($\theta' = 0$), separating (using the Love's nomenclature) different 'bays'.

In-plane secondary bifurcations of the simply supported elastica Let us go back to Fig. 1.46 and note that on each bifurcated branch there is a secondary bifurcation point (marked with a circle), which occurs when the two supports of the rod coincide, namely, when $u_1(l) = -l$, corresponding to $\alpha = 130.7099°$ and different load

[30] The following identities turn out to be useful (Byrd and Friedman, 1954):

$$\mathrm{am}K(k) = \pi/2, \quad \mathrm{am}[(2m+1)K(k)] = (2m+1)\pi/2, \quad \mathrm{E}(\pi/2, k) = \mathrm{E}(k), \quad \mathrm{E}(n\pi/2, k) = n\mathrm{E}(k).$$

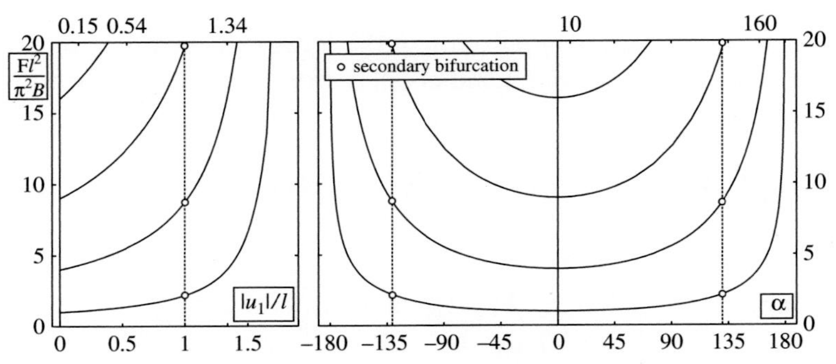

Figure 1.46. Dimensionless load $Fl^2/(\pi^2 B)$ of a doubly supported rod versus dimensionless displacement u_1/l (*left*) and rotation α of the end of the rod (*right*). The first four principal bifurcation points and branches are reported together with the first six secondary bifurcation points. All equilibrium configurations on the second, third and fourth branches are unstable. The first branch become unstable after the secondary bifurcation point.

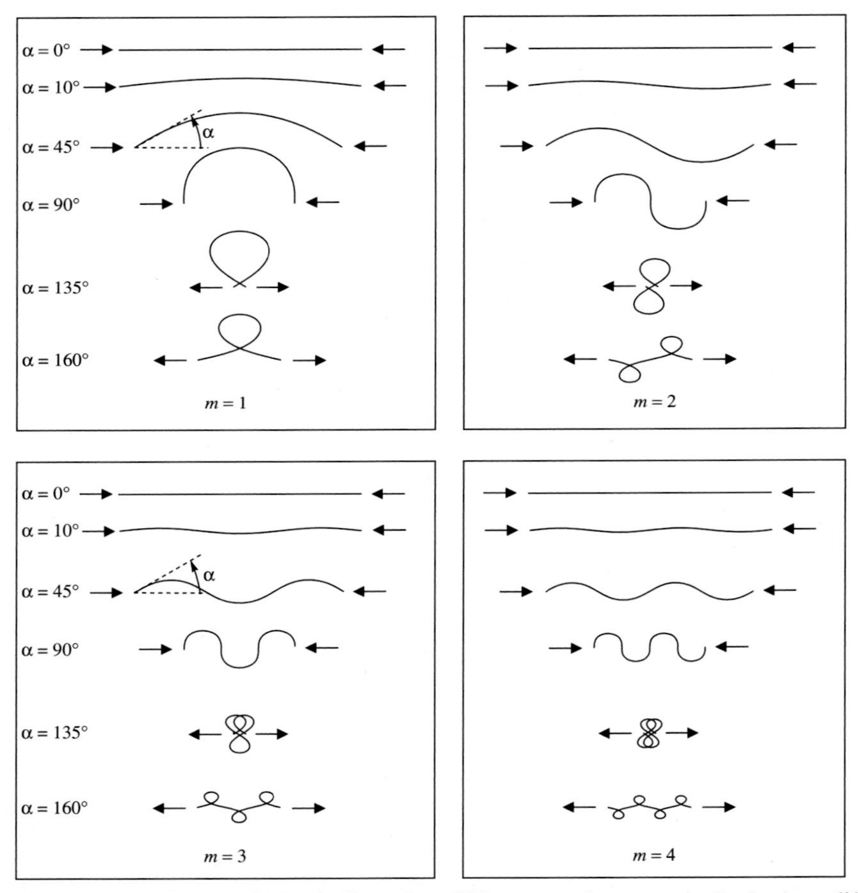

Figure 1.47. Deformed elastic lines for different modes $m = 1, 2, 3, 4$ at different values of parameters setting the deformation: The initial inflexion angle $\alpha = \{0, 10°, 45°, 90°, 135°, 160°\}$ and corresponding dimensionless displacement of the end of the rod $u_1/l = \{0, 0.008, 0.149, 0.543, 1.049, 1.340\}$. The deformed shapes of the elastica represent the post-critical behaviour of the structure.

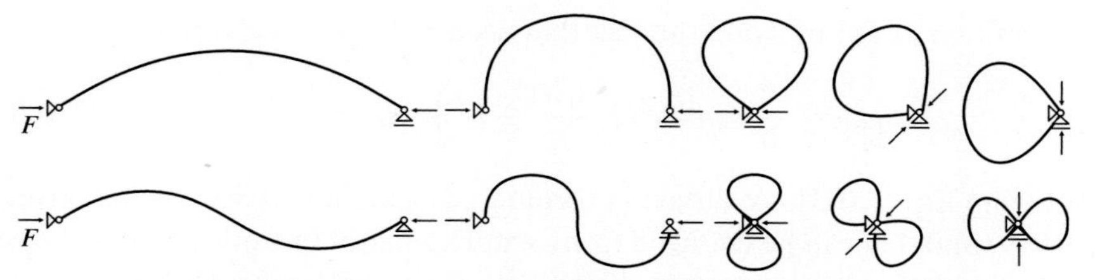

Figure 1.48. Sketch of the in-plane secondary bifurcation modes of the elastica, referred to the first (*top*) and second (*bottom*) modes. When the two supports coincide, the structure can suffer a rigid-body rotation. During this rotation, the horizontal load drops to zero (value reached when the elastica is rotated at 90°), so the force is maintained by the vertical reaction of the support. For rotation angles greater than 90° (not reported), the force changes sign. For imposed horizontal load, the structure becomes unstable when the two supports coincide and snaps to the configuration $u_1 = -2l$, where it is in equilibrium with a tensile load.

values: $Fl^2/(\pi B) = 2.1833$ for the first mode, $Fl^2/(\pi B) = 8.7335$ for the second mode, $Fl^2/(\pi B) = 19.6504$ for the third mode and so on.

These secondary bifurcation modes, which passed unnoticed until Maddocks (1984) (see also Kuznetsov and Levyakov, 2002; Sachkov and Levyakov, 2010), have a simple explanation. In fact, *when the two supports of the rod momentarily coincide during deformation along the bifurcation path, the structure can rigidly rotate about the pin.* During the rigid-body rotation, vertical reactions of the supports are generated, so the horizontal load drops until, when the structure is rotated at 90°, the horizontal load is reduced to zero, and finally, further rotation requires a negative force. The situation is sketched in Fig. 1.48 with reference to the first two modes. For imposed horizontal load, the situation in which the two supports coincide marks an instability point in the sense that (1) at this point the structure rigidly rotates and snaps to the configuration $u_1 = -2l$, where it is subject to a tensile load and (2) equilibrium configurations belonging to the post-critical path $m = 1$ and $\alpha > 130.7099°$ are unstable, as shown in Fig. 1.2c (see also Fig. 1.51). The treatment of instability follows.

Stability of the elastica In order to judge the stability of the various configurations of the elastica, let us now analyse the total potential energy P.

The total potential energy P of the elastica is expressed by the sum of the elastic energy

$$B \int_0^l \frac{(\theta')^2}{2} ds$$

plus the work made by the load F with reversed sign, so we obtain

$$P = B \int_0^l \frac{(\theta')^2}{2} ds - F(l - d), \tag{1.105}$$

where d is given by Eq. (1.56).

It is possible now to calculate the total potential energy for the previously obtained bifurcation solutions. In fact, inserting Eq. (1.81) into Eq. (1.105), and using the definition (1.56) of d and the fact that $\cos\alpha = 1 - 2k^2$, we obtain

$$\frac{P}{B} = -4\lambda^2 l\left(\frac{u_1(l)}{l} - k^2\right). \qquad (1.106)$$

Thus, employing Eq. (1.103), we obtain [a formula coincident with Eq. (3.26a) given by Reiss (1969) and that can be deduced from a more general formula given by Love (1927), his last equation reported on p. 407]

$$\frac{Pl}{B} = 16m^2 \mathrm{K}(k)\left\{2[\mathrm{E}(k) - \mathrm{K}(k)] + k^2\mathrm{K}(k)\right\}, \qquad m = 1,2,... \qquad (1.107)$$

which is always negative for every $k \in [0,1)$. Since the rectilinear configuration corresponds to null potential energy, all non-trivial solutions correspond to a smaller (negative) potential energy. At a fixed load F, the parameter λ is also fixed, so it can be shown that the function

$$\frac{Pl}{B} = 8\frac{\mathrm{E}(k)}{\mathrm{K}(k)} - 8 + k^2 \qquad (1.108)$$

is a negative and strictly decreasing function of k. Therefore, the total potential energy remains negative but increases for m increasing.

Therefore, since the first bifurcated path, $m = 1$, is that corresponding to the lower potential energy, this will be 'preferred' with respect to the others ($m > 1$). However, this statement means nothing about the stability of the bifurcated modes. For instance, the path corresponding to the second mode ($m = 2$) *could* be locally stable, though characterised by a potential energy higher than the first mode. Moreover, the first bifurcation mode could result unstable (and, in fact, it is for $\alpha \geq 130.7099°$).

We will see that *(if the rod is not constrained) the modes m = 2, 3, 4 are unstable, whereas the first mode m = 1 is stable, but only up to the point at which the two supports coincide*; after this configuration, the bifurcation path becomes unstable, a situation reported in Fig. 1.2*c*.

Instability of the configuration reached when the two supports coincide already has been explained qualitatively (Fig. 1.48). However, to decide stability (instability) in a rigorous way, we have to prove that the deformed configurations correspond to minima (maxima) of the total potential energy. Therefore, let us go back to the expression (1.105) of the potential energy and use Eq. (1.56) for d to obtain

$$P = B\int_0^l \frac{(\theta')^2}{2} ds - Fl + F\int_0^l \cos\theta ds, \qquad (1.109)$$

an equation holding for all deformed configurations characterised by a rotation $\theta(s)$, $s \in [0,l]$ satisfying

$$\theta'(0) = \theta'(l) = 0, \qquad (1.110)$$

together with the condition stating that every deformed line has to correspond to null difference of displacement u_2 between the two simply supported ends of the rod

$$u_2(l) - u_2(0) = 0,$$

namely, from Eq. $(1.55)_1$,

$$\int_0^l \sin\theta\, ds = 0. \qquad (1.111)$$

Let us consider the functional

$$V(\theta) = \int_0^l \frac{(\theta')^2}{2} ds - \lambda^2 l + \lambda^2 \int_0^l \cos\theta ds - \frac{R}{B} \int_0^l \sin\theta\, ds, \tag{1.112}$$

where the definition (1.73) has been employed and R is a Lagrangean multiplier (representing—as will be shown below—the vertical reactions of the supports). Functional (1.109) is defined over the set of kinematically admissible deformed lines, meaning every twice-differentiable displacement field $\boldsymbol{u}$ corresponding to a rotation field θ and curvature θ' related via the constitutive Eq. (1.66) to the bending moment.

Considering a variation $\tilde{\theta}$ of θ satisfying the boundary conditions $\tilde{\theta}'(0) = \tilde{\theta}'(l) = 0$, we can evaluate the first variation δV of functional V, which, keeping into account integration by parts,

$$\int_0^l \theta' \tilde{\theta}' = -\int_0^l \theta'' \tilde{\theta},$$

becomes

$$\delta V = -\int_0^l \left[\theta'' + \lambda^2 \sin\theta + \frac{R}{B} \cos\theta \right] \tilde{\theta}\, ds, \tag{1.113}$$

holding for all admissible perturbing fields $\tilde{\theta}$.

Imposing the vanishing of the first variation (1.113) yields the equilibrium equation (1.69) of the elastica, making transparent the fact that R represents the vertical reaction of the support, always null, except in the special case (which will not be addressed for simplicity) in which the two supports coincide.

To decide about stability of the equilibrium configuration, we have to consider the second variation of the functional V with respect to variations $\tilde{\theta}$ satisfying boundary conditions $\tilde{\theta}'(0) = \tilde{\theta}'(l) = 0$ plus the supplementary condition

$$\int_0^l \tilde{\theta} \cos\theta\, ds = 0, \tag{1.114}$$

following from the vanishing of the first variation of the constraint represented by Eq. (1.111). Keeping into account Eq. (1.114), the second variation of V can be readily calculated to be

$$\delta^2 V = \int_0^l \left[\left(\tilde{\theta}'\right)^2 - \lambda^2 \tilde{\theta}^2 \cos\theta \right] ds, \tag{1.115}$$

which, using integration by parts,

$$\int_0^l \left(\tilde{\theta}'\right)^2 = -\int_0^l \tilde{\theta}'' \tilde{\theta},$$

yields *the stability criterion*

$$\delta^2 V = -\int_0^l \left[\tilde{\theta}'' + \lambda^2 \tilde{\theta} \cos\theta \right] \tilde{\theta}\, ds \begin{cases} > 0 & \text{stability} \\ < 0 & \text{instability} \end{cases} \tag{1.116}$$

for all admissible $\tilde{\theta}$, subject to the condition (1.114).

In order to treat condition (1.116), let us denote with $\phi_n(s)$ ($s \in [0, l]$) the non-trivial solutions of the following Sturm-Liouville problem

$$\phi_n'' + \gamma_n \lambda^2 \cos\theta\, \phi_n = 0, \qquad \phi_n'(0) = \phi_n'(l) = 0, \tag{1.117}$$

where ϕ_n are the eigenfunctions associated with the eigenvalues γ_n with weight function $\lambda^2 \cos\theta$. It is known (see, e.g., Broman, 1970) that (1) problem (1.117) admits a countably infinite set of eigenvalues γ_n and these can be arranged in an increasing sequence ($\gamma_n < \gamma_{n+1}$ for each integer n), (2) $\gamma_n \longrightarrow \infty$ when $n \longrightarrow \infty$ and (3) the system $\phi_n(s)$, $s \in [0,l]$ is an orthogonal system with the weight function $\lambda^2 \cos\theta(s)$.

Direct integration of $(1.117)_1$ between 0 and l leads to

$$\gamma_n \lambda^2 \int_0^l \phi_n \cos\theta \, ds = 0, \tag{1.118}$$

showing that functions ϕ_n satisfy the supplementary condition (1.114). Moreover, multiplication of the differential Eq. $(1.117)_1$ by ϕ_n and integration between 0 and l yields [keeping into account the boundary conditions $(1.117)_{2,3}$ and using integration by parts]

$$\gamma_n \lambda^2 \int_0^l \phi_n^2 \cos\theta \, ds = \int_0^l (\phi_n')^2 \, ds > 0 \tag{1.119}$$

and the orthogonality condition

$$\int_0^l \phi_n \phi_m \cos\theta \, ds = 0, \qquad n \neq m. \tag{1.120}$$

Condition (1.119) defines a norm and Eq. (1.120) a weighted orthogonality condition for the functions ϕ_n with weight function $\lambda^2 \cos\theta$.

Therefore, system ϕ_n with weight function $\lambda^2 \cos\theta(s)$ can be used to give a Fourier series representation (converging in the mean) to the square-integrable function $\tilde{\theta}$,

$$\tilde{\theta} \sim \sum_{n=1}^{\infty} c_n \phi_n, \tag{1.121}$$

where c_n are the Fourier coefficients.

We do not need to specify coefficients c_n; rather, we can simply substitute the Fourier representation (1.121) into condition (1.116) and keep into consideration Eq. (1.117) to obtain

$$\delta^2 V = \int_0^l \left[\sum_{n=1}^{\infty} (\gamma_n - 1) c_n \phi_n \lambda^2 \cos\theta \right] \left[\sum_{m=1}^{\infty} c_m \phi_m \right] ds \; \begin{cases} > 0 & \text{stability} \\ < 0 & \text{instability} \end{cases} \tag{1.122}$$

which, employing conditions (1.119) and (1.120), finally becomes

$$\delta^2 V = \sum_{n=1}^{\infty} (1 - \frac{1}{\gamma_n}) c_n^2 \int_0^l (\phi_n')^2 \, ds \; \begin{cases} > 0 & \text{stability} \\ < 0 & \text{instability} \end{cases} \tag{1.123}$$

so that we arrive at the stability requirement

$$\begin{cases} \gamma_n \notin [0,1] & \text{stability} \\ \gamma_n \in [0,1] & \text{instability} \end{cases} \tag{1.124}$$

where γ_n are solutions of the Sturm-Liouville problem (1.117) written for a configuration determined by a specific function $\cos\theta$ and load λ^2. The values $\gamma_n = 0$ or $\gamma_n = 1$ represent 'transition' points and thus are called 'critical'.

Let us begin considering the stability of the straight configuration of the rod, $\cos\theta = 1$. In this case, the Sturm-Liouville problem (1.117) becomes

$$\phi_n'' + \gamma_n \lambda^2 \phi_n = 0, \qquad \phi_n'(0) = \phi_n'(l) = 0, \tag{1.125}$$

which has the non-trivial solutions

$$\phi_n = \cos\frac{n\pi s}{l}, \qquad \text{and} \qquad \gamma_n = \frac{F_n^{cr}}{F}, \tag{1.126}$$

where F_n^{cr} are the Euler's critical loads at different modes n, Eq. (1.78), so that when $\gamma_1 < 0$ or $\gamma_1 > 1$ ($0 < \gamma_1 < 1$), the straight configuration is stable (unstable), which corresponds to $F < F_1^{cr}$ ($F > F_1^{cr}$).

To judge stability of the deformed elastica, we can substitute Eq. (1.88) into Eq. (1.97) to obtain

$$\cos\theta = 1 - 2k^2 \operatorname{sn}^2\left[\left(\frac{s}{l}2m + 1\right)K(k), k\right] \tag{1.127}$$

and rewrite with the non-dimensional variable $\tilde{s} = s/l \in [0,1]$ the Sturm-Liouville problem (1.117) so that for a given mode m and inclination of the rod edge $k = \sin(\alpha/2)$, the *smallest eigenvalue* γ_m has to be determined as the solution of

$$\phi_m'' + \gamma_m 4m^2 K^2(k)\left\{1 - 2k^2 \operatorname{sn}^2\left[(2m\tilde{s} + 1)K(k), k\right]\right\}\phi_m = 0, \tag{1.128}$$

(where $'$ denotes differentiation with respect to $\tilde{s}$) subject to the boundary conditions $(1.117)_2$, namely, $\phi_m'(0) = \phi_m'(1) = 0$.

Problem (1.128) can be solved easily with a numerical routine. As noticed by Kuznestov and Levyakov (2002), a numerical procedure can be easily set, solving the differential Eq. (1.128) with the boundary conditions

$$\phi_m(0) = 1 \qquad \text{and} \qquad \phi_m'(0) = 0, \tag{1.129}$$

so that the trivial solution is always eliminated, and iterations on γ_m can be performed to match the condition $\phi_m'(1) = 0$. The iterations have been performed on the basis of a bisection method and the integration made with the 'NDSolve' function of Mathematica (5.2 from Wolfram Research) to produce the graphs reported in Fig. 1.49. The smallest eigenvalues γ_m for the first four modes $m = 1, 2, 3$ and 4 are reported versus the inclination α (in degrees) of the ends of the deformed rod. It is clear that the first mode, $m = 1$, is stable (the eigenvalues range between 1 and 10) until the two supports coincide for $\alpha = 130.7099°$, at which point the eigenvalues become discontinuous and fall to values within $[0,1]$ (in particular, $\gamma_1 = 9.9228$ at $\alpha = 130.7°$ and $\gamma_1 = 0.0059$ at $\alpha = 130.8°$). All modes higher than the first ($m = 2, 3$ and 4) are unstable with eigenvalues belonging to $[0,1]$. Note that for all the considered modes, the eigenvalues for $\alpha \geq 130.7099°$ are all coincident with the values for $m = 1$.

The instability mode associated with the unstable configurations occurring for $m = 1$ and $\alpha > 130.7099°$ [discovered by Maddocks (1984), his fig. 5*a*] and corresponding to 'self-intersecting' elasticas is not easy to be illustrated and understood. Therefore, we have performed an experiment with the beam model shown in Fig. 1.50, made up of an AISI 1095 steel strip ($180 \times 12 \times 0.07$ mm) having a 7 mm wide cut dividing the strip into two parts (one 5 mm wide and the other ∩-shaped with each of the two legs 2.5 mm wide). The model is suspended vertically in a self-intersecting

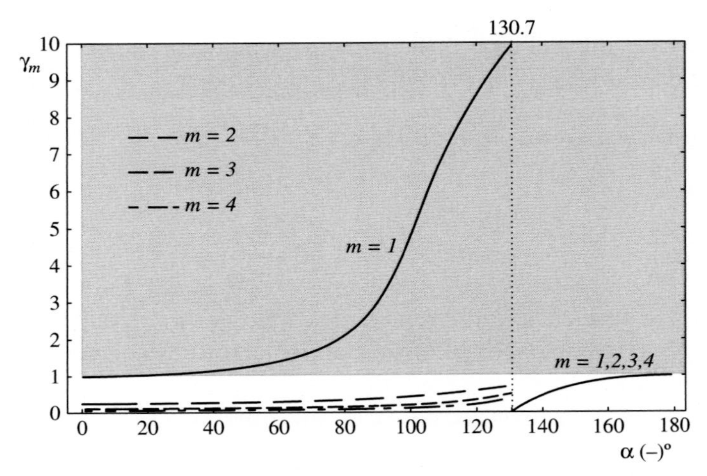

Figure 1.49. The smallest eigenvalues γ_m for the Sturm-Liouville problem (1.128) as functions of the rotation α of the ends of the rod. These determine the stability of the different modes of elastica ($m = 1,2,3$ and 4 are investigated). The grey region corresponds to stability, so only the first mode $m = 1$ is stable and only until the two supports of the rod coincide, a situation corresponding to $\alpha = 130.7099°$.

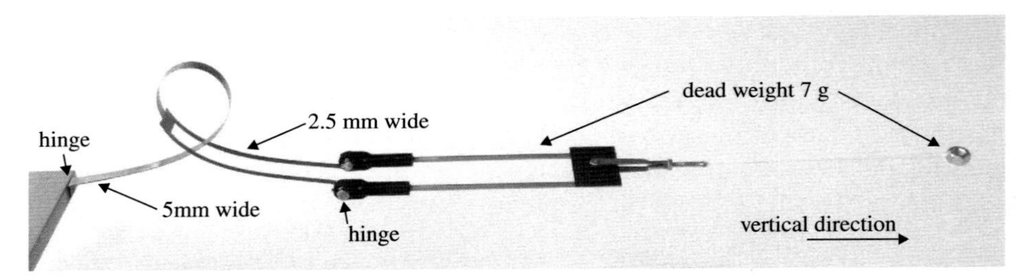

Figure 1.50. An unstable configuration of a beam model used to experimentally check the instability of 'self-intersecting elastica' (cut from a 180 × 12 × 0.07-mm steel strip) suspended vertically and subject to a 7 g weight. The unstable configuration is maintained because of minimal friction at one intersection point. A gentle movement of air is sufficient to break equilibrium, thus generating the motion recorded in Fig. 1.51.

configuration, in equilibrium with a 7-g dead weight. This equilibrium is possible only because of a 'minimal' friction between a contact point internal to the ∩-shaped part of the beam. If this configuration is just touched, the curved loop moves and flips around one hinge so violently that the images shown in Fig. 1.51 are blurred, although taken with 1/500 s exposure time (a Genie HM1400, DALSA Corporation) of a high speed camera (equipped with an 18–35 mm 1:3.5–4.5 D AF Nikkor lens, from Nikon Corporation, at 50 shots per second).

We finally note that with the proposed procedure to check the stability of the elastica configurations, it is not directly possible to conclude that *all* modes $m > 1$ are unstable, though mechanical considerations suggest that this might be the case. Our check of the instability of the modes $m = 2,..,4$ substantiates Love's (1927, p. 412) statement that 'the instability of forms of the elastica with more than the smallest possible number of inflexions between the ends is well known as an experimental fact'.

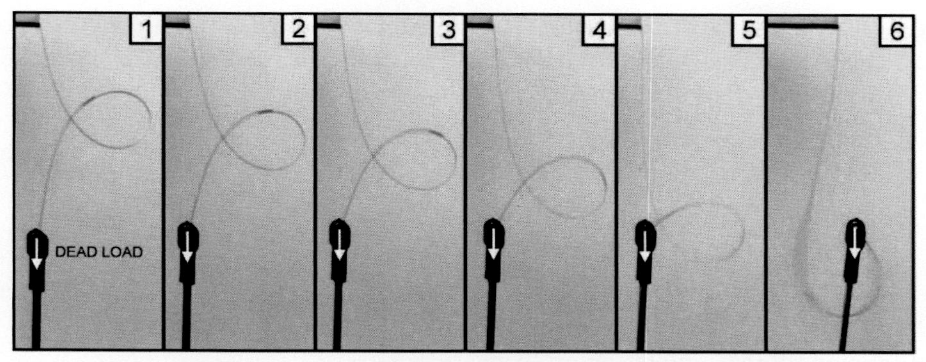

Figure 1.51. Experiment documenting the instability of 'self-intersecting elastica': The curved loop moves vertically towards the hinge and finally flips around this at the bottom of the sample. Photos have been taken at a speed of 50 shots per second, with an exposure time of 1/500 s.

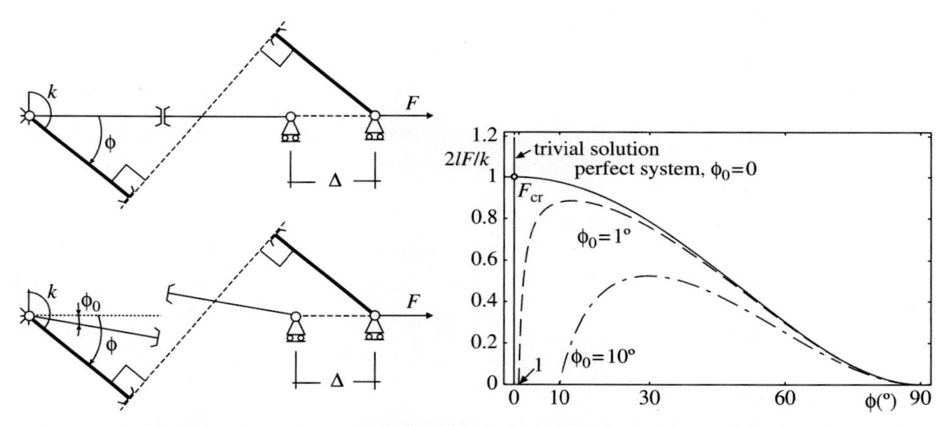

Figure 1.52. Bifurcation of a one degree of freedom elastic system under *tensile dead loading* (the rods of length l are rigid and jointed with a slider; a rotational elastic spring of stiffness k attached at the hinge on the left provides rotational stiffness); an imperfect system, with an initial inclination ϕ_0 of the rods, is reported in the lower part. Note that the bifurcation is 'purely geometrical' and is induced by the constraint in the middle of the beam which transmits rotation but not shear (*left*). The bifurcation diagram, showing bifurcation and softening in tension, is reported on the right, where $\phi_0 = \{1°, 10°\}$.

1.13.2 Exercise: Bifurcation of a structure subject to tensile dead load

Bifurcation of structural elements such as beams, columns and frames under dead loading is normally believed to be associated with compressive loads. It therefore may be interesting to know that bifurcation and instability for tensile dead loading are possible for the one degree of freedom structure shown in Fig. 1.52, where two rigid rods are connected through a 'slider' (a device which imposes the same rotation angle and null shear, leaving the possibility of relative sliding). This structure was invented by Zaccaria et al. (2011), who have also provided examples of systems made up of elastic rods exhibiting instabilities for tensile dead loading (it is a useful exercise to solve the elastica for these systems; see Zaccaria et al., 2011). Bifurcation load and equilibrium paths of the one degree of freedom structure shown in Fig. 1.52 can be calculated by considering the bifurcation mode reported in Fig. 1.52 and defined by the rotation angle ϕ. The elongation of the system is

$$\Delta = 2l\left(\frac{1}{\cos\phi} - 1\right), \tag{1.130}$$

so the potential energy is

$$P(\phi) = \frac{1}{2}k\phi^2 - 2Fl\left(\frac{1}{\cos\phi} - 1\right), \tag{1.131}$$

and solutions of the equilibrium problem are

$$F = \frac{k}{2l}\frac{\phi\cos^2\phi}{\sin\phi} \quad \text{(plus the trivial solution} \quad \phi = 0,\ \forall F). \tag{1.132}$$

Analysis of the second-order derivative of the strain energy reveals that the trivial solution is stable up to the critical load

$$F_{cr} = \frac{k}{2l}, \tag{1.133}$$

whereas the non-trivial path, *evidencing softening*, is unstable. For an imperfect system characterised by an initial angle ϕ_0, we obtain

$$\Delta = 2l\left(\frac{1}{\cos\phi} - \frac{1}{\cos\phi_0}\right), \tag{1.134}$$

and

$$P(\phi,\phi_0) = \frac{1}{2}k(\phi - \phi_0)^2 - 2Fl\left(\frac{1}{\cos\phi} - \frac{1}{\cos\phi_0}\right), \tag{1.135}$$

so equilibrium gives

$$F = \frac{k}{2l}\frac{(\phi - \phi_0)\cos^2\phi}{\sin\phi}. \tag{1.136}$$

The simple structure presented in Fig. 1.52, showing the possibility of a bifurcation under dead load in tension and displaying an overall softening behaviour, can be realised in practice, as shown by the teaching model reported in Fig. 1.53. Note that the rigid rods have been realized in wood, whereas the rotational stiffness of the hinge has been obtained through connection to a thin metal strip. The slider has been realised with two linear bearings (type Easy Rail SN22-80-500-610, purchased from Rollon), commonly used in machine design applications.

Structural systems with diffused elasticity and buckling under tensile dead loading have been analysed theoretically and tested experimentally by Zaccaria et al. (2011).

1.13.3 Exercise: Degrees of freedom and number of critical loads of elastic structures

It is often believed that the number of degrees of freedom of a structural system equals the number of bifurcation modes and loads (although the latter may be coincident). Contrary to this to this common belief, we have seen that there is only one

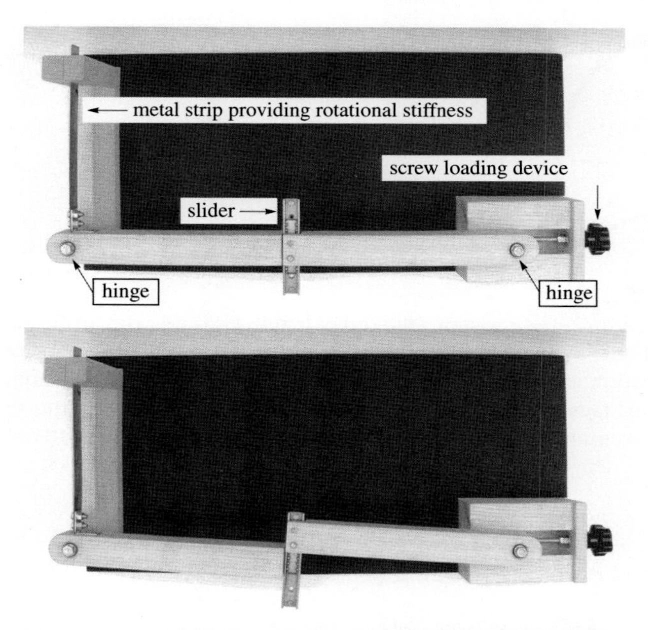

Figure 1.53. A teaching model (*top*: in which a metal strip reproduces the rotational spring and the load is given through a screw loading device to impose horizontal displacement) displaying bifurcation for tensile axial load (*bottom*) and designed to reproduce the model shown in Fig. 1.52 on the left.

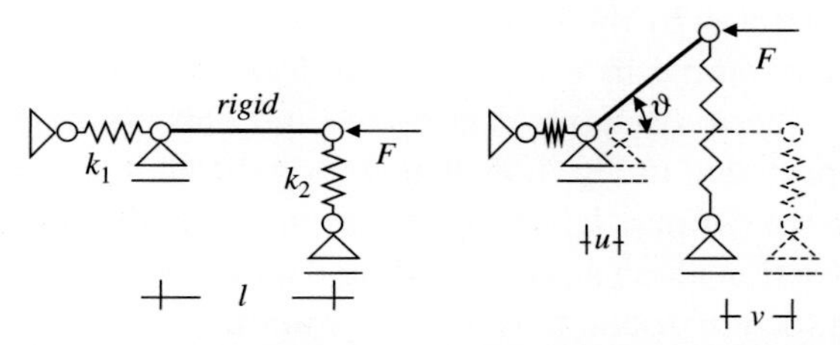

Figure 1.54. A two degree of freedom elastic system (the rod is rigid) with only one bifurcation load $F_{cr} = k_2 l$ for every possible combination of spring stiffnesses k_1 and k_2. *Left*: un-deformed system; *right*: system in a deformed configuration.

bifurcation load for the Shanley rod, when still in the elastic range, and we will see below that there are structures in which the number of critical loads is higher than the number of the degrees of freedom.

The simple structure shown in Fig. 1.54, where $F_{cr} = k_2 l$, provides an example of a structure with two degrees of freedom and only one bifurcation load. The reason for this peculiarity is that the load parameter F enters only one of the two eigenvalues of the Hessian of the potential energy. In particular, the potential energy of the structure is

$$P = \frac{k_1}{2}u^2 + \frac{k_2}{2}l^2 \sin^2\theta - Fl(1-\cos\theta) - Fu, \tag{1.137}$$

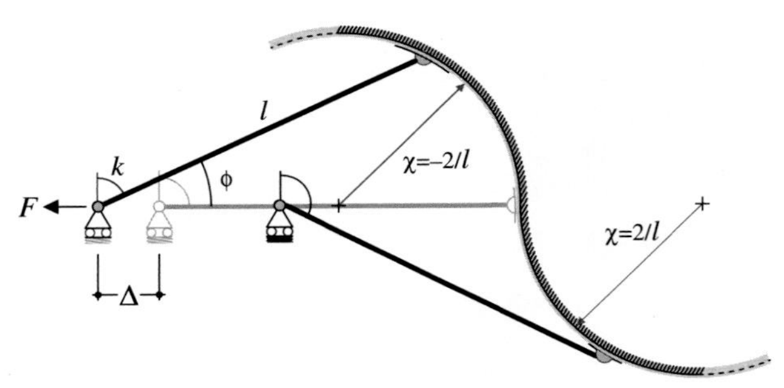

Figure 1.55. A one degree of freedom elastic system (the rod of length l is rigid, with an elastic hinge of stiffness k on the left and a hinge constrained to move on an S-shaped profile on the right) exhibiting two bifurcation loads, one in tension, $F_{cr} = k/l$, and one in compression, $F_{cr} = -k/(3l)$. The two buckling loads are a consequence of the discontinuity in the curvature of the constraint at the point of contact with the straight configuration (shown grey) of trivial equilibrium.

which, truncated at the second order in θ, becomes

$$P^{2nd} = \frac{k_1}{2}u^2 + \frac{k_2}{2}l^2\theta^2 - \frac{Fl}{2}\theta^2 - Fu, \tag{1.138}$$

so the Hessian is

$$\left[\begin{array}{cc} k_1 & 0 \\ 0 & k_2 l^2 - Fl \end{array}\right], \tag{1.139}$$

showing that one eigenvalue is not affected by the load F.

It is more surprising to find an example in which *a single degree of freedom structure has two critical loads, one in tension and one in compression*. This example, invented by Bigoni et al. (2012), is provided in Fig. 1.55, where a rigid rod of length l and with a rotational elastic spring (of stiffness k) on the left is pinned on the right with a hinge that can slide without friction along an S-shaped rigid constraint (made of two semi-circles of radius $l/2$). In the horizontal trivial configuration, the rod is in equilibrium in a position corresponding to a discontinuity in curvature ($\chi = -2/l$ in the upper circle and $\chi = 2/l$ in the lower) of the constraint. The horizontal coordinate x_1 of the constraint (taken null in the trivial configuration) can be expressed as a function of the vertical coordinate x_2 in the form

$$x_1 = f(x_2) = \left\{ \begin{array}{ll} \frac{l}{2} + l\sqrt{1/4 - (x_2/l)^2}, & x_2 \in [0, l/2] \\ \frac{3}{2}l - l\sqrt{1/4 - (x_2/l)^2}, & x_2 \in [0, -l/2] \end{array} \right. \tag{1.140}$$

Thus, since $x_2 = l\sin\phi$, the potential energy of the system can be written as

$$P = \frac{k}{2}\phi^2 - F\left[l\cos\phi - f(l\sin\phi)\right]. \tag{1.141}$$

A derivative of the potential energy (1.141) provides the equilibrium condition

$$F = \frac{k}{l}\frac{\phi}{-\sin\phi \pm \frac{\cos\phi\sin\phi}{\sqrt{1/4 - \sin^2\phi}}}, \tag{1.142}$$

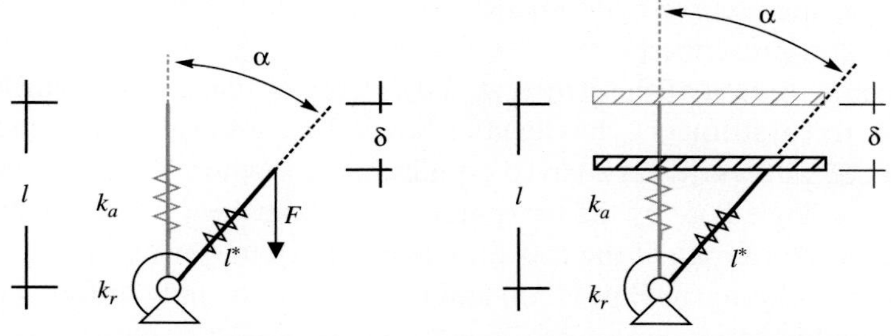

Figure 1.56. A two degree of freedom structure where axial deformation plays an important role. This simple system displays global softening with possible snap-back, cusp and smooth bifurcations. Moreover, the initially stable rectilinear configuration returns to be stable after a certain vertical load has been exceeded (see the experiments shown in Figs. 1.60 and 1.61). The dead force F is prescribed on the left, whereas the vertical displacement δ is prescribed on the right.

(where the + sign refers to the upper circle) yielding the tensile and compressive critical loads

$$F_{cr} = \frac{k}{l} \quad \text{and} \quad F_{cr} = -\frac{k}{3l},$$

respectively.

Stability can be analysed through the second derivative of Eq. (1.141); see Bigoni et al. (2012) for details, where systems such as that shown in Fig. 1.55 but with diffused elasticity (and thus governed by the elastica) also have been considered.

1.13.4 Exercise: A structure with a trivial configuration unstable at a certain load, returning stable at higher load

We reconsider a simplified but equivalent version (Fig. 1.56) of the buckling problem of the structure shown in Fig. 1.17, where the axial deformation plays an important role. This problem has been proposed and analysed by Feodosyev (1977, n. 197).

We will demonstrate the following two surprising features.

- An unstable (stable) vertical configuration of the structure may be stabilised (destabilised) through a reduction (increase) in the axial stiffness. This structure therefore provides a counter-example to the belief that an increase in stiffness cannot destabilise a structure, showing that *instability (stability) can be determined by an increase (decrease) of stiffness of a structural element*, a topic addressed in detail by Tarnai (1980).
- The vertical configuration of this structure, becoming unstable at a certain load, may return to be stable at higher load.

We will show, in addition, a well-known feature, namely, that the condition of stability changes whether the load (Fig. 1.17 *left*) or the displacement (Fig. 1.17, *right*, where the vertical displacement δ is prescribed through a frictionless, rigid constraint) is prescribed at the edge of the structure. In particular, when the load decreases as a function of the vertical displacement ('softening'), the structure is unstable for applied vertical dead force but stable for applied displacement. For a re-entering load/displacement curve ('snap-back'), the structure becomes unstable even if the displacement is prescribed.

The rectilinear configuration of the structure shown grey in Fig. 1.56 (in which the axially deformable rod of stiffness k_a has length l, lies vertically and is constrained by a rotational spring of stiffness k_r) is a trivial equilibrium configuration, whereas other configurations are singled out by the two parameters α and $\eta = l - l^*$, namely, the inclination and axial shortening of the rod. Simple equilibrium consideration of the deformed geometry yields again Eqs. (1.13) and (1.15), with the only difference that now $\tilde{F}$ and $\tilde{k}$ are defined as

$$\tilde{F} = \frac{F}{lk_a} \quad \text{and} \quad \tilde{k} = \frac{k_r}{l^2 k_a} \tag{1.143}$$

so that with the preceding definitions, eq. (1.16) still provides the two critical loads. Therefore, bifurcation curves analogous to those shown in Fig. 1.17 can be found. However, the curves reported in that figure are only those emanating from the smaller bifurcation load [corresponding to the minus sign in Eq. (1.15)], whereas a more complete picture of the behaviour of the structure requires plotting of the branches emanating also from the second bifurcation load and considerations on *stability*. Note that until now, the only difference between the two cases of loading reported in Fig. 1.56 is that in the case on the left, F is given and δ is calculated, whereas in the case on the right, δ is given and F is calculated as the reaction of the smooth constraint. In both cases, the bifurcation conditions and paths remain the same.

Stability under prescribed end force Since stability is different if the load or the displacement at the edge of the structure is prescribed, we initiate from the former case, reported on the left in Fig. 1.56, where the potential energy P (sum of the elastic energy and the potential energy of the external loads) is

$$P = l^2 k_a \left\{ \frac{\tilde{k}}{2}\alpha^2 + \frac{\eta^2}{2l^2} - \tilde{F}\left[1 - \left(1 - \frac{\eta}{l}\right)\cos\alpha\right] \right\}. \tag{1.144}$$

Thus, stationarity again gives the equilibrium condition Eq. (1.15) for non-trivial solutions, whereas stability follows from the properties of the Hessian of P. Written in terms of the non-dimensional quantity $\xi = \eta/l \in (0,1]$ and neglecting an unessential positive multiplier, the Hessian of the potential energy is

$$\begin{bmatrix} \tilde{k} - \tilde{F}(1-\xi)\cos\alpha & \tilde{F}\sin\alpha \\ \tilde{F}\sin\alpha & 1 \end{bmatrix}. \tag{1.145}$$

The eigenvalues of the Hessian (1.145) of the potential energy are proportional (through a factor 1/2) to

$$1 + \tilde{k} - \tilde{F}(1-\xi)\cos\alpha \pm \sqrt{\left[-1 + \tilde{k} - \tilde{F}(1-\xi)\cos\alpha\right]^2 + 4\tilde{F}^2\sin^2\alpha}, \tag{1.146}$$

which, since the equilibrium solutions satisfy

$$\xi = \tilde{F}\cos\alpha, \tag{1.147}$$

can be rewritten as (neglecting an unessential multiplicative factor 1/2)

$$\tilde{k}+\frac{3}{4}+\left(\tilde{F}\cos\alpha-\frac{1}{2}\right)^2\pm\sqrt{\left[\tilde{k}-\frac{5}{4}+\left(\tilde{F}\cos\alpha-\frac{1}{2}\right)^2\right]^2+4\tilde{F}^2\sin^2\alpha}. \qquad (1.148)$$

We see from the expression (1.148) that one of the eigenvalues is always positive, so

$$\text{Stability} \iff \tilde{F}^2\cos(2\alpha)-\tilde{F}\cos\alpha+\tilde{k}>0, \qquad (1.149)$$

and the boundary of stability is determined by

$$\tilde{F}=\frac{\cos\alpha\pm\sqrt{\cos^2\alpha-4\tilde{k}\cos 2\alpha}}{2\cos(2\alpha)}. \qquad (1.150)$$

We may conclude from the condition (1.149) that the vertical configuration $\alpha=0$ is stable for F inferior to the first and *superior to the second* critical load and unstable for loads lying between the first and second critical loads.

Equations (1.143) allow us to rewrite Eq. (1.149) for the vertical configuration $\alpha=0$ as

$$\text{Stability} \iff \left(\frac{lF}{k_r}\right)^2-\frac{l^2k_a}{k_r}\left(\frac{lF}{k_r}-1\right)>0 \qquad (1.151)$$

so that we can consider a fixed lF/k_r for which $lF/k_r>1$. It is clear that stability (instability) always can be obtained by sufficiently decreasing (increasing) the axial stiffness k_a, meaning that certain vertical unstable (stable) configurations can be stabilised (destabilised) by decreasing (increasing) the axial stiffness.

The simple mechanical system analysed in this chapter therefore provides an example of *a structure that can be stabilised (be made unstable) through a decrease (through an increase) in stiffness of one of its components without altering the geometry*, a possibility studied in detail by Tarnai (1980).

Regarding the two non-trivial configurations, we have to check if the path $\tilde{F}$–α satisfies the condition (1.149) or not. This condition defines regions in the $\tilde{F}$–α plane as those sketched in Fig. 1.57, reported for $\tilde{k}=8/48$.

Results of calculations of loads versus displacement δ of the free end of the structure and versus the inclination angle α are reported in Fig. 1.58.

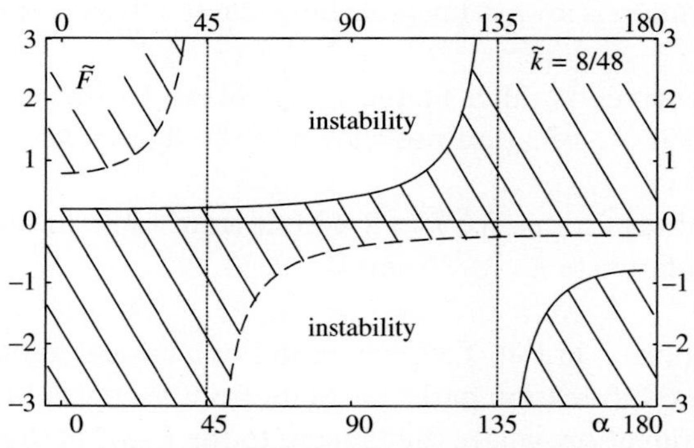

Figure 1.57. Stability and instability regions in the $\tilde{F}$–α plane for the structure subject to a dead force shown in Fig. 1.56 on the left.

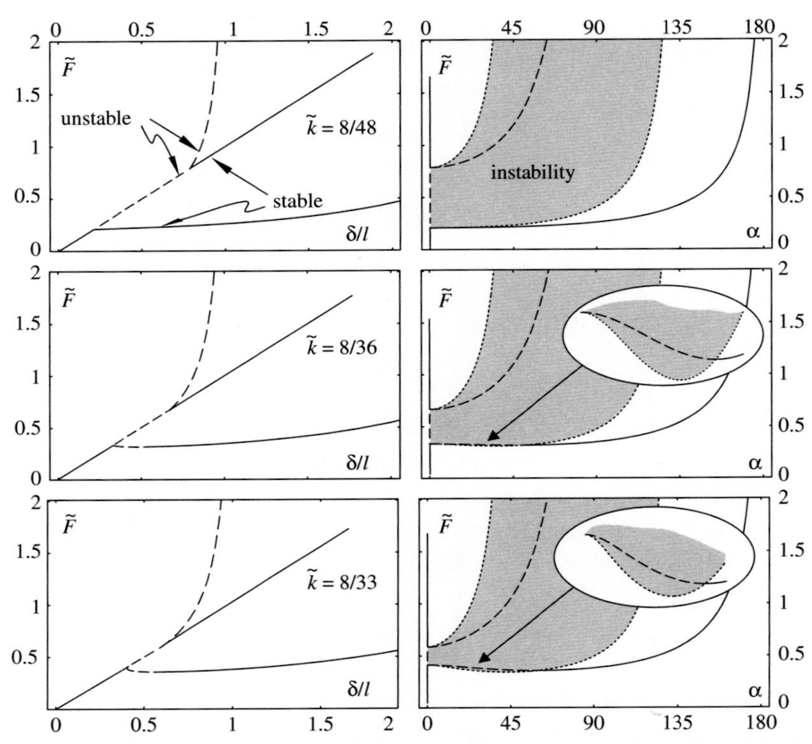

Figure 1.58. Dimensionless load $\tilde{F}$ versus dimensionless displacement δ/l (*left*) and inclination angle α (*right*) of the free edge of the structure shown in Fig. 1.56 on the left for different values of the stiffness ratio $\tilde{k}$.

Note that in the figure, the entire range of variation of $\alpha \in [0, 180°]$ has been explored, whereas δ/l has been limited to 2, to give evidence to the zone near the bifurcation loads. From the figure, we may conclude the following.

- The second bifurcated branch is always unstable and asymptotic to the value $\alpha = \pm\pi/2$, at which point $\eta = 0$ (after this point, the force reverses its sign and a configuration is reached which does not have physical meaning).
- The trivial equilibrium configuration is always unstable between the two critical loads.
- The first bifurcated branch becomes parallel in the $\tilde{F}$–δ/l plane to the line denoting the trivial path at $\alpha = 180°$, a feature not shown in the figure, where $\delta/l \leq 2$.
- The first bifurcated branch is always stable for $\tilde{k} = 8/48$ but is unstable in the descending part of the graphs relative to $\tilde{k} = 8/36$ and $\tilde{k} = 8/33$.

The special case $\tilde{k} = 8/32$ is reported in Fig. 1.59, where the two critical loads coincide. In this special case, the two bifurcation paths emanate from a single (the only) critical point. One is a cusp bifurcation, where the tangent to the $\tilde{F}$–δ/l path is ‘reversed’, whereas the other is a smooth bifurcation, where continuity of the tangent in the $\tilde{F}$–δ/l path is preserved.

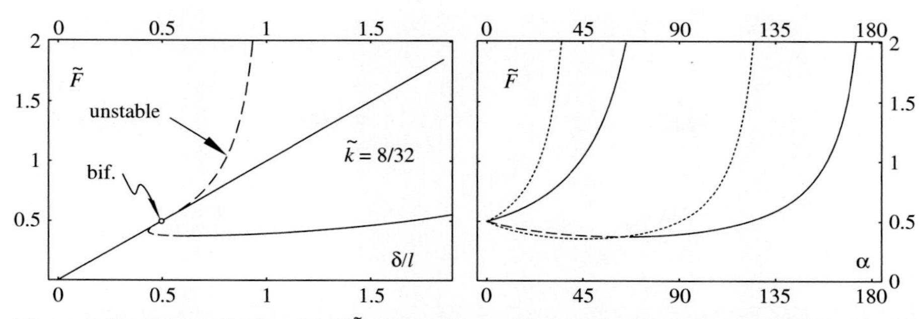

Figure 1.59. Dimensionless load $\tilde{F}$ versus dimensionless displacement δ/l (*left*) and inclination angle α (*right*) of the free edge of the structure shown in Fig. 1.56 on the left, for $\tilde{k} = 8/32$. In this case, there is only one bifurcation point from which a cusp bifurcation and a smooth bifurcation (with continuous tangent to the curve $\tilde{F}$–δ/l) emanate.

Figure 1.60. Teaching model showing the structure sketched in Fig. 1.56 on the left. (*A*) Unloaded configuration. At small load the vertical configuration is stable (*B*), at higher load it becomes unstable (*C*), but it returns stable at an even higher load (*D*). Note that the structure has to be kept momentarily vertical when the load at D is provided.

A curious feature of the analysed structure is that the trivial ($\alpha = 0$) configuration returns to be stable *after* the second bifurcation load. In an experiment on a model structure, owing to the effects of imperfection, the structure will follow a load/displacement behaviour close to the first bifurcation branch, so rotation will occur. However, if the rod is kept vertical and a force F superior to the second bifurcation load is imposed, the structure will remain vertical. This in fact can be observed in the teaching model reported in Figs. 1.60 and 1.61, both representing different views of the same experiment.

Figure 1.61. A different view of the teaching model illustrated in Fig. 1.60 (representing the structure shown in Fig. 1.56 on the left) in which the vertical configuration returns stable (D) for loads higher than those responsible for the first instability (C).

The vertical rod in the teaching model has been realized with a light aluminium tube, and the axial stiffness is provided by three springs. Rotational stiffness is provided by a steel lamina to work in bending, connected to the aluminium rod at one end and clamped (with a device to permit change in lenght and therefore stiffness) at the other end.

Stability under prescribed end displacement If the vertical component of displacement δ of the free edge of the structure is prescribed (Fig. 1.56) and the change in length of the axially deformable rod is

$$\eta = \frac{\delta - l(1-\cos\alpha)}{\cos\alpha}, \tag{1.152}$$

the potential energy is given by

$$W(\alpha) = \frac{l^2 k_a}{2}\left[\tilde{k}\alpha^2 + \left(1 - \frac{1-\delta/l}{\cos\alpha}\right)^2\right] \tag{1.153}$$

and is a function only of the rotation α.

Stationarity of Eq. (1.153) yields the equilibrium configurations, solutions of

$$\tilde{k}\alpha - (1-\delta/l)\left(1 - \frac{1-\delta/l}{\cos\alpha}\right)\frac{\sin\alpha}{\cos^2\alpha} = 0, \tag{1.154}$$

which, though they should be solved for α at given δ/l, become easier when solved for δ/l as a function of α, i.e.,

$$\delta/l = 1 - \frac{\cos\alpha}{2}\left(1 \mp \sqrt{1 - 4\tilde{k}\alpha\cot\alpha}\right). \tag{1.155}$$

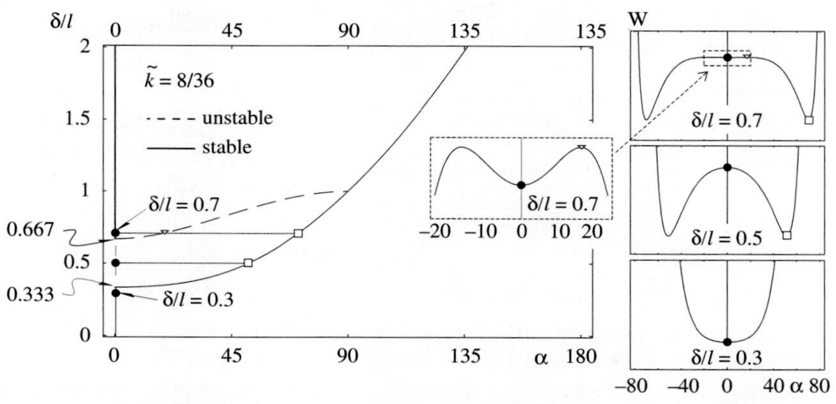

Figure 1.62. Dimensionless displacement δ/l versus rotation α (in degrees) for stable and unstable equilibrium paths of the structure shown in Fig. 1.56 on the right. The dimensionless stiffness $\tilde{k}$ is taken equal to 8/36. Plots of the potential energy as a function of the rotation α are shown on the right at fixed values of δ/l; here we see that the straight configuration is stable and corresponds to an *absolute (or global) minimum* of energy (*bottom*: $\delta/l = 0.3$); it becomes unstable and corresponds to a maximum of energy (*centre*: $\delta/l = 0.5$) and returns stable, though in a *relative (or local) minimum* of energy (*top*: see also the detail, $\delta/l = 0.7$). The two (for positive α) inclined equilibrium configurations for $\delta/l = 0.7$ correspond to a maximum and a relative (or local) minimum.

It can be shown using Eq. (1.152) and transforming η into the reaction force $F = \eta k_a$ that Eqs. (1.155) reduce to Eqs. (1.15), so prescribing the force F or the displacement δ at the edge of the structure gives the same equilibrium paths. Stability is decided on the basis of the sign of the second derivative of the potential energy (1.153), which is

$$\frac{d^2W(\alpha)}{d\alpha^2}\frac{1}{k_a l} = \tilde{k} + \left(1 - \frac{\delta}{l}\right)^2 \frac{1 + 2\sin^2\alpha}{\cos^4\alpha} - \left(1 - \frac{\delta}{l}\right)\frac{1 + \sin^2\alpha}{\cos^3\alpha}, \tag{1.156}$$

so a substitution of $\alpha = 0$ and of solutions (1.155) allows one to decide whether or not the equilibrium paths at prescribed displacement δ are stable. The result for the straight configuration of the rod corresponds to that for prescribed end force: The rectilinear configuration is stable only when the displacement δ falls outside the interval

$$\frac{\delta}{l} = \frac{1 \pm \sqrt{1 - 4\tilde{k}}}{2}, \tag{1.157}$$

corresponding to the same forces (1.16). Note that even in this case, certain unstable (stable) vertical configurations can be stabilised (made unstable) through a sufficiently strong decrease (increase) in axial stiffness.

Results for the dimensionless displacement δ/l versus rotation α (in degrees) are reported in Fig. 1.62 for $\tilde{k} = 8/36$ and in Fig. 1.63 for $\tilde{k} = 8/33$. Note that in both cases there are two non-trivial paths (for positive values of α), each emanating from one of the bifurcation values. Among these, the upper non-trivial path is always unstable and terminates at $\alpha = 90°$ (for higher values of α, this path becomes physically meaningless). The case $\tilde{k} = 8/48$ has not been reported because its results

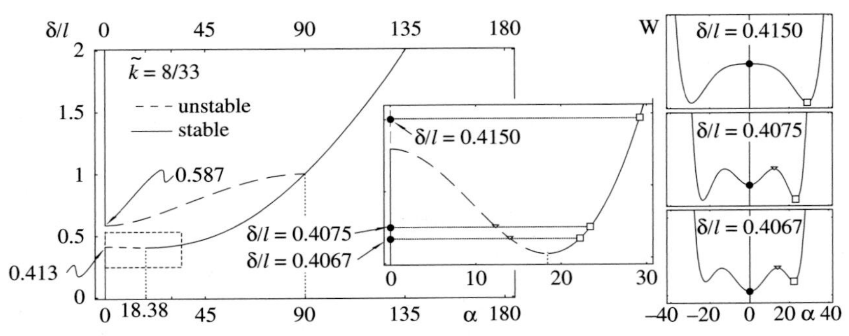

Figure 1.63. Dimensionless displacement δ/l versus rotation α (in degrees) for stable and unstable equilibrium paths of the structure shown in Fig. 1.56 on the right. The dimensionless stiffness $\tilde{k}$ is taken equal to 8/33. Plots of the potential energy as a function of the rotation α are shown on the right for fixed values of δ/l; here, from the lower part to the upper, we see that the straight configuration is stable and corresponds to an *absolute (or global)* (for $\delta/l = 0.4067$) and a *relative (or local)* (for $\delta/l = 0.4075$) minimum of energy, whereas it becomes unstable (corresponding to a maximum of energy) for $\delta/l = 0.4150$. The two (for positive α, in the cases $\delta/l = 0.4067$ and 0.4075) inclined equilibrium configurations are one unstable and one stable.

are qualitatively identical to that corresponding to $\tilde{k} = 8/36$ (Fig. 1.62), so for prescribed displacement δ, there is no difference from the point of view of stability between softening and hardening in the load/displacement curve (see Fig. 1.58). For prescribed displacement, the behaviour becomes unstable in the decreasing portion of the lower equilibrium branch of the displacement versus rotation diagram; see the detail in Fig. 1.63.

In both Figs. 1.62 and 1.63 we have reported plots of the potential energy (1.156) at fixed values of δ/l as a function of α. We see from these graphs that the equilibrium configurations correspond to minima or maxima and that *a minimum of energy can be absolute (also called 'global') or relative (also called 'local')*, so an absolute minimum can, for obvious reasons, be considered 'more stable' than a relative minimum. Using the criterion of the second derivative of the potential energy, we cannot distinguish between local and global minima.

1.13.5 Exercise: Flutter and divergence instability in an elastic structure induced by Coulomb friction

The purpose of this subsection is to illustrate flutter and divergence instability in a structure, the so-called Ziegler column (Ziegler, 1977), as induced by a tangential follower load, and to show that the follower load, and therefore the related instabilities, can be generated by Coulomb friction. We will see that flutter and divergence are dynamical instabilities, the former consisting in an oscillatory vibration of increasing amplitude, whereas the latter in an exponentially growing motion.

The finding that Coulomb friction may provide the follower load necessary for instability is due to Bigoni and Noselli (2011), who have shown how a follower load can be transmitted to a structure through a wheel mounted at the top of it, assuming that the wheel is of negligible mass and can rotate freely about its axis during sliding with Coulomb friction against a rigid support. This finding provides an indisputable

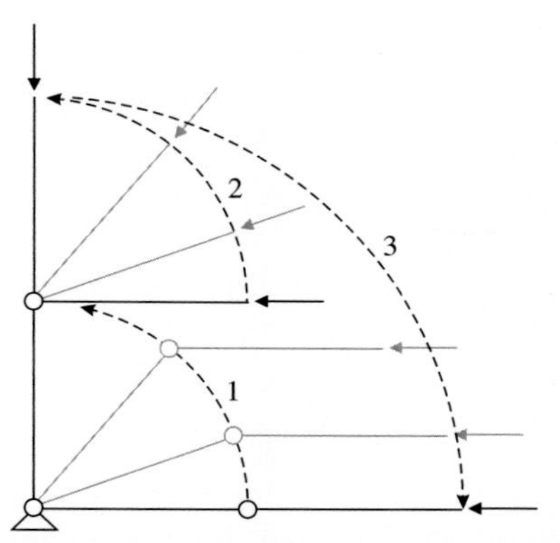

Figure 1.64. A tangential follower force (of the same type that loads the structure shown in Fig. 1.65) is applied at the right end of a two degree of freedom system composed by two hinged rigid bars. The mechanical system is not conservative because in the closed loop $1 \longrightarrow 2 \longrightarrow 3$ (intermediate configurations are represented grey) positive work is produced by the applied force.

and definitive experimental demonstration that flutter instability can be induced by friction. In the following we will limit the presentation to the general concepts and ideas, addressing the interested reader to Bigoni and Noselli (2011) for a thorough treatment with full detail.

A two degree of freedom structure subject to follower load

Flutter and divergence instability may be induced in a structure by a follower tangential force remaining coaxial to the rod to which it is applied and defining an example of a non-conservative load capable of producing positive work in a closed loop (Fig. 1.64). Subject to this load, we consider the two degree of freedom rigid and massless rods system shown in Fig. 1.65, where two rotational springs of stiffnesses k_1 and k_2 provide the elasticity. The generic configuration of the system remains determined by the two Lagrangean parameters α_1 and α_2. The concentrated masses m_1 and m_3 are located at the points D and E, at a distance d from A and h from B, whereas the concentrated mass m_2 is located at C. The tangential follower load $\boldsymbol{P}$, applied at C and taken positive when compressive, maintains the direction parallel to the rod BC. The analysis of a mechanical system similar to that under consideration (in which mass m_3 is not present) can be found in Herrmann (1971), Ziegler (1977) and Nguyen (1995), whereas the akin problem of a clamped elastic rod subjected to a load tangential to its axis at the free end has been solved by Beck (1952) and Pflüger (1955).

A simple static analysis of the structure shown in Fig. 1.65 is sufficient to conclude that only the trivial (straight) configuration satisfies equilibrium (in fact, equilibrium of the rod BC is possible only if $\alpha_1 = \alpha_2$, so equilibrium of the complex ABC requires $\alpha_1 = 0$), so bifurcations are excluded.

Let us now obtain the equations of motion for the system. To this purpose, we start from the position vectors of the concentrated masses m_1, m_2 and m_3

$$
\begin{aligned}
D - A &= d\cos\alpha_1 \boldsymbol{e}_1 + d\sin\alpha_2 \boldsymbol{e}_2,\\
E - A &= (l_1\cos\alpha_1 + h\cos\alpha_2)\,\boldsymbol{e}_1 + (l_1\sin\alpha_1 + h\sin\alpha_2)\,\boldsymbol{e}_2,\\
C - A &= (l_1\cos\alpha_1 + l_2\cos\alpha_2)\,\boldsymbol{e}_1 + (l_1\sin\alpha_1 + l_2\sin\alpha_2)\,\boldsymbol{e}_2,
\end{aligned}
\tag{1.158}
$$

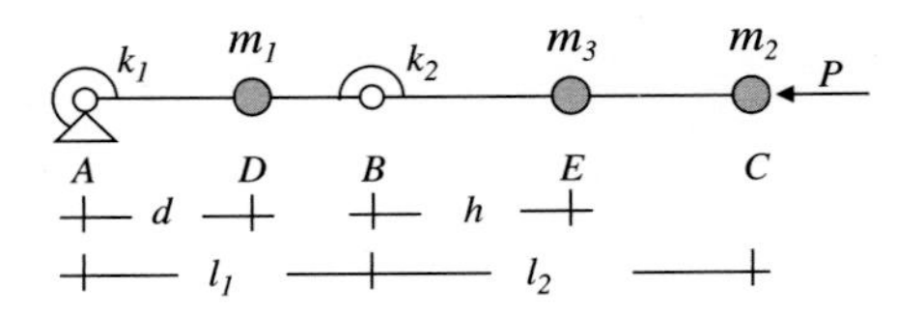

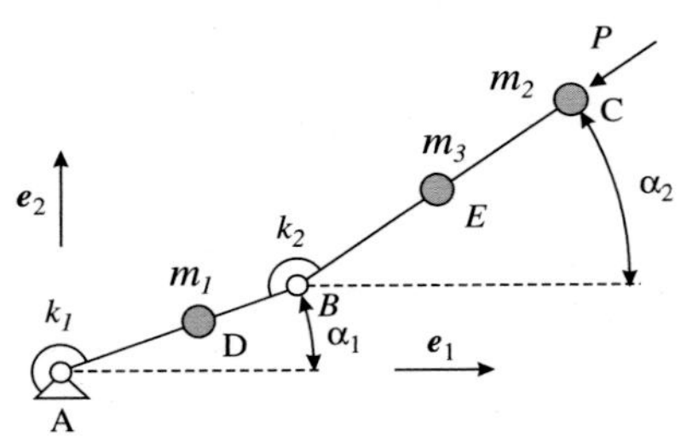

Figure 1.65. A two degree of freedom system subject to follower (non-conservative) load (the force $\boldsymbol{P}$ is applied at C and always remains parallel to rod BC), exhibiting flutter and divergence instability. The rods are rigid and massless and connected with two springs of stiffness k_1 and k_2; three concentrated masses m_1, m_2 and m_3 are present. Note that there is no bifurcation for this system.

where $\boldsymbol{e}_1$ and $\boldsymbol{e}_2$ are the two unit vectors singling out the horizontal and vertical directions, respectively, so that the force $\boldsymbol{P}$, of modulus P, can be expressed as

$$\boldsymbol{P} = -P\cos\alpha_2\boldsymbol{e}_1 - P\sin\alpha_2\boldsymbol{e}_2. \tag{1.159}$$

The velocities of points D, E and C (the time derivative is denoted by a superimposed dot) are

$$\begin{aligned}
\dot{D} &= \frac{\partial(D-A)}{\partial t} = -d\,(\dot{\alpha}_1\sin\alpha_1\boldsymbol{e}_1 - \dot{\alpha}_1\cos\alpha_1\boldsymbol{e}_2),\\
\dot{E} &= \frac{\partial(E-A)}{\partial t} = (-l_1\dot{\alpha}_1\sin\alpha_1 - h\dot{\alpha}_2\sin\alpha_2)\boldsymbol{e}_1 + (l_1\dot{\alpha}_1\cos\alpha_1 + h\dot{\alpha}_2\cos\alpha_2)\boldsymbol{e}_2,\\
\dot{C} &= \frac{\partial(C-A)}{\partial t} = (-l_1\dot{\alpha}_1\sin\alpha_1 - l_2\dot{\alpha}_2\sin\alpha_2)\boldsymbol{e}_1 + (l_1\dot{\alpha}_1\cos\alpha_1 + l_2\dot{\alpha}_2\cos\alpha_2)\boldsymbol{e}_2,
\end{aligned} \tag{1.160}$$

so accelerations $\boldsymbol{a}_1$, $\boldsymbol{a}_2$ and $\boldsymbol{a}_3$ of the masses m_1, m_2 and m_3 are, respectively,

$$\begin{aligned}
\boldsymbol{a}_1 &= \frac{\partial\dot{D}}{\partial t} = -d\left(\dot{\alpha}_1^2\cos\alpha_1 + \ddot{\alpha}_1\sin\alpha_1\right)\boldsymbol{e}_1 - d\left(\dot{\alpha}_1^2\sin\alpha_1 - \ddot{\alpha}_1\cos\alpha_1\right)\boldsymbol{e}_2,\\
\boldsymbol{a}_2 &= \frac{\partial\dot{C}}{\partial t} = \left(-l_1\dot{\alpha}_1^2\cos\alpha_1 - l_1\ddot{\alpha}_1\sin\alpha_1 - l_2\dot{\alpha}_2^2\cos\alpha_2 - l_2\ddot{\alpha}_2\sin\alpha_2\right)\boldsymbol{e}_1\\
&\quad + \left(-l_1\dot{\alpha}_1^2\sin\alpha_1 + l_1\ddot{\alpha}_1\cos\alpha_1 - l_2\dot{\alpha}_2^2\sin\alpha_2 + l_2\ddot{\alpha}_2\cos\alpha_2\right)\boldsymbol{e}_2,\\
\boldsymbol{a}_3 &= \frac{\partial\dot{E}}{\partial t} = \left(-l_1\dot{\alpha}_1^2\cos\alpha_1 - l_1\ddot{\alpha}_1\sin\alpha_1 - h\dot{\alpha}_2^2\cos\alpha_2 - h\ddot{\alpha}_2\sin\alpha_2\right)\boldsymbol{e}_1\\
&\quad + \left(-l_1\dot{\alpha}_1^2\sin\alpha_1 + l_1\ddot{\alpha}_1\cos\alpha_1 - h\dot{\alpha}_2^2\sin\alpha_2 + h\ddot{\alpha}_2\cos\alpha_2\right)\boldsymbol{e}_2.
\end{aligned} \tag{1.161}$$

Noting that the moments transmitted by the rotational springs to the rods are $k_1\alpha_1$ and $k_2(\alpha_2-\alpha_1)$, the principle of virtual power is written as

$$\begin{aligned}
&\boldsymbol{P}\cdot\delta C - m_1\boldsymbol{a}_1\cdot\delta D - m_2\boldsymbol{a}_2\cdot\delta C - m_3\boldsymbol{a}_3\cdot\delta E\\
&\qquad - k_1\alpha_1\delta\alpha_1 - k_2(\alpha_2-\alpha_1)(\delta\alpha_2-\delta\alpha_1) = 0,
\end{aligned} \tag{1.162}$$

where · denotes the scalar product, and the virtual velocities δC, δD and δE have the same expressions (1.160) with the · replaced by δ.

Imposing now condition (1.162) and invoking the arbitrariness of $\delta\alpha_1$ and $\delta\alpha_2$ yields the two equations

$$\left[m_1 d^2+(m_2+m_3)l_1^2\right]\ddot{\alpha}_1+(m_2 l_2+m_3 h)l_1\ddot{\alpha}_2\cos(\alpha_1-\alpha_2) +(m_2 l_2+m_3 h)l_1\dot{\alpha}_2^2\sin(\alpha_1-\alpha_2)-Pl_1\sin(\alpha_1-\alpha_2)+k_1\alpha_1+k_2(\alpha_1-\alpha_2)=0$$

and

$$(m_2 l_2+m_3 h)l_1\ddot{\alpha}_1\cos(\alpha_1-\alpha_2)+(m_2 l_2^2+m_3 h^2)\ddot{\alpha}_2 -(m_2 l_2+m_3 h)\dot{\alpha}_1^2\sin(\alpha_1-\alpha_2)-k_2(\alpha_1-\alpha_2)=0 \quad (1.163)$$

governing the dynamics of the system.

The differential equations (1.163) linearised near the trivial (equilibrium) configuration $\alpha_1=\alpha_2=0$ become

$$\begin{aligned}&\left[m_1 d^2+(m_2+m_3)l_1^2\right]\ddot{\alpha}_1+(m_2 l_2+m_3 h)l_1\ddot{\alpha}_2+(k_2-Pl_1)(\alpha_1-\alpha_2)+k_1\alpha_1=0,\\ &(m_2 l_2+m_3 h)l_1\ddot{\alpha}_1+(m_2 l_2^2+m_3 h^2)\ddot{\alpha}_2-k_2(\alpha_1-\alpha_2)=0.\end{aligned} \quad (1.164)$$

We now look for time-harmonic vibrations near the equilibrium configuration, so the Lagrangean parameters are now assumed to be harmonic functions of time

$$\alpha_j=A_j e^{-i\Omega t}, \qquad j=1,2 \quad (1.165)$$

where A_j are (complex) amplitudes, Ω is the circular frequency and i is the imaginary unit ($i=\sqrt{-1}$), so a substitution of Eq. (1.165) into Eqs. (1.164) yields

$$\begin{bmatrix}(\rho_1 y_1^2+\rho_2+1)\lambda^2\omega^2-1-k+\gamma & \lambda\omega^2(1+\rho_2 y_2)+1-\gamma\\ \lambda\omega^2(1+\rho_2 y_2)+1 & \omega^2(1+\rho_2 y_2^2)-1\end{bmatrix}\begin{bmatrix}A_1\\ A_2\end{bmatrix}=0, \quad (1.166)$$

where

$$\begin{aligned}&\rho_1=\frac{m_1}{m_2}, \quad \rho_2=\frac{m_3}{m_2}, \quad y_1=\frac{d}{l_1}, \quad y_2=\frac{h}{l_2},\\ &\lambda=\frac{l_1}{l_2}, \quad k=\frac{k_1}{k_2}, \quad \omega^2=\frac{m_2 l_2^2\Omega^2}{k_2}, \quad \gamma=\frac{Pl_1}{k_2}.\end{aligned} \quad (1.167)$$

Non-trivial solution of system (1.166) is possible if the determinant of the matrix vanishes, a condition which immediately provides the solutions

$$\omega^2=\frac{b(\gamma)\pm\sqrt{\Delta(\gamma)}}{2a}, \quad (1.168)$$

where

$$\begin{aligned}&a=\lambda^2\left[y_1^2\rho_1+(1-y_2)^2\rho_2+y_1^2 y_2^2\rho_1\rho_2\right]>0,\\ &b(\gamma)=\lambda\left[2+2y_2\rho_2+\lambda(1+y_1^2\rho_1+\rho_2)\right]+(k+1)(1+\rho_2 y_2^2)\\ &\qquad -\gamma\left(1+\lambda+y_2^2\rho_2+y_2\lambda\rho_2\right),\\ &\Delta(\gamma)=b^2(\gamma)-4ka,\end{aligned} \quad (1.169)$$

so

$$\Omega=\frac{1}{l_2}\sqrt{\frac{k_2}{m_2}}\,\omega. \quad (1.170)$$

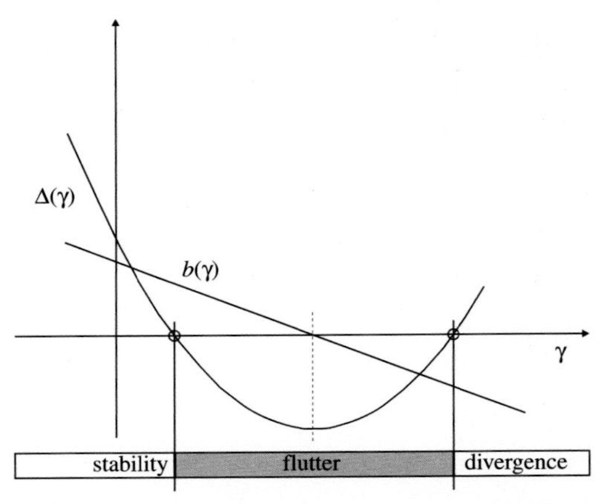

Figure 1.66. Coefficients $b(\gamma)$ and $\Delta(\gamma)$ in Eq. (1.168) (note that the load P is included in the variable γ) determine the stability behaviour of the structure shown in Fig. 1.65.

We note that $\Delta(0) > 0$ because there must be no flutter for the unloaded structure $P = 0$ and that

$$
\begin{aligned}
& b(\gamma) > 0, \Longleftrightarrow \gamma < \gamma_0, \\
& \Delta(\gamma) < 0, \Longleftrightarrow 0 < \gamma_0 - \frac{2\sqrt{ka}}{1+\lambda+y_2^2\rho_2+y_2\lambda\rho_2} < \gamma < \gamma_0 \\
& \qquad + \frac{2\sqrt{ka}}{1+\lambda+y_2^2\rho_2+y_2\lambda\rho_2},
\end{aligned} \tag{1.171}
$$

where

$$
\gamma_0 = \frac{\lambda\left[2+2y_2\rho_2+\lambda(1+y_1^2\rho_1+\rho_2)\right]+(k+1)(1+\rho_2 y_2^2)}{1+\lambda+y_2^2\rho_2+y_2\lambda\rho_2} > 0, \tag{1.172}
$$

a situation sketched in Fig. 1.66.

The following three possibilities only arise:

- Two real and positive values for ω^2 correspond to $b(\gamma) > 0$ and $\Delta(\gamma) > 0$, so vibrations are sinusoidal, a *stable* situation;
- Two complex conjugate values for ω^2 correspond to $\Delta(\gamma) < 0$, so there are four exponential solutions: two blowing up and the other two decaying with time, a situation corresponding to *flutter instability*.
- Two real and negative values for ω^2 correspond to $b(\gamma) < 0$ and $\Delta(\gamma) > 0$, so vibrations are exponential with time (two amplifying and two decaying), a situation corresponding to *divergence instability*.

As a conclusion, *flutter instability* occurs when

$$
0 < \gamma_0 - \frac{2\sqrt{ka}}{1+\lambda+y_2^2\rho_2+y_2\lambda\rho_2} \leq \frac{Pl_1}{k_2} \leq \gamma_0 + \frac{2\sqrt{ka}}{1+\lambda+y_2^2\rho_2+y_2\lambda\rho_2}, \tag{1.173}
$$

and *divergence instability* occurs when

$$
\frac{Pl_1}{k_2} > \gamma_0 + \frac{2\sqrt{ka}}{1+\lambda+y_2^2\rho_2+y_2\lambda\rho_2}. \tag{1.174}
$$

In the particular case in which there are only two masses, $m_3 = 0$, we have $\rho_1 = \rho$, $\rho_2 = 0$ and $y_1 = y$, so the lowest flutter load is obtained in the situation where

$$\sqrt{k} = \lambda y \sqrt{\rho}, \quad \Longleftrightarrow \sqrt{\frac{k_1}{k_2}} = \frac{d}{l_2}\sqrt{\frac{m_1}{m_2}} \tag{1.175}$$

and corresponds to

$$P = k_2\left(\frac{1}{l_1} + \frac{1}{l_2}\right), \tag{1.176}$$

whereas the divergence load becomes

$$P = k_2\left(\frac{1}{l_1} + \frac{1}{l_2}\right) + 4k_1\frac{l_2}{l_1}\left(\frac{1}{l_1 + l_2}\right). \tag{1.177}$$

We finally conclude that *while divergence instability corresponds to a motion growing exponentially in time, flutter instability corresponds to a self-excited oscillation blowing up in time. Note that both these instabilities cannot be detected with a quasi-static analysis.*

The preceding statement is confirmed in Fig. 1.67, where results are reported as solution of the linear differential system (1.164) with the initial conditions $\alpha_1 = \alpha_2 = 0.5°$ ($\alpha_1 = \alpha_2 = -0.5°$ for divergence) and $\dot{\alpha}_1 = \dot{\alpha}_2 = 0$. For the numerical solution, we have selected the following parameters (taken to be representative of the structural model that will be presented in the next chapter):

$$l = 3d = 3h = 100\,\text{mm}, \quad m_1 = 12m_2 = 4m_3 = 552\,\text{g}, \quad k_1 = k_2 = 0.189\,\text{Nm}, \tag{1.178}$$

which correspond from Eqs. (1.173) and (1.174) to a flutter load $P_f \approx 4.8$ N and to a divergence load $P_d \approx 8.8$ N, so we have assumed $P = 6.8$ N ($P = 15.4$ N) for the simulation of flutter (of divergence). A sequence 0.52 (0.2) s long of configurations at different instants of time is reported in Fig. 1.67, where each configuration is drawn at fixed intervals of time (0.04 s). The oscillatory blow-up (exponential growth) of the solution is clearly visible in the case of flutter (of divergence).

The experimental demonstration of flutter induced by Coulomb friction

Experiments on two degree of freedom model structures similar to that shown in Fig. 1.65 have been performed by Herrmann et al. (1966), Herrmann (1971), and Sugiyama et al. (1995, 2000) in which the follower force was obtained either through a fluid flowing from a nozzle at the free end of the structure or through a motor rocket fixed at the free end of the model. Although these experiments have shown that flutter and divergence may occur in reality, the devices employed to obtain the tangential follower force evidence several problems, so follower forces were considered very difficult to be produced and unlikely to be encountered in practice (Koiter, 1996; Elishakoff, 2005).

Bigoni and Noselli (2011) have shown that *a follower tangential force and the related flutter and divergence instabilities may be induced by Coulomb friction in an elastic structure that is stable when the constraints are smooth.* The fact that in the presence of friction a system may become unstable, whereas stability is guaranteed in the absence of it, may appear at a first glance paradoxical because friction is a mere dissipative term, but we will see that this is indeed what happens.

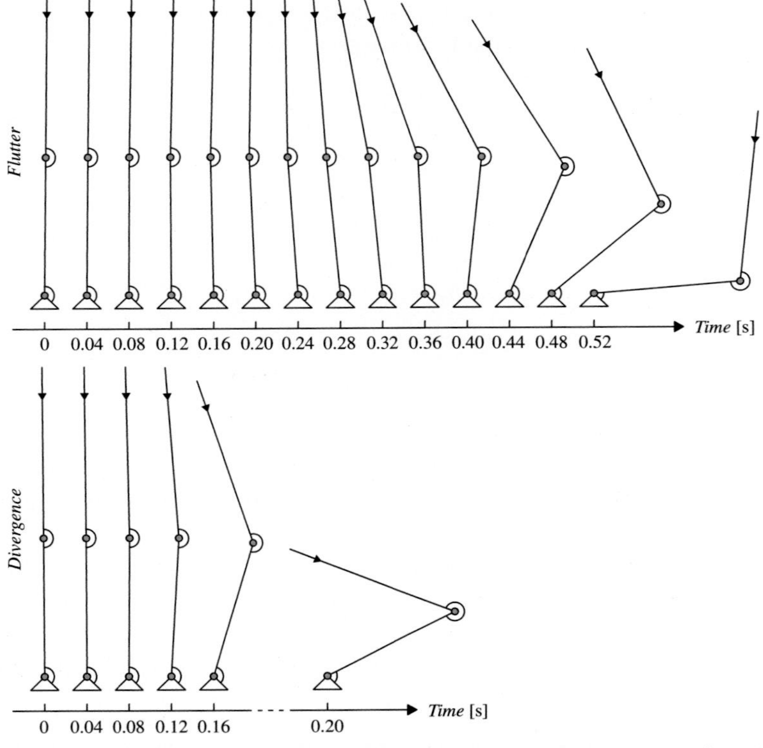

Figure 1.67. A sequence (0.52 s for flutter and to 0.2 s for divergence) of deformed configurations at consecutive time intervals of 0.04 s of the structure sketched in Fig. 1.65 and exhibiting flutter (*top*) and divergence (*bottom*) instability. Results have been obtained through a linearised analysis, Eqs. (1.164), with initial conditions $\alpha_1 = \alpha_2 = 0.5°$ ($\alpha_1 = \alpha_2 = -0.5°$ for divergence) and $\dot{\alpha}_1 = \dot{\alpha}_2 = 0$, at the flutter load $P = 6.8\,\text{N}$ (*top*) and at the divergence load $P = 15.4\,\text{N}$ (*bottom*). The values employed for the analysis are reported in the list (1.178).

The idea of producing a tangential follower load and the related instabilities in a structure through friction is sketched in Fig. 1.68, representing the structure idealised in Fig. 1.65, but where the follower load is obtained as the effect of sliding of a massless wheel (free of rotating about its axis) against a rigid plane. Since the wheel is mounted coaxial with a rigid rod, the frictional force is also transmitted coaxially, whereas any component orthogonal to the wheel cannot develop (because the wheel has negligible mass and is free of rotating). A practical realization of the concept illustrated in Fig. 1.68 is reported in Fig. 1.69, where the so-called second prototype of the structure designed by Bigoni and Noselli (2011) is shown. Note that instead of three concentrated masses, the metal rods now have a diffused mass, a feature that does not change the behaviour of the system much, so for two rods of equal length $l_1 = l_2 = l$ assuming

$$d = h = \frac{l}{3}, \qquad m_1 = 3\rho l, \qquad m_2 = \frac{1}{4}\rho l, \qquad m_3 = \frac{3}{4}\rho l, \tag{1.179}$$

the three-mass system behaves as a system with diffuse masses, in which ρ is the mass per unit length of the 'heavy' rods.

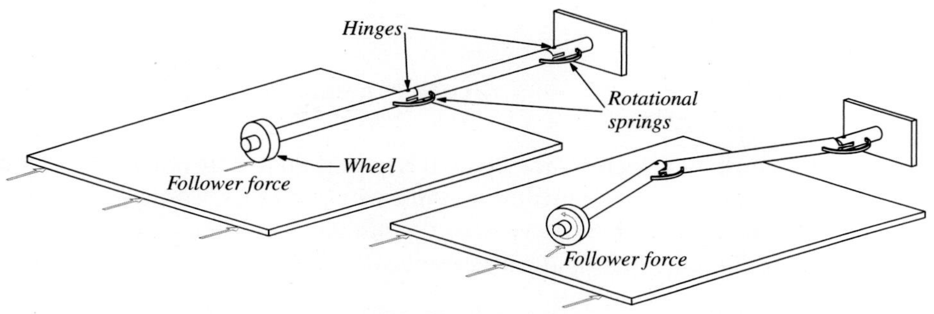

Figure 1.68. The way to produce a force coaxial to a rod through Coulomb friction: A freely rotating massless wheel is mounted at the top of the rod so that this is subject to an axial follower force when a plate is forced to slide against it.

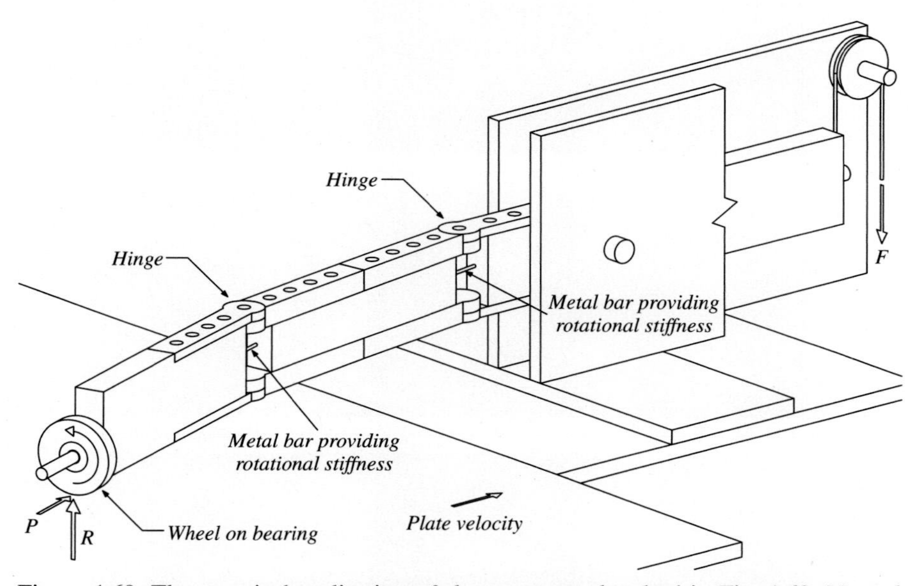

Figure 1.69. The practical realisation of the structure sketched in Fig. 1.68. Note that the structure itself is used as a lever to 'tune' the vertical reaction R through the dead load F. The tangential force P is therefore given by R multiplied by the plate/wheel friction coefficient.

With the choice of parameter (1.179), the mechanical system shown in Fig. 1.69 now behaves as that reported in Fig. 1.65, with the exception that the friction law at the wheel/plate contact, producing the follower load P, is given by the Coulomb rule (similar to that shown in Section 1.9)

$$P = R\mu(\dot{C}_p^r), \qquad \mu(\dot{C}_p^r) = \begin{cases} \mu_d \operatorname{sign}(\dot{C}_p^r), & \text{if } \dot{C}_p^r = \dot{\boldsymbol{C}}_p \cdot \boldsymbol{e}_r \neq 0, \\ [-\mu_s, \mu_s], & \text{if } \dot{C}_p^r = \dot{\boldsymbol{C}}_p \cdot \boldsymbol{e}_r = 0, \end{cases} \tag{1.180}$$

where R is the vertical reaction applied at the wheel (orthogonal to the moving plane), μ_s and μ_d are the static and dynamic friction coefficients (their difference gives the so-called stiction effect, which vanishes taking $\mu_s = \mu_d$) and $\dot{C}_p^r$ is the radial component ($\boldsymbol{e}_r = \cos\alpha_2\, \boldsymbol{e}_1 + \sin\alpha_2\, \boldsymbol{e}_2$) of the velocity of the wheel relative to the plate.

Thus we can write

$$\dot{\boldsymbol{C}}_p = \dot{\boldsymbol{C}} + v_p \boldsymbol{e}_1, \qquad \dot{\boldsymbol{C}}_p \cdot \boldsymbol{e}_r = v_p \cos\alpha_2 - l_1 \sin(\alpha_1 - \alpha_2)\dot{\alpha}_1, \tag{1.181}$$

where v_p is the constant velocity of the plate. It is important to note that the Coulomb law (1.180) introduces an 'incremental non-linearity', in other words, a nonlinear behaviour depending on the relative velocity $\dot{C}_p^r$, similar, in essence, to the piece-wise incremental non-linearity of a rigid perfectly plastic body.

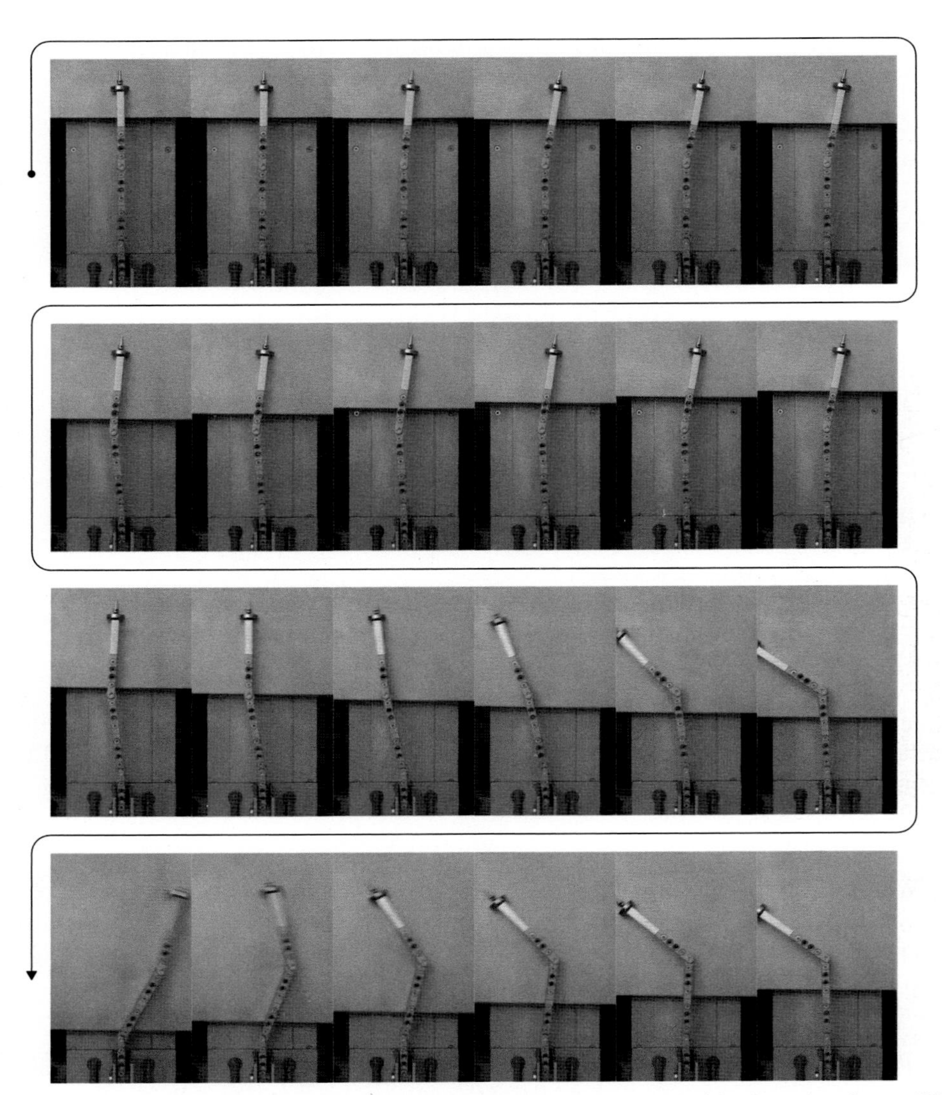

Figure 1.70. A sequence of photos (taken from a movie recorded with a Sony handycam HDR-XR550VE at 25 shots per second) of the structure sketched in Fig. 1.69 and exhibiting flutter instability. Note that the last photos are blurred owing to the increasing velocity of the motion. The whole sequence of photos was recorded in 0.96 s, while the plate was advancing against the wheel. (See color plates section.)

The complication connected to the incremental non-linearity disappears if the plate/wheel sliding condition $\dot{C}_p^r > 0$ is always verified, a situation which certainly holds at the instant of flutter (when $\alpha_1 = \alpha_2 = 0$) and even for a finite interval of time from this instant, for v_p sufficiently high to satisfy the condition

$$v_p > l_1 \sin(\alpha_1 - \alpha_2)\dot{\alpha}_1 / \cos\alpha_2. \tag{1.182}$$

Until the condition (1.182) holds true, the problem becomes identical to that represented by Eqs. (1.164). Thus, since in the experiment the metallic plate under the wheel is moved horizontally with a prescribed, but arbitrary, velocity, this always can be chosen to be high enough to satisfy the condition. At the same time, the structure used as a lever can be loaded (independent of the plate velocity) with a vertical dead load to produce the force on the wheel corresponding to stability, flutter or divergence instability.

It is therefore reasonably expected that flutter and divergence instabilities will be observed at a sufficiently high plate speed v_p, and this expectation is indeed verified in the experiments,[31] as shown in Fig. 1.70, reporting a sequence of photos (taken with a Sony handycam HDR-XR550VE at 25 shots per second) of the structure shown in Fig. 1.69 (mentioned by Bigoni and Noselli, 2011, as 'the second prototype').[32]

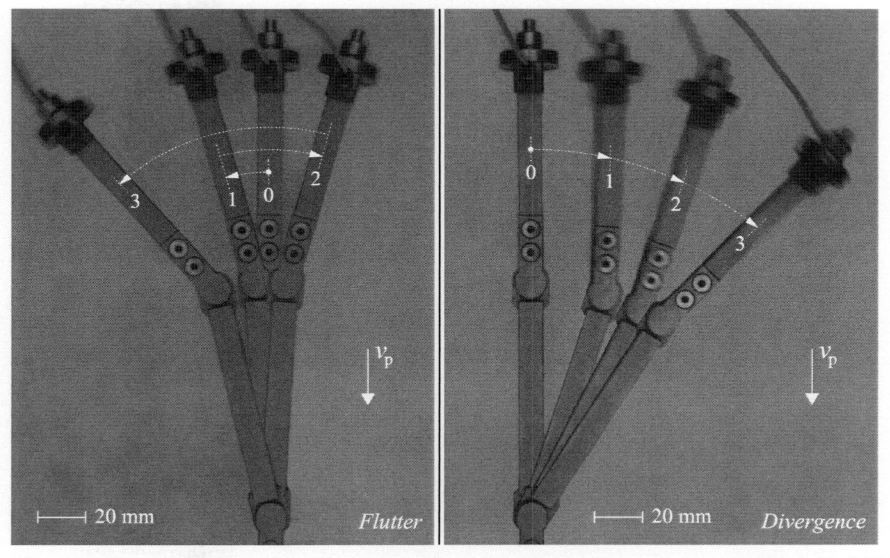

Figure 1.71. A superposition of photos taken at different instants of time [(0, 0.48, 0.72, 1.08) s for flutter and (0, 0.20, 0.28, 0.36) s for divergence] of a two degree of freedom structure (similar to that sketched in Fig. 1.69 and identified as the 'third prototype' by Bigoni and Noselli, 2011) exhibiting flutter (*left*) and divergence (*right*) instability. The velocity of a plate sliding against the wheel v_p has been taken equal to 75 mm/s for flutter and 50 mm/s for divergence.

[31] A fully non-linear analysis of the structure, useful to understand the post-critical behaviour, is reported by Bigoni and Noselli (2011). It shows that the blowing-up vibrational motion corresponding to flutter instability eventually reaches a steady state in which the maximum amplitude of the oscillation depends on the velocity v_p of the plate in a way that the higher v_p is the larger is the amplitude.

[32] The structure (sketched in Fig. 1.69 and shown in Fig. 1.70) has been realised with two AISI 304 steel rods of length 105 mm, rectangular 10 × 30-mm cross section (corresponding to a distributed mass

The sequence of photos has to be compared with the sequence predicted in Fig. 1.67 to conclude that the structure is subject to a typical instability of flutter. Note that the velocity of the system is increasing progressively, so the last photo is clearly blurred owing to the fast movement of the structure.

The structure shown in Fig. 1.70 has broken when we observed divergence instability, which has been investigated systematically by Bigoni and Noselli (2011) with their 'third prototype', a structure characterised by the parameters reported in the list (1.178). Results on this mechanical system are summarised in Fig. 1.71, where photos taken at different instants during flutter (on the left) and divergence (on the right) are superimposed.

Numbers in Fig. 1.71 provide the order in which the photos were taken, so we can clearly detect the oscillatory blow-up of flutter, to be contrasted with the progressive growth of divergence.

We finally note that *in the tested systems, there would be no follower force and therefore no instability in the absence of friction.* The experimental evidence of flutter and divergence shown in Figs. 1.70 and 1.71 is the first definitive proof that these dynamical instabilities can exist in reality in systems where Coulomb friction is present (see Bigoni and Noselli, 2011, for further details).

of 2.355 kg/m), connected with a spring steel wire (1 mm diameter) to obtain rotational springs of 0.148 Nm stiffness. The friction coefficient between the steel wheel and the sandblasted sliding steel plate has been measured to be equal to 0.15. For this system, flutter occurs at follower forces within the interval 3.584 to 6.568 N, whereas divergence occurs for loads higher than 6.568 N. The load for the test has been taken equal to 4.480 N, corresponding to a vertical force of 29.87 N at the wheel (obtained by providing a force $F = 62.72$ N; see Fig. 1.69).

2 Elements of tensor algebra and analysis

A brief review of tensor algebra and calculus is presented for second and fourth-order tensors, with emphasis on concepts that will be useful later and cannot be found easily in the literature, such as, for instance, the representation of non-symmetric and defective tensors. Tensor analysis is also presented, including a brief treatment of cylindrical coordinates. Notions of convexity and quasi-convexity close the presentation of mathematical concepts.

A summary of tensor algebra and analysis is provided below with reference to the three-dimensional Euclidean point space $\mathcal{E}$ and the associated vector space $\mathcal{V}$. Emphasis is given to concepts of interest for our purposes, such as, for instance, those involved in operations with non-symmetric tensors or certain fourth-order tensors. More advanced treatments may be found in the monographs by Gurtin (1972, 1981), Wilkinson (1965) and Bowen and Wang (1976).

We generally refer to Gurtin's notation; in particular, boldface minuscule $(\boldsymbol{a}, \boldsymbol{b}, ...)$ and majuscule $(\boldsymbol{A}, \boldsymbol{B}, ...)$ letters denote vectors (or vector fields) and second-order tensors (or tensor fields), respectively. The space of vectors is denoted by $\mathcal{V}$, the set of second-order tensors by Lin and its symmetric restriction by Sym.

Given two points $\boldsymbol{a}$ and $\boldsymbol{b}$ of the Euclidean point space, their difference

$$\boldsymbol{v} = \boldsymbol{a} - \boldsymbol{b} \tag{2.1}$$

is a vector applied in $\boldsymbol{b}$ (Fig. 2.1*a*). Note that both points and vectors will be indicated with bold lowercase letters. The modulus of a vector $\boldsymbol{v}$ is denoted as $|\boldsymbol{v}|$ and represents the length of the segment joining points $\boldsymbol{a}$ and $\boldsymbol{b}$. Sum between points is meaningless, but sum and difference between vectors may be defined making use of the triangle rule. In particular, as indicated in Fig. 2.1*b*, *c*, we still may operate with points defining the sum of a point and a vector as a new point defined by the extreme of the vector when applied in the point to which the vector is summed. Therefore, for every $\boldsymbol{v} = \boldsymbol{a} - \boldsymbol{b}$ and $\boldsymbol{w} = \boldsymbol{d} - \boldsymbol{a}$, we get

$$\boldsymbol{v} + \boldsymbol{w} = (\boldsymbol{a} - \boldsymbol{b}) + (\boldsymbol{d} - \boldsymbol{a}) = \boldsymbol{d} - \boldsymbol{b}, \quad \boldsymbol{v} - \boldsymbol{w} = (\boldsymbol{a} - \boldsymbol{b}) - (\boldsymbol{c} - \boldsymbol{b}) = \boldsymbol{a} - \boldsymbol{c}. \tag{2.2}$$

The scalar product between two vectors $\boldsymbol{v}$ and $\boldsymbol{w}$ associates the given vectors to the scalar $\boldsymbol{v} \cdot \boldsymbol{w}$, defined as

$$\boldsymbol{v} \cdot \boldsymbol{w} = |\boldsymbol{v}|\,|\boldsymbol{w}| \cos\vartheta, \tag{2.3}$$

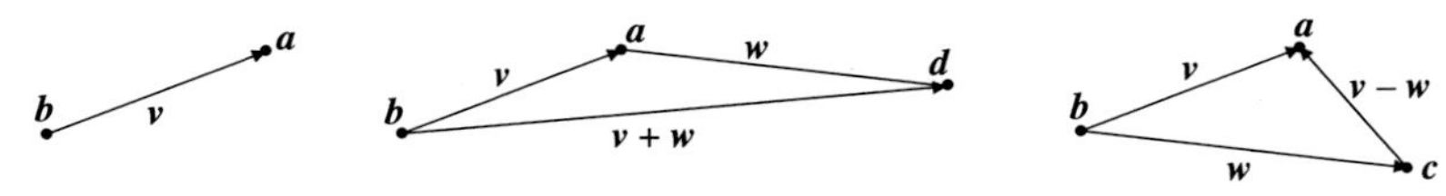

Figure 2.1. Elementary vector algebra: (*a*) Vector $\boldsymbol{v}$ obtained as the difference between two points $\boldsymbol{a}$ and $\boldsymbol{b}$; sum (*b*) and difference (*c*) between vectors.

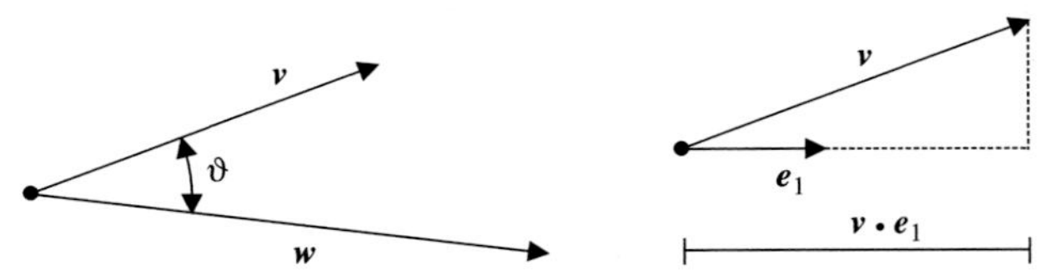

Figure 2.2. Convention used for the angle θ in the definition of scalar product (2.3) between two vectors (*left*). Projection of a vector $\boldsymbol{v}$ on a unit vector $\boldsymbol{e}_1$ (*right*).

where ϑ is the angle between $\boldsymbol{v}$ and $\boldsymbol{w}$ (with reference to their orientation; Fig. 2.2 left). It is worth noting that the scalar product may take any sign and is null when the two vectors are orthogonal. Also, the following identity (holding for every vector) will be useful later

$$|\boldsymbol{w}| = \sqrt{\boldsymbol{w} \cdot \boldsymbol{w}}. \tag{2.4}$$

A 'unit vector' is defined as a vector with unit modulus, so it follows from definition (2.3) that the scalar product between a vector and a unit vector represents the orthogonal projection of the former onto the direction set by the latter (Fig. 2.2 right).

2.1 Components onto an orthonormal basis

Let us consider now three orthogonal unit vectors, $\boldsymbol{e}_1$, $\boldsymbol{e}_2$ and $\boldsymbol{e}_3$. By definition, they satisfy the identity

$$\boldsymbol{e}_i \cdot \boldsymbol{e}_j = \delta_{ij}, \tag{2.5}$$

where the indices i and j range between 1 and 3, and δ_{ij} is the Kronecker[1] delta, equal to one when $i = j$ and equal to zero otherwise. A generic vector $\boldsymbol{v}$ can be represented as (Fig. 2.3)

$$\boldsymbol{v} = v_1\, \boldsymbol{e}_1 + v_2\, \boldsymbol{e}_2 + v_3\, \boldsymbol{e}_3, \tag{2.6}$$

where

$$v_1 = \boldsymbol{v} \cdot \boldsymbol{e}_1, \qquad v_2 = \boldsymbol{v} \cdot \boldsymbol{e}_2, \qquad v_3 = \boldsymbol{v} \cdot \boldsymbol{e}_3 \tag{2.7}$$

are the components of vector $\boldsymbol{v}$ in the orthonormal basis defined by $\boldsymbol{e}_1$, $\boldsymbol{e}_2$ and $\boldsymbol{e}_3$. Note that in the following, the so-called Einstein summation convention is adopted that *repeated indices are summed* between 1 and 3. Thus Eq. (2.6) becomes

$$\boldsymbol{v} = v_i\, \boldsymbol{e}_i. \tag{2.8}$$

[1] Leopold Kronecker (1823–1891) was a German arithmetician and algebraist who retired at the age of 30 from mercantile business to fully devote his activity to mathematics.

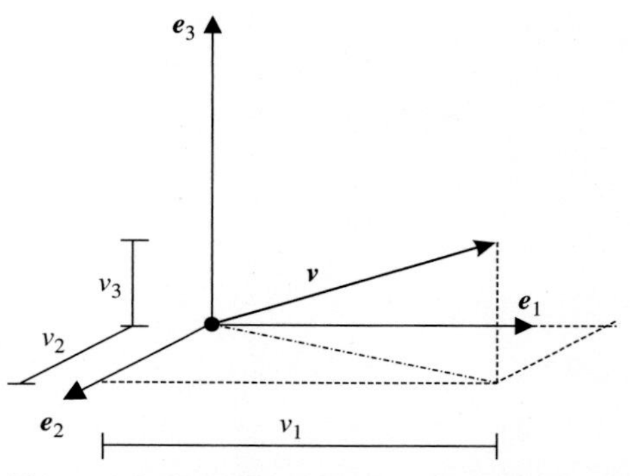

Figure 2.3. Definition of the components of a vector.

The scalar product (2.3) can be written in components as

$$v \cdot w = v_1 w_1 + v_2 w_2 + v_3 w_3 = v_i w_i, \tag{2.9}$$

where w_1, w_2 and w_3 are the components of $\boldsymbol{w}$ in the same basis onto which $\boldsymbol{v}$ is projected, and the repeated index convention has been employed. A consequence of representation (2.6) is that

$$|\boldsymbol{v}| = \sqrt{v_1^2 + v_2^2 + v_3^2} = \sqrt{v_i v_i}. \tag{2.10}$$

We will refer[2] to the following 'cross-product' between vectors, defined with reference to the orthonormal basis $\boldsymbol{e}_i$ as

$$\boldsymbol{a} \times \boldsymbol{b} = (a_2 b_3 - a_3 b_2)\boldsymbol{e}_1 + (a_3 b_1 - a_1 b_3)\boldsymbol{e}_2 + (a_1 b_2 - a_2 b_1)\boldsymbol{e}_3, \tag{2.11}$$

from which the following properties

$$\boldsymbol{a} \times \boldsymbol{b} = -\boldsymbol{b} \times \boldsymbol{a}, \qquad \boldsymbol{a} \times \boldsymbol{a} = \boldsymbol{0} \tag{2.12}$$

can be derived.

Note that the mixed product between three vectors, namely,

$$\boldsymbol{a} \cdot \boldsymbol{b} \times \boldsymbol{c}, \tag{2.13}$$

represents the volume of the parallelepiped having vectors $\boldsymbol{a}$, $\boldsymbol{b}$ and $\boldsymbol{c}$ as edges.

2.2 Dyads

Given two unit vectors $\boldsymbol{m}$ and $\boldsymbol{n}$, let us define the dyad $\boldsymbol{m} \otimes \boldsymbol{n}$, showing the way it operates on a generic vector $\boldsymbol{v}$

$$(\boldsymbol{m} \otimes \boldsymbol{n})\,\boldsymbol{v} = (\boldsymbol{n} \cdot \boldsymbol{v})\,\boldsymbol{m}. \tag{2.14}$$

From definition (2.14), we may conclude that $\boldsymbol{m} \otimes \boldsymbol{n}$ represents a linear operator which associates to every vector $\boldsymbol{v}$ a vector parallel to $\boldsymbol{m}$ having modulus equal to the

[2] Two cross-products can be defined in a three-dimensional vector space.

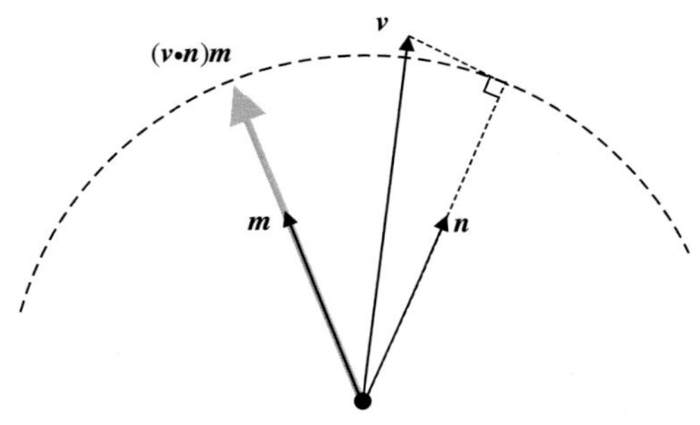

Figure 2.4. A geometrical interpretation of how a dyad $\boldsymbol{m} \otimes \boldsymbol{n}$ transforms a vector $\boldsymbol{v}$. Note that $\boldsymbol{n}$ and $\boldsymbol{m}$ are unit vectors.

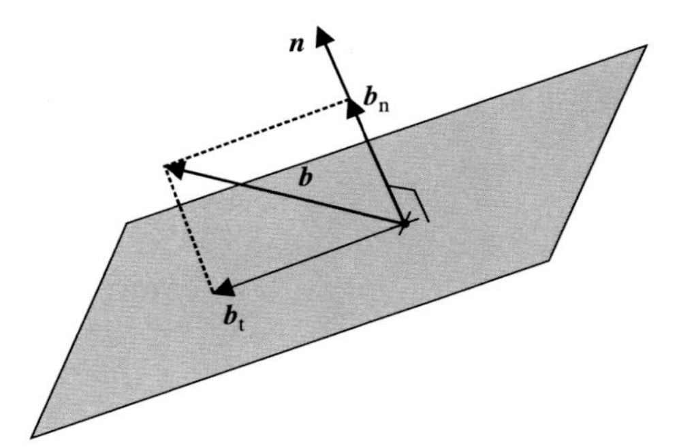

Figure 2.5. Decomposition of a force $\boldsymbol{b}$ acting on a plane of unit normal $\boldsymbol{n}$ into the normal and the tangential components to the plane: $\boldsymbol{b} = \boldsymbol{b}_n + \boldsymbol{b}_t$.

projection of $\boldsymbol{v}$ onto $\boldsymbol{n}$. In other words, the dyad rotates $\boldsymbol{v}$ in the direction of $\boldsymbol{m}$ and reduces the modulus, taking its projection upon $\boldsymbol{n}$ (Fig. 2.4).

As an example, we mention that the dyad allows us to express the normal and tangential vectors to a plane in a very compact form. In particular, let a plane be defined by the unit vector $\boldsymbol{n}$, and let us consider a contact force $\boldsymbol{b}$ acting on it (Fig. 2.5). The two components $\boldsymbol{b}_n$ and $\boldsymbol{b}_t$ of the force normal and tangential to the plane are given, respectively, by

$$\boldsymbol{b}_n = (\boldsymbol{n} \otimes \boldsymbol{n})\,\boldsymbol{b}, \qquad \boldsymbol{b}_t = \boldsymbol{b} - \boldsymbol{b}_n. \tag{2.15}$$

The definition of dyad remains meaningful when generic vectors $\boldsymbol{a}$ and $\boldsymbol{b}$ replace the unit vectors $\boldsymbol{m}$ and $\boldsymbol{n}$, so in the following we will refer to 'dyad' in all cases where we find linear operators such as $\boldsymbol{a} \otimes \boldsymbol{b}$, which is also called a 'dyadic product' between vectors $\boldsymbol{a}$ and $\boldsymbol{b}$.

Since in a given orthonormal reference system $\boldsymbol{e}_1$, $\boldsymbol{e}_2$ and $\boldsymbol{e}_3$ definition (2.9) applies, the scalar product $\boldsymbol{n} \cdot \boldsymbol{v}$ equals $n_i v_i$. Thus the components of a dyad $\boldsymbol{m} \otimes \boldsymbol{n}$ in an orthonormal system of reference $\boldsymbol{e}_1$, $\boldsymbol{e}_2$ and $\boldsymbol{e}_3$ are defined as

$$(\boldsymbol{m} \otimes \boldsymbol{n})_{ij} = \boldsymbol{e}_i \cdot (\boldsymbol{m} \otimes \boldsymbol{n})\boldsymbol{e}_j = m_i n_j. \tag{2.16}$$

Therefore, while a vector is represented as a string of three components in a given reference system, a dyad is represented as a 3×3 matrix in the same reference system.

2.3 Second-order tensors

The dyad $\boldsymbol{n}\otimes\boldsymbol{m}$ defined by Eq. (2.14) defines a linear transformation between vectors and is an example of a *second-order tensor*. More in general, with reference to three orthogonal unit vectors $\boldsymbol{e}_i$, a linear combination of all permutations of dyads $\boldsymbol{e}_i \otimes \boldsymbol{e}_j$, namely,

$$\boldsymbol{A} = A_{ij}\boldsymbol{e}_i \otimes \boldsymbol{e}_j \tag{2.17}$$

(where A_{ij}, with $i,j = 1,2,3$, are coefficients), is a linear operator which transforms a generic vector $\boldsymbol{v}$ into the vector

$$\boldsymbol{A}\boldsymbol{v} = A_{ij}(\boldsymbol{e}_j \cdot \boldsymbol{v})\boldsymbol{e}_i \tag{2.18}$$

and is called a 'second-order tensor'[3] belonging to Lin.

Introducing the components v_j with respect to the basis $\boldsymbol{e}_j$, Eq. (2.18) becomes

$$(\boldsymbol{A}\boldsymbol{v})_i = A_{ij}v_j, \tag{2.19}$$

so the matrix

$$A_{ij} = (\boldsymbol{A})_{ij} = \boldsymbol{e}_i \cdot \boldsymbol{A}\boldsymbol{e}_j \tag{2.20}$$

collects the components of tensor $\boldsymbol{A}$ in the reference system $\boldsymbol{e}_1$, $\boldsymbol{e}_2$ and $\boldsymbol{e}_3$. In general, the components of a tensor $\boldsymbol{A}$ in an orthonormal reference system $\boldsymbol{b}_1$, $\boldsymbol{b}_2$ and $\boldsymbol{b}_3$ remain defined as

$$\hat{A}_{ij} = \boldsymbol{b}_i \cdot \boldsymbol{A}\boldsymbol{b}_j = A_{hk}\left(\boldsymbol{e}_h \cdot \boldsymbol{b}_i\right)\left(\boldsymbol{e}_k \cdot \boldsymbol{b}_j\right), \tag{2.21}$$

providing the transformation rule between the two matrices of components A_{ij} and $\hat{A}_{ij}$ which indeed remains defined through the two matrices $\boldsymbol{e}_h \cdot \boldsymbol{b}_i$ and $\boldsymbol{e}_k \cdot \boldsymbol{b}_j$, collecting the cosine directors of the angles between unit vectors $\boldsymbol{b}_i$ and $\boldsymbol{e}_j$.

The transpose of a tensor $\boldsymbol{A}$ is denoted with a superscript T and is defined as the unique tensor $\boldsymbol{A}^T$ with the property

$$\boldsymbol{a} \cdot \boldsymbol{A}\boldsymbol{b} = \boldsymbol{b} \cdot \boldsymbol{A}^T\boldsymbol{a}, \tag{2.22}$$

holding for every vector $\boldsymbol{a}$ and $\boldsymbol{b}$. It follows that the transpose of a dyad is

$$(\boldsymbol{m}\otimes\boldsymbol{n})^T = \boldsymbol{n}\otimes\boldsymbol{m}. \tag{2.23}$$

As a consequence of Eq. (2.23), a tensor is said to be symmetric (skew-symmetric) when it remains unchanged (with reversed sign) by transposition

$$\boldsymbol{A} = \boldsymbol{A}^T \iff \boldsymbol{A} \in \mathsf{Sym}; \qquad \left(\boldsymbol{A} = -\boldsymbol{A}^T \iff \boldsymbol{A} \in \mathsf{Skw}\right), \tag{2.24}$$

where Sym (Skw) is the subset of Lin collecting symmetric (skew-symmetric) tensors. In every orthogonal reference system, a symmetric (skew-symmetric) tensor is always associated to a symmetric (skew-symmetric) matrix of components, a property that can be deduced from Eq. (2.21).

[3] Although the concept of tensor may be traced back to Georg Friedrich Bernhard Riemann (1826–1866) and Elwin Bruno Christoffel (1829–1900), modern tensor calculus was founded by Gregorio Ricci Curbastro (1853–1925) and Tullio Levi Civita (1873–1941).

A remarkable property of second-order tensors is that they always can be decomposed into a symmetric and a skew-symmetric part, namely,

$$\boldsymbol{A} = \boldsymbol{S} + \boldsymbol{W}, \qquad \boldsymbol{S} = \frac{1}{2}\left(\boldsymbol{A} + \boldsymbol{A}^T\right), \qquad \boldsymbol{W} = \frac{1}{2}\left(\boldsymbol{A} - \boldsymbol{A}^T\right). \tag{2.25}$$

The product between two tensors $\boldsymbol{A}$ and $\boldsymbol{B}$ is defined in the way they act on a generic vector $\boldsymbol{v}$:

$$\boldsymbol{ABv} = \boldsymbol{A}(\boldsymbol{Bv}), \tag{2.26}$$

Thus, applying the definition of components, one gets

$$(\boldsymbol{AB})_{ij} = A_{ik}B_{kj}. \tag{2.27}$$

It is clear from Eq. (2.26) that the product between two tensors defines a tensor; moreover, the product between n identical tensors is immediate:

$$\boldsymbol{A}^n = \underbrace{\boldsymbol{AA}\ldots\boldsymbol{A}}_{n \text{ times}} = \boldsymbol{A}^{n-1}\boldsymbol{A}. \tag{2.28}$$

The identity tensor is defined, for every tensor $\boldsymbol{A}$, as

$$\boldsymbol{IA} = \boldsymbol{AI} = \boldsymbol{A}, \tag{2.29}$$

so the identity can be represented in every orthonormal reference system $\boldsymbol{e}_i$ as

$$\boldsymbol{I} = \boldsymbol{e}_1 \otimes \boldsymbol{e}_1 + \boldsymbol{e}_2 \otimes \boldsymbol{e}_2 + \boldsymbol{e}_3 \otimes \boldsymbol{e}_3 = \boldsymbol{e}_i \otimes \boldsymbol{e}_i. \tag{2.30}$$

Note that the tensor product (2.26) does not in general commute; when it does, the two tensors are called 'coaxial', namely,

$$\boldsymbol{AB} = \boldsymbol{BA} \quad \Longleftrightarrow \quad \boldsymbol{A}, \boldsymbol{B} \qquad \text{are coaxial.} \tag{2.31}$$

A special type of coaxial tensor is two 'parallel' tensors, so $\boldsymbol{A} = \rho\boldsymbol{B}$, where ρ is a scalar. For such a case, the symbol $\propto$ will be used, namely, $\boldsymbol{A} \propto \boldsymbol{B}$.

Let us define the trace of a dyad $\boldsymbol{m} \otimes \boldsymbol{n}$ as

$$\mathrm{tr}(\boldsymbol{m} \otimes \boldsymbol{n}) = \boldsymbol{m} \cdot \boldsymbol{n}. \tag{2.32}$$

From its definition, we may easily note that the trace is a linear operator, so the trace of a generic tensor (2.17) is simply

$$\mathrm{tr}\boldsymbol{A} = A_{11} + A_{22} + A_{33} = A_{ii}. \tag{2.33}$$

By definition, the deviator of a tensor $\boldsymbol{A} \in \mathrm{Lin}$ is

$$\mathrm{dev}\boldsymbol{A} = \boldsymbol{A} - \frac{\mathrm{tr}\boldsymbol{A}}{3}\boldsymbol{I}, \tag{2.34}$$

thus satisfying the property $\mathrm{tr}\,\mathrm{dev}\boldsymbol{A} = 0$.

The scalar product between two tensors $\boldsymbol{A}$ and $\boldsymbol{B}$ associates the scalar $\boldsymbol{A} \cdot \boldsymbol{B}$ to them and is defined as

$$\boldsymbol{A} \cdot \boldsymbol{B} = \mathrm{tr}\left(\boldsymbol{AB}^T\right) = \mathrm{tr}\left(\boldsymbol{A}^T\boldsymbol{B}\right) = A_{ij}B_{ij}. \tag{2.35}$$

The determinant of a second-order tensor is defined as the determinant of the associate matrix in a certain orthonormal reference system:

$$\det \boldsymbol{A} = \det\left[A_{ij}\right]. \tag{2.36}$$

Thus a tensor is called 'singular' when its determinant vanishes. The fact that the determinant is invariant with respect to the choice of reference system can be proved from Eq. (2.21) using the properties

$$\det(\boldsymbol{AB}) = \det \boldsymbol{A} \det \boldsymbol{B}, \qquad \det \boldsymbol{A}^T = \det \boldsymbol{A}. \tag{2.37}$$

An expression which will be useful later gives the determinant of a rank-one modification of the identity $\boldsymbol{I} + \boldsymbol{a} \otimes \boldsymbol{b}$ in the form

$$\det(\boldsymbol{I} + \boldsymbol{a} \otimes \boldsymbol{b}) = 1 + \boldsymbol{a} \cdot \boldsymbol{b}, \tag{2.38}$$

holding for every vector $\boldsymbol{a}$ and $\boldsymbol{b}$.

A tensor $\boldsymbol{A}$ is 'non-singular' when its determinant is non-zero, and in such case, it is invertible with inverse $\boldsymbol{A}^{-1}$ satisfying

$$\boldsymbol{A}\boldsymbol{A}^{-1} = \boldsymbol{A}^{-1}\boldsymbol{A} = \boldsymbol{I}. \tag{2.39}$$

The inverse of a tensor satisfies the following properties:

$$(\boldsymbol{AB})^{-1} = \boldsymbol{B}^{-1}\boldsymbol{A}^{-1}, \qquad \left(\boldsymbol{A}^{-1}\right)^T = \left(\boldsymbol{A}^T\right)^{-1}, \tag{2.40}$$

the latter motivating the notation $\boldsymbol{A}^{-T}$.

Three useful expressions can be derived immediately. One is the identity

$$\boldsymbol{c} \cdot \boldsymbol{a} \times \boldsymbol{b} = \det \begin{bmatrix} c_1 & c_2 & c_3 \\ a_1 & a_2 & a_3 \\ b_1 & b_2 & b_3 \end{bmatrix}, \tag{2.41}$$

holding in every reference system, and the other two are

$$\boldsymbol{Fc} \cdot \boldsymbol{Fa} \times \boldsymbol{Fb} = (\det \boldsymbol{F})\boldsymbol{c} \cdot \boldsymbol{a} \times \boldsymbol{b} \tag{2.42}$$

and

$$\boldsymbol{Fa} \times \boldsymbol{Fb} = (\det \boldsymbol{F})\boldsymbol{F}^{-T}\left(\boldsymbol{a} \times \boldsymbol{b}\right), \tag{2.43}$$

both holding for every non-singular tensor $\boldsymbol{F}$.

Property (2.41) can be proved directly taking the scalar product of Eq. (2.11) with $\boldsymbol{c}$, whereas property (2.42) can be obtained from Eq. (2.41) as follows:

$$\boldsymbol{Fc} \cdot \boldsymbol{Fa} \times \boldsymbol{Fb} = \det \begin{bmatrix} F_{1i}c_i & F_{2i}c_i & F_{3i}c_i \\ F_{1i}a_i & F_{2i}a_i & F_{3i}a_i \\ F_{1i}b_i & F_{2i}b_i & F_{3i}b_i \end{bmatrix} = \det\left(\begin{bmatrix} c_1 & c_2 & c_3 \\ a_1 & a_2 & a_3 \\ b_1 & b_2 & b_3 \end{bmatrix} [\boldsymbol{F}]^T \right)$$

$$= \det \begin{bmatrix} c_1 & c_2 & c_3 \\ a_1 & a_2 & a_3 \\ b_1 & b_2 & b_3 \end{bmatrix} (\det \boldsymbol{F}) = \boldsymbol{c} \cdot \boldsymbol{a} \times \boldsymbol{b}(\det \boldsymbol{F}). \tag{2.44}$$

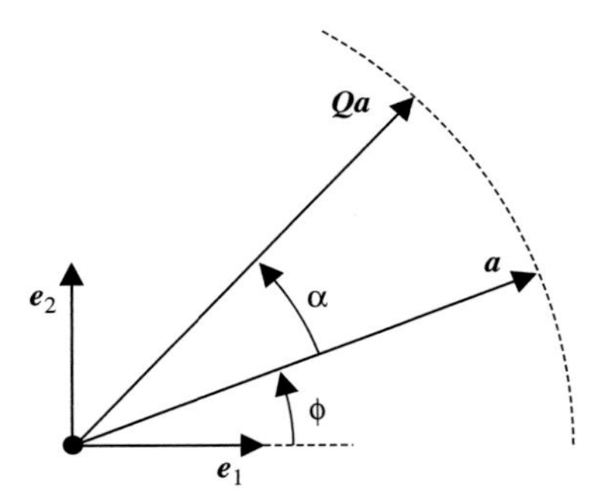

Figure 2.6. The geometrical interpretation of a rotation tensor.

In order to prove Eq. (2.43), let us consider an arbitrary vector $\boldsymbol{x}$, and using Eq. (2.42), we obtain

$$\boldsymbol{x}\cdot\boldsymbol{F}\boldsymbol{a}\times\boldsymbol{F}\boldsymbol{b}=\boldsymbol{F}\left(\boldsymbol{F}^{-1}\boldsymbol{x}\right)\cdot\boldsymbol{F}\boldsymbol{a}\times\boldsymbol{F}\boldsymbol{b}=(\det\boldsymbol{F})\left(\boldsymbol{F}^{-1}\boldsymbol{x}\right)\cdot\boldsymbol{a}\times\boldsymbol{b}. \tag{2.45}$$

Thus, since

$$\left(\boldsymbol{F}^{-1}\boldsymbol{x}\right)\cdot\boldsymbol{a}\times\boldsymbol{b}=\boldsymbol{x}\cdot\boldsymbol{F}^{-T}(\boldsymbol{a}\times\boldsymbol{b}), \tag{2.46}$$

the arbitrariness of $\boldsymbol{x}$ allows us to conclude that Eq. (2.43) holds true.

2.4 Rotation tensors

Given a generic angle α, let us define the tensor

$$\boldsymbol{Q}=\cos\alpha\boldsymbol{e}_1\otimes\boldsymbol{e}_1-\sin\alpha\boldsymbol{e}_1\otimes\boldsymbol{e}_2+\sin\alpha\boldsymbol{e}_2\otimes\boldsymbol{e}_1+\cos\alpha\boldsymbol{e}_2\otimes\boldsymbol{e}_2. \tag{2.47}$$

If we now consider a generic vector in the plane $\boldsymbol{e}_1$–$\boldsymbol{e}_2$ inclined at an angle ϕ with respect to the direction singled out by the unit vector $\boldsymbol{e}_1$ (or inclined at $\pi/2-\phi$ with respect to the unit vector $\boldsymbol{e}_2$ orthogonal to $\boldsymbol{e}_1$), namely,

$$\boldsymbol{a}=|\boldsymbol{a}|(\cos\phi\,\boldsymbol{e}_1+\sin\phi\,\boldsymbol{e}_2), \tag{2.48}$$

it is immediate to conclude that

$$\boldsymbol{Q}\boldsymbol{a}=|\boldsymbol{a}|(\cos(\alpha+\phi)\,\boldsymbol{e}_1+\sin(\alpha+\phi)\,\boldsymbol{e}_2), \tag{2.49}$$

so tensor $\boldsymbol{Q}$ rotates any vector $\boldsymbol{a}$ of an angle α, preserving its modulus (Fig. 2.6). Tensor $\boldsymbol{Q}$ is an example of a rotation tensor.

In general, a tensor $\boldsymbol{Q}$ is called 'orthogonal' when it satisfies the condition

$$\boldsymbol{Q}\boldsymbol{Q}^T=\boldsymbol{Q}^T\boldsymbol{Q}=\boldsymbol{I}, \tag{2.50}$$

and it is said to belong to the subset Orth of Lin.

It follows from Eq. (2.50) that the determinant of an orthogonal tensor is ±1, and in particular, when the determinant equals 1, the orthogonal tensor is called a 'rotation' and belongs to Orth^+ (otherwise, it is called a 'reflection' and belongs to Orth^-).

A necessary and sufficient condition for Eq. (2.50) to be verified is that the orthogonal tensor $\boldsymbol{Q}$ preserves the scalar product between two arbitrary vectors $\boldsymbol{a}$ and $\boldsymbol{b}$, namely,

$$\boldsymbol{Q}\boldsymbol{a} \cdot \boldsymbol{Q}\boldsymbol{b} = \boldsymbol{a} \cdot \boldsymbol{b}. \tag{2.51}$$

2.5 Positive definite second-order tensors, eigenvalues and eigenvectors

A generic tensor $\boldsymbol{A} \in \mathsf{Lin}$ is *positive definite* when

$$\boldsymbol{v} \cdot \boldsymbol{A}\boldsymbol{v} > 0, \tag{2.52}$$

for every non-zero vector $\boldsymbol{v}$. The tensor is positive semi-definite when the preceding $>$ is replaced by $\geq$.

Two unit vectors $\boldsymbol{a}^r$ and $\boldsymbol{a}^l$ are, respectively, a right and a left eigenvector of a tensor $\boldsymbol{A} \in \mathsf{Lin}$ if

$$\boldsymbol{A}\boldsymbol{a}^r = \alpha \boldsymbol{a}^r \qquad \text{and} \qquad \boldsymbol{A}^T \boldsymbol{a}^l = \alpha \boldsymbol{a}^l, \tag{2.53}$$

where α is the eigenvalue corresponding to both $\boldsymbol{a}^r$ and $\boldsymbol{a}^l$. All vectors parallel to all unit left (right) eigenvectors associated to the same eigenvalue α form the left (right) *characteristic space* corresponding to α.

The eigenvalues are the roots of the characteristic equation

$$\det(\boldsymbol{A} - \alpha \boldsymbol{I}) = 0, \tag{2.54}$$

or,[4] equivalently,

$$\alpha^3 - I_1(\boldsymbol{A})\alpha^2 + I_2(\boldsymbol{A})\alpha - I_3(\boldsymbol{A}) = 0, \tag{2.56}$$

where $I_i(\boldsymbol{A})$ are the principal invariants of $\boldsymbol{A}$, defined as

$$I_1(\boldsymbol{A}) = \mathrm{tr}\boldsymbol{A}, \qquad I_2(\boldsymbol{A}) = \frac{1}{2}\left[(\mathrm{tr}\boldsymbol{A})^2 - \mathrm{tr}(\boldsymbol{A}^2)\right], \qquad I_3(\boldsymbol{A}) = \det \boldsymbol{A}, \tag{2.57}$$

and are always real or complex in conjugate pairs. Thus, in a three dimensional space, every tensor always has at least one real eigenvalue. Note that we will for brevity write $\mathrm{tr}\boldsymbol{A}^2$ for $\mathrm{tr}(\boldsymbol{A}^2)$ and $\mathrm{tr}^2\boldsymbol{A}$ for $(\mathrm{tr}\boldsymbol{A})^2$ in the following.

In terms of eigenvalues α_1, α_2 and α_3, the invariants (2.57) become

$$I_1(\boldsymbol{A}) = \alpha_1 + \alpha_2 + \alpha_3, \qquad I_2(\boldsymbol{A}) = \alpha_1\alpha_2 + \alpha_1\alpha_3 + \alpha_2\alpha_3, \qquad I_3(\boldsymbol{A}) = \alpha_1\alpha_2\alpha_3. \tag{2.58}$$

The 'algebraic' multiplicity of an eigenvalue is the multiplicity of the corresponding root of the characteristic equation (2.56), whereas the 'geometrical' multiplicity of an eigenvalue is the dimension of the associated space of eigenvectors (namely, the number of linearly independent associated eigenvectors).

The Cayley-Hamilton theorem states that every tensor $\boldsymbol{A}$ satisfies its own characteristic equation [formally identical to Eq. (2.56) but with α replaced by $\boldsymbol{A}$]

$$\boldsymbol{A}^3 - I_1(\boldsymbol{A})\boldsymbol{A}^2 + I_2(\boldsymbol{A})\boldsymbol{A} - I_3(\boldsymbol{A})\boldsymbol{I} = \boldsymbol{0}, \tag{2.59}$$

[4] As a consequence of the property $(2.37)_2$ of the determinant, the following identity holds:

$$\det(\boldsymbol{A} - \alpha \boldsymbol{I}) = \det(\boldsymbol{A}^T - \alpha \boldsymbol{I}). \tag{2.55}$$

It shows that the eigenvalues of $\boldsymbol{A}$ coincide with the eigenvalues of $\boldsymbol{A}^T$.

which allows one to express $\boldsymbol{A}^3$ as a function of $\boldsymbol{A}^2$ and $\boldsymbol{A}$.

Equations $(2.57)_{1,2}$ and the Cayley-Hamilton theorem show that the principal invariants can be defined as functions of the invariants tr$\boldsymbol{A}$, tr$\boldsymbol{A}^2$ and tr$\boldsymbol{A}^3$. Vice versa, these invariants can be expressed, through the Cayley-Hamilton theorem, as functions of the principal invariants:

$$\text{tr}\boldsymbol{A}^2 = I_1^2(\boldsymbol{A}) - 2I_2(\boldsymbol{A}), \qquad \text{tr}\boldsymbol{A}^3 = I_1^3(\boldsymbol{A}) - 3I_2(\boldsymbol{A})I_1(\boldsymbol{A}) + 3I_3(\boldsymbol{A}). \tag{2.60}$$

Moreover, multiplication of the Cayley-Hamilton formula by $\boldsymbol{A}^{n-3}$ shows that

$$\boldsymbol{A}^n = I_1(\boldsymbol{A})\boldsymbol{A}^{n-1} - I_2(\boldsymbol{A})\boldsymbol{A}^{n-2} + I_3(\boldsymbol{A})\boldsymbol{A}^{n-3}. \tag{2.61}$$

Thus, using Eq. (2.61) recursively (and taking its trace), we can conclude that all tensors $\boldsymbol{A}^n$ (and all invariants tr$\boldsymbol{A}^n$) can be expressed as a function of $\boldsymbol{A}$, $\boldsymbol{A}^2$, $\boldsymbol{A}^3$ (of tr$\boldsymbol{A}$, tr$\boldsymbol{A}^2$ and tr$\boldsymbol{A}^3$ and, equivalently, of the principal invariants of $\boldsymbol{A}$).

The following remarks are important to clarify issues concerning non-symmetric tensors.

- If a tensor is positive definite, the real eigenvalues must be strictly positive, but a tensor may possess all real and positive eigenvalues and be indefinite. Let us explain this concept with an example in two dimensions. Take the matrix

$$\begin{bmatrix} 1 & \epsilon - 10 \\ \epsilon & 9 \end{bmatrix}, \tag{2.62}$$

  which is positive definite with complex eigenvalues when $2 < \epsilon < 8$, whereas it is indefinite with positive eigenvalues for $\epsilon \in [0,2]$.
- A positive definite tensor has the following properties: It is non-singular; all the diagonal components in every reference systems are positive; and the largest component in every reference system lies on the diagonal.
- If a tensor has coincident eigenvalues, it may be *defective*, in which case one or more eigenvectors do not exist (so the geometrical multiplicity of some eigenvalues is inferior to the algebraic multiplicity). To clarify this issue with an example, we mention the Jordan block

$$\begin{bmatrix} 1 & 1 \\ 0 & 1 \end{bmatrix}, \tag{2.63}$$

  which possesses two coincident eigenvalues equal to 1 and only one right eigenvector of components $\{1,0\}$ and left eigenvector $\{0,1\}$. Note that these two eigenvectors are mutually orthogonal; otherwise, the Jordan block would be diagonalisable.
- A left and a right eigenvector, corresponding to different eigenvalues, are mutually orthogonal. In fact,

$$\boldsymbol{A}\boldsymbol{a}_1^r = \alpha_1 \boldsymbol{a}_1^r, \qquad \boldsymbol{A}^T\boldsymbol{a}_2^l = \alpha_2 \boldsymbol{a}_2^l, \quad \rightsquigarrow \quad \alpha_1 \boldsymbol{a}_2^l \cdot \boldsymbol{a}_1^r = \alpha_2 \boldsymbol{a}_1^r \cdot \boldsymbol{a}_2^l. \tag{2.64}$$

  Thus, $\boldsymbol{a}_1^r \cdot \boldsymbol{a}_2^l = 0$ follows.
- If a tensor has three distinct eigenvalues, we may define the eigenvectors such that they form reciprocal (possibly non-orthogonal) bases:

$$\boldsymbol{a}_i^r \cdot \boldsymbol{a}_j^l = \delta_{ij}. \tag{2.65}$$

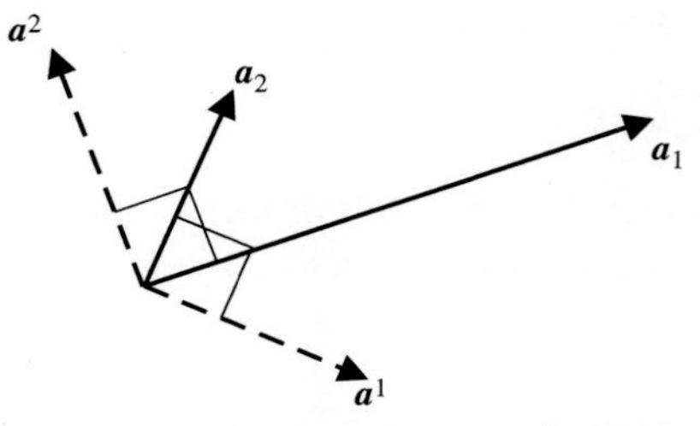

Figure 2.7. Reciprocal bases $\boldsymbol{a}^i$ and $\boldsymbol{a}_j$ in two dimensions.

2.6 Reciprocal bases: Covariant and contravariant components

The concept of reciprocal bases introduced by Eq. (2.65) can be detailed as follows: Let us consider three non-coplanar but otherwise generic vectors $\boldsymbol{a}_i$, $i = 1,2,3$, and define the reciprocal vectors $\boldsymbol{a}^j$, $j = 1,2,3$, by imposing condition (2.65) (see Fig. 2.7 referred for simplicity to two dimensions). Both the triplets of vectors $\boldsymbol{a}_i$ and $\boldsymbol{a}^j$ alternatively can be used to represent any vector; thus

$$\boldsymbol{v} = v_i \boldsymbol{a}^i \qquad \text{or} \qquad \boldsymbol{v} = v^i \boldsymbol{a}_i \tag{2.66}$$

where

$$v_i = \boldsymbol{a}_i \cdot \boldsymbol{v} \qquad \text{and} \qquad v^i = \boldsymbol{a}^i \cdot \boldsymbol{v} \tag{2.67}$$

are[5] the so-called covariant and contravariant components of $\boldsymbol{v}$. Now, the representation $(2.66)_1$ or $(2.66)_2$ or both can be employed to decompose the unit vectors $\boldsymbol{e}_i$ into components

$$\boldsymbol{e}_i = (\boldsymbol{e}_i \cdot \boldsymbol{a}_k)\boldsymbol{a}^k \qquad \text{and} \qquad \boldsymbol{e}_i = (\boldsymbol{e}_i \cdot \boldsymbol{a}^k)\boldsymbol{a}_k \tag{2.68}$$

in Eq. (2.17), so a second-order tensor with orthogonal coordinates A_{ij} in the reference system $\boldsymbol{e}_1$, $\boldsymbol{e}_2$ and $\boldsymbol{e}_3$ equivalently can be represented in four different ways

$$\begin{aligned} \boldsymbol{A} &= A_{ij}(\boldsymbol{e}_i \cdot \boldsymbol{a}_h)(\boldsymbol{e}_j \cdot \boldsymbol{a}_k)\boldsymbol{a}^h \otimes \boldsymbol{a}^k = A_{ij}(\boldsymbol{e}_i \cdot \boldsymbol{a}^h)(\boldsymbol{e}_j \cdot \boldsymbol{a}^k)\boldsymbol{a}_h \otimes \boldsymbol{a}_k \\ &= A_{ij}(\boldsymbol{e}_i \cdot \boldsymbol{a}_h)(\boldsymbol{e}_j \cdot \boldsymbol{a}^k)\boldsymbol{a}^h \otimes \boldsymbol{a}_k = A_{ij}(\boldsymbol{e}_i \cdot \boldsymbol{a}^h)(\boldsymbol{e}_j \cdot \boldsymbol{a}_k)\boldsymbol{a}_h \otimes \boldsymbol{a}^k, \end{aligned} \tag{2.69}$$

therefore defining the *covariant components*

$$A_{hk} = \boldsymbol{a}_h \cdot \boldsymbol{A}\boldsymbol{a}_k = A_{ij}(\boldsymbol{e}_i \cdot \boldsymbol{a}_h)(\boldsymbol{e}_j \cdot \boldsymbol{a}_k), \tag{2.70}$$

the *contravariant components*

$$A^{hk} = \boldsymbol{a}^h \cdot \boldsymbol{A}\boldsymbol{a}^k = A_{ij}(\boldsymbol{e}_i \cdot \boldsymbol{a}^h)(\boldsymbol{e}_j \cdot \boldsymbol{a}^k), \tag{2.71}$$

and the *mixed components*

$$\begin{aligned} A^{h*}_{*k} &= \boldsymbol{a}^h \cdot \boldsymbol{A}\boldsymbol{a}_k = A_{ij}(\boldsymbol{e}_i \cdot \boldsymbol{a}^h)(\boldsymbol{e}_j \cdot \boldsymbol{a}_k), \\ A^{*k}_{h*} &= \boldsymbol{a}_h \cdot \boldsymbol{A}\boldsymbol{a}^k = A_{ij}(\boldsymbol{e}_i \cdot \boldsymbol{a}_h)(\boldsymbol{e}_j \cdot \boldsymbol{a}^k). \end{aligned} \tag{2.72}$$

[5] Throughout this book, an index, for instance, i, at the apex of a vector, for instance, $\boldsymbol{a}^i$, is used to denote the three vectors $\boldsymbol{a}^1$, $\boldsymbol{a}^2$ and $\boldsymbol{a}^3$. Therefore, $\boldsymbol{a}^2$ has no relation with the scalar $\boldsymbol{a} \cdot \boldsymbol{a} = |\boldsymbol{a}|^2$.

In conclusion, the representation in mixed components of a tensor is

$$A = A^{i*}_{*j} a_i \otimes a^j = A^{*j}_{i*} a^i \otimes a_j. \tag{2.73}$$

It is possible now to prove the following *lemma*: A tensor A is defective (so that it cannot be diagonalised) if and only if the right and left eigenvectors corresponding to a multiple eigenvalue with algebraic multiplicity greater than the geometrical are mutually orthogonal.

In fact, if these eigenvectors were not mutually orthogonal, reciprocal bases could be constructed employing these eigenvectors and, projected onto these bases, the component matrix would be diagonal.

As a consequence of the preceding lemma and the mixed representation (2.73), the following theorem can be stated.

2.7 Spectral representation theorem

- A non-defective tensor A always can be represented as

$$A = \alpha_1 \underbrace{a_1}_{\text{right}} \otimes \overbrace{a^1}^{\text{left}} + \alpha_2 \underbrace{a_2}_{\text{right}} \otimes \overbrace{a^2}^{\text{left}} + \alpha_3 \underbrace{a_3}_{\text{right}} \otimes \overbrace{a^3}^{\text{left}}, \tag{2.74}$$

  where a_i and a^j are right and left eigenvectors, respectively, forming reciprocal bases, namely, $a_i \cdot a^j = \delta_{ij}$. Note that two of the eigenvalues α_i can be complex conjugate, and in that case, the corresponding eigenvectors become complex.
- A defective tensor A with

	Multiplicity		Eigenvectors	
Eigenvalues	Algebraic	Geometric	Right	Left
α_1	1	1	b_1,	b^1
α_2	2	1	b_2,	b^2

  (where the eigenvectors are selected such that $b^1 \cdot b_1 = |b_2| = |b^2| = 1$) can be represented as follows: First, we note that $b^1 \cdot b_2 = b_1 \cdot b^2 = b_2 \cdot b^2 = 0$.[6] Therefore, we define the reciprocal bases

$$\begin{aligned} a_1 &= b_1, \quad & a_2 &= b_2, \quad & a_3 &= \frac{b_2 \times b^1}{a^2 \cdot A(b_2 \times b^1)} \\ a^1 &= b^1, \quad & a^2 &= b_2 - (b_1 \cdot b_2) b^1, \quad & a^3 &= \frac{a^2 \cdot A(b_2 \times b^1)}{b^2 \cdot (b_2 \times b^1)} b^2 \end{aligned} \tag{2.75}$$

  such that we obtain the Jordan normal form of tensor A:

$$A = \alpha_1 a_1 \otimes a^1 + \alpha_2 a_2 \otimes a^2 + \alpha_2 a_3 \otimes a^3 + a_2 \otimes a^3. \tag{2.76}$$

  Note that $a^2 \cdot A(b_2 \times b^1) \neq 0$; otherwise, A would be diagonal in a basis similar to (2.75) but rewritten with $a_3 = b_2 \times b^1$. Also, $b^2 \cdot (b_2 \times b^1) \neq 0$; otherwise, b^1, b^2 and b_2 would be co-planar. Note also that if three eigenvalues are coincident

[6] Note that $b_2 \cdot b^2 = 0$; otherwise, A would be diagonalisable and thus non-defective.

with geometrical multiplicity equal to 2, representation (2.76) still holds with $\alpha_1 = \alpha_2$.

- A defective tensor $\boldsymbol{A}$ with

	Multiplicity		Eigenvectors	
Eigenvalues	Algebraic	Geometrical	Right	Left
α	3	1	$\boldsymbol{b}_1$,	$\boldsymbol{b}^1$

(where the eigenvectors are of unit magnitude, $|\boldsymbol{b}_1| = |\boldsymbol{b}^1| = 1$) can be represented as follows: First, we define tensor

$$\boldsymbol{B} = \boldsymbol{A} - \alpha \boldsymbol{I},$$

which has the same eigenvectors of $\boldsymbol{A}$, but all null eigenvalues, so $\boldsymbol{B}^T \boldsymbol{b}^1 = \boldsymbol{B}\boldsymbol{b}_1 = \boldsymbol{0}$. Second, noting that $\boldsymbol{b}^1 \cdot \boldsymbol{b}_1 = 0$, we introduce the reciprocal bases

$$\boldsymbol{a}_1 = \frac{\boldsymbol{b}_1}{\boldsymbol{b}_1 \cdot \boldsymbol{B}\boldsymbol{b}^1 \times \boldsymbol{b}^1}, \qquad \boldsymbol{a}_2 = \frac{\boldsymbol{B}\boldsymbol{b}^1}{\boldsymbol{a}^1 \cdot \boldsymbol{B}^2 \boldsymbol{b}^1}, \qquad \boldsymbol{a}_3 = \frac{\boldsymbol{b}^1}{\boldsymbol{a}^1 \cdot \boldsymbol{B}^2 \boldsymbol{b}^1}$$
$$\boldsymbol{a}^1 = \boldsymbol{B}\boldsymbol{b}^1 \times \boldsymbol{b}^1, \qquad \boldsymbol{a}^2 = \left(\frac{\boldsymbol{a}^1 \cdot \boldsymbol{B}^2 \boldsymbol{b}^1}{\boldsymbol{b}_1 \cdot \boldsymbol{a}^1} \right) \boldsymbol{b}^1 \times \boldsymbol{b}_1, \qquad \boldsymbol{a}^3 = (\boldsymbol{a}^1 \cdot \boldsymbol{B}^2 \boldsymbol{b}^1) \boldsymbol{b}^1, \tag{2.77}$$

using which the Jordan normal form of tensor $\boldsymbol{A}$ is obtained

$$\boldsymbol{A} = \alpha \boldsymbol{I} + \boldsymbol{a}_1 \otimes \boldsymbol{a}^2 + \boldsymbol{a}_2 \otimes \boldsymbol{a}^3. \tag{2.78}$$

Note that $\boldsymbol{B}\boldsymbol{b}^1 \cdot \boldsymbol{b}^1 \times \boldsymbol{b}_1 = \boldsymbol{b}_1 \cdot \boldsymbol{a}^1$ and that $\boldsymbol{a}^1 \cdot \boldsymbol{B}^2 \boldsymbol{b}^1 \neq 0$; otherwise, it would be possible to employ reciprocal bases analogous to (2.77) but with $\boldsymbol{a}^1 \cdot \boldsymbol{B}^2 \boldsymbol{b}^1$ replaced by 1 to obtain a representation of $\boldsymbol{A}$ corresponding to a geometrical multiplicity 2 of the eigenvectors. Note also that the fact that $\boldsymbol{a}^2 \cdot \boldsymbol{B}\boldsymbol{a}_2 = 0$ comes from the observation that otherwise $\boldsymbol{B}$ (and also $\boldsymbol{A}$) would not have coincident eigenvalues [as can be easily checked using the characteristic equation (2.56) and keeping into account that $\boldsymbol{B}$ has all null eigenvalues by assumption].

As a particular case of representation (2.74), we note that for a symmetric tensor $\boldsymbol{A}$, defectivity is excluded, and there is no difference between left and right eigenvectors; thus $\boldsymbol{a}_1$, $\boldsymbol{a}_2$ and $\boldsymbol{a}_3$ can be chosen to form an orthonormal basis, with corresponding real eigenvalues α_1, α_2 and α_3. Therefore, we have

$$\boldsymbol{A} = \alpha_1 \boldsymbol{a}_1 \otimes \boldsymbol{a}_1 + \alpha_2 \boldsymbol{a}_2 \otimes \boldsymbol{a}_2 + \alpha_3 \boldsymbol{a}_3 \otimes \boldsymbol{a}_3, \tag{2.79}$$

and in the particular cases $\alpha_2 = \alpha_3$ and $\alpha_1 = \alpha_2 = \alpha_3$,

$$\boldsymbol{A} = \alpha_1 \boldsymbol{a}_1 \otimes \boldsymbol{a}_1 + \alpha_2 (\boldsymbol{I} - \boldsymbol{a}_1 \otimes \boldsymbol{a}_1) \quad \text{and} \quad \boldsymbol{A} = \alpha_1 \boldsymbol{I}, \tag{2.80}$$

respectively.

2.8 Square root of a tensor

For a symmetric positive definite tensor $\boldsymbol{A}$, with eigenvalues α_i and eigenvectors $\boldsymbol{a}_i$, we define the square-root tensor as

$$\sqrt{\boldsymbol{A}} = \sqrt{\alpha_1}\, \boldsymbol{a}_1 \otimes \boldsymbol{a}_1 + \sqrt{\alpha_2}\, \boldsymbol{a}_2 \otimes \boldsymbol{a}_2 + \sqrt{\alpha_3}\, \boldsymbol{a}_3 \otimes \boldsymbol{a}_3. \tag{2.81}$$

Evidently, $\sqrt{\boldsymbol{A}}$ is positive definite, and it can be proved that it is unique.

2.9 Polar decomposition theorem

For every tensor $\boldsymbol{F}$ such that $\det \boldsymbol{F} > 0$, the following (unique) representations exist

$$\boldsymbol{F} = \boldsymbol{R}\boldsymbol{U} = \boldsymbol{V}\boldsymbol{R}, \tag{2.82}$$

where

$$\boldsymbol{U} = \sqrt{\boldsymbol{F}^T\boldsymbol{F}} \quad \text{and} \quad \boldsymbol{V} = \sqrt{\boldsymbol{F}\boldsymbol{F}^T}, \tag{2.83}$$

where $\boldsymbol{U}$ and $\boldsymbol{V}$ are symmetric positive definite tensors and $\boldsymbol{R}$ is a rotation. The proof of this theorem is standard and is omitted (Gurtin, 1981; Ogden, 1984).

2.10 On coaxiality between second-order tensors

Since coaxiality (2.31) is a property frequently used throughout this book, we feel that it is appropriate to include a detailed discussion. In particular, we prove the following *proposition*, valid for symmetric tensors:

Two symmetric second-order tensors $\boldsymbol{A}$, $\boldsymbol{B}$ commute, that is, $\boldsymbol{A}\boldsymbol{B} = \boldsymbol{B}\boldsymbol{A}$, if and only if they possess at least three common, linearly independent eigenvectors (thus defining a principal reference system).

Proof. The sufficiency is trivial. In fact, let us assume that the two tensors share a principal reference system. Represented in this system, the two tensors evidently commute.

Assume now that the two tensors commute. Represented in the principal reference system of $\boldsymbol{A}$, these two tensors write as

$$\boldsymbol{A} = \sum_{1=1}^{3} \alpha_i \boldsymbol{a}_i \otimes \boldsymbol{a}_i \quad \text{and} \quad \boldsymbol{B} = \beta_{ij} \boldsymbol{a}_i \otimes \boldsymbol{a}_j, \tag{2.84}$$

where $\boldsymbol{a}_i$, $i = 1,2,3$, are the unit eigenvectors of $\boldsymbol{A}$ and α_i the corresponding eigenvalues, β_{ij} are the components of $\boldsymbol{B}$ onto $\boldsymbol{a}_i$, which, owing to symmetry, satisfy $\beta_{ij} = \beta_{ji}$.

Imposing the coaxiality condition $\boldsymbol{A}\boldsymbol{B} = \boldsymbol{B}\boldsymbol{A}$ yields

$$(\alpha_i - \alpha_j)\beta_{ij} = 0, \qquad (i \neq j;\ i,j = 1,2,3). \tag{2.85}$$

Condition (2.85) implies that either $\boldsymbol{a}_i$ and $\boldsymbol{a}_j$ are eigenvectors of $\boldsymbol{B}$, so $\beta_{ij} = 0$ for $i \neq j$, or the characteristic space corresponding to $\boldsymbol{a}_i$ and $\boldsymbol{a}_j$ is a plane, so $\alpha_i = \alpha_j$. In the latter case, it is always possible to choose in this plane a reference system which is principal for $\boldsymbol{B}$. □

Remark 1. The preceding proof is restricted to dimension 3. Its generalisation to dimension n is straightforward.

Remark 2. It is important to realize that the preceding proposition *does not imply* that two coaxial tensors share *all* eigenvectors. Take, for instance, the identity tensor. This is coaxial to every tensor but shares only three eigenvectors with any symmetric second-order tensor with distinct eigenvalues.

2.11 Fourth-order tensors

Fourth-order tensors are denoted by special majuscules, such as, for instance, the elasticity tensor $\mathbb{E}[\cdot]$. They define linear mappings, assigning to each second-order tensor $\boldsymbol{A}$ a second-order tensor $\mathbb{E}[\boldsymbol{A}]$ so that introducing the components $\mathbb{E}_{ijhk}$ in an orthonormal basis $\boldsymbol{e}_i$ [see Itskov (2000) for definitions in non-orthogonal bases]

$$\mathbb{E} = \mathbb{E}_{ijhk}\boldsymbol{e}_i \otimes \boldsymbol{e}_j \otimes \boldsymbol{e}_h \otimes \boldsymbol{e}_k \qquad \text{so that} \qquad (\mathbb{E}[\boldsymbol{A}])_{ij} = \mathbb{E}_{ijhk}A_{hk}. \tag{2.86}$$

The product $\mathbb{E}\mathbb{H}$ of two fourth-order tensors is defined, analogously to second-order tensors, by composition, namely, for every $\boldsymbol{A} \in \mathsf{Lin}$,

$$(\mathbb{E}\mathbb{H})[\boldsymbol{A}] = \mathbb{E}[\mathbb{H}[\boldsymbol{A}]] \qquad \text{or} \qquad (\mathbb{E}\mathbb{H})_{ijhk} = \mathbb{E}_{ijst}\mathbb{H}_{sthk}. \tag{2.87}$$

Three tensorial products will be employed, denoted by symbols $\otimes$, $\boxtimes$ and $\underline{\overline{\otimes}}$. These are defined, for every $\boldsymbol{A}, \boldsymbol{B}, \boldsymbol{C} \in \mathsf{Lin}$, as

$$(\boldsymbol{A} \otimes \boldsymbol{B})[\boldsymbol{C}] = (\boldsymbol{B} \cdot \boldsymbol{C})\boldsymbol{A} \qquad \text{or} \qquad (\boldsymbol{A} \otimes \boldsymbol{B})_{ijhk} = \boldsymbol{A}_{ij}\boldsymbol{B}_{hk},$$

$$(\boldsymbol{A} \boxtimes \boldsymbol{B})[\boldsymbol{C}] = \boldsymbol{A}\boldsymbol{C}\boldsymbol{B}^T \qquad \text{or} \qquad (\boldsymbol{A} \boxtimes \boldsymbol{B})_{ijhk} = \boldsymbol{A}_{ih}\boldsymbol{B}_{jk},$$

$$(\boldsymbol{A} \underline{\overline{\otimes}} \boldsymbol{B})[\boldsymbol{C}] = \frac{1}{2}\left(\boldsymbol{A}\boldsymbol{C}\boldsymbol{B}^T + \boldsymbol{A}\boldsymbol{C}^T\boldsymbol{B}^T\right) \tag{2.88}$$

$$\text{or} \qquad (\boldsymbol{A} \underline{\overline{\otimes}} \boldsymbol{B})_{ijhk} = \frac{1}{2}\left(\boldsymbol{A}_{ih}\boldsymbol{B}_{jk} + \boldsymbol{A}_{ik}\boldsymbol{B}_{jh}\right),$$

Thus

$$(\boldsymbol{A} \underline{\overline{\otimes}} \boldsymbol{B})[\boldsymbol{C}] = \frac{1}{2}(\boldsymbol{A} \boxtimes \boldsymbol{B})[\boldsymbol{C} + \boldsymbol{C}^T]. \tag{2.89}$$

Note that with the preceding definition, $\boldsymbol{I} \boxtimes \boldsymbol{I}$ is the fourth-order identity tensor.

Defining the transpose of a fourth-order tensor (for every $\boldsymbol{A}, \boldsymbol{B} \in \mathsf{Lin}$) as the unique tensor $\mathbb{E}^T$ with the property

$$\boldsymbol{B} \cdot \mathbb{E}^T[\boldsymbol{A}] = \boldsymbol{A} \cdot \mathbb{E}[\boldsymbol{B}], \tag{2.90}$$

we say that

- $\mathbb{E}$ has the major (or diagonal) symmetry whenever $\mathbb{E} = \mathbb{E}^T$;
- $\mathbb{E}$ has the left minor symmetry whenever, for every $\boldsymbol{A}$, $\mathbb{E}[\boldsymbol{A}] \in \mathsf{Sym}$;
- $\mathbb{E}$ has the right minor symmetry whenever, for every $\boldsymbol{A}$, $\mathbb{E}^T[\boldsymbol{A}] \in \mathsf{Sym}$.

We may note now that

$$(\boldsymbol{A} \otimes \boldsymbol{B})^T = \boldsymbol{B} \otimes \boldsymbol{A}, \qquad (\boldsymbol{A} \boxtimes \boldsymbol{B})^T = \boldsymbol{A}^T \boxtimes \boldsymbol{B}^T, \qquad (\boldsymbol{A} \underline{\overline{\otimes}} \boldsymbol{B})^T = \boldsymbol{A}^T \underline{\overline{\otimes}} \boldsymbol{B}^T. \tag{2.91}$$

Positive definiteness and invertibility of fourth-order tensors are defined analogously to second-order tensors.

The fourth-order tensor $\boldsymbol{I} \underline{\overline{\otimes}} \boldsymbol{I}$ is the symmetrising operator, which associates to every second-order tensor $\boldsymbol{X}$ its symmetric part:

$$\boldsymbol{I} \underline{\overline{\otimes}} \boldsymbol{I}[\boldsymbol{X}] = \frac{1}{2}\left(\boldsymbol{X} + \boldsymbol{X}^T\right). \tag{2.92}$$

Obviously $\boldsymbol{I}\,\underline{\overline{\otimes}}\,\boldsymbol{I}$ is singular (because it associates the null tensor to every skew-symmetric tensor), but its restriction to Sym is invertible, and the inverse is the tensor itself.

Finally, since

$$\begin{aligned}(\boldsymbol{A}\boxtimes\boldsymbol{B})(\boldsymbol{C}\boxtimes\boldsymbol{D}) &= \boldsymbol{AC}\boxtimes\boldsymbol{BD},\\ (\boldsymbol{A}\,\underline{\overline{\otimes}}\,\boldsymbol{B})(\boldsymbol{C}\,\underline{\overline{\otimes}}\,\boldsymbol{D}) &= \boldsymbol{AC}\,\underline{\overline{\otimes}}\,\boldsymbol{BD}+\boldsymbol{AD}\,\underline{\overline{\otimes}}\,\boldsymbol{BC},\end{aligned} \tag{2.93}$$

we have

$$(\boldsymbol{A}\boxtimes\boldsymbol{B})^{-1}=\boldsymbol{A}^{-1}\boxtimes\boldsymbol{B}^{-1} \quad \text{and} \quad (\boldsymbol{A}\,\underline{\overline{\otimes}}\,\boldsymbol{A})^{-1}=\boldsymbol{A}^{-1}\,\underline{\overline{\otimes}}\,\boldsymbol{A}^{-1}, \tag{2.94}$$

where the inverses are defined in such a way that

$$(\boldsymbol{A}\boxtimes\boldsymbol{B})(\boldsymbol{A}^{-1}\boxtimes\boldsymbol{B}^{-1})=\boldsymbol{I}\boxtimes\boldsymbol{I} \quad \text{and} \quad (\boldsymbol{A}\,\underline{\overline{\otimes}}\,\boldsymbol{A})(\boldsymbol{A}^{-1}\,\underline{\overline{\otimes}}\,\boldsymbol{A}^{-1})=\boldsymbol{I}\,\underline{\overline{\otimes}}\,\boldsymbol{I}. \tag{2.95}$$

2.12 On the metric induced by semi–positive definite tensors

This section is instrumental to a proof that will be given in Chapter 11 devoted to local criteria for uniqueness and stability.

Let us consider the positive scalar–valued function f defined over the space of nth-order tensors as

$$f(\boldsymbol{A})=\sqrt{\boldsymbol{A}\cdot\mathbb{H}[\boldsymbol{A}]}\geq 0, \tag{2.96}$$

where $\boldsymbol{A}$ is a generic nth-order tensor and $\mathbb{H}$ is any semi–positive definite $(2\times n)$th-order tensor (symmetries are not required). In particular, the cases in which $\mathbb{H}$ is a fourth-order ($n=2$) or a second-order ($n=1$) tensor are particularly relevant here. We begin with a proof of the Cauchy-Schwarz inequality. Take any real λ and consider, for every nth-order tensors $\boldsymbol{A}$ and $\boldsymbol{B}$,

$$0\leq[f(\lambda\boldsymbol{A}+\boldsymbol{B})]^2=\lambda^2\boldsymbol{A}\cdot\mathbb{H}[\boldsymbol{A}]+\lambda(\boldsymbol{A}\cdot\mathbb{H}[\boldsymbol{B}]+\boldsymbol{B}\cdot\mathbb{H}[\boldsymbol{A}])+\boldsymbol{B}\cdot\mathbb{H}[\boldsymbol{B}]. \tag{2.97}$$

It may be observed from Eq. (2.97) that the discriminant of the second-order polynomial in λ must be negative or null. This condition yields the Cauchy-Schwarz inequality:

$$\frac{|\boldsymbol{A}\cdot\mathbb{H}[\boldsymbol{B}]+\boldsymbol{B}\cdot\mathbb{H}[\boldsymbol{A}]|}{2}\leq\sqrt{\boldsymbol{A}\cdot\mathbb{H}[\boldsymbol{A}]}\sqrt{\boldsymbol{B}\cdot\mathbb{H}[\boldsymbol{B}]}. \tag{2.98}$$

Taking now $\lambda=1$ in Eq. (2.97) and using Eq. (2.98), the triangle inequality is readily obtained:

$$\begin{aligned}(\boldsymbol{A}+\boldsymbol{B})\cdot\mathbb{H}[\boldsymbol{A}+\boldsymbol{B}] &\leq \boldsymbol{A}\cdot\mathbb{H}[\boldsymbol{A}]\\ &\quad+|\boldsymbol{A}\cdot\mathbb{H}[\boldsymbol{B}]+\boldsymbol{B}\cdot\mathbb{H}[\boldsymbol{A}]|+\boldsymbol{B}\cdot\mathbb{H}[\boldsymbol{B}]\\ &\leq\left(\sqrt{\boldsymbol{A}\cdot\mathbb{H}[\boldsymbol{A}]}+\sqrt{\boldsymbol{B}\cdot\mathbb{H}[\boldsymbol{B}]}\right)^2.\end{aligned} \tag{2.99}$$

As a consequence of Eq. (2.99), we note that

$$\begin{aligned}f(\boldsymbol{A}) &= f(\boldsymbol{A}\pm\boldsymbol{B}\mp\boldsymbol{B})\leq f(\boldsymbol{A}\pm\boldsymbol{B})+f(\boldsymbol{B}),\\ f(\boldsymbol{B}) &= f(\boldsymbol{B}\pm\boldsymbol{A}\mp\boldsymbol{A})\leq f(\boldsymbol{A}\pm\boldsymbol{B})+f(\boldsymbol{A}),\end{aligned} \tag{2.100}$$

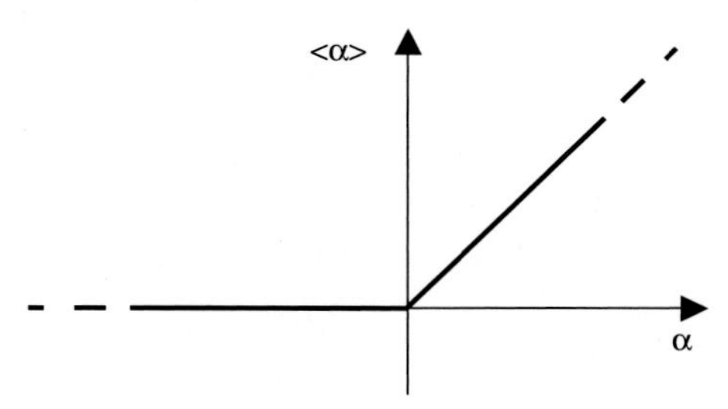

Figure 2.8. Macaulay bracket function $< \alpha >$.

which imply

$$\sqrt{(\boldsymbol{A} \pm \boldsymbol{B}) \cdot \mathbb{H}[\boldsymbol{A} \pm \boldsymbol{B}]} \geq |\sqrt{\boldsymbol{A} \cdot \mathbb{H}[\boldsymbol{A}]} - \sqrt{\boldsymbol{B} \cdot \mathbb{H}[\boldsymbol{B}]}|. \tag{2.101}$$

Finally, we note that for every real λ,

$$f(\lambda \boldsymbol{A}) = |\lambda| f(\boldsymbol{A}). \tag{2.102}$$

Therefore, function f defines a semi-norm on the space of nth-order tensors (a norm when $\mathbb{H}$ is positive definite).

2.13 The Macaulay bracket operator

The Macaulay bracket operator is important because it defines the incremental non-linearity of elastoplasticity. It is defined for every scalar α as (Macaulay, 1919)

$$< \alpha > = \max\{0, \alpha\}, \qquad \text{or, equivalently, as} \qquad < \alpha > = \frac{\alpha + |\alpha|}{2} \tag{2.103}$$

(see the sketch in Fig. 2.8).

2.14 Differential calculus for tensors

We consider here scalar, vector or tensor (of arbitrary order) functions, generically indicated with $\boldsymbol{f}$, of a scalar variable t, defined on an open interval of the real axis. The variable t, for instance, can be identified with a scalar variable as the temperature, or the time, or a time-like parameter governing a deformation process. We denote with a dot over the symbol of the function, namely, $\dot{\boldsymbol{f}}(t)$, the derivative of $\boldsymbol{f}$ at t, so

$$\dot{\boldsymbol{f}}(t) = \lim_{\tau \to 0} \frac{\boldsymbol{f}(t+\tau) - \boldsymbol{f}(t)}{\tau}. \tag{2.104}$$

When $\dot{\boldsymbol{f}}$ exists and is continuous, function $\boldsymbol{f}$ is called 'smooth'.

Standard rules of differentiation can be derived from the definition (2.104), in particular,

$$(\boldsymbol{u} + \boldsymbol{v})^{\cdot} = \dot{\boldsymbol{u}} + \dot{\boldsymbol{v}}, \qquad (\alpha \, \boldsymbol{u} * \boldsymbol{v})^{\cdot} = \dot{\alpha} \, \boldsymbol{u} * \boldsymbol{v} + \alpha \, \dot{\boldsymbol{u}} * \boldsymbol{v} + \alpha \, \boldsymbol{u} * \dot{\boldsymbol{v}}, \tag{2.105}$$

where $\boldsymbol{u}$ and $\boldsymbol{v}$ are vectors or tensors, α is a scalar function and $*$ stands for $\cdot$ or $\otimes$ or $\times$.

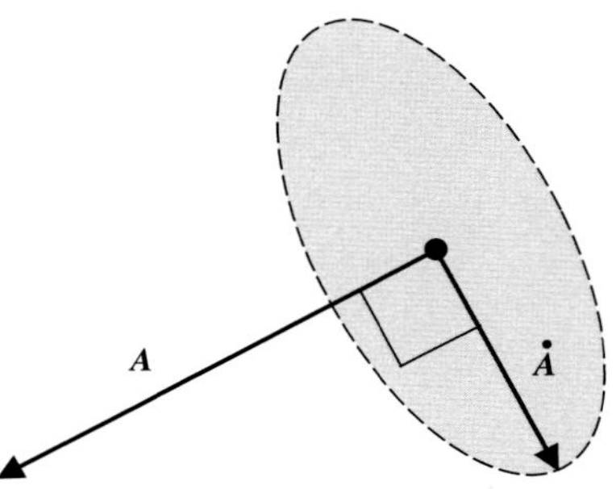

Figure 2.9. When vector $\boldsymbol{A}$ has a constant modulus, $\dot{\boldsymbol{A}}$ results to be orthogonal to $\boldsymbol{A}$. The orthogonality property also holds for tensors.

Note that $\left(\boldsymbol{A}^{-1}\right)^{\boldsymbol{\cdot}} \neq \dot{\boldsymbol{A}}^{-1}$, and the trick for calculating $\left(\boldsymbol{A}^{-1}\right)^{\boldsymbol{\cdot}}$ is the following:

$$\left(\boldsymbol{A}^{-1}\boldsymbol{A}\right)^{\boldsymbol{\cdot}} = (\boldsymbol{I})^{\boldsymbol{\cdot}} = \boldsymbol{0}, \qquad \Longrightarrow \qquad \left(\boldsymbol{A}^{-1}\right)^{\boldsymbol{\cdot}} = -\boldsymbol{A}^{-1}\dot{\boldsymbol{A}}\boldsymbol{A}^{-1}. \tag{2.106}$$

Another useful property occurs when a vector or tensor, say, $\boldsymbol{A}$, has a modulus independent of the variable t. In this case,

$$(\boldsymbol{A}\cdot\boldsymbol{A})^{\boldsymbol{\cdot}} = (const.)^{\boldsymbol{\cdot}} = 0, \qquad \Longrightarrow \qquad \dot{\boldsymbol{A}}\cdot\boldsymbol{A} = 0, \tag{2.107}$$

showing that $\dot{\boldsymbol{A}}$ is orthogonal to $\boldsymbol{A}$ (Fig. 2.9).

2.15 Gradient

We consider now a generic (scalar, vector or tensor of arbitrary order) function $\boldsymbol{f}$ of a vector or tensor variable $\boldsymbol{x}$, defined on an open subset of a finite-dimensional linear normed space. We denote with $\nabla\boldsymbol{f}(\boldsymbol{x})[\boldsymbol{a}]$ the directional derivative of function $\boldsymbol{f}$ at $\boldsymbol{x}$ in the direction (of the vector or tensor) $\boldsymbol{a}$ such that if

$$\nabla\boldsymbol{f}(\boldsymbol{x})[\boldsymbol{a}] = \lim_{\alpha\to 0}\frac{\boldsymbol{f}(\boldsymbol{x}+\alpha\boldsymbol{a}) - \boldsymbol{f}(\boldsymbol{x})}{\alpha} \tag{2.108}$$

exists and is linear for all $\boldsymbol{a}$, then $\nabla\boldsymbol{f}$ is the gradient of $\boldsymbol{f}$, and this function admits the Taylor series expansion

$$\boldsymbol{f}(\boldsymbol{x}+\boldsymbol{a}) = \boldsymbol{f}(\boldsymbol{x}) + \nabla\boldsymbol{f}(\boldsymbol{x})[\boldsymbol{a}] + O\left(|\boldsymbol{a}|^2\right). \tag{2.109}$$

The following notations for the gradient of a (scalar, vector or tensor) field will be employed

$$\nabla\boldsymbol{f} = \frac{\partial\boldsymbol{f}}{\partial\boldsymbol{x}} = \mathsf{grad}\boldsymbol{f} = \mathsf{Grad}\boldsymbol{f}, \tag{2.110}$$

where the difference between symbols grad and Grad will be detailed later (briefly, the different notation refers to differentiation with respect to the spatial, $\boldsymbol{x}$, or the material, $\boldsymbol{x}_0$, placement). When $\nabla\boldsymbol{f}$ is continuous at each point of the definition domain for every $\boldsymbol{a}$, function $\boldsymbol{f}$ is called 'smooth' (or of class C^1).

As a first example of calculation of a gradient, let us consider

$$\boldsymbol{f} = \boldsymbol{X}^n, \tag{2.111}$$

with $\boldsymbol{X} \in \mathsf{Lin}$ and $n \geq 1$. Application of definition (2.108) yields

$$\nabla(\boldsymbol{X}^n)[\boldsymbol{A}] = \lim_{\alpha \to 0} \frac{(\boldsymbol{X}+\alpha\boldsymbol{A})^n - \boldsymbol{X}^n}{\alpha} = \sum_{i=1}^{n} \boldsymbol{X}^{n-i}\boldsymbol{A}\boldsymbol{X}^{i-1} = \sum_{i=1}^{n} (\boldsymbol{X}^{n-i} \boxtimes \boldsymbol{X}^{i-1\,T})[\boldsymbol{A}], \tag{2.112}$$

which, in the particular case of $n = 1$, gives the fourth-order identity tensor.

The derivative of the inverse function

$$\boldsymbol{f}(\boldsymbol{X}) = \boldsymbol{X}^{-1} \tag{2.113}$$

can be obtained in different ways: (1) with a trick similar to that used to obtain Eq. (2.106), namely, by differentiating the identity $\boldsymbol{X}^{-1}\boldsymbol{X} = \boldsymbol{I}$, (2) using the Cayley-Hamilton theorem, Eq. (2.59), written for $\boldsymbol{X}$, multiplied by $\boldsymbol{X}^{-1}$ or finally, (3) just employing the definition (2.108):

$$\nabla(\boldsymbol{X}^{-1})[\boldsymbol{A}] = \lim_{\alpha \to 0} \frac{(\boldsymbol{X}+\alpha\boldsymbol{A})^{-1} - \boldsymbol{X}^{-1}}{\alpha} \tag{2.114}$$

and noting that

$$(\boldsymbol{X}^{-1} - \alpha\boldsymbol{X}^{-1}\boldsymbol{A}\boldsymbol{X}^{-1})(\boldsymbol{X}+\alpha\boldsymbol{A}) = \boldsymbol{I} - \alpha^2\boldsymbol{X}^{-1}\boldsymbol{A}\boldsymbol{X}^{-1}\boldsymbol{A} \tag{2.115}$$

so that

$$(\boldsymbol{X}+\alpha\boldsymbol{A})^{-1} = (\boldsymbol{X}^{-1} - \alpha\boldsymbol{X}^{-1}\boldsymbol{A}\boldsymbol{X}^{-1}) + O(\alpha^2). \tag{2.116}$$

Therefore,

$$\nabla(\boldsymbol{X}^{-1})[\boldsymbol{A}] = -\boldsymbol{X}^{-1}\boldsymbol{A}\boldsymbol{X}^{-1} = -(\boldsymbol{X}^{-1} \boxtimes \boldsymbol{X}^{-T})[\boldsymbol{A}]. \tag{2.117}$$

Considering now the trace invariant function

$$f = \mathsf{tr}\boldsymbol{X}^n, \tag{2.118}$$

it is easy to obtain its gradient

$$\nabla(\mathsf{tr}\boldsymbol{X}^n)[\boldsymbol{A}] = n\,\mathsf{tr}\left(\boldsymbol{X}^{n-1}\boldsymbol{A}\right) = n\boldsymbol{X}^{n-1\,T} \cdot \boldsymbol{A}. \tag{2.119}$$

Thus, taking the trace of the Cayley-Hamilton theorem (2.59), the gradient of the determinant can be computed as

$$\nabla(\det\boldsymbol{X}) = \left(\boldsymbol{X}^2 - I_1(\boldsymbol{X})\boldsymbol{X} + I_2(\boldsymbol{X})\boldsymbol{I}\right)^T = (\det\boldsymbol{X})\boldsymbol{X}^{-T} = \mathsf{Cof}\boldsymbol{X}, \tag{2.120}$$

where we also have introduced the definition of co-factor of a non-singular tensor:

$$\mathsf{Cof}\boldsymbol{A} = (\det\boldsymbol{A})\boldsymbol{A}^{-T}. \tag{2.121}$$

The rule

$$\frac{\partial(\boldsymbol{B}\boldsymbol{X}\boldsymbol{C})}{\partial\boldsymbol{X}} = \boldsymbol{B} \boxtimes \boldsymbol{C}^T \tag{2.122}$$

follows directly from the definition of gradient (2.108), whereas the following formula is immediate:

$$\frac{\partial\mathsf{dev}\boldsymbol{X}}{\partial\boldsymbol{X}} = \boldsymbol{I}\,\underline{\overline{\otimes}}\,\boldsymbol{I} - \frac{1}{3}\boldsymbol{I} \otimes \boldsymbol{I}, \tag{2.123}$$

where $\mathsf{dev}\boldsymbol{X}$ is the deviator of $\boldsymbol{X}$, Eq. (2.34).

The following gradient can be obtained directly from the definition (2.108):

$$\frac{\partial \boldsymbol{K}\cdot\boldsymbol{X}^2}{\partial \boldsymbol{X}} = \boldsymbol{X}^T\boldsymbol{K}+\boldsymbol{K}\boldsymbol{X}^T, \qquad \frac{\partial \mathsf{tr}(\boldsymbol{K}\boldsymbol{X})^2}{\partial \boldsymbol{X}} = (\boldsymbol{K}\boldsymbol{X}\boldsymbol{K})^T+\boldsymbol{K}\boldsymbol{X}\boldsymbol{K}. \tag{2.124}$$

As a finale example, we take the gradient of the characteristic equation (2.56) of a symmetric tensor

$$\boldsymbol{A} = \sum_{i=1}^{3} \alpha_i \boldsymbol{a}_i \otimes \boldsymbol{a}_i \tag{2.125}$$

with respect to the tensor itself to obtain

$$3\alpha_i^2 \frac{\partial \alpha_i}{\partial \boldsymbol{A}} - \alpha_i^2 \boldsymbol{I} - 2\alpha_i I_1(\boldsymbol{A})\frac{\partial \alpha_i}{\partial \boldsymbol{A}} + \alpha_i[I_1(\boldsymbol{A})\boldsymbol{I}-\boldsymbol{A}] + I_2(\boldsymbol{A})\frac{\partial \alpha_i}{\partial \boldsymbol{A}} - (\det\boldsymbol{A})\boldsymbol{A}^{-T} = \boldsymbol{0}, \tag{2.126}$$

where α_i is the ith eigenvalue and the index is not summed. Use of the spectral representation for $\boldsymbol{A}$ [given by Eq. (2.125)] and for $\boldsymbol{A}^{-1}$ and the definition of the invariants (2.58) yields the rule (where the index is not summed)

$$\frac{\partial \alpha_i}{\partial \boldsymbol{A}} = \boldsymbol{a}_i \otimes \boldsymbol{a}_i, \tag{2.127}$$

which easily can be shown to hold for the non-zero eigenvalues even when $\det\boldsymbol{A} = 0$.

2.16 Divergence

For any given smooth vector field $\boldsymbol{a}$, the divergence of $\boldsymbol{a}$ is defined in a unique way as

$$\mathsf{div}\,\boldsymbol{a} = \mathsf{tr}(\nabla\boldsymbol{a}). \tag{2.128}$$

Thus, in rectangular Cartesian components,

$$(\mathsf{grad}\,\boldsymbol{a})_{ij} = a_{i,j} \qquad \text{and} \qquad \mathsf{div}\,\boldsymbol{a} = a_{k,k}, \tag{2.129}$$

where we have used the usual convention of the comma that means differentiation

$$a_{i,j} = \frac{\partial a_i}{\partial x_j}. \tag{2.130}$$

Differently from the divergence of a vector field, the divergence operator of a tensor field $\boldsymbol{A}$ can be given three different definitions. In this book we will follow the definition employed, among others, by Gurtin (1972, 1981), Truesdell and Noll (1965) and Podio Guidugli (2000), in which $\mathsf{div}\boldsymbol{A}$ is the unique vector field which, for every constant vector $\boldsymbol{a}$, satisfies

$$(\mathsf{div}\boldsymbol{A})\cdot\boldsymbol{a} = \mathsf{div}(\boldsymbol{A}^T\boldsymbol{a}). \tag{2.131}$$

In rectangular Cartesian components,

$$(\mathsf{div}\boldsymbol{A})_i = A_{ij,j}. \tag{2.132}$$

Another definition of divergence used, among others, by Hill (1978) and Ogden (1984) is the following:

$$(\mathsf{div}\boldsymbol{A})\cdot\boldsymbol{a} = \mathsf{div}(\boldsymbol{A}\boldsymbol{a}), \tag{2.133}$$

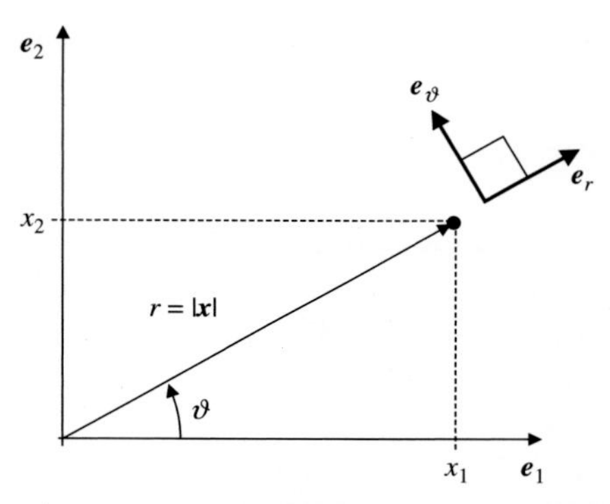

Figure 2.10. Cylindrical reference system.

which becomes coherent with their use of the nominal stress instead of the first Piola-Kirchhoff stress tensor.

The third possibility of defining a divergence, not used in practice, would be to define the divergence of a tensor field as $\nabla(\text{tr}\boldsymbol{A})$.

2.17 Cylindrical coordinates

The use of cylindrical coordinates is very frequent in solid mechanics, and they also will be employed sometimes within this book. Therefore, we present a very concise derivation of the relevant equations.

With reference to the cylindrical reference system sketched in Fig. 2.10, where the out-of-plane coordinates z and 3 coincide (so that the out-of-plane unit vector is $\boldsymbol{e}_3 = \boldsymbol{e}_z$), the point corresponding to the position vector $\boldsymbol{x} = x_i \boldsymbol{e}_i$ is singled out by the two cylindrical coordinates r and θ

$$r = |\boldsymbol{x}|, \qquad \theta = \tan^{-1}\frac{x_2}{x_1} + \begin{cases} 0 & x_1 > 0, x_2 \geq 0, \\ \pi & x_1 < 0, \\ 2\pi & x_1 > 0, x_2 < 0. \end{cases} \tag{2.134}$$

From Eqs. (2.134) it can be observed that

$$\begin{aligned} \frac{\partial r}{\partial x_1} &= \cos\theta, & \frac{\partial r}{\partial x_2} &= \sin\theta, \\ \frac{\partial \theta}{\partial x_1} &= -\frac{\sin\theta}{r}, & \frac{\partial \theta}{\partial x_2} &= \frac{\cos\theta}{r}, \end{aligned} \tag{2.135}$$

so the chain rule of differentiation allows us, using Eqs. (2.135), to relate the partial derivatives of a generic (scalar, vector or tensor) field $\boldsymbol{A}$ taken with respect to x_1, x_2 and x_3 to the partial derivatives taken with respect to r, θ and $z = x_3$ as

$$\frac{\partial \boldsymbol{A}}{\partial x_i} = \left(\frac{\partial \boldsymbol{A}}{\partial r}\cos\theta - \frac{\partial \boldsymbol{A}}{\partial \theta}\frac{\sin\theta}{r}\right)\delta_{1i} + \left(\frac{\partial \boldsymbol{A}}{\partial r}\sin\theta + \frac{\partial \boldsymbol{A}}{\partial \theta}\frac{\cos\theta}{r}\right)\delta_{2i} + \frac{\partial \boldsymbol{A}}{\partial z}\delta_{3i}, \tag{2.136}$$

representing the components of the gradient of $\boldsymbol{A}$ in the rectangular coordinate system x_1–x_2–x_3.

Introducing the unit vectors $\boldsymbol{e}_i$ corresponding to the axes x_i, Eq. (2.136) becomes

$$\nabla \boldsymbol{A} = \left(\frac{\partial \boldsymbol{A}}{\partial r} \cos\theta - \frac{\partial \boldsymbol{A}}{\partial \theta} \frac{\sin\theta}{r} \right) \otimes \boldsymbol{e}_1 + \left(\frac{\partial \boldsymbol{A}}{\partial r} \sin\theta + \frac{\partial \boldsymbol{A}}{\partial \theta} \frac{\cos\theta}{r} \right) \otimes \boldsymbol{e}_2 + \frac{\partial \boldsymbol{A}}{\partial z} \otimes \boldsymbol{e}_3. \quad (2.137)$$

Equation (2.137) holds when $\boldsymbol{A}$ is a scalar, a vector or a second-order tensor. In the case where $\boldsymbol{A}$ is a scalar, the tensor product $\otimes$ has to be replaced by an ordinary product in Eq. (2.137). Some ambiguity may arise in the case in which $\boldsymbol{A}$ is a second-order tensor. In this case, the structure of Eq. (2.137) is

$$\nabla \boldsymbol{A} = \boldsymbol{B}_1 \otimes \boldsymbol{e}_1 + \boldsymbol{B}_2 \otimes \boldsymbol{e}_2 + \boldsymbol{B}_3 \otimes \boldsymbol{e}_3, \quad (2.138)$$

with $\boldsymbol{B}_i$ second-order tensors, so the rule to apply $\nabla \boldsymbol{A}$ to every dyad $\boldsymbol{c} \otimes \boldsymbol{d}$ is

$$\nabla \boldsymbol{A}[\boldsymbol{c} \otimes \boldsymbol{d}] = (\boldsymbol{e}_1 \cdot \boldsymbol{c}) \boldsymbol{B}_1 \boldsymbol{d} + (\boldsymbol{e}_2 \cdot \boldsymbol{c}) \boldsymbol{B}_2 \boldsymbol{d} + (\boldsymbol{e}_3 \cdot \boldsymbol{c}) \boldsymbol{B}_3 \boldsymbol{d}. \quad (2.139)$$

A vector $\boldsymbol{u}$ and a second-order (non-symmetric) tensor $\boldsymbol{S}$ can be represented in cylindrical coordinates as

$$\begin{aligned} \boldsymbol{u} &= u_r \boldsymbol{e}_r + u_\theta \boldsymbol{e}_\theta + u_z \boldsymbol{e}_z, \\ \boldsymbol{S} &= S_{rr} \boldsymbol{e}_r \otimes \boldsymbol{e}_r + S_{r\theta} \boldsymbol{e}_r \otimes \boldsymbol{e}_\theta + S_{rz} \boldsymbol{e}_r \otimes \boldsymbol{e}_z \\ &\quad + S_{\theta r} \boldsymbol{e}_\theta \otimes \boldsymbol{e}_r + S_{\theta\theta} \boldsymbol{e}_\theta \otimes \boldsymbol{e}_\theta + S_{\theta z} \boldsymbol{e}_\theta \otimes \boldsymbol{e}_z \\ &\quad + S_{zr} \boldsymbol{e}_z \otimes \boldsymbol{e}_r + S_{z\theta} \boldsymbol{e}_z \otimes \boldsymbol{e}_\theta + S_{zz} \boldsymbol{e}_z \otimes \boldsymbol{e}_z, \end{aligned} \quad (2.140)$$

where u_r and u_θ and S_{rr}, $S_{\theta\theta}$, $S_{r\theta}$ and $S_{\theta r}$ are the cylindrical components of $\boldsymbol{u}$ and $\boldsymbol{S}$, respectively, and the transformation rule from components (2.140) to Cartesian components u_i is

$$u_i = u_r \boldsymbol{e}_i \cdot \boldsymbol{e}_r + u_\theta \boldsymbol{e}_i \cdot \boldsymbol{e}_\theta + u_z \boldsymbol{e}_i \cdot \boldsymbol{e}_z, \quad (2.141)$$

whereas Cartesian components S_{ij} in the reference system $\boldsymbol{e}_i$ can be obtained immediately using Eq. (2.20), which gives for the ij component in the x_1–x_2 plane

$$\begin{aligned} S_{ij} &= S_{rr} (\boldsymbol{e}_i \cdot \boldsymbol{e}_r)(\boldsymbol{e}_j \cdot \boldsymbol{e}_r) + S_{r\theta} (\boldsymbol{e}_i \cdot \boldsymbol{e}_r)(\boldsymbol{e}_j \cdot \boldsymbol{e}_\theta) \\ &\quad + S_{\theta r} (\boldsymbol{e}_i \cdot \boldsymbol{e}_\theta)(\boldsymbol{e}_j \cdot \boldsymbol{e}_r) + S_{\theta\theta} (\boldsymbol{e}_i \cdot \boldsymbol{e}_\theta)(\boldsymbol{e}_j \cdot \boldsymbol{e}_\theta). \end{aligned} \quad (2.142)$$

Now we note that $\boldsymbol{e}_r$ and $\boldsymbol{e}_\theta$ can be expressed in the reference system $\boldsymbol{e}_1$ and $\boldsymbol{e}_2$ as

$$\boldsymbol{e}_r = \boldsymbol{e}_1 \cos\theta + \boldsymbol{e}_2 \sin\theta, \qquad \boldsymbol{e}_\theta = -\boldsymbol{e}_1 \sin\theta + \boldsymbol{e}_2 \cos\theta, \quad (2.143)$$

so their dependence on θ is made explicit, and we arrive at the formulae

$$\frac{\partial \boldsymbol{e}_r}{\partial \theta} = \boldsymbol{e}_\theta \quad \text{and} \quad \frac{\partial \boldsymbol{e}_\theta}{\partial \theta} = -\boldsymbol{e}_r, \quad (2.144)$$

whereas the derivatives with respect to r are obviously null.

Considering the gradient (2.137) and keeping into account Eqs. (2.143), we can conclude that the formula for the gradient (2.137) in cylindrical coordinates is

$$\nabla \boldsymbol{A} = \frac{\partial \boldsymbol{A}}{\partial r} \otimes \boldsymbol{e}_r + \frac{1}{r} \frac{\partial \boldsymbol{A}}{\partial \theta} \otimes \boldsymbol{e}_\theta + \frac{\partial \boldsymbol{A}}{\partial z} \otimes \boldsymbol{e}_z. \quad (2.145)$$

Equation (2.145) holds when $\boldsymbol{A}$ is a scalar, vector or tensor field. For a vector field $\boldsymbol{u}$, using the representation $(2.140)_1$ and Eqs. (2.144), we obtain the gradient of the vector field $\boldsymbol{u}$ expressed in the cylindrical reference system $\boldsymbol{e}_r$–$\boldsymbol{e}_\theta$–$\boldsymbol{e}_z$:

$$\begin{aligned}\nabla\boldsymbol{u} = {} & (u_{r,r}\boldsymbol{e}_r + u_{\theta,r}\boldsymbol{e}_\theta + u_{z,r}\boldsymbol{e}_z)\otimes\boldsymbol{e}_r \\ & + \frac{1}{r}[(u_{r,\theta} - u_\theta)\boldsymbol{e}_r + (u_r + u_{\theta,\theta})\boldsymbol{e}_\theta + u_{z,\theta}\boldsymbol{e}_z]\otimes\boldsymbol{e}_\theta \\ & + (u_{r,z}\boldsymbol{e}_r + u_{\theta,z}\boldsymbol{e}_\theta + u_{z,z}\boldsymbol{e}_z)\otimes\boldsymbol{e}_z.\end{aligned} \tag{2.146}$$

Thus the trace of Eq. (2.146) gives the divergence of a vector field $\boldsymbol{u}$ in cylindrical coordinates

$$\mathrm{div}\boldsymbol{u} = u_{r,r} + \frac{u_r + u_{\theta,\theta}}{r} + u_{z,z}. \tag{2.147}$$

When field $\boldsymbol{A}$ is a second-order tensor, $\boldsymbol{S} \in \mathsf{Lin}$, we express it in the cylindrical reference system as $(2.140)_2$, so Eq. (2.145) and Eqs. (2.144) provide the gradient of a second-order tensor expressed in the cylindrical reference system $\boldsymbol{e}_r$–$\boldsymbol{e}_\theta$–$\boldsymbol{e}_z$:

$$\begin{aligned}\nabla\boldsymbol{S} = {} & [S_{rr,r}\boldsymbol{e}_r\otimes\boldsymbol{e}_r + S_{r\theta,r}\boldsymbol{e}_r\otimes\boldsymbol{e}_\theta + S_{rz,r}\boldsymbol{e}_r\otimes\boldsymbol{e}_z \\ & + S_{\theta r,r}\boldsymbol{e}_\theta\otimes\boldsymbol{e}_r + S_{\theta\theta,r}\boldsymbol{e}_\theta\otimes\boldsymbol{e}_\theta + S_{\theta z,r}\boldsymbol{e}_\theta\otimes\boldsymbol{e}_z \\ & + S_{zr,r}\boldsymbol{e}_z\otimes\boldsymbol{e}_r + S_{z\theta,r}\boldsymbol{e}_z\otimes\boldsymbol{e}_\theta + S_{zz,r}\boldsymbol{e}_z\otimes\boldsymbol{e}_z]\otimes\boldsymbol{e}_r \\ & + \frac{1}{r}[(S_{rr,\theta} - S_{r\theta} - S_{\theta r})\boldsymbol{e}_r\otimes\boldsymbol{e}_r + (S_{r\theta,\theta} + S_{rr} - S_{\theta\theta})\boldsymbol{e}_r\otimes\boldsymbol{e}_\theta \\ & + (S_{\theta r,\theta} + S_{rr} - S_{\theta\theta})\boldsymbol{e}_\theta\otimes\boldsymbol{e}_r + (S_{\theta\theta,\theta} + S_{r\theta} + S_{\theta r})\boldsymbol{e}_\theta\otimes\boldsymbol{e}_\theta \\ & + (S_{\theta z,\theta} + S_{rz})\boldsymbol{e}_\theta\otimes\boldsymbol{e}_z + (S_{zr,\theta} - S_{z\theta})\boldsymbol{e}_z\otimes\boldsymbol{e}_r + (S_{z\theta,\theta} + S_{zr})\boldsymbol{e}_z\otimes\boldsymbol{e}_\theta \\ & + (S_{rz,\theta} - S_{\theta z})\boldsymbol{e}_r\otimes\boldsymbol{e}_z + S_{zz,\theta}\boldsymbol{e}_z\otimes\boldsymbol{e}_z]\otimes\boldsymbol{e}_\theta \\ & + [S_{rr,z}\boldsymbol{e}_r\otimes\boldsymbol{e}_r + S_{r\theta,z}\boldsymbol{e}_r\otimes\boldsymbol{e}_\theta + S_{rz,z}\boldsymbol{e}_r\otimes\boldsymbol{e}_z \\ & + S_{\theta r,z}\boldsymbol{e}_\theta\otimes\boldsymbol{e}_r + S_{\theta\theta,z}\boldsymbol{e}_\theta\otimes\boldsymbol{e}_\theta + S_{\theta z,z}\boldsymbol{e}_\theta\otimes\boldsymbol{e}_z \\ & + S_{zr,z}\boldsymbol{e}_z\otimes\boldsymbol{e}_r + S_{z\theta,z}\boldsymbol{e}_z\otimes\boldsymbol{e}_\theta + S_{zz,z}\boldsymbol{e}_z\otimes\boldsymbol{e}_z]\otimes\boldsymbol{e}_z.\end{aligned} \tag{2.148}$$

Thus, according to our definition (2.131), the divergence in the cylindrical reference system becomes

$$\begin{aligned}\mathrm{div}\boldsymbol{S} = {} & \left[S_{rr,r} + \frac{1}{r}\left(S_{r\theta,\theta} + S_{rr} - S_{\theta\theta}\right) + S_{rz,z}\right]\boldsymbol{e}_r \\ & + \left[S_{\theta r,r} + \frac{1}{r}\left(S_{\theta\theta,\theta} + S_{r\theta} + S_{\theta r}\right) + S_{\theta z,z}\right]\boldsymbol{e}_\theta \\ & + \left[S_{zr,r} + \frac{1}{r}\left(S_{z\theta,\theta} + S_{zr}\right) + S_{zz,z}\right]\boldsymbol{e}_z.\end{aligned} \tag{2.149}$$

2.18 Divergence theorem

We consider a regular [in other words, a bounded, orientable surface divisible into a finite number of portions possessing continuous normal vectors; see Kellog (1953) for a precise explanation] region B of the Euclidean three-dimensional space with boundary ∂B. If the boundary ∂B has outward unit normal $\boldsymbol{n}$, and $\boldsymbol{v}$ is a smooth vector field, the divergence theorem writes in the usual form (see, e.g., Apostol, 1969):

$$\int_B \mathrm{div}\boldsymbol{v} = \int_{\partial B} \boldsymbol{v} \cdot \boldsymbol{n}, \tag{2.150}$$

where we have omitted the symbols dv and da usually accompanying the volume and surface integrals, respectively; these symbols usually will be omitted throughout this book except in cases where we believe they may help understanding.

Let us now introduce the smooth tensor field $\boldsymbol{T}^T$ and the arbitrary constant vectors $\boldsymbol{a}$ and $\boldsymbol{b}$ and replace vector $\boldsymbol{v}$ in Eq. (2.150) with $(\boldsymbol{v} \cdot \boldsymbol{a})\boldsymbol{T}^T\boldsymbol{b}$, thus obtaining

$$\int_B \mathrm{div}[(\boldsymbol{v} \cdot \boldsymbol{a})\boldsymbol{T}^T\boldsymbol{b}] = \boldsymbol{a} \cdot \left(\int_{\partial B} \boldsymbol{v} \otimes \boldsymbol{T}\boldsymbol{n}\right)\boldsymbol{b}. \tag{2.151}$$

Thus, since

$$\mathrm{div}[(\boldsymbol{v} \cdot \boldsymbol{a})\boldsymbol{T}^T\boldsymbol{b}] = \boldsymbol{a} \cdot \left[(\nabla \boldsymbol{v})\boldsymbol{T}^T + \boldsymbol{v} \otimes \mathrm{div}\boldsymbol{T}\right]\boldsymbol{b}, \tag{2.152}$$

and $\boldsymbol{a}$ and $\boldsymbol{b}$ are arbitrary vectors, we arrive at the integral equation

$$\int_B \left[(\nabla \boldsymbol{v})\boldsymbol{T}^T + \boldsymbol{v} \otimes \mathrm{div}\boldsymbol{T}\right] = \int_{\partial B} \boldsymbol{v} \otimes \boldsymbol{T}\boldsymbol{n}. \tag{2.153}$$

Equation (2.153) gives, considering all constant fields $\boldsymbol{v}$,

$$\int_B \mathrm{div}\boldsymbol{T} = \int_{\partial B} \boldsymbol{T}\boldsymbol{n}, \tag{2.154}$$

whereas, for $\boldsymbol{T} = \boldsymbol{I}$

$$\int_B \nabla \boldsymbol{v} = \int_{\partial B} \boldsymbol{v} \otimes \boldsymbol{n}. \tag{2.155}$$

2.19 Convexity and quasi-convexity

The notions of convexity and quasi-convexity play an important role in the development of yield functions and strain energy functions for elastic materials. In the following we limit ourselves to providing definitions and showing certain equivalences between some of these in a non-formal way, so the interested reader is referred for details to Boyd and Vandenberghe (2004) or Roberts and Varberg (1973).

We begin stating that a *subset U of a linear space is convex* if, for every pair $\boldsymbol{x}$ and $\boldsymbol{y}$ belonging to U and $0 \leq \lambda \leq 1$, all the points

$$\lambda \boldsymbol{x} + (1 - \lambda)\boldsymbol{y} \tag{2.156}$$

belong to U. Note that if the points (2.156) belong to U for all real λ (instead of for $\lambda \in [0,1]$), the U is called 'affine'. A linear space therefore is always both convex and affine.

A function $f(\boldsymbol{x}) : U \longrightarrow \mathbb{R}$, with U being a convex set and $0 < \lambda < 1$,[7] is convex if

$$f(\lambda \boldsymbol{x} + (1 - \lambda)\boldsymbol{y}) \leq \lambda f(\boldsymbol{x}) + (1 - \lambda) f(\boldsymbol{y}) \tag{2.157}$$

for all $\boldsymbol{x}$ and $\boldsymbol{y}$ belonging to U.

[7] The condition $\lambda \in [0,1]$ instead of $\lambda \in (0,1)$ can be used equivalently in the definition of convexity.

Convexity becomes 'strict' when instead of the inequality $\leq$ the strict inequality $<$ holds in definition (2.157) for $\boldsymbol{x} \neq \boldsymbol{y}$. A function f is said to be concave when $-f$ is convex. When f is a continuously differentiable function throughout the open convex set U, definition (2.157) becomes equivalent to

$$f(\boldsymbol{x}) - f(\boldsymbol{y}) \geq (\boldsymbol{x} - \boldsymbol{y}) \cdot \frac{\partial f}{\partial \boldsymbol{y}}(\boldsymbol{y}) \tag{2.158}$$

for all $\boldsymbol{x}$ and $\boldsymbol{y} \in U$. Again, 'strict convexity' means that $\geq$ is replaced by $>$ in Eq. (2.158), which has to hold for every $\boldsymbol{x}$ and $\boldsymbol{y} \in U$ such that $\boldsymbol{x} \neq \boldsymbol{y}$.

We proceed now in a non-formal way, noting that condition (2.158) also has to hold when $\boldsymbol{x}$ and $\boldsymbol{y}$ are interchanged. Thus, summing Eq. (2.158) with the same equation with $\boldsymbol{x}$ and $\boldsymbol{y}$ interchanged, we obtain

$$\left(\frac{\partial f}{\partial \boldsymbol{x}}(\boldsymbol{x}) - \frac{\partial f}{\partial \boldsymbol{x}}(\boldsymbol{y}) \right) \cdot (\boldsymbol{x} - \boldsymbol{y}) \geq 0, \tag{2.159}$$

which, in fact can be proved to be equivalent to condition (2.158) (with $\geq$ replaced by $>$ for strict convexity).

Let us suppose now that the Hessian

$$\mathbb{H} = \nabla(\nabla f) \tag{2.160}$$

exists throughout the open convex subset U. We note that for points belonging to the set U, we may take $\boldsymbol{y}$ near to $\boldsymbol{x}$ in the form

$$\boldsymbol{y} = \boldsymbol{x} + \alpha \boldsymbol{z} \tag{2.161}$$

with small α and expand $\partial f / \partial \boldsymbol{y}$ into a Taylor series near $\alpha = 0$ to obtain the following condition to hold for every $\boldsymbol{z}$:

$$\boldsymbol{z} \cdot \mathbb{H}(\boldsymbol{x})[\boldsymbol{z}] \geq 0. \tag{2.162}$$

Thus positive semi-definiteness of the Hessian is a condition necessary to convexity (2.159). This condition also can be proven to be sufficient (Ogden, 1982) by identifying $\boldsymbol{z}$ with $(\boldsymbol{x} - \boldsymbol{y})dt$, tangent to the straight-line segment $(1-t)\boldsymbol{y} + t\boldsymbol{x}$, and integrating

$$\mathbb{H}[(\boldsymbol{x} - \boldsymbol{y})dt] \cdot (\boldsymbol{x} - \boldsymbol{y}) \tag{2.163}$$

along the line to obtain Eq. (2.158).

Positive semi-definiteness of the Hessian of a continuously differentiable function is a necessary and sufficient condition for convexity of the function. However, positive definiteness of the Hessian implies strict convexity of the function, but the converse is true only with possible exceptions on a nowhere dense subset of U (as shown by Bernstein and Toupin, 1962), so we may write

$$\text{Hessian of } f \text{ positive definite} \quad \underset{\text{'exceptions'}}{\underbrace{\overset{\Longrightarrow}{\Longleftarrow}}} \quad f \text{ strictly convex.} \tag{2.164}$$

After the pioneering work of de Finetti (1949), it became clear that convexity of *every* level set of a function represents its *quasi-convexity*, a property defining a class of functions much broader than the class of convex functions.

In more detail, let us consider a function $f(\mathbf{x}) : U \longrightarrow \mathbb{R}$, with U being a convex set, and its level sets

$$L_\alpha = \{\mathbf{x} \in U \mid f(\mathbf{x}) \leq \alpha\} \tag{2.165}$$

so that (Roberts and Varberg, 1973; Boyd and Vandenberghe, 2004)

f is quasi-convex if the level sets L_α are convex for every $\alpha \in \mathbb{R}$.

Now, this definition of quasi-convexity is equivalent to the definition

$$f[\lambda\mathbf{x} + (1+\lambda)\mathbf{y}] \leq \max\{f(\mathbf{x}), f(\mathbf{y})\}, \qquad \forall \mathbf{x},\mathbf{y} \in U, \qquad \forall \lambda \in [0,1], \tag{2.166}$$

and, if f is continuous and differentiable, to

$$f(\mathbf{y}) \leq f(\mathbf{x}) \;\Rightarrow\; \nabla f(\mathbf{x}) \cdot (\mathbf{x} - \mathbf{y}) \geq 0, \qquad \forall \mathbf{x},\mathbf{y} \in U. \tag{2.167}$$

It is worth noting that the definition (2.166) of quasi-convexity 'in the sense of de Finetti' *is different from the definition of quasi-convexity introduced by Morrey (1952)*, frequently used in non-linear elasticity (Ball, 1977), which for a function $W(\boldsymbol{x}, \boldsymbol{F})$ defined on points $\boldsymbol{x}$ of a closed and bounded subset $\Omega \in \mathbb{R}^3$ and all tensors $\boldsymbol{F} \in \text{Lin}$ writes as

$$\int_D W(\bar{\boldsymbol{x}}_0, \bar{\boldsymbol{F}} + \nabla \boldsymbol{w}(\boldsymbol{x}_0))\, dx_0 \geq m(D)\, W(\bar{\boldsymbol{x}}_0, \bar{\boldsymbol{F}}) \tag{2.168}$$

for all fixed $\bar{\boldsymbol{x}}_0$ in Ω, all constant tensors $\bar{\boldsymbol{F}}$, each bounded open subset D of $\mathbb{R}^3$, and for all $\boldsymbol{w}$ belonging to the set of all infinitely differentiable functions with compact support contained in D. Note that $m(D)$ indicates the volume of D.

For twice continuously differentiable functions W, quasi-convexity (2.168) implies strong ellipticity (see Section 11.3) of the Hessian of W.

2.20 Examples and details

2.20.1 Example: Jordan normal form of a defective tensor with a double eigenvalue

The following matrix

$$\begin{bmatrix} 1 & 1/2 & \sqrt{3}/2 \\ \sqrt{3} & (-5-\sqrt{3})/4 & 3(-1+\sqrt{3})/4 \\ -1 & (1+3\sqrt{3})/4 & (1+\sqrt{3})/4 \end{bmatrix} \tag{2.169}$$

has eigenvalues $\alpha_1 = -2$, $\alpha_2 = 1$ (the latter with algebraic multiplicity 2) and left and right eigenvectors

$$\begin{aligned} &[\boldsymbol{b}_1] = [0, -\sqrt{3}, 1], && [\boldsymbol{b}^1] = [12, -5-9\sqrt{3}, 9-5\sqrt{3}]/36, \\ &[\boldsymbol{b}_2] = [-3, -\sqrt{3}, 1]/\sqrt{13}, && [\boldsymbol{b}^2] = [0, 1, \sqrt{3}]/2. \end{aligned} \tag{2.170}$$

An application of Eq. (2.75) yields

$$[\boldsymbol{a}_1] = [0, -\sqrt{3}, 1], \qquad [\boldsymbol{a}^1] = [12, -5-9\sqrt{3}, 9-5\sqrt{3}]/36,$$
$$[\boldsymbol{a}_2] = [-3, -\sqrt{3}, 1]/\sqrt{13}, \qquad [\boldsymbol{a}^2] = [-39, 5, 5\sqrt{3}]/(9\sqrt{13}),$$
$$[\boldsymbol{a}_3] = [-20, 3(-13+5\sqrt{3}), -3(5+13\sqrt{3})]/(26\sqrt{13}), \qquad [\boldsymbol{a}^3] = [0, -\sqrt{13}, -\sqrt{39}]/6,$$

so the Jordan normal form in the dual bases $\boldsymbol{a}_i$ and $\boldsymbol{a}^i$ $(i = 1,2,3)$ is

$$\begin{bmatrix} -2 & 0 & 0 \\ 0 & 1 & 1 \\ 0 & 0 & 1 \end{bmatrix}. \tag{2.171}$$

2.20.2 Example: Jordan normal form of a defective tensor with a triple eigenvalue

The following matrix

$$\begin{bmatrix} 71+5\sqrt{6} & 5\sqrt{2}+4\sqrt{3} & 2(\sqrt{2}+4\sqrt{3}) \\ -15\sqrt{2}+4\sqrt{3} & 63-5\sqrt{6} & 2(4-\sqrt{6}) \\ -8\sqrt{3} & -8 & 43 \end{bmatrix} \tag{2.172}$$

has one eigenvalue, $\alpha = 59$, with multiplicity 3 and the following left and right eigenvector:

$$[\boldsymbol{b}_1] = [-1/2, \sqrt{3}/2, 0], \quad [\boldsymbol{b}^1] = [\sqrt{3}, 1, 2]/(2\sqrt{2}). \tag{2.173}$$

Using Eqs. (2.77), we can write

$$[\boldsymbol{a}_1] = [1, -\sqrt{3}, 0]/64, \qquad [\boldsymbol{a}^1] = [16-6\sqrt{6}, -6\sqrt{2}-16\sqrt{3}, 12\sqrt{2}],$$
$$[\boldsymbol{a}_2] = [3+2\sqrt{6}, 2\sqrt{2}-3\sqrt{3}, -4\sqrt{2}]/4096, \qquad [\boldsymbol{a}^2] = 128[\sqrt{6}, \sqrt{2}, -2\sqrt{2}],$$
$$[\boldsymbol{a}_3] = [\boldsymbol{b}^1]/16384, \qquad [\boldsymbol{a}^3] = 16384[\boldsymbol{b}^1]. \tag{2.174}$$

Projected onto the preceding reciprocal basis, we arrive at the Jordan normal form of tensor $\boldsymbol{A}$:

$$\begin{bmatrix} 59 & 1 & 0 \\ 0 & 59 & 1 \\ 0 & 0 & 59 \end{bmatrix}. \tag{2.175}$$

2.20.3 Example: Inverse of the acoustic tensor of isotropic elasticity

For every unit vector $\boldsymbol{n}$, the symmetric second-order tensor

$$\boldsymbol{A}(\boldsymbol{n}) = (\lambda+\mu)\,\boldsymbol{n}\otimes\boldsymbol{n} + \mu\boldsymbol{I}, \tag{2.176}$$

where λ and μ are the Lamé constants, is the acoustic tensor of isotropic elasticity (see, e.g., Gurtin, 1972, 1981). It can be verified by direct inspection that the eigenspectrum of the tensor is

Eigenvalue	Multiplicity	Eigenvector
$\lambda + 2\mu$	1	$\boldsymbol{n}$
μ	2	all $\boldsymbol{s}$ such that $\boldsymbol{s}\cdot\boldsymbol{n} = 0$

The inverse of tensor (2.176) therefore is

$$\boldsymbol{A}^{-1}(\boldsymbol{n}) = -\frac{\lambda+\mu}{\mu(\lambda+2\mu)}\boldsymbol{n}\otimes\boldsymbol{n} + \frac{1}{\mu}\boldsymbol{I}, \tag{2.177}$$

and it can be checked easily by taking the product with tensor (2.176). Within the context of isotropic linear elasticity, the conditions that $\boldsymbol{A}(\boldsymbol{n})$ is positive definite and invertible are the so-called strong ellipticity and ellipticity conditions, respectively, which can be obtained from Eq. (2.177) to be

$$\begin{array}{llllll} \boldsymbol{A}(\boldsymbol{n}) & \text{positive definite} & \Longleftrightarrow & \mu > 0, & \lambda+2\mu > 0: & \text{strong ellipticity,} \\ \boldsymbol{A}(\boldsymbol{n}) & \text{non-singular} & \Longleftrightarrow & \mu \neq 0, & \lambda+2\mu \neq 0: & \text{ellipticity.} \end{array}$$

2.20.4 Example: Inverse of the acoustic tensor for a particular class of anisotropic elasticity

Adopting the anisotropic elastic model with positive definite fabric tensor $\boldsymbol{B} \in \mathrm{Sym}$ proposed by Valanis (1990) and Zysset and Curnier (1995), the acoustic tensor results in the form (Bigoni and Loret, 1999)

$$\boldsymbol{A}(\boldsymbol{n}) = (\lambda+\mu)\boldsymbol{B}\boldsymbol{n}\otimes\boldsymbol{B}\boldsymbol{n} + \mu\,(\boldsymbol{n}\cdot\boldsymbol{B}\boldsymbol{n})\boldsymbol{B}, \tag{2.178}$$

where λ and μ play a role similar to the Lamé constants of isotropic elasticity. The inverse is

$$\boldsymbol{A}^{-1}(\boldsymbol{n}) = -\frac{\lambda+\mu}{\mu(\lambda+2\mu)}\frac{\boldsymbol{n}\otimes\boldsymbol{n}}{(\boldsymbol{n}\cdot\boldsymbol{B}\boldsymbol{n})^2} + \frac{1}{\mu}\frac{\boldsymbol{B}^{-1}}{\boldsymbol{n}\cdot\boldsymbol{B}\boldsymbol{n}}, \tag{2.179}$$

and the conditions of strong ellipticity and ellipticity remain the same as for the acoustic tensor (2.176).

2.20.5 Example: A representation for the square root of a tensor

Let us consider a symmetric positive definite tensor $\boldsymbol{B}$ with eigenvalues λ_1^2, λ_2^2 and λ_3^2 and its square root tensor $\sqrt{\boldsymbol{B}}$ (with eigenvalues λ_1, λ_2 and λ_3). From the Cayley-Hamilton theorem (2.59) written for $\sqrt{\boldsymbol{B}}$, we obtain

$$\begin{aligned} \boldsymbol{B}^{-1/2} &= \frac{1}{I_3(\sqrt{\boldsymbol{B}})}\left[\boldsymbol{B} - I_1(\sqrt{\boldsymbol{B}})\sqrt{\boldsymbol{B}} + I_2(\sqrt{\boldsymbol{B}})\boldsymbol{I}\right], \\ \boldsymbol{B}^{-1} &= \frac{1}{I_3(\sqrt{\boldsymbol{B}})}\left[\sqrt{\boldsymbol{B}} - I_1(\sqrt{\boldsymbol{B}})\boldsymbol{I} + I_2(\sqrt{\boldsymbol{B}})\boldsymbol{B}^{(-1/2)}\right]. \end{aligned} \tag{2.180}$$

A substitution of Eq. $(2.180)_1$ into Eq. $(2.180)_2$ allows us to obtain $\sqrt{\boldsymbol{B}}$ as the following function of $\boldsymbol{B}$ and $\boldsymbol{B}^{-1}$ and of the invariants of $\sqrt{\boldsymbol{B}}$ and $\sqrt{I_3(\boldsymbol{B})} = I_3(\sqrt{\boldsymbol{B}})$:

$$\sqrt{\boldsymbol{B}} = \frac{I_3(\boldsymbol{B})}{\sqrt{I_3(\boldsymbol{B})} - I_1(\sqrt{\boldsymbol{B}})I_2(\sqrt{\boldsymbol{B}})}\left[\boldsymbol{B}^{-1} - \frac{I_2(\sqrt{\boldsymbol{B}})}{I_3(\boldsymbol{B})}\boldsymbol{B} + \frac{I_3(\sqrt{\boldsymbol{B}})I_1(\sqrt{\boldsymbol{B}}) - I_2^2(\sqrt{\boldsymbol{B}})}{I_3(\boldsymbol{B})}\boldsymbol{I}\right]. \tag{2.181}$$

Note that the two relations

$$I_2(\boldsymbol{B}) = I_2^2(\sqrt{\boldsymbol{B}}) - 2I_1(\sqrt{\boldsymbol{B}})\sqrt{I_3(\boldsymbol{B})} \quad \text{and} \quad I_1(\boldsymbol{B}) = I_1^2(\sqrt{\boldsymbol{B}}) - 2I_2(\sqrt{\boldsymbol{B}}) \tag{2.182}$$

can be derived easily, whereas the first and second invariants of $\sqrt{\boldsymbol{B}}$ are very complicated functions of the invariants of $\boldsymbol{B}$.

2.20.6 Proof of a property of the scalar product between two symmetric tensors

We provide the proof of a property of the scalar product of two symmetric tensors, which is used often (among others, by Ogden, 1984). The property has been noticed by Hill (1968), who did not provide a complete proof, which is only sketched in a footnote, perhaps because of a lack of space. We were not able to find a detailed proof of the following proposition.

Let $\boldsymbol{A},\boldsymbol{B}$ be two symmetric tensors. Then, denoting by $\alpha_1,\alpha_2,\alpha_3$ and β_1,β_2,β_3 the eigenvalues of $\boldsymbol{A}$ and $\boldsymbol{B}$, respectively,

$$\boldsymbol{A}\cdot\boldsymbol{B} \leq \alpha_1\beta_1+\alpha_2\beta_2+\alpha_3\beta_3, \tag{2.183}$$

given that the eigenvalues of the two tensors are numbered in the same algebraic order.

Proof. Given the eigenvalues of the two tensors, we keep the eigenvectors $\boldsymbol{a}_1,\boldsymbol{a}_2,\boldsymbol{a}_3$ of $\boldsymbol{A}$ fixed and seek for the maximum of $\boldsymbol{A}\cdot\boldsymbol{B}$ as the eigenvectors $\boldsymbol{b}_1,\boldsymbol{b}_2,\boldsymbol{b}_3$ of $\boldsymbol{B}$ rotate with respect to $\boldsymbol{a}_1,\boldsymbol{a}_2,\boldsymbol{a}_3$. Therefore, the problem can be formulated in terms of the following optimisation problem:

$$\max_{\boldsymbol{b}_1,\boldsymbol{b}_2,\boldsymbol{b}_3} \boldsymbol{A}\cdot\boldsymbol{B}, \tag{2.184}$$

with the constraint that $(\boldsymbol{b}_1,\boldsymbol{b}_2,\boldsymbol{b}_3)$ is an orthonormal basis,

$$\boldsymbol{b}_M\cdot\boldsymbol{b}_N=\delta_{MN}, \tag{2.185}$$

where δ_{MN} is the Kronecker symbol.

This optimisation problem can be solved using the Lagrangean multipliers Λ_i $(i=1,...,6)$. Thus we can maximise

$$\begin{aligned}\boldsymbol{A}\cdot\boldsymbol{B} = {} & \alpha_1\beta_1(\boldsymbol{a}_1\cdot\boldsymbol{b}_1)^2+\alpha_1\beta_2(\boldsymbol{a}_1\cdot\boldsymbol{b}_2)^2+\alpha_1\beta_3(\boldsymbol{a}_1\cdot\boldsymbol{b}_3)^2\\ & +\alpha_2\beta_1(\boldsymbol{a}_2\cdot\boldsymbol{b}_1)^2+\alpha_2\beta_2(\boldsymbol{a}_2\cdot\boldsymbol{b}_2)^2+\alpha_2\beta_3(\boldsymbol{a}_2\cdot\boldsymbol{b}_3)^2\\ & +\alpha_3\beta_1(\boldsymbol{a}_3\cdot\boldsymbol{b}_1)^2+\alpha_3\beta_2(\boldsymbol{a}_3\cdot\boldsymbol{b}_2)^2+\alpha_3\beta_3(\boldsymbol{a}_3\cdot\boldsymbol{b}_3)^2\\ & +\Lambda_1(\boldsymbol{b}_1\cdot\boldsymbol{b}_1-1)+\Lambda_2(\boldsymbol{b}_2\cdot\boldsymbol{b}_2-1)+\Lambda_3(\boldsymbol{b}_3\cdot\boldsymbol{b}_3-1)\\ & +\Lambda_4(\boldsymbol{b}_1\cdot\boldsymbol{b}_2)+\Lambda_5(\boldsymbol{b}_2\cdot\boldsymbol{b}_3)+\Lambda_6(\boldsymbol{b}_3\cdot\boldsymbol{b}_1)\end{aligned} \tag{2.186}$$

as a function of $\boldsymbol{b}_1,\boldsymbol{b}_2,\boldsymbol{b}_3$, obtaining

$$\begin{aligned}\frac{\partial\boldsymbol{A}\cdot\boldsymbol{B}}{\partial\boldsymbol{b}_1} &= 2\beta_1\boldsymbol{A}\boldsymbol{b}_1+2\Lambda_1\boldsymbol{b}_1+\Lambda_4\boldsymbol{b}_2+\Lambda_6\boldsymbol{b}_3=\boldsymbol{0},\\ \frac{\partial\boldsymbol{A}\cdot\boldsymbol{B}}{\partial\boldsymbol{b}_2} &= 2\beta_2\boldsymbol{A}\boldsymbol{b}_2+2\Lambda_2\boldsymbol{b}_2+\Lambda_4\boldsymbol{b}_1+\Lambda_5\boldsymbol{b}_3=\boldsymbol{0},\\ \frac{\partial\boldsymbol{A}\cdot\boldsymbol{B}}{\partial\boldsymbol{b}_3} &= 2\beta_3\boldsymbol{A}\boldsymbol{b}_3+2\Lambda_3\boldsymbol{b}_3+\Lambda_5\boldsymbol{b}_2+\Lambda_6\boldsymbol{b}_1=\boldsymbol{0},\end{aligned} \tag{2.187}$$

together with the constraints (2.185).

In the case of distinct eigenvalues β_1,β_2,β_3, the system (2.187) is satisfied if and only if

$$\boldsymbol{b}_M\cdot\boldsymbol{A}\boldsymbol{b}_N=0, \qquad \text{for} \quad M\neq N, \tag{2.188}$$

and thus if and only if $\boldsymbol{b}_1, \boldsymbol{b}_2, \boldsymbol{b}_3$ are eigenvectors of $\boldsymbol{A}$. This proves that the extreme values of $\boldsymbol{A} \cdot \boldsymbol{B}$ are attained when the two tensors are coaxial. The maximum then is selected from six possibilities.

In the case $\beta_1 = \beta_2 \neq \beta_3$, the same line of thought used earlier allows us to conclude that the extreme values of $\boldsymbol{A} \cdot \boldsymbol{B}$ are attained when $\boldsymbol{b}_3$ is an eigenvector of $\boldsymbol{A}$, in which case the two tensors $\boldsymbol{A}$ and $\boldsymbol{B}$ are coaxial and, choosing $\boldsymbol{b}_3 \equiv \boldsymbol{a}_3$, $\boldsymbol{A} \cdot \boldsymbol{B} = (\alpha_1 + \alpha_2)\beta_1 + \alpha_3\beta_3$.

The case $\beta_1 = \beta_2 = \beta_3$ is trivial. The two tensors $\boldsymbol{A}$ and $\boldsymbol{B}$ are coaxial, and the scalar product is $\boldsymbol{A} \cdot \boldsymbol{B} = (\alpha_1 + \alpha_2 + \alpha_3)\beta_1$. □

2.20.7 Example: Inverse and positive definiteness of the fourth-order tensor defining linear isotropic elasticity

The constitutive equations of linear isotropic elasticity are a relation between the Cauchy stress $\boldsymbol{T}$ and the infinitesimal strain $\boldsymbol{E}$:

$$\boldsymbol{T} = \mathbb{E}[\boldsymbol{E}] \tag{2.189}$$

through the isotropic fourth-order tensor

$$\mathbb{E} = \lambda \boldsymbol{I} \otimes \boldsymbol{I} + 2\mu \boldsymbol{I} \underline{\overline{\otimes}} \boldsymbol{I}, \tag{2.190}$$

where λ and μ are the Lamé constants.

Tensor $\mathbb{E}$ is positive definite when, for every $\boldsymbol{X} \in \mathsf{Sym}$ (note that the positive definiteness can be examined only within the class of symmetric tensors because $\mathbb{E}$ has the minor symmetries),

$$\boldsymbol{X} \cdot \mathbb{E}[\boldsymbol{X}] = \lambda(\text{tr}\boldsymbol{X})^2 + 2\mu \boldsymbol{X} \cdot \boldsymbol{X} > 0. \tag{2.191}$$

Thus, considering traceless tensors $\text{tr}\boldsymbol{X} = 0$ and isotropic tensors $\boldsymbol{X} = \boldsymbol{I}$, we can immediately conclude that

$$\mu > 0 \qquad \text{and} \qquad \lambda + \frac{2}{3}\mu > 0 \tag{2.192}$$

are necessary conditions for positive definiteness of $\mathbb{E}$. Using the Cauchy-Schwarz inequality

$$(\text{tr}\boldsymbol{X})^2 \leq 3\boldsymbol{X} \cdot \boldsymbol{X}, \tag{2.193}$$

we conclude that

$$\boldsymbol{X} \cdot \mathbb{E}[\boldsymbol{X}] \geq \begin{cases} (\lambda + \frac{2}{3}\mu)(\text{tr}\boldsymbol{X})^2 > 0, & \text{if } \text{tr}\boldsymbol{X} \neq 0, \\ 2\mu > 0, & \text{if } \text{tr}\boldsymbol{X} = 0, \end{cases} \tag{2.194}$$

so conditions (2.192) are also sufficient.

To invert the isotropic elastic fourth-order tensor, Eq. (2.190), we note the following product:

$$\mathbb{E}\left(\alpha \boldsymbol{I} \otimes \boldsymbol{I} + \frac{1}{2\mu}\boldsymbol{I} \underline{\overline{\otimes}} \boldsymbol{I}\right) = \boldsymbol{I} \underline{\overline{\otimes}} \boldsymbol{I} + \left(2\mu\alpha + 3\alpha\lambda + \frac{\lambda}{2\mu}\right)\boldsymbol{I} \otimes \boldsymbol{I}. \tag{2.195}$$

Thus, selecting α in an appropriate way, we can eliminate the term $\boldsymbol{I} \otimes \boldsymbol{I}$ and arrive at

$$\mathbb{E}^{-1} = -\frac{\lambda}{2\mu(3\lambda + 2\mu)}\boldsymbol{I} \otimes \boldsymbol{I} + \frac{1}{2\mu}\boldsymbol{I} \underline{\overline{\otimes}} \boldsymbol{I}, \tag{2.196}$$

which is the inverse of $\mathbb{E}$ restricted to the space of symmetric tensors.

2.20.8 Example: Inverse and positive definiteness of a fourth-order tensor defining a special anisotropic linear elasticity

Instead of the fourth-order isotropic tensor (2.190), we introduce now the tensor (Valanis, 1990; Zysset and Curnier, 1995; Bigoni and Loret, 1999)

$$\mathbb{E} = \lambda \boldsymbol{B} \otimes \boldsymbol{B} + 2\mu \boldsymbol{B} \underline{\overline{\otimes}} \boldsymbol{B}, \tag{2.197}$$

where λ and μ play a role similar to the Lamé constants, and $\boldsymbol{B}$ is a positive definite fabric tensor satisfying the constraint

$$\text{tr}\boldsymbol{B} = 3. \tag{2.198}$$

The same arguments used in Example 2.20.7 can be employed now to show that the conditions for positive definiteness of (2.197) still remain (2.192), and

$$\mathbb{E}^{-1} = -\frac{\lambda}{2\mu(3\lambda + 2\mu)} \boldsymbol{B}^{-1} \otimes \boldsymbol{B}^{-1} + \frac{1}{2\mu} \boldsymbol{B}^{-1} \underline{\overline{\otimes}} \boldsymbol{B}^{-1} \tag{2.199}$$

is the inverse of $\mathbb{E}$ restricted to the space of symmetric tensors.

2.20.9 Example: Inverse of the elastoplastic fourth-order tangent tensor

Given the scalars λ and μ and the symmetric second-order tensors $\boldsymbol{M}$ and $\boldsymbol{N}$, the fourth-order tensor

$$\mathbb{C} = \lambda \boldsymbol{I} \otimes \boldsymbol{I} + 2\mu \boldsymbol{I} \underline{\overline{\otimes}} \boldsymbol{I} + \boldsymbol{M} \otimes \boldsymbol{N} \tag{2.200}$$

corresponds to the loading branch of the elastoplastic constitutive operator (8.101) with an isotropic tensor $\mathbb{G}$ (the scalar $1/g$ has been included in the dyad $\boldsymbol{M} \otimes \boldsymbol{N}$) and referred to small strain.

In the following, we assume invertibility of $\mathbb{C}$ and find the inverse (restricted to the space of symmetric tensors), a result shown by Bigoni and Piccolroaz in an unpublished note.

For all tensors $\boldsymbol{X} \in \text{Sym}$, we have

$$\mathbb{C}[\boldsymbol{X}] = \lambda\, (\text{tr}\boldsymbol{X})\boldsymbol{I} + 2\mu \boldsymbol{X} + (\boldsymbol{N} \cdot \boldsymbol{X})\boldsymbol{M}, \tag{2.201}$$

so

$$\boldsymbol{X} = \lambda\, (\text{tr}\boldsymbol{X})\mathbb{C}^{-1}[\boldsymbol{I}] + 2\mu\, \mathbb{C}^{-1}[\boldsymbol{X}] + (\boldsymbol{N} \cdot \boldsymbol{X})\mathbb{C}^{-1}[\boldsymbol{M}], \tag{2.202}$$

from which we obtain $\mathbb{C}^{-1}[\boldsymbol{X}]$ in the form

$$\mathbb{C}^{-1}[\boldsymbol{X}] = \frac{1}{2\mu} \left\{ \boldsymbol{X} - \lambda\, (\text{tr}\boldsymbol{X})\mathbb{C}^{-1}[\boldsymbol{I}] - (\boldsymbol{N} \cdot \boldsymbol{X})\mathbb{C}^{-1}[\boldsymbol{M}] \right\}. \tag{2.203}$$

Using Eq. (2.203) with $\boldsymbol{X} = \boldsymbol{I}$ and $\boldsymbol{X} = \boldsymbol{M}$, we arrive at the following linear system for $\mathbb{C}^{-1}[\boldsymbol{I}]$ and $\mathbb{C}^{-1}[\boldsymbol{M}]$

$$\begin{aligned} \mathbb{C}^{-1}[\boldsymbol{I}] &= \frac{1}{2\mu} \left\{ \boldsymbol{I} - 3\lambda\, \mathbb{C}^{-1}[\boldsymbol{I}] - (\text{tr}\boldsymbol{N})\mathbb{C}^{-1}[\boldsymbol{M}] \right\}, \\ \mathbb{C}^{-1}[\boldsymbol{M}] &= \frac{1}{2\mu} \left\{ \boldsymbol{M} - \lambda\, (\text{tr}\boldsymbol{M})\mathbb{C}^{-1}[\boldsymbol{I}] - (\boldsymbol{N} \cdot \boldsymbol{M})\mathbb{C}^{-1}[\boldsymbol{M}] \right\}, \end{aligned} \tag{2.204}$$

which can be solved to yield

$$\mathbb{C}^{-1}[\boldsymbol{I}] = \frac{2\mu + \boldsymbol{M}\cdot\boldsymbol{N}}{D}\boldsymbol{I} - \frac{\text{tr}\boldsymbol{N}}{D}\boldsymbol{M} \quad \text{and} \quad \mathbb{C}^{-1}[\boldsymbol{M}] = \frac{3\lambda + 2\mu}{D}\boldsymbol{M} - \frac{\lambda\,\text{tr}\boldsymbol{M}}{D}\boldsymbol{I}, \tag{2.205}$$

where

$$D = (3\lambda + 2\mu)2\mu - \lambda(\text{tr}\boldsymbol{M})(\text{tr}\boldsymbol{N}) + (3\lambda + 2\mu)(\boldsymbol{M}\cdot\boldsymbol{N}), \tag{2.206}$$

a quantity which is equal or proportional to the determinant of $\mathbb{C}$ (see Section 2.20.10). A substitution of Eqs. (2.205) into Eq. (2.203) yields the inverse (restricted to the space of symmetric tensors) of the fourth-order tensor $\mathbb{C}$:

$$\boxed{\begin{aligned}\mathbb{C}^{-1} = &\frac{1}{2\mu}\boldsymbol{I}\underline{\overline{\otimes}}\boldsymbol{I} - \frac{\lambda(2\mu + \boldsymbol{M}\cdot\boldsymbol{N})}{2\mu D}\boldsymbol{I}\otimes\boldsymbol{I} + \frac{\lambda\text{tr}\boldsymbol{N}}{2\mu D}\boldsymbol{M}\otimes\boldsymbol{I}\\ &+\frac{\lambda\text{tr}\boldsymbol{M}}{2\mu D}\boldsymbol{I}\otimes\boldsymbol{N} - \frac{3\lambda + 2\mu}{2\mu D}\boldsymbol{M}\otimes\boldsymbol{N}.\end{aligned}} \tag{2.207}$$

2.20.10 Example: Spectral representation of the elastoplastic fourth-order tangent tensor

We provide a complete spectral analysis of the tensor $\mathbb{C}$ in Eq. (2.200), representing the loading branch of the elastoplastic constitutive operator (8.101) based on isotropic elasticity and small strain. The results presented below have been obtained by Bigoni and Piccolroaz in an unpublished note and complement the analysis given by Bigoni and Zaccaria (1994b). We rewrite the fourth-order tensor $\mathbb{C}$ splitting tensors $\boldsymbol{M}$ and $\boldsymbol{N}$ into their deviatoric and isotropic components as

$$\mathbb{C} = \left(\lambda + \frac{\text{tr}\boldsymbol{M}\text{tr}\boldsymbol{N}}{9}\right)\boldsymbol{I}\otimes\boldsymbol{I} + 2\mu\boldsymbol{I}\underline{\overline{\otimes}}\boldsymbol{I} + \frac{\text{tr}\boldsymbol{N}}{3}\text{dev}\boldsymbol{M}\otimes\boldsymbol{I} + \frac{\text{tr}\boldsymbol{M}}{3}\boldsymbol{I}\otimes\text{dev}\boldsymbol{N} + \text{dev}\boldsymbol{M}\otimes\text{dev}\boldsymbol{N}, \tag{2.208}$$

from which it follows that

$$\mathbb{C}[\boldsymbol{R}] = 2\mu\boldsymbol{R} \quad \text{and} \quad \mathbb{C}^T[\boldsymbol{L}] = 2\mu\boldsymbol{L} \tag{2.209}$$

for all $\boldsymbol{R}$ satisfying

$$\text{tr}\boldsymbol{R} = 0 \quad \text{and} \quad \boldsymbol{R}\cdot\text{dev}\boldsymbol{N} = 0 \tag{2.210}$$

and all $\boldsymbol{L}$ satisfying

$$\text{tr}\boldsymbol{L} = 0 \quad \text{and} \quad \boldsymbol{L}\cdot\text{dev}\boldsymbol{M} = 0. \tag{2.211}$$

Note that the tensors defined by Eqs. (2.210) and (2.211) define a space of dimension 8. We conclude that 2μ is an eigenvalue with multiplicity 4, and all tensors satisfying conditions (2.210) [(2.211)] are right (left) eigenvectors.

The other two right eigenvectors are in the form

$$\omega_1\boldsymbol{I} + \omega_2\text{dev}\boldsymbol{M} \tag{2.212}$$

and can be found by solving the eigenvalue problem

$$\mathbb{C}[\omega_1\boldsymbol{I} + \omega_2\text{dev}\boldsymbol{M}] = \Lambda\left(\omega_1\boldsymbol{I} + \omega_2\text{dev}\boldsymbol{M}\right), \tag{2.213}$$

which leads to the following linear system:

$$\begin{cases} \left(3\lambda + \dfrac{\text{tr}\boldsymbol{M}\text{tr}\boldsymbol{N}}{3} + 2\mu - \Lambda\right)\omega_1 + \dfrac{\text{tr}\boldsymbol{M}}{3}(\text{dev}\boldsymbol{M}\cdot\text{dev}\boldsymbol{N})\,\omega_2 = 0, \\ \text{tr}\boldsymbol{N}\omega_1 + (2\mu + \text{dev}\boldsymbol{M}\cdot\text{dev}\boldsymbol{N} - \Lambda)\,\omega_2 = 0. \end{cases} \tag{2.214}$$

Introducing the notation

$$\Delta = \sqrt{(\boldsymbol{M}\cdot\boldsymbol{N} - 3\lambda)^2 + 4\lambda\text{tr}\boldsymbol{M}\text{tr}\boldsymbol{N}}, \tag{2.215}$$

non-trivial solutions of system (2.214) are given by

$$\Lambda_{5,6} = \frac{3\lambda + 4\mu + \boldsymbol{M}\cdot\boldsymbol{N} \pm \Delta}{2}, \tag{2.216}$$

with the corresponding right eigenvectors

$$\boldsymbol{R}^{(5),(6)} = \frac{3(\boldsymbol{M}\cdot\boldsymbol{N} \mp \Delta + 3\lambda)}{2\text{dev}\boldsymbol{M}\cdot\text{dev}\boldsymbol{N}}\boldsymbol{I} + \frac{3(3\boldsymbol{M}\cdot\boldsymbol{N} \pm 3\Delta - 9\lambda - 2\text{tr}\boldsymbol{M}\text{tr}\boldsymbol{N})}{2\text{tr}\boldsymbol{M}(\text{dev}\boldsymbol{M}\cdot\text{dev}\boldsymbol{N})}\boldsymbol{M}, \tag{2.217}$$

or, when $\text{tr}\boldsymbol{M}(\text{dev}\boldsymbol{M}\cdot\text{dev}\boldsymbol{N}) = 0$ but $\text{tr}\boldsymbol{N} \neq 0$, we have

$$\boldsymbol{R}^{(5),(6)} = \frac{3\lambda - \boldsymbol{M}\cdot\boldsymbol{N} \pm \Delta}{2\text{tr}\boldsymbol{N}}\boldsymbol{I} + \boldsymbol{M}. \tag{2.218}$$

The two left eigenvectors corresponding to the eigenvalues (2.216) are in the form

$$\omega_1\boldsymbol{I} + \omega_2\text{dev}\boldsymbol{N}, \tag{2.219}$$

so by solving $\mathbb{C}^T[\omega_1\boldsymbol{I} + \omega_2\text{dev}\boldsymbol{N}] = \Lambda\,(\omega_1\boldsymbol{I} + \omega_2\text{dev}\boldsymbol{N})$, we find

$$\boldsymbol{L}^{(5),(6)} = \frac{3(\boldsymbol{M}\cdot\boldsymbol{N} \mp \Delta + 3\lambda)}{2\text{dev}\boldsymbol{M}\cdot\text{dev}\boldsymbol{N}}\boldsymbol{I} + \frac{3(3\boldsymbol{M}\cdot\boldsymbol{N} \pm 3\Delta - 9\lambda - 2\text{tr}\boldsymbol{M}\text{tr}\boldsymbol{N})}{2\text{tr}\boldsymbol{N}(\text{dev}\boldsymbol{M}\cdot\text{dev}\boldsymbol{N})}\boldsymbol{N}, \tag{2.220}$$

or, when $\text{tr}\boldsymbol{N}(\text{dev}\boldsymbol{M}\cdot\text{dev}\boldsymbol{N}) = 0$ but $\text{tr}\boldsymbol{M} \neq 0$, we have

$$\boldsymbol{L}^{(5),(6)} = \frac{3\lambda - \boldsymbol{M}\cdot\boldsymbol{N} \pm \Delta}{2\text{tr}\boldsymbol{M}}\boldsymbol{I} + \boldsymbol{N}. \tag{2.221}$$

The case in which $\text{tr}\boldsymbol{M}(\text{dev}\boldsymbol{M}\cdot\text{dev}\boldsymbol{N}) = \text{tr}\boldsymbol{N}(\text{dev}\boldsymbol{M}\cdot\text{dev}\boldsymbol{N}) = 0$ is trivial.

Note that when $\Delta < 0$, $\Lambda_{5,6}$ become complex conjugate eigenvalues.

In conclusion, we have found the spectral representation of the fourth-order tensor (2.200) without introducing any assumption on $\boldsymbol{M}$ and $\boldsymbol{N}$, except their symmetry. This is summarised in Table 2.1, where the eigenvalues coincide with those derived in a different way by Bigoni and Zaccaria (1994b).

Finally, in the case when tensors $\boldsymbol{M}$ and $\boldsymbol{N}$ are coaxial, we can explicitly write the four eigenvectors corresponding to the eigenvalue 2μ. To this purpose, we assume the spectral representation of $\boldsymbol{M}$ and $\boldsymbol{N}$ as

$$\boldsymbol{M} = M_1\boldsymbol{m}_1\otimes\boldsymbol{m}_1 + M_2\boldsymbol{m}_2\otimes\boldsymbol{m}_2 + M_3\boldsymbol{m}_3\otimes\boldsymbol{m}_3, \tag{2.222}$$

$$\boldsymbol{N} = N_1\boldsymbol{m}_1\otimes\boldsymbol{m}_1 + N_2\boldsymbol{m}_2\otimes\boldsymbol{m}_2 + N_3\boldsymbol{m}_3\otimes\boldsymbol{m}_3, \tag{2.223}$$

Table 2.1. *Eigensystem of the fourth-order tensor (2.200)*

Eigenvalues	Multiplicity	Left eigenvectors	Right eigenvectors
2μ	4	$\forall \boldsymbol{L} \in \mathsf{Sym}$: $\operatorname{tr}\boldsymbol{L} = 0, \boldsymbol{L} \cdot \operatorname{dev}\boldsymbol{M} = 0$	$\forall \boldsymbol{R} \in \mathsf{Sym}$: $\operatorname{tr}\boldsymbol{R} = 0, \boldsymbol{R} \cdot \operatorname{dev}\boldsymbol{N} = 0$
$\Lambda_5 = \dfrac{3\lambda + 4\mu + \boldsymbol{M} \cdot \boldsymbol{N} + \Delta}{2}$	1	$\boldsymbol{L}^{(5)}$, Eq. (2.221)	$\boldsymbol{R}^{(5)}$, Eq. (2.218)
$\Lambda_6 = \dfrac{3\lambda + 4\mu + \boldsymbol{M} \cdot \boldsymbol{N} - \Delta}{2}$	1	$\boldsymbol{L}^{(6)}$, Eq. (2.221)	$\boldsymbol{R}^{(6)}$, Eq. (2.218)

so the spectral representation of $\mathbb{C}$ in the case of coaxial $\boldsymbol{M}$ and $\boldsymbol{N}$ is given by

$$\begin{aligned}\mathbb{C} = 2\mu\left(\boldsymbol{R}^{(12)} \otimes \boldsymbol{L}^{(12)} + \boldsymbol{R}^{(13)} \otimes \boldsymbol{L}^{(13)} + \boldsymbol{R}^{(23)} \otimes \boldsymbol{L}^{(23)} + \tilde{\boldsymbol{R}} \otimes \tilde{\boldsymbol{L}}\right) \\ + \Lambda_5 \boldsymbol{R}^{(5)} \otimes \boldsymbol{L}^{(5)} + \Lambda_6 \boldsymbol{R}^{(6)} \otimes \boldsymbol{L}^{(6)},\end{aligned} \tag{2.224}$$

where

$$\boldsymbol{R}^{(ij)} = \boldsymbol{L}^{(ij)} = \frac{\boldsymbol{m}_i \otimes \boldsymbol{m}_j + \boldsymbol{m}_j \otimes \boldsymbol{m}_i}{\sqrt{2}}, \qquad i,j = 1,2,3, \tag{2.225}$$

$$\tilde{\boldsymbol{R}} = \frac{(N_2 - N_3)\boldsymbol{m}_1 \otimes \boldsymbol{m}_1 + (N_3 - N_1)\boldsymbol{m}_2 \otimes \boldsymbol{m}_2 + (N_1 - N_2)\boldsymbol{m}_3 \otimes \boldsymbol{m}_3}{\sqrt{\operatorname{dev}\boldsymbol{M} \cdot \operatorname{dev}\boldsymbol{N}}}, \tag{2.226}$$

$$\tilde{\boldsymbol{L}} = \frac{(M_2 - M_3)\boldsymbol{m}_1 \otimes \boldsymbol{m}_1 + (M_3 - M_1)\boldsymbol{m}_2 \otimes \boldsymbol{m}_2 + (M_1 - M_2)\boldsymbol{m}_3 \otimes \boldsymbol{m}_3}{\sqrt{\operatorname{dev}\boldsymbol{M} \cdot \operatorname{dev}\boldsymbol{N}}}. \tag{2.227}$$

2.20.11 Example: Strict convexity of the strain energy defining linear isotropic elasticity

The strain energy of linear isotropic elasticity can be written for a strain $\boldsymbol{X} \in \mathsf{Sym}$ as

$$\frac{1}{2}\boldsymbol{X} \cdot \mathbb{E}[\boldsymbol{X}] = \frac{\lambda}{2}(\operatorname{tr}\boldsymbol{X})^2 + \mu \boldsymbol{X} \cdot \boldsymbol{X}. \tag{2.228}$$

Thus, applying directly the definition (2.157) of strict convexity, we have to prove that for every $\gamma \in (0,1)$ and every $\boldsymbol{X}$ and $\boldsymbol{Y} \in \mathsf{Sym}$,

$$\begin{aligned}\frac{\lambda}{2}[\gamma \operatorname{tr}\boldsymbol{X} + (1-\lambda)\operatorname{tr}\boldsymbol{Y}]^2 + \mu(\gamma\boldsymbol{X} + (1-\gamma)\boldsymbol{Y}) \cdot (\gamma\boldsymbol{X} + (1-\gamma)\boldsymbol{Y}) \\ < \gamma\frac{\lambda}{2}\operatorname{tr}^2\boldsymbol{X} + \gamma\mu\boldsymbol{X} \cdot \boldsymbol{X} + (1-\gamma)\frac{\lambda}{2}\operatorname{tr}^2\boldsymbol{Y} + (1-\gamma)\mu\boldsymbol{Y} \cdot \boldsymbol{Y}.\end{aligned} \tag{2.229}$$

Condition (2.229) can be developed to yield

$$-\gamma(1-\gamma)\frac{\lambda}{2}(\operatorname{tr}\boldsymbol{X} - \operatorname{tr}\boldsymbol{Y})^2 - \gamma(1-\gamma)\mu(\boldsymbol{X} - \boldsymbol{Y}) \cdot (\boldsymbol{X} - \boldsymbol{Y}) < 0, \tag{2.230}$$

so the factor $-\gamma(1-\gamma)$ can be simplified, and introduction of the deviator of $\boldsymbol{X} - \boldsymbol{Y}$ allows us to arrive at

$$\left(\frac{\lambda}{2} + \frac{\mu}{3}\right)(\operatorname{tr}\boldsymbol{X} - \operatorname{tr}\boldsymbol{Y})^2 + \mu \operatorname{dev}(\boldsymbol{X} - \boldsymbol{Y}) \cdot \operatorname{dev}(\boldsymbol{X} - \boldsymbol{Y}) > 0, \tag{2.231}$$

an equation that immediately yields the condition of positive definiteness of the elastic tensor [Eq. (2.192)]. On the other hand, conditions (2.192) are exactly the conditions of positive definiteness of the Hessian of the strain energy, from which strict convexity could have been deduced immediately.

3 Solid mechanics at finite strains

Kinematics and motion of a solid body are introduced. Mass balance and the concept of force and stress are provided, with emphasis on the notion of work-conjugated stress and strain measures, fundamental in the constitutive description of materials. Rules governing the changes of field quantities for rigid-body rotations of the reference and current configurations are given evidence to clarify the concept of spatial and material fields.

The description of the motion, deformation and stress of a solid body subject to external actions is the focus of solid mechanics, a science that was initiated more than four centuries ago by G. Galilei (1564–1642). Solid mechanics is articulated into five main parts: (1) kinematics and the concept of deformation, (2) mass conservation, (3) forces and stress, (4) the constitutive equations and (5) the setting of the boundary value problem. We will be concerned in this chapter with the preceding points (1) through (3), whereas constitutive equations and the setting of the boundary value problem will be deferred to chapters 4 and 6 through 9. As a complement to the material that will be presented in this chapter, we suggest the exhaustive treatments by Truesdell and Noll (1965), Truesdell (1966), Chadwick (1976), Gurtin (1981), Ogden (1984), and Podio Guidugli (2000).

3.1 Kinematics

Bodies occupy configurations, which are regions of the three-dimensional Euclidean point space. Obviously, a body should not be confused with its configuration, for the same reason that the center-line of a cantilever beam should not be confused with the points occupied by the elastica. However, a one-to-one correspondence can be set between points of a body and the points they occupy in the Euclidean space. Thus, we will treat a body as a closed set of points occupying a regular[1] region B, called the 'current' configuration, of the three-dimensional Euclidean point space. The displacement $\boldsymbol{u}$ of these points may be measured only with respect to a reference[2] configuration, which will be denoted with B_0. Note that *the choice of the reference*

[1] 'Regular' has here the same meaning detailed in the setting of the divergence theorem, Section 2.18.
[2] We do not distinguish here between material and referential descriptions, such as, for instance, in Truesdell (1966), so these nomenclatures will be used with the same meaning.

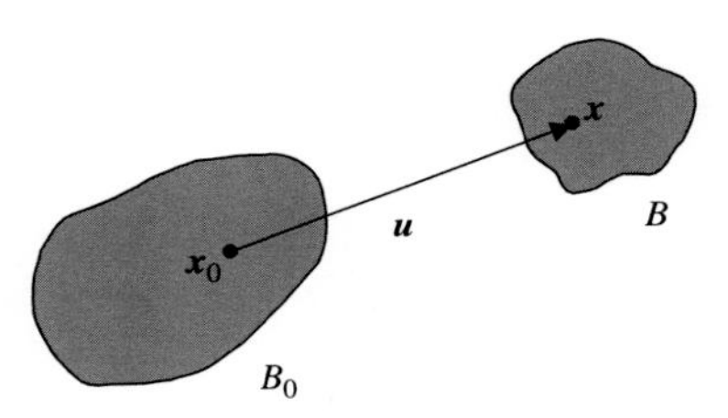

Figure 3.1. Deformation of a body. Vector $\boldsymbol{u}$ denotes the displacement of material points $\boldsymbol{x}_0$.

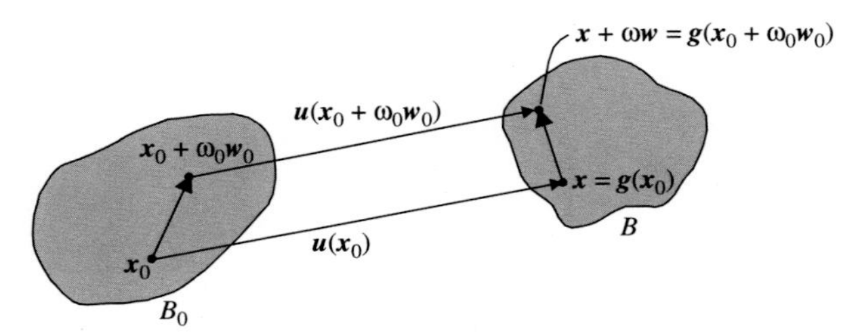

Figure 3.2. Deformation of an embedded oriented line element.

configuration is arbitrary, so even a configuration which has never been occupied by the body can be chosen as reference. Obviously, the resulting description of all fields must be independent of this choice, a statement which will be helpful later. Points $\boldsymbol{x}$ ($\boldsymbol{x}_0$) defined in B (B_0) are called 'spatial' (material). Therefore, with reference to Fig. 3.1, the displacement $\boldsymbol{u}$ is defined as

$$\boldsymbol{u} = \boldsymbol{x} - \boldsymbol{x}_0 \tag{3.1}$$

or

$$\boldsymbol{u} = \boldsymbol{g}(\boldsymbol{x}_0) - \boldsymbol{x}_0, \tag{3.2}$$

where the function $\boldsymbol{g} : B_0 \to B$ is the *deformation*, a bijection mapping[3] relating material points to spatial points. This function is invertible, with inverse $\boldsymbol{g}^{-1}$, so

$$\boldsymbol{x} = \boldsymbol{g}(\boldsymbol{x}_0), \quad \boldsymbol{x}_0 = \boldsymbol{g}^{-1}(\boldsymbol{x}). \tag{3.3}$$

Let us now consider two material points $\boldsymbol{x}_0$ and $\boldsymbol{x}_0 + \omega_0 \boldsymbol{w}_0$, the latter obtained by adding to $\boldsymbol{x}_0$ the unit vector $\boldsymbol{w}_0$ multiplied by the scalar ω_0. The two points are mapped forward to

$$\boldsymbol{x} = \boldsymbol{g}(\boldsymbol{x}_0) \qquad \text{and} \qquad \boldsymbol{x} + \omega \boldsymbol{w} = \boldsymbol{g}(\boldsymbol{x}_0 + \omega_0 \boldsymbol{w}_0) \tag{3.4}$$

(where $\boldsymbol{w}$ is a unit vector) in the spatial configuration (Fig. 3.2), so the embedded vector $\omega_0 \boldsymbol{w}_0$ transforms to $\omega \boldsymbol{w}$ (note that ω_0 and ω can be restricted to be positive). The Taylor series expansions of $\boldsymbol{g}$ and $\boldsymbol{g}^{-1}$ around $\boldsymbol{x}_0$ and $\boldsymbol{x}$, respectively, yield

$$\begin{aligned} \omega \boldsymbol{w} &= \boldsymbol{g}(\boldsymbol{x}_0 + \omega_0 \boldsymbol{w}_0) - \boldsymbol{g}(\boldsymbol{x}_0) = \boldsymbol{F}(\omega_0 \boldsymbol{w}_0) + \boldsymbol{O}(\omega_0^2), \\ \omega_0 \boldsymbol{w}_0 &= \boldsymbol{g}^{-1}(\boldsymbol{x} + \omega \boldsymbol{w}) - \boldsymbol{g}^{-1}(\boldsymbol{x}) = \boldsymbol{F}^{-1}(\omega \boldsymbol{w}) + \boldsymbol{O}(\omega^2), \end{aligned} \tag{3.5}$$

[3] The deformation is assumed for the moment to be twice-continuously differentiable, although there will be circumstances in which the assumption must be relaxed, for instance, across a moving discontinuity surface (Chapter 9).

where $\boldsymbol{F}$ is the deformation gradient and $\boldsymbol{F}^{-1}$ its inverse, $\boldsymbol{F}^{-1}\boldsymbol{F} = \boldsymbol{F}\boldsymbol{F}^{-1} = \boldsymbol{I}$. It is worth noting that $\boldsymbol{F}$ represents a gradient of a material field, whereas $\boldsymbol{F}^{-1}$ is the gradient of a spatial field, namely,

$$\boldsymbol{F} = \frac{\partial \boldsymbol{g}(\boldsymbol{x}_0)}{\partial \boldsymbol{x}_0} \quad \text{and} \quad \boldsymbol{F}^{-1} = \frac{\partial \boldsymbol{g}^{-1}(\boldsymbol{x})}{\partial \boldsymbol{x}}. \tag{3.6}$$

Note that tensor $\boldsymbol{F}$ is a *two-point tensor* in the sense that $\boldsymbol{w}_0$ and $\boldsymbol{w}$ are, respectively, defined in the material and spatial configurations. In order to highlight that the two gradients are taken with respect to different fields, the following convention will be used:

$$\boldsymbol{F} = \mathsf{Grad}\boldsymbol{g} \quad \text{and} \quad \boldsymbol{F}^{-1} = \mathsf{grad}\boldsymbol{g}^{-1} \tag{3.7}$$

to rewrite Eqs. (3.6) in an alternative way. More in general, when an operator will be written with an initial majuscule (minuscule) letter, it will be referred to a material (spatial) field.

3.1.1 Transformation of oriented line elements

Equations (3.5) make transparent the fundamental property of the deformation gradient $\boldsymbol{F}$ ($\boldsymbol{F}^{-1}$): It transforms the embedded material (spatial) oriented line element into a corresponding embedded spatial (material) oriented line element. In particular, Eqs. (3.5) yield, when truncated at the first order:

- The change in the modulus of the embedded oriented line element:

$$\lambda(\boldsymbol{x}_0, \boldsymbol{w}_0) = \frac{\omega}{\omega_0} = |\boldsymbol{F}\boldsymbol{w}_0| \qquad \frac{1}{\lambda(\boldsymbol{x}, \boldsymbol{w})} = \frac{\omega_0}{\omega} = |\boldsymbol{F}^{-1}\boldsymbol{w}|, \tag{3.8}$$

  where λ is the *stretch* at the material point $\boldsymbol{x}_0$ (or, equivalently, at the spatial point $\boldsymbol{x}$) relative to the material direction $\boldsymbol{w}_0$ (or, equivalently, spatial direction $\boldsymbol{w}$). Note that if $\boldsymbol{F}$ were singular, there would be fibres transformed into points, a situation clearly non admissible. Therefore, it is assumed[4] that det $\boldsymbol{F} > 0$, so the stretch always results strictly positive: greater (smaller) than one when the fibre elongates (contracts) deforming from B_0 to B.
- The change in the orientation of an embedded oriented line element:

$$\boldsymbol{w}_0 = \frac{\boldsymbol{F}^{-1}\boldsymbol{w}}{|\boldsymbol{F}^{-1}\boldsymbol{w}|} = \lambda(\boldsymbol{x}, \boldsymbol{w})\, \boldsymbol{F}^{-1}\boldsymbol{w} \qquad \boldsymbol{w} = \frac{\boldsymbol{F}\boldsymbol{w}_0}{|\boldsymbol{F}\boldsymbol{w}_0|} = \frac{\boldsymbol{F}\boldsymbol{w}_0}{\lambda(\boldsymbol{x}_0, \boldsymbol{w}_0)}. \tag{3.9}$$

It is expedient now to introduce the *right and left Cauchy-Green (symmetric and positive definite*[5]*) deformation tensors*

$$\boldsymbol{C} = \boldsymbol{F}^T\boldsymbol{F} \quad \text{and} \quad \boldsymbol{B} = \boldsymbol{F}\boldsymbol{F}^T, \tag{3.10}$$

so Eqs. (3.8) become

$$\lambda(\boldsymbol{x}_0, \boldsymbol{w}_0) = \sqrt{\boldsymbol{w}_0 \cdot \boldsymbol{C}\boldsymbol{w}_0} \quad \text{and} \quad \frac{1}{\lambda(\boldsymbol{x}, \boldsymbol{w})} = \sqrt{\boldsymbol{w} \cdot \boldsymbol{B}^{-1}\boldsymbol{w}}. \tag{3.11}$$

[4] The assumption $\det \boldsymbol{F} \neq 0$ would be sufficient, but then, since $\det \boldsymbol{F} = 1$ when $\boldsymbol{F} = \boldsymbol{I}$ (occurring when $B \equiv B_0$), continuity of deformation imposes positiveness, det $\boldsymbol{F} > 0$.

[5] Positive definiteness of $\boldsymbol{B}$ can be proven noting that for every non-null vector $\boldsymbol{x}$,

$$\boldsymbol{x} \cdot \boldsymbol{B}\boldsymbol{x} = \boldsymbol{x} \cdot \boldsymbol{F}\boldsymbol{F}^T\boldsymbol{x} = \boldsymbol{F}^T\boldsymbol{x} \cdot \boldsymbol{F}^T\boldsymbol{x}$$

is always strictly positive because $\det \boldsymbol{F} > 0$. The same property for $\boldsymbol{C}$ can be proven analogously.

Since the stretch is always strictly positive and, consistently, the two Cauchy-Green deformation tensors are positive definite, we may use the definition of square root of a tensor [Eq. (2.81)] to define the *right and left stretch tensors*

$$\boldsymbol{U} = \boldsymbol{C}^{1/2} \qquad \text{and} \qquad \boldsymbol{V} = \boldsymbol{B}^{1/2}, \tag{3.12}$$

which are symmetric and positive definite tensors admitting the spectral representation

$$\begin{aligned} \boldsymbol{U} &= \lambda_1 \boldsymbol{u}_1 \otimes \boldsymbol{u}_1 + \lambda_2 \boldsymbol{u}_2 \otimes \boldsymbol{u}_2 + \lambda_3 \boldsymbol{u}_3 \otimes \boldsymbol{u}_3, \\ \boldsymbol{V} &= \lambda_1 \boldsymbol{v}_1 \otimes \boldsymbol{v}_1 + \lambda_2 \boldsymbol{v}_2 \otimes \boldsymbol{v}_2 + \lambda_3 \boldsymbol{v}_3 \otimes \boldsymbol{v}_3, \end{aligned} \tag{3.13}$$

where λ_i, $\boldsymbol{u}_i$ and $\boldsymbol{v}_i$ $(i = 1,2,3)$ are, respectively, the *principal stretches* (the stretches in the direction of the eigenvectors of $\boldsymbol{C}$ and $\boldsymbol{B}$), and the *Lagrangean and Eulerian principal axes*. It is clear from definition (3.13) that $\boldsymbol{U}$ ($\boldsymbol{V}^{-1}$) stretches material (spatial) fibres into their spatial (material) length. It is therefore reasonable to understand that the effect of $\boldsymbol{F}$ ($\boldsymbol{F}^{-1}$) on a material (spatial) fibre results in a stretch given by $\boldsymbol{U}$ ($\boldsymbol{V}^{-1}$) and a rotation $\boldsymbol{R}$. Defining this rotation as

$$\boldsymbol{v}_i = \boldsymbol{R}\boldsymbol{u}_i, \tag{3.14}$$

we find the polar representation theorem (2.82) that we rewrite here

$$\boldsymbol{F} = \boldsymbol{R}\boldsymbol{U} = \boldsymbol{V}\boldsymbol{R}, \tag{3.15}$$

where $\boldsymbol{R} \in \mathsf{Orth}^+$, $\boldsymbol{U} \in \mathsf{Sym}^+$ and $\boldsymbol{V} \in \mathsf{Sym}^+$ are uniquely defined by Eqs. (3.13).[6]

From Eqs. (3.13) to (3.15), we may immediately obtain the following representation for the deformation gradient:

$$\boldsymbol{F} = \lambda_1 \boldsymbol{v}_1 \otimes \boldsymbol{u}_1 + \lambda_2 \boldsymbol{v}_2 \otimes \boldsymbol{u}_2 + \lambda_3 \boldsymbol{v}_3 \otimes \boldsymbol{u}_3 = \sum_{i=1}^{3} \lambda_i \underbrace{\boldsymbol{v}_i}_{\text{Eulerian}} \otimes \overbrace{\boldsymbol{u}_i}^{\text{Lagrangean}}, \tag{3.16}$$

and its inverse

$$\boldsymbol{F}^{-1} = \frac{1}{\lambda_1} \boldsymbol{u}_1 \otimes \boldsymbol{v}_1 + \frac{1}{\lambda_2} \boldsymbol{u}_2 \otimes \boldsymbol{v}_2 + \frac{1}{\lambda_3} \boldsymbol{u}_3 \otimes \boldsymbol{v}_3 = \sum_{i=1}^{3} \frac{1}{\lambda_i} \overbrace{\boldsymbol{u}_i}^{\text{Lagrangian}} \otimes \underbrace{\boldsymbol{v}_i}_{\text{Eulerian}}, \tag{3.17}$$

two equations which should not be confused with spectral representations because $\boldsymbol{v}_i$ ($\boldsymbol{u}_i$) are not right (left) eigenvectors of $\boldsymbol{F}$ (see Section 2.7) but highlight the fact that $\boldsymbol{F}$ ($\boldsymbol{F}^{-1}$) is a two-point tensor, transforming Lagrangean (Eluerian) elements into Eulerian (Lagrangean).

[6] Note that since $\boldsymbol{R} = \boldsymbol{F}\boldsymbol{U}^{-1}$, the relations

$$\boldsymbol{R}^T\boldsymbol{R} = \boldsymbol{U}^{-1}\boldsymbol{F}^T\boldsymbol{F}\boldsymbol{U}^{-1} = \boldsymbol{I} \qquad \text{and} \qquad \det \boldsymbol{R} = \det \boldsymbol{F} / \det \boldsymbol{U} > 0$$

confirm that $\boldsymbol{R} \in \mathsf{Orth}^+$.

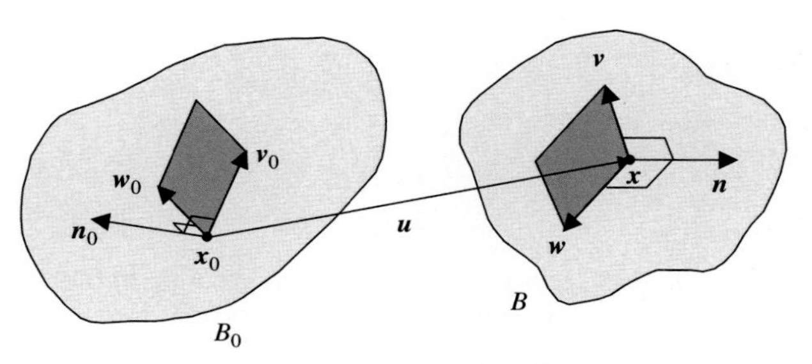

Figure 3.3. Transformation of an embedded oriented area element.

3.1.2 Transformation of oriented area elements

Let us consider an element area A_0 with unit normal $\boldsymbol{n}_0$ in the material configuration (Fig. 3.3)

$$A_0\boldsymbol{n}_0 = \boldsymbol{v}_0 \times \boldsymbol{w}_0, \tag{3.18}$$

where $\boldsymbol{v}_0$ and $\boldsymbol{w}_0$ are two vectors directed along the edges of the element area. The area element (3.18) is transformed in the spatial configuration as

$$A\boldsymbol{n} = \boldsymbol{F}\boldsymbol{v}_0 \times \boldsymbol{F}\boldsymbol{w}_0. \tag{3.19}$$

It is expedient now to use property (2.43) rewritten here for $\boldsymbol{F}$, $\boldsymbol{v}_0$ and $\boldsymbol{w}_0$ as

$$\boldsymbol{F}\boldsymbol{v}_0 \times \boldsymbol{F}\boldsymbol{w}_0 = (\det \boldsymbol{F})\boldsymbol{F}^{-T}(\boldsymbol{v}_0 \times \boldsymbol{w}_0) \tag{3.20}$$

so that Eq. (3.19) becomes *Nanson's rule of area transformation*:

$$A\boldsymbol{n} = A_0 J \boldsymbol{F}^{-T}\boldsymbol{n}_0, \tag{3.21}$$

where $J = \det \boldsymbol{F}$.

Equation (3.21) provides

- The transformation law of unit normal to embedded surfaces (Figs. 3.4 and 3.5):

$$\boldsymbol{n} = \frac{\boldsymbol{F}^{-T}\boldsymbol{n}_0}{|\boldsymbol{F}^{-T}\boldsymbol{n}_0|} \quad \text{and} \quad \boldsymbol{n}_0 = \frac{\boldsymbol{F}^{T}\boldsymbol{n}}{|\boldsymbol{F}^{T}\boldsymbol{n}|}, \tag{3.22}$$

  which makes clear that the normal to an embedded area element is not an embedded vector.
- The transformation law of area elements (Fig. 3.4)

$$A = J|\boldsymbol{F}^{-T}\boldsymbol{n}_0|A_0 \quad \text{and} \quad A_0 = J^{-1}|\boldsymbol{F}^{T}\boldsymbol{n}|A. \tag{3.23}$$

3.1.3 Transformation of volume elements

Since line elements transform according to Eq. (3.5), it may be easily understood that a volume element V_0 at a generic material point is transformed by the deformation as (Fig. 3.6)

$$V_0 = \boldsymbol{t}_0 \cdot \boldsymbol{v}_0 \times \boldsymbol{w}_0 \xrightarrow{\boldsymbol{F}} V = \boldsymbol{F}\boldsymbol{t}_0 \cdot \boldsymbol{F}\boldsymbol{v}_0 \times \boldsymbol{F}\boldsymbol{w}_0 \tag{3.24}$$

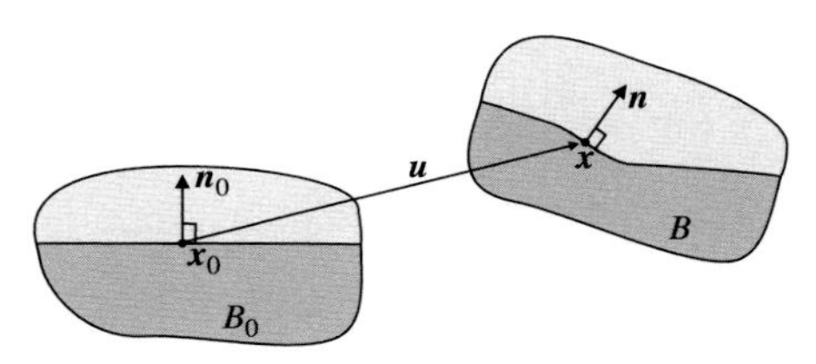

Figure 3.4. Transformation of a normal $\boldsymbol{n}_0$ to a surface from the reference configuration B_0 to the current configuration B. The example is referred to a layer in a composite.

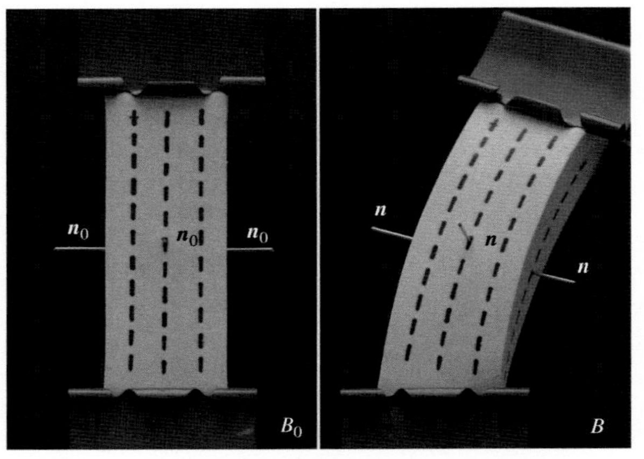

Figure 3.5. Transformation of the normal $\boldsymbol{n}_0$ to the external surfaces of a rubber block from the reference (unloaded) configuration B_0 to the current configuration B. Line elements (parallel in the reference configuration) are also drawn on the surfaces.

(where $\boldsymbol{t}_0$, $\boldsymbol{v}_0$ and $\boldsymbol{w}_0$ are material line elements), so an application of Eq. (3.20) yields

$$V = J\,V_0. \tag{3.25}$$

Equation (3.25) makes explicit the fact that

$$\text{Incompressibility} \iff J = \lambda_1\lambda_2\lambda_3 = 1. \tag{3.26}$$

3.1.4 Angular changes

Employing again Eq. (3.5), we obtain (Fig. 3.7)

$$\theta_0 = \arccos\frac{\boldsymbol{v}_0\cdot\boldsymbol{w}_0}{|\boldsymbol{v}_0|\,|\boldsymbol{w}_0|} \xrightarrow{\ \boldsymbol{F}\ } \theta = \arccos\frac{\boldsymbol{F}\boldsymbol{v}_0\cdot\boldsymbol{F}\boldsymbol{w}_0}{|\boldsymbol{F}\boldsymbol{v}_0|\,|\boldsymbol{F}\boldsymbol{w}_0|}. \tag{3.27}$$

These, using definition (3.12) of stretch tensors, become

$$\begin{aligned} \theta_0 &= \arccos\frac{\boldsymbol{v}\cdot\boldsymbol{V}^{-2}\boldsymbol{w}}{\sqrt{\boldsymbol{v}\cdot\boldsymbol{V}^{-2}\boldsymbol{v}}\sqrt{\boldsymbol{w}\cdot\boldsymbol{V}^{-2}\boldsymbol{w}}}, \\ \theta &= \arccos\frac{\boldsymbol{v}_0\cdot\boldsymbol{U}^{2}\boldsymbol{w}_0}{\sqrt{\boldsymbol{v}_0\cdot\boldsymbol{U}^{2}\boldsymbol{v}_0}\sqrt{\boldsymbol{w}_0\cdot\boldsymbol{U}^{2}\boldsymbol{w}_0}}. \end{aligned} \tag{3.28}$$

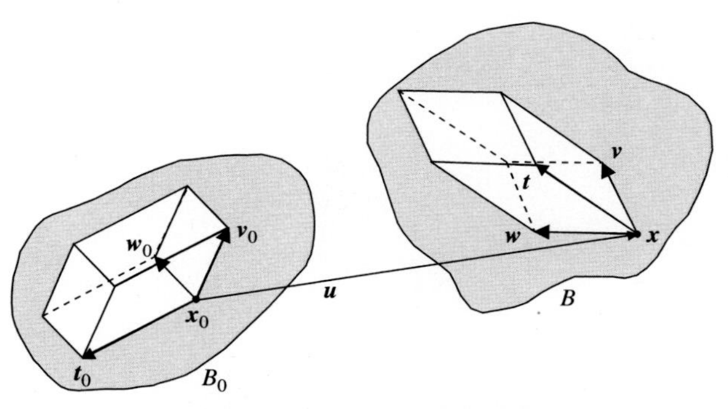

Figure 3.6. Transformation of a volume element.

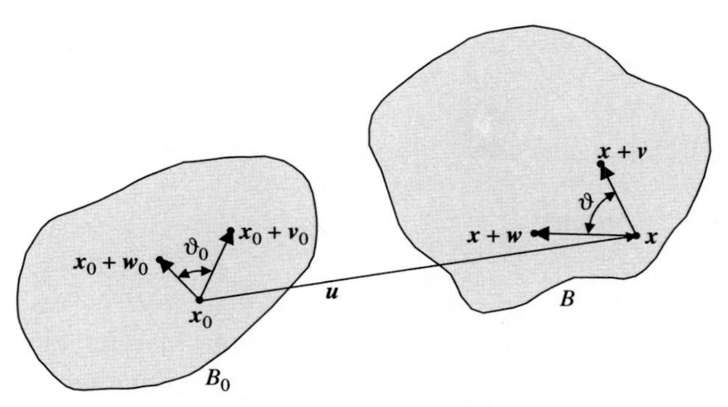

Figure 3.7. Transformation of the angle between two embedded fibres.

Since $\boldsymbol{B} = \boldsymbol{V}^2$ and $\boldsymbol{C} = \boldsymbol{U}^2$, we may collect results (3.11) and (3.28) to represent the Cauchy-Green deformation tensors in the material and spatial reference systems $\boldsymbol{e}_i^0$ and $\boldsymbol{e}_i$ $(i = 1,2,3)$ as (Fig. 3.8)

$$[\boldsymbol{C}]_{\{\boldsymbol{e}_1^0,\boldsymbol{e}_2^0,\boldsymbol{e}_3^0\}} = \begin{bmatrix} \lambda^2(\boldsymbol{e}_1^0) & \lambda(\boldsymbol{e}_1^0)\lambda(\boldsymbol{e}_2^0)\cos\theta^{(12)} & \lambda(\boldsymbol{e}_1^0)\lambda(\boldsymbol{e}_3^0)\cos\theta^{(13)} \\ \lambda(\boldsymbol{e}_2^0)\lambda(\boldsymbol{e}_1^0)\cos\theta^{(21)} & \lambda^2(\boldsymbol{e}_2^0) & \lambda(\boldsymbol{e}_2^0)\lambda(\boldsymbol{e}_3^0)\cos\theta^{(23)} \\ \lambda(\boldsymbol{e}_3^0)\lambda(\boldsymbol{e}_1^0)\cos\theta^{(31)} & \lambda(\boldsymbol{e}_3^0)\lambda(\boldsymbol{e}_2^0)\cos\theta^{(32)} & \lambda^2(\boldsymbol{e}_3^0) \end{bmatrix}, \tag{3.29}$$

and

$$[\boldsymbol{B}^{-1}]_{\{\boldsymbol{e}_1,\boldsymbol{e}_2,\boldsymbol{e}_3\}} = \begin{bmatrix} \lambda^{-2}(\boldsymbol{e}_1) & \dfrac{\cos\theta_0^{(12)}}{\lambda(\boldsymbol{e}_1)\lambda(\boldsymbol{e}_2)} & \dfrac{\cos\theta_0^{(13)}}{\lambda(\boldsymbol{e}_1)\lambda(\boldsymbol{e}_3)} \\ \dfrac{\cos\theta_0^{(21)}}{\lambda(\boldsymbol{e}_2)\lambda(\boldsymbol{e}_1)} & \lambda^{-2}(\boldsymbol{e}_2) & \dfrac{\cos\theta_0^{(23)}}{\lambda(\boldsymbol{e}_2)\lambda(\boldsymbol{e}_3)} \\ \dfrac{\cos\theta_0^{(31)}}{\lambda(\boldsymbol{e}_3)\lambda(\boldsymbol{e}_1)} & \dfrac{\cos\theta_0^{(32)}}{\lambda(\boldsymbol{e}_3)\lambda(\boldsymbol{e}_2)} & \lambda^{-2}(\boldsymbol{e}_3) \end{bmatrix}. \tag{3.30}$$

3.1.5 Measures of strain

It is clear from Eqs. $(3.12)_1$ and (3.28) that $\boldsymbol{U}$ or $\boldsymbol{V}$ provides a local measure of deformation because their principal components represent the stretch of the

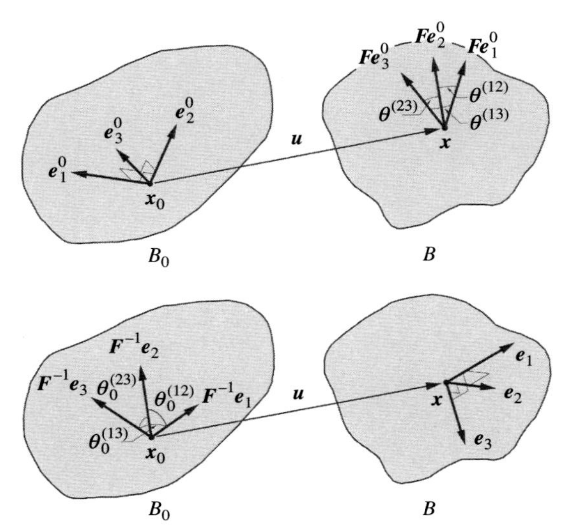

Figure 3.8. Transformation from the three embedded fibers e_1^0, e_2^0 and e_3^0 (e_1, e_2 and e_3) defined and orthogonal in B_0 (in B) to the three fibers defined in B (in B_0).

three orthogonal fibres (aligned with the eigenvectors). In particular, $\boldsymbol{U}$ ($\boldsymbol{V}$) is a Lagrangean (Eulerian) measure of deformation because it transforms material (spatial) quantities into spatial (material). In the absence of strain, the stretch tensors reduce to the identity $\boldsymbol{U} = \boldsymbol{V} = \boldsymbol{I}$. All the tensors $\boldsymbol{C}$, $\boldsymbol{B}$, $\boldsymbol{U}$ and $\boldsymbol{V}$ can be chosen to quantify the strain, and more in general, we may conclude that there are infinite possibilities of choice. It therefore may be convenient to introduce the Lagrangean and Eulerian strain measures

$$\boldsymbol{E}^{(m)} = \begin{cases} \dfrac{\boldsymbol{U}^m - \boldsymbol{I}}{m} & m \neq 0 \\ \log \boldsymbol{U} & m = 0 \end{cases}, \qquad \boldsymbol{G}^{(m)} = \begin{cases} \dfrac{\boldsymbol{V}^m - \boldsymbol{I}}{m} & m \neq 0 \\ \log \boldsymbol{V} & m = 0 \end{cases}, \tag{3.31}$$

defined for every (positive, negative or null) integer m so that they vanish in an undeformed situation. The logarithm of a tensor is defined for symmetric positive definite tensors taking the logarithm of the eigenvalues in its spectral decomposition so that

$$\log \boldsymbol{U} = \sum_{i=1}^{3} \varepsilon_i \boldsymbol{u}_i \otimes \boldsymbol{u}_i \qquad \text{and} \qquad \log \boldsymbol{V} = \sum_{i=1}^{3} \varepsilon_i \boldsymbol{v}_i \otimes \boldsymbol{v}_i, \tag{3.32}$$

where

$$\varepsilon_i = \log \lambda_i \tag{3.33}$$

are the so-called logarithmic strains.

The logarithmic strain ε may be introduced with reference to a fibre of initial length l_0 stretched until it reaches the final length l_f, so the stretch is $\lambda_f = l_f/l_0$. At a certain stage of the process of straining, the fibre has a length l, so the increment in deformation $d\varepsilon$ for an increment dl in length is

$$d\varepsilon = \frac{dl}{l},$$

which, integrated between l_0 and l_f, provides

$$\varepsilon = \log \frac{l_f}{l_0} = \log \lambda_f, \tag{3.34}$$

an expression corresponding to the definition (3.33).

Note that the two notations

$$\boldsymbol{T}^m = \underbrace{\boldsymbol{T}\boldsymbol{T}...\boldsymbol{T}}_{m \text{ times}} \qquad \text{and} \qquad \boldsymbol{T}^{(m)} \tag{3.35}$$

[the former of which is defined by Eq. (2.28)] should not be confused.

When $m = 2$, we obtain the Green-Lagrange strains

$$\boldsymbol{E}^{(2)} = \frac{1}{2}(\boldsymbol{C} - \boldsymbol{I}) \qquad \text{and} \qquad \boldsymbol{G}^{(2)} = \frac{1}{2}(\boldsymbol{B} - \boldsymbol{I}). \tag{3.36}$$

Equations (3.36) can be rewritten in terms of displacement gradient as

$$\boldsymbol{E}^{(2)} = \frac{1}{2}(\nabla \boldsymbol{u} + \nabla \boldsymbol{u}^T + \nabla \boldsymbol{u} \nabla \boldsymbol{u}^T) \qquad \text{and} \qquad \boldsymbol{G}^{(2)} = \frac{1}{2}(\nabla \boldsymbol{u} + \nabla \boldsymbol{u}^T + \nabla \boldsymbol{u}^T \nabla \boldsymbol{u}), \tag{3.37}$$

so they both reduce to the infinitesimal strain $\boldsymbol{E}$ when second-order terms in displacement gradient are neglected

$$\boldsymbol{E} = \frac{1}{2}(\nabla \boldsymbol{u} + \nabla \boldsymbol{u}^T) = \boldsymbol{E}^{(2)} + \boldsymbol{O}(|\nabla \boldsymbol{u}|^2) = \boldsymbol{G}^{(2)} + \boldsymbol{O}(|\nabla \boldsymbol{u}|^2). \tag{3.38}$$

Example: Rigid-body deformation As an example of *homogeneous deformation* occurring with constant gradient, a rigid-body rotation about the point $\boldsymbol{p}$ and a rigid-body translation of displacement $\boldsymbol{v}$ is illustrated in Fig. 3.9. The rotation about point $\boldsymbol{p}$ moves the solid from the reference configuration B_0 to the 'intermediate' configuration B_i, so point $\boldsymbol{x}_0$ is transformed into point $\boldsymbol{x}_i$ as

$$\boldsymbol{x}_i = \boldsymbol{p} + \boldsymbol{Q}(\boldsymbol{x}_0 - \boldsymbol{p}), \tag{3.39}$$

where $\boldsymbol{Q} \in \mathsf{Orth}^+$ describes the rigid-body rotation, constant for all material points in B_0. Finally, point $\boldsymbol{x}_i$ is translated to $\boldsymbol{x}$ simply by adding a vector $\boldsymbol{v}$, constant for all points in B_i. The result is

$$\boldsymbol{x} = \boldsymbol{q} + \boldsymbol{Q}(\boldsymbol{x}_0 - \boldsymbol{p}), \tag{3.40}$$

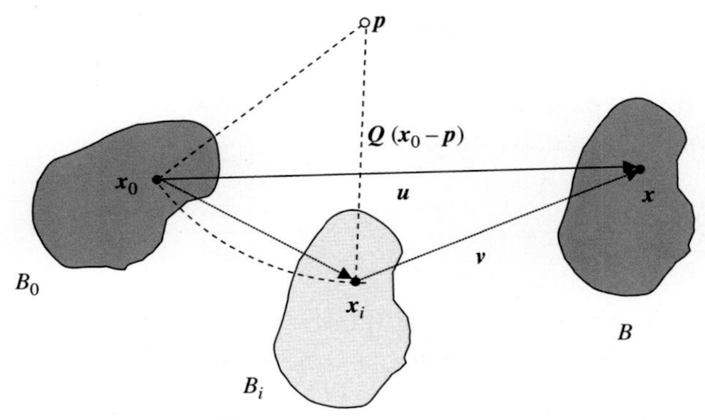

Figure 3.9. Rigid-body deformation.

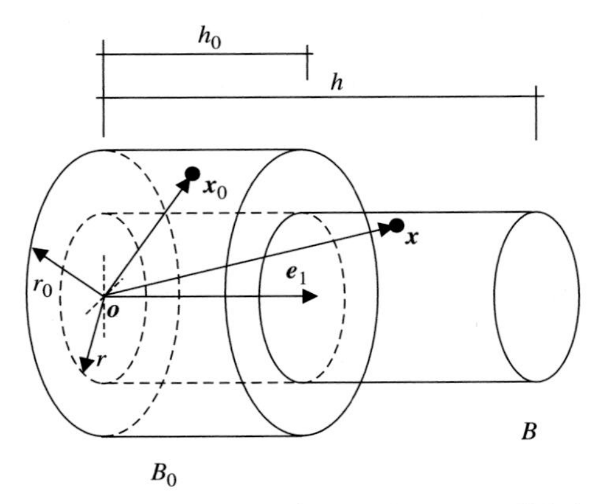

Figure 3.10. Uniform extension of a solid circular cylinder.

where $\boldsymbol{q} = \boldsymbol{p} + \boldsymbol{v}$, so the deformation gradient for a rigid-body rotation and translation is $\boldsymbol{F} = \boldsymbol{Q}$. The deformation described by Eq. (3.40) is rigid because the distance between every pair of points in B_0 is preserved, as can be easily checked.

Example: Uniform extension of a circular cylinder An example of homogeneous deformation with no rigid-body components is the uniform extension of a circular cylindrical solid, as sketched in Fig. 3.10.

Introducing the origin $\boldsymbol{o}$ and the unit vector $\boldsymbol{e}_1$ defining the axis of the cylinder, it can be envisaged easily that the material points $\boldsymbol{x}_0 \in B_0$ are mapped into the final configuration $\boldsymbol{x} \in B$ by the deformation

$$\boldsymbol{x} - \boldsymbol{o} = \left[\frac{h}{h_0} \boldsymbol{e}_1 \otimes \boldsymbol{e}_1 + \frac{r}{r_0} (\boldsymbol{I} - \boldsymbol{e}_1 \otimes \boldsymbol{e}_1) \right] (\boldsymbol{x}_0 - \boldsymbol{o}), \tag{3.41}$$

where r_0 and h_0 are the radius and height of the cylinder in the reference configuration B_0, assuming the values r and h in the deformed configuration B. It is immediate to conclude from Eq. (3.41) that in this case

$$\boldsymbol{F} = \boldsymbol{U} = \boldsymbol{V} = \frac{h}{h_0} \boldsymbol{e}_1 \otimes \boldsymbol{e}_1 + \frac{r}{r_0} (\boldsymbol{I} - \boldsymbol{e}_1 \otimes \boldsymbol{e}_1), \tag{3.42}$$

so there is no rotation, $\boldsymbol{R} = \boldsymbol{I}$, and the axis of the cylinder is simultaneously a principal Lagrangian and Eulerian axis, whereas the other two can be chosen arbitrarily in the plane orthogonal to it. Note that expanding the logarithmic strains into a Taylor series near the undeformed configuration, we have

$$\varepsilon_1 = \frac{h - h_0}{h_0} + O\left(\frac{(h - h_0)^2}{(h_0)^2} \right), \qquad \varepsilon_2 = \varepsilon_3 = \frac{r - r_0}{r_0} + O\left(\frac{(r - r_0)^2}{(r_0)^2} \right), \tag{3.43}$$

showing that when the strain is small, the logarithmic strains reduce to the components of the infinitesimal strain tensor, which is therefore given by

$$\boldsymbol{E} = \frac{h - h_0}{h_0} \boldsymbol{e}_1 \otimes \boldsymbol{e}_1 + \frac{r - r_0}{r_0} (\boldsymbol{I} - \boldsymbol{e}_1 \otimes \boldsymbol{e}_1). \tag{3.44}$$

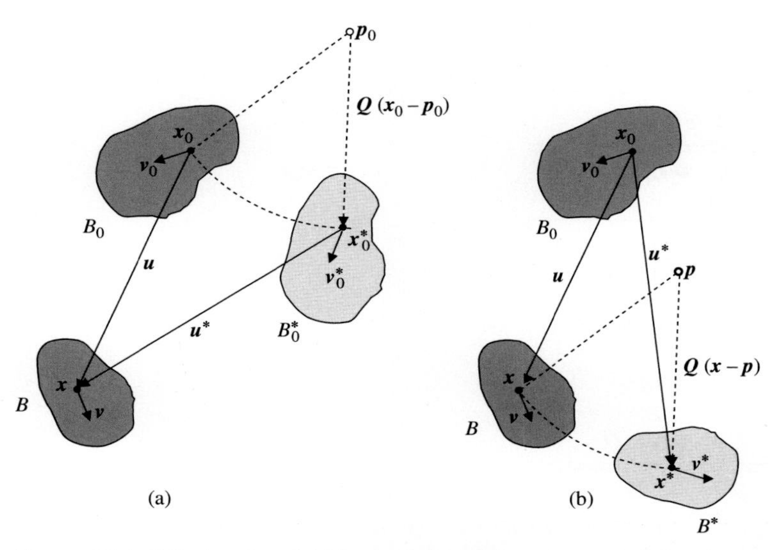

Figure 3.11. Effects of a rigid-body deformation: two rotated reference configurations to describe the same current configuration (*a*) and two rotated final configurations corresponding to the same reference configuration (*b*).

In the case of an incompressible material, Eq. (3.42) becomes

$$\boldsymbol{F} = \boldsymbol{U} = \boldsymbol{V} = \frac{h}{h_0}\boldsymbol{e}_1 \otimes \boldsymbol{e}_1 + \sqrt{\frac{h_0}{h}}\,(\boldsymbol{I} - \boldsymbol{e}_1 \otimes \boldsymbol{e}_1), \tag{3.45}$$

so the incompressibility constraint, $\lambda_1\lambda_2\lambda_3 = 1$, is satisfied.

3.2 On material and spatial strain measures

We now consider the effects of rigid-body rotations, represented by a tensor $\boldsymbol{Q} \in \mathsf{Orth}^+$ and applied on the reference configuration or on the spatial configuration.

3.2.1 Rigid-body rotation of the reference configuration

Since the choice of the reference configuration is arbitrary, two reference configurations, B_0 and B_0^*, are sketched in Fig. 3.11*a*, both transformed into the same final configuration B. The difference between B_0 and B_0^* lies in a rigid-body rotation.

The transformation of oriented line elements provides the following equations:

$$\boldsymbol{v} = \boldsymbol{F}\boldsymbol{v}_0 \qquad \text{and} \qquad \boldsymbol{v} = \boldsymbol{F}^*\boldsymbol{v}_0^*, \tag{3.46}$$

where $\boldsymbol{F}$ and $\boldsymbol{F}^*$ are the two deformation gradients relative to the two reference configurations B_0 and B_0^*, respectively. Since Eq. (3.46) must hold true for every vector $\boldsymbol{v}_0$, and $\boldsymbol{v}_0^* = \boldsymbol{Q}\boldsymbol{v}_0$, it may be concluded that

$$\boldsymbol{F}^* = \boldsymbol{F}\boldsymbol{Q}^T, \tag{3.47}$$

showing how a change in the reference configuration by a rigid-body transformation affects the deformation gradient.

Table 3.1. *Changes of strain and deformation measures for rigid-body rotations of the reference configuration B_0 and of the spatial configuration B*

Rigid-body rotation of B_0 (Fig. 3.11*a*)	Rigid-body rotation of B (Fig. 3.11*b*)
$\boldsymbol{F}\boldsymbol{Q}^T$	$\boldsymbol{Q}\boldsymbol{F}$
$\boldsymbol{R}\boldsymbol{Q}^T$	$\boldsymbol{Q}\boldsymbol{R}$
$\boldsymbol{Q}\boldsymbol{U}\boldsymbol{Q}^T$	$\boldsymbol{U}$
$\boldsymbol{V}$	$\boldsymbol{Q}\boldsymbol{V}\boldsymbol{Q}^T$
$\boldsymbol{Q}\boldsymbol{C}\boldsymbol{Q}^T$	$\boldsymbol{C}$
$\boldsymbol{B}$	$\boldsymbol{Q}\boldsymbol{B}\boldsymbol{Q}^T$
$\boldsymbol{Q}\boldsymbol{E}^{(m)}\boldsymbol{Q}^T$	$\boldsymbol{E}^{(m)}$
$\boldsymbol{G}^{(m)}$	$\boldsymbol{Q}\boldsymbol{G}^{(m)}\boldsymbol{Q}^T$

3.2.2 Rigid-body rotation of the current configuration

Two spatial configurations B and B^* differing in a rigid-body deformation and both corresponding to the same reference configuration B_0 are shown in Fig. 3.11*b*. Now, the two spatial vectors $\boldsymbol{v}$ and $\boldsymbol{v}^*$ are rotated of $\boldsymbol{Q}$, so $\boldsymbol{v}^* = \boldsymbol{Q}\boldsymbol{v}$ and

$$\boldsymbol{v} = \boldsymbol{F}\boldsymbol{v}_0 \qquad \text{and} \qquad \boldsymbol{v}^* = \boldsymbol{F}^*\boldsymbol{v}_0, \tag{3.48}$$

where $\boldsymbol{F}$ and $\boldsymbol{F}^*$ are the two deformation gradients relative to the two configurations B and B^*, respectively. Given the fact that $\boldsymbol{v}_0$ is arbitrary, we obtain

$$\boldsymbol{F}^* = \boldsymbol{Q}\boldsymbol{F}, \tag{3.49}$$

showing how a change in the spatial configuration by a rigid-body transformation affects the deformation gradient.

Since Eqs. (3.47) and (3.49) make explicit the variation of the deformation gradient when rigid-body transformations in the material or spatial configurations are applied, the tensors $\boldsymbol{C}^*$, $\boldsymbol{U}^*$, $\boldsymbol{E}^{(m)*}$, $\boldsymbol{B}^*$, $\boldsymbol{V}^*$ and $\boldsymbol{G}^{(m)*}$ can be obtained directly from their definitions. The result is reported in Table 3.1, where the changes of the strain and deformation measures are shown both for a rigid-body rotation of the reference configuration (in the left column) and for a rigid-body rotation of the spatial configuration (in the right column). Although both transformations do not affect the stretch, they influence the measures of strain. In particular, it may be concluded from Table 3.1 that tensors $\boldsymbol{C}$, $\boldsymbol{U}$ and $\boldsymbol{E}^{(m)}$ are purely material, whereas tensors $\boldsymbol{B}$, $\boldsymbol{V}$ and $\boldsymbol{G}^{(m)}$ are purely spatial, in the sense that the former are unchanged by a rigid-body transformation of the spatial configuration, whereas the latter are left unchanged by transformations in the reference configuration. The deformation gradient $\boldsymbol{F}$ is indeed always affected, so it is both material and spatial; in other words, it is a two-point tensor.

In standard textbooks such as Truesdell and Noll (1965) and Gurtin (1981), Eulerian fields are called 'objective', whereas Hill (in all his papers quoted in this book) defines 'objective' the Lagrangean fields, a circumstance that certainly does not improve clarity. Ogden (1984) 'reconciles' these two views by introducing the nomenclature 'Eulerian objective' and 'Lagrangean objective' so that, for instance, the right and left Cauchy-Green deformation tensors are, respectively, Lagrangean and Eulerian objectives. In addition to these objectivities, Holzapfel (2000) introduces a 'two-point objectivity' so that, for instance, the deformation gradient also conquers his own status of objectivity.

Our view is that the adjective 'objective' implies some positive value, so most humans would prefer belonging to the 'club of objectivity'. On the other hand, the deformation gradient $\boldsymbol{F}$ is 'not worse' than, for instance, the left Cauchy-Green deformation tensor $\boldsymbol{B}$, so it does not make sense to reward $\boldsymbol{B}$ with 'admission to the club of objectivity' and exclude at the same time $\boldsymbol{F}$. Therefore, our conclusion is simple: We will avoid use of the word 'objective'.

3.3 Motion of a deformable body

A 'motion' is an ordered sequence of mappings of a reference configuration into current configurations, or in other words, a 'motion' is a smooth one-parameter family of deformations, ordered by the time t, so that

$$\boldsymbol{x} = \boldsymbol{g}(\boldsymbol{x}_0, t) \qquad \text{and} \qquad \boldsymbol{x}_0 = \boldsymbol{g}^{-1}(\boldsymbol{x}, t) . \tag{3.50}$$

We introduce the material $\dot{\boldsymbol{x}}$ and spatial $\boldsymbol{v}$ descriptions of the velocity

$$\dot{\boldsymbol{x}}(\boldsymbol{x}_0, t) = \frac{\partial \boldsymbol{g}(\boldsymbol{x}_0, t)}{\partial t} \qquad \text{and} \qquad \boldsymbol{v}(\boldsymbol{x}, t) = \dot{\boldsymbol{x}}\left(\boldsymbol{g}^{-1}(\boldsymbol{x}, t), t\right) \tag{3.51}$$

and the material $\ddot{\boldsymbol{x}}$ and spatial $\boldsymbol{a}$ descriptions of the acceleration

$$\ddot{\boldsymbol{x}}(\boldsymbol{x}_0, t) = \frac{\partial^2 \boldsymbol{g}(\boldsymbol{x}_0, t)}{\partial^2 t} \qquad \text{and} \qquad \boldsymbol{a}(\boldsymbol{x}, t) = \ddot{\boldsymbol{x}}\left(\boldsymbol{g}^{-1}(\boldsymbol{x}, t), t\right). \tag{3.52}$$

It is expedient now to introduce three different time derivatives:

- The *material time derivative of a material field* $\boldsymbol{A}(\boldsymbol{x}_0, t)$ as

$$\dot{\boldsymbol{A}}(\boldsymbol{x}_0, t) = \frac{\partial \boldsymbol{A}(\boldsymbol{x}_0, t)}{\partial t}, \tag{3.53}$$

  so $\boldsymbol{x}_0$ is held fixed.
- The *material time derivative of a spatial field* $\boldsymbol{A}(\boldsymbol{x}, t)$ as

$$\dot{\boldsymbol{A}}(\boldsymbol{x}, t) = \left.\frac{\partial \boldsymbol{A}(\boldsymbol{g}(\boldsymbol{x}_0, t), t)}{\partial t}\right|_{\boldsymbol{x}_0 = \boldsymbol{g}^{-1}(\boldsymbol{x}, t)}, \tag{3.54}$$

  so the representation of $\boldsymbol{A}$ is first pulled back to the material description, then a material time derivative is taken and finally, the resulting expression is pushed forward again in the spatial description.
- The *spatial time derivative of a spatial field* $\boldsymbol{A}(\boldsymbol{x}, t)$ as

$$\boldsymbol{A}'(\boldsymbol{x}, t) = \frac{\partial \boldsymbol{A}(\boldsymbol{x}, t)}{\partial t}, \tag{3.55}$$

  so $\boldsymbol{x}$ is now held fixed.

Among these three types of time derivatives, the first and second will be employed frequently here (the third is particularly useful in fluid mechanics), and we will briefly refer to these as 'material time derivatives', meaning the first (second) in the preceding list when the field is material (spatial).

According to Eq. (3.53), the gradient of the material description of the velocity is

$$\mathsf{Grad}\,\dot{\boldsymbol{x}}(\boldsymbol{x}_0,t)=\dot{\boldsymbol{F}}, \tag{3.56}$$

whereas, by definition, the gradient of the spatial description of velocity is denoted by $\boldsymbol{L}$:

$$\boldsymbol{L}(\boldsymbol{x},t)=\mathsf{grad}\,\boldsymbol{v} \tag{3.57}$$

and is decomposed into its symmetric and skew-symmetric components

$$\boldsymbol{L}=\boldsymbol{D}+\boldsymbol{W}, \tag{3.58}$$

where, using the notion introduced by Eqs. $(2.88)_{2,3}$,

$$\boldsymbol{D}=\boldsymbol{I}\,\underline{\overline{\otimes}}\,\boldsymbol{I}[\boldsymbol{L}] \qquad \text{and} \qquad \boldsymbol{W}=(\boldsymbol{I}\boxtimes\boldsymbol{I}-\boldsymbol{I}\,\underline{\overline{\otimes}}\,\boldsymbol{I})[\boldsymbol{L}] \tag{3.59}$$

are the Eulerian strain rate and spin, respectively.

The connection between Eqs. (3.56) and (3.57) is given by observing that

$$\dot{\boldsymbol{x}}\,(\boldsymbol{x}_0,t)=\boldsymbol{v}\,(\boldsymbol{g}(\boldsymbol{x}_0,t),t)\,, \tag{3.60}$$

which, taking the gradient with respect $\boldsymbol{x}_0$ and using the chain rule of differentiation, yields

$$\dot{\boldsymbol{F}}=\boldsymbol{L}\boldsymbol{F}. \tag{3.61}$$

It may be important to observe that for every material fibre $\omega_0\boldsymbol{w}_0$, employing Eq. (3.61) and remembering that $(\omega_0\boldsymbol{w}_0)^{\boldsymbol{\cdot}}=\boldsymbol{0}$ because $\omega_0\boldsymbol{w}_0$ is fixed, we may obtain

$$(\omega\boldsymbol{w})^{\boldsymbol{\cdot}}=\dot{\boldsymbol{F}}(\omega_0\boldsymbol{w}_0)=\boldsymbol{L}(\omega\boldsymbol{w}) \tag{3.62}$$

so that $\boldsymbol{L}$ is a 'dynamical' counterpart of $\boldsymbol{F}$, in the sense that it provides the rate of change of the line element $\omega\boldsymbol{w}$.

Applying Eq. (2.106) to the deformation gradient, we obtain

$$\left(\boldsymbol{F}^{-1}\right)^{\boldsymbol{\cdot}}=-\boldsymbol{F}^{-1}\dot{\boldsymbol{F}}\boldsymbol{F}^{-1}, \tag{3.63}$$

which can be employed together with the material-time derivative of Eq. $(3.9)_2$ to arrive at

$$\left(\frac{\boldsymbol{F}^{-1}\boldsymbol{w}}{|\boldsymbol{F}^{-1}\boldsymbol{w}|}\right)^{\boldsymbol{\cdot}}=\boldsymbol{0}, \tag{3.64}$$

where the unit vector $\boldsymbol{w}$ is the transformed oriented direction of any unit material vector $\boldsymbol{w}_0$ satisfying the condition $\dot{\boldsymbol{w}}_0=\boldsymbol{0}$. Since for any unit vector the property (2.107) holds, $\dot{\boldsymbol{w}}\cdot\boldsymbol{w}=0$, a manipulation of Eq. (3.64) yields

$$\dot{\boldsymbol{w}}=(\boldsymbol{I}-\boldsymbol{w}\otimes\boldsymbol{w})\boldsymbol{L}\boldsymbol{w}, \tag{3.65}$$

providing the rate of change of a spatial oriented direction $\boldsymbol{w}$. Equation (3.65) was obtained by Hill [1978, his Eq. (3.13)][7]; it also can be derived directly from Eq. (3.62).

[7] Rodney Hill, June 11, 1921–February 2, 2011.

Analogously to the preceding derivation, we may take the material time derivative of Eq. (3.22)$_2$ to arrive at

$$\dot{\boldsymbol{n}} = -(\boldsymbol{I} - \boldsymbol{n} \otimes \boldsymbol{n})\boldsymbol{L}^T \boldsymbol{n}, \tag{3.66}$$

providing the rate of change of the unit normal to an embedded area element.

The relation

$$\dot{J} = J \operatorname{tr} \boldsymbol{L} = J \operatorname{div} \boldsymbol{v} \tag{3.67}$$

provides the rate of change of the determinant of the deformation gradient. Equation (3.67) may be proved by noting that

$$\dot{J} = (\lambda_1 \lambda_2 \lambda_3)^{\cdot} = J \left(\frac{\dot{\lambda}_1}{\lambda_1} + \frac{\dot{\lambda}_2}{\lambda_2} + \frac{\dot{\lambda}_3}{\lambda_3} \right) = J \operatorname{tr} \left(\dot{\boldsymbol{U}} \boldsymbol{U}^{-1} \right) \tag{3.68}$$

and

$$\dot{\boldsymbol{U}} = \boldsymbol{R}^T \dot{\boldsymbol{F}} - \boldsymbol{R}^T \dot{\boldsymbol{R}} \boldsymbol{U}, \quad \boldsymbol{R} \cdot \dot{\boldsymbol{R}} = 0. \tag{3.69}$$

We are now in a position to note that taking the material time derivative of $\boldsymbol{F}$ expressed through Eq. (3.16), that is,

$$\dot{\boldsymbol{F}} = \sum_{i=1}^{3} \dot{\lambda}_i \boldsymbol{v}_i \otimes \boldsymbol{u}_i + \sum_{i=1}^{3} \lambda_i \left(\dot{\boldsymbol{v}}_i \otimes \boldsymbol{u}_i + \boldsymbol{v}_i \otimes \dot{\boldsymbol{u}}_i \right), \tag{3.70}$$

using Eq. (3.61) and multiplying by $\boldsymbol{F}^{-1}$ given by Eq. (3.17) yields the following representation for the velocity gradient:

$$\begin{aligned} \boldsymbol{L} = & \sum_{i=1}^{3} \frac{\dot{\lambda}_i}{\lambda_i} \boldsymbol{v}_i \otimes \boldsymbol{v}_i + \sum_{i=1}^{3} \dot{\boldsymbol{v}}_i \otimes \boldsymbol{v}_i \\ & + \frac{\lambda_2}{\lambda_1} (\dot{\boldsymbol{u}}_2 \cdot \boldsymbol{u}_1) \boldsymbol{v}_2 \otimes \boldsymbol{v}_1 + \frac{\lambda_3}{\lambda_1} (\dot{\boldsymbol{u}}_3 \cdot \boldsymbol{u}_1) \boldsymbol{v}_3 \otimes \boldsymbol{v}_1 + \frac{\lambda_1}{\lambda_2} (\dot{\boldsymbol{u}}_1 \cdot \boldsymbol{u}_2) \boldsymbol{v}_1 \otimes \boldsymbol{v}_2 \\ & + \frac{\lambda_3}{\lambda_2} (\dot{\boldsymbol{u}}_3 \cdot \boldsymbol{u}_2) \boldsymbol{v}_3 \otimes \boldsymbol{v}_2 + \frac{\lambda_1}{\lambda_3} (\dot{\boldsymbol{u}}_1 \cdot \boldsymbol{u}_3) \boldsymbol{v}_1 \otimes \boldsymbol{v}_3 + \frac{\lambda_2}{\lambda_3} (\dot{\boldsymbol{u}}_2 \cdot \boldsymbol{u}_3) \boldsymbol{v}_2 \otimes \boldsymbol{v}_3. \end{aligned} \tag{3.71}$$

To obtain Eq. (3.71), the property (2.107) has been used. Now the same property allows us to obtain from Eq. (3.71) the diagonal terms of the velocity gradient (equal to the diagonal components of the Eulerian strain rate) in the Eulerian reference system $\boldsymbol{v}_1$, $\boldsymbol{v}_2$ and $\boldsymbol{v}_3$ as

$$D_{11} = \frac{\dot{\lambda}_1}{\lambda_1} = \dot{\varepsilon}_1, \qquad D_{22} = \frac{\dot{\lambda}_2}{\lambda_2} = \dot{\varepsilon}_2, \qquad D_{33} = \frac{\dot{\lambda}_3}{\lambda_3} = \dot{\varepsilon}_3, \tag{3.72}$$

where the definition (3.33) of logarithmic strain has been employed.

A simple example: Motion with uniform velocity gradient Let us consider the motion [analysed by Bigoni and Petryk (2002) and useful to discuss material instability (Chapter 14)]

$$\boldsymbol{x} = (\boldsymbol{I} + t\boldsymbol{L}_c)\boldsymbol{x}_0, \tag{3.73}$$

where t is the time and $\boldsymbol{L}_c$ is a second-order tensor independent of the place $\boldsymbol{x}_0$. The deformation gradient results in the form

$$\boldsymbol{F} = \boldsymbol{I} + t\boldsymbol{L}_c, \tag{3.74}$$

showing that $\boldsymbol{L}_c$ cannot be arbitrary but must satisfy the condition

$$J = \det(\boldsymbol{I} + t\boldsymbol{L}_c) > 0 \tag{3.75}$$

at every instant of time $t \in [0,\infty)$, ensuring that the deformation is invertible into

$$\boldsymbol{x}_0 = (\boldsymbol{I} + t\boldsymbol{L}_c)^{-1}\boldsymbol{x}. \tag{3.76}$$

The displacement can be expressed as a material or spatial field, respectively,

$$\boldsymbol{u}(\boldsymbol{x}_0,t) = t\,\boldsymbol{L}_c\boldsymbol{x}_0 \qquad \text{and} \qquad \boldsymbol{u}(\boldsymbol{x},t) = t\boldsymbol{L}_c(\boldsymbol{I} + t\,\boldsymbol{L}_c)^{-1}\boldsymbol{x}. \tag{3.77}$$

The material description of velocity and the material time derivative of the deformation gradient are

$$\dot{\boldsymbol{x}} = \boldsymbol{L}_c\boldsymbol{x}_0 \qquad \text{and} \qquad \dot{\boldsymbol{F}} = \boldsymbol{L}_c. \tag{3.78}$$

Now, the spatial description of velocity can be calculated from Eq. $(3.78)_1$ through substitution of Eq. (3.76), yielding

$$\boldsymbol{v}(\boldsymbol{x},t) = \boldsymbol{L}_c\,(\boldsymbol{I} + t\boldsymbol{L}_c)^{-1}\boldsymbol{x}, \tag{3.79}$$

from which the velocity gradient becomes

$$\boldsymbol{L} = \boldsymbol{L}_c\,(\boldsymbol{I} + t\boldsymbol{L}_c)^{-1}. \tag{3.80}$$

Expression (3.80) also could be obtained employing Eqs. (3.61) and $(3.78)_2$, and the spatial velocity $\boldsymbol{v}$ could be obtained taking the material derivative of the spatial description of displacement (recall that the spatial description of $\dot{\boldsymbol{x}}$ is $\boldsymbol{v}$):

$$\begin{aligned} \boldsymbol{v} &= \left[t\boldsymbol{L}_c(\boldsymbol{I} + t\boldsymbol{L}_c)^{-1}\boldsymbol{x}\right]^{\bullet} \\ &= t\boldsymbol{L}_c(\boldsymbol{I} + t\boldsymbol{L}_c)^{-1}\boldsymbol{v} + \boldsymbol{L}_c(\boldsymbol{I} + t\boldsymbol{L}_c)^{-1}\boldsymbol{x} + t\boldsymbol{L}_c\left[(\boldsymbol{I} + t\boldsymbol{L}_c)^{-1}\right]^{\bullet}\boldsymbol{x} \end{aligned} \tag{3.81}$$

and noting that Eq. (3.63) implies

$$\left[(\boldsymbol{I} + t\boldsymbol{L}_c)^{-1}\right]^{\bullet} = -(\boldsymbol{I} + t\boldsymbol{L}_c)^{-1}\boldsymbol{L}_c(\boldsymbol{I} + t\boldsymbol{L}_c)^{-1}, \tag{3.82}$$

so Eq. (3.81) can be solved for $\boldsymbol{v}$, thus providing expression (3.79). Note that Eq. (3.81) is the spatial time derivative (3.55) of the spatial representation of the displacement. This can be checked by employing the following general relation:

$$\dot{\boldsymbol{q}}(\boldsymbol{x},t) = (\mathsf{grad}\boldsymbol{q})\boldsymbol{v} + \boldsymbol{q}', \tag{3.83}$$

holding true for any spatial vector field $\boldsymbol{q}$.

The version of Eq. (3.83) for a scalar spatial field Θ can be obtained using the chain rule of differentiation as

$$\dot{\Theta}(\boldsymbol{x},t) = \boldsymbol{v}\cdot\mathsf{grad}\Theta + \Theta' = \Theta\,\mathsf{div}\boldsymbol{v} - \mathsf{div}(\Theta\boldsymbol{v}) + \Theta'. \tag{3.84}$$

3.4 Mass conservation

A continuous body possesses a mass, so for a generic part P of this body,

$$\text{mass}(P) = \int_P \rho \, dv, \tag{3.85}$$

where ρ is the mass density in the current configuration. Since the part P is the image in the spatial configuration of P_0 defined in the reference configuration, the *principle of conservation of the mass* is expressed by the condition

$$\text{mass}(P) = \text{mass}(P_0) \iff \int_P \rho \, dv = \int_{P_0} \rho_0 \, dv_0, \tag{3.86}$$

for every part $P_0 \subseteq B_0$ and corresponding $P \subseteq B$. Since Eq. (3.86) holds for every part of the body, the following local condition can be deduced:

$$\rho \, dv = \rho_0 \, dv_0, \tag{3.87}$$

which, using the transformation rule of volume elements [Eq. (3.25)], becomes the *local condition of mass conservation*:

$$\rho J = \rho_0. \tag{3.88}$$

It can be noted that for a continuous spatial field $\phi(\boldsymbol{x},t)$, defined on a part P, the following identity can be directly derived from Eq. (3.87):

$$\int_P \phi(\boldsymbol{x},t)\rho \, dv = \int_{P_0} \phi_0(\boldsymbol{x}_0,t)\rho_0 \, dv_0, \tag{3.89}$$

where P_0 is the image of P in the reference configuration and ϕ_0 is the material description of ϕ, that is,

$$\phi_0(\boldsymbol{x}_0,t) = \phi(\boldsymbol{g}(\boldsymbol{x}_0,t),t). \tag{3.90}$$

To take the material time derivative of the left hand side of Eq. (3.89) is not an immediate operation because P depends on time, a problem which does not exist with the right-hand side of the same equation because P_0 is fixed in time. Therefore, taking the material time derivative of the right-hand side of Eq. (3.89) and re-transforming the result to the spatial setting, we obtain

$$\frac{d}{dt}\int_P \phi(\boldsymbol{x},t)\rho \, dv = \int_P \dot{\phi}(\boldsymbol{x},t)\rho \, dv, \tag{3.91}$$

an equation which is certainly not intuitive at first glance.

Considering now Eq. (3.91) in the particular case in which $\phi = 1$, we obtain the *mass balance in integral form*

$$\frac{d}{dt}\int_P \rho \, dv = 0, \tag{3.92}$$

whereas taking the material time derivative of Eq. (3.88), we arrive at the *local rate equation of mass conservation*

$$\dot{\rho} + \rho \, \mathsf{div}\boldsymbol{v} = 0. \tag{3.93}$$

3.5 Stress, dynamic forces

Introduction of the concept of force requires specification that we will refer to an inertial frame. In the current configuration, a body interacts with the environment through forces, which may act on the surface, denoted by $\boldsymbol{\sigma}$ (defined per unit surface) or, at interior points, denoted by $\boldsymbol{b}$ (defined per unit volume). Owing to the presence of these forces, it is postulated that another contact force field exists inside the body, acting between every adjacent part of it: the stress vector or traction $\boldsymbol{s}$ (defined per unit area). In particular, with reference to a generic part P of the body in the spatial configuration (Fig. 3.12), the so-called Cauchy hypothesis is assumed, namely, that traction $\boldsymbol{s}$ depends only on the position $\boldsymbol{x}$ and on the unit normal $\boldsymbol{n}$ to the surface of P at $\boldsymbol{x}$, namely,

$$\boldsymbol{s} = \boldsymbol{s}(\boldsymbol{x}, \boldsymbol{n}). \tag{3.94}$$

It is assumed that the generic part P is in equilibrium under the action of the tractions $\boldsymbol{s}$ on its surface and of the generalised body force

$$\boldsymbol{b}^* = \boldsymbol{b} - \rho\,\dot{\boldsymbol{v}}, \tag{3.95}$$

sum of the volume forces $\boldsymbol{b}$ and the inertial body force $-\rho\dot{\boldsymbol{v}}$.

Since equilibrium holds with the generalised body forces $\boldsymbol{b}^*$, the momentum balance laws must be satisfied for the part P. These are

$$\begin{aligned} &\int_{\partial P} \boldsymbol{s} + \int_P \boldsymbol{b}^* = \boldsymbol{0}, \\ &\int_{\partial P} (\boldsymbol{x} - \boldsymbol{o}) \times \boldsymbol{s} + \int_P (\boldsymbol{x} - \boldsymbol{o}) \times \boldsymbol{b}^* = \boldsymbol{0}, \end{aligned} \tag{3.96}$$

holding for any origin $\boldsymbol{o}$ and expressing the vanishing of the resultant force and the resultant moment over P. Using Eq. (3.91), Eqs. (3.96) can be rewritten in the following equivalent form:

$$\begin{aligned} &\int_{\partial P} \boldsymbol{s} + \int_P \boldsymbol{b} = \frac{d}{dt}\int_P \rho\,\boldsymbol{v}, \\ &\int_{\partial P} (\boldsymbol{x} - \boldsymbol{o}) \times \boldsymbol{s} + \int_P (\boldsymbol{x} - \boldsymbol{o}) \times \boldsymbol{b} = \frac{d}{dt}\int_P \rho\,(\boldsymbol{x} - \boldsymbol{o}) \times \boldsymbol{v}. \end{aligned} \tag{3.97}$$

The Cauchy theorem states that if $\boldsymbol{\sigma}$ and $\boldsymbol{b}^*$ are a system of forces for B during a motion, a necessary and sufficient condition for Eqs. (3.96) to hold for any part $P \in B$

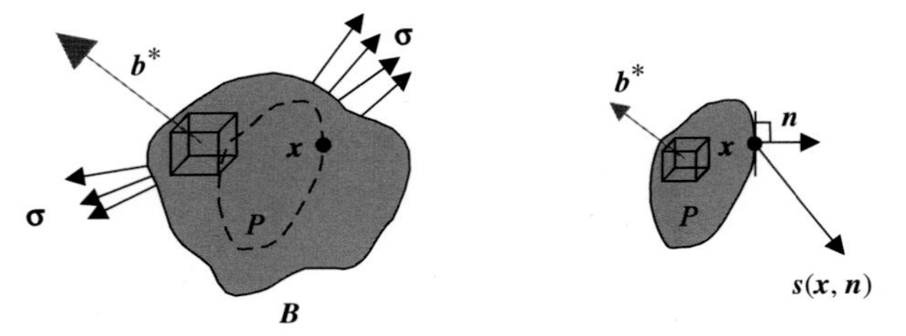

Figure 3.12. Forces acting on a body in the current configuration B and its part P. $\boldsymbol{\sigma}$ is a force acting on the external surface of B, $\boldsymbol{b}^*$ is the sum of body and inertia forces and $\boldsymbol{s}$ is the traction.

is the existence of a spatial tensorial field $\boldsymbol{T}$ (the so-called Cauchy stress tensor) such that

$\mathcal{A}1$. The traction is a linear function of the unit normal $\boldsymbol{n}$ through the Cauchy stress:

$$\boldsymbol{s}(\boldsymbol{n}) = \boldsymbol{T}\boldsymbol{n}, \tag{3.98}$$

transforming the spatial unit normal to the spatial vector $\boldsymbol{s}$.

$\mathcal{A}2$. The Cauchy stress is symmetric:

$$\boldsymbol{T} \in \mathrm{Sym}. \tag{3.99}$$

$\mathcal{A}3$. $\boldsymbol{T}$ satisfies the local equations of motion

$$\mathrm{div}\boldsymbol{T} + \boldsymbol{b}^* = \boldsymbol{0}. \tag{3.100}$$

In other words, Cauchy's theorem states that the local counterpart of Eq. $(3.96)_1$ is $\mathcal{A}1$ and $\mathcal{A}3$, whereas the local counterpart of Eq. $(3.96)_2$ is $\mathcal{A}2$. Since the concepts of force and stress are naturally defined in the spatial configuration, where they have a full physical meaning, the proof of the Cauchy theorem is the same as in the infinitesimal theory and can be found in many textbooks (e.g., Gurtin, 1972, 1981). In particular, the well-known proof of necessity [showing that Euler's axioms (3.96) imply $\mathcal{A}1$–$\mathcal{A}3$] is based on the imposition of equilibrium of the small tetrahedron of material shown in Fig. 3.13, where body forces become higher order with respect to the surface tractions when the dimensions of the tetrahedron are made vanishing small. In particular, since the faces of the Cauchy tetrahedron obey the geometrical relation

$$\Omega_{\boldsymbol{e}_i} = (\boldsymbol{n} \cdot \boldsymbol{e}_i)\, \Omega_{\boldsymbol{n}}, \tag{3.101}$$

where $\Omega_{\boldsymbol{e}_i}$ and $\Omega_{\boldsymbol{n}}$ denote the areas of the faces of unit normal $\boldsymbol{e}_i$ and $\boldsymbol{n}$, respectively, translational equilibrium, in which body forces are neglected (because they are of higher order when the volume of the element tends to zero), implies

$$\boldsymbol{s}(\boldsymbol{n}) = (\boldsymbol{n} \cdot \boldsymbol{e}_1)\, \boldsymbol{s}(\boldsymbol{e}_1) + (\boldsymbol{n} \cdot \boldsymbol{e}_2)\, \boldsymbol{s}(\boldsymbol{e}_2) + (\boldsymbol{n} \cdot \boldsymbol{e}_3)\, \boldsymbol{s}(\boldsymbol{e}_3), \tag{3.102}$$

showing that knowledge of the stress vectors $\boldsymbol{s}(\boldsymbol{e}_i)$ applied on three mutually orthogonal surface elements (of unit normals $\boldsymbol{e}_i$) is sufficient to determine the stress vector

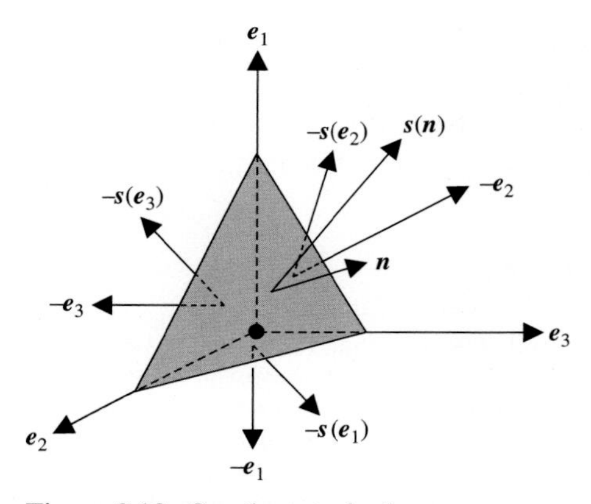

Figure 3.13. Cauchy tetrahedron.

$s(n)$ applied to every surface element of unit normal n. In other words, we can define the Cauchy stress tensor as

$$T = s(e_1) \otimes e_1 + s(e_2) \otimes e_2 + s(e_3) \otimes e_3 \qquad (3.103)$$

so that $\mathcal{A}1$ follows.

The rotational equilibrium of the Cauchy tetrahedron allows us to demonstrate $\mathcal{A}2$, whereas $\mathcal{A}3$ can be obtained from localization of the first of the momentum balance equations $(3.96)_1$, in which $\mathcal{A}1$ and the divergence theorem (2.154) are employed to write

$$\int_{\partial P} s = \int_{\partial P} Tn = \int_P \text{div}T. \qquad (3.104)$$

The sufficiency of the Cauchy theorem, namely, that $\mathcal{A}1$, $\mathcal{A}2$ and $\mathcal{A}3$ imply Eq. $(3.96)_1$, can be obtained by integrating $\mathcal{A}3$ over P, applying the divergence theorem and using $\mathcal{A}1$. With a similar (but harder) proof, Eq. $(3.96)_2$ also can be obtained (see Gurtin, 1981).

Stress is the internal counterpart to forces applied by the environment to the body in its actual configuration: It intrinsically represents, therefore, a *spatial* quantity. However, the following identity may be derived easily from Nanson's formula (3.21):

$$Tnda = Sn_0 da_0, \qquad (3.105)$$

where

$$S = JTF^{-T} \qquad (3.106)$$

is the *first Piola-Kirchhoff stress tensor*,[8] and Sn_0 is the nominal traction. Tensor S is in general un-symmetric ($\in$ Lin) but satisfies

$$SF^T = FS^T, \qquad (3.107)$$

because T is symmetric.

Note that by introducing the so-called Kirchhoff stress,

$$K = JT, \qquad (3.108)$$

Eq. (3.106) can be rewritten in the short form as

$$S = KF^{-T}. \qquad (3.109)$$

The nomenclature 'Kirchhoff stress' to denote (3.108) was used by Hill (in all his papers quoted in this book).

The first Piola-Kirchhoff stress tensor is the transpose of the nominal stress tensor (note that it is denoted by a bold minuscule letter, but it is a second-order un-symmetric tensor), that is,

$$t = S^T = F^{-1}K, \qquad (3.110)$$

[8] From Gabrio Piola (Milano 1794–Giussano 1850) and G. R. Kirchhoff (1824–1887). According to Truesdell and Toupin (1960, section 210), Kirchhoff had nothing to do with the definition of the stress measure (3.106) and also with the so-called second Piola-Kirchhoff stress — which will be introduced later [Eq. (3.132)] — so it would be more appropriate to refer to the former as to the 'Piola stress' and to the latter as the 'Cosserat stress', a nomenclature used by Podio Guidugli (2000). It may be interesting to note that Truesdell himself (see Truesdell and Noll, 1966, and Truesdell, 1966) did not contribute increasing nomenclature, continuing to refer to the first and second Piola-Kirchhoff stresses, which is also our choice.

used, among others, by Hill (1978) and Ogden (1984). We have used both notations, so the nominal stress is employed instead of the first Piola-Kirchhoff stress in Bigoni and Capuani (2002, 2005), Bigoni et al. (2008), Radi et al. (2002) Brun et al. (2003a, 2003b) and Dal Corso et al. (2008). Note also that in all these works another definition of Div is employed, namely, that reported in Eq. (2.133).

Note that the nominal traction is $\boldsymbol{S}\boldsymbol{n}_0 = \boldsymbol{t}^T\boldsymbol{n}_0$, so it remains the same with the two notations.

The peculiarity of the first Piola-Kirchhoff stress is that $\boldsymbol{S}\boldsymbol{n}_0$ is a measure of the surface force per unit area in the reference configuration. Therefore, the first Piola-Kirchhoff stress satisfies

$$\int_{\partial P} \boldsymbol{T}\boldsymbol{n} = \int_{\partial P_0} \boldsymbol{S}\boldsymbol{n}_0, \tag{3.111}$$

Thus, introducing the body force in the reference configuration

$$\boldsymbol{b}_0 = J\boldsymbol{b}, \tag{3.112}$$

and noting that

$$\int_{\partial P} \dot{\boldsymbol{v}}\rho = \int_{\partial P_0} \ddot{\boldsymbol{x}}\rho_0, \tag{3.113}$$

the balance laws (3.96) become

$$\begin{aligned} &\int_{\partial P_0} \boldsymbol{S}\boldsymbol{n}_0 + \int_{P_0} \boldsymbol{b}_0 = \int_{P_0} \ddot{\boldsymbol{x}}\rho_0, \\ &\int_{\partial P_0} (\boldsymbol{x}-\boldsymbol{o}) \times \boldsymbol{S}\boldsymbol{n}_0 + \int_{P_0} (\boldsymbol{x}-\boldsymbol{o}) \times \boldsymbol{b}_0 = \int_{P_0} (\boldsymbol{x}-\boldsymbol{o}) \times \ddot{\boldsymbol{x}}\rho_0. \end{aligned} \tag{3.114}$$

Equation $(3.114)_1$ localises into a local balance expressed in the reference configuration

$$\mathrm{Div}\boldsymbol{S} + \boldsymbol{b}_0 = \rho_0\ddot{\boldsymbol{x}}, \tag{3.115}$$

representing, when the inertia is neglected, translational equilibrium of a spatial element imposed on the reference configuration. However, symmetry of $\boldsymbol{S}$ does not follow from Eq. $(3.114)_2$, which localises into condition (3.107), equivalent to the symmetry of the Kirchhoff (and thus Cauchy) stress.

To prove that Eq. $(3.114)_2$ localises into Eq. (3.107), which means the symmetry of the Kirchhoff stress $\boldsymbol{K}$, we can proceed as follows. For every constant vector $\boldsymbol{a}$, we may write

$$\boldsymbol{a} \cdot (\boldsymbol{x}-\boldsymbol{o}) \times \boldsymbol{S}\boldsymbol{n}_0 = -\left[\boldsymbol{S}^T((\boldsymbol{x}-\boldsymbol{o}) \times \boldsymbol{a})\right] \cdot \boldsymbol{n}_0, \tag{3.116}$$

so the divergence theorem (2.154) gives

$$\int_{\partial P_0} \boldsymbol{a} \cdot (\boldsymbol{x}-\boldsymbol{o}) \times \boldsymbol{S}\boldsymbol{n}_0 = -\int_{P_0} \mathrm{Div}\left[\boldsymbol{S}^T((\boldsymbol{x}-\boldsymbol{o}) \times \boldsymbol{a})\right]. \tag{3.117}$$

Using, now the property

$$\mathrm{Div}\left[\boldsymbol{S}^T((\boldsymbol{x}-\boldsymbol{o}) \times \boldsymbol{a})\right] = \boldsymbol{S} \cdot \mathrm{Grad}[(\boldsymbol{x}-\boldsymbol{o}) \times \boldsymbol{a}] + [(\boldsymbol{x}-\boldsymbol{o}) \times \boldsymbol{a}] \cdot \mathrm{Div}\boldsymbol{S} \tag{3.118}$$

and the mixed product equality

$$((\boldsymbol{x}-\boldsymbol{o}) \times \boldsymbol{a}) \cdot \mathrm{Div}\boldsymbol{S} = -\boldsymbol{a} \cdot (\boldsymbol{x}-\boldsymbol{o}) \times \mathrm{Div}\boldsymbol{S}, \tag{3.119}$$

we can rewrite the scalar product of Eq. $(3.114)_2$ with $\boldsymbol{a}$ as

$$\int_{P_0} \boldsymbol{S} \cdot \mathrm{Grad}\,((\boldsymbol{x}-\boldsymbol{o}) \times \boldsymbol{a}) = -\boldsymbol{a} \cdot \int_{P_0} (\boldsymbol{x}-\boldsymbol{o}) \times [\mathrm{Div}\boldsymbol{S} + \boldsymbol{b}_0 - \ddot{\boldsymbol{x}}\,\rho_0] \tag{3.120}$$

so that through Eq. (3.115) we arrive at

$$\int_{P_0} \boldsymbol{S} \cdot \mathrm{Grad}\,[(\boldsymbol{x}-\boldsymbol{o}) \times \boldsymbol{a}] = 0, \tag{3.121}$$

which, holding for every part P_0 (and vector $\boldsymbol{a}$), implies the integrand to be null, which may be rewritten [using the definition of the first Piola-Kirchhoff stress, Eq. (3.106), the Kirchhoff stress, Eq. (3.108), and the polar decomposition of $\boldsymbol{F}$, Eq. (3.15)], as

$$\mathrm{Grad}\,[(\boldsymbol{x}-\boldsymbol{o}) \times \boldsymbol{a}]\,\boldsymbol{U}^{-1} \cdot \boldsymbol{K}\boldsymbol{R} = 0. \tag{3.122}$$

Now, since $\boldsymbol{x} = \boldsymbol{g}(\boldsymbol{x}_0)$ and for all vectors $\boldsymbol{u}$

$$[\mathrm{Grad}\,((\boldsymbol{x}-\boldsymbol{o}) \times \boldsymbol{a})]\,\boldsymbol{u} = \lim_{\beta \to 0} \frac{\boldsymbol{g}(\boldsymbol{x}_0 + \beta\boldsymbol{u}) - \boldsymbol{g}(\boldsymbol{x}_0)}{\beta} \times \boldsymbol{a} = -\boldsymbol{a} \times \boldsymbol{F}\boldsymbol{u}, \tag{3.123}$$

the spectral representation of $\boldsymbol{U}^{-1}$ [recall the spectral representations of the right and left stretch tensors (3.13)] allows us to write

$$\mathrm{tr}\,[(\boldsymbol{a} \times \boldsymbol{v}_1) \otimes \boldsymbol{K}\boldsymbol{v}_1 + (\boldsymbol{a} \times \boldsymbol{v}_2) \otimes \boldsymbol{K}\boldsymbol{v}_2 + (\boldsymbol{a} \times \boldsymbol{v}_3) \otimes \boldsymbol{K}\boldsymbol{v}_3] = 0, \tag{3.124}$$

which developing the scalar product and invoking arbitrariness of $\boldsymbol{a}$ yields

$$\boldsymbol{v}_1 \times \boldsymbol{K}\boldsymbol{v}_1 + \boldsymbol{v}_2 \times \boldsymbol{K}\boldsymbol{v}_2 + \boldsymbol{v}_3 \times \boldsymbol{K}\boldsymbol{v}_3 = \boldsymbol{0}. \tag{3.125}$$

Thus, using the representation of $\boldsymbol{K}$ in the reference system $\boldsymbol{v}_i$, that is,

$$\boldsymbol{K} = \sum_{i,j=1}^{3} K_{ij}\boldsymbol{v}_i \otimes \boldsymbol{v}_j, \tag{3.126}$$

we finally conclude the symmetry of $\boldsymbol{K}$ (note that the symmetry of a tensor in one reference system implies symmetry in every reference system):

$$(K_{12} - K_{21})\,\boldsymbol{v}_2 \times \boldsymbol{v}_1 + (K_{13} - K_{31})\,\boldsymbol{v}_3 \times \boldsymbol{v}_1 + (K_{23} - K_{32})\,\boldsymbol{v}_3 \times \boldsymbol{v}_2 = \boldsymbol{0}. \tag{3.127}$$

3.6 Power expended and work-conjugate stress/strain measures

Taking the scalar product of $\mathcal{A}1$ [of Eq. (3.115)] with the spatial (material) velocity $\boldsymbol{v}$ ($\dot{\boldsymbol{x}}$), the subsequent application of the divergence theorem yields the *theorem of power expended* in the spatial description

$$\underbrace{\int_{\partial P} \boldsymbol{s}(\boldsymbol{n}) \cdot \boldsymbol{v} + \int_{P} \boldsymbol{b} \cdot \boldsymbol{v}}_{\text{external power}} = \underbrace{\int_{P} \boldsymbol{T} \cdot \boldsymbol{D}}_{\text{stress power}} + \frac{d}{dt} \underbrace{\int_{P} \frac{\rho}{2} \boldsymbol{v}^2}_{\text{kinetic energy}}, \tag{3.128}$$

and referential description

$$\underbrace{\int_{\partial P_0} \boldsymbol{S}\boldsymbol{n}_0 \cdot \dot{\boldsymbol{x}} + \int_{P_0} \boldsymbol{b}_0 \cdot \dot{\boldsymbol{x}}}_{\text{external power}} = \underbrace{\int_{P_0} \boldsymbol{S} \cdot \dot{\boldsymbol{F}}}_{\text{stress power}} + \frac{d}{dt} \underbrace{\int_{P_0} \frac{\rho_0}{2} \dot{\boldsymbol{x}}^2}_{\text{kinetic energy}}. \tag{3.129}$$

which, in other words, represent the equation of energy balance for isothermal deformation (when temperature effects are kept into account, the internal energy, the heat flux and supply come into play; see Chadwick, 1976).

Note that the stress power *per unit volume* in the reference configuration is

$$\boldsymbol{K}\cdot\boldsymbol{D}=\boldsymbol{S}\cdot\dot{\boldsymbol{F}}, \tag{3.130}$$

but it also can be expressed as[9]

$$\boldsymbol{T}^{(2)}\cdot\dot{\boldsymbol{E}}^{(2)}=\boldsymbol{S}\cdot\dot{\boldsymbol{F}}, \tag{3.131}$$

where

$$\boldsymbol{T}^{(2)}=\boldsymbol{F}^{-1}\boldsymbol{K}\boldsymbol{F}^{-T}=\boldsymbol{F}^{-1}\boldsymbol{S} \tag{3.132}$$

is the *second Piola-Kirchhoff (symmetric) stress tensor*. Since its scalar product with the Green-Lagrange rate of strain gives the stress power per unit volume (in B_0), the second Piola-Kirchhoff stress and the Green-Lagrange strain tensor are said to be 'work conjugate' (Hill, 1968, 1978). More in general, we can define the Lagrangean and Eulerian stress measures $\boldsymbol{T}^{(m)}$ and $\boldsymbol{Z}^{(m)}$ work-conjugate to the strain measures (3.31) so that

$$\boldsymbol{K}\cdot\boldsymbol{D}=\boldsymbol{Z}^{(m)}\cdot\dot{\boldsymbol{G}}^{(m)}=\boldsymbol{S}\cdot\dot{\boldsymbol{F}}=\boldsymbol{T}^{(m)}\cdot\dot{\boldsymbol{E}}^{(m)}. \tag{3.133}$$

A conjugate pair of stress and strain that will become useful later is formed by the *Biot*[10] *stress tensor* $\boldsymbol{T}^{(1)}$ and the *right stretch-strain tensor* $\boldsymbol{E}^{(1)}$, defined as

$$\boldsymbol{E}^{(1)}=\boldsymbol{U}-\boldsymbol{I} \qquad \text{conjugate to} \qquad \boldsymbol{T}^{(1)}=\frac{1}{2}\left(\boldsymbol{T}^{(2)}\boldsymbol{U}+\boldsymbol{U}\boldsymbol{T}^{(2)}\right), \tag{3.134}$$

whereas another pair is the *Almansi*[11] *strain* $\boldsymbol{E}^{(-2)}$ and its conjugate stress $\boldsymbol{T}^{(-2)}$, defined as

$$\boldsymbol{E}^{(-2)}=-\frac{1}{2}(\boldsymbol{U}^{-2}-\boldsymbol{I}) \qquad \text{conjugate to} \qquad \boldsymbol{T}^{(-2)}=\boldsymbol{F}^T\boldsymbol{K}\boldsymbol{F}, \tag{3.135}$$

where $\boldsymbol{U}^{-2}=\boldsymbol{C}^{-1}=\boldsymbol{F}^{-1}\boldsymbol{F}^{-T}$.

It is well known, however, that it is not always the easy task of the preceding examples to obtain the stress measure conjugated to a given strain of the form (3.31). For instance, the conjugate of the logarithmic strain $\boldsymbol{E}^{(0)}$ has a very complex form (Hoger, 1987), which simplifies to the so-called rotated stress $\boldsymbol{T}^{(0)}$ only when the two measures result coaxial, namely,

$$\boldsymbol{E}^{(0)}=\log\boldsymbol{U} \qquad \text{conjugate to} \qquad \boldsymbol{T}^{(0)}=\boldsymbol{R}^T\boldsymbol{K}\boldsymbol{R} \tag{3.136}$$

[9] Since

$$\dot{\boldsymbol{E}}^{(2)}=(\dot{\boldsymbol{F}}^T\boldsymbol{F}+\boldsymbol{F}^T\dot{\boldsymbol{F}})/2,$$

it follows that

$$\begin{aligned}\dot{\boldsymbol{E}}^{(2)}\cdot\boldsymbol{T}^{(2)}&=(\dot{\boldsymbol{F}}^T\boldsymbol{F}\cdot\boldsymbol{F}^{-1}\boldsymbol{K}\boldsymbol{F}^{-T}+\boldsymbol{F}^T\dot{\boldsymbol{F}}\cdot\boldsymbol{F}^{-1}\boldsymbol{K}\boldsymbol{F}^{-T})/2\\&=\mathrm{tr}(\dot{\boldsymbol{F}}^T\boldsymbol{F}\boldsymbol{F}^{-1}\boldsymbol{K}\boldsymbol{F}^{-T}+\boldsymbol{F}^T\dot{\boldsymbol{F}}\boldsymbol{F}^{-1}\boldsymbol{K}\boldsymbol{F}^{-T})/2=\boldsymbol{S}\cdot\dot{\boldsymbol{F}}.\end{aligned}$$

[10] Maurice Anthony Biot, Antwerp, 25 May 1905--New York City, 12 September 1985.

[11] Emilio Almansi, Firenze, 15 April 1869–Firenze, 10 August 1948.

if and only if the following coaxiality condition holds true:

$$\boldsymbol{E}^{(0)}\boldsymbol{T}^{(0)} = \boldsymbol{T}^{(0)}\boldsymbol{E}^{(0)} \iff (\log \boldsymbol{V})\boldsymbol{K} = \boldsymbol{K} \log \boldsymbol{V}, \tag{3.137}$$

where $\boldsymbol{V}$ is the left stretch tensor, so $\boldsymbol{F} = \boldsymbol{R}\boldsymbol{U} = \boldsymbol{V}\boldsymbol{R}$ (note that the preceding equivalence is an immediate consequence of the fact that the logarithmic function is isotropic). Condition (3.137) is satisfied for isotropic elasticity but may not be in more general contexts, such as, for instance, elastoplasticity (Sansour, 2001).

It may be instructive for subsequent considerations to demonstrate that the rotated stress $\boldsymbol{R}^T\boldsymbol{K}\boldsymbol{R}$ and $\boldsymbol{E}^{(0)}$ are work conjugate under the coaxiality assumptions and to find their spatial counterparts.

When the coaxiality condition (3.137) holds true, the Eulerian principal axes $\boldsymbol{v}_i$ $(i = 1,2,3)$ determine a principal reference system for $\boldsymbol{K}$ that can be written as

$$\boldsymbol{K} = \sum_{i=1}^{3} K_i \boldsymbol{v}_i \otimes \boldsymbol{v}_i, \tag{3.138}$$

Thus Eq. (3.14) allows us to write the spectral representation for the rotated stress as

$$\boldsymbol{R}^T\boldsymbol{K}\boldsymbol{R} = \sum_{i=1}^{3} K_i \boldsymbol{u}_i \otimes \boldsymbol{u}_i, \tag{3.139}$$

where K_i are the principal Kirchhoff stresses. Equations (3.138) and (3.139) can be used together with the material time derivative of the Lagrangean and Eulerian logarithmic strains, written employing Eq. (3.72), as

$$\begin{aligned} (\log \boldsymbol{U})^{\boldsymbol{\cdot}} &= \sum_{i=1}^{3} D_{ii}\boldsymbol{u}_i \otimes \boldsymbol{u}_i + \sum_{i=1}^{3} \varepsilon_i \left(\dot{\boldsymbol{u}}_i \otimes \boldsymbol{u}_i + \boldsymbol{u}_i \otimes \dot{\boldsymbol{u}}_i\right), \\ (\log \boldsymbol{V})^{\boldsymbol{\cdot}} &= \sum_{i=1}^{3} D_{ii}\boldsymbol{v}_i \otimes \boldsymbol{v}_i + \sum_{i=1}^{3} \varepsilon_i \left(\dot{\boldsymbol{v}}_i \otimes \boldsymbol{v}_i + \boldsymbol{v}_i \otimes \dot{\boldsymbol{v}}_i\right), \end{aligned} \tag{3.140}$$

to prove the following relation

$$\boldsymbol{T}^{(0)} \cdot \dot{\boldsymbol{E}}^{(0)} = \boldsymbol{K} \cdot (\log \boldsymbol{V})^{\boldsymbol{\cdot}} = \boldsymbol{K} \cdot \boldsymbol{D}, \tag{3.141}$$

which shows that the Lagrangian (Eulerian) stress and strain measures $\boldsymbol{T}^{(0)} = \boldsymbol{R}^T\boldsymbol{K}\boldsymbol{R}$ and $\boldsymbol{E}^{(0)}$ ($\boldsymbol{K}$ and $\log \boldsymbol{V}$) are work conjugate.

To better understand the material or spatial nature of the stress measures that we have introduced, let us consider the effects of a rigid-body rotation of the current configuration (Fig. 3.14), represented by the rotation tensor $\boldsymbol{Q} \in \mathsf{Orth}^+$. Since the unit normal $\boldsymbol{n}^*$ and the stress vector $\boldsymbol{s}^*$ in the rotated configuration B^* are related to the corresponding vectors in B via

$$\boldsymbol{n}^* = \boldsymbol{Q}\boldsymbol{n}, \quad \boldsymbol{s}^* = \boldsymbol{Q}\boldsymbol{s}, \tag{3.142}$$

it is concluded from (3.98) that the Cauchy stress transforms as an Eulerian quantity

$$\boldsymbol{T}^* = \boldsymbol{Q}\boldsymbol{T}\boldsymbol{Q}^T, \tag{3.143}$$

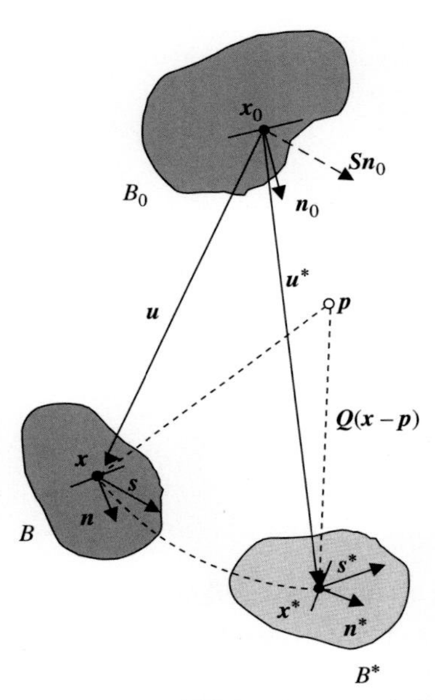

Figure 3.14. Effects of a rigid-body rotation on the stress vector: two final configurations B (with stress vector $\boldsymbol{s}$) and B^* (with stress vector $\boldsymbol{s}^*$). Note that $\boldsymbol{S}\boldsymbol{n}_0$ is parallel to $\boldsymbol{s}$.

expressing the fact that the two stress vectors $\boldsymbol{T}\boldsymbol{n}$ and $\boldsymbol{T}^*\boldsymbol{n}^*$ are related by

$$\boldsymbol{T}^*\boldsymbol{n}^* = \boldsymbol{Q}(\boldsymbol{T}\boldsymbol{n}). \tag{3.144}$$

The transformation rules of kinematic fields for rigid-body rotations of the current configuration are reported in Table 3.1 in the right column, and we can find the transformation laws of the first and second Piola-Kirchhoff tensors

$$\boldsymbol{Q}\boldsymbol{S} \quad \text{and} \quad \boldsymbol{T}^{(2)}, \tag{3.145}$$

showing that the latter remains unchanged as $\boldsymbol{E}^{(2)}$, whereas the former behaves as $\boldsymbol{F}$.

Let us now consider a rigid-body rotation of the reference configuration such as that sketched in Fig. 3.11*a*, whereas the current configuration remains fixed. The Cauchy stress does not change, and all kinematic fields transform according to the rules reported in Table 3.1 in the left column. Therefore, the first and the second Piola-Kirchhoff tensors transform as

$$\boldsymbol{S}\boldsymbol{Q}^T \quad \text{and} \quad \boldsymbol{Q}\boldsymbol{T}^{(2)}\boldsymbol{Q}^T, \tag{3.146}$$

showing that they behave again as $\boldsymbol{F}$ and $\boldsymbol{E}^{(2)}$, respectively.

Finally, since the stress power (3.130) is independent of the rigid-body rotations of the reference and current configurations, it can be concluded that the stress measures are affected by these in the way listed in Table 3.2, where transformations for rigid-body rotations of the reference (spatial) configuration are reported in the left (right) column. It can be noted that $\boldsymbol{S}$ behaves as the deformation gradient, whereas $\boldsymbol{T}^{(m)}$ as $\boldsymbol{E}^{(m)}$, so these are material fields, and $\boldsymbol{Z}^{(m)}$ behaves as $\boldsymbol{G}^{(m)}$, so these are spatial fields.

Table 3.2. *Changes of stress measures for rigid-body rotations of the reference configuration B_0 and of the spatial configuration B*

Rigid-body rotation of B_0 (Fig. 3.11*a*)	Rigid-body rotation of B (Fig. 3.11*b*)
$\boldsymbol{S}\boldsymbol{Q}^T$	$\boldsymbol{Q}\boldsymbol{S}$
$\boldsymbol{T}$	$\boldsymbol{Q}\boldsymbol{T}\boldsymbol{Q}^T$
$\boldsymbol{Q}\boldsymbol{T}^{(m)}\boldsymbol{Q}^T$	$\boldsymbol{T}^{(m)}$
$\boldsymbol{Z}^{(m)}$	$\boldsymbol{Q}\boldsymbol{Z}^{(m)}\boldsymbol{Q}^T$

3.7 Changes of fields for a superimposed rigid-body motion

The way in which spatial, material and two-point tensor fields change when the current configuration suffers a rigid-body rotation has been shown in Tables 3.1 and 3.2. Two observers viewing the same body record the same motion modulo a rigid motion. Two motions $\boldsymbol{x}$ and $\boldsymbol{x}^*$ are related by a rigid-body motion when

$$\boldsymbol{x}^*(\boldsymbol{x}_0,t) = \boldsymbol{c}(t) + \boldsymbol{Q}(t)\,(\boldsymbol{x}(\boldsymbol{x}_0,t) - \boldsymbol{o}), \tag{3.147}$$

where $\boldsymbol{c}$ is a spatial point, $\boldsymbol{o}$ is a spatial fixed origin and $\boldsymbol{Q}$ is a rotation tensor.

Taking the material gradient of Eq. (3.147), we find the same transformation rule $\boldsymbol{F}^* = \boldsymbol{Q}\boldsymbol{F}$ shown in Table 3.1 (right column). From this rule, it is immediate to conclude that all the spatial, material and two-point fields transform with the rules shown in the right columns of Tables 3.1 and 3.2. However, there are important fields not shown in the two tables. We derive these fields below.

Taking the material time derivative of Eq. (3.147), we obtain the transformation rules of velocity in a material description

$$\dot{\boldsymbol{x}}^*(\boldsymbol{x}_0,t) = \dot{\boldsymbol{c}}(t) + \boldsymbol{Q}(t)\dot{\boldsymbol{x}}(\boldsymbol{x}_0,t) + \dot{\boldsymbol{Q}}(t)\,(\boldsymbol{x}(\boldsymbol{x}_0,t) - \boldsymbol{o}), \tag{3.148}$$

and in a spatial description

$$\boldsymbol{v}^* = \dot{\boldsymbol{c}}(t) + \boldsymbol{Q}(t)\,\boldsymbol{v} + \dot{\boldsymbol{Q}}(t)\,(\boldsymbol{x} - \boldsymbol{o}). \tag{3.149}$$

Recalling now Eq. (3.61), we obtain

$$\boldsymbol{L}^* = \dot{\boldsymbol{F}}^*(\boldsymbol{F}^{-1})^* = (\boldsymbol{Q}\boldsymbol{F})^{\boldsymbol{\cdot}}\boldsymbol{F}^{-1}\boldsymbol{Q}^T = \dot{\boldsymbol{Q}}\boldsymbol{Q}^T + \boldsymbol{Q}\dot{\boldsymbol{F}}\boldsymbol{F}^{-1}\boldsymbol{Q}^T, \tag{3.150}$$

so the transformation rule of the velocity gradient becomes

$$\boldsymbol{L}^* = \boldsymbol{Q}\boldsymbol{L}\boldsymbol{Q}^T + \dot{\boldsymbol{Q}}\boldsymbol{Q}^T. \tag{3.151}$$

We note that tensor $\dot{\boldsymbol{Q}}\boldsymbol{Q}^T$ is skew-symmetric because

$$\left(\boldsymbol{Q}\boldsymbol{Q}^T\right)^{\boldsymbol{\cdot}} = \boldsymbol{0} \iff \dot{\boldsymbol{Q}}\boldsymbol{Q}^T = -\boldsymbol{Q}\dot{\boldsymbol{Q}}^T. \tag{3.152}$$

Therefore, by isolating the symmetric and skew-symmetric parts of Eq. (3.151), we obtain the transformation rules

$$\boldsymbol{D}^* = \boldsymbol{Q}\boldsymbol{D}\boldsymbol{Q}^T \qquad \text{and} \qquad \boldsymbol{W}^* = \boldsymbol{Q}\boldsymbol{W}\boldsymbol{Q}^T + \dot{\boldsymbol{Q}}\boldsymbol{Q}^T. \tag{3.153}$$

Let us consider the Kirchhoff stress $\boldsymbol{K}$, which transforms under a superimposed rigid motion as

$$\boldsymbol{K}^* = \boldsymbol{Q}\boldsymbol{K}\boldsymbol{Q}^T. \tag{3.154}$$

Taking the material time derivative on both sides of Eq. (3.154), we find

$$\dot{\boldsymbol{K}}^* = \left(\boldsymbol{Q}\boldsymbol{K}\boldsymbol{Q}^T\right)^{\boldsymbol{\cdot}} = \boldsymbol{Q}\dot{\boldsymbol{K}}\boldsymbol{Q}^T + \dot{\boldsymbol{Q}}\boldsymbol{K}\boldsymbol{Q}^T + \boldsymbol{Q}\boldsymbol{K}\dot{\boldsymbol{Q}}^T. \tag{3.155}$$

Note that Eq. (3.155) holds for every spatial field replacing $\boldsymbol{K}$ and following the rule (3.154), for instance, the Cauchy stress $\boldsymbol{T}$. It is concluded that the material time derivative of a spatial field transforming like (3.154) does not in general transform following the same rule.

Let us now consider the following tensor fields

$$\begin{aligned}
\overset{\triangledown}{\boldsymbol{K}} &= \dot{\boldsymbol{K}} - \boldsymbol{W}\boldsymbol{K} + \boldsymbol{K}\boldsymbol{W}, \\
\overset{\circ}{\boldsymbol{K}} &= \dot{\boldsymbol{K}} - \boldsymbol{L}\boldsymbol{K} - \boldsymbol{K}\boldsymbol{L}^T, \\
\overset{\diamond}{\boldsymbol{K}} &= \dot{\boldsymbol{K}} + \boldsymbol{L}^T\boldsymbol{K} + \boldsymbol{K}\boldsymbol{L},
\end{aligned} \tag{3.156}$$

defining, respectively, the Zaremba-Jaumann or co-rotational (Zaremba, 1903; Jaumann, 1905), Oldroyd (1950), and Cotter-Rivlin (or convected) rates of the Kirchhoff stress. It is a simple exercise to prove that all the preceding fields transform as

$$\overset{*}{\boldsymbol{K}} = \boldsymbol{Q}\overset{*}{\boldsymbol{K}}\boldsymbol{Q}^T, \tag{3.157}$$

where $*$ stands for $\triangledown$, or $\circ$, or $\diamond$. Obviously, there are infinite possible definitions of derivatives (usually called 'objective', but not by us) which transform as Eq. (3.157); see Truesdell and Toupin (1960, sections 148–151).

Finally, the following are useful relations between the rates (3.156) and the material derivative of the first Piola-Kirchhoff stress:

$$\begin{aligned}
\overset{\triangledown}{\boldsymbol{K}} &= \dot{\boldsymbol{S}}\boldsymbol{F}^T - \boldsymbol{W}\boldsymbol{K} + \boldsymbol{K}\boldsymbol{D}, \\
\overset{\circ}{\boldsymbol{K}} &= \dot{\boldsymbol{S}}\boldsymbol{F}^T - \boldsymbol{L}\boldsymbol{K}, \\
\overset{\diamond}{\boldsymbol{K}} &= \dot{\boldsymbol{S}}\boldsymbol{F}^T + 2\boldsymbol{K}\boldsymbol{D} + \boldsymbol{L}^T\boldsymbol{K}.
\end{aligned} \tag{3.158}$$

4 Isotropic non-linear hyperelasticity

We introduce in a simple way the constitutive equations for compressible and incompressible hyperelastic materials, isotropic in their unstressed configuration. Several specific constitutive models are given: the so-called Kirchhoff–Saint Venant, the neo-Hookean and the Mooney-Rivlin materials, as well as the J_2-deformation theory of plasticity and the GBG material.

Elastic behaviour is characterised by the immediate reversibility of the deformation on release of the stress. Though common at small strain, this behaviour becomes 'rare' for materials subjected to large strain. In fact, elastic strain is limited to 1% for crystalline materials and amorphous materials in their rigid state and decreases to 0.1% and less for steel and to 0.001% for granular materials. In practice, the only materials behaving elastically at large strain are rubber, where extensibility can reach 500% to 1000%, and biological soft tissues[1]. However, our interest in elastic modelling is not only limited to materials really behaving elastically; rather, it is also important to describe the loading branch of the constitutive behaviour of elastoplastic materials (roughly speaking, the behaviour exhibited when unloading is never involved). We will see, in fact, that bifurcation and instability analyses for elastoplastic materials are usually reduced to the analysis of so-called elastic comparison solids (Chapter 10).

The objective of this chapter is to introduce the constitutive equations for elastic solids *isotropic in their unloaded configuration* in the simplest way, deferring the detailed treatment of elastic anisotropy and of a general constitutive framework to Chapter 6. We present the specific constitutive equations for several elastic materials that will be used in future applications. These are the so-called Kirchhoff–Saint Venant, neo-Hookean, Mooney-Rivlin, J_2-deformation theory of plasticity, and GBG materials. We will see that each of these models has its own 'merits'. For instance, the Kirchhoff–Saint Venant and neo-Hookean materials are very simple, the J_2-deformation theory material is tailored to describe the loading branch of metals whereas the GBG material has been invented to model the loading branch of quasi-brittle materials, such as, for instance, rock, soil and concrete.

[1] Soft biological tissues usually are fluid saturated, so their behaviour is complicated by the hydrated nature of the material; see, e.g., Franceschini et al. (2006).

A material is called 'hyperelastic' or 'Green elastic' if a strain energy density W exists, function of a symmetric strain measure $\boldsymbol{E}^{(m)}$, such that the stress measure $\boldsymbol{T}^{(m)}$ can be obtained as

$$\boldsymbol{T}^{(m)} = \frac{\partial W(\boldsymbol{E}^{(m)})}{\partial \boldsymbol{E}^{(m)}}. \tag{4.1}$$

Therefore, a strain energy function relates work conjugate stress/strain measures so that, if such a function exists, the stress power (3.129) can be expressed, employing Eq. (3.133), as the material time derivative of the strain energy

$$\int_{P_0} \boldsymbol{S} \cdot \dot{\boldsymbol{F}} = \int_{P_0} \boldsymbol{T}^{(m)} \cdot \dot{\boldsymbol{E}}^{(m)} = \int_{P_0} \frac{\partial W(\boldsymbol{E}^{(m)})}{\partial \boldsymbol{E}^{(m)}} \cdot \dot{\boldsymbol{E}}^{(m)} = \frac{d}{dt} \int_{P_0} W. \tag{4.2}$$

As a conclusion, the *theorem of power expended (3.129) becomes for a hyperelastic material* (in the referential version)

$$\int_{\partial P_0} \dot{\boldsymbol{x}} \cdot \boldsymbol{S} \boldsymbol{n}_0 + \int_{P_0} \boldsymbol{b}_0 \cdot \dot{\boldsymbol{x}} = \frac{d}{dt} \int_{P_0} \left(W + \frac{\rho_0}{2} \dot{\boldsymbol{x}}^2 \right). \tag{4.3}$$

Note that Eq. (3.133) implies that we also can define the energy function directly to obtain the first Piola-Kirchhoff stress tensor upon derivation with respect to the deformation gradient as

$$\boldsymbol{S} = \frac{\partial \tilde{W}(\boldsymbol{F})}{\partial \boldsymbol{F}}. \tag{4.4}$$

The relation between the strain energies (4.1) and (4.4) can be obtained for a given m. For instance, for $m = 2$ and using the chain rule of differentiation, we may write

$$\boldsymbol{S} = \frac{\partial \tilde{W}(\boldsymbol{F})}{\partial \boldsymbol{F}} = \left(\frac{\partial \boldsymbol{E}^{(2)}}{\partial \boldsymbol{F}} \right)^T \left[\frac{\partial W(\boldsymbol{E}^{(2)})}{\partial \boldsymbol{E}^{(2)}} \right] = \boldsymbol{F} \frac{\partial W(\boldsymbol{E}^{(2)})}{\partial \boldsymbol{E}^{(2)}} = \boldsymbol{F} \boldsymbol{T}^{(2)}. \tag{4.5}$$

4.1 Isotropic compressible hyperelastic material

All choices of strain measure $\boldsymbol{E}^{(m)}$ *are equivalent* in defining an elastic potential, Eq. (4.1), but it is simpler to proceed referring to the logarithmic strain $\boldsymbol{E}^{(0)}$ so that

$$\boldsymbol{T}^{(0)} = \frac{\partial W(\boldsymbol{E}^{(0)})}{\partial \boldsymbol{E}^{(0)}}. \tag{4.6}$$

Note that every symmetric tensor, and therefore also $\boldsymbol{E}^{(0)}$, is defined by six quantities: its three eigenvalues and the Euler angles, singling out its three eigenvectors. For a *material isotropic in the reference, unloaded, configuration*, it becomes spontaneous to postulate that the strain energy function W, defined through Eq. (4.6), cannot depend on 'directional properties' of $\boldsymbol{E}^{(0)}$ but only on its eigenvalues, the principal logarithmic strains. Thus

$$W = W(\varepsilon_1, \varepsilon_2, \varepsilon_3) \tag{4.7}$$

a function that has to be indifferent to every permutation of ε_1, ε_2 and ε_3 (otherwise isotropy would be violated). Note that Eq. (4.7) is indifferent to rigid-body rotations of the reference or of the current configuration.

Combining Eq. (4.6) with Eq. (4.7) and using the chain rule of differentiation, we obtain

$$\boldsymbol{T}^{(0)} = \frac{\partial W}{\partial \varepsilon_1}\frac{\partial \varepsilon_1}{\partial \boldsymbol{E}^{(0)}} + \frac{\partial W}{\partial \varepsilon_2}\frac{\partial \varepsilon_2}{\partial \boldsymbol{E}^{(0)}} + \frac{\partial W}{\partial \varepsilon_3}\frac{\partial \varepsilon_3}{\partial \boldsymbol{E}^{(0)}}, \tag{4.8}$$

which, since ε_i are the eigenvalues of $\boldsymbol{E}^{(0)}$, employing the differentiation rule of eigenvalues (2.127), yields

$$\boldsymbol{T}^{(0)} = \frac{\partial W}{\partial \varepsilon_1}\boldsymbol{u}_1 \otimes \boldsymbol{u}_1 + \frac{\partial W}{\partial \varepsilon_2}\boldsymbol{u}_2 \otimes \boldsymbol{u}_2 + \frac{\partial W}{\partial \varepsilon_3}\boldsymbol{u}_3 \otimes \boldsymbol{u}_3, \tag{4.9}$$

where the strain potential alternatively can be expressed as a function of the principal stretches $\hat{W}(\lambda_1,\lambda_2,\lambda_3)$. Thus

$$\frac{\partial W}{\partial \varepsilon_i} = \lambda_i \frac{\partial \hat{W}}{\partial \lambda_i}, \qquad i=1,2,3 \quad \text{not summed.} \tag{4.10}$$

Equation (4.9) shows that $\boldsymbol{T}^{(0)}$ is coaxial with $\boldsymbol{E}^{(0)}$, so $\boldsymbol{T}^{(0)}$ can be identified with the so-called rotated stress, and Eq. (3.136) applies, yielding the expression for the Kirchhoff stress [Eq. (3.108)] as

$$\boldsymbol{K} = \frac{\partial W}{\partial \varepsilon_1}\boldsymbol{v}_1 \otimes \boldsymbol{v}_1 + \frac{\partial W}{\partial \varepsilon_2}\boldsymbol{v}_2 \otimes \boldsymbol{v}_2 + \frac{\partial W}{\partial \varepsilon_3}\boldsymbol{v}_3 \otimes \boldsymbol{v}_3. \tag{4.11}$$

On application of the definition of first Piola-Kirchhoff stress, Eq. (3.106), with the representation (3.17) for $\boldsymbol{F}^{-T}$, allows us to write

$$\boldsymbol{S} = \frac{1}{\lambda_1}\frac{\partial W}{\partial \varepsilon_1}\boldsymbol{v}_1 \otimes \boldsymbol{u}_1 + \frac{1}{\lambda_2}\frac{\partial W}{\partial \varepsilon_2}\boldsymbol{v}_2 \otimes \boldsymbol{u}_2 + \frac{1}{\lambda_3}\frac{\partial W}{\partial \varepsilon_3}\boldsymbol{v}_3 \otimes \boldsymbol{u}_3, \tag{4.12}$$

showing again that the first Piola-Kirchhoff stress is a two-point tensor. Finally, we can employ the definition of the second Piola-Kirchhoff stress and the representation (3.16) to obtain

$$\boldsymbol{T}^{(2)} = \frac{1}{\lambda_1^2}\frac{\partial W}{\partial \varepsilon_1}\boldsymbol{u}_1 \otimes \boldsymbol{u}_1 + \frac{1}{\lambda_2^2}\frac{\partial W}{\partial \varepsilon_2}\boldsymbol{u}_2 \otimes \boldsymbol{u}_2 + \frac{1}{\lambda_3^2}\frac{\partial W}{\partial \varepsilon_3}\boldsymbol{u}_3 \otimes \boldsymbol{u}_3. \tag{4.13}$$

4.1.1 Kirchhoff–Saint Venant material

An example of an isotropic elastic material is provided by the so-called Kirchhoff–Saint Venant model (Truesdell and Noll, 1965, see also Holzapfel, 2000), in which the strain energy is defined as

$$W(\boldsymbol{E}^{(2)}) = \frac{\lambda}{2}\mathrm{tr}^2\boldsymbol{E}^{(2)} + \mu \boldsymbol{E}^{(2)} \cdot \boldsymbol{E}^{(2)}, \tag{4.14}$$

a function of the Green-Lagrange strain, so the second Piola-Kirchhoff stress is given by

$$\boldsymbol{T}^{(2)} = \frac{\partial W}{\partial \boldsymbol{E}^{(2)}} = \lambda \left(\mathrm{tr}\boldsymbol{E}^{(2)}\right)\boldsymbol{I} + 2\mu \boldsymbol{E}^{(2)}, \tag{4.15}$$

where λ and μ are material constants playing a role similar to the Lamé constants of the linear elasticity theory, which can be assumed for simplicity to be subject to the same restrictions valid in that theory ($\mu > 0$, and $3\lambda + 2\mu > 0$, see Section 2.20.7).

Employing the definitions of second and first Piola-Kirchhoff stress tensors [Eqs. (3.106) and (3.132)], we may transform Eq. (4.15) into the equivalent versions

$$\begin{aligned} \boldsymbol{S} &= \frac{\lambda}{2}(\boldsymbol{F}\cdot\boldsymbol{F}-3)\boldsymbol{F}+\mu\left(\boldsymbol{F}\boldsymbol{F}^T\boldsymbol{F}-\boldsymbol{F}\right), \\ \boldsymbol{K} &= \frac{\lambda}{2}(\mathrm{tr}\boldsymbol{B}-3)\boldsymbol{B}+\mu\left(\boldsymbol{B}^2-\boldsymbol{B}\right). \end{aligned} \tag{4.16}$$

Using the spectral representation of $\boldsymbol{B}$, Eq. $(4.16)_2$ can be rewritten as

$$\boldsymbol{K}=\sum_{i=1}^{3}\left[\frac{\lambda}{2}\left(\lambda_1^2+\lambda_2^2+\lambda_3^2-3\right)+\mu\left(\lambda_i^2-1\right)\right]\lambda_i^2\,\boldsymbol{v}_i\otimes\boldsymbol{v}_i, \tag{4.17}$$

so the strain energy potential $W(\lambda_i)$ results in the form

$$W(\lambda_i)=\frac{\lambda}{4}\left[\frac{1}{2}\left(\lambda_1^2+\lambda_2^2+\lambda_3^2\right)^2-3\left(\lambda_1^2+\lambda_2^2+\lambda_3^2\right)+\frac{9}{2}\right]+\frac{\mu}{4}\sum_{i=1}^{3}\left(\lambda_i^2-1\right)^2, \tag{4.18}$$

from which we can immediately obtain representation (4.7) by using the transformation $\lambda_i=\exp(\varepsilon_i)$.

4.2 Incompressible isotropic elasticity

Incompressible elasticity is an important model for finitely strained materials. In fact, we mention that

- Rubber and biological soft tissues, which are important materials capable of large elastic strain, are nearly incompressible.
- Finite strain elasticity is often used as a model for the plastic branch of severely deformed ductile metals. For these materials, the plastic deformation is incompressible and dominates the elastic deformation (which remains small), so incompressible elasticity again can be used as a sound model (however, with such a model, it evidently will be impossible to analyse situations involving unloading).
- The equations governing deformations of incompressible elasticity are in some cases easier to handle for obtaining analytical solutions [although from a numerical point of view, incompressibility is a problem, which can be addressed using various techniques, for instance, mixed finite elements (Auricchio et al., 2004) or boundary element methods (Brun et al., 2003*a*, 2003*b*; Bigoni et al., 2007)].

The incompressibility constraint in terms of logarithmic strains

$$\varepsilon_1+\varepsilon_2+\varepsilon_3=0 \tag{4.19}$$

enters the formulation of the constitutive equation as a Lagrangean multiplier, so Eq. (4.6) is replaced by

$$\boldsymbol{T}^{(0)}=-\pi\frac{\partial(\varepsilon_1+\varepsilon_2+\varepsilon_3)}{\partial\boldsymbol{E}^{(0)}}+\frac{\partial W(\boldsymbol{E}^{(0)})}{\partial\boldsymbol{E}^{(0)}}, \tag{4.20}$$

where π is an arbitrary Lagrangean multiplier, playing the role of the mean pressure, which remains 'constitutively indeterminate' in an incompressible material. The rule of differentiation of the eigenvalues of a tensor with respect to the tensor itself (2.127) allows us to calculate

$$\frac{\partial(\varepsilon_1+\varepsilon_2+\varepsilon_3)}{\partial \boldsymbol{E}^{(0)}} = \sum_{i=1}^{3} \boldsymbol{u}_i \otimes \boldsymbol{u}_i = \boldsymbol{I}, \tag{4.21}$$

so Eq. (4.20) can be rewritten as

$$\boldsymbol{T}^{(0)} = -\pi \boldsymbol{I} + \frac{\partial W(\boldsymbol{E}^{(0)})}{\partial \boldsymbol{E}^{(0)}}, \qquad \sum_{i=1}^{3} \varepsilon_i = 0. \tag{4.22}$$

Invoking isotropy of the response, we can proceed as for the compressible case to obtain (note that now, owing to incompressibility, $J = 1$, and there is no difference between Kirchhoff and Cauchy stresses, i.e., $\boldsymbol{T} = \boldsymbol{K}$) the following expressions:

$$\begin{aligned}
\boldsymbol{T}^{(0)} &= -\pi \boldsymbol{I} + \frac{\partial W}{\partial \varepsilon_1}\boldsymbol{u}_1 \otimes \boldsymbol{u}_1 + \frac{\partial W}{\partial \varepsilon_2}\boldsymbol{u}_2 \otimes \boldsymbol{u}_2 + \frac{\partial W}{\partial \varepsilon_3}\boldsymbol{u}_3 \otimes \boldsymbol{u}_3,\\
\boldsymbol{T} &= -\pi \boldsymbol{I} + \frac{\partial W}{\partial \varepsilon_1}\boldsymbol{v}_1 \otimes \boldsymbol{v}_1 + \frac{\partial W}{\partial \varepsilon_2}\boldsymbol{v}_2 \otimes \boldsymbol{v}_2 + \frac{\partial W}{\partial \varepsilon_3}\boldsymbol{v}_3 \otimes \boldsymbol{v}_3,\\
\boldsymbol{S} &= -\pi \boldsymbol{F}^{-T} + \frac{1}{\lambda_1}\frac{\partial W}{\partial \varepsilon_1}\boldsymbol{v}_1 \otimes \boldsymbol{u}_1 + \frac{1}{\lambda_2}\frac{\partial W}{\partial \varepsilon_2}\boldsymbol{v}_2 \otimes \boldsymbol{u}_2 + \frac{1}{\lambda_3}\frac{\partial W}{\partial \varepsilon_3}\boldsymbol{v}_3 \otimes \boldsymbol{u}_3,\\
\boldsymbol{T}^{(2)} &= -\pi \boldsymbol{U}^{-2} + \frac{1}{\lambda_1^2}\frac{\partial W}{\partial \varepsilon_1}\boldsymbol{u}_1 \otimes \boldsymbol{u}_1 + \frac{1}{\lambda_2^2}\frac{\partial W}{\partial \varepsilon_2}\boldsymbol{u}_2 \otimes \boldsymbol{u}_2 + \frac{1}{\lambda_3^2}\frac{\partial W}{\partial \varepsilon_3}\boldsymbol{u}_3 \otimes \boldsymbol{u}_3,
\end{aligned} \tag{4.23}$$

to be complemented by the incompressibility constraint (4.19).

We remark, again, that instead of expressing the strain energy as a function of the principal logarithmic strains, we can express it as a function of the principal stretches [Eq. (4.10)] to obtain

$$T_i = -\pi + \lambda_i \frac{\partial \hat{W}(\lambda_1,\lambda_2,\lambda_3)}{\partial \lambda_i}, \quad i = 1,2,3 \quad \text{not summed}, \quad \lambda_1\lambda_2\lambda_3 = 1. \tag{4.24}$$

Moreover, we can eliminate the Lagrangean multiplier π by providing the differences in the principal stresses. In this way, we find the following often employed form of constitutive equations (Mooney, 1940):

$$T_i - T_j = \lambda_i \frac{\partial \hat{W}(\lambda_1,\lambda_2,\lambda_3)}{\partial \lambda_i} - \lambda_j \frac{\partial \hat{W}(\lambda_1,\lambda_2,\lambda_3)}{\partial \lambda_j} \quad i,j = 1,2,3 \quad \text{not summed}. \tag{4.25}$$

Several models can be given as particular cases of the above-formulated constitutive equations. The following constitutive models are particularly important for later developments.

4.2.1 Mooney-Rivlin elasticity

The Mooney-Rivlin material, proposed by Mooney (1940), is defined through the strain energy density function

$$W(\varepsilon_i) = \frac{\mu_1}{2}\left(e^{2\varepsilon_1} + e^{2\varepsilon_2} + e^{2\varepsilon_3} - 3\right) - \frac{\mu_2}{2}\left(e^{-2\varepsilon_1} + e^{-2\varepsilon_2} + e^{-2\varepsilon_3} - 3\right)$$

or

$$W(\lambda_i) = \frac{\mu_1}{2}\left(\lambda_1^2+\lambda_2^2+\lambda_3^2-3\right) - \frac{\mu_2}{2}\left(\lambda_1^{-2}+\lambda_2^{-2}+\lambda_3^{-2}-3\right), \tag{4.26}$$

where μ_1 and μ_2 represent two *constant* moduli, and we have to add the incompressibility constraint $\varepsilon_1+\varepsilon_2+\varepsilon_3=0$ or $\lambda_1\lambda_2\lambda_3=1$, complementing Eqs. (4.26).

Applying Eq. $(4.23)_2$ and the definition of left stretch tensor (3.12), we obtain

$$\boldsymbol{T} = -\pi\boldsymbol{I} + \mu_1\boldsymbol{B} + \mu_2\boldsymbol{B}^{-1}. \tag{4.27}$$

We note that from Eq. (4.27) and the definition of first Piola-Kirchhoff stress (3.106), we may obtain

$$\boldsymbol{S} = -\pi\boldsymbol{F}^{-T} + \mu_1\boldsymbol{F} + \mu_2\boldsymbol{F}^{-T}\boldsymbol{F}^{-1}\boldsymbol{F}^{-T}. \tag{4.28}$$

Thus, since

$$\frac{\partial \boldsymbol{F}^{-1}\boldsymbol{\cdot}\boldsymbol{F}^{-1}}{\partial \boldsymbol{F}} = -2\boldsymbol{F}^{-T}\boldsymbol{F}^{-1}\boldsymbol{F}^{-T} \quad \text{and} \quad \frac{\partial \boldsymbol{F}\boldsymbol{\cdot}\boldsymbol{F}}{\partial \boldsymbol{F}} = 2\boldsymbol{F}, \tag{4.29}$$

we arrive at the expression of the strain energy written as a function of the deformation gradient

$$\tilde{W}(\boldsymbol{F}) = \frac{\mu_1}{2}\boldsymbol{F}\boldsymbol{\cdot}\boldsymbol{F} - \frac{\mu_2}{2}\boldsymbol{F}^{-1}\boldsymbol{\cdot}\boldsymbol{F}^{-1}, \tag{4.30}$$

which allows us to write

$$\boldsymbol{S} = -\pi\boldsymbol{F}^{-T} + \frac{\partial \tilde{W}(\boldsymbol{F})}{\partial \boldsymbol{F}}. \tag{4.31}$$

An infinitesimal shear deformation of amplitude γ, parallel to axes $\boldsymbol{e}_1$ and $\boldsymbol{e}_2$, and applied to an unloaded solid, can be expressed in terms of left Cauchy-Green deformation tensor and its inverse as (see Section 5.4 for details)

$$\boldsymbol{B} = \boldsymbol{I} + \gamma(\boldsymbol{e}_1\otimes\boldsymbol{e}_2 + \boldsymbol{e}_2\otimes\boldsymbol{e}_1) \quad \text{and} \quad \boldsymbol{B}^{-1} \approx \boldsymbol{I} - \gamma(\boldsymbol{e}_1\otimes\boldsymbol{e}_2 + \boldsymbol{e}_2\otimes\boldsymbol{e}_1), \tag{4.32}$$

which substituted into Eq. (4.27) yields for the shear stress T_{12}

$$T_{12} = \gamma(\mu_1 - \mu_2), \tag{4.33}$$

so we conclude that $\mu_1-\mu_2$ is the shear modulus in the undeformed configuration, which has to be strictly positive

$$\mu_1 - \mu_2 > 0. \tag{4.34}$$

A substitution of the stretches $\lambda_1=\lambda_2=\lambda$ and $\lambda_3 = 1/\lambda^2$, representing an incompressible deformation, into Eq. $(4.26)_2$, that is,

$$W(\lambda) = \frac{\mu_1}{2}\left(\frac{2\lambda^6-3\lambda^4+1}{\lambda^4}\right) - \frac{\mu_2}{2}\left(\frac{\lambda^8-3\lambda^4+2\lambda^2}{\lambda^4}\right), \tag{4.35}$$

yields the conclusion that for $\lambda \longrightarrow 0$, the strain energy W has the sign of μ_1, whereas for $\lambda \longrightarrow \infty$, the strain energy W has the sign of $-\mu_2$. Therefore, if we want the energy (4.26) to be definite positive, we have to conclude that $\mu_1 \geq 0$ and $\mu_2 \leq 0$. As an example, we report that the values

$$\mu_1 = 3 \text{ bar} \quad \text{and} \quad \mu_2 = -0.3 \text{ bar} \tag{4.36}$$

provide an excellent description of the behaviour of rubber at room temperature (Müller and Strehlow, 2004).

4.2.2 Neo-Hookean elasticity

The neo-Hookean material, proposed by Rivlin (1948a), is a special case of the Mooney-Rivlin material obtained by setting $\mu_1 = \mu_0$ and $\mu_2 = 0$ (and modifying the definition of the Lagrangean multiplier π), yielding

$$\boldsymbol{T} = -\pi \boldsymbol{I} + \mu_0 (\boldsymbol{B} - \boldsymbol{I}), \tag{4.37}$$

where μ_0 is a constant representing the shear stiffness in the unstressed configuration.

Note that for plane strain isochoric deformation (with out-of-plane direction singled out by the unit vector $\boldsymbol{v}_3$), the left Cauchy-Green strain tensor $\boldsymbol{B}$ takes the form

$$\boldsymbol{B} = \lambda^2 \boldsymbol{v}_1 \otimes \boldsymbol{v}_1 + \frac{1}{\lambda^2} \boldsymbol{v}_2 \otimes \boldsymbol{v}_2 + \boldsymbol{v}_3 \otimes \boldsymbol{v}_3, \tag{4.38}$$

so Eqs. (4.27) and (4.37), employing the arbitrariness of π, can be expressed in the same form. Therefore, *there is no difference between Mooney-Rivlin and neo-Hookean constitutive modelling under the plane strain constraint.*

4.2.3 J_2-Deformation theory of plasticity

The J_2-deformation theory of plasticity was introduced by Hutchinson and Neale (1979) (see also Hutchinson and Tvergaard, 1980; Neale, 1981) to model the plastic response of metals subject to monotonically increasing loading.

The constitutive law for the J_2-deformation theory of plasticity can be expressed as

$$T_i = \frac{2}{3} E_s \varepsilon_i + p; \qquad i = 1,2,3, \quad \varepsilon_1 + \varepsilon_2 + \varepsilon_3 = 0, \tag{4.39}$$

where $\varepsilon_i = \log \lambda_i$ are the logarithmic strains (3.33) and $p = \mathrm{tr}\boldsymbol{T}/3$. In Eq. (4.39), E_s is the secant modulus to the curve representing the effective stress σ_e versus effective strain ε_e:

$$\sigma_e = \sqrt{\frac{3}{2} (\mathrm{dev}T_1^2 + \mathrm{dev}T_2^2 + \mathrm{dev}T_3^2)}, \qquad \varepsilon_e = \sqrt{\frac{2}{3} (\varepsilon_1^2 + \varepsilon_2^2 + \varepsilon_3^2)}, \tag{4.40}$$

where $\mathrm{dev}T_i$ are the principal components of deviatoric Cauchy stress (2.34)

$$\mathrm{dev}T_i = T_i - \frac{T_1 + T_2 + T_3}{3}. \tag{4.41}$$

The σ_e–ε_e curve is assumed to be determined by

$$E_s = K \varepsilon_e^{N-1}, \tag{4.42}$$

where $N \in (0,1]$ is a strain hardening exponent, and K is a positive constitutive parameter with the dimension of stress. The strain energy function results therefore are to be given by

$$W = \frac{K}{N+1} \varepsilon_e^{N+1}. \tag{4.43}$$

Note that the strain energy (4.43) is a convex function (see Section 2.19) of the logarithmic strains ε_i $(i = 1,2,3)$.

In the plane strain version of the model (used, among others, by Bigoni et al., 1997; Radi et al., 2002; Bigoni et al., 2007), $\varepsilon_3 = 0$, Eq. (4.25) gives

$$T_1 - T_2 = \frac{dW}{d\varepsilon_e}\left(\lambda_1 \frac{\partial \varepsilon_e}{\partial \varepsilon_1}\frac{\partial \varepsilon_1}{\partial \lambda_1} - \lambda_2 \frac{\partial \varepsilon_e}{\partial \varepsilon_2}\frac{\partial \varepsilon_2}{\partial \lambda_2}\right), \tag{4.44}$$

which can be transformed to

$$T_1 - T_2 = \frac{2(\varepsilon_1 - \varepsilon_2)}{3\varepsilon_e}\frac{dW}{d\varepsilon_e}. \tag{4.45}$$

Keeping into account the plane strain and incompressibility constraints ($\varepsilon_3 = 0$, $\varepsilon_1 = -\varepsilon_2$) and $\lambda_1 = 1/\lambda_2 = \lambda$, the effective stress and strain reduce to

$$\sigma_e = \frac{\sqrt{3}}{2}(T_1 - T_2) \qquad \text{and} \qquad \varepsilon_e = |\log\lambda|\frac{2}{\sqrt{3}}, \tag{4.46}$$

whereas Eq. (4.45) becomes

$$T_1 - T_2 = \frac{2\varepsilon_1}{\sqrt{3}|\varepsilon_1|}\frac{\partial W}{\partial \varepsilon_e}. \tag{4.47}$$

Finally, we arrive at

$$T_1 - T_2 = K\left(\frac{2}{\sqrt{3}}\right)^{N+1} |\log\lambda|^{N-1}\log\lambda, \tag{4.48}$$

from which the behaviour in uniaxial plane strain tension/compression is determined by taking $T_2 = 0$ (see Chapter 9).

Note that the stress response of the J_2-deformation theory material is singular at the unloaded state $\lambda = 1$, where the tangent stiffness blows up to infinity. For this reason, although the material is isotropic in its unloaded state, this isotropy is immediately lost in a discontinuous way during every loading program starting from the unloaded state (a fact that will become clear in Section 12.2.1, where it will be shown that the incremental anisotropy parameter ξ does not tend to 1—the value corresponding to isotropy—when the material is unloaded, $k = 0$).

4.2.4 The GBG model

The model proposed by Gei, Bigoni and Guicciardi (2004), referred as the 'GBG model' in the following, is similar to the J_2-deformation theory of plasticity but is tailored to produce a uniaxial Cauchy stress curve with a stiff behaviour ending in a peak followed by a 'tunable' softening. The model has been proposed to be employed for frictional-cohesive materials, evidencing softening behaviour (see also Chapter 5).

With reference to the effective logarithmic strain ε_e [Eq. (4.40)$_2$], the strain energy is defined as

$$W = Kc\,\varepsilon_0\left[\frac{1}{c+1}\exp\left(-\frac{\varepsilon_e}{\varepsilon_0}\right) - 1\right]\exp\left(-\frac{\varepsilon_e}{c\,\varepsilon_0}\right), \tag{4.49}$$

where the values $K > 0$, $c > 0$ and $\varepsilon_0 > 0$ are positive material parameters.[2]

[2] For instance, the values $K = 1680$ MPa, $c = 42$, and $\varepsilon_0 = 0.0045$ are suited to represent the behaviour of silicon nitride at 1,200°C analysed by Gei et al. (2004), see Fig. 1.32.

Note that, differently from the J_2-deformation theory material, the potential energy (4.49) is not a convex function (see Section 2.19) of the logarithmic strains ε_i $(i = 1,2,3)$.

The principal components of Cauchy stress can be obtained from Eq. (4.25), which in this case still takes the form (4.44) or (4.45). For plane strain and incompressibility ($\varepsilon_3 = 0$, $\varepsilon_1 = -\varepsilon_2$, $\lambda_1 = 1/\lambda_2 = \lambda$), the effective stress again is given by Eqs. $(4.46)_2$ and (4.47) still hold, so we arrive at the expression for the principal stress difference in plane strain:

$$T_1 - T_2 = \frac{2}{\sqrt{3}} K \operatorname{sign}(\varepsilon_1) \left[1 - \exp\left(-\frac{2|\varepsilon_1|}{\sqrt{3}\varepsilon_0} \right) \right] \exp\left(-\frac{2|\varepsilon_1|}{c\sqrt{3}\varepsilon_0} \right). \tag{4.50}$$

In the particular (but important for our subsequent calculations) case of uniaxial tension/compression of a cylinder, the states of strain and stress are axial-symmetric. In a cylindrical coordinate system (r, θ, z) with z coincident with the axis of the cylinder, the Eulerian logarithmic strain $\boldsymbol{G}^{(0)}$ [defined by Eq. (3.31) with $m = 0$] and the Cauchy stress $\boldsymbol{T}$ are represented as

$$\boldsymbol{G}^{(0)} = -\varepsilon/2\,(\boldsymbol{e}_r \otimes \boldsymbol{e}_r + \boldsymbol{e}_\theta \otimes \boldsymbol{e}_\theta) + \varepsilon \boldsymbol{e}_z \otimes \boldsymbol{e}_z, \qquad \boldsymbol{T} = T_z \boldsymbol{e}_z \otimes \boldsymbol{e}_z, \tag{4.51}$$

where $\varepsilon < 0$ (> 0) in compression (in tension) and $\{\boldsymbol{e}_r, \boldsymbol{e}_\theta, \boldsymbol{e}_z\}$ are the orthonormal basis associated with the cylindrical system just introduced.

In the present setting, the form (4.51) of $\boldsymbol{G}^{(0)}$ leads to

$$\varepsilon_e = |\varepsilon|, \tag{4.52}$$

so for non-linear incompressible isotropic materials undergoing an axial-symmetric deformation, the Cauchy stress T_z simply can be derived from the relationship (4.25), which now turns out to be given by

$$T_z - T_r = \frac{\partial W}{\partial \varepsilon} = \frac{dW}{d\varepsilon_e} \frac{\varepsilon}{\varepsilon_e}, \tag{4.53}$$

where the lateral stress T_r vanishes in this case. Therefore, we obtain from Eq. (4.49)

$$T_z = K \frac{\varepsilon}{\varepsilon_e} \left(1 - e^{-\varepsilon_e/\varepsilon_0} \right) e^{-\varepsilon_e/(c\varepsilon_0)}, \tag{4.54}$$

or, in term of nominal stress S_z,

$$S_z = T_z e^{-\varepsilon}. \tag{4.55}$$

Note that taking the limit for ε tending to zero in Eq. (4.54), we obtain

$$\frac{T_z}{\varepsilon} = \frac{K}{\varepsilon_0}, \tag{4.56}$$

playing the role of an initial Young modulus.

The behaviour in uniaxial compression turns out to be identical to that in tension, a circumstance not fully consistent with experimental observations on frictional cohesive materials but yielding a substantial simplification. Moreover, we can note that the peaks in the nominal stress versus logarithmic strain curve in uniaxial tension or compression are given by

$$\varepsilon_{\text{peak nominal}} = \pm \varepsilon_0 \log\left(1 + \frac{c}{1 \pm c\varepsilon_0} \right), \qquad \begin{cases} + & \text{for tensile stress,} \\ - & \text{for compressive stress,} \end{cases} \tag{4.57}$$

which always has a solution for tension but can have no solution for uniaxial compression. Therefore, the nominal stress/logarithmic strain curve may not present a maximum (occurring when the argument of the logarithm is negative, $1+c/(1-c\varepsilon_0)<0$). On the other hand, a peak in the Cauchy (true) stress versus logarithmic strain is always attained at

$$\varepsilon_{\text{peak true}} = \pm\varepsilon_0 \log(1+c)\,, \qquad \begin{cases} + & \text{for tensile stress,} \\ - & \text{for compressive stress,} \end{cases} \tag{4.58}$$

so $\varepsilon_{\text{peak nominal}} < \varepsilon_{\text{peak true}}$ for tension and compression (note that the values of strain are negative in compression). Examples of uniaxial stress/strain behaviour are postponed to Section 5.3.

5 Solutions of simple problems in finitely deformed non-linear elastic solids

Simple elastic quasi-static problems involving large deformations are solved, which will become useful later for bifurcation analyses. These are (1) uniaxial tension/compression of a rectangular elastic block deformed in plane strain, (2) uniaxial tension/compression of an incompressible elastic cylinder (with circular cross section), (3) simple shear of an elastic block, and (4) plane strain bending of an elastic block.

We address a few simple problems of large deformation of elastic materials with the aim of providing simple examples in which the non-linear behaviour related both to the non-linearity introduced in the constitutive description and in the treatment of large deformation effects can be appreciated and easily observed. The solutions of plane strain and axial-symmetric uniaxial tension/compression and simple shear are fundamental for comparisons with experiments, whereas the solution of plane strain flexure of a block is interesting in view of several engineering applications. These solutions also will be used in Chapters 12 and 13 as reference states for incremental bifurcation analyses.

5.1 Uniaxial plane strain tension and compression of an incompressible elastic block

With reference to Fig. 5.1, we consider a plane strain problem of an incompressible elastic block subject to uniaxial tensile or compressive stress in the direction $\boldsymbol{e}_1$, the unit vector defining the x_1 axis.

Kinematics We begin with the kinematics, which is an elongation parallel and a contraction orthogonal to the $\boldsymbol{e}_1$ axis, so the deformation is given by

$$\boldsymbol{x}-\boldsymbol{o}=\left(\frac{h}{h_0}\boldsymbol{e}_1\otimes\boldsymbol{e}_1+\frac{h_0}{h}\boldsymbol{e}_2\otimes\boldsymbol{e}_2+\boldsymbol{e}_3\otimes\boldsymbol{e}_3\right)(\boldsymbol{x}_0-\boldsymbol{o}), \tag{5.1}$$

and its gradient is

$$\boldsymbol{F}=\boldsymbol{U}=\boldsymbol{V}=\frac{h}{h_0}\boldsymbol{e}_1\otimes\boldsymbol{e}_1+\frac{h_0}{h}\boldsymbol{e}_2\otimes\boldsymbol{e}_2+\boldsymbol{e}_3\otimes\boldsymbol{e}_3, \qquad \boldsymbol{R}=\boldsymbol{I}. \tag{5.2}$$

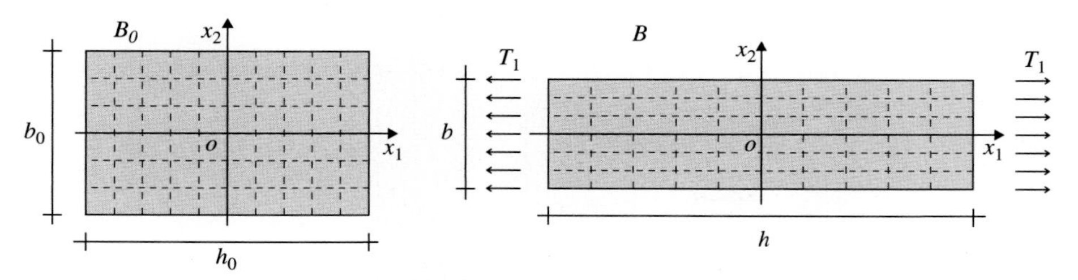

Figure 5.1. Plane strain elongation (contraction) in the x_1 (x_2) direction of an incompressible elastic material.

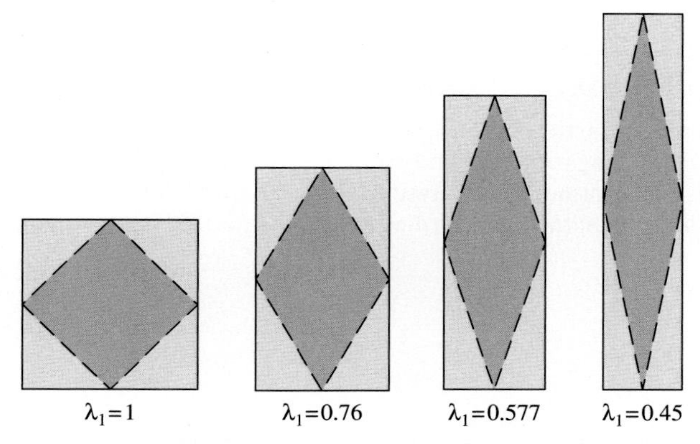

Figure 5.2. Equivalence for an incompressible material of elongation (contraction) in the x_2 (x_1) direction and pure shear in a reference system rotated 45° with respect to the x_1–x_2 reference system.

We can easily calculate the Cauchy-Green deformation tensors

$$\boldsymbol{B}=\boldsymbol{C}=\frac{h^2}{h_0^2}\boldsymbol{e}_1\otimes\boldsymbol{e}_1+\frac{h_0^2}{h^2}\boldsymbol{e}_2\otimes\boldsymbol{e}_2+\boldsymbol{e}_3\otimes\boldsymbol{e}_3, \tag{5.3}$$

describing a homogeneous strain in which there is no difference between Lagrangean and Eulerian axes (coincident with $\boldsymbol{e}_1$ and $\boldsymbol{e}_2$). We can note that uniaxial extension in the x_1 direction for an incompressible material is equivalent to a pure shear in a reference system rotated of 45° with respect to the x_1–x_2 reference system, as sketched in Fig. 5.2.

It may be interesting to note that during elongation of the block, a circle in the reference configuration transforms into an ellipse, with axes aligned parallel and orthogonal to the principal axes of $\boldsymbol{B}$ and $\boldsymbol{C}$ (see Fig. 5.3), where $\lambda_1 = h/h_0$.

Let us consider a fibre inclined at an angle ϑ_0 with respect to the x_1 axis in the reference configuration. Weissenberg (1935, 1949) has shown that such a fibre (with the exceptions of $\vartheta_0 = 0$, $\pi/4$ and $\pi/2$) may be transformed (by imposing an appropriate deformation) into a fibre with identical length (but different orientation, say, ϑ). In fact, the stretch of a fibre initially inclined at ϑ_0 can be calculated from Eq. $(3.8)_1$ to be given by

$$\lambda(\vartheta_0)=|\boldsymbol{F}(\cos\vartheta_0\boldsymbol{e}_1+\sin\vartheta_0\boldsymbol{e}_2)|=\frac{|\cos\vartheta_0|}{\lambda_1}\sqrt{\lambda_1^4+\tan^2\vartheta_0}, \tag{5.4}$$

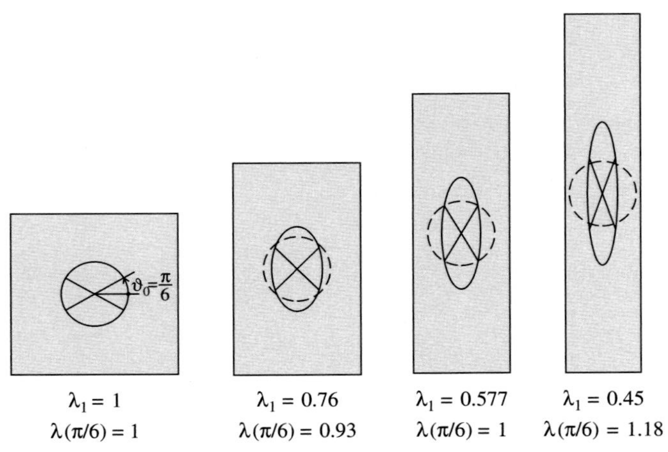

Figure 5.3. Plane strain elongation (contraction) in the x_2 (x_1) direction of an incompressible elastic block, where $\lambda_1 = h/h_0$: zero elongation line and strain ellipse. Note that the initial circle has been reported dashed in the deformed configurations for comparison. The two fibres marked by segments are inclined at $\pm\pi/6$ in the un-deformed state; for $\lambda_1 = 0.577$, the original length of the fibre is recovered.

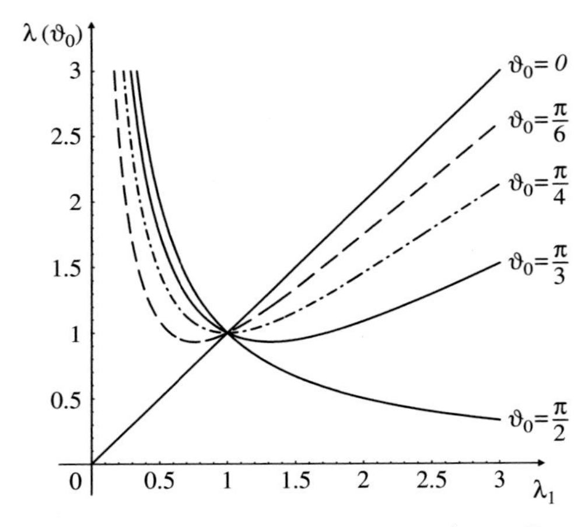

Figure 5.4. Plane strain elongation ($\lambda_1 > 1$) and contraction ($\lambda_1 < 1$) in the x_1 direction of an incompressible elastic block: stretch of fibres inclined at ϑ_0 with respect to the horizontal axis in the reference configuration as a function of principal stretch $\lambda_1 = h/h_0$.

a relationship plotted in Fig. 5.4. We may observe from Figs. 5.3 and 5.4, referred to a contraction parallel to the x_1 axis, $\lambda_1 < 1$ (elongation parallel to the x_2 axis), that a fibre inclined at $\pi/6$ suffers a contraction followed by an elongation so that for $h/h_0 = \lambda_1 = 0.577$, the original length of the fibre is recovered. Obviously, there are no fibres that always can be held at the original length during a continuous, finite elongation. Note that for $\vartheta_0 = \pi/4$, the tangent of the graph is horizontal, so there is no elongation for a deformation increment. Moreover, the minima of the curves reported in Fig. 5.4 correspond to

$$\lambda_1^{\min} = \sqrt{\tan\vartheta_0} \qquad \text{and} \qquad \lambda^{\min} = \sqrt{\sin 2\vartheta_0}. \tag{5.5}$$

Stress For an incompressible Mooney-Rivlin or neo-Hookean material (there are no differences between the models in plane strain; see Section 4.2.2), the principal Cauchy stress components can be evaluated from Eqs. (4.27) or (4.37) and (5.3) in the form

$$T_1 = -\pi + \mu_1 \frac{h^2}{h_0^2} + \mu_2 \frac{h_0^2}{h^2}, \qquad T_2 = -\pi + \mu_1 \frac{h_0^2}{h^2} + \mu_2 \frac{h^2}{h_0^2}, \qquad T_3 = -\pi + \mu_1 + \mu_2 \quad (5.6)$$

for Mooney-Rivlin or

$$T_1 = -\pi + \mu_0 \left(\frac{h^2}{h_0^2} - 1 \right), \qquad T_2 = -\pi + \mu_0 \left(\frac{h_0^2}{h^2} - 1 \right), \qquad T_3 = -\pi \quad (5.7)$$

for neo-Hookean material. Imposing that the stress is uniaxial, $T_2 = 0$, allows determination of π, so the plane strain uniaxial stress state becomes

$$T_1 = \mu_0 \left(\frac{h^2}{h_0^2} - \frac{h_0^2}{h^2} \right), \quad (5.8)$$

where $\mu_0 = \mu_1 - \mu_2$ for Mooney-Rivlin material, showing that the in-plane stress components are the same in the two materials. The out-of-plane stress T_3 remains different for the two materials, so for Mooney-Rivlin,

$$T_3 = \mu_1 \left(1 - \frac{h_0^2}{h^2} \right) + \mu_2 \left(1 - \frac{h^2}{h_0^2} \right), \quad (5.9)$$

whereas for neo-Hookean,

$$T_3 = \mu_0 \left(1 - \frac{h_0^2}{h^2} \right). \quad (5.10)$$

For the Mooney-Rivlin material, in terms of logarithmic strain $\varepsilon = \log(h/h_0)$, we may write

$$T_1 = 2\mu_0 \sinh 2\varepsilon \qquad \text{and} \qquad T_3 = \mu_1(1 - e^{-2\varepsilon}) + \mu_2(1 - e^{2\varepsilon}), \quad (5.11)$$

so the stress state (5.11) corresponds for small ε to

$$T_1 \sim 4\mu_0\varepsilon \qquad \text{and} \qquad T_3 \sim 2\mu_0\varepsilon, \quad (5.12)$$

which, compared to the plane strain equations of incompressible infinitesimal elasticity, shows that μ_0 corresponds to the initial shear modulus.

Finally, it may be interesting to evaluate the first and second Piola-Kirchhoff stresses for Mooney-Rivlin ($\mu_0 = \mu_1 - \mu_2$) and neo-Hookean materials:

$$\boldsymbol{S} = \mu_0 \left(\frac{h}{h_0} - \frac{h_0^3}{h^3} \right) \boldsymbol{e}_1 \otimes \boldsymbol{e}_1 + T_3 \boldsymbol{e}_3 \otimes \boldsymbol{e}_3,$$
$$\boldsymbol{T}^{(2)} = \mu_0 \left(1 - \frac{h_0^4}{h^4} \right) \boldsymbol{e}_1 \otimes \boldsymbol{e}_1 + T_3 \boldsymbol{e}_3 \otimes \boldsymbol{e}_3. \quad (5.13)$$

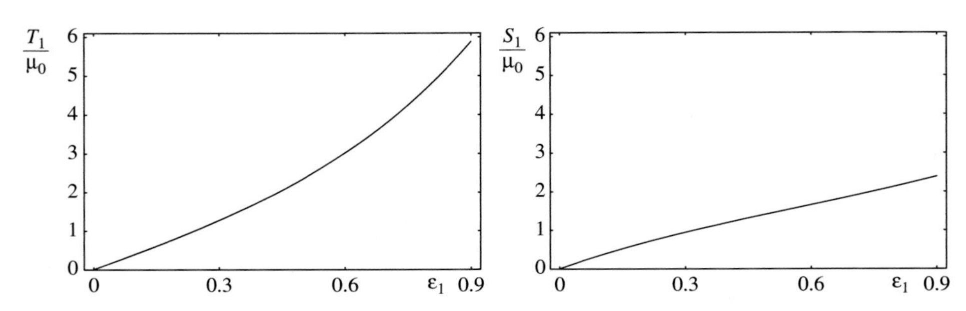

Figure 5.5. Uniaxial plane strain tension of a Mooney-Rivlin incompressible elastic material. The Cauchy and nominal stresses (normalised through division by the shear modulus at the unloaded state μ_0) are reported on the left and on the right, respectively, as functions of the logarithmic strain.

Let us compare now the response given by Eq. (5.11) with the corresponding response to an uniaxial elongation for a J_2-deformation theory material [Eq. (4.48)]:

$$T_1 = K\left(\frac{2}{\sqrt{3}}\right)^{N+1} |\varepsilon|^{N-1}\varepsilon, \qquad T_2 = 0 \qquad \text{and} \qquad T_3 = \frac{T_1}{2} \tag{5.14}$$

and with the response of a GBG material [Eq. (4.50)]:

$$\begin{aligned} T_1 &= \frac{2}{\sqrt{3}} K \text{sign}[\varepsilon_1]\left[1 - \exp\left(-\frac{2|\varepsilon_1|}{\sqrt{3}\varepsilon_0}\right)\right]\exp\left(-\frac{2|\varepsilon_1|}{c\sqrt{3}\varepsilon_0}\right), \\ T_2 &= 0, \\ T_3 &= \frac{T_1}{2}. \end{aligned} \tag{5.15}$$

The Cauchy stress T_1 and the nominal stress S_1 responses versus logarithmic strain for the Mooney-Rivlin, J_2-deformation theory and GBG materials are reported in Figs. 5.5 through 5.7. In the figures pertaining to the J_2-deformation theory and GBG material, we have reported the thresholds for loss of positive definiteness (PD) and ellipticity loss (E). These are thresholds for material instabilities that will be explained in Chapters 11 and 12 and have been calculated for the J_2-deformation theory material using Eqs. (12.7) and (12.61) and for the GBG material employing Eqs. (12.6) and (12.51). Note that the PD condition, which will be introduced in Chapter 11, corresponds to the maximum of the nominal stress curve, which for uniaxial stress ($T_2 = 0$) can be calculated from

$$S_1 = \frac{T_1}{\lambda} \qquad \text{and} \qquad \frac{dS_1}{d\lambda} = \frac{1}{\lambda}\frac{dT_1}{d\lambda} - \frac{T_1}{\lambda^2} = \frac{4\mu_* - T_1}{\lambda^2}. \tag{5.16}$$

Results reported in Fig. 5.7, relative to the GBG material, have been obtained by employing the following parameter sets:

$$\begin{aligned} c^A &= 42, \quad & \varepsilon_0^A &= 0.0045, \quad & K^A &= 1680 \text{ MPa}; \\ c^B &= 10, \quad & \varepsilon_0^B &= 0.0080, \quad & K^B &= 2100 \text{ MPa}; \\ c^C &= 5, \quad & \varepsilon_0^C &= 0.4000, \quad & K^C &= 2600 \text{ MPa}. \end{aligned} \tag{5.17}$$

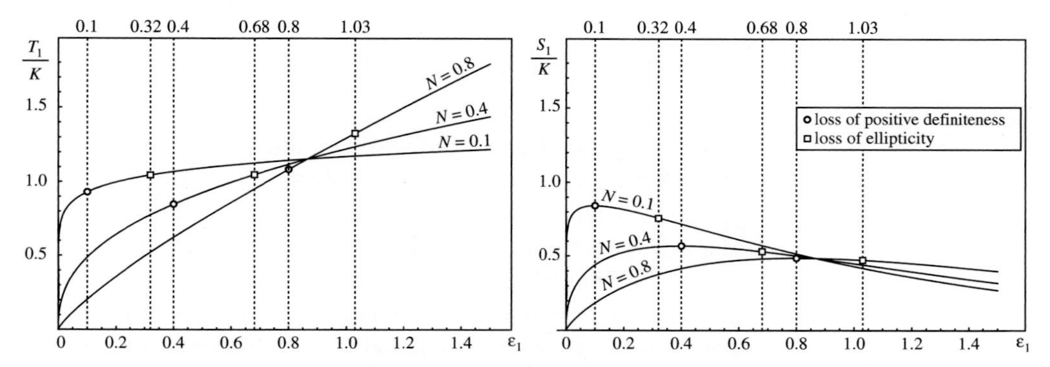

Figure 5.6. Uniaxial plane strain tension of a J_2-deformation theory incompressible elastic material for values of the hardening exponent $N = \{0.1, 0.4, 0.8\}$. The Cauchy and nominal stresses (normalised through division by K) are reported on the left and on the right, respectively, as functions of the logarithmic strain. The points corresponding to loss of positive definiteness and ellipticity [explained in Chapters 11 and 12 and calculated for the J_2-deformation theory material with Eqs. (12.7) and (12.61)] of the elasticity tensor are reported on the curves. Note that loss of positive definiteness always occurs at the peak of the nominal stress curves.

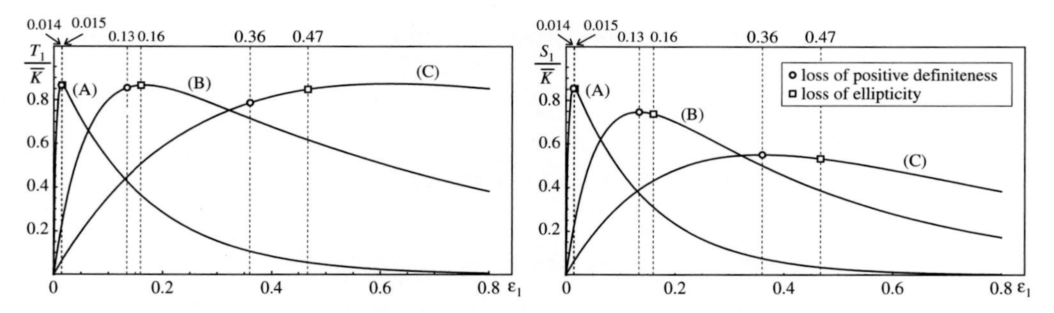

Figure 5.7. Uniaxial plane strain tension of a GBG material for various values of parameters c, K and ε_0, see the list (5.17). The Cauchy and nominal stresses (normalised through division by $\overline{K} = 2000$ MPa) are reported on the left and on the right, respectively, as functions of the logarithmic strain. The points corresponding to loss of positive definiteness and ellipticity [explained in Chapters 11 and 12 and calculated for the GBG material employing Eqs. (12.6) and (12.51)] of the elasticity tensor are reported on the curves. Note that loss of positive definiteness always occurs at the peak of the nominal stress curves.

We can observe from Figs. 5.5 through 5.7 that (1) the behaviours are identical in tension and compression, so tension has been reported only, (2) the neo-Hookean material has a locking behaviour, so stiffness increases at increasing strain, a feature evidenced by rubbers and biological materials, and (3) the J_2-deformation theory material is tailored to mimic the behaviour of a metal, with a very stiff initial behaviour, followed by a strain hardening remaining always positive; the GBG material has been invented to provide a description of rock-like materials, so we note strain softening in the Cauchy stress versus logarithmic strain representation, a feature not present in both the neo-Hookean and the J_2-deformation theory materials.

5.2 Uniaxial plane strain tension and compression of Kirchhoff–Saint Venant material

The so-called Kirchhoff–Saint Venant material has been introduced by Eq. (4.15). Although in this material model the relation between second Piola-Kirchhoff stress and Green-Lagrange strain tensor is linear, the material response is non-linear because $\boldsymbol{E}^{(2)}$ contains quadratic terms in the displacement gradient. This constitutive model is reasonably simple and has been used in many stability problems. However, as pointed out by Ciarlet (1988), the associated energy function is not 'polyconvex', a mathematical statement with interesting 'practical' consequences. To investigate these, we consider uniaxial plane strain tension/compression of the material (4.15).

We consider an elastic rectangular block of initial dimensions $b_0 \times h_0$ deformed into a rectangular $b \times h$ block (as in Fig. 5.1, except that now the material is compressible). For uniaxial plane strain tension/compression aligned parallel to $\boldsymbol{e}_1$, the deformation gradient is similar to Eq. (5.2), except that now the two stretches are unrelated, that is,

$$\boldsymbol{F} = \boldsymbol{U} = \boldsymbol{V} = \frac{h}{h_0}\boldsymbol{e}_1 \otimes \boldsymbol{e}_1 + \frac{b}{b_0}\boldsymbol{e}_2 \otimes \boldsymbol{e}_2 + \boldsymbol{e}_3 \otimes \boldsymbol{e}_3, \tag{5.18}$$

and the determinant of the deformation gradient is

$$J = \frac{hb}{h_0 b_0}, \tag{5.19}$$

Thus the Green-Lagrange strain tensor becomes

$$\boldsymbol{E}^{(2)} = \frac{1}{2}\left(\frac{h^2}{h_0^2}\boldsymbol{e}_1 \otimes \boldsymbol{e}_1 + \frac{b^2}{b_0^2}\boldsymbol{e}_2 \otimes \boldsymbol{e}_2 + \boldsymbol{e}_3 \otimes \boldsymbol{e}_3 - \boldsymbol{I}\right). \tag{5.20}$$

Using the constitutive equation (4.15), we may calculate the stress in terms of second Piola-Kirchhoff tensor

$$\boldsymbol{T}^{(2)} = \frac{\lambda}{2}\left(\frac{h^2}{h_0^2} + \frac{b^2}{b_0^2} - 2\right)\boldsymbol{I} + \mu\left(\frac{h^2}{h_0^2}\boldsymbol{e}_1 \otimes \boldsymbol{e}_1 + \frac{b^2}{b_0^2}\boldsymbol{e}_2 \otimes \boldsymbol{e}_2 + \boldsymbol{e}_3 \otimes \boldsymbol{e}_3 - \boldsymbol{I}\right), \tag{5.21}$$

from which we may obtain the Kirchhoff stress as

$$\boldsymbol{K} = \frac{\lambda}{2}\left(\frac{h^2}{h_0^2} + \frac{b^2}{b_0^2} - 2\right)\boldsymbol{F}\boldsymbol{F}^T + \mu\left(\frac{h^2}{h_0^2}\boldsymbol{F}\boldsymbol{e}_1 \otimes \boldsymbol{F}\boldsymbol{e}_1 + \frac{b^2}{b_0^2}\boldsymbol{F}\boldsymbol{e}_2 \otimes \boldsymbol{F}\boldsymbol{e}_2 + \boldsymbol{F}\boldsymbol{e}_3 \otimes \boldsymbol{F}\boldsymbol{e}_3 - \boldsymbol{F}\boldsymbol{F}^T\right), \tag{5.22}$$

where $\boldsymbol{F}$ is given by Eq. (5.18) and J by Eq. (5.19). Thus

$$\begin{aligned}\boldsymbol{K} = &\left[\frac{\lambda}{2}\left(\frac{h^2}{h_0^2} + \frac{b^2}{b_0^2} - 2\right) - \mu\right]\left(\frac{h^2}{h_0^2}\boldsymbol{e}_1 \otimes \boldsymbol{e}_1 + \frac{b^2}{b_0^2}\boldsymbol{e}_2 \otimes \boldsymbol{e}_2 + \boldsymbol{e}_3 \otimes \boldsymbol{e}_3\right) \\ &+ \mu\left(\frac{h^4}{h_0^4}\boldsymbol{e}_1 \otimes \boldsymbol{e}_1 + \frac{b^4}{b_0^4}\boldsymbol{e}_2 \otimes \boldsymbol{e}_2 + \boldsymbol{e}_3 \otimes \boldsymbol{e}_3\right).\end{aligned} \tag{5.23}$$

Until now, we have not imposed that the stress is uniaxial, namely, that $T_2 = K_2 = 0$. This condition can be expressed as

$$JT_2 = \left[\frac{\lambda}{2}\left(\frac{h^2}{h_0^2} + \frac{b^2}{b_0^2} - 2\right) - \mu\right]\frac{b^2}{b_0^2} + \mu\frac{b^4}{b_0^4} = 0, \tag{5.24}$$

an equation that (excluding for the moment that $b/b_0 = 0$) permits calculation of the transversal stretch b/b_0 as a function of the longitudinal stretch

$$\left(\frac{b}{b_0}\right)^2 = 1 + \frac{\nu}{1-\nu}\left[1 - \left(\frac{h}{h_0}\right)^2\right], \tag{5.25}$$

where ν has been introduced in analogy with Poisson's ratio of the infinitesimal theory as

$$\nu = \frac{\lambda}{2(\lambda+\mu)}. \tag{5.26}$$

Equation (5.25) reveals a 'deficiency' of the Saint Venant–Kirchhoff model. In particular, for a positive ν, a longitudinal stretch exists which makes the transversal stretch vanish (e.g., for $\nu = 1/2$, $b/b_0 = 0$ when $h/h_0 = \sqrt{2}$). This is a limit stretch, which cannot be exceeded, a fact in clear contrast with the real behaviour of materials.

The stress T_1 parallel to the direction $\boldsymbol{e}_1$ is given by

$$T_1 = \frac{\mu}{\sqrt{1-\nu}}\lambda_1\left(\lambda_1^2 - 1\right)\sqrt{\frac{1}{1-\nu\lambda_1^2}}, \tag{5.27}$$

where $\lambda_1 = h/h_0$ and in terms of logarithmic strain,

$$T_1 = \frac{\mu}{\sqrt{1-\nu}}\exp\varepsilon_1\left(\exp 2\varepsilon_1 - 1\right)\sqrt{\frac{1}{1-\nu\exp 2\varepsilon_1}}, \tag{5.28}$$

so the nominal stress is given by

$$S_1 = \frac{\mu}{\sqrt{1-\nu}}\left(\exp 2\varepsilon_1 - 1\right)\sqrt{\frac{1}{1-\nu\exp 2\varepsilon_1}}. \tag{5.29}$$

The response to uniaxial tension/compression is reported in Fig. 5.8 in terms of uniaxial Cauchy and nominal stresses (divided by μ) versus logarithmic strain ε_1 for $\nu = 0$ and $\nu = 0.49$. The behaviour in tension is similar to a Mooney-Rivlin material, but for $\nu \neq 0$, there is an asymptote (for $\lambda_1 = \sqrt{1/\nu}$), whereas in compression the material exhibits softening (and ellipticity, which will be introduced in Chapter 11, is lost soon after the peak).

We finally note that a linearisation through Taylor series expansion of Eq. (5.28) or Eq. (5.29) about $\varepsilon_1 = 0$ provides

$$T_1 = S_1 = \frac{2\mu}{1-\nu^2}\varepsilon_1, \tag{5.30}$$

which coincides with the uniaxial plane strain behaviour of linear isotropic elasticity.

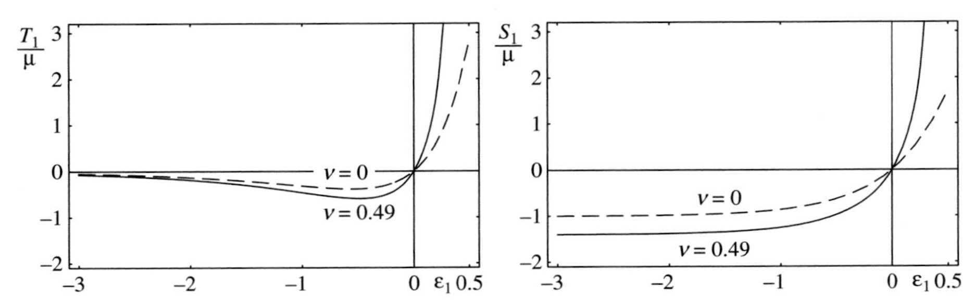

Figure 5.8. Uniaxial plane strain tension and compression of a Kirchhoff–Saint Venant material. Axial Cauchy T_1 and nominal S_1 stress (normalised through division by μ) versus the logarithmic strain ε_1. Two values of Poisson's ratio have been considered, namely, $\nu = 0$ and $\nu = 0.49$. In tension, the material becomes progressively stiff, and the stress becomes infinite when the transversal stretch tends to zero. In compression, the material exhibits softening.

5.3 Uniaxial tension and compression of an incompressible elastic cylinder

The kinematics of axial extension and compression of a cylinder was explained in Chapter 3, where the homogeneous deformation gradient was given for [compressible, Eq. (3.42) and] incompressible materials [Eq. (3.45)].

Referring to an incompressible neo-Hookean response [Eq. (4.37)], we obtain for the axial stress

$$T_1 = -\pi + \mu_0 \left(\frac{h^2}{h_0^2} - 1 \right), \tag{5.31}$$

whereas, since the radial stress must be zero to match the boundary conditions,

$$T_r = -\pi + \mu_0 \left(\frac{h_0}{h} - 1 \right) = 0, \tag{5.32}$$

we can calculate π and substitute in Eq. (5.31) to obtain

$$T_1 = \mu_0 \left(\frac{h^2}{h_0^2} - \frac{h_0}{h} \right). \tag{5.33}$$

In terms of first Piola-Kirchhoff (or nominal) stress, Eq. (5.33) can be written as

$$S_1 = \mu_0 \left(\frac{h}{h_0} - \frac{h_0^2}{h^2} \right). \tag{5.34}$$

For the J_2-deformation theory, Eq. (4.39) can be used to calculate T_1 and $T_2 = T_3$. Setting the latter to zero, we determine p, so finally we arrive at

$$T_1 = K|\varepsilon_1|^{N-1}\varepsilon_1, \tag{5.35}$$

whereas for the GBG model, Eq. (4.54) gives the uniaxial tension/compression behaviour.

The responses of a neo-Hookean incompressible material (5.33), of the J_2-deformation theory material (5.35) and of the GBG model (4.54) to uniaxial tension

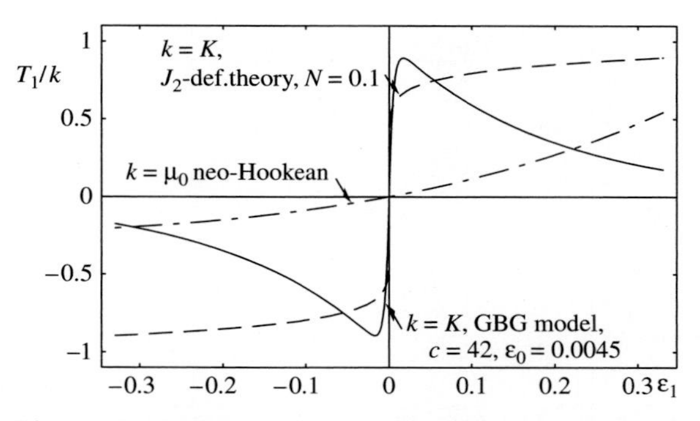

Figure 5.9. Uniaxial tension and compression of an incompressible elastic cylinder: neo-Hookean, J_2-deformation theory of plasticity and GBG model (Gei et al., 2004) are reported in terms of Cauchy stress versus logarithmic strain.

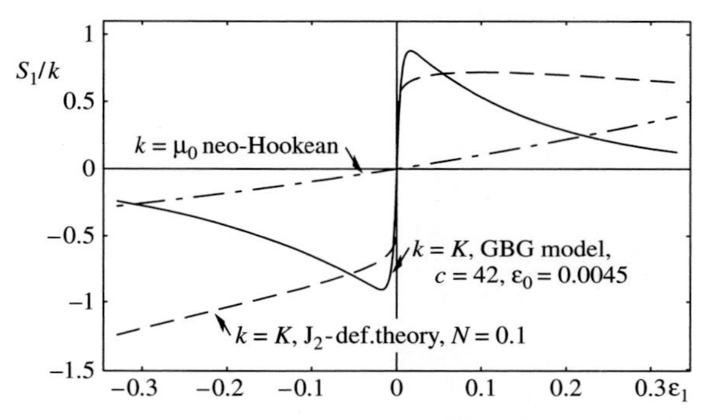

Figure 5.10. Uniaxial tension and compression of an incompressible elastic cylinder: neo-Hookean, J_2-deformation theory of plasticity, and GBG model (Gei et al., 2004) are reported in terms of nominal stress versus logarithmic strain.

and compression are plotted in Fig. 5.9 in terms of Cauchy stress versus logarithmic strain.

We clearly see from the figure that (1) J_2-deformation theory and GBG material models yield an identical response in tension and in compression and (2) unlike the neo-Hookean and J_2-deformation theory materials, the GBG model (plotted with $c = 42$ and $\varepsilon_0 = 0.0045$) has been proposed to mimic the behaviour of quasi-brittle materials such as ceramics under high temperature or hard soils so that it describes a softening response occurring after a peak in the Cauchy stress/logarithmic strain curve. Hardening and softening behaviour depends on the stress measure reported. The nominal stress versus logarithmic strain is reported in Fig. 5.10, where we see that now the behaviour changes from tension to compression, and while softening now occurs for J_2-deformation theory material in tension, the same material displays a strong hardening in compression. On the contrary, the behaviour remains softening for the GBG model, for $c = 42$ and $\varepsilon_0 = 0.0045$.

To better elucidate the features of the GBG model, we report additional results in Fig. 5.11, limited to positive logarithmic strains (the compression/tension behaviour is identical for Cauchy stress versus logarithmic strain). In Fig. 5.11 on the left, parameter ε_0 is varied, at fixed $c = 42$, whereas on the right, parameter c is varied

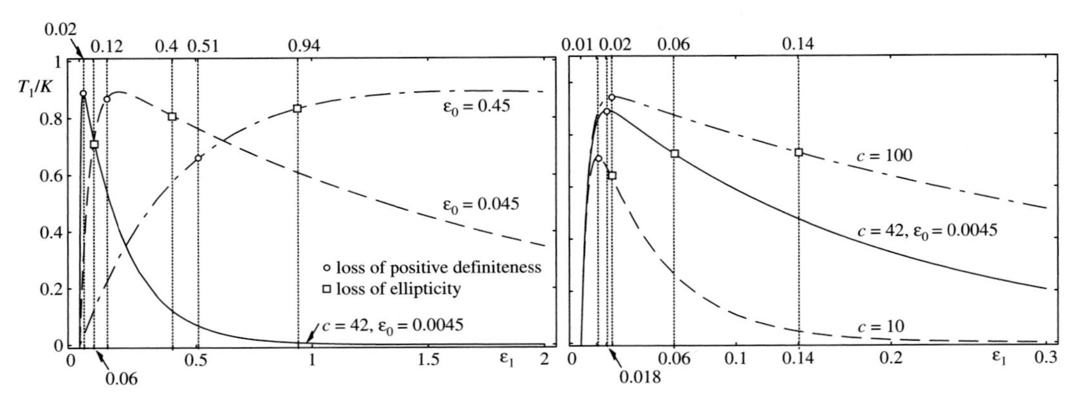

Figure 5.11. Uniaxial tension of an incompressible elastic cylinder made up of GBG material: Cauchy stress versus logarithmic strain for different values of material parameters. Note that loss of positive definiteness (PD) and ellipticity loss (E) of the tangent constitutive tensor (see Chapter 11 and Section 12.3) are marked with spots on the curves.

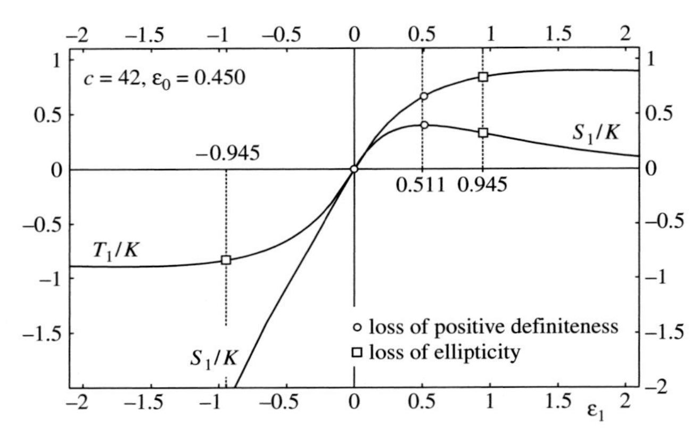

Figure 5.12. Uniaxial tension and compression of an incompressible elastic cylinder made up of GBG material is reported in terms of Cauchy (true) T_1 and nominal stress S_1 (both normalised through division by K) versus logarithmic strain. Note that loss of positive definiteness and of ellipticity of the tangent constitutive tensor (see Chapter 11 and Section 12.3) are marked with spots on the curves.

at fixed $\varepsilon_0 = 0.0045$. We may appreciate from the figure that the parameters ε_0 and c control the 'inclination' of the softening and the position of the peak of the curve. The points corresponding to loss of positive definiteness and of ellipticity of the tangent constitutive tensor (concepts that will be developed later in Chapters 11 and 12.3) are marked with spots on the curves in Fig. 5.11. We note that loss of positive definiteness and of ellipticity occur in the hardening for $c = 42$ and $\varepsilon_0 = 0.45$ for the Cauchy (true) stress versus logarithmic strain representation. It is instructive to compare the Cauchy stress with the nominal stress representations (Fig. 5.12), so that we can see that loss of positive definiteness occurs in tension at the peak of the nominal stress curve and that the loss of ellipticity is insensitive to the sign of the strain (while positive definiteness is lost when $\varepsilon_1 \leq 0$).

Figure 5.13. A collection of blocks representing homogeneous and heterogeneous simple shear deformations forming the artwork *Muraglia, giallo Mori e verde Alpi*, by Pietro Consagra (displayed at Castelvecchio, Verona, December 2006).

We highlighted in the Introduction that the softening behaviour observed during a test of a material element usually should be considered as a structural effect related to the inhomogeneities (bifurcations, necking, strain localization or crack nucleation and growth) developing near failure. However, it may be important in many circumstances to ignore these structural effects and to analyse the consequences of softening within a region of material, assuming softening as a constitutive response. Therefore, this constitutive response has been implemented in the GBG model, which has been tailored to reproduce a Cauchy stress/logarithmic strain curve evidencing the peak and softening characteristic of the behaviour near failure of several engineering materials.

5.4 Simple shear of an elastic block

We analyse simple shear of an elastic block, the deformation of the blocks which are collected in an incompatible way to form the artpiece shown in Fig. 5.13.

Kinematics With reference to Fig. 5.14, a block of a material is subject to a simple shear deformation (Truesdell and Toupin, 1960, section 45) when two displacement components are null (along axes $\hat{x}_2$ and $\hat{x}_3$; see the figure) and the other component, $\hat{u}_1$, depends linearly only on $\hat{x}_2^0$, namely,[1]

$$\hat{u}_1(\hat{x}_2^0) = \gamma \hat{x}_2^0 \qquad \text{and} \qquad \hat{u}_2 = \hat{u}_3 = 0, \tag{5.36}$$

so if a point at $\hat{x}_2^0 = h$ horizontally displaces s, we can determine the dimensionless parameter $\gamma = s/h$, controlling the amplitude of shear deformation. The deformation is defined by

$$\hat{\boldsymbol{x}} = \hat{\boldsymbol{x}}_0 + \gamma (\hat{\boldsymbol{x}}_0 \cdot \hat{\boldsymbol{e}}_2)\hat{\boldsymbol{e}}_1, \tag{5.37}$$

[1] We use $\hat{x}_i$, $(i = 1,2,3)$ instead of x_i to give evidence to the fact that axes $\hat{x}_i$ are parallel to the shear but not principal.

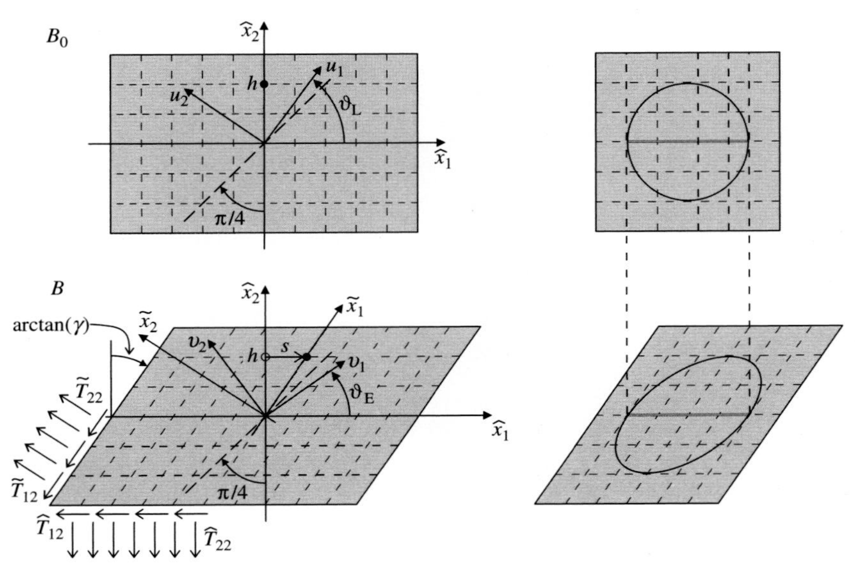

Figure 5.14. Simple shear deformation of an elastic block. *Top*: Reference configuration; *bottom*: configuration at a shear $\gamma = s/h$, with indicated Eulerian $\{\boldsymbol{v}_1, \boldsymbol{v}_2\}$ and Lagrangean $\{\boldsymbol{u}_1, \boldsymbol{u}_2\}$ principal axes, inclined at ϑ_E and ϑ_L, respectively. The $\tilde{x}_1$ and $\tilde{x}_2$ axes are parallel and orthogonal to the fibres initially vertical in the un-deformed configuration. Note the circle in the reference configuration (*top*, *right*) that becomes a strain ellipse (*bottom*, *right*). In this sketch, $\arctan\gamma = 34°$, so $\vartheta_E \approx 35.682°$ and $\vartheta_L \approx 54.318°$.

so the deformation gradient is homogeneous and given by

$$\boldsymbol{F} = \boldsymbol{I} + \gamma\, \hat{\boldsymbol{e}}_1 \otimes \hat{\boldsymbol{e}}_2, \tag{5.38}$$

and the left and right Cauchy-Green deformation tensors are

$$\begin{aligned} \boldsymbol{B} &= \boldsymbol{I} + \gamma \left(\hat{\boldsymbol{e}}_1 \otimes \hat{\boldsymbol{e}}_2 + \hat{\boldsymbol{e}}_2 \otimes \hat{\boldsymbol{e}}_1\right) + \gamma^2 \hat{\boldsymbol{e}}_1 \otimes \hat{\boldsymbol{e}}_1, \\ \boldsymbol{C} &= \boldsymbol{I} + \gamma \left(\hat{\boldsymbol{e}}_1 \otimes \hat{\boldsymbol{e}}_2 + \hat{\boldsymbol{e}}_2 \otimes \hat{\boldsymbol{e}}_1\right) + \gamma^2 \hat{\boldsymbol{e}}_2 \otimes \hat{\boldsymbol{e}}_2. \end{aligned} \tag{5.39}$$

Note that $\det \boldsymbol{F} = 1$, so the simple shear deformation is isochoric.

The eigenvectors of $\boldsymbol{B}$ and $\boldsymbol{C}$ define the principal Eulerian and Lagrangean axes, so in the $\hat{x}_1$–$\hat{x}_2$ plane these are

$$\begin{aligned} \{\mathbf{v}_j\} &= \frac{1}{\sqrt{4 + \left[\gamma - (-1)^j \sqrt{4+\gamma^2}\right]^2}} \left\{\gamma - (-1)^j \sqrt{4+\gamma^2},\ 2\right\}, \\ \{\mathbf{u}_j\} &= \frac{1}{\sqrt{4 + \left[\gamma + (-1)^j \sqrt{4+\gamma^2}\right]^2}} \left\{-\gamma - (-1)^j \sqrt{4+\gamma^2},\ 2\right\}, \end{aligned} \tag{5.40}$$

(where $j = 1, 2$), respectively, whereas the principal stretches result to be given by

$$\lambda_j = -(-1)^j \frac{\gamma}{2} + \sqrt{1 + \left(\frac{\gamma}{2}\right)^2}. \tag{5.41}$$

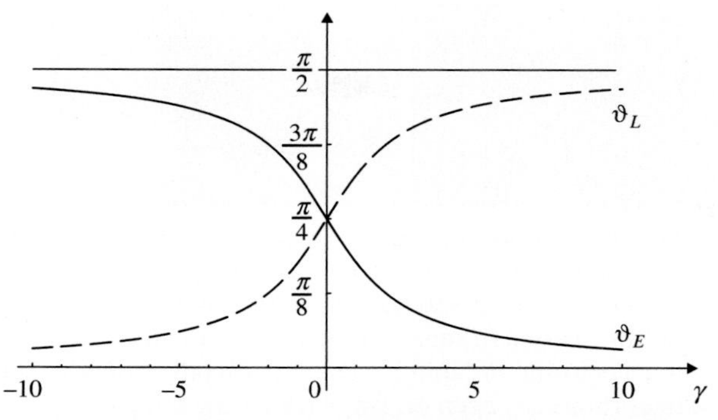

Figure 5.15. Simple shear deformation of an elastic block: Angles of inclination of Eulerian, ϑ_E, and Lagrangean, ϑ_L, principal axes during simple shear deformation, as functions of the amount of shear γ.

Note that

$$\lambda_1 - \lambda_2 = \gamma, \tag{5.42}$$

so if $\gamma > 0$ (< 0) then $\lambda_1 > \lambda_2$ ($\lambda_1 < \lambda_2$).

The eigenvectors (5.40) yield (through an elementary use of trigonometry) the inclination of the Eulerian, ϑ_E, and Lagrangean, ϑ_L, axes with respect to the $\hat{x}_1$ axis, that is,

$$\vartheta_E = \frac{1}{2}\arctan\left(\frac{2}{\gamma}\right) \qquad \text{and} \qquad \vartheta_L = \frac{\pi}{2} - \vartheta_E, \tag{5.43}$$

plotted in Fig. 5.15 as functions of the amount of shear γ. For $\gamma > 0$, we have $0 \leq \vartheta_E \leq \pi/4$ and $\pi/4 \leq \vartheta_L \leq \pi/2$, so in the reference configuration, at null strain, $\vartheta_E = \vartheta_L = \pi/4$, and the Lagrangean and Eulerian axes coincide, whereas at infinite strain, $\vartheta_E = 0$ and $\vartheta_L = \pi/2$.

With reference to the Cauchy-Green deformation tensors (5.39), we note that a fibre parallel to the $\hat{x}_1$ axis, singled out by the unit vector $\hat{\boldsymbol{e}}_1$, remains unstretched because

$$\hat{\boldsymbol{e}}_1 \cdot \boldsymbol{C}\hat{\boldsymbol{e}}_1 = \hat{\boldsymbol{e}}_1 \cdot \boldsymbol{B}^{-1}\hat{\boldsymbol{e}}_1 = 1, \tag{5.44}$$

so the fibres parallel to the $\hat{x}_1$ axis in Fig. 5.14 are the so-called zero elongation lines (Weissenberg, 1949). Note that a rigid line inclusion lying parallel to a zero elongation line is 'neutral', so it does not perturb the stress state generated during finite shear. This feature has been tested experimentally using transmission photoelasticity on a two-component epoxy resin sample containing a thin (0.1 mm thick) steel lamina (20 mm long) enclosed in a rectangle drawn on the sample's surface (with a black pen) (see Fig. 5.16). The rectangle has been drawn to provide a graphical measure of the shear strain. Note that the lamina is embedded at the centre of a 260 mm (edge parallel to $\hat{x}_1$ axis) $\times$ 100 mm (edge parallel to $\hat{x}_2$ axis) $\times$ 5 mm (out-of-plane thickness) resin sample subjected to simple shear conditions on its boundary. The maximum shear deformation has been limited to 10% in the experiment to avoid shear buckling (see Noselli et al., 2010, for further details). Since the steel lamina lies along a zero elongation line, it should leave the simple shear stress state almost ('almost' because the inclusion has a small, though finite, thickness) unperturbed. The experiment

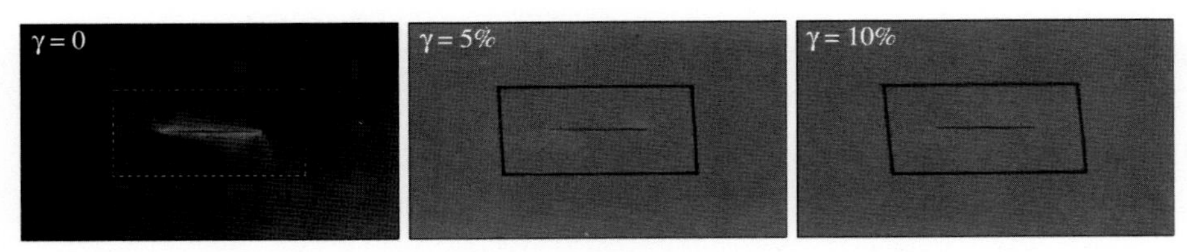

Figure 5.16. Photoelastic experiment demonstrating neutrality of a thin rigid inclusion (made of steel, 0.1 mm thick, 20 mm long) embedded in an elastic (two-component epoxy resin) matrix subject to simple shear parallel to it. Homogeneous colours denote uniform stress. The sample is unloaded on the left, whereas the amount of shear γ is equal to 5% in the centre and 10% on the right. Note that a rectangle enclosing the thin inclusion (with edges parallel and orthogonal to it in the unloaded configuration) has been drawn black on the sample's surface to measure shear deformation. (See color plates section.)

shows in fact that the colour fringes are nearly uniform, demonstrating that the lamina does not disturb the stress state produced by simple shear.

The left and right stretch tensors are

$$\boldsymbol{V} = \frac{1}{\sqrt{4+\gamma^2}}\left[(2+\gamma^2)\,\hat{\boldsymbol{e}}_1\otimes\hat{\boldsymbol{e}}_1 + \gamma\,(\hat{\boldsymbol{e}}_1\otimes\hat{\boldsymbol{e}}_2 + \hat{\boldsymbol{e}}_2\otimes\hat{\boldsymbol{e}}_1) + 2\,\hat{\boldsymbol{e}}_2\otimes\hat{\boldsymbol{e}}_2\right] + \boldsymbol{e}_3\otimes\boldsymbol{e}_3,$$
$$\boldsymbol{U} = \frac{1}{\sqrt{4+\gamma^2}}\left[2\,\hat{\boldsymbol{e}}_1\otimes\hat{\boldsymbol{e}}_1 + \gamma\,(\hat{\boldsymbol{e}}_1\otimes\hat{\boldsymbol{e}}_2 + \hat{\boldsymbol{e}}_2\otimes\hat{\boldsymbol{e}}_1) + (2+\gamma^2)\,\hat{\boldsymbol{e}}_2\otimes\hat{\boldsymbol{e}}_2\right] + \boldsymbol{e}_3\otimes\boldsymbol{e}_3, \tag{5.45}$$

and the rotation tensor is

$$\boldsymbol{R} = \frac{1}{\sqrt{4+\gamma^2}}\left[2\,(\hat{\boldsymbol{e}}_1\otimes\hat{\boldsymbol{e}}_1 + \hat{\boldsymbol{e}}_2\otimes\hat{\boldsymbol{e}}_2) + \gamma\,(\hat{\boldsymbol{e}}_1\otimes\hat{\boldsymbol{e}}_2 - \hat{\boldsymbol{e}}_2\otimes\hat{\boldsymbol{e}}_1)\right] + \boldsymbol{e}_3\otimes\boldsymbol{e}_3, \tag{5.46}$$

so there is an in-plane rigid-body rotation of an angle (taken positive when anti-clockwise) $\arctan(-\gamma/2)$.

Stress For a material isotropic in the reference configuration, the Cauchy stress is coaxial with the left Cauchy-Green tensor $\boldsymbol{V}$ $(5.45)_1$, so in the Eulerian principal reference system we have the spectral representation

$$\boldsymbol{T} = T_1\boldsymbol{v}_1\otimes\boldsymbol{v}_1 + T_2\boldsymbol{v}_2\otimes\boldsymbol{v}_2 + T_3\boldsymbol{e}_3\otimes\boldsymbol{e}_3, \tag{5.47}$$

where, in the $\hat{x}_1$–$\hat{x}_2$ reference system, the unit vectors $\boldsymbol{v}_1$ and $\boldsymbol{v}_2$ are defined by the components

$$\{\boldsymbol{v}_1\} = \{\cos\vartheta_E, \sin\vartheta_E\} \qquad \text{and} \qquad \{\boldsymbol{v}_2\} = \{-\sin\vartheta_E, \cos\vartheta_E\}. \tag{5.48}$$

Therefore, in the reference system $\hat{\boldsymbol{e}}_1$, $\hat{\boldsymbol{e}}_2$, we have

$$\begin{aligned}
\widehat{T}_{11} &= T_1\cos^2\vartheta_E + T_2\sin^2\vartheta_E,\\
\widehat{T}_{22} &= T_1\sin^2\vartheta_E + T_2\cos^2\vartheta_E,\\
\widehat{T}_{12} &= (T_1-T_2)\cos\vartheta_E\sin\vartheta_E = \frac{T_1-T_2}{2}\sin 2\vartheta_E,
\end{aligned} \tag{5.49}$$

which, using simple trigonometric identities, yield

$$\left.\begin{array}{l}\widehat{T}_{11}\\ \\ \widehat{T}_{22}\end{array}\right\} = p \pm \frac{T_1 - T_2}{2}\frac{\gamma}{\sqrt{4+\gamma^2}}, \qquad \widehat{T}_{12} = \frac{T_1 - T_2}{\sqrt{4+\gamma^2}}, \qquad T_3 = p, \tag{5.50}$$

where p is the in-plane mean stress, that is,

$$p = \frac{T_1 + T_2}{2}. \tag{5.51}$$

From Eqs. (5.49) we obtain

$$\widehat{T}_{11} - \widehat{T}_{22} = \frac{2}{\tan 2\vartheta_E}\widehat{T}_{12} = \gamma\,\widehat{T}_{12}, \tag{5.52}$$

which is independent of the specific constitutive equation and therefore is an example of a so-called universal relation. Note that for $\gamma < 1$ ($\gamma > 1$), $\widehat{T}_{11} - \widehat{T}_{22} < \widehat{T}_{12}$ ($\widehat{T}_{11} - \widehat{T}_{22} > \widehat{T}_{12}$), and in particular, the deviatoric stress $\widehat{T}_{11} - \widehat{T}_{22}$ is of higher order in γ than $\widehat{T}_{12}$.

The stress on the inclined faces of the deformed block (sketched in the lower part of Fig. 5.15) can be obtained from the components of the unit tangential and normal vector

$$\{\tilde{\boldsymbol{e}}_1\} = \left\{\frac{\gamma}{\sqrt{1+\gamma^2}}, \frac{1}{\sqrt{1+\gamma^2}}\right\}, \quad \{\tilde{\boldsymbol{e}}_2\} = \left\{-\frac{1}{\sqrt{1+\gamma^2}}, \frac{\gamma}{\sqrt{1+\gamma^2}}\right\}, \tag{5.53}$$

in the form

$$\left.\begin{array}{l}\tilde{T}_{11}\\ \\ \tilde{T}_{22}\end{array}\right\} = p \pm \frac{T_1 - T_2}{2}\frac{\gamma(3+\gamma^2)}{(1+\gamma^2)\sqrt{4+\gamma^2}}, \qquad \tilde{T}_{12} = \frac{T_1 - T_2}{(1+\gamma^2)\sqrt{4+\gamma^2}}. \tag{5.54}$$

Assuming a neo-Hookean response [Eq. (4.37)], and using Eq. $(5.39)_1$, we obtain

$$\left.\begin{array}{l}\widehat{T}_{11}\\ \\ \widehat{T}_{22}\end{array}\right\} = p \pm \frac{\mu_0\gamma^2}{2}, \qquad \widehat{T}_{12} = \mu_0\gamma, \qquad T_3 = -\pi, \tag{5.55}$$

in which π, and consequently the in-plane mean stress p, remains undetermined.

In the case of a J_2-deformation theory material (Section 4.2.3), we obtain

$$\left.\begin{array}{l}\widehat{T}_{11}\\ \\ \widehat{T}_{22}\end{array}\right\} = p \pm \frac{\mu\gamma}{\sqrt{4+\gamma^2}\coth 2\varepsilon}, \qquad \widehat{T}_{12} = \frac{4\mu_*\varepsilon}{N\sqrt{4+\gamma^2}}, \qquad T_3 = p, \tag{5.56}$$

where $\varepsilon = \varepsilon_1 = \log\lambda_1$ is the logarithmic strain, function of γ through Eq. (5.41), p is the in-plane mean stress, $N \in (0,1]$ is the strain hardening exponent and μ and μ_* are incremental shear moduli and depend on the current stretch in the following way[2]

$$\mu = \overline{\mu}\,2\varepsilon\,|\varepsilon|^{N-1}\coth 2\varepsilon \qquad \text{and} \qquad \mu_* = N\overline{\mu}|\varepsilon|^{N-1}, \tag{5.57}$$

[2] Note that the J_2-deformation theory material is a non-linear elastic material isotropic in the unloaded state. Owing to the fact that the incremental shear moduli μ and μ_* tend to infinity when ε tends to zero (and $N < 1$), the ratio μ_*/μ tends to N (and therefore is different from 1).

where

$$\overline{\mu} = \frac{K}{3}\left(\frac{2}{\sqrt{3}}\right)^{N-1}, \tag{5.58}$$

in which K is a positive stiffness parameter.

Considering small shear amplitude γ, the constitutive relations (5.56) become

$$\left.\begin{array}{c}\widehat{T}_{11}\\ \\ \widehat{T}_{22}\end{array}\right\} = p \pm \frac{\overline{\mu}}{2^N}\,|\gamma|^{N+1} \qquad \text{and} \qquad \widehat{T}_{12} = \text{sign}(\gamma)\frac{\overline{\mu}}{2^{N-1}}\,|\gamma|^N. \tag{5.59}$$

The deviatoric stress and the shear stress, made dimensionless through division by μ_0 for a neo-Hookean material and for K for a J_2–deformation theory material, are reported in Figs. 5.17 and 5.18, respectively [in the latter figure, the curve relative to $N = 0.1$ coincides with that plotted by Harren et al. (1989), their fig. 2].

Note that it is evident from Eq. (5.59) that for $\gamma = 0$, the curve representing the deviatoric stress has an horizontal tangent, and the curve representing the shear stress has a vertical tangent. Although the vertical tangent is not visible in Fig. 5.18 (*right*), it is easy to check its existence; in fact, the incremental equations corresponding to

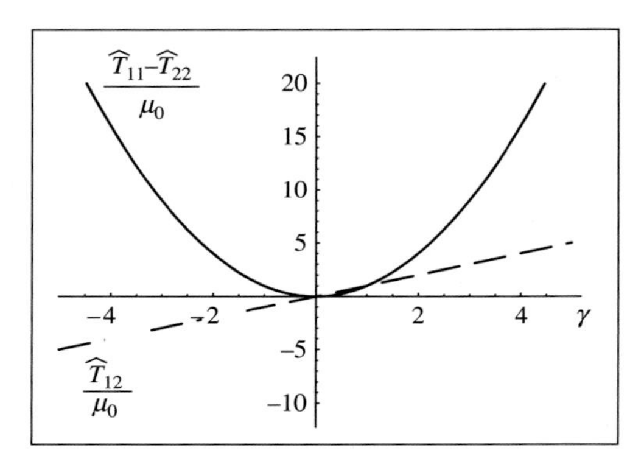

Figure 5.17. Simple shear of an elastic incompressible Mooney-Rivlin block: In-plane deviatoric stress response for finite shear amplitude γ.

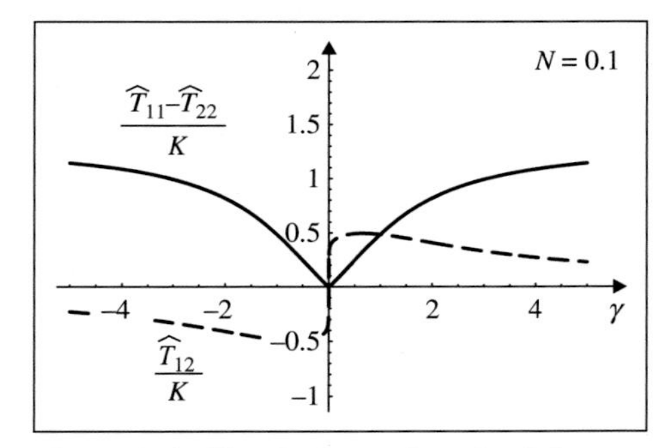

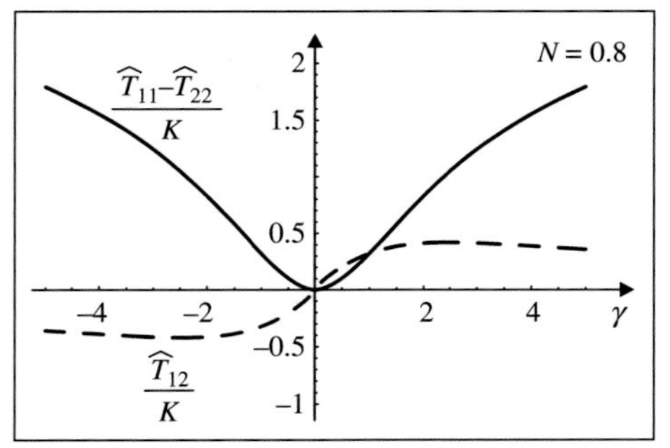

Figure 5.18. Simple shear of an elastic incompressible block made up of J_2-deformation theory material: In-plane deviatoric stress response for finite shear amplitude γ at low $N = 0.1$ and high $N = 0.8$ strain hardening.

Eq. (5.56) become

$$\dot{T}_{11} = \dot{T}_{22} = \dot{p} \qquad \text{and} \qquad \dot{T}_{12} = \mu_* \dot{\gamma}, \tag{5.60}$$

where $\dot{\gamma}$ ($\dot{p}$) represents the shear (the mean stress) increment and μ_* tends to infinity when γ tends to zero.

The following features may be observed from the above-obtained equations, holding for both neo-Hookean and J_2-deformation theory materials:

1. The coefficient μ represents a shear modulus.
2. A normal stress is generated through a simple shear deformation; this is the so-called Kelvin effect.
3. The two normal stress components during simple shear are not equal, that is, $\widehat{T}_{11} \neq \widehat{T}_{22}$, a feature representing the so-called Poynting effect.
4. The linear theory can be recovered by neglecting terms on the order of γ^2 or higher in a Taylor series expansion of the equations. Doing this, we obtain the following linear approximations:

$$\begin{aligned}
&\lambda_j \sim -(-1)^j \frac{\gamma}{2} + 1, \\
&\vartheta_E \sim \frac{\pi - \gamma}{4}, \\
&\vartheta_L \sim \frac{\pi + \gamma}{4}, \\
&\{\mathbf{v}_j\} \sim \frac{1}{\sqrt{2}} \left\{ -(-1)^j + \frac{\gamma}{4},\ 1 + (-1)^j \frac{\gamma}{4} \right\}, \\
&\{\mathbf{u}_j\} \sim \frac{1}{\sqrt{2}} \left\{ -(-1)^j - \frac{\gamma}{4},\ 1 - (-1)^j \frac{\gamma}{4} \right\},
\end{aligned} \tag{5.61}$$

where $j = 1, 2$. For a Mooney-Rivlin material,

$$\widehat{T}_{11} \sim \widehat{T}_{22} \sim p \qquad \text{and} \qquad \widehat{T}_{12} \sim \mu_0 \gamma, \tag{5.62}$$

whereas for a J_2-deformation theory material, a linearisation cannot be obtained near the unloaded state [see Eqs. (5.59)].

5.5 Finite bending of an incompressible elastic block

We consider in this section the problem of finite bending of an incompressible elastic block deformed in plane strain (Fig. 5.19). This problem has been solved by Rivlin (1949; see also Green and Zerna, 1968; Lurie, 2005; Ogden, 1984; Truesdell and Toupin, 1960, section 50) and is of considerable interest for engineering applications. Moreover, it will be the basis for the bifurcation analysis performed in Section 12.4.

Kinematics Plane strain flexure of an incompressible elastic block, rectangular with dimensions $l_0 \times h_0$ in the undeformed configuration (Fig. 5.20), is considered, which is deformed into a sector of a circular tube. In the reference, unloaded, configuration, the Cartesian coordinates are defined so that

$$x_1^0 \in [-h_0/2, h_0/2] \qquad \text{and} \qquad x_2^0 \in [-l_0/2, l_0/2], \tag{5.63}$$

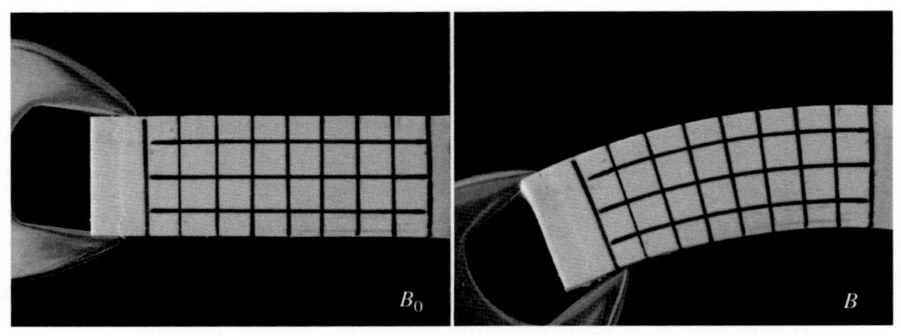

Figure 5.19. Finite bending of a rubber block. *Left*: Reference, unloaded configuration B_0 (an orthogonal grid has been drawn on the surface of the block). *Right*: Deformed configuration B.

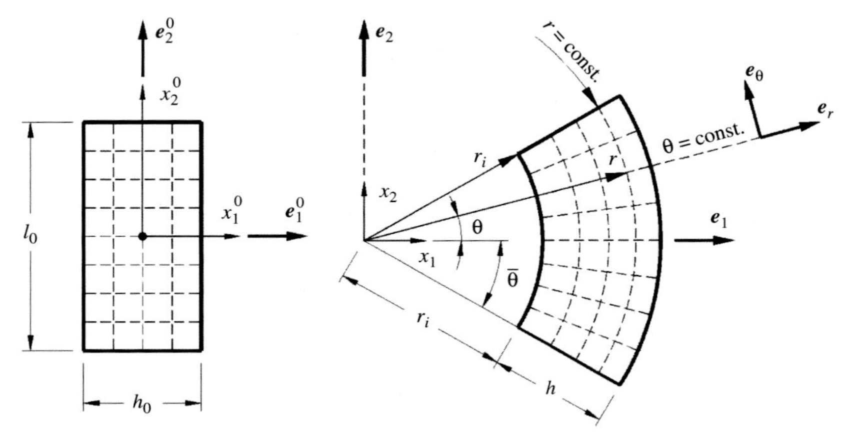

Figure 5.20. Plane strain bending of a rectangular incompressible elastic block. *Left*: Reference configuration; right: deformed configuration, with associated cylindrical coordinate system. Owing to the incompressibility constraint, the deformed configuration is fixed by two of the three parameters $\bar{\theta}$, h and r_i.

(note that we have shifted to the apex the index 0 denoting the reference configuration), and x_3^0 is the out-of-plane coordinate. We denote with $\boldsymbol{e}_i^0$ $(i = 1,2)$ the unit vectors defining the x_i^0 axes so that

$$\boldsymbol{x}^0 = x_1^0 \boldsymbol{e}_1^0 + x_2^0 \boldsymbol{e}_2^0. \tag{5.64}$$

The deformed configuration is a portion of a cylindrical tube of semi-angle $\bar{\theta}$ and thickness h (Fig. 5.20), so it is instrumental to use cylindrical coordinates $r \in [r_i, r_i + h]$, $\theta \in [-\bar{\theta}, \bar{\theta}]$, and z.

The deformation is prescribed so that planes at constant x_1^0 transform to sectors of cylindrical surface at constant r, whereas planes at constant x_2^0 become planes at constant θ. The out-of-plane strain is simply null, so $x_3^0 = z$. We distinguish between the unit vectors $\boldsymbol{e}_i^0$ in the reference configuration and $\boldsymbol{e}_i$ in the current configuration in Fig. 5.20. However, these unit vectors are identical from a practical point of view, so they will be used without discrimination in the subsequent calculations.

The incompressibility constraint requires the deformed area of the body to be identical to the initial area $l_0 \times h_0$; thus

$$r_i = \frac{l_0 h_0}{2\bar{\theta} h} - \frac{h}{2}. \tag{5.65}$$

Note that for a given initial configuration $l_0 \times h_0$, there are ∞^2 possible deformed configurations, determined by two of the parameters $\bar{\theta}$, h and r_i.

In these conditions, the deformation is described by in-plane cylindrical coordinates as

$$r = f(x_1^0) \qquad \text{and} \qquad \theta = g(x_2^0), \tag{5.66}$$

where the two functions f and g are for the moment unknown. Introducing now the two cylindrical unit vectors

$$\boldsymbol{e}_r = \boldsymbol{e}_1^0 \cos\theta + \boldsymbol{e}_2^0 \sin\theta \qquad \text{and} \qquad \boldsymbol{e}_\theta = -\boldsymbol{e}_1^0 \sin\theta + \boldsymbol{e}_2^0 \cos\theta, \tag{5.67}$$

the deformation is

$$\boldsymbol{x} = r\,\boldsymbol{e}_r = f(x_1^0)\left[\boldsymbol{e}_1^0 \cos g\left(x_2^0\right) + \boldsymbol{e}_2^0 \sin g\left(x_2^0\right)\right], \tag{5.68}$$

and therefore, the deformation gradient takes the form

$$\boldsymbol{F} = f'\,\boldsymbol{e}_r \otimes \boldsymbol{e}_1^0 + f\,g'\,\boldsymbol{e}_\theta \otimes \boldsymbol{e}_2^0, \tag{5.69}$$

where a prime denotes derivative with respect to the relevant variable. Note that the two-point, material (Cartesian) and spatial (cylindrical) bases of the deformation gradient are explicit in Eq. (5.69).

It becomes immediate now to evaluate—from their definitions (3.10)—the right and left Cauchy-Green deformation tensors

$$\begin{aligned} \boldsymbol{C} &= f'^2\,\boldsymbol{e}_1^0 \otimes \boldsymbol{e}_1^0 + (f\,g')^2\,\boldsymbol{e}_2^0 \otimes \boldsymbol{e}_2^0, \\ \boldsymbol{B} &= f'^2\,\boldsymbol{e}_r \otimes \boldsymbol{e}_r + (f\,g')^2\,\boldsymbol{e}_\theta \otimes \boldsymbol{e}_\theta, \end{aligned} \tag{5.70}$$

the right and left stretch tensors

$$\begin{aligned} \boldsymbol{U} &= f'\,\boldsymbol{e}_1^0 \otimes \boldsymbol{e}_1^0 + f\,g'\,\boldsymbol{e}_2^0 \otimes \boldsymbol{e}_2^0, \\ \boldsymbol{V} &= f'\,\boldsymbol{e}_r \otimes \boldsymbol{e}_r + f\,g'\,\boldsymbol{e}_\theta \otimes \boldsymbol{e}_\theta, \end{aligned} \tag{5.71}$$

and the rotation tensor (appearing in the polar decomposition theorem)

$$\boldsymbol{R} = \boldsymbol{e}_r \otimes \boldsymbol{e}_1^0 + \boldsymbol{e}_\theta \otimes \boldsymbol{e}_2^0. \tag{5.72}$$

The stretches therefore are

$$\lambda_1 = f'(x_1^0) \qquad \text{and} \qquad \lambda_2 = f(x_1^0) g'(x_2^0), \qquad \lambda_3 = 1, \tag{5.73}$$

so incompressibility requires that the following differential equation be satisfied:

$$f'(x_1^0) f(x_1^0) g'(x_2^0) = 1, \tag{5.74}$$

from which functions f and g remain determined, except for the three constants A, B and C in the form

$$f(x_1^0) = \sqrt{\frac{2}{A} x_1^0 + C} \qquad \text{and} \qquad g(x_2^0) = A x_2^0 + B, \tag{5.75}$$

showing that the strain is not homogeneous.

Symmetry of the deformed configuration about axis x_1, that is,

$$\theta(-x_2^0) = g(-x_2^0) = -\theta(x_2^0) = -g(x_2^0), \tag{5.76}$$

implies that $B = 0$, so we conclude that the deformed configuration depends on two parameters A and C. These can be rewritten in terms of the more intuitive parameters $\bar{\theta}$, h and r_i [related by Eq. (5.65)] imposing that

- At $x_2^0 = \pm l_0/2$, $\theta = \pm\bar{\theta}$, namely,

$$\pm\bar{\theta} = g(\pm l_0/2), \tag{5.77}$$

  yielding $A = 2\bar{\theta}/l_0$, and
- At $x_1^0 = -h_0/2$, $r = r_i$, namely,

$$r_i = f(-h_0/2), \tag{5.78}$$

  yielding $C = r_i^2 + l_0 h_0/(2\bar{\theta})$. Note that the imposition of the condition $r_e = f(h_0/2)$ leads, through Eq. (5.65), to the same expression just found for C.

  We therefore can conclude that

$$\theta = g(x_0^2) = \frac{2\bar{\theta}}{l_0} x_2^0 \qquad \text{and} \qquad r = f(x_1^0) = \sqrt{\frac{l_0}{\bar{\theta}} x_1^0 + r_i^2 + \frac{l_0 h_0}{2\bar{\theta}}}, \tag{5.79}$$

and a calculation of the derivatives f' and g' yields

$$\lambda_1 = \frac{1}{\lambda_2} = \frac{l_0}{2\bar{\theta}\, r}, \tag{5.80}$$

which shows that the stretches are mere functions of r in the Eulerian description and depend only on x_1^0 in the Lagrangean description:

$$\frac{1}{\lambda_1} = \lambda_2 = \frac{2\bar{\theta}}{l_0} \sqrt{\frac{l_0}{\bar{\theta}} x_1^0 + r_i^2 + \frac{l_0 h_0}{2\bar{\theta}}}. \tag{5.81}$$

Stress Assuming an incompressible neo-Hookean response [Eq. (4.37)], we obtain

$$\boldsymbol{T} = -\pi \boldsymbol{I} + \mu_0 \left(\frac{l_0^2}{4\bar{\theta}^2 r^2} \boldsymbol{e}_r \otimes \boldsymbol{e}_r + \frac{4\bar{\theta}^2}{l_0^2} r^2 \boldsymbol{e}_\theta \otimes \boldsymbol{e}_\theta - \boldsymbol{I} \right), \tag{5.82}$$

or, in terms of the first Piola-Kirchhoff stress,

$$\boldsymbol{S} = \left(-\frac{\pi + \mu_0}{\lambda_1} + \mu_0 \lambda_1 \right) \boldsymbol{e}_r \otimes \boldsymbol{e}_1^0 + \left(-\frac{\pi + \mu_0}{\lambda_2} + \mu_0 \lambda_2 \right) \boldsymbol{e}_\theta \otimes \boldsymbol{e}_2^0. \tag{5.83}$$

Note that the Eulerian nature of the Cauchy stress is explicitly visible in Eq. (5.82), where the Eulerian unit vectors $\boldsymbol{e}_r$ and $\boldsymbol{e}_\theta$ appear, whereas the fact that the first Piola-Kirchhoff stress is a two-point tensor emerges from the expression of the dyads in Eq. (5.83).

The state of stress in not homogeneous, so the local equilibrium equations have to be imposed. This can be done either in the deformed configuration (in terms of

Cauchy stress) or in the reference configuration (in terms of first Piola-Kirchhoff stress). In the former case, the local equilibrium equations are needed in cylindrical form [Eq. (2.149)]. Since there are only two non-null principal stresses, T_r and T_θ, equilibrium becomes

$$\frac{\partial T_r}{\partial r} + \frac{T_r - T_\theta}{r} = 0 \qquad \text{and} \qquad \frac{\partial T_\theta}{\partial \theta} = 0. \tag{5.84}$$

Using Eq. (5.82), we obtain the following partial differential equations for π:

$$\frac{\partial \pi}{\partial r} + \mu_0 \left(\frac{l_0^2}{4\bar{\theta}^2 r^3} + \frac{4\bar{\theta}^2}{l_0^2} r \right) = 0 \qquad \text{and} \qquad \frac{\partial \pi}{\partial \theta} = 0, \tag{5.85}$$

from which we can determine π as the following function of r:

$$\pi = \mu_0 \frac{l_0^2}{8\bar{\theta}^2 r^2} - \mu_0 \frac{2\bar{\theta}^2}{l_0^2} r^2 + D, \tag{5.86}$$

where D is an integration constant, so the stress components become

$$\begin{aligned} T_r(r) &= \mu_0 \left(\frac{l_0^2}{8\bar{\theta}^2 r^2} + \frac{2\bar{\theta}^2}{l_0^2} r^2 - 1 \right) - D, \\ T_\theta(r) &= \mu_0 \left(-\frac{l_0^2}{8\bar{\theta}^2 r^2} + \frac{6\bar{\theta}^2}{l_0^2} r^2 - 1 \right) - D. \end{aligned} \tag{5.87}$$

We are in a position now to impose the boundary conditions, corresponding to null tractions on the external, curved sides:

$$T_r(r_i) = 0 \qquad \text{and} \qquad T_r(r_i + h) = 0, \tag{5.88}$$

with $T_r(r)$ given by Eq. $(5.87)_1$. Equations (5.88) are a non-linear system of equations which fortunately can be solved, thus providing

$$D = \frac{\mu_0}{2} \left[\frac{l_0^2}{4\bar{\theta}^2 (r_i + h)^2} + \frac{4\bar{\theta}^2 (r_i + h)^2}{l_0^2} - 2 \right] \tag{5.89}$$

and

$$h = \frac{l_0}{\bar{\theta}\sqrt{2}} \sqrt{-1 + \sqrt{1 + 4\bar{\theta}^2 \frac{h_0^2}{l_0^2}}}. \tag{5.90}$$

Therefore, the solution now can be obtained in the following steps:

- Chose the initial geometry, l_0 and h_0 and shear modulus μ_0, and fix $\bar{\theta}$.
- h can be calculated from Eq. (5.90), so the deformed geometry becomes known, and r_i can be obtained from Eq. (5.65) and the stretches from Eqs. (5.80).
- D can be calculated from Eq. (5.89), so the stresses T_r and T_θ are obtained from Eqs. (5.87).

Solutions are plotted in Fig. 5.21 for an elastic block having aspect ratio $h_0/l_0 = 1/2$ and for various angles $\bar{\theta} = \{\pi/20, \pi/4, \pi/2, \pi - \pi/20\}$. The corresponding values of h/h_0 have been calculated to be $\{0.997, 0.938, 0.836, 0.694\}$, whereas

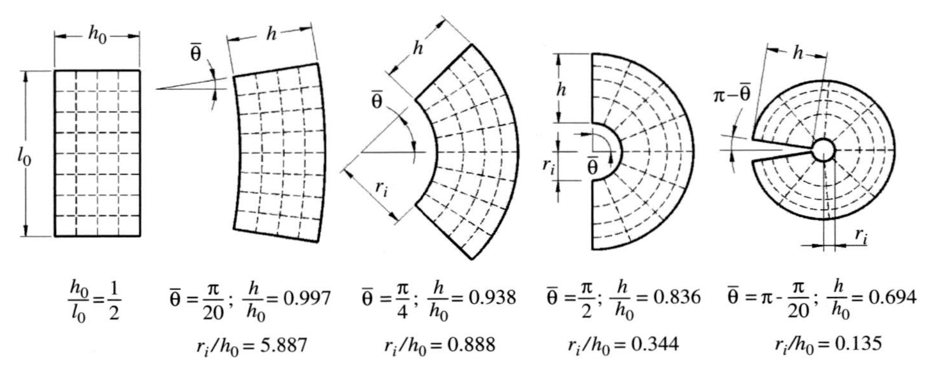

Figure 5.21. Plane strain bending of a rectangular incompressible elastic neo-Hookean block of initial aspect ratio 2 (*left*). *From left to right*: Un-deformed configuration, deformed configurations corresponding to: small ($\bar{\theta}=\pi/20$), moderate ($\bar{\theta}=\pi/4$), large ($\bar{\theta}=\pi/2$), and extreme ($\bar{\theta}=\pi-\pi/20$) bending. Fixing the un-deformed slender ratio h_0/l_0 and the flexure angle $\bar{\theta}$ determines the deformed configuration through parameters r_i/h_0 and h/h_0.

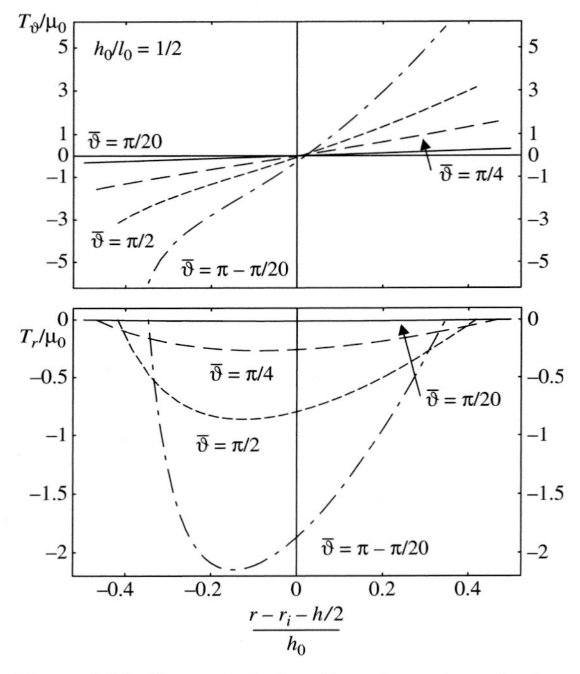

Figure 5.22. Plane strain bending of a rectangular incompressible elastic neo-Hookean block of initial aspect ratio 2. Radial and circumferential Cauchy stress components T_r and T_θ divided by μ_0, corresponding to the deformations shown in Fig. 5.21. Note that the radial stress is small but not null for $\bar{\theta}=\pi/20$.

$r_i/h_0=\{5.887, 0.888, 0.344, 0.135\}$. The stress components T_r and T_θ are plotted in Fig. 5.22 for $r\in[r_i, r_i+h]$ referred to the variable $r-r_i-h/2$, which ranges between $-h/2$ and $h/2$ and is normalized through division by h_0. Interestingly, the circumferential stress remains nearly linear, and its maximum is an order of magnitude greater

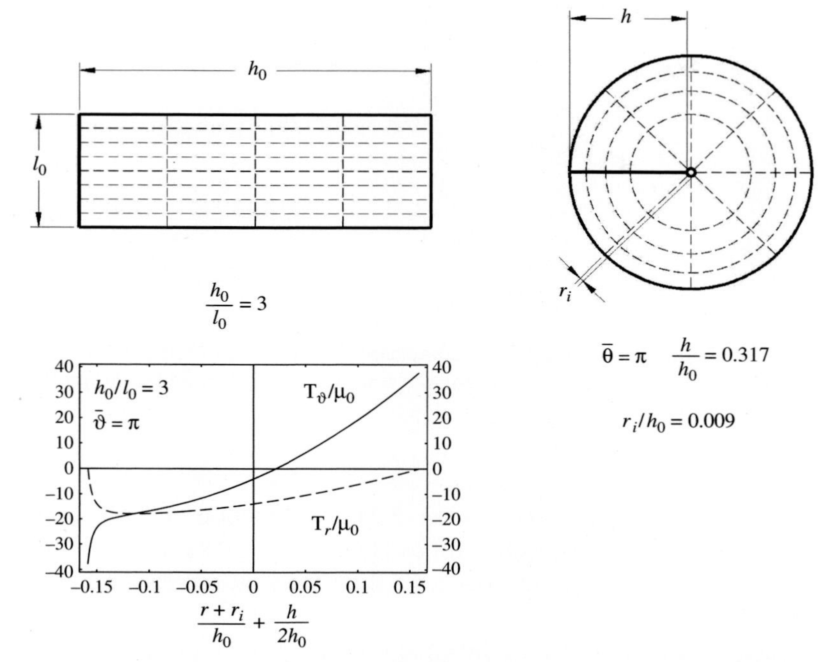

Figure 5.23. Extreme plane strain bending of a rectangular incompressible elastic neo-Hookean block of initial aspect ratio 1/3 (*top*, *left*: reference unloaded configuration; *right*: deformed configuration with $\bar{\theta} = \pi$, $r_i/h_0 = 0.009$, and $h/h_0 = 0.317$). Radial and circumferential Cauchy stress components T_r and T_θ divided by μ_0 (*below*); note that both the radial and circumferential stresses blow up near $r = r_i$, a feature similar to the rise of stress near the tip of a fracture.

than the maximum radial stress at least until large deformations are attained. This circumstance fully confirms the validity of the linear elastic beam theory.

Another interesting feature is that for extreme deformations, the circumferential stress near the maximum curvature, $r = r_i$, tends to blow up when r_i approaches zero (Fig. 5.23), confirming the singular behaviour known in linear elastic fracture mechanics.

The stress resultants at the planar external edges of the deformed configurations can be calculated by integrating T_θ and its moment. The result is that the normal stress resultant N is null, that is,

$$N = \int_{r_i}^{r_i+h} T_\theta(r)\, dr = 0, \tag{5.91}$$

but its moment M is not, that is,

$$\begin{aligned} M &= \int_{r_i}^{r_i+h} r\, T_\theta(r)\, dr \\ &= \mu_0 \frac{h(r_e + r_i)[l_0^2(1 + r_e^4) - 6r_e^2(r_e^2 + r_i^2)\bar{\theta}^2]}{4 l_0^2 r_e^2} + \frac{\mu_0 l_0^2}{8\bar{\theta}^2} \log \frac{r_e}{r_i}, \end{aligned} \tag{5.92}$$

where $r_e = r_i + h$.

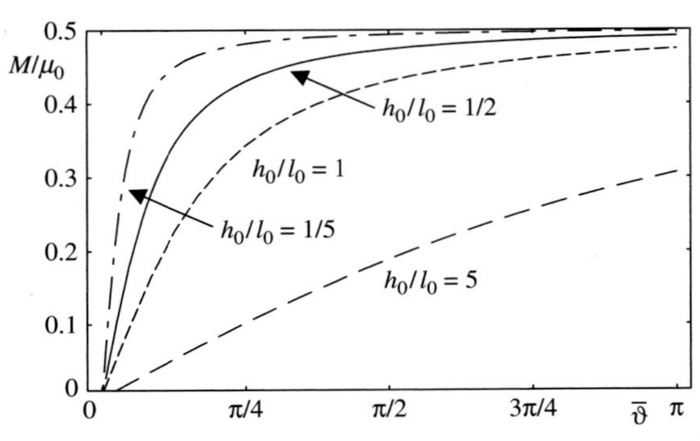

Figure 5.24. Plane strain bending of rectangular incompressible elastic neo-Hookean blocks of initial aspect ratios $\{1/5, 1/2, 1, 5\}$. Bending moment as a function of the bending semi-angle $\bar{\theta}$.

The bending moment M is reported in Fig. 5.24 as a function of the bending angle $\bar{\theta}$ for an elastic block of initial aspect ratios $\{1/5, 1/2, 1, 5\}$. We may note that the bending moment decreases and approaches a limit value at increasing flexure angle $\bar{\theta}$.

Before to conclude, we want to show how the preceding stress problem, solved in a Eulerian description, can be managed using the reference configuration and Lagrangean description.

The components of the first Piola-Kirchhoff stress tensor (5.83) are

$$\begin{aligned} S_{r1} &= -\frac{\pi + \mu_0}{\lambda_1(x_1^0)} + \mu_0 \lambda_1(x_1^0), \\ S_{r2} &= S_{\theta 1} = 0, \\ S_{\theta 2} &= -\frac{\pi + \mu_0}{\lambda_2(x_1^0)} + \mu_0 \lambda_2(x_1^0), \end{aligned} \tag{5.93}$$

so imposing equilibrium,

$$\mathrm{Div}\boldsymbol{S} = \frac{\partial}{\partial x_1^0} S_{r1} \boldsymbol{e}_r + \frac{\partial}{\partial x_2^0} S_{\theta 2} \boldsymbol{e}_\theta = \boldsymbol{0}, \tag{5.94}$$

we obtain

$$\frac{\partial \pi}{\partial x_2^0} = 0 \quad \text{and} \quad \frac{\partial \pi}{\partial x_1^0} = -\frac{\mu_0 l_0}{2\bar{\theta}(h_0 + 2\bar{\theta} r_i^2 / l_0 + 2x_1^0)} - \mu_0 \frac{2\bar{\theta}}{l_0}, \tag{5.95}$$

where we have used the relations

$$\frac{\partial \boldsymbol{e}_r}{\partial x_1^0} = \boldsymbol{0} \quad \text{and} \quad \frac{\partial \boldsymbol{e}_\theta}{\partial x_2^0} = -\boldsymbol{e}_r \frac{\partial \theta}{\partial x_2^0}. \tag{5.96}$$

Note that the same result (5.95) can be obtained by taking the divergence $S_{ij,j} = 0$ using the Cartesian components of the first Piola-Kirchhoff stress:

$$
\begin{aligned}
S_{11} &= \left[-\frac{\pi + \mu_0}{\lambda_1(x_1^0)} + \mu_0 \lambda_1(x_1^0) \right] \cos \frac{2\bar{\theta}}{l_0} x_2^0, \\
S_{12} &= \left[\frac{\pi + \mu_0}{\lambda_2(x_1^0)} - \mu_0 \lambda_2(x_1^0) \right] \sin \frac{2\bar{\theta}}{l_0} x_2^0, \\
S_{21} &= \left[-\frac{\pi + \mu_0}{\lambda_1(x_1^0)} + \mu_0 \lambda_1(x_1^0) \right] \sin \frac{2\bar{\theta}}{l_0} x_2^0, \\
S_{22} &= \left[-\frac{\pi + \mu_0}{\lambda_2(x_1^0)} + \mu_0 \lambda_2(x_1^0) \right] \cos \frac{2\bar{\theta}}{l_0} x_2^0.
\end{aligned}
\tag{5.97}
$$

The differential equations (5.95) can be integrated to give

$$
\pi(x_1^0) = -\frac{\mu_0 l_0}{4\bar{\theta}(h_0 + 2\bar{\theta} r_i^2 / l_0 + 2x_1^0)} - \mu_0 \frac{2\bar{\theta}}{l_0} x_1^0 + E, \tag{5.98}
$$

where E is an integration constant.

Now the stress state, in terms of the first Piola-Kirchhoff tensor, is known (except for the constant E). The boundary conditions are

$$
S_{11}\left(x_1^0 = \pm \frac{h_0}{2} \right) = 0 \quad \text{and} \quad S_{21}\left(x_1^0 = \pm \frac{h_0}{2} \right) = 0. \tag{5.99}
$$

or, in equivalent form,

$$
S_{r1}\left(x_1^0 = \pm \frac{h_0}{2} \right) = 0. \tag{5.100}
$$

The three conditions (5.99) and (5.100) are all satisfied when

$$
\pi\left(x_1^0 = \pm \frac{h_0}{2} \right) + \mu_0 - \mu_0 \lambda_1^2 \left(x_1^0 = \pm \frac{h_0}{2} \right) = 0, \tag{5.101}
$$

which are identical to conditions (5.88), and they now provide E and h.

6 Constitutive equations and anisotropic elasticity

Introduction of (1) material frame indifference, (2) indifference with respect to rigid-body rotations of the reference configuration, and (3) material symmetry classification guides the development of non-linear (both Cauchy and hyper-) elastic constitutive laws for isotropic and anisotropic behaviour (for materials with a micro-structure). Incremental elasticity is introduced with reference to the relative Lagrangean description, and the Biot framework for incompressible elasticity is derived. Hypo-elasticity is also briefly included.

Broadly speaking, constitutive laws set a bridge between strain and stress, so they have to keep into account the behaviour of the specific material under consideration. This behaviour can be fully reversible and described by elasticity (e.g., in the case of rubber) or can be partially irreversible and described by elastoplasticity (e.g., in the case of mild steel), or it may be time-dependent and described by viscous laws (e.g., in the case of a Newtonian fluid). We exclude for the moment the elastoplastic behaviour, and we present constitutive laws in this chapter for anisotropic elastic materials and for incrementally deformed solids (elastoplasticity is deferred to Chapter 8).

The introduction of constitutive laws for materials subject to large strains requires preliminary statements of general principles, such as the so-called material frame indifference and indifference with respect to rigid-body rotations of the reference configuration. When these concepts are specified, the development of constitutive equations to model different materials is greatly simplified. After introduction of these general principles, we conclude the present chapter with the formulation of elastic constitutive laws, generalising concepts proposed in Chapter 4, now to include micro-structural anisotropy. Moreover, we introduce the incremental and rate forms of constitutive equations and the relative Lagrangean description.

6.1 Constitutive equations: General concepts

With the aim of introducing constitutive equations through simple examples, we begin stating that *a constitutive equation for*

- An *elastic material* relates two work-conjugate stress and strain measures (in a linear or non-linear way).

- A *Stokesian fluid* relates the Cauchy stress to the Eulerian strain rate, 'often'[1] in a linear way;
- An *elastoplastic material* relates appropriate rates of work conjugate stress and strain measures in a non-linear way.

In these materials,[2] anisotropy may be inherent or induced (e.g., by stress or strain), and elastoplastic materials have to be considered with reference to a loaded configuration because plasticity may develop only after a certain threshold stress has been reached.

We refer to materials which may possess an intrinsic or induced anisotropy, determined by a referential or spatial micro-structure, expressible in terms of vectors or tensors of arbitrary order.

To retain generality, we define now 'constitutive equation' as a relation in which a — finite or a rate of a — stress measure (briefly, 'stress') is a function of a — finite or a rate of a — strain measure (briefly, 'strain') and of the appropriate micro-structure descriptors (briefly, 'm.d.'), namely,

$$\text{Stress}\,(\text{strain, m.d.})\,. \tag{6.1}$$

Our aim is to introduce and explain with simple examples the general requirements that a constitutive equation must satisfy. These are

- Material frame indifference or invariance of material response with respect to a change in observer
- Indifference with respect to rigid-body rotations of the reference configuration
- Material symmetries, defining the type of anisotropy

6.1.1 Change in observer and related principle of invariance of material response

In order to introduce the principle of invariance of material response under a change of observer, we start considering the following ideal experiment borrowed from Gurtin (1981) and sketched in Fig. 6.1: A weight attached to a spring does not move in one situation, so it elongates the spring under the effect of gravity, whereas the same weight in another situation is rotated at a constant speed, precisely selected to produce the same elongation in the spring as in the static situation. We intuitively are prone to believe that if the elongation in the spring is the same in the two situations, then the internal force is also the same. Note that the two motions of the spring differ in a change in observer, so the ideal experiment can be rationalised in the principle of invariance of material response under a change in observer as follows (Noll, 1958; Truesdell and Noll, 1965; see also Gurtin, 1981):

Two observers *adopting the same reference configuration* B_0 record the same motion, modulo a rigid-body displacement of the current configuration (so they will view the two configurations B and B^* as in Fig. 3.11b), which represents the

[1] A remarkable exception is the case of the so-called Reiner-Rivlin fluid, where the Cauchy stress is a function of $\boldsymbol{D}$ and $\boldsymbol{D}^2$ (Reiner, 1945; Rivlin, 1948b).

[2] As already remarked, the classification is far from exhaustive. For instance, the Cauchy stress may depend on the strain and the Eulerian strain rate in viscoelastic materials.

Figure 6.1. Material frame indifference seen from the point of view of an ant sitting on a weight hanging on a deformable spring. The ant sees the same elongation in the spring when the weight is at rest and when it is rotating at a certain constant angular velocity. The ant therefore believes the force generated in the spring to be the same in the two situations.

movement of one observer with respect to the other. The situation at a given instant is that sketched in Fig. 3.11*b*, so kinematic quantities transform from B to B^* as specified in Table 3.1, right column, and stress measures as specified in Table 3.2, right column. If micro-structures are embedded in the continuum to describe an intrinsic or induced anisotropy, these also will transform with their specific rules. We denote through the three operators

$$\mathbf{Q}^B_{\text{stress}}[\cdot], \qquad \mathbf{Q}^B_{\text{strain}}[\cdot] \qquad \mathbf{Q}^B_{\text{micro}}[\cdot], \tag{6.2}$$

(where the superscript B reminds us that the spatial configuration B is rotated at fixed B_0) the appropriate transformations of stress, strain measures, and micro-structure descriptors consequent to a rotation of the spatial configuration with fixed reference configuration.

With reference to the constitutive equation (6.1), *the principle of invariance of material response under a change in observer writes as*

$$\mathbf{Q}^B_{\text{stress}}[\text{stress (strain, m.d.)}] = \text{stress}\left(\mathbf{Q}^B_{\text{strain}}[\text{strain}], \mathbf{Q}^B_{\text{micro}}[\text{m.d.}]\right), \tag{6.3}$$

holding for all rotations $\boldsymbol{Q} \in \text{Orth}^+$.

For instance, let us consider the constitutive relations between stress and deformation measures listed below and the way they transform under a change in observer.

Type	Constitutive eqn.	Rigid-body rotation of B
a.	$\boldsymbol{T} = -p\boldsymbol{I} + 2\eta_T\boldsymbol{D}$,	$\boldsymbol{QTQ}^T = -p\boldsymbol{I} + 2\eta_T\boldsymbol{QDQ}^T$,
b.	$\boldsymbol{S} = -\pi\boldsymbol{F}^{-T} + \mu\left(\boldsymbol{F} - \boldsymbol{F}^{-T}\right)$,	$\boldsymbol{QS} = -\pi\boldsymbol{QF}^{-T} + \mu\left(\boldsymbol{QF} - \boldsymbol{QF}^{-T}\right)$,
c.	$\boldsymbol{T}^{(2)} = \lambda\text{tr}\boldsymbol{E}^{(2)}\boldsymbol{I} + 2\mu\boldsymbol{E}^{(2)}$,	$\boldsymbol{T}^{(2)} = \lambda\text{tr}\boldsymbol{E}^{(2)}\boldsymbol{I} + 2\mu\boldsymbol{E}^{(2)}$.

(6.4)

Therefore, Eq. (6.3) is understood in three forms without micro-structure, so $\mathbf{Q}^B_{\text{micro}}$ is not defined, and with the following definitions.

$$\begin{array}{llll} \text{Type} & \text{Stress} & \text{Strain} & \text{Rotation tensors in Eq. (6.3)} \\ a. & \boldsymbol{T} & \boldsymbol{D} & \mathbf{Q}^B_{\text{stress}} = \mathbf{Q}^B_{\text{strain}} = \boldsymbol{Q}\boxtimes\boldsymbol{Q}, \\ b. & \boldsymbol{S} & \boldsymbol{F} & \mathbf{Q}^B_{\text{stress}} = \mathbf{Q}^B_{\text{strain}} = \boldsymbol{Q}\boxtimes\boldsymbol{I}, \\ c. & \boldsymbol{T}^{(2)} & \boldsymbol{E}^{(2)} & \mathbf{Q}^B_{\text{stress}} = \mathbf{Q}^B_{\text{strain}} = \boldsymbol{I}\boxtimes\boldsymbol{I}, \end{array} \tag{6.5}$$

The constitutive equation $(6.4)_1$ [where p represents the incompressibility constraint, η_T is a viscosity coefficient and $\boldsymbol{D}$ the Eulerian strain rate (3.59)] defines an incompressible Newtonian fluid. The constitutive equation $(6.4)_2$ (where π defines the incompressibility constraint, μ is a shear modulus and $\boldsymbol{S}$ and $\boldsymbol{F}$ are the first Piola-Kirchhoff and the deformation gradient, respectively) defines the neo-Hookean incompressible elastic material [Eq. (4.37)]. The constitutive equation $(6.4)_3$ (where λ and μ are constants) defines the Kirchhoff–Saint Venant elastic material [Eq. (4.15)]. Condition (6.3) is satisfied for all constitutive equations (6.4), so these are material frame indifferent.

Note that all the examples (6.4) do not involve any intrinsic anisotropy.

The following relation between the material time derivative of the Cauchy stress and Eulerian strain rate

$$\dot{\boldsymbol{T}} = (\lambda \text{tr}\boldsymbol{D})\boldsymbol{I} + 2\mu\boldsymbol{D} - \frac{\boldsymbol{D}\cdot\text{dev}\boldsymbol{T}}{g}\text{dev}\boldsymbol{T}, \tag{6.6}$$

where $\text{dev}\boldsymbol{T} = \boldsymbol{T} - \text{tr}\boldsymbol{T}/3$ is the deviator of the Cauchy stress, λ and μ are the elastic Lamé constants and g a plastic modulus, defines the plastic branch of a von Mises elastoplastic model. Note that there is an anisotropy in the model induced by the deviatoric Cauchy stress $\text{dev}\boldsymbol{T}$, so stress $\longrightarrow \dot{\boldsymbol{T}}$, strain $\longrightarrow \boldsymbol{D}$ and micro $\longrightarrow \text{dev}\boldsymbol{T}$ and

$$\mathbf{Q}^B_{\text{strain}} = \mathbf{Q}^B_{\text{micro}} = \boldsymbol{Q}\boxtimes\boldsymbol{Q}, \tag{6.7}$$

but

$$\mathbf{Q}^B_{\text{stress}}[\dot{\boldsymbol{T}}] = \boldsymbol{Q}\dot{\boldsymbol{T}}\boldsymbol{Q}^T + \dot{\boldsymbol{Q}}\boldsymbol{T}\boldsymbol{Q}^T + \boldsymbol{Q}\boldsymbol{T}\dot{\boldsymbol{Q}}^T. \tag{6.8}$$

Therefore, the model (6.6) is properly defined only in a geometrically linear theory because it tranforms under a change of observer as

$$\boldsymbol{Q}\dot{\boldsymbol{T}}\boldsymbol{Q}^T + \dot{\boldsymbol{Q}}\boldsymbol{T}\boldsymbol{Q}^T + \boldsymbol{Q}\boldsymbol{T}\dot{\boldsymbol{Q}}^T = (\lambda\text{tr}\boldsymbol{D})\boldsymbol{I} + 2\mu\boldsymbol{Q}\boldsymbol{D}\boldsymbol{Q}^T - \frac{\boldsymbol{D}\cdot\text{dev}\boldsymbol{T}}{g}\boldsymbol{Q}(\text{dev}\boldsymbol{T})\boldsymbol{Q}^T \tag{6.9}$$

and therefore *violates the principle of material frame indifference* [Eq. (6.3)], so Eq. (6.6) cannot be used for finite strain analyses (although it has been used sometimes). The constitutive equation (6.6) can be 'fixed' from the point of view of material frame indifference by replacing $\dot{\boldsymbol{T}}$ with one of the stress derivatives (3.156), for instance, the Jaumann derivative of Kirchhoff stress, thus obtaining in this case the elastoplastic model suggested by Key and Krieg (1982):

$$\overset{\triangledown}{\boldsymbol{K}} = (\lambda\text{tr}\boldsymbol{D})\boldsymbol{I} + 2\mu\boldsymbol{D} - \frac{\text{dev}\boldsymbol{T}\cdot\boldsymbol{D}}{g}\text{dev}\boldsymbol{T} \tag{6.10}$$

so that here stress $\longrightarrow \overset{\nabla}{\boldsymbol{K}}$ and now

$$\mathbf{Q}^B_{\text{stress}} = \mathbf{Q}^B_{\text{strain}} = \mathbf{Q}^B_{\text{micro}} = \boldsymbol{Q} \boxtimes \boldsymbol{Q}. \tag{6.11}$$

The constitutive Eqs. (6.6) and (6.10) are in rate form and incorporate a stress-induced anisotropy (through the terms $\text{dev}\boldsymbol{T}$).

Let us consider now a fibre-reinforced incompressible fluid, inextensible in the fibre direction, as defined by Rogers (1989) and Spencer (1997, 2004)

$$\boldsymbol{T} = -p\boldsymbol{I} + T\boldsymbol{a} \otimes \boldsymbol{a} + 2\eta_T \boldsymbol{D} + 2(\eta_L - \eta_T)(\boldsymbol{a} \otimes \boldsymbol{D}\boldsymbol{a} + \boldsymbol{D}\boldsymbol{a} \otimes \boldsymbol{a}), \tag{6.12}$$

where p and T, respectively, the pressure and the tension in the fibre direction, define the so-called reaction stress, η_T and η_L are viscosity coefficients, $\boldsymbol{a}$ is the (spatial) unit vector defining the direction of fibres embedded in the fluid and $\boldsymbol{D}$ is the Eulerian strain rate (3.59). The constitutive law (6.12) transforms in the following way under a rigid-body rotation of the spatial configuration

$$\boldsymbol{Q}\boldsymbol{T}\boldsymbol{Q}^T = -p\boldsymbol{I} + T\boldsymbol{Q}\boldsymbol{a} \otimes \boldsymbol{Q}\boldsymbol{a} + 2\eta_T \boldsymbol{Q}\boldsymbol{D}\boldsymbol{Q}^T + 2(\eta_L - \eta_T)(\boldsymbol{Q}\boldsymbol{a} \otimes \boldsymbol{Q}\boldsymbol{D}\boldsymbol{a} + \boldsymbol{Q}\boldsymbol{D}\boldsymbol{a} \otimes \boldsymbol{Q}\boldsymbol{a}), \tag{6.13}$$

showing that the constitutive law is frame indifferent. Note that the spatial vector $\boldsymbol{a}$ defining the micro-structure transforms as $\boldsymbol{Q}\boldsymbol{a}$, so

$$\mathbf{Q}^B_{\text{stress}}[\cdot] = \mathbf{Q}^B_{\text{strain}}[\cdot] = \boldsymbol{Q} \boxtimes \boldsymbol{Q}[\cdot], \quad \mathbf{Q}^B_{\text{micro}}[\cdot] = \boldsymbol{Q}(\cdot), \tag{6.14}$$

where stress $\longrightarrow \boldsymbol{T}$, strain $\longrightarrow \boldsymbol{D}$ and micro $\longrightarrow \boldsymbol{a}$.

Finally, we consider a particular case of the transversely isotropic Kirchhoff–Saint Venant model proposed by Bonet and Burton (1998). Here the anisotropy is intrinsic and described by the *referential* unit vector $\boldsymbol{a}_0$, so we obtain

$$\boldsymbol{T}^{(2)} = \lambda \left(\text{tr}\boldsymbol{E}^{(2)}\right)\boldsymbol{I} + 2\mu \boldsymbol{E}^{(2)} + \gamma \left(\boldsymbol{a}_0 \cdot \boldsymbol{E}^{(2)}\boldsymbol{a}_0\right) \boldsymbol{a}_0 \otimes \boldsymbol{a}_0 \tag{6.15}$$

(where λ, μ and γ are constants), a Kirchhoff–Saint Venant material reinforced by fibres parallel to $\boldsymbol{a}_0$ (see also Section 6.1.5). Under a rigid-body rotation of the spatial configuration, constitutive Eq. (6.15) remains unchanged, so

$$\mathbf{Q}^B_{\text{stress}}[\cdot] = \mathbf{Q}^B_{\text{strain}}[\cdot] = \boldsymbol{I} \boxtimes \boldsymbol{I}[\cdot], \quad \mathbf{Q}^B_{\text{micro}}[\cdot] = \boldsymbol{I}(\cdot) \tag{6.16}$$

where stress $\longrightarrow \boldsymbol{T}^{(2)}$, strain $\longrightarrow \boldsymbol{E}^{(2)}$ and micro $\longrightarrow \boldsymbol{a}_0 \otimes \boldsymbol{a}_0$, and the constitutive equation satisfies material frame indifference.

6.1.2 Indifference with respect to rigid-body rotation of the reference configuration

In the principle of material frame indifference, the two observers employ the same reference configuration, and the spatial configurations differ in a rigid-body rotation, according with Tables 3.1 and 3.2, right columns (and also micro-structures have to be rotated accordingly). Our aim now is to introduce a 'material invariance', that is, an indifference with respect to a rotation of the reference configuration, following the rules reported in Tables 3.1 and 3.2, left columns. This indifference is inherent to the fact that the choice of the reference configuration is arbitrary, so a material

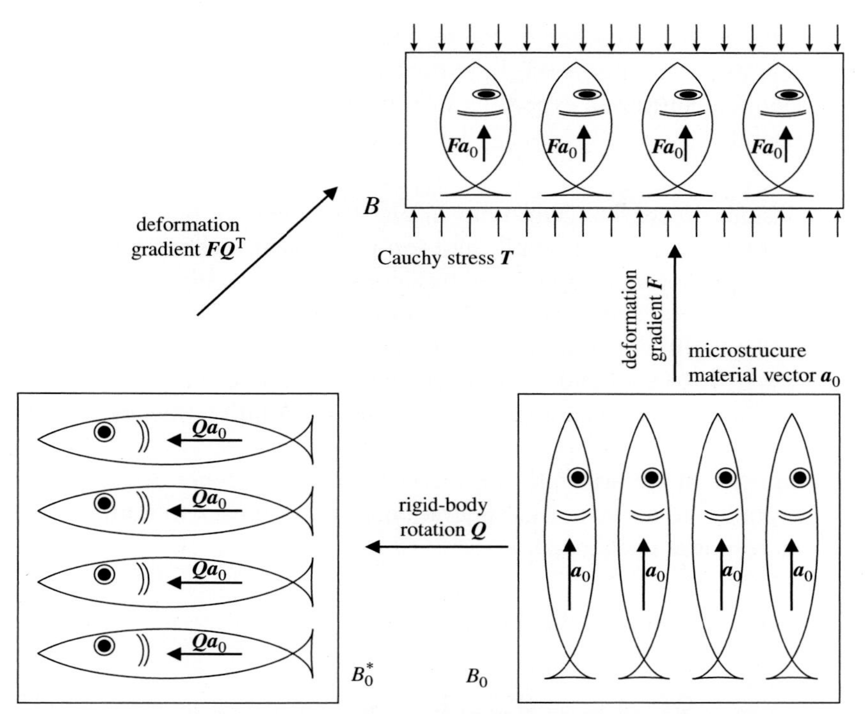

Figure 6.2. Invariance with respect to a rigid-body rotation of the reference configuration. The fish is a symbol to represent a micro-structure, defined by referential vector $\boldsymbol{a}_0$, so the material in this example is orthotropic. Applied to the unloaded configuration B_0^* (rotated $\pi/2$ anti-clockwise with respect to B_0, so vector $\boldsymbol{a}_0$ embedded in B_0 is rotated to $\boldsymbol{Q}\boldsymbol{a}_0$), a deformation gradient FQ^T yields the same final configuration B obtained by applying $\boldsymbol{F}$ (taken uniform for simplicity and corresponding to a uniform Cauchy stress) to B_0.

response cannot be affected by this choice (Truesdell, 1966). This statement becomes particularly simple when two reference configurations are considered, differing in a rigid-body rotation.

Let us therefore consider the following two gedanken experiments referred for simplicity to an elastic material possessing an unloaded initial configuration (see the sketch in Fig. 6.2):

- *Experiment 1.* A body is deformed from the initial unloaded configuration B_0 to B through an arbitrary deformation gradient $\boldsymbol{F}$.
- *Experiment 2.* The same body is considered, and the initial unloaded configuration, B_0 is first rotated through the rigid-body rotation $\boldsymbol{Q}$ to the new initial unloaded configuration B_0^*, assumed as a new reference configuration. In this rigid-body transformation, all material micro-structures are rotated, so, for instance, if a micro-structure is represented by a material vector $\boldsymbol{a}_0$ (or by a material tensor $\boldsymbol{A}_0$), this is rotated to $\boldsymbol{Q}\boldsymbol{a}_0$ (to $\boldsymbol{Q}\boldsymbol{A}_0\boldsymbol{Q}^T$). At this point, B_0^* is deformed in a way to obtain *the same spatial configuration* reached in experiment 1, namely, with the deformation gradient used in experiment 1 but rotated with a rigid-body rotation $\boldsymbol{Q}$ as in Table 3.1, left column; in other words, B_0^* is subject to $\boldsymbol{F}\boldsymbol{Q}^T$.

It is clear that the situation reached in the second experiment is exactly the same as that reached in the first experiment. In particular, it is expected that the Cauchy stress and all spatial quantities in experiment 2 remain the same as in experiment 1.

More in general:

- We now consider the two configurations B_0 and B of a generic body. At the two points $\boldsymbol{x}_0$ and $\boldsymbol{x}$, all referential, spatial and two-point quantities are defined, and some of them appropriately are related by a constitutive equation.
- If we now rigidly rotate *both* B_0 *and* B, with the same rigid-body rotation, to B_0^* and B^*, all quantities simultaneously rotate, combining together the left and right columns of Tables 3.1 and 3.2. For instance, both $\boldsymbol{C}$ and $\boldsymbol{B}$ rotate to $\boldsymbol{QCQ}^T$ and $\boldsymbol{QBQ}^T$. This rotation changes nothing from the point of view of material response.
- As a consequence of the principle of material frame indifference, we can give a rigid-body rotation to the spatial configuration, bringing it back to B but leaving the material response unchanged.

As a conclusion, denoting by the three operators

$$\mathbf{Q}^{B_0}_{\text{stress}}[\cdot], \qquad \mathbf{Q}^{B_0}_{\text{strain}}[\cdot], \qquad \mathbf{Q}^{B_0}_{\text{micro}}[\cdot], \tag{6.17}$$

the appropriate transformations of stress and strain measures and micro-structure descriptors consequent to a rigid-body rotation of the referential configuration B_0 at fixed B, reminded by the superscript B_0 in Eq. (6.17), we conclude that a constitutive equation must satisfy *the invariance with respect a rigid-body rotation of the reference configuration*:

$$\mathbf{Q}^{B_0}_{\text{stress}}[\text{stress}\,(\text{strain, m.d.})] = \text{stress}\left(\mathbf{Q}^{B_0}_{\text{strain}}[\text{strain}], \mathbf{Q}^{B_0}_{\text{micro}}[\text{m.d.}]\right), \tag{6.18}$$

holding for all rotations $\boldsymbol{Q} \in \mathsf{Orth}^+$.

Condition (6.18) has been used by Boehler (1987), Ogden (2001) and Spencer (1984).

To provide some simple examples of application of the indifference with respect to rigid-body rotations of the reference configuration, let us consider once more the constitutive relations (6.4); the way they transform under a rigid-body rotation of the reference configuration is shown on the right.

Type	Constitutive Equation	Rigid-body rotation of B_0
a.	$\boldsymbol{T} = -p\boldsymbol{I} + 2\eta_T\boldsymbol{D}$,	$\boldsymbol{T} = -p\boldsymbol{I} + 2\eta_T\boldsymbol{D}$,
b.	$\boldsymbol{S} = -\pi\boldsymbol{F}^{-T} + \mu\left(\boldsymbol{F} - \boldsymbol{F}^{-T}\right)$,	$\boldsymbol{SQ}^T = -\pi\boldsymbol{F}^{-T}\boldsymbol{Q}^T + \mu\left(\boldsymbol{FQ}^T - \boldsymbol{F}^{-T}\boldsymbol{Q}^T\right)$,
c.	$\boldsymbol{T}^{(2)} = \lambda\text{tr}\boldsymbol{E}^{(2)}\boldsymbol{I} + 2\mu\boldsymbol{E}^{(2)}$,	$\boldsymbol{QT}^{(2)}\boldsymbol{Q}^T = \left(\lambda\text{tr}\boldsymbol{E}^{(2)}\right)\boldsymbol{I} + 2\mu\boldsymbol{QE}^{(2)}\boldsymbol{Q}^T$.

(6.19)

Therefore, Eq. (6.18) is understood in three forms without micro-structure, so $\mathbf{Q}_{\text{micro}}^{B_0}$ is not defined, and with the following definitions:

$$\begin{array}{llll} \text{Type} & \text{Stress} & \text{Strain} & \text{Rotation tensors in Eq. (6.18)} \\ a. & \boldsymbol{T} & \boldsymbol{D} & \mathbf{Q}_{\text{stress}}^{B_0} = \mathbf{Q}_{\text{strain}}^{B_0} = \boldsymbol{I} \boxtimes \boldsymbol{I}, \\ b. & \boldsymbol{S} & \boldsymbol{F} & \mathbf{Q}_{\text{stress}}^{B_0} = \mathbf{Q}_{\text{strain}}^{B_0} = \boldsymbol{I} \boxtimes \boldsymbol{Q}, \\ c. & \boldsymbol{T}^{(2)} & \boldsymbol{E}^{(2)} & \mathbf{Q}_{\text{stress}}^{B_0} = \mathbf{Q}_{\text{strain}}^{B_0} = \boldsymbol{Q} \boxtimes \boldsymbol{Q}. \end{array} \tag{6.20}$$

Finally, the constitutive equation (6.12) remains unchanged under a rigid-body rotation of the reference confguration, so in this case

$$\mathbf{Q}_{\text{stress}}^{B_0}[\cdot] = \mathbf{Q}_{\text{strain}}^{B_0}[\cdot] = \boldsymbol{I} \boxtimes \boldsymbol{I}[\cdot] \quad \text{and} \quad \mathbf{Q}_{\text{micro}}^{B_0}[\cdot] = \boldsymbol{I}(\cdot), \tag{6.21}$$

where stress $\longrightarrow \boldsymbol{T}$, strain $\longrightarrow \boldsymbol{D}$ and micro $\longrightarrow \boldsymbol{a}$, while all quantities (including $\boldsymbol{a}_0$) are transformed according to Table 3.1 left column, for the constitutive Eq. (6.15) so in this case

$$\mathbf{Q}_{\text{stress}}^{B_0}[\cdot] = \mathbf{Q}_{\text{strain}}^{B_0}[\cdot] = \boldsymbol{Q} \boxtimes \boldsymbol{Q}[\cdot] \quad \text{and} \quad \mathbf{Q}_{\text{micro}}^{B_0}[\cdot] = \boldsymbol{Q}(\cdot), \tag{6.22}$$

where stress $\longrightarrow \boldsymbol{T}^{(2)}$, strain $\longrightarrow \boldsymbol{E}^{(2)}$ and micro $\longrightarrow \boldsymbol{a}_0$.

As a conclusion, since for all constitutive equations (6.4), (6.12), and (6.15), condition (6.18) is satisfied, these are indifferent with respect to arbitrary rotations of the reference configuration.

6.1.3 Material symmetries

The principle of material frame indifference (Section 6.1.1) and the invariance under rigid-body rotation of the reference configuration (Section 6.1.2) set restrictions which are independent of the specific material under consideration. These two conditions do not allow discrimination between isotropy or anisotropy of material response. This discrimination is our next subject.

We follow here Boehler (1987) and start considering the following two gedanken experiments (sketched in Fig. 6.3 with reference for simplicity to elasticity):

- *Experiment 1.* In a first situation, a homogeneous stress field in a homogeneous body is produced by a homogeneous deformation agent $\boldsymbol{F}$.
- *Experiment 2.* In a second situation, the body considered in experiment 1 but with the micro-structure rotated by $\boldsymbol{Q}$ is subject to the homogeneous deformation agent $\boldsymbol{F}$ employed in experiment 1.

If the stress responses at the end of the two experiments are the same, $\boldsymbol{Q}$ belongs to the symmetry class of the material. In mathematical terms and denoting with $\mathbf{Q}_{\text{micro}}^{B_0 \text{ or } B}$ the appropriate rotation of the micro-structure, if

$$\text{Stress (strain, m.d.)} = \text{stress}\left(\text{strain}, \mathbf{Q}_{\text{micro}}^{B_0 \text{ or } B}[\text{m.d.}]\right), \tag{6.23}$$

then $\mathbf{Q}_{\text{micro}}^{B_0 \text{ or } B}$ belongs to the *symmetry class of the material.*

If condition (6.23) holds for every rotation, the material is *isotropic*, which is equivalent to saying that there are no micro-structures providing anisotropy.

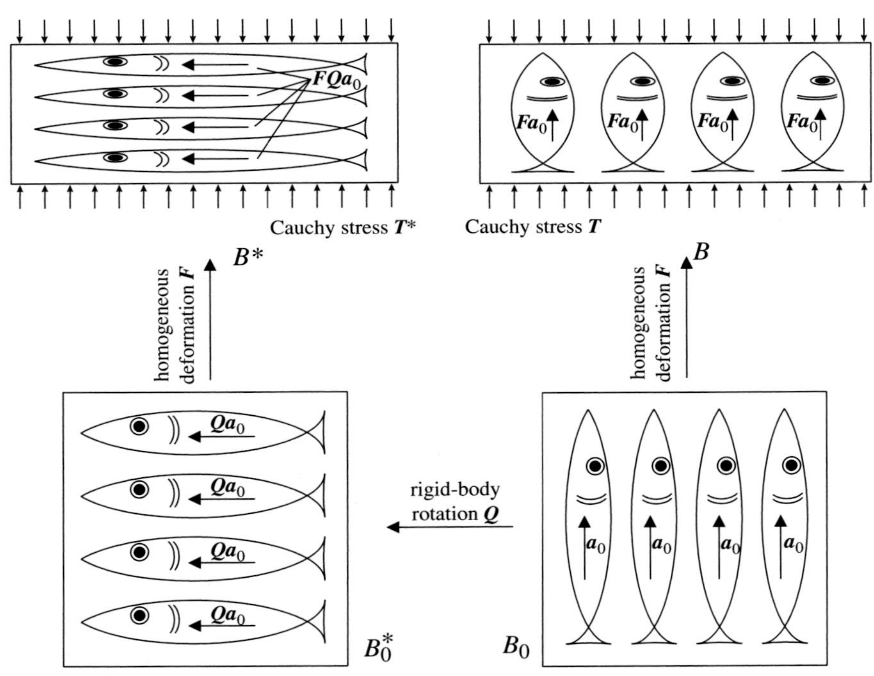

Figure 6.3. Sketch of material symmetry concept with reference to an elastic material in which for simplicity configurations B_0 and B_0^* are both unloaded. The fish is a symbol to represent a micro-structure, which makes the two responses different and the material of this example orthotropic. The unloaded configuration B_0^* is obtained through a $\pi/2$ rigid-body anti-clockwise rotation of B_0. B and B^* are obtained from B_0 and B_0^* with the same deformation gradient $\boldsymbol{F}$ (taken uniform for simplicity and corresponding to a uniform stress). In particular, the two Cauchy stresses obtained with the two experiments, $\boldsymbol{T}$ and $\boldsymbol{T}^*$, are in general different, except when $\boldsymbol{Q}$ belongs to the symmetry class of the material, in this example, for instance, when the rotation of B is of a π angle.

Employing condition (6.23), we can conclude immediately that constitutive equations (6.4) describe isotropic material responses (in fact, there is no micro-structure to be rotated).

Both constitutive equations (6.12) and (6.15) describe an anisotropic behaviour. In particular, the following tensor $\boldsymbol{Q}^B$ is a spatial rotation α around $\boldsymbol{a}$, that is,

$$\boldsymbol{Q}^B = (1-\cos\alpha)\,\boldsymbol{a}\otimes\boldsymbol{a}+\cos\alpha\,\boldsymbol{I}+\sin\alpha(\boldsymbol{t}\otimes\boldsymbol{a}\times\boldsymbol{t}-\boldsymbol{a}\times\boldsymbol{t}\otimes\boldsymbol{t}), \tag{6.24}$$

where $\boldsymbol{t}$ is a vector orthogonal to $\boldsymbol{a}$ ($\boldsymbol{a}\cdot\boldsymbol{t}=0$). This rotation leaves unchanged the response (6.12) because $\boldsymbol{Q}^B\boldsymbol{a}=\boldsymbol{a}$. The material version of Eq. (6.24) is

$$\boldsymbol{Q}^{B_0} = (1-\cos\alpha)\,\boldsymbol{a}_0\otimes\boldsymbol{a}_0+\cos\alpha\,\boldsymbol{I}+\sin\alpha(\boldsymbol{t}_0\otimes\boldsymbol{a}_0\times\boldsymbol{t}_0-\boldsymbol{a}_0\times\boldsymbol{t}_0\otimes\boldsymbol{t}_0), \tag{6.25}$$

where $\boldsymbol{t}_0$ is a vector orthogonal to $\boldsymbol{a}_0$ ($\boldsymbol{a}_0\cdot\boldsymbol{t}_0=0$), so $\boldsymbol{Q}^{B_0}$ is a referential rotation α around $\boldsymbol{a}_0$, leaving unchanged the response (6.15) because $\boldsymbol{Q}^{B_0}\boldsymbol{a}_0=\boldsymbol{a}_0$.

As a conclusion, both materials [(6.12) and (6.15)] are examples of transversely isotropic materials.

Following Gurtin (1981), material symmetry can be introduced in elasticity with reference to the following gedanken experiments (see Fig. 6.4).

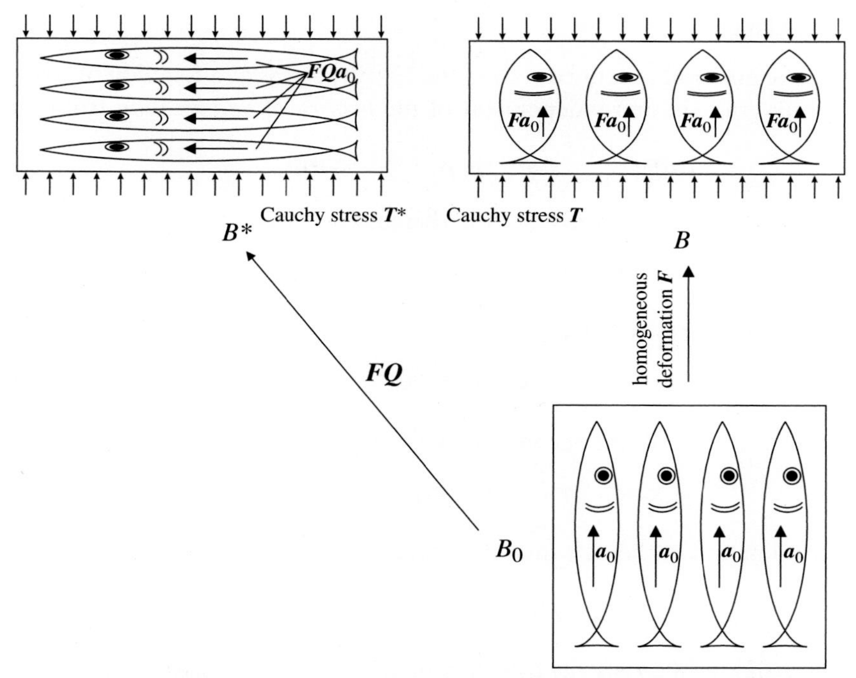

Figure 6.4. Sketch of material symmetry concept for elastic materials. B and B^* are both obtained from the same unloaded configuration B_0 using the two deformation gradients $\boldsymbol{F}$ and $\boldsymbol{FQ}$, respectively. The two Cauchy stresses generated with $\boldsymbol{F}$ and $\boldsymbol{FQ}$ are in general different, except when $\boldsymbol{Q}$ belongs to the symmetry class of the material. In the sketch, $\boldsymbol{Q}$ is a $\pi/2$ rigid-body anti-clockwise rotation of B_0, and the fish is a symbol to represent a micro-structure, which makes the two responses different and the material of this example orthotropic.

- In a first situation, a body in the reference configuration B_0 is deformed to B via a homogeneous deformation $\boldsymbol{F}$.
- In a second situation, an identical body in the reference configuration B_0 is subject to a deformation gradient $\boldsymbol{FQ}$ rigidly rotating B_0 to B_0^* and 'subsequently' providing the same homogeneous deformation as in the first situation. Note that the micro-structure is left fixed in this experiment.

If the two Cauchy stresses at the end of the two experiments are equal, then the rotation belongs to a symmetry class of the material, namely, assuming a constitutive equation in the form

$$\boldsymbol{T} = \hat{\boldsymbol{T}}(\boldsymbol{F}, \text{m.d.}), \tag{6.26}$$

If (note that 'm.d.' is kept fixed)

$$\hat{\boldsymbol{T}}(\boldsymbol{F}, \text{m.d.}) = \hat{\boldsymbol{T}}(\boldsymbol{FQ}, \text{m.d.}), \tag{6.27}$$

$\boldsymbol{Q}$ belongs to the symmetry class of the material. Owing to the invariance under rigid-body rotation of the material configuration [Eq. (6.18)], conditions (6.27) and (6.23) are equivalent.

6.1.4 Cauchy elasticity

A Cauchy elastic material is defined setting the Cauchy stress as a function of solely the deformation gradient and independent of the *history of deformation* from the reference state:

$$\boldsymbol{T} = \hat{\boldsymbol{T}}(\boldsymbol{F}), \tag{6.28}$$

which can be transformed into the first Piola-Kirchhoff stress

$$\boldsymbol{S} = J\hat{\boldsymbol{T}}(\boldsymbol{F})\boldsymbol{F}^{-T}, \tag{6.29}$$

which therefore can be rewritten as

$$\boldsymbol{S} = \hat{\boldsymbol{S}}(\boldsymbol{F}) \tag{6.30}$$

and can be transformed into the second Piola-Kirchhoff stress

$$\boldsymbol{T}^{(2)} = \boldsymbol{F}^{-1}\hat{\boldsymbol{S}}(\boldsymbol{F}), \tag{6.31}$$

which is also a function of the deformation gradient, namely,

$$\boldsymbol{T}^{(2)} = \hat{\boldsymbol{T}}^{(2)}(\boldsymbol{F}). \tag{6.32}$$

Function (6.32) must satisfy the principle of material frame indifference (6.3)

$$\boldsymbol{T}^{(2)} = \hat{\boldsymbol{T}}^{(2)}(\boldsymbol{QF}), \qquad \forall \boldsymbol{Q} \in \mathsf{Orth}^+. \tag{6.33}$$

Moreover, the constitutive equation (6.33) must hold for every deformation gradient $\boldsymbol{F}$, which, using the polar decomposition theorem (3.15), can be written as $\boldsymbol{F} = \boldsymbol{RU}$. Therefore, it can be concluded that it is always possible to express $\boldsymbol{T}^{(2)}$ as a function of $\boldsymbol{U}$. Finally, $\boldsymbol{U}$ can be written as a function of the Green-Lagrange strain tensor $\boldsymbol{E}^{(2)}$, so we arrive at the expression

$$\boldsymbol{T}^{(2)} = \tilde{\boldsymbol{T}}^{(2)}(\boldsymbol{E}^{(2)}). \tag{6.34}$$

More in general than (6.34), we define an elastic Cauchy law as a relation between two conjugate stress and strain measures in the form

$$\boldsymbol{T}^{(m)} = \tilde{\boldsymbol{T}}^{(m)}(\boldsymbol{E}^{(m)}). \tag{6.35}$$

Imposing Eq. (6.35) to satisfy invariance with respect to rigid-body rotations of the reference configuration (6.18),

$$\boldsymbol{Q}\boldsymbol{T}^{(m)}\boldsymbol{Q}^T = \hat{\boldsymbol{T}}^{(m)}(\boldsymbol{Q}\boldsymbol{E}^{(m)}\boldsymbol{Q}^T), \qquad \forall \boldsymbol{Q} \in \mathsf{Orth}^+. \tag{6.36}$$

Thus, using our notation, *we are implicitly assuming an isotropic material behaviour* because we are not giving evidence to any micro-structure.

Equation (6.36) expresses the fact that function $\boldsymbol{T}^{(m)}$ is a so-called isotropic function of the tensor $\boldsymbol{E}^{(m)}$. For an isotropic function of a symmetric tensor, there exists a representation theorem (Truesdell and Noll, 1965; Wang, 1970; Zheng, 1994) stating that function (6.36) can be represented, retaining full generality, as

$$\boldsymbol{T}^{(m)} = f_1(l_1, l_2, l_3)\boldsymbol{I} + f_2(l_1, l_2, l_3)\boldsymbol{E}^{(m)} + f_3(l_1, l_2, l_3)\boldsymbol{E}^{(m)2}, \tag{6.37}$$

where the three coefficients f_i are arbitrary functions of the principal invariants l_1, l_2 and l_3 of $\boldsymbol{E}^{(m)}$ [see Eqs. (2.57)], or, equivalently, of the invariants

$$\mathrm{tr}\boldsymbol{E}^{(m)}, \qquad \mathrm{tr}\boldsymbol{E}^{(m)2}, \qquad \mathrm{tr}\boldsymbol{E}^{(m)3}. \tag{6.38}$$

It is immediate to conclude from Eq. (6.37) that

for isotropic elasticity, $\boldsymbol{T}^{(m)}$ and $\boldsymbol{E}^{(m)}$ are coaxial [see Eq. (2.31)], so they commute

$$\boldsymbol{T}^{(m)}\boldsymbol{E}^{(m)} = \boldsymbol{E}^{(m)}\boldsymbol{T}^{(m)} \quad \Longleftarrow \quad \text{isotropy}. \tag{6.39}$$

The difference between a Cauchy elastic material and a Green elastic or hyperelastic material is that a strain energy function does not necessarily exist in the former case. Therefore, Green elastic materials are a subset of Cauchy elastic.

The fact that the constitutive Equation (6.37) satisfies condition (6.36) can be proved immediately. The difficult part in the representation theorems is the converse, namely, that Eq. (6.37) is the most general representation of the isotropic tensorial function satisfying (6.36).

The fact that the constitutive Equation (6.37) does not necessarily follow from a potential is made transparent from the following example. Let us consider a constitutive equation referred for simplicity to the small strain theory, where the Cauchy stress $\boldsymbol{T}$ is non-linearly related to the infinitesimal strain $\boldsymbol{E}$ [Eq. (3.38)], as

$$\boldsymbol{T} = \lambda(\mathrm{tr}\boldsymbol{E})\boldsymbol{I} + 2\mu(\mathrm{tr}\boldsymbol{E}+1)\boldsymbol{E}, \tag{6.40}$$

where λ and μ are two constants resembling the Lamé constants. Now, Eq. (6.40) is in the form of Eq. (6.37) with $\boldsymbol{T}$ and $\boldsymbol{E}$ replacing $\boldsymbol{T}^{(m)}$ and $\boldsymbol{E}^{(m)}$ and $f_3 = 0, f_1 = \lambda\mathrm{tr}\boldsymbol{E}$ and $f_2 = 2\mu(\mathrm{tr}\boldsymbol{E}+1)$. Assume that the constitutive equation (6.40) can be derived from a potential. To obtain the form of Eq. (6.40), the potential must be a function of invariants $\mathrm{tr}\boldsymbol{E}$ and $\mathrm{tr}\boldsymbol{E}^2$, so

$$\boldsymbol{T} = \frac{\partial W}{\partial \boldsymbol{E}} = \frac{\partial W}{\partial \mathrm{tr}\boldsymbol{E}}\boldsymbol{I} + 2\frac{\partial W}{\partial \mathrm{tr}\boldsymbol{E}^2}\boldsymbol{E}, \tag{6.41}$$

but since

$$0 = \frac{\partial\, \lambda(\mathrm{tr}\boldsymbol{E})}{\partial \mathrm{tr}\boldsymbol{E}^2} \neq \frac{\partial\, 2\mu(\mathrm{tr}\boldsymbol{E}+1)}{\partial \mathrm{tr}\boldsymbol{E}} = 2\mu, \tag{6.42}$$

that potential W does not exist.

The Cayley-Hamilton theorem (2.59) allows us to express $\boldsymbol{E}^{(2)2}$ as a function of of $\boldsymbol{E}^{(2)}$, its invariants and $\boldsymbol{E}^{(2)-1}$, so Eq. (6.37) can be rewritten as

$$\boldsymbol{T}^{(2)} = g_1\boldsymbol{I} + g_2\boldsymbol{E}^{(2)} + g_3\boldsymbol{E}^{(2)-1}, \tag{6.43}$$

where the coefficients g_i are related to coefficients f_i.

It is easy to check that the Kirchhoff–Saint Venant material [Eqs. (4.15) and (6.15)], is a particular form of Eq. (6.37).

The constitutive equation (6.37) can be transformed into a relation between the first Piola-Kirchhoff stress and the deformation gradient

$$\boldsymbol{S} = f_1\boldsymbol{F} + f_2\frac{1}{2}(\boldsymbol{F}\boldsymbol{C} - \boldsymbol{F}) + f_3\frac{1}{4}\left(\boldsymbol{F}\boldsymbol{C}^2 - 2\boldsymbol{F}\boldsymbol{C} + \boldsymbol{F}\right) \tag{6.44}$$

or between the Kirchhoff stress and the left Cauchy-Green deformation tensor

$$\boldsymbol{K} = f_1 \boldsymbol{B} + f_2 \frac{1}{2}\left(\boldsymbol{B}^2 - \boldsymbol{B}\right) + f_3 \frac{1}{4}\left(\boldsymbol{B}^3 - 2\boldsymbol{B}^2 + \boldsymbol{B}\right). \tag{6.45}$$

In a slightly more general approach and including micro-structures, we define as

> *elastic constitutive equation* a one-to-one relation between work-conjugate stress and strain measures, independent of the history of deformation.

Although not necessary, we will refer for simplicity to a relation between Lagrangean symmetric work-conjugate stress $\boldsymbol{T}^{(m)}$ and strain $\boldsymbol{E}^{(m)}$ measures as

$$\boldsymbol{T}^{(m)} = \hat{\boldsymbol{T}}^{(m)}(\boldsymbol{E}^{(m)}, \boldsymbol{H}_0), \tag{6.46}$$

where the response function $\hat{\boldsymbol{T}}^{(m)}$ is a symmetric tensor-valued function of the strain measure $\boldsymbol{E}^{(m)}$ and also of the micro-structural descriptors $\boldsymbol{H}_0$, denoting a collection of referential structural tensorial quantities describing directional properties of $\hat{\boldsymbol{T}}^{(m)}$. For simplicity, $\boldsymbol{H}_0$ will be identified with a *symmetric second-order 'fabric' tensor*, although the generalisation to arbitrary tensorial quantities is not difficult.

The principle of invariance of material response under a change in observer (6.3) requires nothing on Eq. (6.46), which therefore 'automatically' is material frame invariant. However, invariance for rigid-body rotation of the reference configuration imposes the following relation:

$$\boldsymbol{Q}\boldsymbol{T}^{(m)}\boldsymbol{Q}^T = \hat{\boldsymbol{T}}^{(m)}(\boldsymbol{Q}\boldsymbol{E}^{(m)}\boldsymbol{Q}^T, \boldsymbol{Q}\boldsymbol{H}_0\boldsymbol{Q}^T), \qquad \forall \boldsymbol{Q} \in \mathsf{Orth}^+, \tag{6.47}$$

so function $\boldsymbol{T}^{(m)}$ is called an 'isotropic function of the two tensors $\boldsymbol{E}^{(m)}$ and $\boldsymbol{H}_0$', a nomenclature which does not mean that the material response is isotropic! For isotropic functions of arbitrary tensors, we may refer to general representation theorems already available (Truesdell and Noll, 1965; Wang, 1970; Zheng, 1994). In our case, for an isotropic tensorial function of two second-order tensors $\hat{\boldsymbol{T}}^{(m)}$, the most general representation is

$$\begin{aligned} \boldsymbol{T}^{(m)} &= f_1 \boldsymbol{I} + f_2 \boldsymbol{E}^{(m)} + f_3 \boldsymbol{E}^{(m)2} + f_4 \boldsymbol{H}_0 + f_5 \boldsymbol{H}_0^2 \\ &\quad + f_6\left(\boldsymbol{E}^{(m)}\boldsymbol{H}_0 + \boldsymbol{H}_0\boldsymbol{E}^{(m)}\right) + f_7\left(\boldsymbol{H}_0^2\boldsymbol{E}^{(m)} + \boldsymbol{E}^{(m)}\boldsymbol{H}_0^2\right) \\ &\quad + f_8\left(\boldsymbol{E}^{(m)2}\boldsymbol{H}_0^2 + \boldsymbol{H}_0^2\boldsymbol{E}^{(m)2}\right), \end{aligned} \tag{6.48}$$

where the eight coefficients f_i are arbitrary functions of the invariants of $\boldsymbol{E}^{(m)}$ and $\boldsymbol{H}_0$

$$\mathrm{tr}\boldsymbol{E}^{(m)}, \quad \mathrm{tr}\boldsymbol{E}^{(m)2}, \quad \mathrm{tr}\boldsymbol{E}^{(m)3}, \quad \mathrm{tr}\boldsymbol{H}_0, \quad \mathrm{tr}\boldsymbol{H}_0^2, \quad \mathrm{tr}\boldsymbol{H}_0^3, \tag{6.49}$$

and of the mixed invariants

$$I_4 = \mathrm{tr}\left(\boldsymbol{H}_0\boldsymbol{E}^{(m)}\right), \quad I_5 = \mathrm{tr}\left(\boldsymbol{H}_0\boldsymbol{E}^{(m)2}\right), \quad I_6 = \mathrm{tr}\left(\boldsymbol{H}_0^2\boldsymbol{E}^{(m)}\right), \quad I_7 = \mathrm{tr}\left(\boldsymbol{H}_0^2\boldsymbol{E}^{(m)2}\right), \tag{6.50}$$

of $\boldsymbol{E}^{(m)}$ and $\boldsymbol{H}_0$.

Alternatively to the list of mixed invariants (6.50), Zysset and Curnier (1995) have shown that it may be more convenient to replace the invariant I_7 by the following

$$\bar{I}_7 = \mathrm{tr}(\boldsymbol{H}_0\boldsymbol{E}^{(m)})^2. \tag{6.51}$$

If the material response (6.46) has a class of symmetry, the following condition holds, in addition to Eq. (6.47):

$$\boldsymbol{T}^{(m)}(\boldsymbol{E}^{(m)},\boldsymbol{H}_0) = \hat{\boldsymbol{T}}^{(m)}(\boldsymbol{E}^{(m)},\boldsymbol{Q}\boldsymbol{H}_0\boldsymbol{Q}^T), \tag{6.52}$$

for every rotation $\boldsymbol{Q}$ belonging to the symmetry class.

6.1.5 Green elastic or hyperelastic materials

A material is hyperelastic if a strain energy density function W exists such that the stress measure $\boldsymbol{T}^{(m)}$ can be obtained as expressed by Eq. (4.1). In this case, the theorem of power expended can be written in the form expressed by Eq. (4.3).

The case of isotropy already has been addressed (for compressible and incompressible materials) in Chapter 4. We concentrate now on the inclusion of intrinsic anisotropy.

The strain energy density function for an elastic material with a micro-structural anisotropy described by the fabric tensor $\boldsymbol{H}_0$ can be written as

$$W = W(\boldsymbol{E}^{(m)},\boldsymbol{H}_0), \tag{6.53}$$

which automatically satisfies the principle of material frame indifference, but the invariance with respect to rigid-body rotation of the reference configuration requires

$$W(\boldsymbol{E}^{(m)},\boldsymbol{H}_0) = W(\boldsymbol{Q}\boldsymbol{E}^{(m)}\boldsymbol{Q}^T,\boldsymbol{Q}\boldsymbol{H}_0\boldsymbol{Q}^T), \qquad \forall \boldsymbol{Q} \in \mathrm{Orth}^+. \tag{6.54}$$

The scalar function of tensorial argument W therefore is isotropic (again, this does not mean that the material response is), and it is therefore possible to apply a representation theorem stating that W must be a function of the invariants (6.49) and mixed (6.50) of $\boldsymbol{E}^{(m)}$ and $\boldsymbol{H}_0$

$$W = W(I_1,I_2,I_3,\hat{I}_1,\hat{I}_2,\hat{I}_3,I_4,I_5,I_6,I_7), \tag{6.55}$$

where I_i and $\hat{I}_i$, $i = 1,2,3$, denote the invariants of $\boldsymbol{E}^{(m)}$ and $\boldsymbol{H}_0$, respectively, and I_i, $i = 4,5,6,7$, the mixed invariants (6.50).

Employing Eq. (4.1), we can write the stress measure $\boldsymbol{T}^{(m)}$ as

$$\begin{aligned}\boldsymbol{T}^{(m)} &= W_1\boldsymbol{I} + 2W_2\boldsymbol{E}^{(m)} + 3W_3\boldsymbol{E}^{(m)2} + W_4\boldsymbol{H}_0\\ &\quad + W_5(\boldsymbol{H}_0\boldsymbol{E}^{(m)} + \boldsymbol{E}^{(m)}\boldsymbol{H}_0) + W_6\boldsymbol{H}_0^2 + W_7(\boldsymbol{H}_0^2\boldsymbol{E}^{(m)} + \boldsymbol{E}^{(m)}\boldsymbol{H}_0^2),\end{aligned} \tag{6.56}$$

or, alternatively, employing Eq. (6.51) instead of Eq. (6.50), as

$$\boxed{\begin{aligned}\boldsymbol{T}^{(m)} &= W_1\boldsymbol{I} + 2W_2\boldsymbol{E}^{(m)} + 3W_3\boldsymbol{E}^{(m)2} + W_4\boldsymbol{H}_0\\ &\quad + W_5(\boldsymbol{H}_0\boldsymbol{E}^{(m)} + \boldsymbol{E}^{(m)}\boldsymbol{H}_0) + W_6\boldsymbol{H}_0^2 + 2W_7\boldsymbol{H}_0\boldsymbol{E}^{(m)}\boldsymbol{H}_0,\end{aligned}} \tag{6.57}$$

where we have introduced the notation

$$W_i = \frac{\partial W}{\partial I_i}, \quad i = 1,\dots,6 \qquad \text{and} \qquad W_7 = \frac{\partial W}{\partial I_7} \quad \text{or} \quad W_7 = \frac{\partial W}{\partial \bar{I}_7}. \tag{6.58}$$

Equation (6.57) is the constitutive equation for a Green elastic material with an intrinsic anisotropy defined by a material fabric tensor $\boldsymbol{H}_0$.

Note that in the particular case of isotropy, $\boldsymbol{H}_0 = \boldsymbol{0}$ (or, equivalently, $\boldsymbol{H}_0 = \boldsymbol{I}$), Eq. (6.57) shows that $\boldsymbol{T}^{(m)}$ and $\boldsymbol{E}^{(m)}$ are coaxial; see Eq. (6.39).

Exercise: Uniaxial plane strain tension and compression of a fibre-reinforced version of the Kirchhoff–Saint Venant material

The Kirchhoff–Saint Venant material was introduced by Eq. (4.15) and analysed in plane strain tension/compression in Section 5.2.

We consider now uniaxial plane strain tension/compression of the fibre-reinforced material defined by Eq. (6.15), which is a particularisation of a model proposed by Bonet and Burton (1998) and reduces to the Kirchhoff–Saint Venant model (4.15) when the Lagrangean vector $\boldsymbol{a}_0$ specifying the direction of the fibres is taken to be zero.

First, we note that constitutive equation (6.15) satisfies the principle of material frame indifference because a change in the spatial configuration leaves it unchanged.

As in Section 5.2, we consider an elastic rectangular block of initial dimensions $b_0 \times h_0$ deformed into a rectangular $b \times h$ block (as in Fig. 5.1, except that now the material is compressible). We assume that the fibres are all parallel to the edge h_0, so $\boldsymbol{a}_0 = \boldsymbol{e}_1$ and to highlight this, we introduce the notation $\boldsymbol{a}_1 = \boldsymbol{e}_1 = \boldsymbol{a}_0$. For uniaxial plane strain tension/compression aligned parallel to $\boldsymbol{e}_1, \boldsymbol{F}, J$ and $\boldsymbol{E}^{(2)}$ remain given by Eqs. (5.18) to (5.20). Thus, using the constitutive equation (6.15), we may calculate the stress in terms of the second Piola-Kirchhoff tensor

$$\begin{aligned}\boldsymbol{T}^{(2)} &= \frac{\lambda}{2}\left(\frac{h^2}{h_0^2}+\frac{b^2}{b_0^2}-2\right)\boldsymbol{I}+\mu\left(\frac{h^2}{h_0^2}\boldsymbol{e}_1\otimes\boldsymbol{e}_1+\frac{b^2}{b_0^2}\boldsymbol{e}_2\otimes\boldsymbol{e}_2+\boldsymbol{e}_3\otimes\boldsymbol{e}_3-\boldsymbol{I}\right)\\ &\quad+\frac{\gamma}{2}\left(\frac{h^2}{h_0^2}-1\right)\boldsymbol{a}_1\otimes\boldsymbol{a}_1,\end{aligned} \tag{6.59}$$

from which we may obtain the Kirchhoff stress as

$$\begin{aligned}\boldsymbol{K} &= \frac{\lambda}{2}\left(\frac{h^2}{h_0^2}+\frac{b^2}{b_0^2}-2\right)\boldsymbol{F}\boldsymbol{F}^T+\mu\left(\frac{h^2}{h_0^2}\boldsymbol{F}\boldsymbol{e}_1\otimes\boldsymbol{F}\boldsymbol{e}_1+\frac{b^2}{b_0^2}\boldsymbol{F}\boldsymbol{e}_2\otimes\boldsymbol{F}\boldsymbol{e}_2+\boldsymbol{F}\boldsymbol{e}_3\otimes\boldsymbol{F}\boldsymbol{e}_3-\boldsymbol{F}\boldsymbol{F}^T\right)\\ &\quad+\frac{\gamma}{2}\left(\frac{h^2}{h_0^2}-1\right)\boldsymbol{F}\boldsymbol{a}_1\otimes\boldsymbol{F}\boldsymbol{a}_1,\end{aligned} \tag{6.60}$$

where $\boldsymbol{F}$ is given by Eq. (5.18) and J by Eq. (5.19). Thus

$$\begin{aligned}\boldsymbol{K} &= \left[\frac{\lambda}{2}\left(\frac{h^2}{h_0^2}+\frac{b^2}{b_0^2}-2\right)-\mu\right]\left(\frac{h^2}{h_0^2}\boldsymbol{e}_1\otimes\boldsymbol{e}_1+\frac{b^2}{b_0^2}\boldsymbol{e}_2\otimes\boldsymbol{e}_2+\boldsymbol{e}_3\otimes\boldsymbol{e}_3\right)\\ &\quad+\mu\left(\frac{h^4}{h_0^4}\boldsymbol{e}_1\otimes\boldsymbol{e}_1+\frac{b^4}{b_0^4}\boldsymbol{e}_2\otimes\boldsymbol{e}_2+\boldsymbol{e}_3\otimes\boldsymbol{e}_3\right)+\frac{\gamma}{2}\left(\frac{h^2}{h_0^2}-1\right)\frac{h^2}{h_0^2}\boldsymbol{a}_1\otimes\boldsymbol{a}_1.\end{aligned} \tag{6.61}$$

The imposition of the condition that the stress is uniaxial, $T_2 = 0$, yields again Eqs. (5.24) and (5.25), which do not involve the fibre $\boldsymbol{a}_1$, so they hold also for an isotropic material, and the fibre reinforced material suffers the same limitations as its isotropic counterpart.

If now $\boldsymbol{a}_0$ is selected in the direction $\boldsymbol{e}_2$ and defined as $\boldsymbol{a}_2 = \boldsymbol{e}_2 = \boldsymbol{a}_0$, we obtain

$$\begin{aligned}\boldsymbol{K} = &\left[\frac{\lambda}{2}\left(\frac{h^2}{h_0^2}+\frac{b^2}{b_0^2}-2\right)-\mu\right]\left(\frac{h^2}{h_0^2}\boldsymbol{e}_1\otimes\boldsymbol{e}_1+\frac{b^2}{b_0^2}\boldsymbol{e}_2\otimes\boldsymbol{e}_2+\boldsymbol{e}_3\otimes\boldsymbol{e}_3\right)\\ &+\mu\left(\frac{h^4}{h_0^4}\boldsymbol{e}_1\otimes\boldsymbol{e}_1+\frac{b^4}{b_0^4}\boldsymbol{e}_2\otimes\boldsymbol{e}_2+\boldsymbol{e}_3\otimes\boldsymbol{e}_3\right)+\frac{\gamma}{2}\left(\frac{b^2}{b_0^2}-1\right)\frac{b^2}{b_0^2}\boldsymbol{a}_2\otimes\boldsymbol{a}_2.\end{aligned} \tag{6.62}$$

Note that since the material is now orthotropic, the stress given by Eq. (6.61) is different from that given by Eq. (6.62).

Let us now go back to the case $\boldsymbol{a}_1 = \boldsymbol{e}_1 = \boldsymbol{a}_0$ and prescribe first an anti-clockwise rigid-body rotation of $\pi/2$ and, second, again the deformation $\boldsymbol{F}$ given by Eq. (5.18). This is equivalent to the deformation gradient $\boldsymbol{FQ}$, where

$$\boldsymbol{Q} = -\boldsymbol{e}_1\otimes\boldsymbol{e}_2+\boldsymbol{e}_2\otimes\boldsymbol{e}_1+\boldsymbol{e}_3\otimes\boldsymbol{e}_3, \tag{6.63}$$

so the Green-Lagrange tensor tranforms like $\boldsymbol{Q}^T\boldsymbol{E}^{(2)}\boldsymbol{Q}$, whereas $\boldsymbol{a}_1$ is not affected. Moreover, we can write the conditions

$$\boldsymbol{a}_1\cdot\boldsymbol{Q}^T\boldsymbol{E}^{(2)}\boldsymbol{Q}\boldsymbol{a}_1 = \boldsymbol{Q}\boldsymbol{a}_1\cdot\boldsymbol{E}^{(2)}\boldsymbol{Q}\boldsymbol{a}_1 = \boldsymbol{e}_2\cdot\boldsymbol{E}^{(2)}\boldsymbol{e}_2 = \frac{b^2}{b_0^2}, \tag{6.64}$$

revealing that the stress response can be written as

$$\begin{aligned}\boldsymbol{T}^{(2)} = &\frac{\lambda}{2}\left(\frac{h^2}{h_0^2}+\frac{b^2}{b_0^2}-2\right)\boldsymbol{I}+\mu\left(\frac{h^2}{h_0^2}\boldsymbol{e}_2\otimes\boldsymbol{e}_2+\frac{b^2}{b_0^2}\boldsymbol{e}_1\otimes\boldsymbol{e}_1+\boldsymbol{e}_3\otimes\boldsymbol{e}_3\right)\\ &+\frac{\gamma}{2}\left(\frac{b^2}{b_0^2}-1\right)\boldsymbol{a}_1\otimes\boldsymbol{a}_1\end{aligned} \tag{6.65}$$

and the Kirchhoff stress as

$$\begin{aligned}\boldsymbol{K} = &\left[\frac{\lambda}{2}\left(\frac{h^2}{h_0^2}+\frac{b^2}{b_0^2}-2\right)-\mu\right]\left(\frac{h^2}{h_0^2}\boldsymbol{e}_1\otimes\boldsymbol{e}_1+\frac{b^2}{b_0^2}\boldsymbol{e}_2\otimes\boldsymbol{e}_2+\boldsymbol{e}_3\otimes\boldsymbol{e}_3\right)\\ &+\mu\left(\frac{h^4}{h_0^4}\boldsymbol{e}_1\otimes\boldsymbol{e}_1+\frac{b^4}{b_0^4}\boldsymbol{e}_2\otimes\boldsymbol{e}_2+\boldsymbol{e}_3\otimes\boldsymbol{e}_3\right)+\frac{\gamma}{2}\left(\frac{b^2}{b_0^2}-1\right)\frac{b^2}{b_0^2}\boldsymbol{e}_2\otimes\boldsymbol{e}_2.\end{aligned} \tag{6.66}$$

Note that the stress response associated by Eq. (6.66) to the deformation gradient $\boldsymbol{FQ}$ is different from the response to $\boldsymbol{F}$ with $\boldsymbol{a}_0 = \boldsymbol{e}_1$ given by Eq. (6.61), but it is equal to the response (6.62) obtained with $\boldsymbol{a}_0 = \boldsymbol{e}_2$.

A different situation arises if a rigid-body rotation is given to the body, and this configuration is taken as a reference configuration to which the constitutive equation is referred. In this case, the unit vector $\boldsymbol{a}_0$ transforms to $Q\boldsymbol{a}_0$, the deformation gradient to $\boldsymbol{FQ}$ and the Green-Lagrange tensor and the second Piola-Kirchhoff transform to $\boldsymbol{QE}^{(2)}\boldsymbol{Q}$ and $\boldsymbol{QT}^{(2)}\boldsymbol{Q}$. In this case, it is easy to show that the stress response in terms of Cauchy stress does not change, a property holding true for every rotation tensor $\boldsymbol{Q}$.

6.1.6 Incompressible hyperelasticity and constrained materials

For incompressible materials (see also Section 4.2), the incompressibility constraint is

$$J-1=0, \tag{6.67}$$

which, from the polar decomposition theorem (3.15), is equivalent to

$$\det \boldsymbol{U}^m - 1 = 0. \tag{6.68}$$

The constraint [(6.67) or (6.68)] enters the constitutive equation through an arbitrary Lagrangean multiplier π. Thus, assuming the existence of a strain energy potential as in Eq. (6.53), we may write

$$\boldsymbol{T}^{(m)} = -\frac{\pi}{m}\frac{\partial[\det \boldsymbol{U}^m - 1]}{\partial \boldsymbol{E}^{(m)}} + \frac{\partial W(\boldsymbol{E}^{(m)},\boldsymbol{H}_0)}{\partial \boldsymbol{E}^{(m)}}, \tag{6.69}$$

where π has been divided by m (an inconsequent operation because π is arbitrary) to simplify subsequent calculations.

Employing the definition (3.31) of $\boldsymbol{E}^{(m)}$ and the gradient rule of a determinant [Eq. (2.120)], we may note the following identity

$$\frac{\partial \det \boldsymbol{U}^m}{\partial \boldsymbol{U}^m}\frac{\partial \boldsymbol{U}^m}{\partial \boldsymbol{E}^{(m)}} = m\boldsymbol{U}^{-m}. \tag{6.70}$$

Thus we arrive at the *constitutive law for an incompressible elastic material with an anisotropy determined by tensor* $\boldsymbol{H}_0$:

$$\boxed{\boldsymbol{T}^{(m)} = -\pi\, \boldsymbol{U}^{-m} + \frac{\partial W(\boldsymbol{E}^{(m)},\boldsymbol{H}_0)}{\partial \boldsymbol{E}^{(m)}}.} \tag{6.71}$$

As far as the strain energy density is concerned, invariance for rigid-body rotation of the reference configuration (6.18) allows once more the use of representation theorems, so we arrive again at an expression similar to Eq. (6.57), to which the constraint $-\pi\boldsymbol{U}^{-m}$ has to be added.

To make this statement precise, we note that the incompressibility constraint reduces the functional dependence of the strain energy W on the invariants of $\boldsymbol{E}^{(m)}$. In particular, assuming $m \neq 0$, the expressions

$$(\boldsymbol{E}^{(m)})^2 = m^{-2}\left(\boldsymbol{U}^{2m} - 2\boldsymbol{U}^m + \boldsymbol{I}\right), \tag{6.72}$$

and

$$(\boldsymbol{E}^{(m)})^3 = m^{-3}\left(\boldsymbol{U}^{3m} - 3\boldsymbol{U}^{2m} + 3\boldsymbol{U}^m - \boldsymbol{I}\right)$$

show that the strain energy equivalently may be viewed as a function of the invariants of $\boldsymbol{E}^{(m)}$ or $\boldsymbol{U}^m$. Thus, using the Cayley-Hamilton theorem (2.59) for $\boldsymbol{U}^m$ and the incompressibility constraint $\det \boldsymbol{U}^m = 1$, we find that the independent invariants reduce to two or, in other words, that $\mathrm{tr}\boldsymbol{U}^{3m}$ can be made a function of the other two invariants

$$\mathrm{tr}\boldsymbol{U}^{3m} = \frac{3}{2}\mathrm{tr}\boldsymbol{U}^m\mathrm{tr}\boldsymbol{U}^{2m} - \frac{1}{2}\mathrm{tr}^3\boldsymbol{U}^m + 3. \tag{6.73}$$

From Eqs. (6.73) and (6.72) we conclude that in the incompressible limit, the strain energy function depends only on the two invariants I_1 and I_2 of $\boldsymbol{E}^{(m)}$, so we arrive at the representation

$$\boxed{\begin{aligned}\boldsymbol{T}^{(m)} &= -\pi\boldsymbol{U}^{-m} + W_1\boldsymbol{I} + 2W_2\boldsymbol{E}^{(m)} + W_4\boldsymbol{H}_0\\ &\quad + W_5(\boldsymbol{H}_0\boldsymbol{E}^{(m)} + \boldsymbol{E}^{(m)}\boldsymbol{H}_0) + W_6\boldsymbol{H}_0^2 + 2W_7\boldsymbol{H}_0\boldsymbol{E}^{(m)}\boldsymbol{H}_0.\end{aligned}} \tag{6.74}$$

Isotropic incompressible hyperelasticity For isotropic incompressible elasticity, $\boldsymbol{H}_0 = \boldsymbol{0}$ (or, equivalently, $\boldsymbol{H}_0 = \boldsymbol{I}$) (it may be instructive to compare with Section 4.2), we may select $m = 1$ (Biot stress and its conjugate strain measure) and represent the strain energy W directly as a function of $\boldsymbol{U}$ to obtain

$$\boldsymbol{T}^{(1)} = -\pi \boldsymbol{U}^{-1} + W_1 \boldsymbol{I} + 2 W_2 \boldsymbol{U}, \qquad \det \boldsymbol{U} = 1, \tag{6.75}$$

which using Eq. (3.134)$_2$ yields

$$\boldsymbol{T}^{(2)} = -\pi \boldsymbol{U}^{-2} + W_1 \boldsymbol{U}^{-1} + 2 W_2 \boldsymbol{I}, \qquad \det \boldsymbol{U} = 1. \tag{6.76}$$

Thus, employing the definitions of second Piola-Kirchhoff stress [Eq. (3.132)], we arrive at the representation

$$\boldsymbol{T} = -\pi \boldsymbol{I} + W_1 \boldsymbol{V} + 2 W_2 \boldsymbol{V}^2, \qquad \det \boldsymbol{V} = 1. \tag{6.77}$$

Tensors $\boldsymbol{V}^2 = \boldsymbol{B}$ and $\boldsymbol{V}$ also, can through Eq. (2.181), be expressed as a function of $\boldsymbol{B}, \boldsymbol{B}^{-1}$ and of their invariants. Therefore, Eq. (6.77) can be expressed alternatively as

$$\boldsymbol{T} = -\pi \boldsymbol{I} + \beta_0 \boldsymbol{B} + \beta_1 \boldsymbol{B}^{-1}, \qquad \det \boldsymbol{B} = 1, \tag{6.78}$$

where π is a Lagrangean multiplier different from that employed in Eq. (6.77), and β_0 and β_1 are functions of the two invariants I_1 and I_2 of $\boldsymbol{B}$. When β_0 and β_1 are constants, the Mooney-Rivlin material [Eq. (4.27)] is recovered.

Note 1. For plane strain deformation $\lambda_2 = 1/\lambda_1$ and $\lambda_3 = 1$, the in-plane mean stress, calculated with Eq. (6.78), results

$$\frac{T_1 + T_2}{2} = -\pi + \frac{(\beta_0 + \beta_1)(1 + \lambda_1^4)}{2\lambda_1^2}, \tag{6.79}$$

whereas the three-dimensional mean stress (obtained summing up the out-of-plane stress) is

$$\frac{T_1 + T_2 + T_3}{3} = -\pi + \frac{(\beta_0 + \beta_1)(1 + \lambda_1^2 + \lambda_1^4)}{3\lambda_1^2}, \tag{6.80}$$

so only for particular constitutive models (where $\beta_0 = -\beta_1$, as in the cases of the J_2-deformation theory and GBG materials; see Section 4.2.3), the two mean stresses coincide. Under small strain, the mean stresses coincide independently of the constitutive parameters, as can be easily verified with a Taylor series expansion:

$$\frac{T_1 + T_2}{2} \sim \frac{T_1 + T_2 + T_3}{3} \sim -\pi + \beta_0 + \beta_1 + \left[\frac{d\beta_0}{d\lambda_1} + \frac{d\beta_1}{d\lambda_1}\right]_{\lambda_1 = 1} (\lambda_1 - 1). \tag{6.81}$$

Note 2. Internal constraint may be introduced more in general than the preceding incompressibility constraint. In particular, we may consider a constraint of the form

$$c(\boldsymbol{F}) = 0, \tag{6.82}$$

where c is a scalar function of the deformation gradient. Constraint (6.82) is subject to the objectivity requirement

$$c(\boldsymbol{F}) = c(\boldsymbol{Q}\boldsymbol{F}), \qquad \forall \boldsymbol{Q} \in \mathsf{Orth}^+, \tag{6.83}$$

so from the polar representation theorem (3.15), we conclude that function $c(\boldsymbol{F})$ must be expressible as a function of the right stretch tensor $\boldsymbol{U}$:

$$c(\boldsymbol{F}) = \hat{c}(\boldsymbol{U}). \tag{6.84}$$

The constitutive equation therefore becomes

$$\boldsymbol{T}^{(m)} = -\frac{\pi}{m}\frac{\partial \hat{c}(\boldsymbol{U})}{\partial \boldsymbol{E}^{(m)}} + \frac{\partial W(\boldsymbol{E}^{(m)}, \boldsymbol{H}_0)}{\partial \boldsymbol{E}^{(m)}} \tag{6.85}$$

(where we have inconsequentially divided π through the factor m), so employing the relation between $\boldsymbol{E}^{(m)}$ and $\boldsymbol{U}^m$ [Eq. (3.31)], we arrive at

$$\boldsymbol{T}^{(m)} = -\pi \frac{\partial \hat{c}(\boldsymbol{U})}{\partial \boldsymbol{U}^m} + \frac{\partial W(\boldsymbol{E}^{(m)}, \boldsymbol{H}_0)}{\partial \boldsymbol{E}^{(m)}}. \tag{6.86}$$

In the particular case of incompressible materials, the constraint (6.84) becomes Eq. (6.68), and the constitutive equation (6.71) is recovered.

In the case of *inextensibility in a direction* $\boldsymbol{m}_0$, we have

$$\hat{c}(\boldsymbol{U}) = \boldsymbol{m}_0 \cdot (\boldsymbol{U}^2 \boldsymbol{m}_0) - 1 \tag{6.87}$$

so that considering now the second Piola-Kirchhoff stress tensor $\boldsymbol{T}^{(2)}$ and the Green-Lagrange strain $\boldsymbol{E}^{(2)}$, we have

$$\frac{\partial \hat{c}}{\partial \boldsymbol{U}^2} = \boldsymbol{m}_0 \otimes \boldsymbol{m}_0, \tag{6.88}$$

and therefore, Eq. (6.86) becomes

$$\boldsymbol{T}^{(2)} = -\pi\,(\boldsymbol{m}_0 \otimes \boldsymbol{m}_0) + \frac{\partial W(\boldsymbol{E}^{(2)}, \boldsymbol{m}_0 \otimes \boldsymbol{m}_0)}{\partial \boldsymbol{E}^{(2)}}. \tag{6.89}$$

If we now refer to an incompressible elastic material reinforced by inextensible fibres in the material direction $\boldsymbol{m}_0$, the constitutive equations (6.71) and (6.89) can be combined to yield

$$\boldsymbol{T}^{(2)} = -\pi\,\boldsymbol{U}^{-2} - T\,(\boldsymbol{m}_0 \otimes \boldsymbol{m}_0) + \frac{\partial W(\boldsymbol{E}^{(2)}, \boldsymbol{m}_0 \otimes \boldsymbol{m}_0)}{\partial \boldsymbol{E}^{(2)}}, \tag{6.90}$$

where π and T are two Lagrangean multipliers.

Employing representation (6.57) in Eq. (6.90) and transforming the result to the spatial setting, we arrive at the form reported by Spencer [1984, his Eq. (68)]

$$\boldsymbol{T} = -\pi \boldsymbol{I} + \hat{T}\boldsymbol{m} \otimes \boldsymbol{m} + 2\left\{\hat{W}_1 \boldsymbol{B} - \hat{W}_2 \boldsymbol{B}^{-1} + \hat{W}_3\,(\boldsymbol{m} \otimes \boldsymbol{B}\boldsymbol{m} + \boldsymbol{B}\boldsymbol{m} \otimes \boldsymbol{m})\right\}, \tag{6.91}$$

where $\boldsymbol{m} = \boldsymbol{F}\boldsymbol{m}_0$ and coefficients $\hat{T}$ and $\hat{W}_i$ $(i = 1,2,3)$ are functions of T, W_i $(i = 1,...,6)$ and W_7.

Another example of constraint is the so-called kinematic Bell constraint (Beatty, 2001), which can be expressed as

$$\mathrm{tr}\boldsymbol{E}^{(1)} = 0, \tag{6.92}$$

where $\boldsymbol{E}^{(1)}$ is the strain work conjugate to the Biot stress [see Eqs. (3.134)].

We do not introduce for simplicity any dependence on $\boldsymbol{H}_0$; thus, since the relation between $\boldsymbol{E}^{(1)}$ and $\boldsymbol{U}$ is linear, Eq. (6.85) leads to

$$\boldsymbol{T}^{(1)} = -\pi \boldsymbol{I} + \frac{\partial W(\boldsymbol{E}^{(1)})}{\partial \boldsymbol{E}^{(1)}}. \tag{6.93}$$

Having assumed an isotropic behaviour, all stress and strain work-conjugate measures commute (6.39), so $\boldsymbol{T}^{(2)}$ commutes with $\boldsymbol{U}$, and it is possible to obtain

$$\boldsymbol{T}^{(1)} = \boldsymbol{U}\boldsymbol{T}^{(2)} = \boldsymbol{T}^{(2)}\boldsymbol{U} \qquad \text{(for isotropic behaviour)} \tag{6.94}$$

from Eq. (3.134). Therefore, we may transform Eq. (6.93) into

$$\boldsymbol{T}^{(2)} = -\pi \boldsymbol{U}^{-1} + \frac{\partial W(\boldsymbol{E}^{(1)})}{\partial \boldsymbol{E}^{(1)}} \boldsymbol{U}^{-1}. \tag{6.95}$$

Employing representation (6.57) with $\boldsymbol{H}_0 = \boldsymbol{0}$ and $m = 1$, and noting that $\boldsymbol{E}^{(1)}$ is a linear function of $\boldsymbol{U}$ [Eq. (3.134)$_1$], we may transform Eq. (6.95) into

$$\boldsymbol{T}^{(2)} = -\hat{\pi}\, \boldsymbol{U}^{-1} + \hat{W}_1 \boldsymbol{I} + \hat{W}_2 \boldsymbol{U}, \tag{6.96}$$

where $\hat{\pi}$ and $\hat{W}_i$ $(i = 1,2)$ are functions of π and W_i [see representation (6.57)].

Recalling the definition of second Piola-Kirchhoff stress (3.132), Eq. (6.96) can be transformed into a relation between the Cauchy stress $\boldsymbol{T}$ and the left stretch tensor $\boldsymbol{V}$, which, noting that

$$\boldsymbol{R}\boldsymbol{U}^3\boldsymbol{R}^T = (\boldsymbol{R}\boldsymbol{U}\boldsymbol{R}^T)^3 = \boldsymbol{V}^3, \tag{6.97}$$

and using the Cayley-Hamilton theorem (2.59) to reduce $\boldsymbol{V}^3$ in terms of $\boldsymbol{V}$ and $\boldsymbol{V}^2$, allows us to rewrite Eq. (6.96) in the same form given by Beatty [2001, his Eq. (4.14)]

$$\boldsymbol{T} = p\boldsymbol{V} + \omega_0 \boldsymbol{I} + \omega_2 \boldsymbol{V}^2, \tag{6.98}$$

where p is the (modified) constraint parameter, and ω_0 and ω_2 are functions of the invariants of $\boldsymbol{V}$, with the exception of $\mathrm{tr}\,\boldsymbol{V}$, because the constraint (6.92) equivalently can be written as $\mathrm{tr}\,\boldsymbol{V} = 3$.

6.2 Rate and incremental elastic constitutive equations

Rate constitutive equations represent the only possibility of describing the behaviour of elastoplastic material, so their study in the simpler case of elasticity is fundamental to our purposes. We will see that when the time is interpreted as a scalar parameter governing loading, treatment of the *rate* form of elastic constitutive equations is essentially analogous to the development of *incremental* constitutive laws. Constitutive equations in rate and incremental form are crucial for the bifurcation and stability analyses that will be performed subsequently (see in particular, Chapters 11, 12 and 16).

6.2.1 Elastic laws in incremental and rate form

The analysis of small elastic deformations superimposed on a given state of finite deformation is the focus of *incremental elasticity*. From the point of view of constitutive equations, the incremental stress $\Delta\boldsymbol{T}^{(m)}$ is obtained as a linearisation (through

Taylor series expansion) of the constitutive equations (6.46) (where the dependence on $\boldsymbol{H}_0$ is for the moment omitted) for a small increment of the conjugate strain measure $\Delta\boldsymbol{E}^{(m)}$. Thus,

$$\Delta\boldsymbol{T}^{(m)} \sim \frac{\partial\hat{\boldsymbol{T}}^{(m)}}{\partial\boldsymbol{E}^{(m)}}[\Delta\boldsymbol{E}^{(m)}] \tag{6.99}$$

becomes a state-dependent, but linear in $\Delta\boldsymbol{E}^{(m)}$, constitutive equation.

Given the assumption of time-independent behaviour, Eq. (6.99) is formally identical to the *rate equations of elasticity*, namely, the equations that can be obtained taking the material-time derivative of the constitutive equations (6.46), or Eq. (4.1) when a strain potential exists, to obtain

$$\dot{\boldsymbol{T}}^{(m)} = \mathbb{E}[\dot{\boldsymbol{E}}^{(m)}], \tag{6.100}$$

where the *elastic tensor* is defined as

$$\mathbb{E} = \frac{\partial\hat{\boldsymbol{T}}^{(m)}(\boldsymbol{E}^{(m)})}{\partial\boldsymbol{E}^{(m)}} \quad \text{or} \quad \mathbb{E} = \frac{\partial^2 W(\boldsymbol{E}^{(m)})}{\partial\boldsymbol{E}^{(m)2}} \tag{6.101}$$

(respectively, for Cauchy elasticity and for hyperelasticity), and when this is positive definite over the set of symmetric tensors, Eq. (6.100) can be inverted to give

$$\dot{\boldsymbol{E}}^{(m)} = \mathbb{M}[\dot{\boldsymbol{T}}^{(m)}], \tag{6.102}$$

where

$$\mathbb{E} = \mathbb{M}^{-1}. \tag{6.103}$$

Note that $\mathbb{E}$ always has the minor symmetries, which are induced by $\boldsymbol{E}^{(m)}$, has the major symmetry for hyperelasticity but may loose this in the case of Cauchy elasticity.

Since rate and incremental equations are formally identical, we will refer only to the former, avoiding the notation Δ.

Equation (6.100) automatically satisfies invariance with respect to a change in observer and indifference with respect to a rigid-body rotation of the reference configuration if Eq. (6.46) or Eq. (4.1) — as expected — does satisfy. In particular, assuming the existence of a strain energy function W and thus employing Eq. (6.57), we obtain

$$\boxed{\begin{aligned}\mathbb{E} = {} & \boldsymbol{I}\otimes\frac{\partial W_1}{\partial\boldsymbol{E}^{(m)}} + 2W_2\boldsymbol{I}\,\underline{\overline{\otimes}}\,\boldsymbol{I} + 2\boldsymbol{E}^{(m)}\otimes\frac{\partial W_2}{\partial\boldsymbol{E}^{(m)}} + 3W_3(\boldsymbol{E}^{(m)}\otimes\boldsymbol{I} + \boldsymbol{I}\otimes\boldsymbol{E}^{(m)}) \\ & + 3\boldsymbol{E}^{(m)2}\otimes\frac{\partial W_3}{\partial\boldsymbol{E}^{(m)}} + \boldsymbol{H}_0\otimes\frac{\partial W_4}{\partial\boldsymbol{E}^{(m)}} + W_5(\boldsymbol{H}_0\,\underline{\overline{\otimes}}\,\boldsymbol{I} + \boldsymbol{I}\,\underline{\overline{\otimes}}\,\boldsymbol{H}_0) \\ & + (\boldsymbol{H}_0\boldsymbol{E}^{(m)} + \boldsymbol{E}^{(m)}\boldsymbol{H}_0)\otimes\frac{\partial W_5}{\partial\boldsymbol{E}^{(m)}} + \boldsymbol{H}_0^2\otimes\frac{\partial W_6}{\partial\boldsymbol{E}^{(m)}} \\ & + 2W_7\boldsymbol{H}_0\,\underline{\overline{\otimes}}\,\boldsymbol{H}_0 + 2\boldsymbol{H}_0\boldsymbol{E}^{(m)}\boldsymbol{H}_0\otimes\frac{\partial W_7}{\partial\boldsymbol{E}^{(m)}},\end{aligned}} \tag{6.104}$$

where

$$\begin{aligned}\frac{\partial W_i}{\partial\boldsymbol{E}^{(m)}} = {} & W_{i1}\boldsymbol{I} + 2W_{i2}\boldsymbol{E}^{(m)} + 3W_{i3}\boldsymbol{E}^{(m)2} + W_{i4}\boldsymbol{H}_0 \\ & + 2W_{i5}\boldsymbol{H}_0\boldsymbol{E}^{(m)} + W_{i6}\boldsymbol{H}_0^2 + 2W_{i7}\boldsymbol{H}_0\boldsymbol{E}^{(m)}\boldsymbol{H}_0,\end{aligned} \tag{6.105}$$

with

$$W_{ij} = \frac{\partial W_i}{\partial I_j}, \quad i,j = 1,...,6, \qquad \text{and} \qquad W_{i7} = \frac{\partial W_i}{\partial \bar{I}_7}, \quad i = 1,...,7. \tag{6.106}$$

Note that since $W_{ij} = W_{ji}$, $\mathbb{E}$ possesses all symmetries for hyperelastic behaviour.

Obviously, it is always possible to transform Eq. (6.100) from the material to the spatial description or to a formulation in terms of material derivatives of the first Piola-Kirchhoff stress and of the deformation gradient, the latter being particularly convenient to our purposes.

For instance, in the case of the second Piola-Kirchhoff stress tensor $\boldsymbol{T}^{(2)}$ and the Green-Lagrange strain tensor $\boldsymbol{E}^{(2)}$, the following relations hold true

$$\dot{\boldsymbol{S}} = \boldsymbol{F}\dot{\boldsymbol{T}}^{(2)} + J\boldsymbol{L}\boldsymbol{T}\boldsymbol{F}^{-T} \qquad \text{and} \qquad \dot{\boldsymbol{E}}^{(2)} = \boldsymbol{F}^T\boldsymbol{D}\boldsymbol{F}, \tag{6.107}$$

where $\boldsymbol{L}$ is the velocity gradient (3.57) and $\boldsymbol{D}$ the Eulerian strain rate (3.59).

The minor symmetries of $\mathbb{E}$ and the relation (3.61) imply

$$\mathbb{E}[\boldsymbol{F}^T\boldsymbol{D}\boldsymbol{F}] = \frac{1}{2}\mathbb{E}[\boldsymbol{F}^T\dot{\boldsymbol{F}} + (\boldsymbol{F}^T\dot{\boldsymbol{F}})^T] = \mathbb{E}[\boldsymbol{F}^T\dot{\boldsymbol{F}}]. \tag{6.108}$$

Moreover, from the equality

$$\boldsymbol{F}^{-T}\boldsymbol{B}\cdot\boldsymbol{F}\boldsymbol{A} = \boldsymbol{B}\cdot\boldsymbol{A}, \tag{6.109}$$

holding for every $\boldsymbol{A},\boldsymbol{B} \in \mathsf{Lin}$, we conclude that the law (6.100) transforms to

$$\dot{\boldsymbol{S}} = \mathbb{G}[\dot{\boldsymbol{F}}], \tag{6.110}$$

where the fourth-order tensors

$$\mathbb{G} = \mathbb{B} + \boldsymbol{I}\boxtimes\boldsymbol{S}^T\boldsymbol{F}^{-T} \qquad \text{and} \qquad \mathbb{B} = (\boldsymbol{F}\boxtimes\boldsymbol{I})\mathbb{E}(\boldsymbol{F}\boxtimes\boldsymbol{I})^T \tag{6.111}$$

have been introduced. It should be noted that neither $\mathbb{B}$ nor $\mathbb{G}$ has the minor symmetries and that both $\mathbb{B}$ and $\mathbb{G}$ have the major symmetry only in the case of Green elasticity, that is, when $\mathbb{E}$ has the major symmetry, too.

Note that

$$\boldsymbol{S}^T\boldsymbol{F}^{-T} = J\boldsymbol{F}^{-1}\boldsymbol{T}\boldsymbol{F}^{-T} = \boldsymbol{F}^{-1}\boldsymbol{S} \in \mathsf{Sym}, \tag{6.112}$$

so

$$\mathbb{G} - \mathbb{B} = \boldsymbol{I}\boxtimes\boldsymbol{S}^T\boldsymbol{F}^{-T}, \tag{6.113}$$

always has the major symmetry.

It is now possible to transform Eq. (6.110) into a purely spatial rate-of-stress/rate-of-strain constitutive equation. To this purpose, let us observe that from definition $(3.156)_2$, we obtain

$$\dot{\boldsymbol{T}}^{(2)} = \boldsymbol{F}^{-1}\overset{\circ}{\boldsymbol{K}}\boldsymbol{F}^{-T} \qquad \text{and} \qquad \dot{\boldsymbol{E}}^{(2)} = \boldsymbol{F}^T\boldsymbol{D}\boldsymbol{F}, \tag{6.114}$$

so the constitutive equation (6.110) becomes

$$\overset{\circ}{\boldsymbol{K}} = \mathbb{H}[\boldsymbol{D}], \tag{6.115}$$

Table 6.1. *Symmetries of the elastic fourth-order tensors* $\mathbb{B}$, $\mathbb{E}$, $\mathbb{G}$ *and* $\mathbb{H}$

Tensor	Minor symmetry	Major symmetry
$\mathbb{B}$	No	Yes, iff $\mathbb{E}$ has the major symmetry
$\mathbb{E}$	Yes	Yes for hyperelasticity
$\mathbb{G}-\mathbb{B}$	No	Yes
$\mathbb{H}$	Yes	Yes, iff $\mathbb{E}$ has the major symmetry

where the fourth-order tensor $\mathbb{H}$ is

$$\mathbb{H} = (\boldsymbol{F}\boxtimes\boldsymbol{F})\mathbb{E}(\boldsymbol{F}\boxtimes\boldsymbol{F})^T, \tag{6.116}$$

which always has the minor symmetries, whereas it has the major symmetry only if $\mathbb{E}$ has.

Note that from Eq. (2.94) we may write

$$\mathbb{H} = (\boldsymbol{I}\boxtimes\boldsymbol{F})\mathbb{B}(\boldsymbol{I}\boxtimes\boldsymbol{F})^T \qquad \text{and} \qquad \mathbb{B} = (\boldsymbol{I}\boxtimes\boldsymbol{F})^{-1}\mathbb{H}(\boldsymbol{I}\boxtimes\boldsymbol{F})^{-T}. \tag{6.117}$$

In summary, the symmetries of the elastic fourth-order tensors $\mathbb{B}$, $\mathbb{E}$, $\mathbb{G}$ and $\mathbb{H}$ are reported in Table 6.1.

6.2.2 Relative Lagrangean description

We have already remarked that the choice of the reference configuration B_0 is arbitrary. However, constitutive equations usually are introduced with respect to *one* 'special' configuration, say, $B_\square$, in which the material possesses special characteristics; for instance, it is isotropic, or unloaded, or it displays certain anisotropy properties.

In incremental and rate problems, it is often instrumental to work in a so-called relative Lagrangean description, where the reference configuration moves and is at each instant identified with the current configuration, $B_0 \equiv B$. This operation is completely legitimate, but the constitutive equation must continue to be referred to its 'special' configuration. Moreover, the treatment of the time derivatives (or increments) needs some care in the relative Lagrangean description. Thus, for instance, although $\boldsymbol{F} = \boldsymbol{I}$ when $B_0 \equiv B$, the rate of the deformation gradient is not null but equal to the velocity gradient.

We therefore have to deal with two problems: One is that the constitutive equations have to be referred to a 'special' configuration $B_\square$, and in another is that in adopting the relative Lagrangean description, we need to calculate the rates of the quantities before to assume the coincidence between B_0 and B. It is possible to proceed in the following way (see Fig. 6.5).

- We define three configurations: $B_\square$, B_0 and B. Therefore, we have the following motion and deformation

$$\boldsymbol{x} = \boldsymbol{g}(\boldsymbol{x}_0, t), \qquad \boldsymbol{x}_0 = \boldsymbol{g}_*(\boldsymbol{x}_\square), \tag{6.118}$$

so

$$\boldsymbol{x} = \boldsymbol{g}(\boldsymbol{g}_*(\boldsymbol{x}_\square), t), \tag{6.119}$$

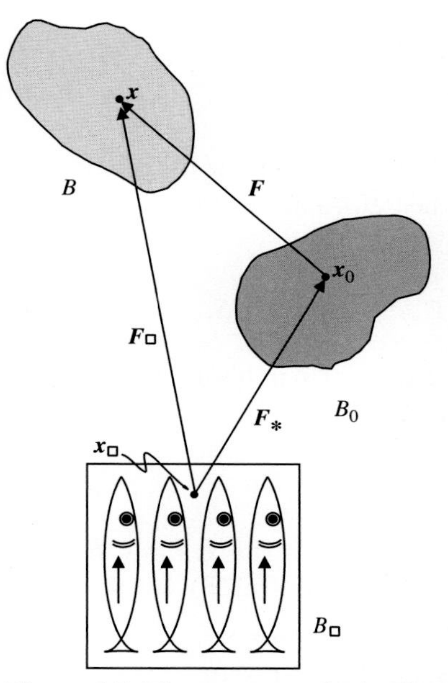

Figure 6.5. The route to obtain the relative Lagrangean description: three configurations are introduced, namely, the 'special' configuration for the constitutive description $B_\square$ (e.g., unloaded and orthotropic), the reference configuration B_0 and the current configuration B. Eventually, B_0 is assumed to coincide with B, so $\boldsymbol{F} = \boldsymbol{I}$, $J = 1$, $\boldsymbol{F}_\square = \boldsymbol{F}_*$ and $J_\square = J_*$.

and we can introduce the three deformation gradients

$$\boldsymbol{F} = \frac{\partial \boldsymbol{g}}{\partial \boldsymbol{x}_0}, \qquad \boldsymbol{F}_* = \frac{\partial \boldsymbol{g}_*}{\partial \boldsymbol{x}_\square}, \qquad \boldsymbol{F}_\square = \frac{\partial \boldsymbol{g}(\boldsymbol{g}_*(\boldsymbol{x}_\square), t)}{\partial \boldsymbol{x}_\square} = \boldsymbol{F}\boldsymbol{F}_*, \tag{6.120}$$

where $\boldsymbol{F}_*$ is independent of time. We refer to a constitutive equation of the type (6.46), defined in the configuration $B_\square$ (and 'producing' the stress in B associated with the deformation gradient $\boldsymbol{F}_\square$ through $\boldsymbol{E}_\square^{(m)}$ and the referential fabric tensor $\boldsymbol{H}_\square$)

$$\boldsymbol{T}_\square^{(m)} = \hat{\boldsymbol{T}}_\square^{(m)}(\boldsymbol{E}_\square^{(m)}, \boldsymbol{H}_\square). \tag{6.121}$$

Assuming $m = 2$ (second Piola-Kirchhoff stress and Green-Lagrange strain), we can express $\boldsymbol{T}_\square^{(2)}$ as

$$\boldsymbol{T}_\square^{(2)} = J_\square \boldsymbol{F}_\square^{-1} \boldsymbol{T} \boldsymbol{F}_\square^{-T} = J_* \boldsymbol{F}_*^{-1} \boldsymbol{T}^{(2)} \boldsymbol{F}_*^{-T}, \tag{6.122}$$

and we can transport the stress (6.122) in B as

$$\boldsymbol{T} = \frac{1}{JJ_*} \boldsymbol{F}\boldsymbol{F}_* \boldsymbol{T}_\square^{(2)} \boldsymbol{F}_*^T \boldsymbol{F}^T. \tag{6.123}$$

- Assuming that B moves in time, we take the material-time derivatives, beginning with Eqs. (6.121) and (6.122). Thus, since

$$\dot{\boldsymbol{F}}_\square = \dot{\boldsymbol{F}} \boldsymbol{F}_*, \tag{6.124}$$

we arrive at

$$\dot{\boldsymbol{T}}^{(2)} = \frac{1}{J_*} (\boldsymbol{F}_* \boxtimes \boldsymbol{F}_*) \mathbb{E}_\square (\boldsymbol{F}_* \boxtimes \boldsymbol{F}_*)^T [\dot{\boldsymbol{E}}^{(2)}], \tag{6.125}$$

where

$$\mathbb{E}_\square = \frac{\partial \hat{\boldsymbol{T}}_\square^{(2)}}{\partial \boldsymbol{E}_\square^{(2)}}. \tag{6.126}$$

Equation (6.125) can be transformed into a relation between the time derivatives of the first Piola-Kirchhoff stress tensor and the deformation gradients using the rule (6.110) and Eq. $(2.93)_1$, that is,

$$\dot{\boldsymbol{S}} = \frac{1}{J_*}(\boldsymbol{F}\boldsymbol{F}_*\boxtimes\boldsymbol{F}_*)\mathbb{E}_\square(\boldsymbol{F}\boldsymbol{F}_*\boxtimes\boldsymbol{F}_*)^T[\dot{\boldsymbol{F}}] + \dot{\boldsymbol{F}}\boldsymbol{F}^{-1}\boldsymbol{S}, \tag{6.127}$$

or into a relation between the Oldroyd derivative of Kirchhoff stress and the Eulerian strain rate using the rule (6.115), that is,

$$\overset{\circ}{\boldsymbol{K}} = \frac{1}{J_*}(\boldsymbol{F}\boldsymbol{F}_*\boxtimes\boldsymbol{F}\boldsymbol{F}_*)\mathbb{E}_\square(\boldsymbol{F}\boldsymbol{F}_*\boxtimes\boldsymbol{F}\boldsymbol{F}_*)^T[\boldsymbol{D}]. \tag{6.128}$$

- The configuration B_0 is now taken to coincide with B, so $\boldsymbol{F} = \boldsymbol{I}$, $J = 1$, $\boldsymbol{F}_\square = \boldsymbol{F}_*$ and $J_\square = J_*$ can be substituted into Eqs. (6.123) through (6.128). It follows that Eq. (6.125) remains unchanged, but we arrive at

$$\boldsymbol{T} = \frac{1}{J_\square}\boldsymbol{F}_\square\hat{\boldsymbol{T}}_\square^{(m)}(\boldsymbol{E}_\square^{(m)}, \boldsymbol{H}_\square)\boldsymbol{F}_\square^T, \tag{6.129}$$

defining the Cauchy stress in B, the so-called pre-stress, and

$$\dot{\boldsymbol{S}} = \mathbb{G}[\boldsymbol{L}], \qquad \mathbb{G} = \frac{1}{J_\square}(\boldsymbol{F}_\square\boxtimes\boldsymbol{F}_\square)\mathbb{E}_\square(\boldsymbol{F}_\square\boxtimes\boldsymbol{F}_\square)^T + \boldsymbol{I}\boxtimes\boldsymbol{T}, \tag{6.130}$$

defining the rate constitutive equation in terms of first Piola-Kirchhoff stress tensor and velocity gradient, and

$$\overset{\circ}{\boldsymbol{K}} = \mathbb{H}[\boldsymbol{D}], \qquad \mathbb{H} = \frac{1}{J_\square}(\boldsymbol{F}_\square\boxtimes\boldsymbol{F}_\square)\mathbb{E}_\square(\boldsymbol{F}_\square\boxtimes\boldsymbol{F}_\square)^T \tag{6.131}$$

in terms of the Oldroyd derivative of the Kirchhoff stress and Eulerian strain rate.

Incompressible materials For an *incompressible material*, we have to start from Eq. (6.71), which, for $n = 2$ and because $\boldsymbol{U}^{-2} = \boldsymbol{F}^{-1}\boldsymbol{F}^{-T}$, can be rewritten as

$$\boldsymbol{T}_\square^{(2)} = -\pi\,\boldsymbol{F}_\square^{-1}\boldsymbol{F}_\square^{-T} + \frac{\partial W(\boldsymbol{E}_\square^{(2)}, \boldsymbol{H}_\square)}{\partial \boldsymbol{E}_\square^{(2)}}. \tag{6.132}$$

The material time derivative of Eq. (6.132) yields

$$\dot{\boldsymbol{T}}_\square^{(2)} = -\dot{\pi}\,\boldsymbol{F}_\square^{-1}\boldsymbol{F}_\square^{-T} + 2\pi\,\boldsymbol{F}_\square^{-1}\boldsymbol{D}\boldsymbol{F}_\square^{-T} + \mathbb{E}_\square[\dot{\boldsymbol{E}}_\square^{(2)}], \tag{6.133}$$

where

$$\mathbb{E}_\square = \frac{\partial^2 W(\boldsymbol{E}_\square^{(2)}, \boldsymbol{H}_\square)}{\partial \boldsymbol{E}_\square^{(2)\,2}}. \tag{6.134}$$

Taking the material time derivative of the relation between the first and the second Piola-Kirchhoff stresses, $\boldsymbol{S} = \boldsymbol{F}\boldsymbol{T}^{(2)}$, in which $\boldsymbol{T}^{(2)}$ is expressed via Eq. (6.122) as a function of $\boldsymbol{T}_\square^{(2)}$, we obtain (note that $J_* = J_\square = J = 1$ due to incompressibility)

$$\dot{\boldsymbol{S}} = \dot{\boldsymbol{F}}\boldsymbol{F}_*\boldsymbol{T}_\square^{(2)}\boldsymbol{F}_*^T + \boldsymbol{F}_\square\dot{\boldsymbol{T}}_\square^{(2)}\boldsymbol{F}_*^T. \tag{6.135}$$

Using now the relation $\dot{\boldsymbol{F}} = \boldsymbol{L}\boldsymbol{F}$, Eqs. (6.132) through (6.134) and imposing the coincidence of B with B_0 (so that $\boldsymbol{F} = \boldsymbol{I}$ and $\boldsymbol{F}_* = \boldsymbol{F}_\square$), we arrive at

$$\dot{\boldsymbol{S}} = -\dot{\pi}\boldsymbol{I} + \mathbb{G}[\boldsymbol{L}], \tag{6.136}$$

where the fourth-order tensor $\mathbb{G}$ is defined as

$$\mathbb{G} = 2\pi \boldsymbol{I}\,\underline{\overline{\otimes}}\,\boldsymbol{I} + (\boldsymbol{F}_\square \boxtimes \boldsymbol{F}_\square)\mathbb{E}_\square(\boldsymbol{F}_\square \boxtimes \boldsymbol{F}_\square)^T + \boldsymbol{I} \boxtimes \boldsymbol{T}. \tag{6.137}$$

In terms of the Oldroyd derivative of Cauchy stress (owing to incompressibility, there is no distinction between Cauchy and Kirchhoff stresses), Eq. (6.136) becomes

$$\overset{\circ}{\boldsymbol{T}} = -\dot{\pi}\boldsymbol{I} + \mathbb{H}[\boldsymbol{D}], \quad \mathbb{H} = 2\pi \boldsymbol{I}\,\underline{\overline{\otimes}}\,\boldsymbol{I} + (\boldsymbol{F}_\square \boxtimes \boldsymbol{F}_\square)\mathbb{E}_\square(\boldsymbol{F}_\square \boxtimes \boldsymbol{F}_\square)^T, \tag{6.138}$$

a relation which follows immediately from Eq. $(3.158)_2$.

The Biot formulation of constitutive equations for incompressible, plane elasticity

Biot (1965) has provided a general framework for incremental elastic incompressible constitutive equations. Our interest here is in the formulation for plane strain incremental deformations superimposed on a given uniform stress. Following Brun et al. (2003a), we will obtain the Biot constitutive equations starting from the response of a Cauchy *isotropic* elastic solid (6.78); we will comment later that the equations obtained are more general to embrace orthotropic solids with orthotropy axes aligned parallel to the pre-stress state.

We begin from the relation (6.78) between the Cauchy stress $\boldsymbol{T}$ and the left Cauchy-Green strain tensor $\boldsymbol{B}$ for a Cauchy-elastic, incompressible and isotropic solid.

The two particular cases of Mooney-Rivlin and neo-Hookean materials are recovered when β_0 and β_1 are taken constant and, in addition, when $\beta_1 = 0$ in the latter case.

The constitutive equation (6.78) corresponds to Cauchy elasticity, so it describes a class of behaviours broader than hyperelasticity. The requirement of existence of an elastic potential influences the dependence of coefficients β_0 and β_1 on the invariants of $\boldsymbol{B}$. To develop this point, let us consider that the constitutive equation (6.78) implies coaxiality of tensors $\boldsymbol{B}$ and $\boldsymbol{T}$, so these share (at least) one principal reference system — the Eulerian principal axes — where

$$\operatorname{diag}\boldsymbol{B} = (\lambda_1^2, \lambda_2^2, \lambda_3^2) \qquad \text{and} \qquad \operatorname{diag}\boldsymbol{T} = (T_1, T_2, T_3), \tag{6.139}$$

in which $\lambda_i > 0$, $i = 1,2,3$, are the principal stretches, satisfying the incompressibility constraint

$$\lambda_1\lambda_2\lambda_3 = 1. \tag{6.140}$$

Expressing Eq. (6.78) in the Eulerian principal reference system and solving for β_0 and β_1 yields

$$\begin{aligned}
\beta_0 &= \frac{1}{\lambda_1^2 - \lambda_2^2}\left[\frac{(T_1 - T_3)\lambda_1^2}{\lambda_1^2 - \lambda_3^2} - \frac{(T_2 - T_3)\lambda_2^2}{\lambda_2^2 - \lambda_3^2}\right], \\
\beta_1 &= \frac{1}{\lambda_1^2 - \lambda_2^2}\left(\frac{T_1 - T_3}{\lambda_1^2 - \lambda_3^2} - \frac{T_2 - T_3}{\lambda_2^2 - \lambda_3^2}\right),
\end{aligned} \tag{6.141}$$

two equations that can be expressed alternatively employing every permutation of 1, 2 and 3 as indices. It is clear from Eq. (6.141) that the existence of an elastic potential restricts the functional dependence of coefficients β_i on the stretch. In particular, defining the elastic potential as a function of the principal stretches, Eq. (4.25) can be inserted immediately into Eq. (6.141), yielding expressions less general than Eq. (6.141) and depending only on the principal stretches.

Taking the material-time derivative of eqn. (6.78) yields

$$\dot{\boldsymbol{T}} = -\dot{\pi}\boldsymbol{I} + \beta_0\dot{\boldsymbol{B}} + \beta_1\left(\boldsymbol{B}^{-1}\right)^{\cdot} + \dot{\beta}_0\boldsymbol{B} + \dot{\beta}_1\boldsymbol{B}^{-1}, \tag{6.142}$$

where

$$\dot{\boldsymbol{B}} = \boldsymbol{DB} + \boldsymbol{BD} + \boldsymbol{WB} - \boldsymbol{BW} \tag{6.143}$$

and

$$\left(\boldsymbol{B}^{-1}\right)^{\cdot} = -\boldsymbol{B}^{-1}\boldsymbol{D} - \boldsymbol{D}\boldsymbol{B}^{-1} - \boldsymbol{B}^{-1}\boldsymbol{W} + \boldsymbol{W}\boldsymbol{B}^{-1}, \tag{6.144}$$

in which $\boldsymbol{D}$ is the Eulerian strain rate and $\boldsymbol{W}$ the spin tensor [Eqs. (3.59)]. Keeping into account now Eq. (6.78) and the definition of Jaumann derivative $(3.156)_1$, the constitutive equation (6.78) becomes

$$\overset{\nabla}{\boldsymbol{T}} + \dot{\pi}\boldsymbol{I} = \beta_0(\boldsymbol{DB} + \boldsymbol{BD}) - \beta_1(\boldsymbol{B}^{-1}\boldsymbol{D} + \boldsymbol{D}\boldsymbol{B}^{-1}) + \dot{\beta}_0\boldsymbol{B} + \dot{\beta}_1\boldsymbol{B}^{-1}. \tag{6.145}$$

Noting that

$$\dot{\beta}_i = \frac{\partial \beta_i}{\partial I_1}\mathrm{tr}\dot{\boldsymbol{B}} + 2\frac{\partial \beta_i}{\partial I_2}\boldsymbol{B}\cdot\dot{\boldsymbol{B}}, \tag{6.146}$$

where $\beta_i = \beta_i(I_1, I_2)$, with $i = 0, 1$, and

$$\mathrm{tr}\dot{\boldsymbol{B}} = 2\boldsymbol{B}\cdot\boldsymbol{D}, \quad \boldsymbol{B}\cdot\dot{\boldsymbol{B}} = 2\boldsymbol{B}^2\cdot\boldsymbol{D} \tag{6.147}$$

or

$$\dot{\hat{\beta}}_i = \frac{\partial \hat{\beta}_i}{\partial \lambda_1}\dot{\lambda}_1 + \frac{\partial \hat{\beta}_i}{\partial \lambda_2}\dot{\lambda}_2, \tag{6.148}$$

where $\hat{\beta}_i = \hat{\beta}_i(\lambda_1, \lambda_2)$, with $i = 0, 1$, are the coefficients β_0 and β_1 expressed as functions of the principal stretches, we arrive at

$$\begin{aligned}\overset{\nabla}{\boldsymbol{T}} + \dot{\pi}\boldsymbol{I} = {} & \beta_0(\boldsymbol{DB} + \boldsymbol{BD}) - \beta_1(\boldsymbol{B}^{-1}\boldsymbol{D} + \boldsymbol{D}\boldsymbol{B}^{-1}) \\ & + 2\left(\boldsymbol{B}\cdot\boldsymbol{D}\frac{\partial \beta_0}{\partial I_1} + 2\boldsymbol{B}^2\cdot\boldsymbol{D}\frac{\partial \beta_0}{\partial I_2}\right)\boldsymbol{B} + 2\left(\boldsymbol{B}\cdot\boldsymbol{D}\frac{\partial \beta_1}{\partial I_1} + 2\boldsymbol{B}^2\cdot\boldsymbol{D}\frac{\partial \beta_1}{\partial I_2}\right)\boldsymbol{B}^{-1},\end{aligned} \tag{6.149}$$

which is equivalent to

$$\begin{aligned}\overset{\nabla}{\boldsymbol{T}} + \dot{\pi}\boldsymbol{I} = {} & \beta_0(\boldsymbol{DB} + \boldsymbol{BD}) - \beta_1(\boldsymbol{B}^{-1}\boldsymbol{D} + \boldsymbol{D}\boldsymbol{B}^{-1}) \\ & + \left(\frac{\partial \hat{\beta}_0}{\partial \lambda_1}\dot{\lambda}_1 + \frac{\partial \hat{\beta}_0}{\partial \lambda_2}\dot{\lambda}_2\right)\boldsymbol{B} + \left(\frac{\partial \hat{\beta}_1}{\partial \lambda_1}\dot{\lambda}_1 + \frac{\partial \hat{\beta}_1}{\partial \lambda_2}\dot{\lambda}_2\right)\boldsymbol{B}^{-1}.\end{aligned} \tag{6.150}$$

The incremental constitutive equation in the form of Eq. (6.149) or Eq. (6.150) is valid for three-dimensional, incompressible Cauchy elasticity.

We are interested here in the particularisation of Eq. (6.149) to *incremental plane strain deformations* superimposed on a generic state of *homogeneous* deformation. In the Eulerian principal, reference system,

$$\operatorname{diag}\boldsymbol{B} = \left(\lambda_1^2, \lambda_2^2, \frac{1}{\lambda_1^2\lambda_2^2}\right), \qquad D_{i3} = D_{3i} = 0, \quad i = 1,2,3, \tag{6.151}$$

so the out-of-plane stress rate components can be determined as

$$\overset{\nabla}{T}_{3i} = \overset{\nabla}{T}_{i3} = 0, \qquad i = 1,2, \tag{6.152}$$

and

$$\begin{aligned}\overset{\nabla}{T}_{33} = -\dot{\pi} + (\lambda_1^2 - \lambda_2^2)\left\{\frac{1}{\lambda_1^2\lambda_2^2}\left[\frac{\partial\beta_0}{\partial I_1} + 2\left(\lambda_1^2 + \lambda_2^2\right)\frac{\partial\beta_0}{\partial I_2}\right]\right. \\ \left. + \lambda_1^2\lambda_2^2\left[\frac{\partial\beta_1}{\partial I_1} + 2\left(\lambda_1^2 + \lambda_2^2\right)\frac{\partial\beta_1}{\partial I_2}\right]\right\}(D_{11} - D_{22})\end{aligned} \tag{6.153}$$

or

$$\overset{\nabla}{T}_{33} = -\dot{\pi} + \left[\frac{1}{\lambda_1^2\lambda_2^2}\left(\lambda_1\frac{\partial\hat{\beta}_0}{\partial\lambda_1} - \lambda_2\frac{\partial\hat{\beta}_0}{\partial\lambda_2}\right) + \lambda_1^2\lambda_2^2\left(\lambda_1\frac{\partial\hat{\beta}_1}{\partial\lambda_1} - \lambda_2\frac{\partial\hat{\beta}_1}{\partial\lambda_2}\right)\right]\frac{(D_{11} - D_{22})}{2}. \tag{6.154}$$

The rate or incremental constitutive equations for plane strain incremental isochoric deformation superimposed on a given state of stress are (Biot, 1965)

$$\begin{cases} \overset{\nabla}{T}_{12} = 2\mu D_{12}, \\ \overset{\nabla}{T}_{11} - \overset{\nabla}{T}_{22} = 2\mu_* (D_{11} - D_{22}), \\ D_{11} + D_{22} = 0, \end{cases} \tag{6.155}$$

where μ and μ_* are two incremental moduli corresponding, respectively, to shearing parallel to and at 45° to the Eulerian principal axes. Note that constitutive relations of the form (6.155) with generic coefficients μ and μ_* embrace a much broader class of material behaviours than Cauchy isotropic elasticity (including the Mooney-Rivlin and GBG material models) and, in particular, the J_2-deformation theory of plasticity and the alternative model to describe the behaviour of ductile metals proposed by Stören and Rice (1975; see the discussion in Brun et al., 2003a).

The two incremental shear moduli μ and μ_* appearing in Eqs. (6.155) can be deduced by comparing Eqs. (6.155) with Eq. (6.150), writting in components and keeping into account the plane strain constraint. These moduli can be expressed as functions of the invariants of $\boldsymbol{B}$ as

$$\begin{aligned}\mu &= \frac{\lambda_1^2 + \lambda_2^2}{2}\left(\beta_0 - \frac{\beta_1}{\lambda_1^2\lambda_2^2}\right), \\ \mu_* &= \frac{\lambda_1^2 + \lambda_2^2}{2}\beta_0 + \frac{(\lambda_1^2 - \lambda_2^2)^2}{2}\left[\frac{\partial\beta_0}{\partial I_1} + 2\left(\lambda_1^2 + \lambda_2^2\right)\frac{\partial\beta_0}{\partial I_2}\right] \\ &\quad - \frac{1}{\lambda_1^2\lambda_2^2}\left\{\frac{\lambda_1^2 + \lambda_2^2}{2}\beta_1 + \frac{(\lambda_1^2 - \lambda_2^2)^2}{2}\left[\frac{\partial\beta_1}{\partial I_1} + 2\left(\lambda_1^2 + \lambda_2^2\right)\frac{\partial\beta_1}{\partial I_2}\right]\right\},\end{aligned} \tag{6.156}$$

or as functions of the principal stretches as

$$\mu = \frac{\lambda_1^2+\lambda_2^2}{2}\left(\hat{\beta}_0 - \frac{\hat{\beta}_1}{\lambda_1^2\lambda_2^2}\right),$$

$$\mu_* = \frac{\lambda_1^2+\lambda_2^2}{2}\hat{\beta}_0 + \frac{\lambda_1^2-\lambda_2^2}{4}\left(\lambda_1\frac{\partial\hat{\beta}_0}{\partial\lambda_1} - \lambda_2\frac{\partial\hat{\beta}_0}{\partial\lambda_2}\right) - \frac{1}{\lambda_1^2\lambda_2^2}\left[\frac{\lambda_1^2+\lambda_2^2}{2}\hat{\beta}_1 + \frac{\lambda_1^2-\lambda_2^2}{4}\left(\lambda_1\frac{\partial\hat{\beta}_1}{\partial\lambda_1} - \lambda_2\frac{\partial\hat{\beta}_1}{\partial\lambda_2}\right)\right]. \tag{6.157}$$

For an elastic solid isotropic in its reference configuration and deformed under plane strain, $\lambda_3 = 1$ and $\lambda = \lambda_1 = 1/\lambda_2 > 1$, we can express the elastic energy $\hat{W}(\lambda_1,\lambda_2,\lambda_3)$ used in Eq. (4.23) as a function of λ as follows

$$\tilde{W}(\lambda) = \hat{W}(\lambda, 1/\lambda, 1), \tag{6.158}$$

from which we obtain

$$\frac{d\tilde{W}(\lambda)}{d\lambda} = \frac{\partial\hat{W}}{\partial\lambda_1}\frac{\partial\lambda_1}{\partial\lambda} + \frac{\partial\hat{W}}{\partial\lambda_2}\frac{\partial\lambda_2}{\partial\lambda} = \frac{\partial\hat{W}}{\partial\lambda_1} - \frac{1}{\lambda^2}\frac{\partial\hat{W}}{\partial\lambda_2}. \tag{6.159}$$

Since Eq. (4.25) provides

$$T_1 - T_2 = \lambda_1\frac{\partial\hat{W}}{\partial\lambda_1} - \lambda_2\frac{\partial\hat{W}}{\partial\lambda_2} = \lambda\frac{\partial\hat{W}}{\partial\lambda_1} - \frac{1}{\lambda}\frac{\partial\hat{W}}{\partial\lambda_2}, \tag{6.160}$$

and employing Eq. (6.159), we arrive at the useful expression (holding for plane strain incompressible hyperelasticity)

$$T_1 - T_2 = \lambda\frac{d\tilde{W}(\lambda)}{d\lambda}. \tag{6.161}$$

Taking the material-time derivative of the spectral representation of the Cauchy stress, which for *isotropic* material is coaxial to left Cauchy-Green deformation tensor $\boldsymbol{B}$, namely,

$$\boldsymbol{T} = T_1\boldsymbol{v}_1\otimes\boldsymbol{v}_1 + T_2\boldsymbol{v}_2\otimes\boldsymbol{v}_2 + T_3\boldsymbol{v}_3\otimes\boldsymbol{v}_3, \tag{6.162}$$

we obtain (for plane strain $\boldsymbol{v}_3$ remains fixed)

$$\dot{\boldsymbol{T}} = \dot{T}_1\boldsymbol{v}_1\otimes\boldsymbol{v}_1 + \dot{T}_2\boldsymbol{v}_2\otimes\boldsymbol{v}_2 + \dot{T}_3\boldsymbol{v}_3\otimes\boldsymbol{v}_3 + T_1(\dot{\boldsymbol{v}}_1\otimes\boldsymbol{v}_1 + \boldsymbol{v}_1\otimes\dot{\boldsymbol{v}}_1) + T_2(\dot{\boldsymbol{v}}_2\otimes\boldsymbol{v}_2 + \boldsymbol{v}_2\otimes\dot{\boldsymbol{v}}_2). \tag{6.163}$$

Through a representation of the unit vectors $\boldsymbol{v}_1$ and $\boldsymbol{v}_2$ in the 1–2 plane, in terms of the azimuthal angle θ as

$$\{\boldsymbol{v}_1\} = \{\cos\theta, \sin\theta\} \quad \text{and} \quad \{\boldsymbol{v}_2\} = \{-\sin\theta, \cos\theta\}, \tag{6.164}$$

we immediately find that

$$\dot{\boldsymbol{v}}_1 = \dot{\theta}\boldsymbol{v}_2, \qquad \dot{\boldsymbol{v}}_2 = -\dot{\theta}\boldsymbol{v}_1, \qquad \dot{\theta} = |\dot{\boldsymbol{v}}_1| = |\dot{\boldsymbol{v}}_2|, \tag{6.165}$$

so Eq. (6.163) becomes

$$\dot{\boldsymbol{T}} = \dot{T}_1\boldsymbol{v}_1\otimes\boldsymbol{v}_1 + \dot{T}_2\boldsymbol{v}_2\otimes\boldsymbol{v}_2 + \dot{T}_3\boldsymbol{v}_3\otimes\boldsymbol{v}_3 + \dot{\theta}(T_1 - T_2)(\boldsymbol{v}_1\otimes\boldsymbol{v}_2 + \boldsymbol{v}_2\otimes\boldsymbol{v}_1). \tag{6.166}$$

From the definition of Jaumann derivative (3.156)$_1$, in the principal reference system of $\boldsymbol{T}$, we have

$$\overset{\triangledown}{T}_{11}-\overset{\triangledown}{T}_{22}=\dot{T}_{11}-\dot{T}_{22}, \qquad \overset{\triangledown}{T}_{12}=\dot{T}_{12}+(T_1-T_2)\,W_{12}. \tag{6.167}$$

Thus, using Eq. (6.163), we obtain

$$\overset{\triangledown}{T}_{11}-\overset{\triangledown}{T}_{22}=(T_1-T_2)^{\cdot}, \qquad \overset{\triangledown}{T}_{12}=(T_1-T_2)\left(\dot{\theta}+W_{12}\right). \tag{6.168}$$

The material-time derivative of the left Cauchy-Green deformation tensor (6.143) yields

$$\frac{\dot{\lambda}}{\lambda}=\frac{D_{11}-D_{22}}{2}, \quad \dot{\theta}=\frac{\lambda^4+1}{\lambda^4-1}D_{12}-W_{12}, \tag{6.169}$$

the second of which, used into Eq. (6.168)$_2$, provides for these equations the expressions

$$\overset{\triangledown}{T}_{11}-\overset{\triangledown}{T}_{22}=\frac{\lambda}{2}\frac{d\,(T_1-T_2)}{d\lambda}(D_{11}-D_{22}), \qquad \overset{\triangledown}{T}_{12}=(T_1-T_2)\frac{\lambda^4+1}{\lambda^4-1}D_{12}. \tag{6.170}$$

A comparison between Eqs. (6.170) and Eqs. (6.155) yields another useful expression for the two incremental moduli μ and μ_*, namely,

$$\mu=\frac{\lambda^4+1}{\lambda^4-1}\frac{T_1-T_2}{2}, \qquad \mu_*=\frac{\lambda}{4}\frac{d\,(T_1-T_2)}{d\lambda}, \tag{6.171}$$

which, using Eqs. (6.161), become

$$\mu=\frac{\lambda}{2}\frac{\lambda^4+1}{\lambda^4-1}\frac{d\tilde{W}(\lambda)}{d\lambda}, \qquad \mu_*=\frac{\lambda}{4}\left(\frac{d\tilde{W}(\lambda)}{d\lambda}+\lambda\frac{d^2\tilde{W}(\lambda)}{d\lambda^2}\right). \tag{6.172}$$

Assuming now that the stress difference T_1-T_2 can be represented in a Taylor series expansion near the unstressed state $\lambda=1$ (which need not always be true, such as, for instance, in the case of the J_2-deformation theory of plasticity), we obtain

$$T_1-T_2\sim\left.\frac{d\,(T_1-T_2)}{d\lambda}\right|_{\lambda=1}(\lambda-1), \tag{6.173}$$

so a substitution into Eq. (6.171) yields

$$\frac{\mu_*}{\mu}\sim\lambda, \tag{6.174}$$

thus showing that the material is incrementally isotropic at the unstressed state $\lambda=1$ (the ratio between the incremental moduli μ_* and μ tends to 1).

The constitutive equation (6.155) has been used in a series of works of interest in terms of material time derivative of the nominal stress tensor $\boldsymbol{t}=\boldsymbol{S}^T$ [Eq. (3.110)], related to the transpose of the velocity gradient and referred to a Lagrangean formulation of field equations, with the current state taken as reference. Simple transformations show that Eq (6.155) can be rewritten as

$$\dot{S}_{ji}=\dot{t}_{ij}=\mathbb{G}_{ijkl}v_{l,k}+\dot{p}\,\delta_{ij}, \qquad v_{i,i}=0, \tag{6.175}$$

where v_i is the velocity and δ_{ij} is the Kronecker delta, and

$$\dot{p} = \frac{\dot{T}_1 + \dot{T}_2}{2} \tag{6.176}$$

measures the in-plane hydrostatic stress rate (positive in tension) as related to the Cauchy stress rate.

Note that *we have introduced a little abuse of notation* because in Eq. (6.175), $\mathbb{G}_{ijkl}$ transforms the transpose of the gradient of the incremental displacement into the increment of the nominal stress tensor, whereas tensor $\mathbb{G}$ defined by Eq. (6.110) transforms the incremental displacement gradient into the first Piola-Kirchhoff stress; thus the two tensors are different, although they represent the same constitutive equation. We do not want to complicate notation by introducing a different symbol for something analogous, so we will use $\mathbb{G}$ in the following to relate either $\dot{\boldsymbol{S}}$ to $\nabla \boldsymbol{v}$ or $\dot{\boldsymbol{t}} = \dot{\boldsymbol{S}}^T$ to $(\nabla \boldsymbol{v})^T$.

Tensor $\mathbb{G}$ represents the elastic tensor or the material instantaneous moduli and possesses the major symmetry ($\mathbb{G}_{ijkl} = \mathbb{G}_{klij}$), and its components can be written as

$$\begin{aligned}
&\mathbb{G}_{1111} = \mu_* - \frac{\sigma}{2} - p, && \mathbb{G}_{1122} = -\mu_*, && \mathbb{G}_{1112} = \mathbb{G}_{1121} = 0,\\
&\mathbb{G}_{2211} = -\mu_*, && \mathbb{G}_{2222} = \mu_* + \frac{\sigma}{2} - p, && \mathbb{G}_{2212} = \mathbb{G}_{2221} = 0,\\
&\mathbb{G}_{1212} = \mu + \frac{\sigma}{2}, && \mathbb{G}_{1221} = \mathbb{G}_{2112} = \mu - p, && \mathbb{G}_{2121} = \mu - \frac{\sigma}{2},
\end{aligned}$$

with

$$\sigma = T_1 - T_2 \quad \text{and} \quad p = \frac{T_1 + T_2}{2}. \tag{6.177}$$

Finally, we note that constitutive Eqs. (6.175) through (6.177) can be rewritten in the useful form

$$\begin{aligned}
\dot{t}_{11} &= \mu(2\xi - k - \eta)v_{1,1} + \dot{p},\\
\dot{t}_{22} &= \mu(2\xi + k - \eta)v_{2,2} + \dot{p},\\
\dot{t}_{12} &= \mu[(1+k)v_{2,1} + (1-\eta)v_{1,2}],\\
\dot{t}_{21} &= \mu[(1-\eta)v_{2,1} + (1-k)v_{1,2}],
\end{aligned} \tag{6.178}$$

where we have introduced the non-dimensional parameters

$$\xi = \frac{\mu_*}{\mu}, \qquad \eta = \frac{p}{\mu} = \frac{T_1 + T_2}{2\mu}, \qquad k = \frac{T_1 - T_2}{2\mu}, \tag{6.179}$$

the first representing the ratio between incremental shear moduli, the second and third dimensionless measures of the in-plane hydrostatic and deviatoric stresses, respectively. If a material is initially isotropic, we find from Eq. (6.171) that

$$k = \tanh(2\varepsilon_1), \tag{6.180}$$

where ε_1 is the positive in-plane logarithmic strain.

Mooney-Rivlin material In plane strain, the Mooney-Rivlin strain energy function [Eq. (4.27)], takes the form

$$W(I_1, I_2) = \frac{\mu_1}{2}(I_1 - 3) - \frac{\mu_2}{4}\left(I_1^2 - I_2 - 6\right), \tag{6.181}$$

or

$$\hat{W}(\lambda_1,\lambda_2) = \frac{\mu_1}{2}\left(\lambda_1^2+\lambda_2^2+\lambda_1^{-2}\lambda_2^{-2}-3\right) - \frac{\mu_2}{2}\left(\lambda_1^{-2}+\lambda_2^{-2}+\lambda_1^2\lambda_2^2-3\right), \quad (6.182)$$

where μ_1 and μ_2 are material parameters and $\mu_0 = \mu_1 - \mu_2$ represents the shear modulus in the original unstressed state.

With reference to Eq. (6.78), we simply obtain

$$\beta_0 = \mu_1, \qquad \beta_1 = \mu_2,$$
$$\mu = \mu_* = \frac{\mu_0}{2}\left(\lambda_1^2+\lambda_2^2\right) \quad (6.183)$$

and

$$\overset{\nabla}{T}_{33} = -\dot{\pi}, \quad (6.184)$$

where $\lambda_1\lambda_2 = 1$ owing to the incompressibility constraint.

J_2-Deformation theory of plasticity In plane strain, the J_2-deformation theory of plasticity [Eqs. (4.39) through (4.48)] corresponds to the following choice of parameters π, β_0 and β_1 in Eq. (6.78)

$$\begin{aligned}
\pi &= -\mathrm{tr}\boldsymbol{T}/3 + \frac{2}{3}E_s \frac{\left(\lambda_2^4-\lambda_2^{-2}\right)\left(\lambda_1^2+\lambda_3^2\right)\varepsilon_1 - \left(\lambda_1^4-\lambda_1^{-2}\right)\left(\lambda_2^2+\lambda_3^2\right)\varepsilon_2}{\left(\lambda_1^2-\lambda_2^2\right)\left(\lambda_1^2-\lambda_3^2\right)\left(\lambda_2^2-\lambda_3^2\right)},\\
\beta_0 &= \frac{2}{3}E_s\frac{1}{\lambda_1^2-\lambda_2^2}\left[\frac{(\varepsilon_1-\varepsilon_3)\lambda_1^2}{\lambda_1^2-\lambda_3^2} - \frac{(\varepsilon_2-\varepsilon_3)\lambda_2^2}{\lambda_2^2-\lambda_3^2}\right],\\
\beta_1 &= \frac{2}{3}E_s\frac{1}{\lambda_1^2-\lambda_2^2}\left(\frac{\varepsilon_1-\varepsilon_3}{\lambda_1^2-\lambda_3^2} - \frac{\varepsilon_2-\varepsilon_3}{\lambda_2^2-\lambda_3^2}\right),
\end{aligned} \quad (6.185)$$

with $\lambda_3 = 1$, $\varepsilon_3 = 0$, $\lambda_1 = 1/\lambda_2$, $\varepsilon_1 = -\varepsilon_2$, and to the incremental shear moduli:

$$\begin{aligned}
\mu &= \frac{1}{3}E_s(\varepsilon_1-\varepsilon_2)\coth(\varepsilon_1-\varepsilon_2),\\
\mu_* &= \frac{1}{9}\frac{E_s}{\varepsilon_e^2}\left[3(\varepsilon_1+\varepsilon_2)^2 + N(\varepsilon_1-\varepsilon_2)^2\right],
\end{aligned} \quad (6.186)$$

whereas the out-of-plane stress increment is given by

$$\overset{\nabla}{T}_{33} = \mathrm{tr}\dot{\boldsymbol{T}}/3 = \dot{p}. \quad (6.187)$$

Taking $\varepsilon_3 = 0$, so that $\varepsilon_1 = -\varepsilon_2$, parameters ξ and k can be given by

$$\xi = \frac{N}{2\varepsilon_1\coth(2\varepsilon_1)} \quad \text{and} \quad k = \frac{1}{\coth(2\varepsilon_1)}. \quad (6.188)$$

Note that $\xi = N$ in the limit $\varepsilon_1 \longrightarrow 0$.

GBG elasticity In plane strain and with reference to the elastic potential (4.49), we can derive the expressions for the incremental moduli in the form

$$\begin{aligned} \mu_* &= \frac{K}{3c\varepsilon_0}\left[(1+c)\exp\left(-\frac{2|\varepsilon_1|}{\sqrt{3}\varepsilon_0}\right)-1\right]\exp\left(-\frac{2|\varepsilon_1|}{\sqrt{3}c\varepsilon_0}\right), \\ \mu &= \frac{K}{\sqrt{3}}\left[1-\exp\left(-\frac{2|\varepsilon_1|}{\sqrt{3}\varepsilon_0}\right)\right]\exp\left(-\frac{2|\varepsilon_1|}{\sqrt{3}c\varepsilon_0}\right) \end{aligned} \tag{6.189}$$

so that

$$\xi = \frac{c\coth\left(\dfrac{\varepsilon_1}{\sqrt{3}\varepsilon_0}\right) - \mathsf{Sign}[\varepsilon_1](2+c)}{2\sqrt{3}c\varepsilon_0}\tanh(2\varepsilon_1). \tag{6.190}$$

We note the following features:

- For null pre-strain,

$$\text{For } \varepsilon_1 \to 0 \quad \begin{cases} \xi = 1, \mu = \mu_* = \dfrac{K}{3\varepsilon_0}, \\ k = 0, \end{cases} \tag{6.191}$$

  so the initially isotropic behaviour is recovered.
- For infinite pre-strain,

$$\text{For } \varepsilon_1 \to \infty \quad \begin{cases} \xi = -\dfrac{1}{\sqrt{3}c\varepsilon_0} < 0, \\ k = 1, \end{cases} \tag{6.192}$$

  so at 'large' deformation, an incremental shear stiffness becomes negative (for positive $c\varepsilon_0$).

6.2.3 Hypoelasticity

A constitutive law in rate form which is simple and therefore useful for bifurcation analyses is that defined in the case of so-called hypoelastic materials (Truesdell, 1955; Bernstein, 1960; Truesdell and Noll, 1965), where a rate of the type (3.157), for instance, the Oldroyd rate $(3.156)_2$, of a spatial symmetric stress measure, for instance, the Kirchhoff stress (3.108), is related to a Eulerian symmetric strain measure, for instance, the Eulerian strain rate, through a fourth-order constitutive tensor $\mathbb{E}$ as

$$\overset{\circ}{\boldsymbol{K}} = \mathbb{E}[\boldsymbol{D}], \tag{6.193}$$

which is defined on the spatial configuration, so it has to satisfy the principle of material frame indifference (6.3)

$$\mathbb{E}[\boldsymbol{D}] = \boldsymbol{Q}^T\mathbb{E}[\boldsymbol{Q}\boldsymbol{D}\boldsymbol{Q}^T]\boldsymbol{Q}. \tag{6.194}$$

Note the similarity between Eq. (6.193) and the rate form in the spatial setting of an elastic law [see Eq. (6.115)]. However, Eq. (6.193) is more general than Eq. (6.115) because it includes rate equations which are not integrable in time (in other words, incremental constitutive equations which are not a Taylor series expansion of a finite strain law along a strain path), whereas Eq. (6.115) follows from a finite strain law and therefore is always integrable in time. Although $\mathbb{E}$ is usually taken to be a function of the current state of stress $\boldsymbol{T}$, Eq. (6.194) is certainly satisfied when $\mathbb{E}$ is selected in the isotropic form (2.190), namely,

$$\mathbb{E} = \lambda \boldsymbol{I} \otimes \boldsymbol{I} + 2\mu \boldsymbol{I} \underline{\overline{\otimes}} \boldsymbol{I}, \tag{6.195}$$

a choice that provides a specific hypoelastic law (less general than a so-called grade 1 hypoelastic material; Truesdell and Noll, 1965) used, among others, to describe the reversible part of the deformation of an elastic-plastic material by Hill (1962), Hutchinson (1973), Neale (1981), Bigoni (1995) and Bigoni and Zaccaria (1994a).

Hypoelastic models of the type (6.193) have the drawback that they do not in general follow from a potential structure, so they can describe unphysical effects such as, for instance, residual elastic strain and energy 'production' along closed stress and strain paths (Hill, 1959; Zytinski et al., 1978).

The hypoelastic model (6.193) with the isotropic law (6.195) with constant coefficients does not in general determine a hyperelastic behaviour (Simo and Pister, 1984). For the specific model (6.193), based on the Oldroyd derivative, with the isotropic law (6.195), Christoffersen (1991) has shown that hyperelasticity is obtained only if λ and μ are not constants but depend in a certain way on the deformation.

In the way that hypoelasticty has been formulated, the only source of anisotropy comes from the effect of pre-stress, so Truesdell and Noll (1965) write that 'anisotropic elasticity is not included in hypo-elasticity as a special case'. However, as shown by Hill (1959) and Green and McInnis (1967), the theory easily can be generalised to include anisotropic behaviour. We will also employ a hypo-elastic model such as that described by Eq. (6.193) with an anisotropy fourth-order tensor $\mathbb{E}$ generated by a so-called fabric tensor $\boldsymbol{B}$, taken in the form of Eq. (2.197) with the constraint (2.198), suggested by Valanis (1990) and Zysset and Curnier (1995) to describe evolving damage and by Bigoni and Loret (1999), Bigoni et al. (2000) and Piccolroaz et al. (2006a; 2006b) to describe anisotropic elastic behaviour in an elastoplastic framework, namely,

$$\mathbb{E} = \lambda \boldsymbol{B} \otimes \boldsymbol{B} + 2\mu \boldsymbol{B} \underline{\overline{\otimes}} \boldsymbol{B}, \qquad \mathrm{tr}\boldsymbol{B}^2 = 3, \tag{6.196}$$

where $\boldsymbol{B}$ (not to be confused with the left Cauchy-Green deformation tensor) is a positive-definite fabric tensor defined in the *spatial configuration*, and λ and μ are material constants subject to the restrictions of positive definiteness of $\mathbb{E}$ [Eqs. (2.192)]. Note that tensor (6.196) satisfies restriction (6.194) because the fabric tensor $\boldsymbol{B}$ transforms as a spatial quantity.

A simplification of Eq. (6.196) arises in the special case of transverse isotropy about an axis $\boldsymbol{b}$, for which we have

$$\boldsymbol{B} = b_1 \boldsymbol{b} \otimes \boldsymbol{b} + b_2 (\boldsymbol{I} - \boldsymbol{b} \otimes \boldsymbol{b}), \tag{6.197}$$

where b_1 and b_2 are the eigenvalues of $\boldsymbol{B}$, whereas the line spanned by the unit vector $\boldsymbol{b}$ and the plane perpendicular to it are the corresponding eigenspaces. Moreover, for transverse isotropy, the material constants b_1 and b_2 are assumed to depend on a single angular parameter $\hat{b}$, restricted to the range $]0°, 90°[$, to meet the positive definiteness requirement of $\boldsymbol{B}$,

$$b_1 = \sqrt{3}\cos\hat{b}, \qquad b_2 = \sqrt{\frac{3}{2}}\sin\hat{b}, \tag{6.198}$$

so the isotropic behaviour is simply recovered when $b_1 = b_2 = 1$ or $\hat{b} \approx 54.74°$.

7 Yield functions with emphasis on pressure sensitivity

The concept of yield function is introduced to discriminate between states where either the elastic or the elastoplastic behaviours are possible. Special emphasis is given to solids yielding isotropically and possessing so-called pressure-sensitivity, a typical feature of the behaviour of frictional materials. A recently-proposed yield function is presented, from which many important yield criteria are derived as specific cases. General propositions on convexity of yield functions and yield surfaces are provided.

A 'yield function' is a constitutive prescription introduced to discriminate between states of a material in which the elastic deformation is the only possible deformation mechanism and states where plastic flow may occur. This concept is evidently central to the development of elastoplasticity theory. We will limit the presentation to the stress space formulation, in which the state of the material is represented by a stress variable plus a collection $\mathcal{H}$ of so-called hidden or internal state variables (e.g., a plastic strain tensor) describing the inelasticity of the material.[1] These variables can be left for the moment completely arbitrary.

A yield function is a scalar-valued tensor function associating (for a given set of hidden state variables $\mathcal{H}$) negative values to elastic stress states and the null value to stress states for which plastic flow becomes *possible*, namely,

$$\begin{aligned} f(\text{stress}, \mathcal{H}) < 0, \quad & \text{elastic behaviour,} \\ f(\text{stress}, \mathcal{H}) = 0, \quad & \text{possibility of plastic flow.} \end{aligned} \tag{7.1}$$

Note that a stress state for which $f(\text{stress}, \mathcal{H}) > 0$ is by definition impossible. The condition $(7.1)_2$ can be visualised as a *yield surface* in the hyper-space of the variables stress and $\mathcal{H}$, so the elastic states lie internally to this surface, whereas the states for which plastic flow is possible lie on it.

Since there is an evident mathematical analogy between a yield function described by Eqs. (7.1) and a strain energy density function for a material with a micro-structure [Eq. (6.53)], concepts developed for the latter can be used for the former. Therefore, we will avoid repetitions and will reduce the presentation to the essentials.

[1] Employing the constitutive laws, it is possible to transform the stress space description into a strain space formulation; however, we do not insist on this aspect here.

Similarly to statement (6.3), *the principle of invariance of material response under a change in observer* is stated now as

$$f\,(\text{stress},\mathcal{H}) = f\left(\mathbf{Q}^{B}_{\text{stress}}\,[\text{stress}]\,,\mathbf{Q}^{B}_{\mathcal{H}}[\mathcal{H}]\right), \tag{7.2}$$

holding for all rotations $\boldsymbol{Q} \in \mathsf{Orth}^+$ implicit in the operators $\mathbf{Q}^{B}_{\text{stress}}$ and $\mathbf{Q}^{B}_{\mathcal{H}}$, defining the 'appropriate' rotations of the stress measure and of the elements of the set $\mathcal{H}$ for a rotation of the current configuration B.

Similarly to statement (6.18), *the invariance with respect a rigid-body rotation of the reference configuration* is stated now as

$$f\,(\text{stress},\mathcal{H}) = f\left(\mathbf{Q}^{B_0}_{\text{stress}}\,[\text{stress}]\,,\mathbf{Q}^{B_0}_{\mathcal{H}}[\mathcal{H}]\right), \tag{7.3}$$

holding for all rotations $\boldsymbol{Q} \in \mathsf{Orth}^+$, implicit in the operators $\mathbf{Q}^{B_0}_{\text{stress}}$ and $\mathbf{Q}^{B_0}_{\mathcal{H}}$, defining the 'appropriate' rotations of the stress measure and of the elements of the set $\mathcal{H}$ for a rotation of the reference configuration B_0.

The consequences of the invariances [Eqs. (7.2) and (7.3)], depend on the selected stress variable and the set $\mathcal{H}$; for instance, if the variable 'stress' is identified with a Lagrangean stress measure $\boldsymbol{T}^{(m)}$ and the set $\mathcal{H}$ is identified with a symmetric *referential* second-order tensor, say, $\boldsymbol{H}_0$, the invariance expressed by Eq. (7.2) is automatically satisfied, whereas Eq. (7.3) imposes the analogous of Eq. (6.54), namely,

$$f(\boldsymbol{T}^{(m)},\boldsymbol{H}_0) = f(\boldsymbol{Q}\boldsymbol{T}^{(m)}\boldsymbol{Q}^T,\boldsymbol{Q}\boldsymbol{H}_0\boldsymbol{Q}^T), \qquad \forall \boldsymbol{Q} \in \mathsf{Orth}^+, \tag{7.4}$$

so the theorems on isotropic functions of two symmetric tensors can be applied to f. Therefore, an invariant representation similar to Eq. (6.55) can be written for f, and its gradient can be expressed similarly to Eq. (6.56).

Requirements connected to *material symmetries* also need to be stated, which take a form analogous to Eq. (6.23). Thus, denoting with $\mathbf{Q}^{B_0\ or\ B}_{\mathcal{H}}$ the appropriate rotation of the variables $\mathcal{H}$, *if*

$$f\,(\text{stress},\mathcal{H}) = f\left(\text{stress},\mathbf{Q}^{B_0\ or\ B}_{\mathcal{H}}[\mathcal{H}]\right), \tag{7.5}$$

then $\mathbf{Q}^{B_0\ or\ B}_{\mathcal{H}}$ *belongs to the symmetry class of the material.* Obviously, if condition (7.5) holds for every rotation, the yield response is isotropic.

Within this book, specific forms of the yield function are employed only in the examples, so

> *the formulations and results that will be presented are all general and do not require any particular feature of the yield function, except smoothness and satisfaction of the invariances [eqs. (7.2) and (7.3)],*

and therefore, most of the results will be given for yield functions that are required neither to possess any specific mathematical expression nor to be isotropic or even convex.

We address the description of the behaviour of ductile solids, with an emphasis on frictional materials. Several micro-mechanisms (sliding on grain or fracture contacts, pore collapses, nucleation of defects and interaction between inclusions) determine the overall behaviour of these materials and produce the so-called pressure sensitivity of yielding, deeply influencing stability and bifurcation. As illustrated

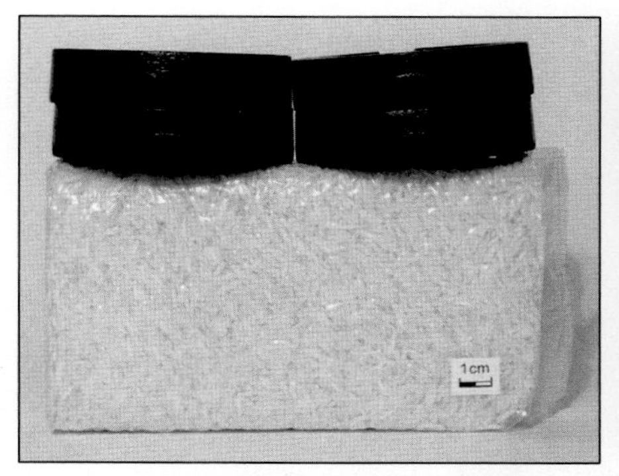

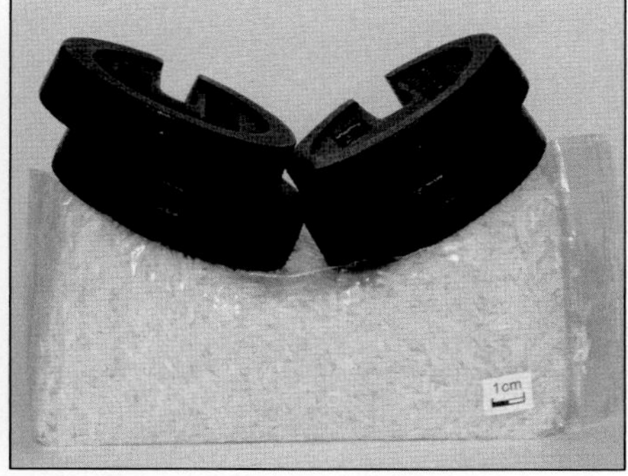

Figure 7.1. A simple explanation (suggested to me by Prof. A. Gajo) of the pressure sensitivity concept in a granular material (rice): a soft-pack under vacuum provides the confining pressure to the material, raising its strength to a value sufficient to support a dead weight (left). The pack has been punctured on the right, so the confining pressure, and with it the shear strength of the material, has dropped to a low value, not sufficient to bear the weight.

in Fig. 7.1, pressure sensitivity means that the yield limit increases with an increase in the confining pressure (the mean value of the stress with reversed sign, $-\mathrm{tr}\,(\text{stress})/3$) so that it is typically connected to Coulomb friction at the micro-scale.

Pressure sensitivity implies the so-called strength-differential (SD) phenomenon, so *yielding in uniaxial tension occurs at a much lower stress than in uniaxial compression*. Examples of pressure-sensitive materials are granular media, geomaterials (soils, rocks, concretes, ceramics), plastics, porous metals and steels exhibiting the SD effect (Hirth and Cohen, 1970; Kalish and Rack, 1972; Drucker, 1973; Spitzig et al., 1976).

We introduce the main concepts following Bigoni and Piccolroaz (2004) and assuming that *the material is isotropic with respect to yielding*, an effect sketched in Fig. 7.2 and implying that the set $\mathcal{H}$ reduces to a collection of *scalar isotropic variables* which can be omitted for conciseness from the formulae.

We will use the Cauchy stress $\boldsymbol{T}$ as a stress measure (but a change of this should not create any problem, so the treatment remains the same with any other material or spatial symmetric stress measure replacing $\boldsymbol{T}$). In these conditions, invariance (7.3) and isotropy (7.5) are automatically satisfied, whereas the principle of indifference with respect to a change in observer [Eq. (7.2)], provides the requirement

$$f(\boldsymbol{T}) = f(\boldsymbol{Q}\boldsymbol{T}\boldsymbol{Q}^T), \qquad \forall \boldsymbol{Q} \in \mathrm{Orth}^+, \tag{7.6}$$

meaning that f is an isotropic function of $\boldsymbol{T}$ and can be rewritten as a function of the invariants of $\boldsymbol{T}$.

7.1 The Haigh-Westergaard representation

The assumption of isotropic behaviour at yielding means that the yield function depends on the stress through its invariants; thus, since the principal values of stress are invariants, we can always express f [Eq. (7.6)] as a function of the Cauchy

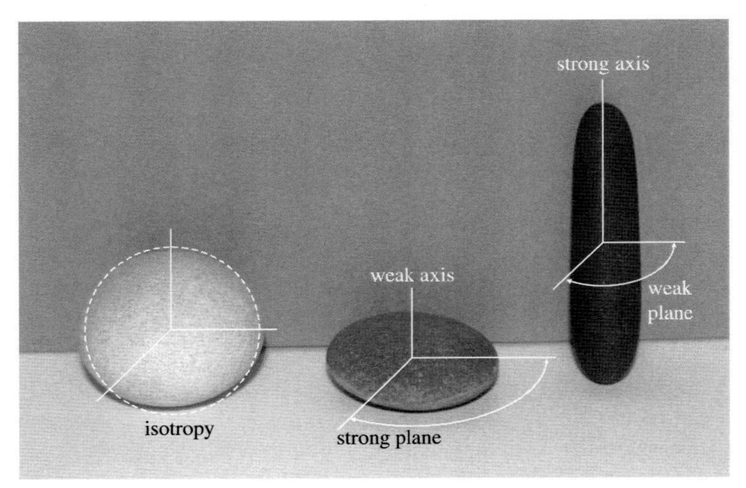

Figure 7.2. Isotropy and anisotropy in the strength of a rock: The round shape of the pebble on the left is an indication of isotropy in the strength of the material, whereas the two pebbles on the right have a form evidencing transverse isotropy (samples have been collected from the shore of Aberystwyth, UK, the round one by Prof. G. Mishuris).

principal stresses T_1, T_2 and T_3. Therefore, the yielding condition becomes

$$f(T_1, T_2, T_3) = 0, \tag{7.7}$$

which has to be indifferent with respect to all possible permutations of T_1, T_2 and T_3.

Equation (7.7) describes a surface in the T_i-space, called 'Haigh-Westergaard stress space' (Haigh, 1920; Westergaard, 1920), such that 'stress points' internal to the surface correspond to elastic behaviour, whereas plastic flow is possible only for stress points lying on the surface, and finally, points external to the surface are 'forbidden' (Hill, 1950a).

Since a stress state is represented in the yield function through its principal values, we note that

$\mathcal{A}1$. A single point in the Haigh-Westergaard space is representative of the infinite (to the power 3) stress tensors having the same principal values.

$\mathcal{A}2$. The Haigh-Westergaard representation preserves the scalar product only between coaxial tensors [we recall Eq. (2.31) for the definition of coxiality].

A stress point $\{T_1, T_2, T_3\}$ can be decomposed into a deviatoric and an isotropic part as (Fig. 7.3, *left*)

$$\{T_1, T_2, T_3\} = \{\text{dev}T_1, \text{dev}T_2, \text{dev}T_3\} + \frac{T_1 + T_2 + T_3}{3}\{1, 1, 1\}, \tag{7.8}$$

so vector $\{1, 1, 1\}$ defines the space diagonal in the Haigh-Westergaard representation, the so-called hydrostatic or octaedral axis. The space diagonal is normal to the so-called deviatoric or π plane, where all vectors $\{\text{dev}T_1, \text{dev}T_2, \text{dev}T_3\}$ lie. The

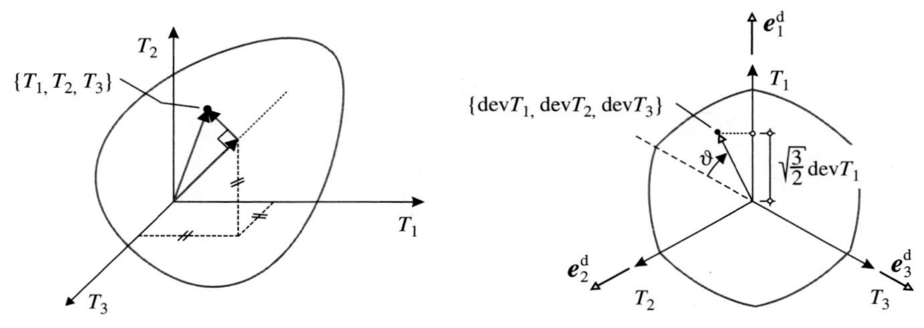

Figure 7.3. Haigh-Westergaard representation of a yield surface for an isotropic material. Three-dimensional representation (*left*); the section with the deviatoric (or π) plane, where θ is the Lode angle (*right*).

projections of the axes T_i on the deviatoric plane are defined by the unit vectors

$$\{e_1^d\} = \frac{\{2,-1,-1\}}{\sqrt{6}}, \qquad \{e_2^d\} = \frac{\{-1,2,-1\}}{\sqrt{6}}, \qquad \{e_3^d\} = \frac{\{-1,-1,2\}}{\sqrt{6}}, \tag{7.9}$$

so the projections of a stress point onto the deviatoric axes (7.9) define segments proportional through the factor $\sqrt{3/2}$ to the deviatoric components of the stress (Fig. 7.3, *right*);

$$\{T_1, T_2, T_3\} \cdot \{e_i^d\} = \sqrt{\frac{3}{2}}\,\mathrm{dev}T_i, \qquad i = 1,2,3. \tag{7.10}$$

From the preceding properties, we can derive the following further observations:

$\mathcal{A}3$. Owing to the arbitrariness in the ordering of the eigenvalues of a tensor, six different points correspond in the Haigh-Westergaard representation to a given stress tensor. As a result, the yield surface results symmetric about the projections of the principal axes on the deviatoric plane (see Fig. 7.4, top, *left*, where the stress states $\{\alpha, \beta, -\gamma\}$ and $\{\alpha, -\gamma, \beta\}$ differ only in a permutation, so they both must be in the same condition with respect to yielding).

$\mathcal{A}4$. The stress points denoted by $\{\alpha, \beta, -\gamma\}$ and $\{\gamma, -\beta, -\alpha\}$ in Fig. 7.4 (top, *right*) differ in sign and in a permutation. Since yielding of ductile metals (we exclude porous metals and metals evidencing the SD effect) is within a good accuracy insensible to the sign of the stress (so the behaviour is the same for tensile or compression tests), the deviatoric section of the yield surface must possess the symmetries sketched in Fig. 7.4 (bottom, *right*).

$\mathcal{A}5$. A convex yield surface—for a material with a given yield strength in compression—must be internal to the two (upper and lower) limit situations shown in Fig. 7.4 (bottom, *left*). Note that the inner bound will be referred as the 'Rankine limit' (Haythornthwaite, 1985).

In the Haigh-Westergaard representation, a stress point can be singled out using cylindrical coordinates by the hydrostatic (note the negative sign in p) and deviatoric

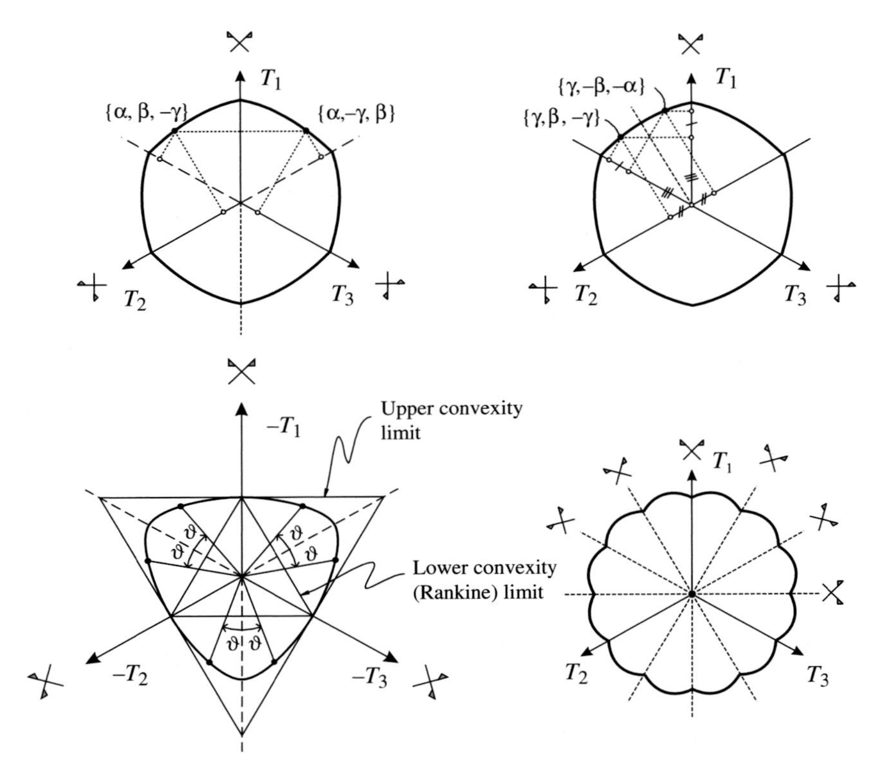

Figure 7.4. Deviatoric section of isotropic yield surfaces: symmetries, definition of the Lode angle θ, lower and upper convexity bounds. Note that the notation $-T_1$, $-T_2$ and $-T_3$ means that the sign of the stress has been reversed.

stress invariants:

$$p = -\frac{\mathrm{tr}\boldsymbol{T}}{3} \quad \text{and} \quad q = \sqrt{3J_2}, \tag{7.11}$$

where[2]

$$J_2 = \frac{1}{2}\mathrm{dev}\boldsymbol{T}\cdot\mathrm{dev}\boldsymbol{T} = \frac{1}{2}|\{\mathrm{dev}T_1, \mathrm{dev}T_2, \mathrm{dev}T_3\}|, \tag{7.12}$$

and by the the Lode (1926) angle θ, defined as

$$\theta = \frac{1}{3}\cos^{-1}\left(\frac{3\sqrt{3}}{2}\frac{J_3}{J_2^{3/2}}\right), \qquad J_3 = \frac{1}{3}\mathrm{tr}(\mathrm{dev}\boldsymbol{T})^3, \tag{7.13}$$

so $\theta \in [0, \pi/3]$. As a consequence of property ($\mathcal{A}2$) of the Haigh-Westergaard representation, a single value of θ corresponds to six different points in the deviatoric plane (Fig. 7.4, bottom, *left*).

[2] As the name anticipates, the J_2-deformation theory of plasticity is based on use of the invariant J_2 both for the stress and for the strain; see Eqs. (4.40) and note that $\sigma_e = \sqrt{3J_2}$.

7.2 The BP yield function

Bigoni and Piccoloraz (2004; see also Piccoloraz and Bigoni, 2009) have proposed a yield function (referred as the BP yield function in the following) which is convex, smooth and extremely flexible so that it results particularly suited to represent the yield surface of pressure sensitive materials. Since this yield function reduces in limiting cases to practically all known models (von Mises, Tresca, Drucker-Prager, modified Tresca, Coulomb-Mohr, Cam-Clay)[3] and provides excellent smooth approximations to singular surfaces, we will introduce this function and derive known cases directly from it.

Generally speaking, smoothness is not necessarily a valuable requisite for a yield function. In fact, the existence of corners in the yield surface and multi-mechanism plasticity often has been advocated as the proper way to approach the behaviour of crystalline metals (Batdorf and Budiansky, 1949; Hill, 1967c; Hutchinson, 1970) and of micro-fissured rocks (Rudnicki and Rice, 1975). However, the experimental detection of yield surface corners is prohibitive (Phillips, 1974), and regarding geomaterials, the situation is already so complicated by many effects that the use of a multi-mechanism or a corner-theory plasticity seems for the moment not motivated enough. Therefore, the availability of a yield function which smoothly approximates the classical Coulomb-Mohr or Tresca yield criterion is important in geomechanics, at least until definitive arguments in favour of more complicated theories will be presented. To better elucidate this point, we note that the original Cam-Clay model (Schofield and Wroth, 1968) has two corners at the intersections of the yield surface with the hydrostatic axis. Employed, for instance, to analyse powder compaction, these corners are encountered in the quite common case of the hydrostatic compression test. There is no evidence at all that the yield surface vertices of the preceding model reflect any peculiarity of the mechanical behaviour of the material, so the incorporation of these in the constitutive description of the material simply represents a useless complication. Another example is the vertex of the Drucker-Prager criterion (Fig. 7.8), which has no mechanical interpretation but is reached during certain stress paths, for instance, in the near-tip fields developed when a crack steadily propagates in a ductile pressure-sensitive material (Bigoni and Radi, 1993).

The BP yield function is a seven-parameter function $F : \mathsf{Sym} \to \mathbb{R} \cup \{+\infty\}$ defined as

$$F(\boldsymbol{T}) = f(p) + \frac{q}{g(\theta)}, \tag{7.14}$$

where the dependence on the stress $\boldsymbol{T}$ is included in the invariants p, q and θ, [Eqs. (7.11)] and (7.13), through the meridian function

$$f(p) = \begin{cases} -Mp_c\sqrt{(\Phi - \Phi^m)\,[2(1-\alpha)\Phi + \alpha]} & \textit{if } \Phi \in [0,1], \\ +\infty & \textit{if } \Phi \notin [0,1], \end{cases} \tag{7.15}$$

[3] Richard von Mises (Lviv, 18 April 1883–Boston, 14 July 1953). Henri-Édouard Tresca (Dunkerque, 12 October 1814–Paris, 21 June 1885). Daniel C. Drucker (New York, 3 June 1918–Gainesville, 1 September 2001). William Prager (Karlsruhe, 23 May 1903–Zurich, 16 March 1980). Charles-Augustin de Coulomb (Angoulême, 14 June 1736–Paris, 23 August 1806). Christian Otto Mohr (Wesselburen, 8 October 1835–Dresden, 2 October 1918).

where

$$\Phi = \frac{p+c}{p_c+c}, \tag{7.16}$$

describing the pressure sensitivity and the deviatoric function[4]

$$g(\theta) = \frac{1}{\cos\left[\beta\frac{\pi}{6} - \frac{1}{3}\cos^{-1}(\gamma\cos 3\theta)\right]}, \tag{7.17}$$

describing the Lode dependence of yielding. The seven, non-negative material parameters

$$\underbrace{M > 0,\ p_c > 0,\ c \geq 0,\ 0 < \alpha < 2,\ m > 1}_{\text{defining } f(p)} \quad \text{and} \quad \underbrace{0 \leq \beta \leq 2,\ 0 \leq \gamma < 1}_{\text{defining } g(\theta)} \tag{7.18}$$

define the shape of the associated yield surface (note that the limitations on the range of parameters ensure convexity and smoothness of the yield surface, as will be detailed later). In particular, we can highlight the following features:

- The dimensionless parameter M controls the pressure sensitivity.
- Parameters p_c and c are the yield strengths under isotropic compression and tension, respectively.
- The dimensionless parameters α and m define the distortion of the meridian section, whereas β and γ model the shape of the deviatoric section.

Note that the deviatoric function describes a piece-wise linear deviatoric surface in the limit $\gamma \longrightarrow 1$.

Note 1: Within the interval [0,2] of β, the yield function is convex independent of the values assumed by parameter γ. Convexity requirements, which will be proved later, impose a broader variation of β than Eq. $(7.18)_6$, but the interval where β may range becomes a function of γ. In particular, *the yield function is convex* when

$$2 - \mathcal{B}(\gamma) \leq \beta \leq \mathcal{B}(\gamma), \tag{7.19}$$

where function $\mathcal{B}(\gamma)$ takes values within the interval $(2,4]$ when γ ranges in $[0,1)$ and is defined as

$$\mathcal{B}(\gamma) = 3 - \frac{6}{\pi}\tan^{-1}\frac{1-2\cos z - 2\cos^2 z}{2\sin z(1-\cos z)}\Bigg|_{z=2/3(\pi-\cos^{-1}\gamma)}. \tag{7.20}$$

Note 2: The meridian function (7.15) can be written in an alternative form by using the Macaulay bracket operator (2.103) and the indicator function $\chi_{[0,1]}(\Phi)$, which takes the value 0 when $\Phi \in [0,1]$ and is equal to $+\infty$ otherwise

$$f(p) = -Mp_c\sqrt{\left(\tilde{\Phi} - \tilde{\Phi}^m\right)\left[2(1-\alpha)\tilde{\Phi} + \alpha\right]} + \chi_{[0,1]}(\Phi), \qquad \tilde{\Phi} = <\Phi> - <\Phi - 1>. \tag{7.21}$$

[4] As indicated to the author by Prof. B. Raniecki (IPPT, Warsaw) in a private communication, the deviatoric yield function (7.17) was introduced by Podgórski (1984, 1985) and rediscovered independently 20 years later (Bigoni and Piccolroaz, 2004).

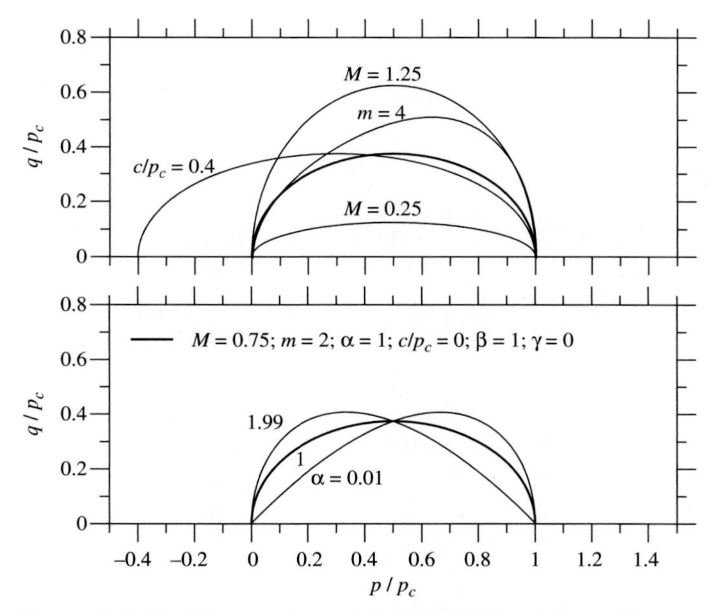

Figure 7.5. Meridian section of the BP yield surface: effects related to the variation of parameters $M, c/p_c, m$ (*top*) and α (*bottom*).

The yield function (7.14) corresponds to the following *yield surface*:

$$q = -f(p)g(\theta), \qquad p \in [-c, p_c], \qquad \theta \in [0, \pi/3], \tag{7.22}$$

which makes explicit the fact that $f(p)$ and $g(\theta)$ define the shape of the meridian and deviatoric sections, respectively.

The yield surface (7.22) is sketched in Fig. 7.5, where the meridian section is reported ($g(\theta) = 1$ has been assumed and non-dimensionalisation is introduced through division by p_c), and in Fig. 7.6, where the deviatoric section is shown, for different values of the seven above-defined material parameters.

As a reference, the case corresponding to the modified Cam-Clay model introduced by Roscoe and Burland (1968) and Schofield and Wroth (1968) and corresponding to

$$\beta = 1, \qquad \gamma = 0, \qquad \alpha = 1, \qquad m = 2, \qquad c = 0$$

is reported in Fig. 7.5 as a solid line for $M = 0.75$ and $c/p_c = 0$. The distortion of meridian section obtained by changing parameters M, m and c/p_c is shown in the upper part of Fig. 7.5, whereas the effects of a variation of parameter α is shown in the lower part of the figure. In particular, note that the shape distortion induced by the variation of parameters m and α is crucial to fit experimental results relative to frictional materials.

A unique feature of the BP model is the possibility of extreme shape distortion of the deviatoric section, which may range between the upper and lower convexity limits and approach the Tresca, von Mises and Coulomb-Mohr criteria. This is sketched in Fig. 7.6, where to simplify the reading of the figure, function $g(\theta)$

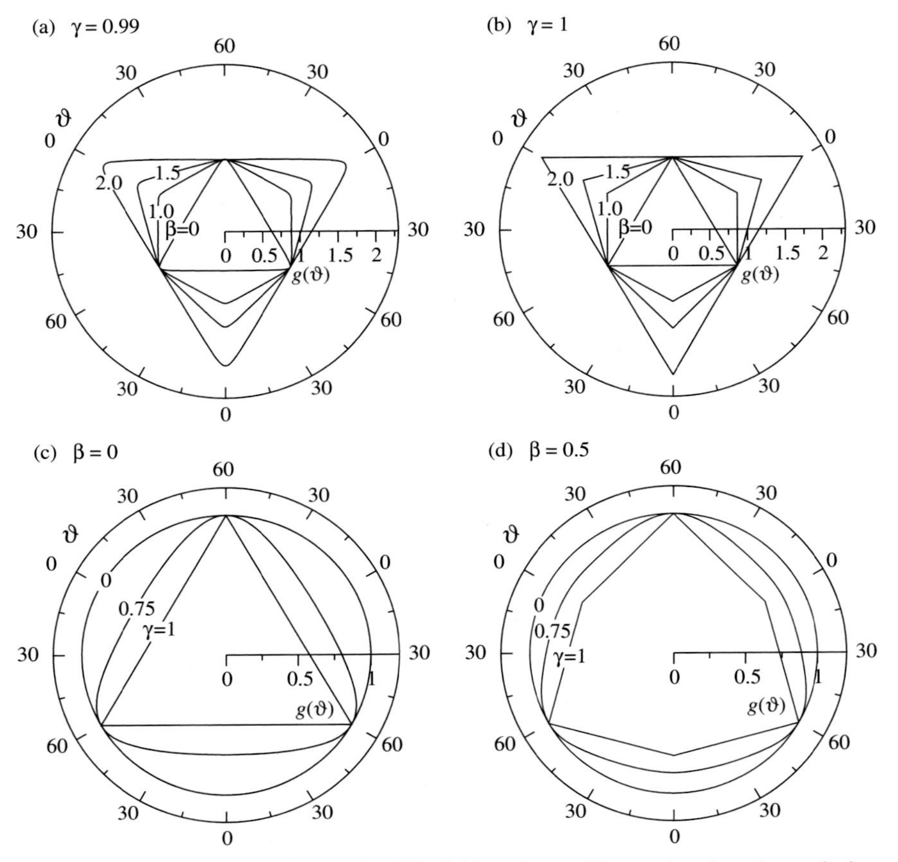

Figure 7.6. Deviatoric section of the BP yield surface: effects related to the variation of β and γ. Variation of $\beta = 0,1,1.5$ and 2 at fixed $\gamma = 0.99$ (*a*) and $\gamma = 1$ (*b*). Variation of $\gamma = 1,0.75$ and 0 at fixed $\beta = 0$ (*c*) and $\beta = 0.5$ (*d*).

has been normalised through division by $g(\pi/3)$ so that all deviatoric sections coincide at the point $\theta = \pi/3$. Use of the BP model therefore may allow one simply to obtain a convex, smooth approximation of several yielding criteria (e.g., Tresca and Coulomb-Mohr). If this is not substantial from a theoretical point of view, it clearly avoids the necessity of introducing independent yielding mechanisms and the related complications.

Parameter γ is kept fixed in Fig. 7.6*a* and *b* and equal to 0.99 and 1, respectively, whereas parameter β is fixed in Fig. 7.6*c* and *d* and equal to 0 and 1/2. Therefore, parts *a* and *b* demonstrate the effect of the variation in β (= 0, 1, 1.5 and 2), which makes possible a distortion of the yield surface from the upper to lower convexity limits going through the Tresca and Coulomb-Mohr shapes. The role played by γ (= 1, 0.75 and 0) is investigated in parts *c* and *d*, from which it becomes evident that γ has a smoothing effect on the corners, emerging in the limit $\gamma = 1$. The von Mises (circular) deviatoric section is obtained when $\gamma = 0$.

The BP yield surface in the plane stress biaxial plane T_1 versus T_2, with $T_3 = 0$, is sketched in Fig. 7.7, where axes are normalised through division by the uniaxial

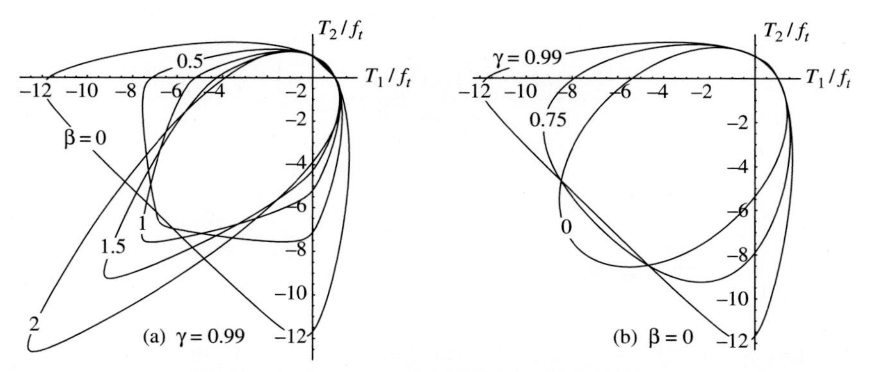

Figure 7.7. The BP yield surface in the plane stress biaxial plane T_1/f_t versus T_2/f_t, with $T_3 = 0$. Variation of $\beta = 0, 0.5, 1, 1.5$ and 2 at fixed $\gamma = 0.99$ (*a*) and variation of $\gamma = 0, 0.75$ and 0.99 at fixed $\beta = 0$ (*b*).

tensile strength f_t. In particular, the figure pertains to $M = 0.75$, $p_c = 50c$, $m = 2$ and $\alpha = 1$, whereas $\gamma = 0.99$ is fixed and β is equal to {0, 0.5, 1, 1.5, 2} in Fig. 7.7*a*, and vice versa, $\beta = 0$ is fixed and γ is equal to {0, 0.75, 0.99} in Fig. 7.7*b*.

7.2.1 Smoothness of the BP yield surface

Smoothness of yield surface (7.22) within the interval of material parameters defined in Eqs. (7.18) and (7.19) can be proved considering the yield function gradient. This can be obtained from the following expressions for the gradients of the invariants, which can be easily obtained from Eqs. (2.119) and (2.123):

$$
\begin{aligned}
&\frac{\partial p}{\partial \boldsymbol{T}} = -\frac{1}{3}\boldsymbol{I}, \qquad \frac{\partial J_2}{\partial \boldsymbol{T}} = \mathsf{dev}\boldsymbol{T}, \qquad \frac{\partial J_3}{\partial \boldsymbol{T}} = (\mathsf{dev}\boldsymbol{T})^2 - \frac{\mathsf{tr}(\mathsf{dev}\boldsymbol{T})^2}{3}\boldsymbol{I}, \\
&\frac{\partial \theta}{\partial \boldsymbol{T}} = -\frac{9}{2q^3 \sin 3\theta}\left((\mathsf{dev}\boldsymbol{T})^2 - \frac{\mathsf{tr}(\mathsf{dev}\boldsymbol{T})^2}{3}\boldsymbol{I} - q\frac{\cos 3\theta}{3}\mathsf{dev}\boldsymbol{T}\right).
\end{aligned} \tag{7.23}
$$

Note that $\partial\theta/\partial\boldsymbol{T}$ is orthogonal to $\boldsymbol{I}$ and to the deviatoric stress $\mathsf{dev}\boldsymbol{T}$.

Therefore, the yield function gradient can be written in the form

$$
\frac{\partial F}{\partial \boldsymbol{T}} = a(p)\boldsymbol{I} + b(\theta)\tilde{\boldsymbol{S}} + c(\theta)\tilde{\boldsymbol{S}}^{\perp}, \tag{7.24}
$$

where

$$
\tilde{\boldsymbol{S}} = \sqrt{\frac{3}{2}}\frac{\mathsf{dev}\boldsymbol{T}}{q}, \qquad \tilde{\boldsymbol{S}}^{\perp} = -\frac{q\sqrt{2}}{\sqrt{3}}\frac{\partial\theta}{\partial\boldsymbol{T}} = \frac{1}{\sin 3\theta}\left[\sqrt{6}\left(\tilde{\boldsymbol{S}}^2 - \frac{1}{3}\boldsymbol{I}\right) - \cos 3\theta\,\tilde{\boldsymbol{S}}\right], \tag{7.25}
$$

and

$$a(p) = -\frac{1}{3}\frac{\partial f(p)}{\partial p} = \frac{Mp_c}{3(p_c+c)}\frac{(1-m\Phi^{m-1})[2(1-\alpha)\Phi+\alpha]+2(1-\alpha)(\Phi-\Phi^m)}{2\sqrt{(\Phi-\Phi^m)[2(1-\alpha)\Phi+\alpha]}},$$

$$b(\theta) = \sqrt{\frac{3}{2}}\frac{1}{g(\theta)}, \tag{7.26}$$

$$c(\theta) = -\frac{\sqrt{3}\gamma\sin 3\theta}{\sqrt{2}\sqrt{1-\gamma^2\cos^2 3\theta}}\sin\left[\beta\frac{\pi}{6}-\frac{1}{3}\cos^{-1}(\gamma\cos 3\theta)\right].$$

It should be noted that $c(0) = c(\pi/3) = 0$ and that $\tilde{\boldsymbol{S}}$ and $\tilde{\boldsymbol{S}}^{\perp}$ are unit norm coaxial tensors, normal to each other.

Coaxiality and orthogonality are immediate properties, whereas the proof that $|\tilde{\boldsymbol{S}}^{\perp}| = 1$ is facilitated when the following identities are kept into account:

$$\tilde{\boldsymbol{S}}^3 - \frac{1}{2}\tilde{\boldsymbol{S}} - \frac{\cos 3\theta}{3\sqrt{6}}\boldsymbol{I} = \boldsymbol{0}, \quad \rightsquigarrow \quad \tilde{\boldsymbol{S}}^2 \cdot \tilde{\boldsymbol{S}}^2 = \frac{1}{2}, \tag{7.27}$$

the former of which is the Cayley-Hamilton theorem (2.59) written for $\tilde{\boldsymbol{S}}$.

Note that $c\tilde{\boldsymbol{S}}^{\perp} = \boldsymbol{0}$ at $\theta = 0, \pi/3$. This can be deduced from the fact that $|\tilde{\boldsymbol{S}}^{\perp}| = 1$ and $c = 0$ for $\theta = 0, \pi/3$.

Finally, the gradient (7.24) can be rewritten to have a unit norm as

$$\mathbf{Q} = \frac{a}{\sqrt{3a^2+b^2+c^2}}\boldsymbol{I} + \frac{b}{\sqrt{3a^2+b^2+c^2}}\tilde{\boldsymbol{S}} + \frac{c}{\sqrt{3a^2+b^2+c^2}}\tilde{\boldsymbol{S}}^{\perp}, \tag{7.28}$$

defining, for stress states satisfying $F(\boldsymbol{T}) = 0$, the unit normal to the yield surface.

The following limits can be calculated easily:

$$\lim_{\Phi\to 0^+}\mathbf{Q} = \frac{1}{\sqrt{3}}\boldsymbol{I} \quad \text{and} \quad \lim_{\Phi\to 1^-}\mathbf{Q} = -\frac{1}{\sqrt{3}}\boldsymbol{I}, \tag{7.29}$$

so the yield surface results to be smooth at the limit points where the hydrostatic axis is met. Moreover, smoothness of the deviatoric section of the yield surface is proved observing that

$$\lim_{\theta\to 0,\pi/3}\mathbf{Q} = \frac{a}{\sqrt{3a^2+b^2}}\boldsymbol{I} + \frac{b}{\sqrt{3a^2+b^2}}\tilde{\boldsymbol{S}}, \tag{7.30}$$

where $\tilde{\boldsymbol{S}}$ and b are evaluated at $\theta = 0$ and $\theta = \pi/3$, and noting that $\tilde{\mathbf{S}}$ and $\tilde{\mathbf{S}}^{\perp}$ are coaxial deviatoric tensors so that they are represented by two orthogonal vectors in the deviatoric plane in the Haigh-Westergaard stress space. We observe, finally, that the limits (7.29) do not hold true when α equals 0 and 2 and that the limits (7.30) do not hold true when $\gamma = 1$. In particular, a corner appears at the intersection of the yield surface with the hydrostatic axis in the former case, and the deviatoric section becomes piece-wise linear in the latter.

7.3 Reduction of the BP yield criterion to known cases

The yield function [Eqs. (7.14) through (7.17)] reduces to almost all 'classical' criteria of yielding, sketched in Fig. 7.8 [the isotropic criterion proposed by Hill (1950b),

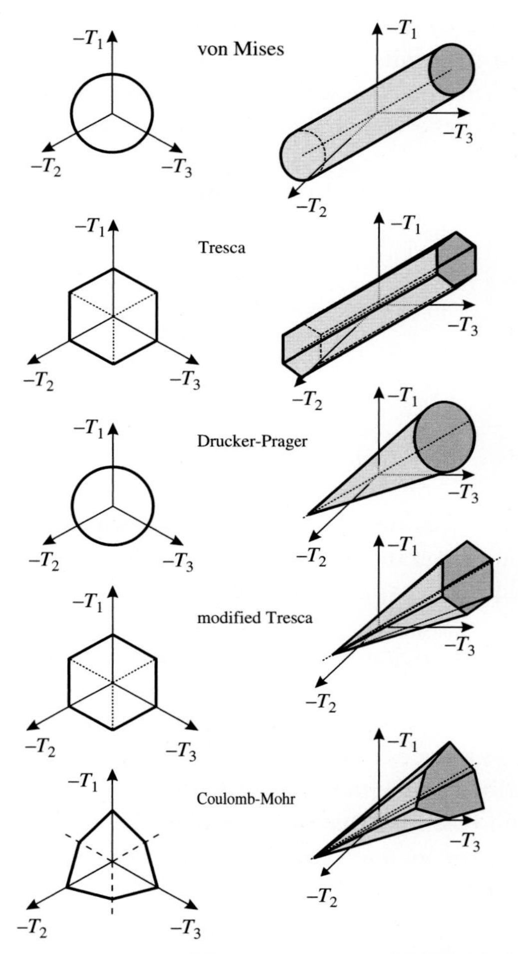

Figure 7.8. The Haigh-Westergaard representation of the von Mises, Tresca, Drucker-Prager, modified Tresca, and Coulomb-Mohr yield surfaces. Note that the notation '$-T_1$', '$-T_2$' and '$-T_3$' means that the sign of the stress has been reversed to highlight compressive states of stress.

corresponding to a Tresca yield surface rotated $\pi/6$ in the deviatoric plane can also be obtained as detailed by Piccolroaz and Bigoni (2009)], where we have also included the so-called modified Tresca criterion introduced by Drucker (1953). Note that the Haigh-Westergaard representation of the Coulomb-Mohr criterion was obtained by Shield (1955).

All the yield criteria reported in Fig. 7.8 can be obtained as limit cases of the BP yield function in the way illustrated in Table 7.1, where parameter r denotes the ratio between the uniaxial strengths in compression (taken positive) and tension, indicated by f_c and f_t, respectively. We note that for usual materials (an exception is wood, as mentioned in the introduction) $r \geq 1$ and that we did not explicitly consider the special cases of no-tension $f_t = 0$ or granular $f_t = f_c = 0$ materials [which anyway can be incorporated easily as limits of Eqs. (7.14) through (7.17)].

We note that the expression of the Tresca criterion which follows from the BP criterion [Eqs. (7.14) through (7.17)] in the limits specified in Table 7.1 also was provided by Bardet (1990) and answers—in a positive way—the question (raised by Salencon, 1974) of whether a proper form of the Tresca yield function exists in terms of stress invariants.[5]

The Coulomb-Mohr limit merits a special mention. In fact, if the following values of the parameters are selected

$$\alpha=0, \qquad c=\frac{f_c\left[\cos\left(\beta\frac{\pi}{6}-\frac{\pi}{3}\right)+\cos\beta\frac{\pi}{6}\right]}{3r\cos\left(\beta\frac{\pi}{6}-\frac{\pi}{3}\right)-3\cos\beta\frac{\pi}{6}}, \qquad M=\frac{3\left[r\cos\left(\beta\frac{\pi}{6}-\frac{\pi}{3}\right)-\cos\beta\frac{\pi}{6}\right]}{\sqrt{2}(r+1)}, \tag{7.31}$$

and then the limits

$$\gamma\longrightarrow 1 \quad \text{and} \quad p_c=f_c m\longrightarrow\infty \tag{7.32}$$

are performed, a three-parameter generalisation of Coulomb-Mohr criterion is obtained which reduces to the latter criterion in the special case where β is selected in the form specified in Table 7.1 (yielding an expression noted also by Chen and Saleeb, 1982).

The cases reported in Table 7.1 refer to situations in which the BP criterion [Eqs. (7.14) through (7.17)], reduces to known yield functions in terms of both meridian function $f(p)$ and deviatoric function $g(\theta)$. It is, however, important to mention that the Lode's dependence function $g(\theta)$ also reduces to well-known cases but in which the pressure sensitivity cannot be described by the meridian function (7.15). These are reported in Table 7.2. It is important to mention that the form of the BP function $g(\theta)$ [Eq. (7.17)] was indeed invented by Bigoni and Piccolroaz as a generalisation of the deviatoric function introduced by Ottosen (1977).

7.3.1 Drucker-Prager and von Mises yield criteria

The Drucker-Prager (1952) and Mises yield criteria will be employed often throughout this book. The values of parameters to reduce the BP criterion to these criteria are listed in Table 7.1. However, owing to their importance, we write these criteria explicitly.

We note that for $\beta=1$ and $\gamma=0$, the function $g(\theta)$ reduces to unity. Moreover, taking the appropriate limits indicated in the table, using the definitions (7.11) of the invariants p and q and dividing by the unessential factor $\sqrt{3}$, we arrive at the following expression for the Drucker-Prager yield criterion in terms of uniaxial compression

[5] The expression for the Tresca yield criterion

$$f(\boldsymbol{T})=4J_2^3-27J_3^2-36k^2J_2^2+96k^4J_2-64k^6,$$

where k is the yield stress under shear (i.e., $k=f_t/2$), reported in several textbooks on plasticity, is definitively wrong. This can be verified easily by taking a stress state belonging to one of the planes defining the Tresca criterion, but outside the yield locus, for instance, the point $\{T_1=0,\ T_2=-2k,\ T_3=2k\}$, corresponding to $J_2=4k^2$ and $J_3=0$. Obviously, the point lies well outside the yield locus but satisfies $f(\boldsymbol{T})=0$ when the preceding, wrong yield function is used.

Table 7.1. *Yield criteria obtained as special cases of the BP yield function, Eqs. (7.14) through (7.17) (Parameter $r = f_c/f_t$ is the ratio between the uniaxial strengths in compression f_c and in tension f_t.)*

Criterion	Meridian function $f(p)$	Deviatoric function $g(\theta)$
von Mises	$\alpha = 1, \quad m = 2, \quad M = \frac{2f_t}{p_c}, \quad c = p_c = \longrightarrow \infty$	$\beta = 1, \gamma = 0$
Drucker-Prager	$\alpha = 0, \quad M = \frac{3(r-1)}{\sqrt{2}(r+1)}, \quad c = \frac{2f_c}{3(r-1)}, \quad p_c = f_c m \longrightarrow \infty$	As for von Mises
Tresca	As for von Mises, except that $M = \frac{\sqrt{3}f_t}{p_c}$	$\beta = 1, \gamma \longrightarrow 1$
Mod. Tresca	As for Drucker-Prager, except that $M = \frac{3\sqrt{3}(r-1)}{2\sqrt{2}(r+1)}$	As for Tresca
Coulomb-Mohr	As for Drucker-Prager, except that $M = \frac{3\left[r\cos\left(\beta\frac{\pi}{6} - \frac{\pi}{3}\right) - \cos\beta\frac{\pi}{6}\right]}{\sqrt{2}(r+1)}$, $c = \frac{f_c\left[\cos\left(\beta\frac{\pi}{6} - \frac{\pi}{3}\right) + \cos\beta\frac{\pi}{6}\right]}{3r\cos\left(\beta\frac{\pi}{6} - \frac{\pi}{3}\right) - 3\cos\beta\frac{\pi}{6}}$	$\beta = \frac{6}{\pi}\tan^{-1}\frac{\sqrt{3}}{2r+1}, \quad \gamma \longrightarrow 1$
Mod. Cam-Clay	$m = 2, \ \alpha = 1, \ c = 0$	As for von Mises

Table 7.2. *Deviatoric yield functions obtained as special cases of the BP yield function, Eq. (7.17)*

Criterion	Deviatoric function $g(\theta)$
Lower convexity (Rankine)	$\beta = 0, \ \gamma \longrightarrow 1$
Upper convexity	$\beta = 2, \ \gamma \longrightarrow 1$
Ottosen	$\beta = 0, \ 0 \leq \gamma < 1$

stress at failure f_c and ratio r between strengths in uniaxial compression f_c and in tension f_t:

$$F_{DP}(\boldsymbol{T}) = \sqrt{J_2} + \text{tr}\boldsymbol{T}\,\frac{r-1}{\sqrt{3}(r+1)} - \frac{2f_c}{\sqrt{3}(r+1)}. \tag{7.33}$$

In the Mohr representation (Den Hartog, 1952), defined by the normal T_{nn} and tangential T_{nt} components of the stress vector $\boldsymbol{Tn}$ relative to the elementary surface of unit normal $\boldsymbol{n}$,

$$T_{nn} = \boldsymbol{n}\cdot\boldsymbol{Tn} \qquad \text{and} \qquad T_{nt} = \sqrt{\boldsymbol{Tn}\cdot\boldsymbol{Tn} - T_{nn}^2}, \tag{7.34}$$

we consider the two 'limit' circles corresponding to the achievement of uniaxial compression and tension strengths (Fig. 7.9). In this representation, the inclination of the lines tangent to the circles defines the so-called internal friction angle, denoted

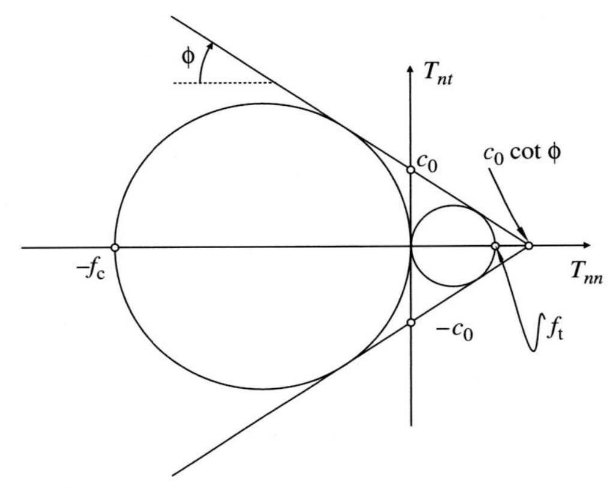

Figure 7.9. Definition of angle of internal friction ϕ and cohesion c_0 in the Mohr representation, with reference to the two Mohr circles representative of uniaxial compression f_c and tensile f_t strengths.

here with ϕ, whereas the intercept of these lines with the T_{nt} axis denotes the so-called cohesion of the material, denoted here with c_0. It is a simple exercise to show that

$$f_c = \frac{2c_0\cos\phi}{1-\sin\phi}, \qquad f_t = \frac{2c_0\cos\phi}{1+\sin\phi}, \qquad r = \frac{1+\sin\phi}{1-\sin\phi}, \tag{7.35}$$

and therefore Eq. (7.33) becomes the *Drucker-Prager criterion expressed in terms of internal friction angle and cohesion*

$$F_{DP}(\boldsymbol{T}) = \sqrt{J_2} + \frac{\sin\phi}{\sqrt{3}}\,\mathrm{tr}\boldsymbol{T} - \frac{2c_0\cos\phi}{\sqrt{3}}. \tag{7.36}$$

The gradient of the yield function (7.36) can be calculated using Eqs. (7.23) as

$$\nabla F_{DP} = \frac{\mathrm{dev}\boldsymbol{T}}{2\sqrt{J_2}} + \frac{\sin\phi}{\sqrt{3}}\boldsymbol{I}, \tag{7.37}$$

which can be rewritten in a unit-norm setting as

$$\boldsymbol{Q} = \cos\phi\frac{\mathrm{dev}\,\boldsymbol{T}}{|\mathrm{dev}\,\boldsymbol{T}|} + \frac{\sin\phi}{\sqrt{3}}\boldsymbol{I}. \tag{7.38}$$

The particular case of the von Mises yield criterion[6] can be obtained from Eqs. (7.36) and (7.38) by setting $\phi = 0$ and $c = f_t/2$ and multiplying by $\sqrt{2}$:

$$F_{VM}(\boldsymbol{T}) = |\mathrm{dev}\boldsymbol{T}| + f_t\sqrt{\frac{2}{3}}, \qquad \boldsymbol{Q} = \frac{\mathrm{dev}\,\boldsymbol{T}}{|\mathrm{dev}\,\boldsymbol{T}|}. \tag{7.39}$$

[6] Maksymilian Tytus Huber (Kroscienku, 4 January 1872–Cracow, 9 December 1950) proposed the criterion in 1904, before von Mises (1913). A few years later, Hencky (1924) (Heinrich Hencky, Ansbach, 2 November 1885–6 July 1951) provided the interpretation of the criterion in terms of elastic energy, an idea which goes back to Beltrami (1885) (Eugenio Beltrami, Cremona, 16 November 1835–Roma, 18 February 1900). Following Bell (1973) and Paul (1968), the von Mises criterion was already known to Maxwell.

7.3.2 A comparison of the BP yield criterion with experimental results

A brief comparison of the BP yield criterion with experimental results referred to several materials is reported (see Bigoni and Piccolroaz; 2004, for more details) to demonstrate the extreme flexibility of the yield surface to fit different material behaviours. In particular, we start considering the meridian section, and we include later a few examples for the deviatoric section (which has a shape so deformable and ranging between well-known forms that fitting experiments is a priori expected) and for the plane-stress biaxial T_1–T_2 representation.

The experimental results reported in Fig. 7.10 refer to the following pressure-sensitive materials.

- *Soils:* Aio dry sand (marked with black spots) and Weald clay (marked with grey spots). The experimental data have been taken, respectively, from Yasufuku et al. (1991, their fig. 10*a*) and Parry (reported by Wood, 1990, their fig. 7.22, where p has been normalised through division by p_e, the equivalent consolidation pressure). Note that in Fig. 7.10*a* the upper plane of the graph refers to triaxial compression ($\theta = \pi/3$), whereas triaxial extension is reported in the lower part ($\theta = 0$).
- *Polymers:* Polymethyl methacrylate (marked with black spots) and an epoxy binder (marked with grey spots). The experimental data have been taken from Ol'khovik (1983, their fig. 5), see also Altenbach and Tushtev (2001, their figs. 2 and 3).
- *Aluminium powder* (Al $D_0 = 0.67$, $D = 0.81$), marked by black spots; and *aluminium powder reinforced by 40 vol%SiC* (Al 40%, SiC $D_0 = 0.66$, $D = 0.82$), marked with grey spots; the experimental data have been taken from Sridhar and Fleck (2000, their fig. 5*b*). Note that the behaviour of the aluminium powder is different from soils and lead-based powders. It results in a meridian section of the yield surface similar to the early version of the Cam-Clay model (Roscoe and Schofield, 1963).
- *Lead powder* (0% steel), marked by black spots; and *lead shot–steel composite powder* (20% steel), marked with grey spots; the experimental data have been taken from Sridhar and Fleck (2000, their fig. 9*c*).
- *Rocks:* Chert (marked with black spots) and dolomite (marked with grey spots). The experimental data have been taken from Hoek and Brown (1980, their pages 143 and 144). Note that owing to the fact that the BP criterion approaches the Coulomb-Mohr limit, the criterion is particularly suited for rocks.
- The Newman and Newman (1971) empirical relationship to experimental results for *concrete* has been considered and marked with black spots. The behaviour at low confining pressure, reported in the insert and marked with grey spots, has been taken from Sfer et al. (2002, their fig. 6).

As far as the deviatoric section and the plane-stress biaxial T_1–T_2 representation are concerned, we report results on the following pressure-sensitive materials (Fig. 7.11):

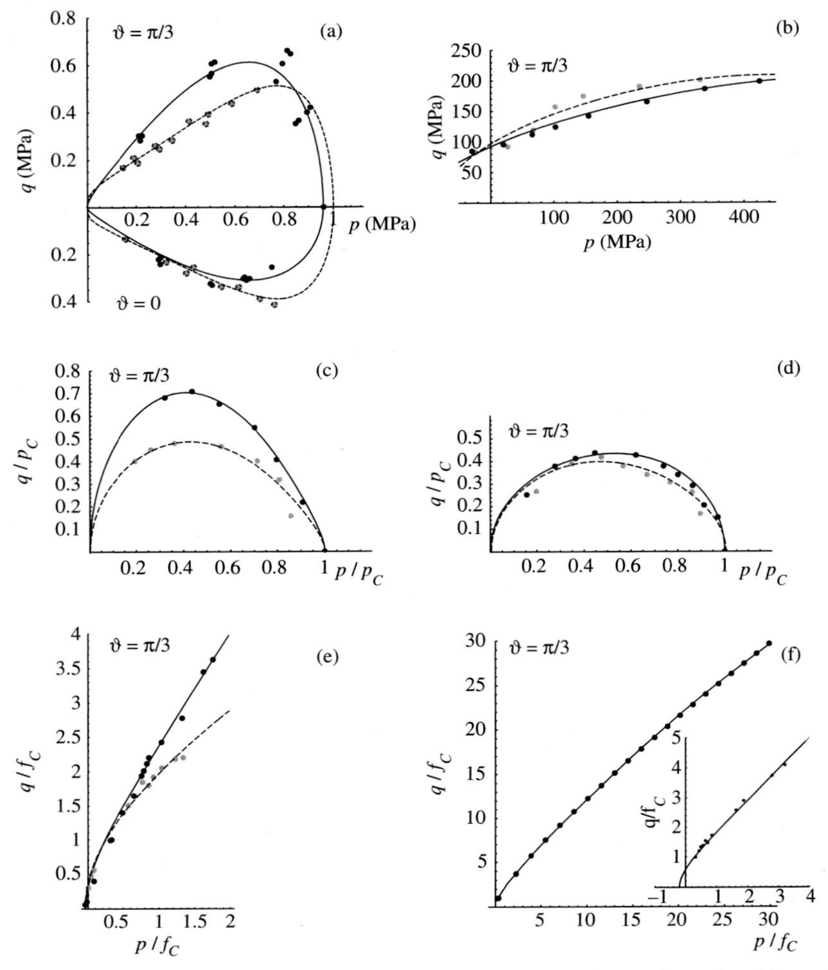

Figure 7.10. Capability of the BP yield surface to fit experimental results: (*a*) sand (black spots; data taken from Yasufuku et al., 1991) and clay (grey spots, data taken from Parry, reported by Wood, 1990); (*b*) polymers: methacrylate (black spots) and an epoxy binder (grey spots; data taken from Ol'khovik, 1983); (*c*) aluminium power (black spots) and aluminium composite powder (grey spots; data taken from Sridhar and Fleck, 2000); (*d*) lead powder (black spots) and lead shot–steel composite (grey spots; data taken from Sridhar and Fleck, 2000); (*e*) rocks: chert (black spots) and dolomite (grey spots; data taken from Hoek and Brown, 1980); (*f*) concrete (black spots; experimental relation proposed by Newman and Newman, 1971) and low confining pressure experiments (grey spots; data taken from Sfer et al., 2002) .

- *Sandstone* (marked with black spots) and *dense sand* (marked with grey spots). Experimental data have been taken from Lade (1997, their figs. 2 and 9*a*).
- *Cast iron* (experimental data taken from Coffin and Schenectady, 1950), marked with black spots, and *concrete* (experimental data taken from Tasuji et al., 1978), marked with grey spots.

The BP yield function has been invented to model pressure-sensitive materials, a class including granular materials, powders, rocks, soils, concrete and cast iron, as well as porous metals and cellular materials and foams.

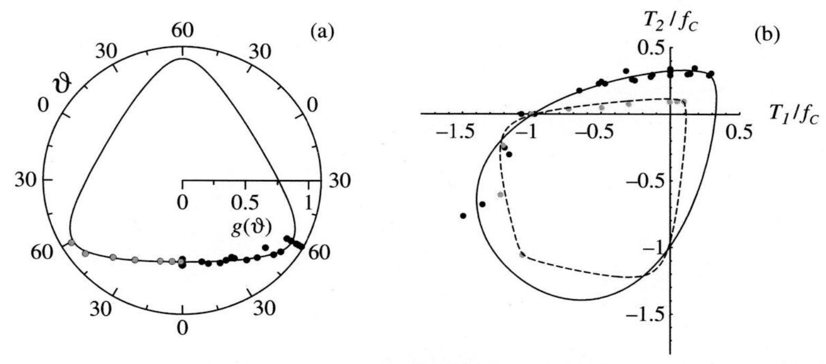

Figure 7.11. Capability of the BP yield surface to fit experimental results: (*a*) deviatoric section for sandstone (black spots) and dense sand (grey spots; data taken from Lade, 1997); (*b*) plane stress biaxial representation for cast iron (black spots; data taken from Coffin and Schenectady, 1950) and concrete (grey spots; data taken from Tasuji et al., 1978).

Regarding metallic foams, we note that the BP yield function describes, with a different expression and for a particular combination of the seven parameters defining it, the same yield surface proposed by Deshpande and Fleck (2000). Regarding the behaviour of porous ductile metals, for appropriate values of the seven parameters defining the criterion, the BP yield surface becomes very close to the yield surface of the Gurson (1977) model; see Bigoni and Piccolroaz (2004) for details.

7.4 Convexity of yield function and yield surface

Convexity of the BP yield function [Eqs. (7.14) through (7.17)] within the range of parameters [Eqs. (7.18) and (7.19)] was until now simply stated.

A proof of convexity of the BP yield function within the parameter range [Eqs. (7.18) and (7.19)] is provided below as an application of a general proposition relating convexity of yield functions and surfaces (recall the notion of convexity from Section 2.19).

We begin by noting that while convexity of the yield function implies convexity of the corresponding yield surface, the converse is usually false; namely, convexity of the level set of a function is unrelated to convexity of the function itself. As an example, let us consider the non-convex yield function

$$f(p,q) = \left(\frac{p}{a}\right)^4 - \left(\frac{p}{a}\right)^2 + \left(\frac{q}{b}\right)^2, \qquad 0 \le \frac{p}{a} \le 1 \tag{7.40}$$

(where a and b are non-null material parameters having the dimension of stress), which corresponds to a convex yield surface $f(p,q) = 0$ (Fig. 7.12).

Convexity of *every* level set of a function represents its *quasi-convexity* [Eqs. (2.166) and (2.167)[7]].

[7] Franchi et al. (1990) noticed that the definition (2.167) is very similar to the so-called Drucker's postulate [Eq. (8.18)]. However, this postulate merely prescribes the so-called normality rule of plastic flow and convexity of yield surface, so quasi-convexity becomes a consequence of Drucker's postulate only in the special case—considered by Franchi et al.—in which convexity of yield surface implies convexity of all level sets of the corresponding function.

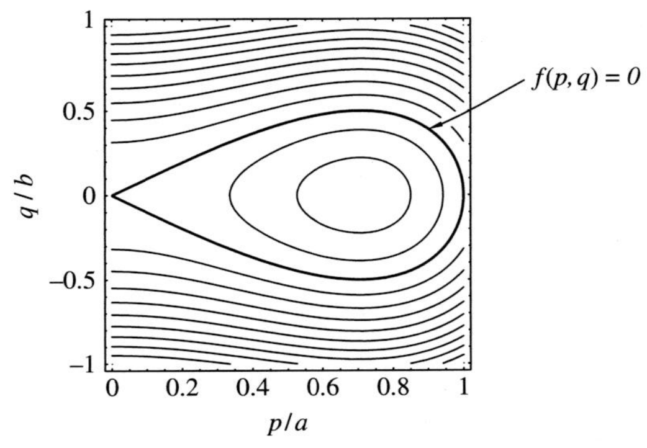

Figure 7.12. Level sets of the function $f(p,q)$ given by Eq. (7.40). Note that while $f(p,q) = 0$ defines a convex yield surface, the function $f(p,q)$ given by Eq. (7.40) lacks quasi-convexity and therefore convexity.

Practically all experiments on yielding (failure) of ductile (brittle) materials demonstrate convexity of the yield (failure) *surface*, which also can be accepted on the basis of engineering argumentations, for instance, the validity of the so-called Drucker's postulate (Drucker, 1956, 1964; see Section 8.1). Obviously, a convex yield locus can be expressed as a level set of a function that generally may lack convexity and even quasi-convexity. For example, the level sets of function (7.40) are given in Fig. 7.12. It can be observed that while $f(p,q) = 0$ may serve perfectly as a convex yield surface, the corresponding yield function even lacks quasi-convexity. Although it is true that, in principle, a convex yield function always can be found to represent a convex yield surface, to find this in a reasonably simple form may be a formidable task. In other words, a number of yield functions that were formulated as an interpolation of experimental results still need a proof of convexity, even in cases where the corresponding yield locus is convex. The propositions that will be given below set a basis to provide these proofs.

7.4.1 A general convexity result for a class of yield functions

The BP yield function [Eqs. (7.14) through (7.17) presented in the preceding section may be viewed as an element of a family of models specified in the generic form of Eq. (7.14). This family includes, among others, the models by Gudheus (1973), Argyris et al. (1974), Willam and Warnke (1975), Eekelen (1980), Lin and Bazant (1986), Bardet (1990), Ehlers (1995), Menétrey and Willam (1995), Christensen (1997) and Christensen et al. (2002).

A general result is provided below showing that for the range of material parameters for which the Haigh-Westergaard representation of a yield surface (7.22) is convex, the function is also convex.

Proposition: Convexity of the yield function (7.14) *assumed smooth* (for simplicity, an extension to non-smooth yield functions has been given by Piccolroaz and Bigoni, 2009) is equivalent to convexity of the meridian and deviatoric sections of the corresponding yield *surface* (7.22) in the Haigh-Westergaard representation. In symbols,[8]

$$\text{Convexity of } F(\boldsymbol{T}) = f(p) + \frac{q}{g(\theta)} \iff \begin{cases} f'' \geq 0, \\ g^2 + 2g'^2 - gg'' \geq 0, & \forall \theta \in (0, \pi/3) \\ g'(0) = g'(\pi/3) = 0. \end{cases} \tag{7.41}$$

where $g(\theta)$ is a positive function.

Proof. It is a well-known theorem of convex analysis (Ekeland and Temam, 1976) that the sum of two convex functions is also a convex function. Since q, p and θ are independent parameters, failure of convexity of $f(p)$ or $q/g(\theta)$ implies failure of convexity of $F(\boldsymbol{T})$, and therefore, convexity of both $f(p)$ and $q/g(\theta)$ are necessary and sufficient conditions for convexity of $F(\boldsymbol{T})$.

Now let us first analyse $f(p)$. The fact that convexity of $f(p)$ as a function of $\boldsymbol{T}$ is equivalent to convexity of the meridian section follows from the linearity of the trace operator in view of the fact that $p = -\text{tr}\boldsymbol{T}/3$.

Second, the fact that convexity of $q/g(\theta)$ as function of $\boldsymbol{T}$ is equivalent to the convexity of the deviatoric section follows from the five lemmas listed below. □

Lemma 7.4.1. Let us consider a scalar isotropic function ϕ of tensorial argument $T_{ij} \in \text{Sym}$ and the corresponding function $\tilde{\phi}$ written with reference to the principal values T_i:

$$\phi(T_{11}, T_{22}, T_{33}, T_{12}, T_{13}, T_{23}) = \tilde{\phi}(T_1, T_2, T_3).$$

Then, owing to isotropy, the following equality holds:

$$\tilde{\phi}(T_1, T_2, T_3) = \phi(T_1, T_2, T_3, 0, 0, 0). \tag{7.42}$$

Thus $\tilde{\phi}$ is the restriction of ϕ to the subdomain of diagonal tensors.

Proof. Property (7.42) is easily proven by the following consideration: The isotropy of $\phi(\boldsymbol{T})$ implies that the function $\phi(\boldsymbol{T})$ is equal to a function $\hat{\phi}$ of the invariants of $\boldsymbol{T}$, that is,

$$\phi(T_{11}, T_{22}, T_{33}, T_{12}, T_{13}, T_{23}) = \hat{\phi}(\text{tr}\boldsymbol{T}, \text{tr}\boldsymbol{T}^2, \text{tr}\boldsymbol{T}^3),$$

and thus

$$\tilde{\phi}(T_1, T_2, T_3) = \hat{\phi}(\text{tr}\boldsymbol{T}, \text{tr}\boldsymbol{T}^2, \text{tr}\boldsymbol{T}^3) = \phi(T_1, T_2, T_3, 0, 0, 0).$$

□

[8] Note that the generalization of Proposition (7.41) to the case of surface corners given by Piccolroaz and Bigoni (2009) remains identical to (7.41) with the conditions

$$g'(0) = g'(\pi/3) = 0$$

simply replaced by the conditions

$$g'(0) \leq 0 \quad \text{and} \quad g'(\pi/3) \geq 0.$$

Note also that condition (7.41) has been obtained by Bigoni and Piccolroaz (2004) under the implicit assumption of a smooth yield function, so conditions $g'(0) = g'(\pi/3) = 0$ have been omitted in their equations. These conditions have been made explicit by Raniecki and Mróz (2008), who also (independently and using a different route) derived condition (7.41).

The following two lemmas find several applications, even beyond the context of yield functions, for instance, to the representation of strain energy functions for finitely deformed materials. In that context, the lemmas have been proved by Hill (1968 and later re-derived by Yang, 1980), whereas Piccolroaz and Bigoni (2009) have provided the generalisation to the non-smooth case (describing, for instance, yield surfaces with corners).

Lemma 7.4.2. Given a convex function of the principal stresses $\tilde{\phi}(T_1, T_2, T_3)$, the algebraic order of components of its gradient (Q_1, Q_2, Q_3) at (T_1, T_2, T_3) is the same as (T_1, T_2, T_3).

Proof . From the strict convexity of $\tilde{\phi}$, it follows that

$$\sum_{i=1}^{3}(Q_i - Q_i^0)(T_i - T_i^0) > 0, \tag{7.43}$$

where Q_i^0 are the components of the gradient of $\tilde{\phi}$ calculated at T_i^0. Choosing $(T_2^0, T_1^0, T_3^0) = (T_1, T_2, T_3)$ and *taking into account isotropy*, it follows that

$$(Q_1 - Q_2)(T_1 - T_2) > 0, \tag{7.44}$$

and similarly for each of the other pairs. It follows that the vector (Q_1, Q_2, Q_3) is ordered in the same algebraic order as (T_1, T_2, T_3), a property which remains true also assuming convexity $\geq$ instead of strict convexity $>$. □

Lemma 7.4.3. Convexity of an isotropic smooth function (for a generalisation to the non-smooth case, see Piccolroaz and Bigoni, 2009) of a symmetric (stress) tensor $\boldsymbol{T}$ is equivalent to convexity of the corresponding function of the principal (stress) values T_i $(i = 1, 2, 3)$. In symbols, given

$$\phi(\boldsymbol{T}) = \tilde{\phi}(T_1, T_2, T_3), \tag{7.45}$$

then $\forall \boldsymbol{T}, \boldsymbol{T}' \in \text{Sym}$:

$$\phi(\boldsymbol{T}') - \phi(\boldsymbol{T}) \geq \boldsymbol{Q} \cdot (\boldsymbol{T}' - \boldsymbol{T}) \tag{7.46}$$

$$\Updownarrow$$

$$\tilde{\phi}(T_1', T_2', T_3') - \tilde{\phi}(T_1, T_2, T_3) \geq \sum_{i=1}^{3} Q_i \,(T_i' - T_i) \tag{7.47}$$

where $\boldsymbol{Q}$ and Q_i are the gradients of ϕ and $\tilde{\phi}$, that is,

$$\boldsymbol{Q} = \frac{\partial \phi}{\partial \boldsymbol{T}} \quad \text{and} \quad Q_i = \frac{\partial \tilde{\phi}}{\partial T_i}. \tag{7.48}$$

Proof . The proof that (7.46) $\Longrightarrow$ (7.47) follows immediately from property (7.42). The converse, that (7.47) $\Longrightarrow$ (7.46), is not trivial and is proven in the following:

We denote by (T_1, T_2, T_3) the principal values of a given $\boldsymbol{T}$. Assuming that (T_1', T_2', T_3') are numbered in the same algebraic order as (T_1, T_2, T_3), and because from the Lemma 7.4.2 we know that the algebraic order of (Q_1, Q_2, Q_3) is also the same as (T_1, T_2, T_3), the auxiliary property of the scalar product (proven in the Introduction, Section 2.20.6) implies that

$$\sum_{i=1}^{3} Q_i \left(T_i' - T_i\right) \geq \boldsymbol{Q} \cdot (\boldsymbol{T}' - \boldsymbol{T}). \qquad (7.49)$$

Since, by hypothesis, the following equation holds true

$$\phi(\boldsymbol{T}') - \phi(\boldsymbol{T}) = \tilde{\phi}(T_1', T_2', T_3') - \tilde{\phi}(T_1, T_2, T_3) \geq \sum_{i=1}^{3} Q_i \left(T_i' - T_i\right), \qquad (7.50)$$

Eq. (7.49) guarantees that $\phi(\boldsymbol{T})$ is convex. □

Lemma 7.4.4. Given a generic isotropic function ϕ of the stress that can be expressed as

$$\phi(T_1, T_2, T_3) = \tilde{\phi}(\mathsf{dev}T_1, \mathsf{dev}T_2), \qquad (7.51)$$

where $\mathsf{dev}T_1$ and $\mathsf{dev}T_2$ are two of the principal components of deviatoric stress, that is,

$$\mathsf{dev}T_1 = \frac{1}{3}(2T_1 - T_2 - T_3) \quad \text{and} \quad \mathsf{dev}T_2 = \frac{1}{3}(-T_1 + 2T_2 - T_3), \qquad (7.52)$$

convexity of $\phi(T_1, T_2, T_3)$ is equivalent to convexity of $\tilde{\phi}(\mathsf{dev}T_1, \mathsf{dev}T_2)$.

Proof. The proof follows immediately from the observation that the relation (7.52) between $\{\mathsf{dev}T_1, \mathsf{dev}T_2\}$ and $\{T_1, T_2, T_3\}$ is linear. □

Lemma 7.4.5. Convexity of

$$\frac{q}{g(\theta)} \qquad (7.53)$$

as a function of $\mathsf{dev}T_1, \mathsf{dev}T_2$ is equivalent to the convexity of the deviatoric section in the Haigh-Westergaard space:

$$g^2 + 2g'^2 - gg'' \geq 0. \qquad (7.54)$$

Proof. The Hessian of Eq. (7.53) is

$$\frac{\partial^2 q/g(\theta)}{\partial \mathsf{dev}T_i \partial \mathsf{dev}T_j} = \frac{1}{g^3}\left[g^2 \frac{\partial^2 q}{\partial \mathsf{dev}T_i \partial \mathsf{dev}T_j} + q(2g'^2 - gg'')\frac{\partial \theta}{\partial \mathsf{dev}T_i}\frac{\partial \theta}{\partial \mathsf{dev}T_j}\right.$$
$$\left. - gg'\left(\frac{\partial q}{\partial \mathsf{dev}T_i}\frac{\partial \theta}{\partial \mathsf{dev}T_j} + \frac{\partial q}{\partial \mathsf{dev}T_j}\frac{\partial \theta}{\partial \mathsf{dev}T_i} + q\frac{\partial^2 \theta}{\partial \mathsf{dev}T_i \partial \mathsf{dev}T_j}\right)\right], \qquad (7.55)$$

where i and j range between 1 and 2, and all functions q and θ are to be understood as functions of $\mathsf{dev}T_1$ and $\mathsf{dev}T_2$ only. Derivatives of q may be easily calculated to be[9]

$$\frac{\partial q}{\partial \mathsf{dev}T_i} = \frac{3}{2q}\left[2\mathsf{dev}T_i - (-1)^i m_i\right], \qquad \frac{\partial^2 q}{\partial \mathsf{dev}T_i \partial \mathsf{dev}T_j} = \frac{27}{4q^3} m_i m_j, \qquad (7.56)$$

[9] Inconsequential misprints in Eqs. $(39)_1$ and $(43)_1$ of Bigoni and Piccolroaz (2004) have been fixed in Eqs. $(7.56)_1$ and $(7.60)_1$, respectively.

where indices are not summed, and vector m_i has the components

$$m_i = \{\mathrm{dev}T_2, -\mathrm{dev}T_1\}. \tag{7.57}$$

The derivatives of θ can be performed through $\cos 3\theta$ [Eq. (7.13)$_1$], noting that

$$\frac{\partial\theta}{\partial \mathrm{dev}T_i} = \frac{-1}{3\sin 3\theta}\frac{\partial\cos 3\theta}{\partial \mathrm{dev}T_i} \quad \text{and}$$
$$\frac{\partial^2\theta}{\partial \mathrm{dev}T_i \partial \mathrm{dev}T_j} = \frac{-1}{3\sin 3\theta}\left(\frac{\cos 3\theta}{\sin^2 3\theta}\frac{\partial\cos 3\theta}{\partial \mathrm{dev}T_i}\frac{\partial\cos 3\theta}{\partial \mathrm{dev}T_j} + \frac{\partial^2\cos 3\theta}{\partial \mathrm{dev}T_i \partial \mathrm{dev}T_j}\right) \tag{7.58}$$

so that

$$\frac{\partial q}{\partial \mathrm{dev}T_i}\frac{\partial\theta}{\partial \mathrm{dev}T_j} + \frac{\partial q}{\partial \mathrm{dev}T_j}\frac{\partial\theta}{\partial \mathrm{dev}T_i} + q\frac{\partial^2\theta}{\partial \mathrm{dev}T_i \partial \mathrm{dev}T_j}$$
$$= \frac{-1}{\sin 3\theta}\left[\frac{\partial^2 q\cos 3\theta}{\partial \mathrm{dev}T_i \partial \mathrm{dev}T_j} - \cos 3\theta\frac{\partial^2 q}{\partial \mathrm{dev}T_i \partial \mathrm{dev}T_j} + q\frac{\cos 3\theta}{\sin^2 3\theta}\frac{\partial\cos 3\theta}{\partial \mathrm{dev}T_i}\frac{\partial\cos 3\theta}{\partial \mathrm{dev}T_j}\right], \tag{7.59}$$

where

$$\frac{\partial\cos 3\theta}{\partial \mathrm{dev}T_i} = \frac{9\sqrt{3}\sin 3\theta}{2q^2} f(\mathrm{dev}T_1, \mathrm{dev}T_2)\, m_i \quad \text{and} \quad \frac{\partial^2 q\cos 3\theta}{\partial \mathrm{dev}T_i \partial \mathrm{dev}T_j} = -27^2\frac{J_3}{q^6} m_i m_j. \tag{7.60}$$

A substitution of Eqs. (7.60) into Eq. (7.59) yields

$$\frac{\partial q}{\partial \mathrm{dev}T_i}\frac{\partial\theta}{\partial \mathrm{dev}T_j} + \frac{\partial q}{\partial \mathrm{dev}T_j}\frac{\partial\theta}{\partial \mathrm{dev}T_i} + q\frac{\partial^2\theta}{\partial \mathrm{dev}T_i \partial \mathrm{dev}T_j} = 0, \tag{7.61}$$

so we may conclude that the Hessian (7.55) can be written as

$$\frac{\partial^2 q/g(\theta)}{\partial \mathrm{dev}T_i \partial \mathrm{dev}T_j} = \frac{27}{4}\frac{(g^2 + 2g'^2 - gg'')}{q^3 g^3} m_i m_j. \tag{7.62}$$

Positive semi-definiteness of the Hessian (7.62) is condition (7.54), which, in turn, represents non-negativeness of the curvature (and thus convexity) of deviatoric section.

□

7.4.2 Convexity of the BP yield function

Now we are in a position to prove the convexity of the BP yield function [Eqs. (7.14) through (7.17)] within the range [Eqs. (7.18) and (7.19)] of material parameters.

First, we show that $f(p)$ [Eq. (7.14)] is a convex function of p (so that the meridian section is convex) and, second, that the deviatoric section described by $g(\theta)$ [Eq. (7.17)] is convex, for the range of material parameters listed in Eqs. (7.18) and (7.19). Therefore, as a conclusion from Proposition (7.41), function $F(\boldsymbol{T})$ is convex.

A well-known result of convex analysis (Ekeland and Temam, 1976) states that function $f(p)$ is convex if and only if the restriction to its effective domain (i.e., $\Phi \in [0,1]$) is convex. Moreover, the function Φ appearing in Eq. (7.16)$_1$ is a linear function of p, so convexity of $f(p)$ can be inferred from convexity of the corresponding function, say, $\tilde{f}$, of Φ. Introducing for simplicity the function

$$h(\Phi) = (\Phi - \Phi^m)\,[2(1-\alpha)\Phi + \alpha], \tag{7.63}$$

the convexity of function $\tilde{f}(\Phi)$ reduces to the condition

$$\left[h'(\Phi)\right]^2 - 2h''(\Phi)h(\Phi) \geq 0, \tag{7.64}$$

where

$$\begin{aligned} h'(\Phi) &= \left(1 - m\Phi^{m-1}\right)[2(1-\alpha)\Phi + \alpha] + 2(1-\alpha)\left(\Phi - \Phi^m\right), \\ h''(\Phi) &= -m(m-1)\Phi^{m-2}[2(1-\alpha)\Phi + \alpha] + 4(1-\alpha)\left(1 - m\Phi^{m-1}\right). \end{aligned} \tag{7.65}$$

Fulfilment of Eq. (7.64) now can be proven easily considering the inequality

$$h''(\Phi) \leq 4(1-\alpha)\left(1 - m\Phi^{m-1}\right), \qquad \forall\, \Phi \in [0,1]. \tag{7.66}$$

It remains now to show the convexity of $q/g(\theta)$. To this purpose, Proposition (7.41) can be employed, through substitution of Eq. (7.17) into the convexity condition [Eq. (7.54)], thus yielding

$$\frac{1}{g(\theta)} + \frac{3\gamma\cos 3\theta}{\sqrt{1-\gamma^2\cos^2 3\theta}} \sin\left[\beta\frac{\pi}{6} - \frac{1}{3}\cos^{-1}(\gamma\cos 3\theta)\right] \geq 0, \tag{7.67}$$

where $\theta \in [0,\pi/3]$ and $g(\theta)$ is given by Eq. (7.17). For values of γ belonging to the interval specified in Eq. $(7.18)_7$, condition (7.67) can be transformed into

$$\sin\left(\frac{\beta\pi}{6} - \frac{4}{3}x\right) + 2\sin\left(\frac{\beta\pi}{6} + \frac{2}{3}x\right) \geq 0 \tag{7.68}$$

with $x \in [\cos^{-1}\gamma,\ \pi - \cos^{-1}\gamma]$ and then into

$$\frac{-1 + 2\cos z + 2\cos^2 z}{2\sin z(1-\cos z)} \sin\beta\frac{\pi}{6} + \cos\beta\frac{\pi}{6} \geq 0 \tag{7.69}$$

with $z \in [2/3\cos^{-1}\gamma,\ 2/3(\pi - \cos^{-1}\gamma)]$, an inequality that can be shown to be verified within the interval of β specified in Eq. (7.19) and thus also within its subinterval [Eq. $(7.18)_6$].

7.4.3 Generating convex yield functions

We show in this subsection that Proposition (7.41) is constructive, in the sense that it can be employed easily to 'generate' convex yield functions within the class described by Eq. (7.14).

The simplest possibility is to maintain $f(p)$ in the form of Eq. (7.15) and change the deviatoric function [Eq. (7.17)]. As a first proposal, we can introduce the following function:

$$g(\theta) = [1 + \beta\,(1 + \cos 3\theta)]^{-1/n}, \tag{7.70}$$

instead of Eq. (7.17). This describes a smooth deviatoric section approaching (without reaching) the triangular (Rankine) shape when parameters $n > 0$ and $\beta \geq 0$ are varied.

We show that the deviatoric section described by Eq. (7.70) is convex, within the range of parameters reported in Table 7.3. To this purpose, a substitution of Eq. (7.70) into condition (7.54) yields

$$a\cos^2(3\theta) + b\cos(3\theta) + c \geq 0, \quad \theta \in [0,\pi/3], \tag{7.71}$$

Table 7.3. *Conditions for the convexity of the deviatoric yield function (7.70)*

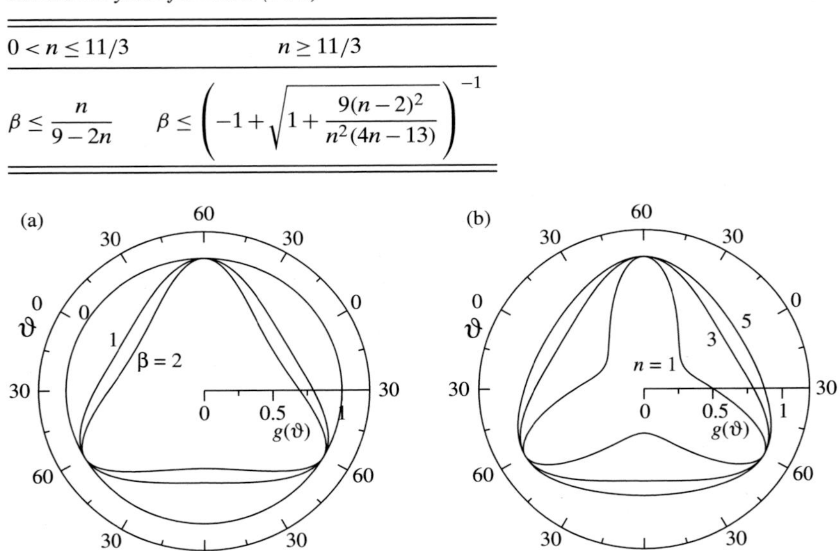

$0 < n \leq 11/3$	$n \geq 11/3$
$\beta \leq \dfrac{n}{9-2n}$	$\beta \leq \left(-1+\sqrt{1+\dfrac{9(n-2)^2}{n^2(4n-13)}}\right)^{-1}$

Figure 7.13. Deviatoric section of the yield surface (7.70): effects related to the variation of β (*a*) and n (*b*).

where the coefficients a, b and c are

$$a = \beta^2\left(n^2 - 9\right), \tag{7.72}$$

$$b = \beta n(1+\beta)(9-2n), \tag{7.73}$$

$$c = n^2(1+\beta)^2 + 9\beta^2(1-n). \tag{7.74}$$

Since condition (7.71) is bounded by a parabola, it suffices to analyse the position of its vertex to obtain the values of the parameters reported in Table 7.3.

The yield function defined by Eqs (7.15) and (7.70) does not possess the extreme deformability of Eqs. (7.15) and (7.17) and does not admit Mohr-Coulomb and Tresca as limits but results in a simple expression. The performance of the deviatoric shape of the yield surface is analysed in Fig. 7.13, where the limit of convexity corresponds to $\beta = 1$ and $n = 3$. The curves reported in Fig. 7.13*a* are relative to the values of $\beta = 0, 1, 2$, whereas for Fig. 7.13*b*, n takes the values $\{1, 3, 5\}$.

A limitation of the yield surface described by Eqs (7.15) and (7.70) is that the deviatoric section cannot be stretched until the Rankine limit. This can be easily amended, assuming for $g(\theta)$ the BP expression (7.17) or that proposed by Willam and Warnke (1975) (see also Menétrey and Willam, 1995), that is,

$$g(\theta) = \frac{2(1-e^2)\cos\theta + (2e-1)\left[4(1-e^2)\cos^2\theta + 5e^2 - 4e\right]^{0.5}}{4(1-e^2)\cos^2\theta + (2e-1)^2}, \tag{7.75}$$

where $e \in]0.5, 1]$ is a material parameter yielding in the limit $e \longrightarrow 0.5$ the Rankine criterion and the von Mises criterion when $e = 1$.

It is already known that the deviatoric section of the yield surface corresponding to Eq (7.75) remains convex for any value of the parameter e ranging within the interval (0.5,1], so—from Proposition (7.41)—function (7.14) equipped with definition (7.75) of function $g(\theta)$ is also convex.

As a final example, we can employ function $g(\theta)$ defined by the expression proposed by Gudheus (1973) and Argyris et al. (1974):

$$g(\theta) = \frac{2k}{1+k+(1-k)\cos 3\theta}, \tag{7.76}$$

where $k \in (0.777, 1]$ is a material parameter.

Otherwise, we can act on the meridian function. For instance, we can modify a Drucker-Prager criterion—which again fits in the framework described by eqn. (7.14)—obtaining a non-circular deviatoric section described by Eq. (7.17):

$$F(\boldsymbol{T}) = -\Gamma(p+c) + q\cos\left[\beta\frac{\pi}{6} - \frac{1}{3}\cos^{-1}(\gamma\cos 3\theta)\right], \tag{7.77}$$

where c is the yield strength under isotropic tension and Γ is a material parameter, or by Eq. (7.70)

$$F(\boldsymbol{T}) = -\Gamma(p+c) + q[1+\beta(1+\cos 3\theta)]^{1/n}, \tag{7.78}$$

or by the Gudheus-Argyris condition (7.76)

$$F(\boldsymbol{T}) = -\Gamma(p+c) + \frac{q}{2k}(1+k+(1-k)\cos 3\theta). \tag{7.79}$$

It may be noted that the yield criterion (7.79) has been employed by Laroussi et al. (2002) to describe the behaviour of foams. In all the preceding cases, Proposition (7.41) ensures that for the range of parameters in which the Haigh-Westergaard representation of the yield surface is convex, the yield function is also convex.

The behaviour of concrete and a generalisation of Proposition (7.41) In the modelling of concrete, there is some experimental evidence that the deviatoric section starts close to the Rankine limit for low hydrostatic stress components and tends to approach a circle when confinement increases. This effect has been described by Ottosen (1977) through a model which does not fit the general framework specified by Eq. (7.14) and can be written in our notation in the form

$$F(\boldsymbol{T}) = Aq^2 + B\frac{q}{g(\theta)} + C - p, \tag{7.80}$$

where $A > 0$, $B \geq 0$ and $C \leq 0$ are constants and $g(\theta)$ is in the form of Eq. (7.17) with $\beta = 0$. The criterion therefore is defined by four parameters.

The preceding expression (7.80) of the yield function suggests the following generalisation of Proposition (7.41):

Proposition: Convexity of the yield function

$$F(\boldsymbol{T}) = Aq^2 + B\frac{q}{g(\theta)} + f(p), \tag{7.81}$$

where A and B are positive constants, is equivalent to

$$f'' \geq 0, \qquad g^2 + 2g'^2 - gg'' \geq 0, \qquad g'(0) = g'(\pi/3) = 0, \tag{7.82}$$

which, in turn, is equivalent to the convexity of the surface

$$B\frac{q}{g(\theta)} + f(p) = 0 \tag{7.83}$$

in the Haigh-Westergaard stress space.

Proof. Let us begin assuming that Eq. (7.82) holds true. Under this condition, Proposition (7.41) ensures that $f(p)$ and $q/g(\theta)$ are convex functions of $\boldsymbol{T}$, so Eq. (7.81) produces the sum of three convex functions, and its convexity follows. Vice versa, failure of convexity of $f(p)$ immediately implies failure of the convexity of Eq. (7.81) because p is independent of θ and q. Finally, let us assume that condition $(7.82)_2$ is violated, for a certain value, say, $\tilde{\theta}$, of θ. The Hessian of

$$Aq^2 + Bq/g(\theta) \tag{7.84}$$

as a function of two components of deviatoric stress $\mathsf{dev}T_1$ and $\mathsf{dev}T_2$ is given by the matrix represented by Eq. (7.62), summed to a constant and positive definite matrix to yield

$$3A\begin{bmatrix} 2 & 1 \\ 1 & 2 \end{bmatrix} + B\frac{27}{4}\frac{(g^2 + 2g'^2 - gg'')}{q^3 g^3}\begin{bmatrix} \mathsf{dev}T_2^2 & -\mathsf{dev}T_1\mathsf{dev}T_2 \\ -\mathsf{dev}T_1\mathsf{dev}T_2 & \mathsf{dev}T_1^2 \end{bmatrix}. \tag{7.85}$$

Considering now the Haigh-Westergaard representation, it is easy to understand that we can keep $\theta = \tilde{\theta}$ fixed and change $\mathsf{dev}T_1$ and consequently $\mathsf{dev}T_2$ so that $\mathsf{dev}T_1/\mathsf{dev}T_2$ remains constant. In this situation, while g and its derivatives remain fixed, the quantity

$$\frac{\mathsf{dev}T_1^2}{q^3} \tag{7.86}$$

in matrix (7.85) tends to $+\infty$ when $\mathsf{dev}T_1$ tends to zero. Therefore, violation of Eq. $(7.82)_2$ cannot be compensated by a constant term, and function (7.81) is not convex. □

The above-stated proposition provides the conditions for convexity of the Ottosen criterion. Moreover, the same proposition allows us to generalise the BP yield function [Eqs. (7.14) through (7.17)], adding a q^2 term as in the Ottosen criterion. This leads immediately to

$$F(\boldsymbol{T}) = A\,q^2 + B\,q\cos\left[\beta\frac{\pi}{6} - \frac{1}{3}\cos^{-1}(\gamma\cos 3\theta)\right] + f(p), \tag{7.87}$$

where $f(p)$ is given by Eq. (7.15).

8 Elastoplastic constitutive equations

After a presentation of elastoplasticity at small strain, a general framework for elastoplasticity is derived under the assumptions of smoothness of the yield surface and independence of the plastic flow mode tensor on the strain rate. Therefore, restrictions are not introduced on: (1) the type of elastic and plastic strain decomposition, (2) the hardening rule, (3) isotropy of the behaviour, (4) convexity of the yield function, and (5) the existence of an elastic potential. As particular cases of the general theory, a constitutive model is presented for describing the behaviour of metallic materials at large strain, together with a small strain derivation of elastoplastic coupling, useful in the constitutive description of geomaterials.

When a ductile material such as, for instance, mild steel (Fig. 8.1) is deformed in a sufficiently severe way, irreversible or, in other words, 'plastic' strain occurs.

In the case of steel, the irreversible deformation is the 'global effect' of dislocation activity which initiates at a certain threshold stress. More in general, plastic flow is always related to the activation of some irreversible micro-mechanism, such as micro-cracking in rock and concrete and sliding between grains in granular matter. From the point of view of constitutive modelling, the 'activation stress' is decided on the basis of a suitable yield function of the type (7.1), which is the first 'building block' of a plasticity or damage theory. The other 'blocks' are (1) the law relating the stress to the elastic deformation, (2) the plastic flow rule, setting the tensorial character of the plastic rate of deformation, (3) the hardening rule, allowing us to describe hardening, softening or perfectly plastic behaviour, (4) the decomposition of elastic and plastic deformation.

8.1 The theory of elastoplasticity at small strain

Before embarking on the analysis of large strain elastoplasticity, it is instructive to briefly develop the theory at small strain (see also Hill, 1950a; Kachanov, 1971; Lubliner, 1990). Let us begin by referring to an ideal uniaxial stress test on a ductile metal sample evidencing a linear elastic behaviour until the stress reaches the value σ_S. We find two idealisations of such a behaviour in Fig. 8.2, the so-called linear isotropic hardening on the left and a non-linear hardening law with Bauschinger effect on the right. The so-called Bauschinger effect consists of the decrease in the yield stress in compression σ'_S owing to previous plastic deformation in tension,

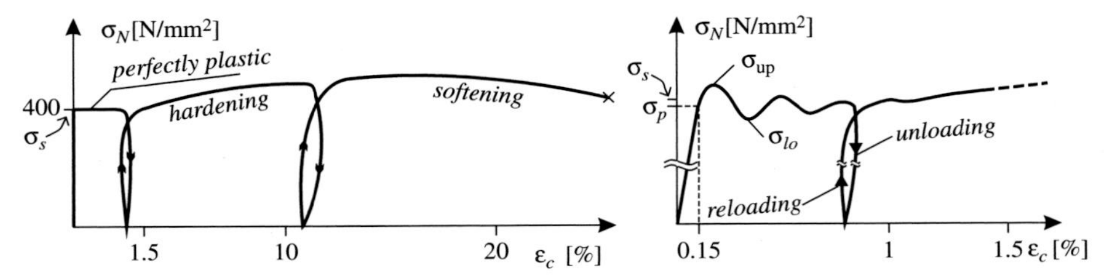

Figure 8.1. Uniaxial nominal stress σ_N versus conventional strain ε_C behaviour of a ductile metal evidencing initial elasticity, followed by perfect plasticity, hardening and final softening. In the detail reported on the right, the upper σ_{up} and lower σ_{lo} yield limits have been indicated, together with the so-called proportionality stress σ_P and the yield stress σ_S. Note that the hysteresis loops and the stress oscillations in the perfectly plastic curve have been magnified.

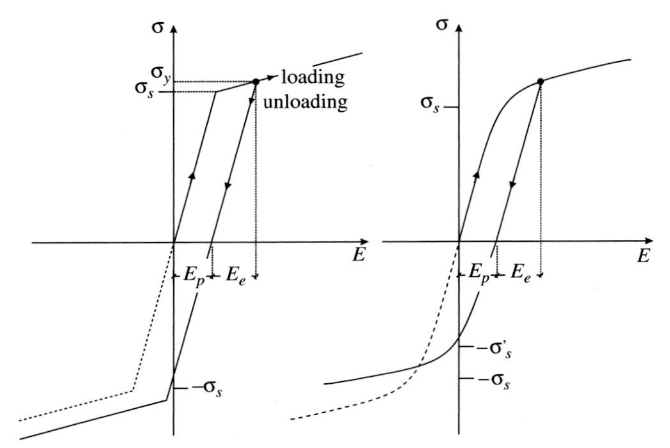

Figure 8.2. Uniaxial stress strain behaviour of a ductile metal evidencing linear (*left*) and non-linear (*right*) strain hardening. In the case on the left, the hardening is isotropic, whereas the Bauschinger effect is visible on the right (this effect consists of the fact that though the initial yield stress for virgin loading in tension and in compression is the same, say, σ_S, after unloading from the plastic range, a new yield stress is found, say, σ'_S, that is different from σ_S). Note the additive decomposition of the strain in the elastic E_e and plastic E_p components, namely, $E = E_e + E_p$.

whereas 'virgin' samples would have the same yield stress σ_S both in tension and in compression.

Uniaxial model of plasticity For both models sketched in Fig. 8.2, the plastic threshold, say, σ_y, though starting from the value σ_S, varies during plastic deformation according to a more or less complicate law. Therefore, the yield function is expressed by the condition

$$|\sigma| - \sigma_y(E_p) \leq 0, \tag{8.1}$$

in which E_p is the irreversible, or plastic, strain, measurable only on unloading.

We can see from the figure that on complete unloading the strain E can be interpreted as the sum of an elastic strain E_e and a plastic strain E_p and that only the former determines the stress through an elastic modulus Y. It therefore becomes spontaneous to introduce the following laws:

$$
\begin{aligned}
&\dot{\sigma} = Y\dot{E}_e && \text{stress rate determined by elastic strain rate,}\\
&\dot{E}_p = \dot{\Lambda}\,\mathsf{sign}(\sigma) && \text{plastic flow rule,}\\
&\dot{\sigma}_y = \dot{\Lambda}\,h && \text{hardening rule,}\\
&\dot{E} = \dot{E}_e + \dot{E}_p && \text{additive decomposition of elastic and plastic strain rates,}
\end{aligned}
\tag{8.2}
$$

where superimposed dots denote derivatives with respect to a time-like parameter governing the load, $\dot{\Lambda} \geq 0$ is the plastic multiplier, positive when plastic flow occurs and otherwise null, and h, the hardening modulus, is a constant for linear (a function of E_p for non-linear) hardening.

A chain substitution of Eq. $(8.2)_2$ into Eq. $(8.2)_4$ and into Eq. $(8.2)_1$ yields

$$
\dot{\sigma} = Y\dot{E} - Y\mathsf{sign}(\sigma)\,\dot{\Lambda}. \tag{8.3}
$$

When plastic loading may occur, the yield condition [Eq. (8.1)], holds with the $=$ sign, and in this situation, the possibility of plastic loading or elastic unloading can be decided from a rate form condition, which can be obtained by differentiating the yield function (8.1), taken with the $=$ sign, with respect to the time-like parameter governing the loading. The differentiation yields

$$
\begin{aligned}
&|\sigma| = \sigma_y \quad \text{and} \quad \mathsf{sign}(\sigma)\,\dot{\sigma} - \dot{\sigma}_y = 0 \quad \text{plastic loading,}\\
&|\sigma| = \sigma_y \quad \text{and} \quad \mathsf{sign}(\sigma)\,\dot{\sigma} < 0 \quad \text{elastic unloading,}
\end{aligned}
\tag{8.4}
$$

where the first of these conditions is the so-called Prager consistency condition. From conditions (8.4), using Eqs. $(8.2)_3$ and (8.3), we can derive the plastic multiplier in the two forms

$$
\begin{aligned}
&\dot{\Lambda} = \frac{Y < \mathsf{sign}(\sigma)\,\dot{E} >}{h+Y} && \text{if } |\sigma| = \sigma_y \text{ and } h+Y > 0,\\
&\dot{\Lambda} = \frac{< \mathsf{sign}(\sigma)\,\dot{\sigma} >}{h} && \text{if } |\sigma| = \sigma_y \text{ and } h > 0,
\end{aligned}
\tag{8.5}
$$

where the operator $< \cdot >$ denotes the Macaulay brackets defined by Eq. (2.103). Note that condition $(8.5)_1$ holds when $h+Y > 0$, whereas condition $(8.5)_2$ holds when $h > 0$, so the former is more general than the latter, and they become equivalent when $h > 0$.

Using Eq. (8.3), we can write the constitutive equations in rate (direct and inverse) form as

$$
\begin{aligned}
&\dot{\sigma} = Y\dot{E} - \frac{Y^2 < \mathsf{sign}(\sigma)\,\dot{E} >}{h+Y}\,\mathsf{sign}(\sigma) && \text{if } |\sigma| = \sigma_y \text{ and } h+Y > 0,\\
&\dot{E} = \frac{\dot{\sigma}}{Y} + \frac{< \mathsf{sign}(\sigma)\,\dot{\sigma} >}{h}\,\mathsf{sign}(\sigma) && \text{if } |\sigma| = \sigma_y \text{ and } h > 0.
\end{aligned}
\tag{8.6}
$$

Note that restricting our attention to the condition of plastic flow, $\mathsf{sign}(\sigma)\,\dot{\sigma} > 0$, the quantity

$$
\frac{Yh}{h+Y} \tag{8.7}
$$

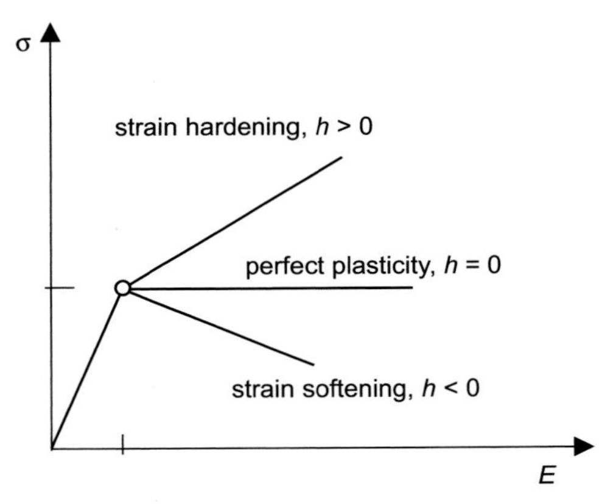

Figure 8.3. Hardening ($h > 0$), softening ($h < 0$) and ideally plastic ($h = 0$) behaviours in the uniaxial stress σ–strain E diagram.

provides the local tangent to the σ–E diagram in the plastic range. We can finally observe that for $h = 0$ ($h \longrightarrow \infty$), perfectly plastic (elastic) behaviour is obtained; moreover, $h > 0$ corresponds to strain hardening, whereas $h < 0$ corresponds to strain softening (Fig. 8.3).

Elastoplasticity at small strain A generalization of the preceding theory to an elastoplastic solid can be developed quickly on the basis of a yield function of the type described by Eq. (7.1), namely,

$$f(\boldsymbol{T}, \mathcal{H}) \leq 0,$$

function of the Cauchy stress $\boldsymbol{T}$ and of a set of internal variables $\mathcal{H}$ describing the inelastic behaviour of the material and assuming the following generalisation of the laws (8.2).

- Rate of stress function of the elastic strain rate

$$\dot{\boldsymbol{T}} = \mathbb{E}[\dot{\boldsymbol{E}}_e], \tag{8.8}$$

  where $\boldsymbol{T}$ is the Cauchy stress, $\mathbb{E}$ is the elastic fourth-order tensor and $\boldsymbol{E}_e$ is the elastic infinitesimal strain.
- Plastic flow rule

$$\dot{\boldsymbol{E}}_p = \dot{\Lambda} \boldsymbol{P}, \tag{8.9}$$

  where $\boldsymbol{E}_p$ is the plastic infinitesimal strain, $\boldsymbol{P} \in \mathsf{Sym}$ is the plastic flow mode tensor (setting the 'direction' of plastic strain rate and therefore replacing the $\mathsf{sign}(\sigma)$ of the unidimensional model) and $\dot{\Lambda} \geq 0$ is the (non-negative) plastic multiplier, positive when plastic flow occurs and null otherwise. It is important to mention that in the special case $\boldsymbol{P} = \boldsymbol{Q}$, the flow rule is called 'associative' (also called the 'normality rule'), whereas in general, when $\boldsymbol{P} \neq \boldsymbol{Q}$, the flow rule is said to be 'non-associative'.
- Hardening rule

$$\dot{\Lambda} h = -\frac{\partial f}{\partial \mathcal{H}} \cdot \dot{\mathcal{H}}, \tag{8.10}$$

where h is the hardening modulus [positive for strain hardening, null for perfectly plastic behaviour and negative for strain softening (Fig. 8.3; see also Hill, 1967b; Maier, 1967) and $\mathcal{H}$ is the set of variables describing the plastic behaviour. Note that h can be a constant for linear hardening but is in general a function of the internal variables.

- Additive decomposition of strain rates

$$\dot{\boldsymbol{E}} = \dot{\boldsymbol{E}}_e + \dot{\boldsymbol{E}}_p. \tag{8.11}$$

A chain substitution of Eqs. (8.9) and (8.11) into Eq. (8.8) yields

$$\dot{\boldsymbol{T}} = \mathbb{E}[\dot{\boldsymbol{E}}] - \dot{\Lambda}\mathbb{E}[\boldsymbol{P}], \tag{8.12}$$

in which the plastic multiplier $\dot{\Lambda}$ is for the moment unknown, and it represents the 'on-off switch' for plastic flow and related evolution of internal variables.

Assuming that the stress satisfies yielding, $f = 0$, the plastic multiplier can be calculated noting that in the case of elastic behaviour, $\dot{f} < 0$, whereas it is $\dot{f} = 0$ in the cases of the plastic behaviour and in the special case of the so-called neutral loading that will be detailed later. The two conditions (the latter is the so-called Prager consistency) lead to

$$\begin{aligned}
&f=0 \quad \text{and} \quad \dot{f}=\boldsymbol{Q}\cdot\dot{\boldsymbol{T}}<0 && \text{for elastic unloading}\\
&f=0 \quad \text{and} \quad \dot{f}=\boldsymbol{Q}\cdot\dot{\boldsymbol{T}}+\frac{\partial f}{\partial \mathcal{H}}\cdot\dot{\mathcal{H}}=0 && \text{for plastic loading or neutral loading}
\end{aligned} \tag{8.13}$$

where

$$\boldsymbol{Q} = \frac{\partial f}{\partial \boldsymbol{T}}, \tag{8.14}$$

is the yield function gradient, and the condition of neutral loading corresponds to the special case in which both conditions

$$\boldsymbol{Q}\cdot\dot{\boldsymbol{T}} = 0 \quad \text{and} \quad \frac{\partial f}{\partial \mathcal{H}}\cdot\dot{\mathcal{H}} = 0 \quad \text{neutral loading}$$

are verified, so there is neither plastic flow nor elastic unloading (because the stress remains tangent to the yield surface).

Equations (8.13) and (8.10) allow us to calculate the plastic multiplier as

$$\begin{aligned}
\dot{\Lambda} &= \frac{<\boldsymbol{Q}\cdot\mathbb{E}[\dot{\boldsymbol{E}}]>}{h+\boldsymbol{Q}\cdot\mathbb{E}[\boldsymbol{P}]} && \text{if } f(\boldsymbol{T},\mathcal{H})=0 \text{ and } h+\boldsymbol{Q}\cdot\mathbb{E}[\boldsymbol{P}]>0,\\
\dot{\Lambda} &= \frac{<\dot{\boldsymbol{T}}\cdot\boldsymbol{Q}>}{h} && \text{if } f(\boldsymbol{T},\mathcal{H})=0 \text{ and } h>0,
\end{aligned} \tag{8.15}$$

where the operator $<\cdot>$ denotes the Macaulay brackets defined by Eq. (2.103), and $h+\boldsymbol{Q}\cdot\mathbb{E}[\boldsymbol{P}]>0$ is the so-called plastic modulus.

Finally, we arrive at the elastoplastic constitutive equations in rate form

$$\dot{\boldsymbol{T}} = \begin{cases} \mathbb{E}[\dot{\boldsymbol{E}}] - \dfrac{<\boldsymbol{Q}\cdot\mathbb{E}[\dot{\boldsymbol{E}}]>}{h+\boldsymbol{Q}\cdot\mathbb{E}[\boldsymbol{P}]}\mathbb{E}[\boldsymbol{P}] & \text{if } f(\boldsymbol{T},\mathcal{H})=0,\\ \mathbb{E}[\dot{\boldsymbol{E}}] & \text{if } f(\boldsymbol{T},\mathcal{H})<0, \end{cases} \tag{8.16}$$

holding under the hypothesis of positive plastic modulus $h+\boldsymbol{Q}\cdot\mathbb{E}[\boldsymbol{P}]>0$.

Equations (8.16) can be inverted under the hypothesis $h > 0$ and rewritten in the form

$$\dot{\boldsymbol{E}} = \begin{cases} \mathbb{E}^{-1}[\dot{\boldsymbol{T}}] + \dfrac{< \boldsymbol{Q} \cdot \dot{\boldsymbol{T}} >}{h} \boldsymbol{P} & \text{if } f(\boldsymbol{T}, \mathcal{H}) = 0, \\ \mathbb{E}^{-1}[\dot{\boldsymbol{T}}] & \text{if } f(\boldsymbol{T}, \mathcal{H}) < 0, \end{cases} \tag{8.17}$$

so the inverse form (8.17) holds only for a material exhibiting positive hardening. Thus it is less general than the direct form (8.16), holding under the sole hypothesis of positive plastic modulus and thus also valid for strain softening.

We finally observe that the rate equations [Eqs. (8.16) and (8.17)] are valid for every smooth yield function f, every hardening rule and a general class of non-associative flow rule.

Drucker postulate, normality rule and plastic dilatancy We can immediately deduce from the plastic flow rule [Eq. (8.9)], that the plastic strain rate is coaxial to $\boldsymbol{P}$. The special case of associative flow rule, in which the plastic flow mode tensor becomes coaxial to the yield function gradient, $\boldsymbol{P} = \boldsymbol{Q}$, can be derived from the so-called Drucker postulate (Drucker, 1956, 1964). This can be expressed considering every possible stress $\boldsymbol{T}$ at yielding, so $f(\boldsymbol{T}) = 0$, and imposing that the plastic strain rate $\dot{\boldsymbol{E}}_p$ which may develop as associated to $\boldsymbol{T}$ satisfies

$$\dot{\boldsymbol{E}}_p \cdot (\boldsymbol{T} - \boldsymbol{T}^*) \geq 0, \quad \forall \boldsymbol{T}^* \qquad \text{such that} \qquad f(\boldsymbol{T}^*) \leq 0. \tag{8.18}$$

Retaining for simplicity the assumption of smoothness of the yield surface and employing our previous assumptions and Eqs. (8.16) and (8.17), we can deduce from the Drucker postulate (8.18) the following:

- *Associative flow rule.* Let us consider now every stress $\boldsymbol{T}^*$ infinitely close to $\boldsymbol{T}$ but satisfying $f(\boldsymbol{T}^*) \leq 0$. The vectors $\boldsymbol{T} - \boldsymbol{T}^*$ represent all infinitesimal vectors pointing outside the plane tangent to the yield surface at $\boldsymbol{T}$ so condition (8.18) implies $\dot{\boldsymbol{E}}_p$ coaxial to the normal $\boldsymbol{Q}$ at the yield surface at $\boldsymbol{T}$, meaning the associative flow rule, $\boldsymbol{P} = \boldsymbol{Q}$ (Fig. 8.4, *left*).
- *Convexity of the yield surface.* Since $\dot{\boldsymbol{E}}_p$ has to be coaxial to the normal $\boldsymbol{Q}$ at the yield surface at $\boldsymbol{T}$, condition (8.18) implies convexity of the yield surface because it is immediate to show that for a non-convex yield surface a stress $\boldsymbol{T}^*$ can be found to violate (8.18) (Fig. 8.4, *right*).
- *Non-negativeness of the second-order plastic work.* Since all stress rates directed inside the yield function, $\dot{\boldsymbol{T}} \cdot \boldsymbol{Q} < 0$, are associated to elastic unloading for which $\dot{\boldsymbol{E}}_p = \boldsymbol{0}$, and since $\dot{\boldsymbol{E}}_p$ is coaxial with $\boldsymbol{Q}$ for all stress rates directed outside the yield function, we have

$$\dot{\boldsymbol{E}}_p \cdot \dot{\boldsymbol{T}} \geq 0, \tag{8.19}$$

  representing the non-negativeness of second-order plastic work.
- *Non-negativeness of the hardening modulus h.* Since $\dot{\boldsymbol{E}}_p = \dot{\lambda} \boldsymbol{Q}$, taking the scalar product of this with $\dot{\boldsymbol{T}}$ given by Eq. (8.16), condition (8.19) implies $h \geq 0$.
- *Positivity of the second-order work.* Since $\boldsymbol{P} = \boldsymbol{Q}$ and $h > 0$, assuming positive definiteness of $\mathbb{E}$, Eq. (8.17) provides

$$\dot{\boldsymbol{T}} \cdot \dot{\boldsymbol{E}} > 0, \tag{8.20}$$

  representing the positivity of second-order work.

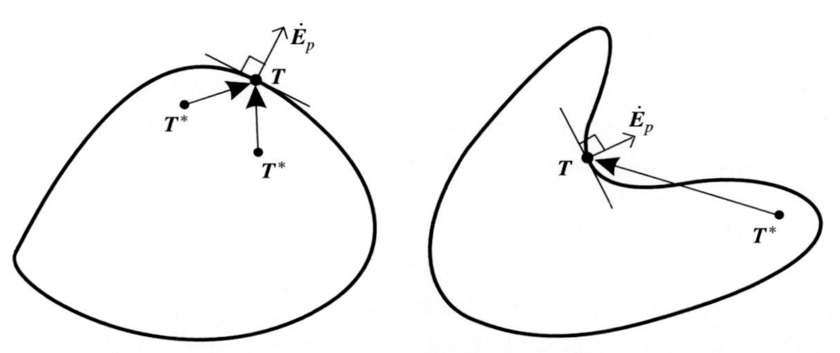

Figure 8.4. A graphical proof that the Drucker postulate (8.18) implies (1) the normality (or associated) rule for the plastic strain rate (*left*) and (2) the convexity of the yield function (*right*).

As a conclusion, we can point out that the assumption of the Drucker postulate leads to severe restrictions, often too restrictive, on the constitutive laws. In particular, the assumption of associativity of the flow rule does not allow us to correctly describe the behaviour of frictional materials, which, as dominated by Coulomb friction occurring at the micro-scale, are inherently non-associative (as shown in Section 1.9). More in general, assuming the normality rule for pressure-sensitive materials usually yields a volumetric plastic dilatancy that is much higher than that the experiments predict. To exemplify this concept with an example, we consider the Drucker-Prager yield function (7.36); thus, assuming the associative flow rule, the plastic strain rate is given from Eq. (7.38) by

$$\dot{\boldsymbol{E}}_p = \dot{\lambda}\left(\cos\phi\frac{\mathrm{dev}\,\boldsymbol{T}}{|\mathrm{dev}\,\boldsymbol{T}|} + \frac{\sin\phi}{\sqrt{3}}\boldsymbol{I}\right), \tag{8.21}$$

and a plastic volumetric dilatancy is predicted to be

$$\mathrm{tr}\dot{\boldsymbol{E}}_p = \dot{\lambda}\left(\sqrt{3}\sin\phi\right), \tag{8.22}$$

positive for positive internal friction ϕ.

8.2 The essential structure of rate elastoplastic constitutive equations at large strain

Although the small strain theory of time-independent elastoplasticity has reached the status of a 'mature' formulation, the state of the art on the large strain theory shows that there is still a lack of consensus on many issues. We would avoid building a general theory of bifurcation and stability on specific constitutive laws which sooner or later could be abandoned; rather, we are interested in considering a general constitutive framework to provide results that are valid over a broad context. Therefore, our main purpose in this section is to show that rate elastoplastic constitutive equations for inviscid materials at large strains can be developed under extremely weak hypotheses (thus assuming *neither* any choice of stress and strain measures *nor* any specific laws for strain decomposition, isotropy, convexity, existence of rate potentials, elastic law, type of hardening and form of yield function) and demonstrate that the rate equations retain a sort of 'universal structure' (which is, for instance, invariant to a change in the decomposition rule of deformation into elastic and plastic parts). It is precisely that universal structure of the elastoplastic constitutive laws which enters the treatment of bifurcation and instability, so many results are general and can be applied within a broad class of different constitutive laws.

We will slightly generalise Hill and Rice (1973) (see also Hill, 1978), avoiding the introduction of elastic potentials, and derive an abstract structure of constitutive equations which does not involve any choice of stress and strain measures or the decomposition rule of elastic and plastic deformations and also does not impose any requirement on the nature of elastic part of deformation (which can be hypoelastic, anisotropic, even non-symmetric), hardening rule (which can describe ideal plasticity, constant or variable hardening and softening), yield function (which can be pressure sensitive, closed or open, but has to be smooth), and plastic flow rule (associative or non-associative, with every possible kind of non-associativity and non-coaxiality). All our treatment is devoted to time independent and isothermal behaviour.[1] As a result, the formulation is so general that most of the time-independent large and small strain elastoplastic or elasto-damaging models are included as particular cases. We will finally present examples of elastoplastic formulations, namely, the theory developed by Simo (1988, based on the multiplicative decomposition of the deformation gradient and the von Mises yield function with kinematic hardening), and the infinitesimal theory of elastoplastic coupling (useful in the constitutive description of geomaterials; Hueckel, 1975, 1976; Hueckel and Maier, 1977; Maier and Hueckel, 1979, Dougill, 1976; Dafalias, 1977; Capurso, 1979; Gajo and Bigoni, 2008).

Elastoplastic, isothermal and time independent material behaviour of solids subject to large strain is described, which may at any stage of deformation exhibit a purely elastic response for appropriate loading. For these materials, elastic response is assumed to be a one-to-one relation between $\boldsymbol{T}^{(m)}$ and $\boldsymbol{E}^{(m)}$ depending on the prior inelastic history 'condensed' in the set of internal (or hidden) variables symbolically denoted with $\mathcal{K}$, that is,

$$\boldsymbol{T}^{(m)} = \hat{\boldsymbol{T}}^{(m)}(\boldsymbol{E}^{(m)}, \mathcal{K}), \qquad \boldsymbol{E}^{(m)} = \hat{\boldsymbol{E}}^{(m)}(\boldsymbol{T}^{(m)}, \mathcal{K}). \tag{8.23}$$

In particular, functionals $\hat{\boldsymbol{T}}^{(m)}$ and $\hat{\boldsymbol{E}}^{(m)}$ depend on the prior history of inelastic deformation through the unspecified set $\mathcal{K}$ of internal variables of generic tensorial nature (thus embracing second-order tensors and scalars), which may play the role of micro-structure descriptors, in a similar vein as the 'm.d.' of Eq. (6.1). Functionals (8.23) have to satisfy the general requirements of material frame indifference and indifference with respect to a rigid-body rotation of the reference configuration (see Chapter 6), but these restrictions need not be imposed at this stage.

For a purely elastic deformation rate, the set $\mathcal{K}$ remains fixed, and we can differentiate with respect to a time-like parameter governing the load to obtain

$$\dot{\boldsymbol{T}}^{(m)} = \mathbb{E}[\dot{\boldsymbol{E}}^{(m)}], \qquad \dot{\boldsymbol{E}}^{(m)} = \mathbb{M}[\dot{\boldsymbol{T}}^{(m)}], \tag{8.24}$$

[1] Adiabatic constitutive law easily may be included in our general treatment; see, for instance, Benallal and Bigoni (2004).

where

$$\mathbb{E}(\boldsymbol{E}^{(m)},\mathcal{K}) = \frac{\partial \hat{\boldsymbol{T}}^{(m)}}{\partial \boldsymbol{E}^{(m)}}, \qquad \mathbb{M}(\boldsymbol{T}^{(m)},\mathcal{K}) = \frac{\partial \hat{\boldsymbol{E}}^{(m)}}{\partial \boldsymbol{T}^{(m)}}, \tag{8.25}$$

and obviously

$$\mathbb{E} = \mathbb{M}^{-1}. \tag{8.26}$$

For a rate involving elastic and inelastic strain, we may write

$$\dot{\boldsymbol{T}}^{(m)} = \mathbb{E}[\dot{\boldsymbol{E}}^{(m)}] - \dot{\Lambda}\mathbb{E}[\boldsymbol{P}], \qquad \dot{\boldsymbol{E}}^{(m)} = \mathbb{M}[\dot{\boldsymbol{T}}^{(m)}] + \dot{\Lambda}\boldsymbol{P}, \tag{8.27}$$

where $\boldsymbol{P} \in \mathsf{Sym}$ is the *flow mode tensor* assumed to be independent of the strain rate and defined in such a way that

$$\dot{\Lambda}\boldsymbol{P} = -\mathbb{M}\left[\frac{\partial \hat{\boldsymbol{T}}^{(m)}}{\partial \mathcal{K}}[\dot{\mathcal{K}}]\right] = \frac{\partial \hat{\boldsymbol{E}}^{(m)}}{\partial \mathcal{K}}[\dot{\mathcal{K}}], \tag{8.28}$$

and the scalar $\dot{\Lambda} \geq 0$, called the *plastic multiplier*, is null for purely elastic rates when $\dot{\mathcal{K}} = 0$. We note that the plastic multiplier is the 'switch' to allow or not the possibility of plastic flow and related evolution of internal variables.

Note that $\boldsymbol{P}$ has been defined with reference to the $\boldsymbol{T}^{(m)}$ stress space, whereas $\mathbb{E}[\boldsymbol{P}]$ is its counterpart in the deformation space.

A yield function is assumed at each $\mathcal{K}$. This may be expressed alternatively in the stress space formulation (see Chapter 7)

$$f_{\boldsymbol{T}^{(m)}}(\boldsymbol{T}^{(m)},\mathcal{H}) \leq 0 \tag{8.29}$$

or in the strain space formulation (included now to retain the maximum of generality)

$$f_{\boldsymbol{E}^{(m)}}(\boldsymbol{E}^{(m)},\tilde{\mathcal{H}}) \leq 0, \tag{8.30}$$

where $\mathcal{H}$ and $\tilde{\mathcal{H}}$ symbolically denote generic collections of internal (or hidden) state variables ($\mathcal{H}$ and $\tilde{\mathcal{H}}$ often may coincide) of arbitrary tensorial nature, thus defining regions of the $\boldsymbol{T}^{(m)}$ or $\boldsymbol{E}^{(m)}$ space, respectively, within which the response is elastic. We always can switch from a stress state representation (8.29) to a strain state representation (8.30), and vice versa, using Eqs. (8.23), namely,

$$f_{\boldsymbol{T}^{(m)}}(\boldsymbol{T}^{(m)},\mathcal{H}) = f_{\boldsymbol{E}^{(m)}}(\hat{\boldsymbol{E}}^{(m)}(\boldsymbol{T}^{(m)},\mathcal{K}),\tilde{\mathcal{H}}), \tag{8.31}$$

Thus, introducing the (symmetric) *gradients of the yield functions in stress and strain spaces*, respectively, as

$$\boldsymbol{Q}_{\boldsymbol{T}^{(m)}} = \frac{\partial f_{\boldsymbol{T}^{(m)}}}{\partial \boldsymbol{T}^{(m)}}, \qquad \boldsymbol{Q}_{\boldsymbol{E}^{(m)}} = \frac{\partial f_{\boldsymbol{E}^{(m)}}}{\partial \boldsymbol{E}^{(m)}}, \tag{8.32}$$

we can apply the chain rule to Eq. (8.31) to obtain the connections between yield function gradients

$$\boldsymbol{Q}_{\boldsymbol{T}^{(m)}} = \mathbb{M}^T[\boldsymbol{Q}_{\boldsymbol{E}^{(m)}}], \qquad \boldsymbol{Q}_{\boldsymbol{E}^{(m)}} = \mathbb{E}^T[\boldsymbol{Q}_{\boldsymbol{T}^{(m)}}], \tag{8.33}$$

two expressions which will become useful later.

Note that Eqs. (8.32) make explicit the assumption that will be used throughout this book of *smooth yield surface*, so the corners are not included.

Since in our developments we will use the stress space version of the yield function gradient $\boldsymbol{Q}_{\boldsymbol{T}^{(m)}}$ more often than the strain space version $\boldsymbol{Q}_{\boldsymbol{E}^{(m)}}$, we usually will drop the subscript $\boldsymbol{T}^{(m)}$ from it, so we introduce the symbol $\boldsymbol{Q} \in \mathsf{Sym}$ with the meaning

$$\boldsymbol{Q} = \boldsymbol{Q}_{\boldsymbol{T}^{(m)}}, \tag{8.34}$$

a notation also giving evidence to the fact that both $\boldsymbol{P}$ and $\boldsymbol{Q}$ 'live' in the same stress space description.

Since the material response is elastic for states satisfying $f_{\boldsymbol{T}^{(m)}} < 0$ or $f_{\boldsymbol{E}^{(m)}} < 0$, we may write the so-called Kuhn-Tucker conditions

$$\dot{\Lambda} f = 0, \qquad \dot{\Lambda} \geq 0, \qquad f \leq 0, \tag{8.35}$$

where f is now the yield function defined in stress or strain space.

Roughly speaking, plastic loading may only occur when the stress (or strain) state 'remains' on the yield surface, while this is changing as a consequence of hardening evolution. More precisely, for a stress (or strain) state at yield, namely, satisfying $f = 0$, the Prager (1949) consistency condition requires that the stress point lie on the yield surface when the inelastic strain rate is different from zero, namely,

$$\dot{f}_{\boldsymbol{T}^{(m)}} = 0, \qquad \dot{f}_{\boldsymbol{E}^{(m)}} = 0 \qquad \text{and} \qquad \dot{\Lambda} > 0 \qquad \text{plastic loading,} \tag{8.36}$$

or the strain rate is purely elastic when the stress (or strain) point moves inside the yield surface, namely,

$$\dot{f}_{\boldsymbol{T}^{(m)}} < 0, \qquad \dot{f}_{\boldsymbol{E}^{(m)}} < 0 \qquad \text{and} \qquad \dot{\Lambda} = 0 \qquad \text{elastic unloading;} \tag{8.37}$$

Finally, there is also a thrid possibility, 'intermediate' to conditions (8.36) and (8.37), namely, in case of the so-called neutral loading, where

$$\dot{f}_{\boldsymbol{T}^{(m)}} = \dot{f}_{\boldsymbol{E}^{(m)}} = 0 \qquad \text{and} \qquad \dot{\Lambda} = 0, \qquad \text{neutral loading.} \tag{8.38}$$

In all these cases, the Kuhn-Tucker conditions hold

$$\dot{\Lambda} \dot{f}_{\boldsymbol{T}^{(m)}} = 0, \qquad \dot{\Lambda} \dot{f}_{\boldsymbol{E}^{(m)}} = 0, \qquad \dot{\Lambda} \geq 0, \qquad \dot{f}_{\boldsymbol{T}^{(m)}} \leq 0, \qquad \dot{f}_{\boldsymbol{E}^{(m)}} \leq 0. \tag{8.39}$$

Now, in the case of plastic loading [Eqs. (8.36)], we obtain

$$\begin{aligned} \dot{f}_{\boldsymbol{T}^{(m)}} &= \boldsymbol{Q} \cdot \dot{\boldsymbol{T}}^{(m)} + \frac{\partial f_{\boldsymbol{T}^{(m)}}}{\partial \mathcal{H}} \cdot \dot{\mathcal{H}} = 0, \\ \dot{f}_{\boldsymbol{E}^{(m)}} &= \boldsymbol{Q}_{\boldsymbol{E}^{(m)}} \cdot \dot{\boldsymbol{E}}^{(m)} + \frac{\partial f_{\boldsymbol{E}^{(m)}}}{\partial \tilde{\mathcal{H}}} \cdot \dot{\tilde{\mathcal{H}}} = 0. \end{aligned} \tag{8.40}$$

Introducing the *hardening modulus* h and the *plastic modulus* g as

$$\dot{\Lambda} h = -\frac{\partial f_{\boldsymbol{T}^{(m)}}}{\partial \mathcal{H}} \cdot \dot{\mathcal{H}} \qquad \text{and} \qquad \dot{\Lambda} g = -\frac{\partial f_{\boldsymbol{E}^{(m)}}}{\partial \tilde{\mathcal{H}}} \cdot \dot{\tilde{\mathcal{H}}}, \tag{8.41}$$

we note that Eqs. (8.33) and (8.27) permit us to find the relation between the two quantities

$$g = h + \boldsymbol{Q} \cdot \mathbb{E}[\boldsymbol{P}]. \tag{8.42}$$

In the following, *we will always assume the plastic modulus g to be strictly positive*. A null value of g would correspond to snap-back in the stress-strain curve, and negative values would correspond to the so-called subcritical behaviour not investigated here.[2]

Equations (8.40), through a substitution of Eq. $(8.27)_1$ or Eq. $(8.27)_2$, allow one to obtain the two expressions for the plastic multiplier:

$$\dot{\Lambda} = \frac{\boldsymbol{Q}_{\boldsymbol{E}^{(m)}} \cdot \dot{\boldsymbol{E}}^{(m)}}{g} = \frac{\boldsymbol{Q} \cdot \mathbb{E}[\dot{\boldsymbol{E}}^{(m)}]}{g} \tag{8.43}$$

or

$$\dot{\Lambda} = \frac{\boldsymbol{Q} \cdot \dot{\boldsymbol{T}}^{(m)}}{h} = \frac{\boldsymbol{Q}_{\boldsymbol{E}^{(m)}} \cdot \mathbb{M}[\dot{\boldsymbol{T}}^{(m)}]}{h}, \tag{8.44}$$

where the former, holding for $g > 0$, is more general than the latter, holding for $h > 0$.

We note that in the Hill (1967b) notation, the hardening modulus h in Eq. (8.41) describes *hardening* when positive, *softening* when negative and *perfect plasticity* when null. As Hill (1967b) remarks, hardening and softening are not measure-invariant concepts, in the sense that h depends on the choice of $\boldsymbol{T}^{(m)}$ and $\boldsymbol{E}^{(m)}$. Therefore, the nomenclature is, to some extent, arbitrary, except in the case of the infinitesimal theory. Moreover, we remark that in addition to h, also $\boldsymbol{Q}$, $\boldsymbol{P}$ and $\mathbb{E}$ are measure-dependent. On the contrary, the plastic modulus g can be shown to be measure-independent (Hill, 1967b; Petryk, 2000).

Analysing now elastic unloading, we may conclude from conditions (8.37) and (8.38) that

$$\boldsymbol{Q} \cdot \mathbb{E}[\dot{\boldsymbol{E}}^{(m)}] \leq 0, \quad \Longleftrightarrow \quad \dot{\Lambda} = 0 \tag{8.45}$$

so that Eqs. (8.43) and (8.46) yield

$$\dot{\Lambda} = \frac{< \boldsymbol{Q} \cdot \mathbb{E}[\dot{\boldsymbol{E}}^{(m)}] >}{g}, \tag{8.46}$$

where the operator $< \cdot >$ denotes the Macaulay brackets [Eqs. (2.103)].

A substitution of Eq. (8.46) into Eq. (8.27) yields the elastoplastic rate constitutive equations in the stress space representation,

$$\dot{\boldsymbol{T}}^{(m)} = \begin{cases} \mathbb{E}[\dot{\boldsymbol{E}}^{(m)}] & - \dfrac{1}{g} < \boldsymbol{Q} \cdot \mathbb{E}[\dot{\boldsymbol{E}}^{(m)}] > \mathbb{E}[\boldsymbol{P}] \quad \text{if } f_{\boldsymbol{T}^{(m)}}(\boldsymbol{T}^{(m)}, \mathcal{H}) = 0, \\ \mathbb{E}[\dot{\boldsymbol{E}}^{(m)}] & \quad \text{if } f_{\boldsymbol{T}^{(m)}}(\boldsymbol{T}^{(m)}, \mathcal{H}) < 0, \end{cases} \tag{8.47}$$

defining an incrementally non-linear rate law with two branches, the elastic and the elastoplastic.

We say that the Eqs. (8.47) follow the 'associative flow law' or the 'normality rule' when the operator associated with Eq. (8.47) always has the major symmetry, which means that (1) the elastic tensor $\mathbb{E}$ is hyperelastic (or Green elastic), so it possesses the major symmetry, and (2) the flow rule is associative $\boldsymbol{P} = \boldsymbol{Q}$.

Note that all quantities appearing in the rate equations (8.47) may fully depend on the entire path of deformation reckoned from some ground state.

[2] It has been proven by Bigoni and Zaccaria (1992a;1992b), and it will be shown in Chapter 11 that subcritical behaviour excludes strong ellipticity.

The scalar product of the first equation in Eqs. (8.47) with $\boldsymbol{Q}$ gives

$$\boldsymbol{Q}\cdot\dot{\boldsymbol{T}}^{(m)} = \boldsymbol{Q}\cdot\mathbb{E}[\dot{\boldsymbol{E}}^{(m)}] - \frac{\boldsymbol{Q}\cdot\mathbb{E}[\boldsymbol{P}]}{g} < \boldsymbol{Q}\cdot\mathbb{E}[\dot{\boldsymbol{E}}^{(m)}] > . \tag{8.48}$$

In the case where $h > 0$, we note that

$$\text{sign}(\boldsymbol{Q}\cdot\mathbb{E}[\dot{\boldsymbol{E}}^{(m)}]) = \text{sign}(\boldsymbol{Q}\cdot\dot{\boldsymbol{T}}^{(m)}).$$

Therefore, *assuming positive hardening*, $h > 0$, and using Eq. (8.48), we obtain the inverse rate constitutive equations

$$\dot{\boldsymbol{E}}^{(m)} = \begin{cases} \mathbb{M}[\dot{\boldsymbol{T}}^{(m)}] + \dfrac{1}{h} < \boldsymbol{Q}\cdot\dot{\boldsymbol{T}}^{(m)} > \boldsymbol{P} & \text{if } f(\boldsymbol{T}^{(m)},\mathcal{H}) = 0, \\ \mathbb{M}[\dot{\boldsymbol{T}}^{(m)}] & \text{if } f(\boldsymbol{T}^{(m)},\mathcal{H}) < 0. \end{cases} \tag{8.49}$$

It may be important to remark that *all possible choices of $\boldsymbol{T}^{(m)}$ and $\boldsymbol{E}^{(m)}$ in Eqs. (8.47) or Eqs. (8.49) are equivalent* and that *all resulting constitutive equations respect the requirement of material frame indifference* (6.3).

It is particularly convenient to write the constitutive equation (8.47) in terms of the material time derivative of the first Piola-Kirchhoff stress and of the deformation gradient. This can be done for any choice of $\boldsymbol{T}^{(m)}$ and $\boldsymbol{E}^{(m)}$. Considering once more the second Piola-Kirchhoff stress tensor $\boldsymbol{T}^{(2)}$ and the Green-Lagrange strain tensor $\boldsymbol{E}^{(2)}$, and keeping into account the form of the elastic law (6.110), we conclude that Eqs. (8.47) can be written as

$$\dot{\boldsymbol{S}} = \begin{cases} \mathbb{G}[\dot{\boldsymbol{F}}] - \dfrac{1}{g} < (\boldsymbol{F}^{-T}\boldsymbol{Q})\cdot\mathbb{B}[\dot{\boldsymbol{F}}] > \mathbb{B}[\boldsymbol{F}^{-T}\boldsymbol{P}] & \text{if } f_{\boldsymbol{S}}(\boldsymbol{S},\mathcal{H}) = 0, \\ \mathbb{G}[\dot{\boldsymbol{F}}] & \text{if } f_{\boldsymbol{S}}(\boldsymbol{S},\mathcal{H}) < 0, \end{cases} \tag{8.50}$$

where $\mathbb{B}$ and $\mathbb{G}$ are given by Eqs. (6.111), and the yield function $f_{\boldsymbol{S}}(\boldsymbol{S},\mathcal{H})$ in Eqs. (8.50) has been expressed in terms of the first Piola-Kirchhoff stress tensor, so its gradient with respect to $\boldsymbol{S}$ is $\boldsymbol{F}^{-T}\boldsymbol{Q}$. This follows from the chain rule of differentiation

$$\frac{\partial f_{\boldsymbol{S}}}{\partial \boldsymbol{S}}\cdot\boldsymbol{A} = \frac{\partial f_{\boldsymbol{T}^{(2)}}}{\partial \boldsymbol{T}^{(2)}}\cdot\frac{\partial \boldsymbol{T}^{(2)}}{\partial \boldsymbol{S}}[\boldsymbol{A}] = \boldsymbol{F}^{-T}\boldsymbol{Q}\cdot\boldsymbol{A} \tag{8.51}$$

for every $\boldsymbol{A} \in \mathsf{Lin}$. When the yield criterion is satisfied, $f_{\boldsymbol{S}}(\boldsymbol{S},\mathcal{H}) = 0$, the constitutive equations (8.50) define a piece-wise linear function of $\dot{\boldsymbol{F}}$

$$\dot{\boldsymbol{S}} = \mathcal{C}(\dot{\boldsymbol{F}}), \tag{8.52}$$

which is *an incrementally non-linear constitutive equation with two branches corresponding to plastic loading and elastic unloading*. This constitutive equation may be written in the following form, useful for subsequent analysis:

$$\dot{\boldsymbol{S}} = \mathbb{G}[\dot{\boldsymbol{F}}] - \frac{1}{g} < \boldsymbol{N}\cdot\dot{\boldsymbol{F}} > \boldsymbol{M}, \tag{8.53}$$

where

$$\boldsymbol{N} = \mathbb{B}^T[\boldsymbol{F}^{-T}\boldsymbol{Q}] \in \mathsf{Lin} \qquad \text{and} \qquad \boldsymbol{M} = \mathbb{B}[\boldsymbol{F}^{-T}\boldsymbol{P}] \in \mathsf{Lin} \tag{8.54}$$

are the yield surface and plastic potential normals, respectively, in the strain space.[3] In particular, expressing the yield function in terms of the deformation gradient, $f_{\boldsymbol{F}}(\boldsymbol{F},\mathcal{H})$, its gradient with respect to $\boldsymbol{F}$ is exactly $\boldsymbol{N}$; in fact,

$$\frac{\partial f_{\boldsymbol{F}}}{\partial \boldsymbol{F}} \cdot \boldsymbol{A} = \frac{\partial f_{\boldsymbol{T}^{(2)}}}{\partial \boldsymbol{T}^{(2)}} \cdot \frac{\partial \boldsymbol{T}^{(2)}}{\partial \boldsymbol{E}^{(2)}} \left[\frac{\partial \boldsymbol{E}^{(2)}}{\partial \boldsymbol{F}} [\boldsymbol{A}] \right] = \mathbb{B}^T [\boldsymbol{F}^{-T} \boldsymbol{Q}] \cdot \boldsymbol{A} \tag{8.55}$$

for every $\boldsymbol{A} \in \mathsf{Lin}$. Note also that $\boldsymbol{P} = \boldsymbol{Q}$ is equivalent to $\boldsymbol{M} = \boldsymbol{N}$ only when $\mathbb{B}$ has the major symmetry, that is, for Green elasticity. In other words, simultaneous normality in stress and strain spaces is not assured for Cauchy elasticity (Hill, 1978).

In the following, we will refer to *associative flow rule* or *normality* when $\boldsymbol{P} = \boldsymbol{Q}$ *and* $\mathbb{B}$ *has the major symmetry* so that normality is preserved in both the strain and the stress spaces and $\boldsymbol{M} = \boldsymbol{N}$.

The constitutive equations (8.50) can be transformed in a spatial setting, transforming the yield function as a function of the Kirchhoff stress

$$f_{\boldsymbol{T}^{(2)}}(\boldsymbol{F}^{-1}\boldsymbol{K}\boldsymbol{F}^{-T},\mathcal{H}) = f_{\boldsymbol{K}}(\boldsymbol{K},\mathcal{H}), \tag{8.56}$$

noting that the chain rule applied to the yield function gives

$$\boldsymbol{Q}_{\boldsymbol{K}} = (\boldsymbol{F} \boxtimes \boldsymbol{F})^{-T} \boldsymbol{Q}, \qquad \boldsymbol{Q}_{\boldsymbol{K}} = \frac{\partial f_{\boldsymbol{K}}(\boldsymbol{K},\mathcal{H})}{\partial \boldsymbol{K}} \tag{8.57}$$

and employing Eqs. (6.115) and (6.116) to finally obtain

$$\overset{\circ}{\boldsymbol{K}} = \begin{cases} \mathbb{H}[\boldsymbol{D}] - \dfrac{1}{g} < (\boldsymbol{Q}_{\boldsymbol{K}}) \cdot \mathbb{H}[\boldsymbol{D}] > \mathbb{H}[\boldsymbol{P}_{\boldsymbol{K}}] & \text{if } f_{\boldsymbol{K}}(\boldsymbol{K},\mathcal{H}) = 0, \\ \mathbb{H}[\boldsymbol{D}] & \text{if } f_{\boldsymbol{K}}(\boldsymbol{K},\mathcal{H}) < 0, \end{cases} \tag{8.58}$$

where $\overset{\circ}{\boldsymbol{K}}$ is the Oldroyd derivative of the Kirchhoff stress $(3.156)_2$, the fourth-order elastic tensor $\mathbb{H}$ is defined by Eq. (6.116) and

$$\boldsymbol{P}_{\boldsymbol{K}} = (\boldsymbol{F} \boxtimes \boldsymbol{F})^{-T} \boldsymbol{P}. \tag{8.59}$$

As far as the choice of $\boldsymbol{P}$ as related to $\boldsymbol{Q}$ is concerned, this is to some extent arbitrary. However, experiments show that many materials exhibit a peculiar kind of non-associativity involving only the volumetric part of plastic deformation. This case of special interest corresponds to so-called deviatoric associativity, where the deviatoric parts of $\mathbf{P}$ and $\mathbf{Q}$ are aligned. This may be defined generically as

$$\mathbf{P} = \chi_1 \hat{\mathbf{S}} + \frac{\chi_2}{3} \boldsymbol{I} \quad \text{and} \quad \mathbf{Q} = \psi_1 \hat{\mathbf{S}} + \frac{\psi_2}{3} \boldsymbol{I}, \tag{8.60}$$

where $\hat{\mathbf{S}} \in \mathsf{Sym}$ is traceless, and χ_1 and ψ_1 are assumed strictly positive. The parameters ψ_2 and χ_2, respectively, describe the *pressure sensitivity* and the *dilatancy* (when

[3] The 'plastic potential' is defined as a smooth potential providing the flow mode tensor with its gradient. If, on the one hand, this notion need not to be introduced, it is often useful to 'visualize' $\boldsymbol{P}$ as the normal to a potential in the same way that $\boldsymbol{Q}$ can be represented as the normal to the yield surface. Obviously, the plastic potential coincides with the yield function for associative flow rules.

$\chi_2 > 0$) or *contractility* (when $\chi_2 < 0$) of the material. In the case of the Drucker-Prager model (7.36) with the gradient (7.38) and the flow used rule non-associativity, among many others, by Bigoni and Loret (1999), we have

$$\chi_1 = \cos\chi, \qquad \chi_2 = \sqrt{3}\sin\chi, \qquad \psi_1 = \cos\phi, \qquad \psi_2 = \sqrt{3}\sin\phi, \qquad \hat{\mathsf{S}} = \frac{\mathsf{dev}\boldsymbol{T}}{|\mathsf{dev}\boldsymbol{T}|}, \tag{8.61}$$

whereas in the *equivalent* but slightly different formulation[4] by Rudnicki and Rice (1975) we have

$$\chi_1 = \psi_1 = \frac{1}{2\sqrt{J_2}}, \qquad \hat{\mathsf{S}} = \mathsf{dev}\boldsymbol{T}, \tag{8.62}$$

where J_2 is the second invariant of deviatoric stress [Eq. (7.12)]. Note that the parameters ψ_1 and ψ_2 and χ_1 and χ_2 are expressed in terms of *friction angle* ϕ and *dilatancy angle* χ. The non-associativity of the flow rule allows us to introduce a key ingredient of the behaviour of granular materials, namely, that the friction angle is usually different from the dilatancy angle. The reason for the term 'dilatancy' is that the trace of the flow mode tensor

$$\mathsf{tr}\boldsymbol{P} = \chi_2 = \sqrt{3}\sin\chi \tag{8.63}$$

determines the amount of the volumetric deformation of the material, so its sign is positive for dilatant behaviour and negative for contractant behaviour. Finally, $\mathsf{tr}\boldsymbol{P} = 0$ for isochoric plastic flow.

Note finally that deviatoric associativity [Eqs. (8.60)], implies that the product $\boldsymbol{QP}$ commutes, so $\boldsymbol{Q}$ and $\boldsymbol{P}$ are coaxial [Eq. (2.31)].

8.2.1 The small strain theory recovered

In the special case of the infinitesimal theory, Eqs. (8.23) simplify to

$$\boldsymbol{T} = \mathbb{E}[\boldsymbol{E} - \boldsymbol{E}_p] \qquad \text{and} \qquad \boldsymbol{E} = \mathbb{M}[\boldsymbol{T}] + \boldsymbol{E}_p, \tag{8.64}$$

where $\boldsymbol{E} = \frac{1}{2}[\mathsf{grad}\,\boldsymbol{u} + (\mathsf{grad}\,\boldsymbol{u})^T]$ is the infinitesimal strain tensor, with $\boldsymbol{E}_p$ being its plastic part. The fourth-order tensor $\mathbb{E} = \mathbb{M}^{-1}$ is now the elasticity tensor of the small strain theory,[5] which in the well-known case of isotropy is

$$\mathbb{E} = \lambda \boldsymbol{I} \otimes \boldsymbol{I} + 2\mu \boldsymbol{I} \underline{\overline{\otimes}} \boldsymbol{I}, \tag{8.65}$$

where $\boldsymbol{I} \underline{\overline{\otimes}} \boldsymbol{I}$ is symmetrising fourth-order tensor (2.92), and λ and μ are the Lamé constants. In the stress space formulation, the yield function can be written as $f(\boldsymbol{T}, \mathcal{H}) \leq 0$, where $\mathcal{H}$ is a generic set of internal variables, possibly depending on $\boldsymbol{E}_p$. Constitutive equations (8.52) therefore become [in the case $f(\boldsymbol{T}, \mathcal{H}) = 0$] in the form (8.16)

$$\dot{\boldsymbol{T}} = \mathbb{E}[\boldsymbol{D}] - \frac{1}{g} < \boldsymbol{Q} \cdot \mathbb{E}[\boldsymbol{D}] > \mathbb{E}[\boldsymbol{P}], \tag{8.66}$$

[4] An inconsequential modification in the definition of hardening modulus is sufficient to show the equivalency between the two models.

[5] Usually $\mathbb{E}$ is constant. In the case of elastoplastic coupling, which will be introduced later, $\mathbb{E}$ is assumed to depend on $\boldsymbol{E}_p$ (Hueckel, 1976). For the moment, we take it constant.

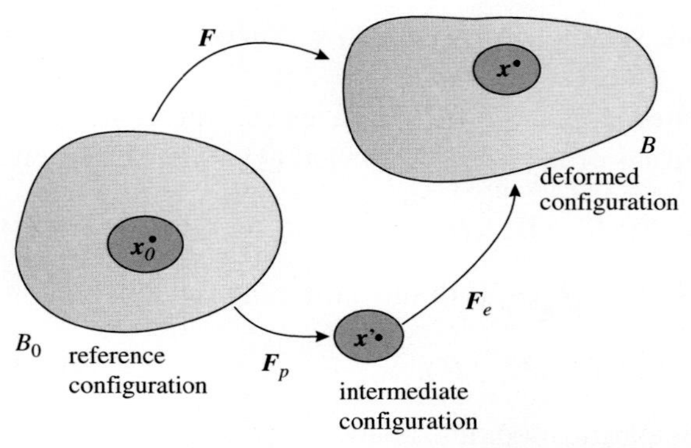

Figure 8.5. Reference, current and intermediate configurations employed when the multiplicative decomposition of the deformation gradient into an elastic and a plastic part is employed.

where the plastic modulus g has been introduced, assumed strictly positive and defined by Eq. (8.42), in which h is the hardening modulus, positive for strain hardening, null for ideal plastic behaviour and negative for strain softening (see Fig. 8.3).

Note that $\boldsymbol{Q}\cdot\mathbb{E}[\boldsymbol{D}]=\boldsymbol{D}\cdot\mathbb{E}^T[\boldsymbol{Q}]$; therefore, the tangent operator is symmetric for the associative flow rule if and only if $\mathbb{E}$ has the major symmetry. Equations (8.66) are identical to Eqs. (8.16) because $\dot{\boldsymbol{E}}$ has the same meaning of $\boldsymbol{D}$ under the small strain assumption.

8.2.2 A theory of elastoplasticity based on multiplicative decomposition of the deformation gradient

It may be instructive to note that the elastoplastic formulation at finite strain proposed by Simo (1988) falls within the given constitutive framework. In particular, Simo (1988) assumed the multiplicative decomposition of the deformation gradient (Lee, 1969; Willis, 1969; Fig. 8.5), that is,

$$\boldsymbol{F}=\boldsymbol{F}_e\boldsymbol{F}_p, \tag{8.67}$$

where $\boldsymbol{F}_e$ and $\boldsymbol{F}_p$ denote the elastic and plastic parts, respectively. The plastic Green-Lagrange strain tensor [Eq. $(3.36)_1$] is defined accordingly:

$$\boldsymbol{E}_p^{(2)}=\frac{1}{2}\left(\boldsymbol{F}_p^T\boldsymbol{F}_p-\boldsymbol{I}\right), \tag{8.68}$$

and a strain energy function is assumed, providing the second Piola-Kirchhoff stress as

$$\boldsymbol{T}^{(2)}=\frac{\partial W(\boldsymbol{E}^{(2)},\boldsymbol{E}_p^{(2)})}{\partial \boldsymbol{E}^{(2)}}, \tag{8.69}$$

which in our notation means that the set $\mathcal{K}$ reduces to the tensor $\boldsymbol{E}_p^{(2)}$.

The elastic fourth-order tensor therefore becomes

$$\mathbb{E}=\frac{\partial^2 W(\boldsymbol{E}^{(2)},\boldsymbol{E}_p^{(2)})}{\partial \boldsymbol{E}^{(2)}\,\partial \boldsymbol{E}^{(2)}}, \tag{8.70}$$

endowed with the minor and major symmetries, so we do not need to distinguish between $\mathbb{E}$ and $\mathbb{E}^T$.

Simo (1988) used a yield function defined in the strain space, where the set $\mathcal{H}$ reduces to the tensor $\boldsymbol{E}_p^{(2)}$ and a second-order tensor $\boldsymbol{H}$ (called $\boldsymbol{Q}$ by Simo), namely,

$$f_{\boldsymbol{E}^{(2)}} = f(\boldsymbol{E}^{(2)}, \boldsymbol{E}_p^{(2)}, \boldsymbol{H}), \tag{8.71}$$

and uses the associative flow rule $\mathbb{E}[\boldsymbol{P}] = \boldsymbol{Q}_{\boldsymbol{E}^{(2)}}$, with the hardening law

$$\dot{\boldsymbol{H}} = \dot{\Lambda}\,\hat{\boldsymbol{H}}(\boldsymbol{E}^{(2)}, \boldsymbol{E}_p^{(2)}, \boldsymbol{H}). \tag{8.72}$$

Thus, following our framework, the plastic modulus satisfies

$$\dot{\Lambda} g = \dot{\Lambda}\,\frac{\partial f_{\boldsymbol{E}^{(2)}}}{\partial \boldsymbol{H}} \cdot \hat{\boldsymbol{H}}(\boldsymbol{E}^{(2)}, \boldsymbol{E}_p^{(2)}, \boldsymbol{H}) + \frac{\partial f_{\boldsymbol{E}^{(2)}}}{\partial \boldsymbol{E}_p^{(2)}} \cdot \dot{\boldsymbol{E}}_p^{(2)}. \tag{8.73}$$

From Eq. $(8.27)_1$ we note that

$$-\dot{\Lambda}\,\boldsymbol{Q}_{\boldsymbol{E}^{(2)}} = \frac{\partial W(\boldsymbol{E}^{(2)}, \boldsymbol{E}_p^{(2)})}{\partial \boldsymbol{E}^{(2)}\,\partial \boldsymbol{E}_p^{(2)}}[\dot{\boldsymbol{E}}_p^{(2)}], \tag{8.74}$$

and so we finally obtain the plastic modulus in the form

$$g = \frac{\partial f_{\boldsymbol{E}^{(2)}}}{\partial \boldsymbol{H}} \cdot \hat{\boldsymbol{H}}(\boldsymbol{E}^{(2)}, \boldsymbol{E}_p^{(2)}, \boldsymbol{H}) - \frac{\partial f_{\boldsymbol{E}^{(2)}}}{\partial \boldsymbol{E}_p^{(2)}} \cdot \left(\frac{\partial W(\boldsymbol{E}^{(2)}, \boldsymbol{E}_p^{(2)})}{\partial \boldsymbol{E}^{(2)}\,\partial \boldsymbol{E}_p^{(2)}} \right)^{-1} [\boldsymbol{Q}_{\boldsymbol{E}^{(2)}}], \tag{8.75}$$

the same expression provided by Simo (1988).

As a conclusion, the model fits in our constitutive framework, so we can directly write down the specific form of rate constitutive Eqs. (8.47) based on the previous assumptions and formulated in the strain space, that is,

$$\dot{\boldsymbol{T}}^{(2)} = \begin{cases} \mathbb{E}[\dot{\boldsymbol{E}}^{(2)}] - \dfrac{1}{g} < \boldsymbol{Q}_{\boldsymbol{E}^{(2)}} \cdot \dot{\boldsymbol{E}}^{(2)} > \boldsymbol{Q}_{\boldsymbol{E}^{(2)}} & \text{if } f(\boldsymbol{E}^{(2)}, \boldsymbol{E}_p^{(2)}, \boldsymbol{H}) = 0, \\ \mathbb{E}[\dot{\boldsymbol{E}}^{(2)}] & \text{if } f(\boldsymbol{E}^{(2)}, \boldsymbol{E}_p^{(2)}, \boldsymbol{H}) < 0, \end{cases} \tag{8.76}$$

which provide the same tangent operator obtained by Simo [1988, his Eq. (1.37)].

At this point, we can re-formulate Eq. (8.76) in terms of material time derivative of the first Piola-Kirchhoff stress, through a direct use of Eq. (8.50) or go directly to the spatial representation (8.58). Following the latter way, we note that Simo (1988) used the so-called Finger strain tensor, in other words, the Eulerian strain measure that in our notation is denoted by $\boldsymbol{G}^{-2}$ [see Eq. (3.31)], namely,

$$\boldsymbol{G}^{(-2)} = -\frac{1}{2}(\boldsymbol{V}^{-2} - \boldsymbol{I}). \tag{8.77}$$

The plastic strain tensor $\boldsymbol{G}_p^{(-2)}$ and the elastic strain tensor $\boldsymbol{G}_e^{(-2)}$ are *defined* as

$$\boldsymbol{G}_p^{(-2)} = \boldsymbol{G}^{(-2)} - \boldsymbol{G}_e^{(-2)} \quad \text{and} \quad \boldsymbol{G}_e^{(-2)} = \frac{1}{2}\left(\boldsymbol{I} - \boldsymbol{B}_e^{-1}\right), \tag{8.78}$$

so the following relations hold

$$\boldsymbol{E}_p^{(2)} = \boldsymbol{F}^T \boldsymbol{G}_p^{(-2)} \boldsymbol{F} \quad \text{and} \quad \boldsymbol{E}^{(2)} = \boldsymbol{F}^T \boldsymbol{G}^{(-2)} \boldsymbol{F}. \tag{8.79}$$

Employing Eqs. (8.79), we note that the strain energy and the yield function may be transformed as

$$W(\boldsymbol{E}^{(2)},\boldsymbol{E}_p^{(2)}) = W(\boldsymbol{F}^T\boldsymbol{G}^{(-2)}\boldsymbol{F},\boldsymbol{F}^T\boldsymbol{G}_p^{(-2)}\boldsymbol{F}) = \hat{W}(\boldsymbol{G}^{(-2)},\boldsymbol{G}_p^{(-2)},\boldsymbol{F}),$$

$$f(\boldsymbol{E}^{(2)},\boldsymbol{E}_p^{(2)}) = f(\boldsymbol{F}^T\boldsymbol{G}^{(-2)}\boldsymbol{F},\boldsymbol{F}^T\boldsymbol{G}_p^{(-2)}\boldsymbol{F}) = \hat{f}(\boldsymbol{G}^{(-2)},\boldsymbol{G}_p^{(-2)},\boldsymbol{F}), \tag{8.80}$$

so applying the chain rule and employing the definition of Kirchhoff stress (3.108), we obtain

$$\boldsymbol{K} = \frac{\partial \hat{W}}{\partial \boldsymbol{G}^{(-2)}}, \qquad \mathbb{H} = \frac{\partial^2 \hat{W}}{\partial \boldsymbol{G}^{(-2)}\,\partial \boldsymbol{G}^{(-2)}}, \qquad \boldsymbol{Q}_{\boldsymbol{G}^{(-2)}} = \frac{\partial \hat{f}}{\partial \boldsymbol{G}^{(-2)}} = \boldsymbol{F}\boldsymbol{Q}_{\boldsymbol{E}^{(2)}}\boldsymbol{F}^T, \tag{8.81}$$

where $\mathbb{H}$ and $\mathbb{E}$ given by Eq. (8.70) are related through Eq. (6.116).

The constitutive equation (8.58) therefore becomes

$$\overset{\circ}{\boldsymbol{K}} = \begin{cases} \mathbb{H}[\boldsymbol{D}] - \dfrac{1}{g} < \boldsymbol{D}\cdot\boldsymbol{Q}_{\boldsymbol{G}^{(-2)}} > \boldsymbol{Q}_{\boldsymbol{G}^{(-2)}} & \text{if } \hat{f}(\boldsymbol{G}^{(-2)},\boldsymbol{G}_p^{(-2)},\boldsymbol{F}) = 0, \\ \mathbb{H}[\boldsymbol{D}] & \text{if } \hat{f}(\boldsymbol{G}^{(-2)},\boldsymbol{G}_p^{(-2)},\boldsymbol{F}) < 0, \end{cases} \tag{8.82}$$

where g is the transformation of expression (8.75) to the spatial setting, thus providing the same tangent operator obtained by Simo (1988, his table 2).

8.2.3 A simple constitutive model for granular materials evidencing flutter instability

We refer to the model proposed by Bigoni and Petryk (2002) as a large strain version of that employed by Bigoni and Loret (1999) [see also Bigoni (1995) and Bigoni and Zaccaria (1994a)]. The model is the piece-wise linear elastoplastic constitutive equation (8.58), which can be written in the form (8.50) or in the form (8.47) with $m = 2$, but with simplified fourth-order elastic tensors $\mathbb{H}$ and $\mathbb{B}$. In particular, we refer to a relative Lagrangean description of the type (8.50), so the elastic part of the constitutive equation is given by Eq. (6.130). We consider two versions of the model (8.50) that, when the stress satisfies the yield condition, can both be written in the form

$$\dot{\boldsymbol{S}} = \mathbb{E}[\boldsymbol{D}] + \boldsymbol{L}\boldsymbol{T} - \frac{1}{g} < \boldsymbol{D}\cdot\mathbb{E}[\boldsymbol{Q}] > \mathbb{E}[\boldsymbol{P}], \tag{8.83}$$

where the Drucker-Prager yield function (7.36) is assumed, so its gradient $\boldsymbol{Q}$ and the plastic mode tensor $\boldsymbol{P}$ are obtained from Eqs. (8.60) with one of the equivalent choices represented by Eqs. (8.61) and (8.62), depending on the specific convenience.

In one version of the model, the elastic fourth-order tensor $\mathbb{E}$ assumes the isotropic form (6.195) of the infinitesimal isotropic theory (with Lamé constants λ and μ), whereas in the other version, the elastic fourth-order tensor $\mathbb{E}$ takes the particular anisotropic elastic law (6.196) used in the flutter and strain localisation analyses by Bigoni and Loret (1999), Bigoni et al. (2000) and Piccolroaz et al. (2006a, 2006b).

Note that Rudnicki and Rice (1975), Rice (1977) and Rice and Rudnicki (1980) have employed a model similar to Eqs. (8.58), with an isotropic fourth-order elastic tensor (6.195), but with the Oldroyd derivative $(3.156)_2$ of the Kirchhoff stress

$\overset{\circ}{\boldsymbol{K}}$ replaced by the Jaumann derivative $(3.156)_1$ of the Cauchy stress $\overset{\nabla}{\boldsymbol{T}}$. The elastoplastic branch of the resulting constitutive equation again can be written in a form similar to Eq. (8.83), involving the material time derivative of the first Piola-Kirchhoff stress and of the deformation gradient, which need not be reported here.

8.2.4 Elastoplastic coupling in the modelling of granular materials and geomaterials

Key points in the constitutive description of the behaviour of granular materials and geomaterials are that the elastic stiffness of the material does not remain constant and that there is an elastic anisotropy evolving during strain. Both these effects are partially related to the development of plastic strain and partially to the evolution of the stress state.

To explain this point with an example, we refer to the ready-to-press alumina powder modeled by Piccolroaz et al. (2006a, 2006b). When subject to increasing uniaxial strain, the grains of this material initially re-arrange their positions, (Fig. 8.6*a*), but after a certain threshold pressure has been overcome, they begin to deform plastically (Fig. 8.6*b*). Induced anisotropy begins to develop even in early stages of the compaction process (because contact between grains is promoted orthogonal to the direction of the principal compressive stresses), but it becomes particularly evident when grains suffer plastic flow. It is in fact easy to conclude that the densely packed structure shown in the inset of Fig. 8.6*b* must be related to an increased elastic stiffness and induced elastic orthotropy of the material when compared with the inset of Fig. 8.6*a*. The increase in stiffness with the development of plastic strain is confirmed by the load versus displacement curve referred to an uniaxial deformation test and reported in Fig. 8.6*c*. In particular, the increase in elastic stiffness is related to the inclination of the hysteresis loops at unloading/reloading, becoming steeper at increasing strain.

The mechanism of progressive stiffening and anisotropisation owing to plastic deformation of grains is particularly easy to be envisaged in the preceding example. However, this mechanism does not exist at low confining pressure for granular materials with stiff grains (e.g., sand) because the grains do not deform plastically. Nevertheless, these materials still present a strong stress- and plastic strain–induced stiffening and anisotropisation. In addition, they exhibit at relatively high stress a reverse phenomenon of stiffening degradation (see Gajo and Bigoni, 2008). It has been shown by Gajo et al. (2004, 2007) that induced anisotropy is a main ingredient in the constitutive description of granular materials to quantitatively explain strain localisation phenomena. For this reason, we feel that it is appropriate to present a brief introduction to explain how to incorporate stress and plastic strain dependence of elastic properties, namely, by introducing the concept of elastoplastic coupling. The main idea is that the plastic deformation may influence the elastic properties of the material, a feature already implicit in Eq. (8.23), which has been applied to geomaterials by Hueckel (Hueckel, 1975, 1976; Hueckel and Maier, 1977; Maier and Hueckel, 1979; see also Dougill, 1976; Dafalias, 1977; Capurso, 1979) and recently re-formulated by Gajo and Bigoni (2008).

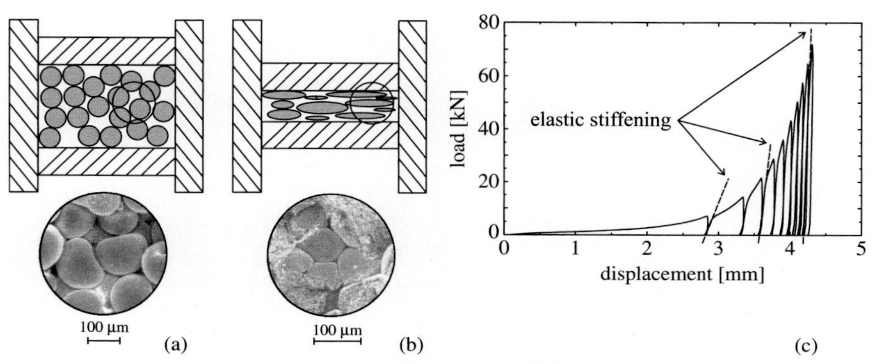

Figure 8.6. Uniaxial strain compaction of a ready-to-press alumina powder. After a certain pressure has been overcome, plastic deformation of the grains begins, so the densely packed structure shown in inset (*b*) is obtained from the randomly distributed and loose arrangement of grains shown in inset (*a*). The densely packed structure is elastically orthotropic and stiffer than the powder in the loose state. The load versus displacement curve (*c*) displays an increasing slope of the loading/reloading cycles, an indication of the progressive elastic stiffening connected with the development of plastic strain.

We will refer for simplicity to dry granular materials (where effects related to interstitial fluids often present in geomaterials are not a concern), and we will limit the presentation to the infinitesimal theory so that all large strain effects will be neglected (a model in which elastoplastic coupling is generalised to large strain has been presented by Piccolroaz et al., 2006b).

We begin by assuming an additive decomposition of infinitesimal strain into an elastic and plastic part

$$\boldsymbol{E} = \boldsymbol{E}_e + \boldsymbol{E}_p, \tag{8.84}$$

and we remark that now plastic strain is defined only on compete unloading, whereas incrementally we will distinguish between a plastic rate $\dot{\boldsymbol{E}}_p$ and an inelastic rate $\dot{\boldsymbol{E}}_i$ to be defined later.

Using Eq. (8.84), we may rewrite Eq. (8.23) in terms of plastic and elastic strains as

$$\boldsymbol{T} = \hat{\boldsymbol{T}}\left(\boldsymbol{E}_p, \boldsymbol{E}_e\right) \qquad \text{and} \qquad \boldsymbol{E}_e = \hat{\boldsymbol{E}}_e\left(\boldsymbol{E}_p, \boldsymbol{T}\right), \tag{8.85}$$

postulating that the Cauchy stress $\boldsymbol{T}$ (or the elastic strain $\boldsymbol{E}_e$) is a prescribed function of the plastic deformation $\boldsymbol{E}_p$ and of the elastic strain $\boldsymbol{E}_e$ (or of the stress $\boldsymbol{T}$).

Taking the rate of Eqs. (8.85) and using the rate form of Eq. (8.84), we obtain

$$\dot{\boldsymbol{T}} = \frac{\partial \hat{\boldsymbol{T}}}{\partial \boldsymbol{E}_e}[\dot{\boldsymbol{E}}] - \left(\frac{\partial \hat{\boldsymbol{T}}}{\partial \boldsymbol{E}_e} - \frac{\partial \hat{\boldsymbol{T}}}{\partial \boldsymbol{E}_p}\right)[\dot{\boldsymbol{E}}_p] \qquad \text{and} \qquad \dot{\boldsymbol{E}} = \frac{\partial \hat{\boldsymbol{E}}_e}{\partial \boldsymbol{T}}[\dot{\boldsymbol{T}}] + \left(\frac{\partial \hat{\boldsymbol{E}}_e}{\partial \boldsymbol{E}_p} + \boldsymbol{I}\underline{\otimes}\boldsymbol{I}\right)[\dot{\boldsymbol{E}}_p]. \tag{8.86}$$

Note that a substitution of Eq. $(8.85)_1$ in Eq. $(8.85)_2$ yields the condition

$$\boldsymbol{E}_e = \hat{\boldsymbol{E}}_e\Big(\boldsymbol{E}_p, \hat{\boldsymbol{T}}\left(\boldsymbol{E}_p, \boldsymbol{E}_e\right)\Big), \tag{8.87}$$

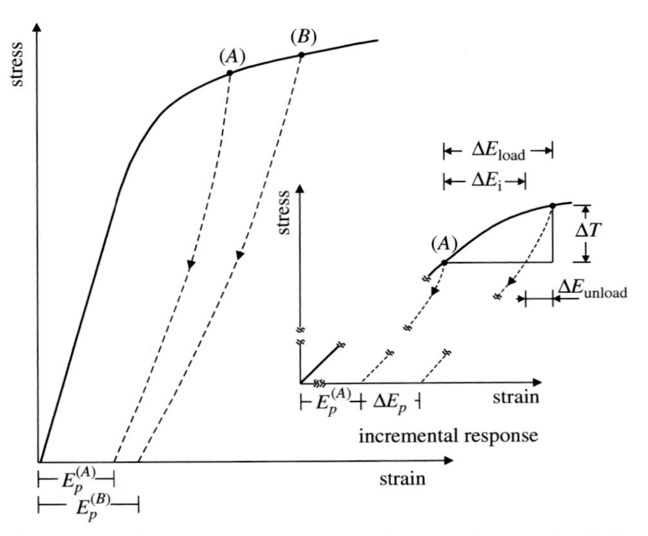

Figure 8.7. Elastoplastic coupling with a non-linear elastic law.

which, differentiated with respect to $\boldsymbol{E}_e$ and $\boldsymbol{E}_p$, and through use of the chain rule, provides the condition

$$\frac{\partial \hat{\boldsymbol{E}}_e}{\partial \hat{\boldsymbol{T}}} \frac{\partial \hat{\boldsymbol{T}}}{\partial \boldsymbol{E}_e} = \boldsymbol{I} \underline{\otimes} \boldsymbol{I} \qquad \text{and} \qquad \frac{\partial \hat{\boldsymbol{E}}_e}{\partial \boldsymbol{E}_p} = -\frac{\partial \hat{\boldsymbol{E}}_e}{\partial \hat{\boldsymbol{T}}} \frac{\partial \hat{\boldsymbol{T}}}{\partial \boldsymbol{E}_p}, \tag{8.88}$$

showing that the two expressions $(8.86)_1$ and $(8.86)_2$ are equivalent.

Let us consider a small stress cycle of loading and unloading from a state at yield, defined by $\{\boldsymbol{T}, \boldsymbol{E}_p\}$, in which the stress is first incremented and subsequently decremented by $\Delta \boldsymbol{T}$ (Fig. 8.7), thus producing an increment $\Delta \boldsymbol{E}_p$ in the plastic deformation. Expanding Eq. $(8.85)_2$ for loading and unloading, we obtain respectively:

$$\begin{aligned} \Delta \boldsymbol{E}_{\text{load}} &\sim \left.\frac{\partial \hat{\boldsymbol{E}}_e}{\partial \boldsymbol{T}}\right|_{\{\boldsymbol{T},\boldsymbol{E}_p\}} [\Delta \boldsymbol{T}] + \left(\left.\frac{\partial \hat{\boldsymbol{E}}_e}{\partial \boldsymbol{E}_p}\right|_{\{\boldsymbol{T},\boldsymbol{E}_p\}} + \boldsymbol{I} \underline{\otimes} \boldsymbol{I} \right) [\Delta \boldsymbol{E}_p], \\ \Delta \boldsymbol{E}_{\text{unload}} &\sim \left.\frac{\partial \hat{\boldsymbol{E}}_e}{\partial \boldsymbol{T}}\right|_{\{\boldsymbol{T},\boldsymbol{E}_p\}} [-\Delta \boldsymbol{T}], \end{aligned} \tag{8.89}$$

showing that *for an infinitesimal loading/unloading cycle, the irreversible strain rate* $\dot{\boldsymbol{E}}_i$ *differs from the plastic strain rate* $\dot{\boldsymbol{E}}_p$ *and follows the rule*

$$\dot{\boldsymbol{E}}_i = \mathbb{P}[\dot{\boldsymbol{E}}_p], \tag{8.90}$$

where

$$\mathbb{P} = \boldsymbol{I} \underline{\otimes} \boldsymbol{I} - \mathbb{E}^{-1} \frac{\partial \hat{\boldsymbol{T}}}{\partial \boldsymbol{E}_p} = \boldsymbol{I} \underline{\otimes} \boldsymbol{I} + \frac{\partial \hat{\boldsymbol{E}}_e}{\partial \boldsymbol{E}_p}, \tag{8.91}$$

in which tensor $\mathbb{E}$ is defined, keeping into account Eq. (8.88), as

$$\mathbb{E} = \frac{\partial \hat{\boldsymbol{T}}}{\partial \boldsymbol{E}_e} = \left(\frac{\partial \hat{\boldsymbol{E}}_e}{\partial \boldsymbol{T}} \right)^{-1}. \tag{8.92}$$

Going back now to Eqs. (8.86) and following our previous treatment, Eqs. (8.24) through (8.46), we introduce the flow rule, defining the rate of irreversible strain as

$$\dot{\boldsymbol{E}}_i = \dot{\Lambda} \boldsymbol{P}, \tag{8.93}$$

where now tensor $\boldsymbol{P}$ defines the 'direction' of the *irreversible* strain rate and $\dot{\Lambda} \geq 0$ its modulus so that $\dot{\Lambda} = 0$ for unloading.

As noted by Hill and Rice (1973) and Hill (1978), we remark that

for associative flow rule $\boldsymbol{P} = \boldsymbol{Q}$, the irreversible strain rate $\dot{\boldsymbol{E}}_i$, rather than the plastic strain rate $\dot{\boldsymbol{E}}_p$, is proportional to the yield function gradient, so the so-called normality rule applies to $\dot{\boldsymbol{E}}_i$ instead of to $\dot{\boldsymbol{E}}_p$.

Assuming now the existence of a yield function, Eq. (8.29), where $\boldsymbol{T}^{(m)}$ is now simply the Cauchy stress $\boldsymbol{T}$, defining a hardening rule as in Eq. (8.41) and imposing Prager consistency yields the rate equations exactly in the form (8.66) but where $\mathbb{E}$ is now not constant; rather, it depends on the plastic strain and possibly on the stress state.

So far the formulation of elastoplastic coupling is so general that even the existence of a free-energy density is, strictly speaking, not needed. However, the introduction of hyperelastic behaviour is important because a hypoelastic law can introduce an additional irreversible strain that would be difficult to discriminate from the plastic one. It is therefore expedient to operate now with a strain energy density $\varphi(\boldsymbol{E}_p, \boldsymbol{E}_e)$ or a complementary strain energy density $\psi(\boldsymbol{E}_p, \boldsymbol{T})$ so that Eqs. (8.85) are obtained as

$$\boldsymbol{T} = \frac{\partial \varphi(\boldsymbol{E}_p, \boldsymbol{E}_e)}{\partial \boldsymbol{E}_e} \quad \text{and} \quad \boldsymbol{E}_e = \frac{\partial \psi(\boldsymbol{E}_p, \boldsymbol{T})}{\partial \boldsymbol{T}}. \tag{8.94}$$

The preceding treatment provides the framework for describing a stress- and plastic strain–dependent elastic anisotropy, a type of induced anisotropy important to describe the behaviour of granular materials.

To obtain a specific model, for instance, for sand, as in Gajo and Bigoni (2008), the following 'ingredients' still have to be prescribed:

- A form for the strain energy density φ (or the complementary strain energy density ψ), incorporating
  - A non-linear dependence on elastic strain (or stress in a representation based on complementary strain energy) reproducing (1) the mean stress dependence of the elastic tangent stiffness, Eq. (8.92), and (2) the stress induced elastic anisotropy.
  - A second-order fabric tensor depending on deviatoric plastic strain reproducing the fraction of plastically induced elastic anisotropy;
  - A scalar dependence on plastic strain describing the elastic degradation and void ratio dependency (see the experiments performed by Hoque and Tatsuoka, 2004).

- A plastic constitutive framework, incorporating
  - A yield function (reproducing either the mean stress and density dependence of shear strength for granular materials or the effects of stress history for cohesive soils).
  - A flow rule (reproducing the observed irreversible dilatancy).
  - A hardening rule (reproducing the progressive decrease of stiffness typical of geomaterials).

A strain energy density describing a non-linear dependence on elastic strain and reproducing the mean stress dependence of the elastic tangent stiffness and the stress induced elastic anisotropy typical of granular materials has been suggested by Gajo and Bigoni (2008) in the form

$$\varphi(\boldsymbol{E}_e,\boldsymbol{E}_p) = \alpha d\left(-\mathrm{tr}(\boldsymbol{B}\boldsymbol{E}_e)\right)^n + \beta d\left(\mathrm{tr}(\boldsymbol{B}\boldsymbol{E}_e)^2\right)^l \tag{8.95}$$

where $\boldsymbol{B}$ is a symmetric second-order positive definite fabric tensor [playing a role similar to that introduced in Eq. (6.196)] describing fabric anisotropy and depending on the plastic strain $\boldsymbol{E}_p$ (Valanis, 1990; Zysset and Curnier, 1995; Bigoni and Loret, 1999), α, β, l and n are material parameters and $d = d(\boldsymbol{E}_p)$ is a function of the plastic strain introduced to describe the elastic stiffness degradation and void ratio dependency typical of granular materials.

When elastic stiffness degradation and void ratio dependency are neglected so that d is constant, $n = 2$ and $l = 1$, the linear elastic anisotropy (6.196) is recovered so that, eventually, when $\boldsymbol{B} = \boldsymbol{I}$, linear isotropic elasticity is recovered.

From the elastic potential (8.95), the stress $\boldsymbol{T}$ can be derived immediately in the form $(8.85)_1$

$$\boldsymbol{T} = \frac{\partial \varphi}{\partial \boldsymbol{E}_e} = -\alpha d n\left(-\mathrm{tr}(\boldsymbol{B}\boldsymbol{E}_e)\right)^{n-1}\boldsymbol{B} + 2\beta d l\left(\mathrm{tr}(\boldsymbol{B}\boldsymbol{E}_e)^2\right)^{l-1}\boldsymbol{B}\boldsymbol{E}_e\boldsymbol{B}. \tag{8.96}$$

Equivalently, the stress increment $\dot{\boldsymbol{T}}$ in the form $(8.86)_1$ can be obtained by taking the rates of Eq. (8.96), that is,

$$\dot{\boldsymbol{T}} = \mathbb{E}[\dot{\boldsymbol{E}}_e] + \frac{\partial^2 \varphi}{\partial \boldsymbol{E}_e\, \partial \boldsymbol{B}}[\dot{\boldsymbol{B}}] + \frac{\partial^2 \varphi}{\partial \boldsymbol{E}_e\, \partial d}\left(\frac{\partial d}{\partial \boldsymbol{E}_p}\cdot \dot{\boldsymbol{E}}_p\right), \tag{8.97}$$

where

$$\dot{\boldsymbol{B}} = \frac{\partial \boldsymbol{B}}{\partial \boldsymbol{E}_p}[\dot{\boldsymbol{E}}_p]. \tag{8.98}$$

The tangent elastic stiffness $\mathbb{E}$ tensors appearing in Eq. (8.97) therefore result to be given by

$$\mathbb{E} = k_1 \boldsymbol{B}\otimes\boldsymbol{B} + k_2 \boldsymbol{B}\boldsymbol{E}_e\boldsymbol{B}\otimes\boldsymbol{B}\boldsymbol{E}_e\boldsymbol{B} + k_3 \boldsymbol{B}\,\underline{\overline{\otimes}}\,\boldsymbol{B}, \tag{8.99}$$

where for an accurate modelling $\boldsymbol{B}$ becomes a function of the plastic strain (see Gajo et al., 2004) and eventually also of the stress state (see Gajo and Bigoni, 2008).

8.3 A summary on rate constitutive equations

In conclusion, we have to highlight that

> *in the development of bifurcation and stability theory, only 'general' features of the constitutive equations will be needed.*

In particular, the rate constitutive equations expressed in terms of material-time derivative of the first Piola-Kirchhoff and of the deformation gradient will play a crucial role. We will refer to the following rate or incremental constitutive laws:

- Constitutive equation for elastic or hypoelastic rate (or incremental) behaviour

$$\dot{\boldsymbol{S}} = \mathbb{G}[\dot{\boldsymbol{F}}], \tag{8.100}$$

  a linear equation where tensor $\mathbb{G}$, which does not possess the minor symmetries, and for non-hyperelastic behaviour does not possess the major either, can depend arbitrarily on the current state (through pre-stress, pre-strain or both). Examples of the constitutive equations (8.100) are provided by Eqs. (6.110), (6.127) and (6.175).
- Constitutive equation for (piece-wise linear) elastoplastic incremental behaviour

$$\dot{\boldsymbol{S}} = \mathbb{G}[\dot{\boldsymbol{F}}] - \frac{1}{g} < \boldsymbol{N} \cdot \dot{\boldsymbol{F}} > \boldsymbol{M}. \tag{8.101}$$

  In this equation, nothing is required on $\mathbb{G}$ (except that it does not depend on rate quantities), which may lack all symmetries, on $\boldsymbol{N}$ and $\boldsymbol{M}$, which are generic tensors, and on g, which, together with the other quantities, may be a generic function, although continuous, of the state. Examples of the constitutive equations (8.101) are provided by Eqs. (8.50) and (8.66) (for small strain, $\dot{\boldsymbol{S}}$ and $\dot{\boldsymbol{F}}$ simply become $\dot{\boldsymbol{T}}$ and $\dot{\boldsymbol{E}}$).

  Note that damaging materials (Benallal et al., 1993) and adiabatic thermoplastic constitutive laws (Raniecki, 1979; Raniecki and Bruhns, 1981; Benallal and Bigoni, 2004) are characterised by a tangent constitutive operator which again can be cast in a form identical to Eq. (8.101).

As an important comment to the preceding constitutive equations, we remark that

> *two sources of non-linearity characterise plastic flow. One is related to the incremental non-linearity of the tangent constitutive operator and the other to its dependence on the current state. The latter arises also in non-linear elasticity or in hypo-elasticity, whereas the former is typical of the plastic behaviour.*

When needed to give full evidence to the preceding non-linearities, we will occasionally employ the symbol $\boldsymbol{\mathcal{C}}(\cdot)$, reserved for the non-linear constitutive *operator*

$$\dot{\boldsymbol{S}} = \boldsymbol{\mathcal{C}}(\dot{\boldsymbol{F}}), \tag{8.102}$$

positively homogeneous of degree one with respect to its argument, relating the material time derivative of the first Piola-Kirchhoff stress to the material time derivative of the deformation gradient.

Note that in our subsequent treatment, the following constitutive equations will *not* be addressed:

- Corner theories and multi-mechanism plasticity (Koiter, 1960; Maier, 1970b; Capurso, 1972)
- Fully non-linear rate constitutive equations and hypo-plasticity (Christofferson and Hutchinson, 1979; Petryk, 1992; Petryk and Thermann, 2002; Gudehus, 2004)
- Thermal or viscous behaviour (Benallal and Bigoni, 2004).

Plate 1.4. Severe folding of metamorphic rock layers (so-called accomodation structures) initiated as buckling owing to compression stresses (Trearddur Bay, Holyhead, N. Wales, UK; the coin in the photos is a pound).

Plate 1.13. Fault active during the strongest inland earthquake in the history of Japan (October 28, 1891). Photo taken at the Neo Valley Earthquake Fault Line Observatory, Midori Neo-mura Motosu-gun (Neomidori Motosu-shi), Gifu, Japan.

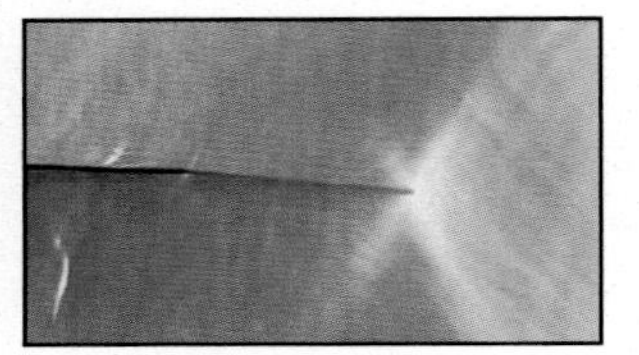

Plate 1.16. Shear bands emerging from the tip of a stiffener (a 0.3 mm thick 44×18-mm aluminium platelet) embedded in a two-component epoxy resin prismatic ($100 \times 100 \times 18$-mm) sample. The sample has been loaded uniaxially in vertical compression (photo taken in bright light at a 50 MPa compressive stress). The material exhibited a ductile behaviour and suffered an out-of plane buckling. Light reflection evidences strain localisation in the form of shear bands at the end of the platelet (clearly visible in the detail on the right).

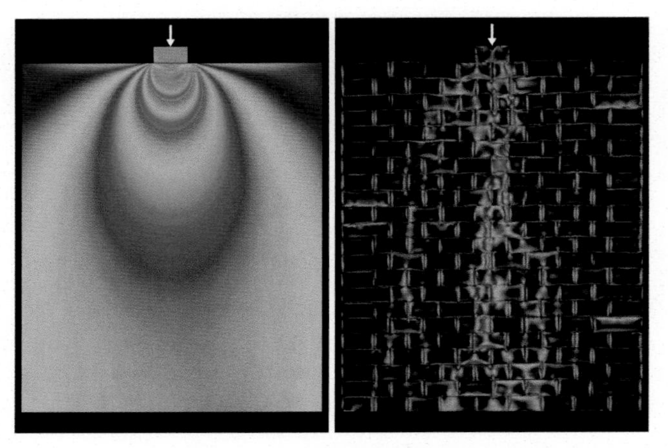

Plate 1.37. Isocromatic photoelastic fringes (detected with a circular transmission polariscope at white light) of a two-component resin platelet subject to a vertical load, denoted with a white arrow (*left*), and of a model of dry masonry (*right*). Vertical load is 500 N for the sample on the left and 125 N for the sample on the right. While the load diffuses within the material according to the Flamant solution in the case reported on the left, there is a strong vertically localised stress percolation in the case reported on the right, according with the Lekhnitskii solution (1.42) at high orthotropy contrast; see Fig. 1.36.

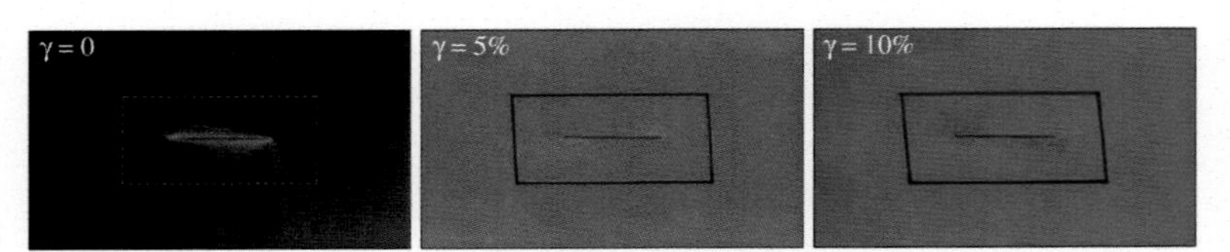

Plate 5.16. Photoelastic experiment demonstrating neutrality of a thin rigid inclusion (made of steel, 0.1 mm thick, 20 mm long) embedded in an elastic (two-component epoxy resin) matrix subject to simple shear parallel to it. Homogeneous colours denote uniform stress. The sample is unloaded on the left, whereas the amount of shear γ is equal to 5% in the centre and 10% on the right. Note that a rectangle enclosing the thin inclusion (with edges parallel and orthogonal to it in the unloaded configuration) has been drawn black on the sample's surface to measure shear deformation.

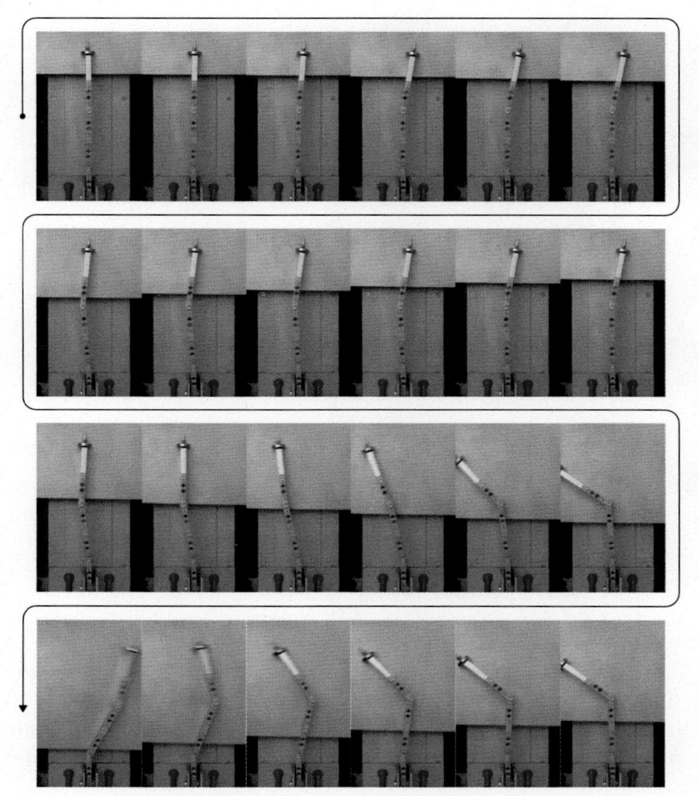

Plate 1.70. A sequence of photos (taken from a movie recorded with a Sony handycam HDR-XR550VE at 25 shots per second) of the structure sketched in Fig. 1.69 and exhibiting flutter instability. Note that the last photos are blurred owing to the increasing velocity of the motion. The whole sequence of photos was recorded in 0.96 s, while the plate was advancing against the wheel.

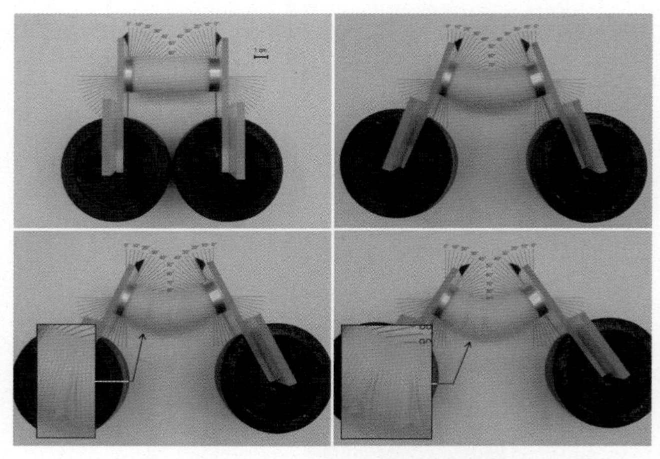

Plate 12.21. Behaviour of a corrugated plastic tube subjected to finite bending. *Top*: Undeformed configuration (*left*); bent configuration (before bifurcation) at an opening semi-angle of 20° (*right*). *Bottom*: First (on the left, at an opening semi-angle of 25°) and second (*right*, at an opening semi-angle of 35°) localisation of deformation, which may be interpreted as a bifurcation of the homogeneous bending solution.

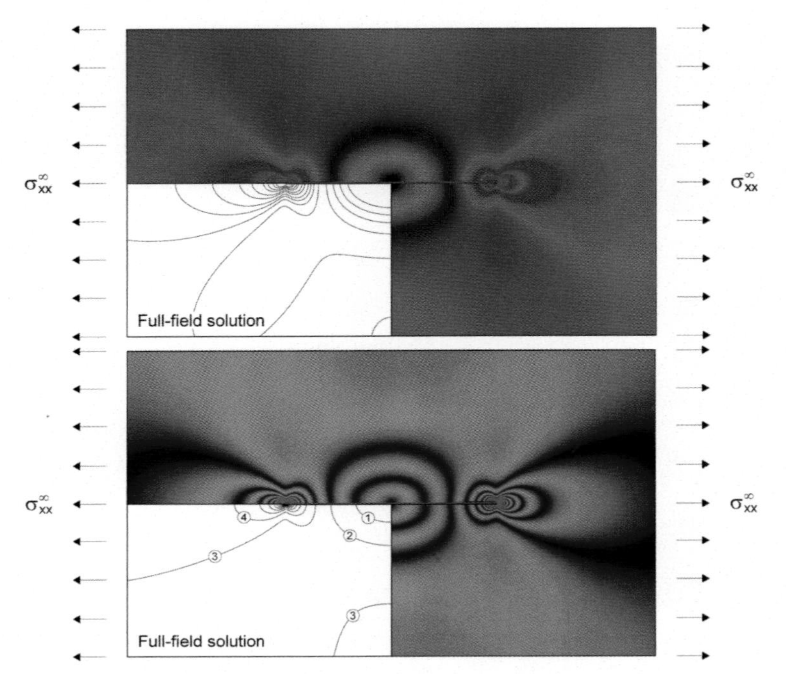

Plate 16.18. Photoelastic (isochromatic, *top*, and monochromatic, *bottom*) fringes revealing the stress field near a thin line inclusion (0.1-mm-thick steel platelet) embedded in an elastic matrix (a two-component 'soft' epoxy resin) and subjected to mode I uniform loading σ_{xx}^{∞}. A comparison is made with the elastic solution in plane strain, with Poisson's ratio equal to 0.45.

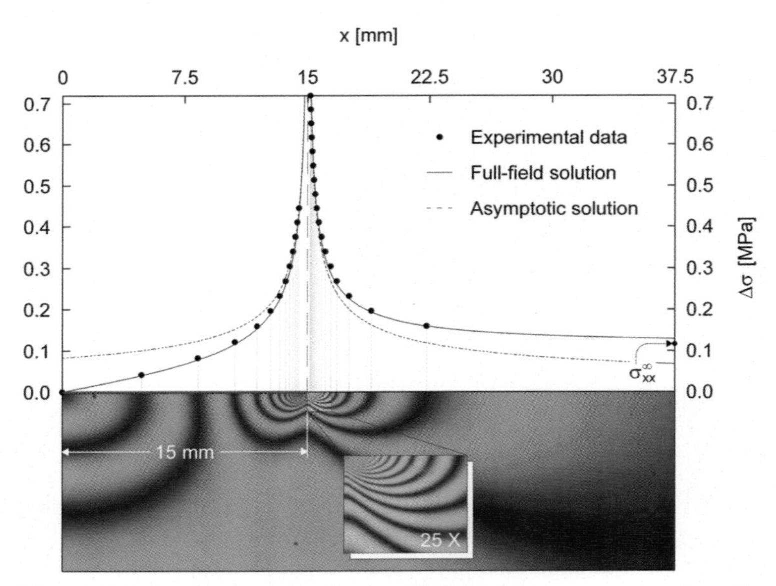

Plate 16.19. In-plane principal stress difference $\Delta\sigma = \sigma_1 - \sigma_2$ along the stiffener line: Comparison between the photoelastic experiment with monochromatic light and the full-field and asymptotic elastic solutions. A detail of the region near to the (*right*) stiffener tip has been captured using an optical microscope (photo reported in the insert), so a great number of fringes (20) have been detected, until a distance from the tip of the same order of magnitude as the stiffener thickness (0.1 mm). A stress concentration of 7 is visible.

9 Moving discontinuities and boundary value problems

Rules governing propagating and stationary discontinuity surfaces in solids are given in view of applications to shear banding and acceleration waves. The finite, rate and incremental boundary value problems are set for solids loaded by prescribed controlled nominal tractions on the boundary.

Analysis of kinematics and balance laws was given in Chapter 3, whereas constitutive equations were detailed in Chapters 4, 6 and 8, with a digression on yield functions presented in Chapter 7. We are now in a position to 'collect the equations' and set boundary value problems in finite and rate forms for solids loaded on the boundary. However, we have until now assumed certain hypotheses of regularity that we want to relax, so before setting boundary value problems, a digression on moving singularities in solids becomes instrumental. This also will be useful in the development of acceleration waves and shear band analysis.

9.1 Moving discontinuities in solids

Until now, we have more or less tacitly assumed that all the fields are 'sufficiently' regular, which is to say smooth. However, there are many situations in which smoothness or even continuity is lost. For instance, displacement, deformations or stresses can suffer jumps across a fracture or a so-called imperfect interface (such as those considered by Bigoni et al., 1997, 1998), or across a rigid thin inclusion (such as that considered by Dal Corso et al., 2008, and Dal Corso and Bigoni, 2009), or simply at an interface separating two different solids, for instance, in a multilaminated material. Although treatment of the compatibility conditions that arise in different situations is an important and fascinating branch of solid mechanics (Hadamard, 1903; Prager, 1954; Truesdell and Toupin, 1960; Thomas, 1961; Hill, 1961, 1962; Chadwick and Powdrill, 1965; Chadwick, 1976; see also the excellent review by Scheidler, 1984), our interest here is limited to describing the following two simple situations.

- A surface of discontinuity 'spontaneously' emerges as an incremental or a rate solution from a smooth field in a quasi-static deformation process, a problem corresponding to the formation of a shear band.
- Second-order discontinuities in the displacement field occur across a surface which propagates in a deformed material, a problem corresponding to a so-called acceleration wave.

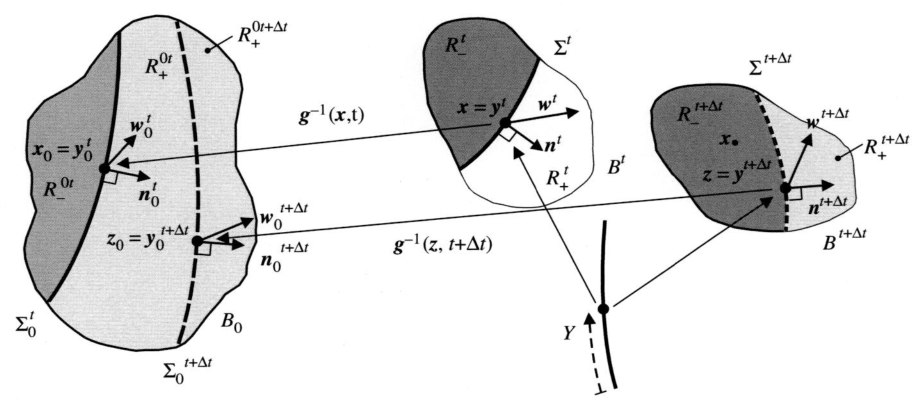

Figure 9.1. Discontinuity surface Σ^t moving with speed $\boldsymbol{w}^t$ with normal component w_n. The unit vector $\boldsymbol{n}^t$ defines the normal at each point of the discontinuity surface.

We will see that the former problem can be obtained as a particular case of the latter, in the limit when the propagation speed of the discontinuity surface vanishes.[1] Therefore, we begin presenting the latter problem, namely, the propagation conditions of acceleration waves.

Assuming that the motion is continuous across a propagating discontinuous surface, this is classified according to the type of jump it carries. In particular,

- *Order 1.* The velocity and the deformation gradient are discontinuous across the surface. The propagating singular surface is called
  - A 'shock wave' if it carries a jump in the normal velocity
  - A 'vortex sheet' if the discontinuity is only tangential
- *Order 2.* The velocity and the deformation gradient are continuous across the surface, but the acceleration is not. The propagating singular surface is called an 'acceleration wave'.
- *Order n* (> 2). The velocity and the deformation gradient are continuous across the surface, together with the $(n-1)$th-order derivatives of all relevant fields, but some nth-order derivatives of certain fields suffer jump discontinuities.

In the following we will limit the presentation to cases in which *the motion and the deformation gradient remain continuous across the propagating singular surface*, so we will exclude a treatment of shock waves and vortex sheets.

9.1.1 Local jump conditions for propagating discontinuity surfaces

Let us denote by σ^t a discontinuous (smooth and orientable) surface with unit normal $\boldsymbol{n}^t$ moving with speed $\boldsymbol{w}^t$ with respect to the material into a region denoted by + and leaving another region denoted by − (Fig. 9.1).

[1] The fact that a stationary discontinuity coincides with an acceleration wave of vanishing speed becomes not straightforward when thermal effects are accounted for (see Benallal and Bigoni, 2004).

The surface 'sweeps through' the spatial configuration of the solid, which is itself moving. Therefore, if at the instant t the surface point $\boldsymbol{y}^t$ is superimposed on the point $\boldsymbol{x}$ of the configuration B^t, at the instant, say, $t+\Delta t$, the same point of the surface occupies the place $\boldsymbol{y}^{t+\Delta t}$, superimposed on a different spatial point $\boldsymbol{z}$ of $B^{t+\Delta t}$. At a generic instant, the surface divides the body into two parts so that $B^t = R^t_- \cup R^t_+$, whereas at each instant the surface is superimposed on points $\Sigma^t = \sigma^t \cap B^t$. Henceforth, the apex t is systematically omitted to simplify notation.

Roughly speaking, the movement of the surface can be described in a way similar to the description of the motion of a continuum presented in Chapter 3, but with the reference configuration replaced by two surface parameters $\boldsymbol{Y}$ so that the surface at each instant t, which points are denoted by $\boldsymbol{y}$ can be described by introducing a mapping such that

$$\boldsymbol{y} = \boldsymbol{y}(\boldsymbol{Y}, t). \tag{9.1}$$

The velocity $\boldsymbol{w}$ of the point $\boldsymbol{Y}$ of the surface can be calculated by taking the time derivative of Eq. (9.1) at fixed $\boldsymbol{Y}$ and mapping the result forward to σ^t by using the inverse of function (9.1):

$$\boldsymbol{w}(\boldsymbol{y}, t) = \left.\frac{\partial \boldsymbol{y}(\boldsymbol{Y}, t)}{\partial t}\right|_{\boldsymbol{Y}=\boldsymbol{y}^{-1}(\boldsymbol{y},t)}. \tag{9.2}$$

The use of parameters $\boldsymbol{Y}$ can be avoided, and Eq. (9.1) can be rewritten as

$$f(\boldsymbol{y}, t) = 0, \tag{9.3}$$

with the unit normal

$$\boldsymbol{n} = \frac{\mathrm{grad} f}{|\mathrm{grad} f|} \tag{9.4}$$

at each time t.

Equation (9.3) holds at every time t, so a differentiation with respect to time yields

$$\dot{f} = 0, \iff (\mathrm{grad} f) \cdot \boldsymbol{w} + \frac{\partial f}{\partial t} = 0. \tag{9.5}$$

Thus, using Eq. (9.4), we find that the partial derivative of f with respect to time is related to the normal velocity of propagation $w_n = \boldsymbol{w} \cdot \boldsymbol{n}$ as

$$w_n = -\frac{1}{|\mathrm{grad} f|} \frac{\partial f}{\partial t}. \tag{9.6}$$

At every instant of time, the moving discontinuity surface Σ is superimposed on points of the spatial configuration B which can be mapped back to the reference configuration B_0 through the inverse of the deformation of the motion of B [Eq. (3.50)$_2$], so $\boldsymbol{g}^{-1}$ 'projects an image' of Σ on the reference configuration that we denote by Σ_0. The points occupied by Σ_0 in B_0 are mapped back to Σ in B, and vice versa, respectively, by functions (3.50)$_2$ and (3.50)$_1$, where $\boldsymbol{x}$ is replaced by $\boldsymbol{y}$ [Eq. (9.1)], namely,

$$\boldsymbol{y}_0 = \boldsymbol{g}^{-1}(\boldsymbol{y}, t), \qquad \boldsymbol{y} = \boldsymbol{g}(\boldsymbol{y}_0, t). \tag{9.7}$$

Using Eq. $(9.7)_2$ in Eq. (9.3), we obtain the referential description f_0 of the surface and its normal

$$f_0(\boldsymbol{y}_0,t)=f(\boldsymbol{g}(\boldsymbol{y}_0,t),t)=0, \qquad \boldsymbol{n}_0=\frac{\mathrm{Grad} f_0}{|\mathrm{Grad} f_0|}=\frac{\boldsymbol{F}^T\boldsymbol{n}}{|\boldsymbol{F}^T\boldsymbol{n}|}, \tag{9.8}$$

so the normals $\boldsymbol{n}_0$ and $\boldsymbol{n}$ transform according to the rule of normals to embedded surfaces [Eq. (3.22)].

Since Σ occupies different points in the various configurations B, its 'image' Σ_0 *travels through* B_0. The speed $\boldsymbol{w}^0$ of Σ_0 on B_0 can be calculated by taking the time derivative of Eq. $(9.7)_1$ at fixed $\boldsymbol{Y}$, that is,

$$\boldsymbol{w}^0(\boldsymbol{y}_0)=\left.\frac{\partial \boldsymbol{g}^{-1}(\boldsymbol{y}(\boldsymbol{Y},t),t)}{\partial t}\right|_{\boldsymbol{Y}=\boldsymbol{y}^{-1}(\boldsymbol{g}(\boldsymbol{y}_0,t),t)}, \tag{9.9}$$

which using the chain rule[2] provides the relation between the velocity $\boldsymbol{w}$ of the discontinuity surface in B and the velocity $\boldsymbol{w}^0$ of its image in B_0, namely,

$$\boldsymbol{w}=\boldsymbol{v}+\boldsymbol{F}\boldsymbol{w}^0. \tag{9.10}$$

Kinematical compatibility conditions Let us consider a generic spatial (or material) field $\boldsymbol{\Theta}$ discontinuous across Σ (or Σ_0) corresponding to the fields $\boldsymbol{\Theta}^+$ and $\boldsymbol{\Theta}^-$, defined and smooth on R^+ and R^- (or on R_0^+ and R_0^-). These fields take finite limit values on Σ (or on Σ_0), where they need not be defined. We denote the jump of $\boldsymbol{\Theta}$ as the difference between the limits approached by $\boldsymbol{\Theta}^\pm$ when approaching the discontinuity surface, that is,

$$[\![\boldsymbol{\Theta}]\!]=\boldsymbol{\Theta}^+(\boldsymbol{y}^+)-\boldsymbol{\Theta}^-(\boldsymbol{y}^-), \tag{9.11}$$

where $\boldsymbol{y}^\pm$ denote the limit points $\boldsymbol{y}$ of Σ (or $\boldsymbol{y}_0$ of Σ_0).

For a generic scalar, vectorial or tensorial spatial or material field $\boldsymbol{\Theta}=\boldsymbol{\Theta}(\boldsymbol{z},t)$ which is a continuous function of place $\boldsymbol{z}=\boldsymbol{x}$ (or $\boldsymbol{z}=\boldsymbol{x}_0$ for material fields) so that $[\![\boldsymbol{\Theta}(\boldsymbol{z},t)]\!]=\boldsymbol{0}$, continuity does not restrict the normal derivative along $\boldsymbol{n}$ (or $\boldsymbol{n}_0$ for material fields), so $[\![\nabla\boldsymbol{\Theta}(\boldsymbol{z},t)]\!]\boldsymbol{m}$ (with $\boldsymbol{m}=\boldsymbol{n}$ or $\boldsymbol{m}=\boldsymbol{n}_0$) remains unrestricted, but the derivative orthogonal to $\boldsymbol{m}$ must vanish:

$$[\![\boldsymbol{\Theta}(\boldsymbol{x},t)]\!]=\boldsymbol{0} \quad\Longrightarrow\quad [\![\nabla\boldsymbol{\Theta}(\boldsymbol{x},t)]\!]\boldsymbol{t}=\boldsymbol{0}, \tag{9.12}$$

for every unit vector $\boldsymbol{t}$ orthogonal to $\boldsymbol{m}$ at each fixed time t. This condition implies that $[\![\boldsymbol{\Theta}]\!]$ has only the normal derivative (i.e., the directional derivative in the direction $\boldsymbol{m}$) different from zero, so defining a vector $\boldsymbol{g}$ as the normal derivative of $[\![\boldsymbol{\Theta}]\!]$ along $\boldsymbol{m}$, we immediately obtain the *Maxwell compatibility conditions* in the material (we use the symbol $\boldsymbol{\Theta}_0$ to highlight referential description) and spatial versions, respectively,

$$[\![\mathrm{Grad}\boldsymbol{\Theta}_0]\!]=\boldsymbol{g}_0\otimes\boldsymbol{n}_0, \qquad [\![\mathrm{grad}\boldsymbol{\Theta}]\!]=\boldsymbol{g}\otimes\boldsymbol{n}, \tag{9.13}$$

[2] Taking the material-time derivative of the identity

$$\boldsymbol{x}_0=\boldsymbol{g}^{-1}(\boldsymbol{g}(\boldsymbol{x}_0,t),t),$$

we can derive the relation

$$\left(\boldsymbol{g}^{-1}(\boldsymbol{x},t)\right)'=-\boldsymbol{F}^{-1}\boldsymbol{v},$$

(where $'$ denotes the spatial time derivative [Eq. (3.55)]) which is useful to obtain Eq. (9.10).

where

$$\boldsymbol{g}_0 = [\text{Grad}\boldsymbol{\Theta}_0]\boldsymbol{n}_0, \qquad \boldsymbol{g} = [\text{grad}\boldsymbol{\Theta}]\boldsymbol{n}. \tag{9.14}$$

Note that the Maxwell compatibility conditions also hold when functions $\boldsymbol{\Theta}_0$ and $\boldsymbol{\Theta}$ are scalar-valued, in which case the preceding $\boldsymbol{g}$ and $\boldsymbol{g}_0$ become scalars and the tensor product $\otimes$ a product between a scalar and a vector.

We note that the moving surface Σ (or Σ_0) always can be viewed as hypersurface $f(\boldsymbol{x},t)$ [or $f_0(\boldsymbol{x}_0,t)$] in a space-time domain, so the normal at $(\boldsymbol{x},t)$ [or $(\boldsymbol{x}_0,t)$] to this is parallel to the gradient in this space and is given by

$$\tilde{\boldsymbol{m}} = (\boldsymbol{m},-c) = \nabla_{(4)}f = \frac{1}{|\nabla f|}\left(\nabla f, \frac{\partial f}{\partial t}\right), \tag{9.15}$$

where $\boldsymbol{m} = \boldsymbol{n}$, $c = \boldsymbol{w}\cdot\boldsymbol{n} = w_n$ and $\nabla = \text{grad}$ in the spatial setting, whereas $\boldsymbol{m} = \boldsymbol{n}_0$, $c = \boldsymbol{w}^0\cdot\boldsymbol{n}_0 = w^0_{n_0}$ and $\nabla = \text{Grad}$, and f is replaced by f_0 in the material setting.

For a scalar function Θ (or Θ_0) continuous across the hypersurface $f(\boldsymbol{x},t)$ [or $f_0(\boldsymbol{x}_0,t))$], the Maxwell compatibility conditions hold with reference to the four-dimensional gradient $\nabla_{(4)}$ introduced in Eqs. (9.15), so

$$[\nabla_{(4)}\Theta] = \lambda\tilde{\boldsymbol{m}}, \quad \Longleftrightarrow \quad [\nabla\Theta] = \lambda\boldsymbol{m}, \qquad [\frac{\partial\Theta}{\partial t}] = -\lambda c, \tag{9.16}$$

where λ is a scalar defining the modulus of the jump. The scalar λ can be calculated from Eq. $(9.16)_2$ and substituted into Eq. $(9.16)_3$ to obtain, using Eq. (3.84) — which relates the partial derivative with respect to time to the material time derivative — the following expressions (given in the material and spatial versions)

$$[\dot{\Theta}_0] = -w^0_{n_0}[\text{Grad}\Theta_0]\cdot\boldsymbol{n}_0, \qquad [\dot{\Theta} - \boldsymbol{v}\cdot\text{grad}\Theta] = -w_n[\text{grad}\Theta]\cdot\boldsymbol{n}, \tag{9.17}$$

Using Eq. $(9.13)_2$ in Eq. $(9.17)_2$ yields the *kinematic conditions of compatibility for the material-time derivative of a scalar field* in the material and spatial versions

$$[\dot{\Theta}_0] = -w^0_{n_0}[\text{Grad}\Theta_0]\cdot\boldsymbol{n}_0, \qquad [\dot{\Theta}] = -\omega_n[\text{grad}\Theta]\cdot\boldsymbol{n}, \tag{9.18}$$

where

$$\omega_n = w_n - \boldsymbol{v}\cdot\boldsymbol{n} \tag{9.19}$$

is the velocity of the discontinuity surface relative to the point $\boldsymbol{x}$ of B superimposed on $\boldsymbol{y}$. Note that taking the scalar product of Eq. (9.10) with $\boldsymbol{n}$ or with $\boldsymbol{n}_0$ and using the law of transformation of normals [Eq. (3.22)], we obtain the relations

$$w^0_{n_0} = \frac{\omega_n}{|\boldsymbol{F}^T\boldsymbol{n}|} \qquad \text{and} \qquad \frac{w^0_{n_0}}{|\boldsymbol{F}^{-T}\boldsymbol{n}_0|} = \omega_n, \tag{9.20}$$

consistent with Eqs. (3.23).

If $\boldsymbol{\Theta}$ is a vector function instead of a scalar, we can multiply it by an arbitrary constant vector $\boldsymbol{a}$ and apply Eqs. (9.17) to obtain the *kinematic conditions of compatibility for the material-time derivative of a vector field* in the material and spatial versions:

$$[\dot{\boldsymbol{\Theta}}_0] = -w^0_{n_0}[\text{Grad}\boldsymbol{\Theta}_0]\boldsymbol{n}_0, \qquad [\dot{\boldsymbol{\Theta}}] = -\omega_n[\text{grad}\boldsymbol{\Theta}]\boldsymbol{n}. \tag{9.21}$$

9.1.2 Balance equations for regions containing a moving discontinuity surface

We consider scalar fields $\Theta^{\pm}$ defined on varying domains. It is instrumental to use a result which is a three-dimensional generalisation of the Leibniz differentiation rule of definite integrals whose limits depend on the differential variable or, in other words, a generalisation of the Reynolds' transport theorem for domains with moving boundaries *in the material description*:

$$\frac{d}{dt}\int_{R_0^+} \Theta^+ \, dv_0 = \int_{R_0^+} \dot{\Theta}^+ \, dv_0 - \int_{\Sigma_0} \Theta^+ w_{n_0}^0 \, da_0, \tag{9.22}$$

where Θ^+ is either a material field or a spatial field, in both cases written in the material description.

Equation (9.22) also can be written for the region R_0^-, thus obtaining a similar equation, but with a positive sign in front of the surface integral (because $\boldsymbol{n}_0$ points inward to R_0^+ but outward to R_0^-). Summing up the resulting equation to Eq. (9.22), we obtain

$$\frac{d}{dt}\int_{B_0} \Theta \, dv_0 = \int_{B_0} \dot{\Theta} \, dv_0 - \int_{\Sigma_0} [\Theta] w_{n_0}^0 \, da_0, \tag{9.23}$$

where B_0 is fixed but contains a moving surface of discontinuity so that the surface integral on Σ_0 (carrying the jump) arises in the preceding equation. Equation (9.23) represents the material version of a generalisation of the Reynolds' transport theorem for domains containing moving singular surfaces.

A formal proof of theorem (9.22) can be found in Marsden and Hughes (1983, chapter 2) and is not repeated here. However, to obtain Eq. (9.22) in a non-formal way is an interesting and simple exercise. In particular, we refer to a fictitious motion $x = \boldsymbol{g}_f(\boldsymbol{x}_0, t)$ in which a solid with reference configuration B_0 is mapped into the region R_0^+ (Fig. 9.2). For this deformation, we define a deformation gradient $\boldsymbol{F}_f$ and its determinant J_f. The kinematics mapping B_0 into R_0^+ corresponds to a velocity $\boldsymbol{v}_f$ that is null on ∂R_0^+ (because the part ∂R_0^+ of the boundary of B_0 does not change) and is equal to $\boldsymbol{w}^0$ on Σ_0. Moreover, note that the field Θ^+, which is defined in R_0^+, is interpreted now as a spatial field. Therefore, we can write

$$\frac{d}{dt}\int_{R_0^+} \Theta^+ \, dv_0 = \frac{d}{dt}\int_{B_0} \Theta^+ J_f \, dv_f = \int_{B_0} \frac{d}{dt}\left[\Theta^+(\boldsymbol{g}_f(\boldsymbol{x}_0, t), t) J_f\right] dv_f, \tag{9.24}$$

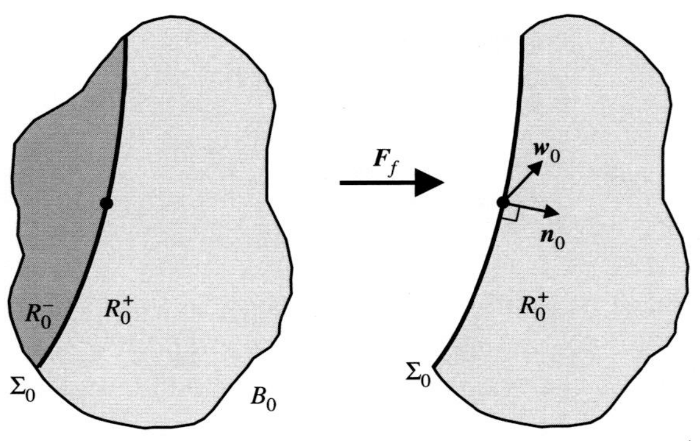

Figure 9.2. Fictitious deformation mapping B_0 into R_0^+ used to prove theorem (9.22).

and remembering the rule (3.84), we obtain

$$\frac{d}{dt}\left[\Theta^+(g_f(x_0,t),t)J_f\right]=\left[\frac{\partial\Theta^+(x,t)}{\partial t}+\mathrm{div}\left(\Theta^+ v\right)\right]J_f, \tag{9.25}$$

where the partial derivative of Θ^+ with respect to time is what we have called the material-time derivative $\partial\Theta^+(x,t)/\partial t=\dot{\Theta}^+$, so the divergence theorem yields

$$\int_{B_0}\left[\frac{\partial\Theta^+(x,t)}{\partial t}+\mathrm{div}\left(\Theta^+ v\right)\right]J_f=\int_{R_0^+}\dot{\Theta}^+ +\int_{\partial R_0^+}\dot{\Theta}^+ v\cdot n, \tag{9.26}$$

where n is the outward unit normal to ∂R_0^+. The last integral in Eq. (9.26) coincides with the last integral in Eq. (9.22), because $v\cdot n=-w_{n_0}^0$ on Σ_0 and $v\cdot n=0$ in the rest of the boundary ∂R_0^+.

Our goal is to transform Eqs. (9.22) and (9.23) into their spatial versions. This can be done in different ways, one is as follows. Write Eq. (9.23) with $J\Theta$ replacing Θ so that the first integral immediately transforms to the spatial version, namely,

$$\frac{d}{dt}\int_{B_0}\Theta J\,dv_0=\frac{d}{dt}\int_B\Theta\,dv, \tag{9.27}$$

whereas

$$\begin{aligned}\int_{B_0}(J\Theta)^{\cdot}\,dv_0 &=\int_B(\Theta'+\mathrm{div}(\Theta v))\,dv\\ &=\int_B\Theta'\,dv+\int_{\partial B}\Theta v\cdot n\,da-\int_\Sigma[\Theta]v\cdot n\,da.\end{aligned} \tag{9.28}$$

Finally, the surface integral in Eq. (9.23) transforms with the help of Nanson's rule (3.21) to

$$\int_{\Sigma_0}[\Theta]w^0\cdot n_0\,da_0=\int_\Sigma[\Theta F w^0]\cdot n\,da. \tag{9.29}$$

Now, by assembling Eqs. (9.27) through (9.29) and keeping into account the relation between material and spatial velocities of the moving surface [Eq. (9.10)], we obtain the spatial version of Reynolds' theorem for a body containing a surface of discontinuity:

$$\frac{d}{dt}\int_{B^t}\Theta\,dv=\int_{B^t}\frac{\partial\Theta}{\partial t}\,dv+\int_{\partial B^t}\Theta(v\cdot n)\,da-\int_{\Sigma^t}w_n[\Theta]da, \tag{9.30}$$

where $w_n=w\cdot n$ is the normal component of the speed of the discontinuity surface (note that we have re-introduced the apex t to highlight the dependence on time). Note that Eq. (9.30) has been obtained with reference to a scalar field Θ, but it is not difficult to see that it holds true even when the scalar field Θ is replaced by a vector field u, that is,

$$\frac{d}{dt}\int_{B^t}u\,dv=\int_{B^t}\frac{\partial u}{\partial t}\,dv+\int_{\partial B^t}u(v\cdot n)\,da-\int_{\Sigma^t}w_n[u]da. \tag{9.31}$$

If we now consider the spatial versions of the mass balance (3.92), the momentum balance law (3.97), and the power expended (3.128), we conclude that they are all expressed in a similar way, say, as

$$\frac{d}{dt}\int_{B^t}\rho\xi\,dv=\int_{B^t}\beta\,dv+\int_{\partial B^t}\chi(n)\,da, \tag{9.32}$$

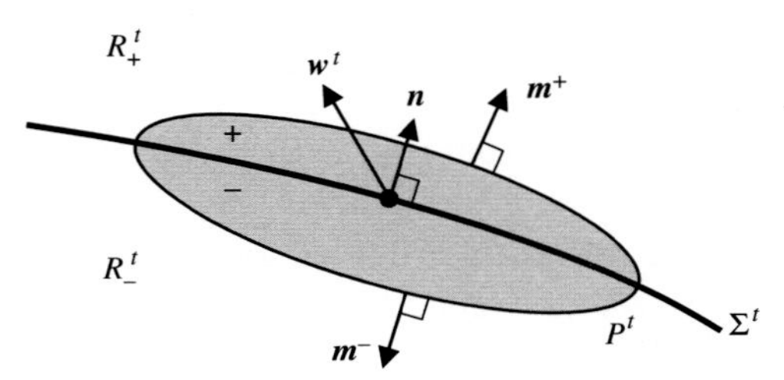

Figure 9.3. Localisation of the integral equation (9.33). Note that the outward unit normal to the part P_t has been denoted by $\boldsymbol{m}^{\pm}$, so when the element P_t is shrunk to Σ^t, $\boldsymbol{m}^+$ tends to $\boldsymbol{n}$ and $\boldsymbol{m}^-$ to $-\boldsymbol{n}$.

where

- $\xi = 1$, $\beta = \chi(\boldsymbol{n}) = 0$ in the case of the mass balance [Eq. (3.92)].
- $\xi = \boldsymbol{v}$ [or $\xi = (\boldsymbol{o} - \boldsymbol{x}) \times \boldsymbol{v}$], $\beta = \boldsymbol{b}$ [or $\beta = (\boldsymbol{o} - \boldsymbol{x}) \times \boldsymbol{b}$] and $\chi(\boldsymbol{n}) = \boldsymbol{s}(\boldsymbol{n})$ [or $\chi(\boldsymbol{n}) = (\boldsymbol{o} - \boldsymbol{x}) \times \boldsymbol{s}(\boldsymbol{n})$] in the case of the momentum balance [Eqs. (3.97)].
- $\xi = v^2/2$, $\beta = \boldsymbol{b} \cdot \boldsymbol{v} - \boldsymbol{T} \cdot \boldsymbol{D}$ and $\chi(\boldsymbol{n}) = \boldsymbol{s}(\boldsymbol{n}) \cdot \boldsymbol{v}$ in the case of the power expended [Eq. (3.128)].

The derivative of the integral in Eq. (9.32) can be transformed using Eq. (9.30) [or the vectorial version Eq. (9.31)] with reference to an arbitrary portion P_t of B_t as

$$\int_{P^t} \left(\frac{\partial \rho \xi}{\partial t} - \beta \right) dv + \int_{\partial P^t} (\rho \xi \, \boldsymbol{v} \cdot \boldsymbol{n} - \chi(\boldsymbol{n})) \, da - \int_{\Sigma^t} w_n [\![\rho \xi]\!] da = 0. \tag{9.33}$$

Since Eq. (9.33) is valid for every part P_t of B_t, it can be localised first on the surface Σ^t (Fig. 9.3) so that it becomes

$$\int_{\Sigma^t} ([\![\rho \xi \, (\boldsymbol{v} \cdot \boldsymbol{n}) - \chi(\boldsymbol{n})]\!] - w_n [\![\rho \xi]\!]) \, da = 0, \tag{9.34}$$

which holds for every surface element of Σ^t and therefore can be further localised into a jump condition holding point-wise on Σ^t

$$[\![\rho \omega_n \xi + \chi(\boldsymbol{n})]\!] = 0, \tag{9.35}$$

where ω_n is the normal velocity of the singular surface relative to the spatial points defined earlier [Eq. (9.19)].

Equation (9.35) provides the local jump conditions across Σ^t; however, all the preceding treatment can be repeated with reference to the material configuration, by considering the integral conditions[3] (3.114) and (3.129), noting that they can be written in the form (9.32), using the transport theorem for domains with moving discontinuities in the material setting [Eq. (9.23)] and thus obtaining an equation similar to Eq. (9.33) but in the material setting, which can be localized to corresponding Eq. (9.35).

[3] Mass balance in the reference configuration reduces to the trivial integral identity

$$\frac{d}{dt} \int_{P_0} \rho_0 = 0.$$

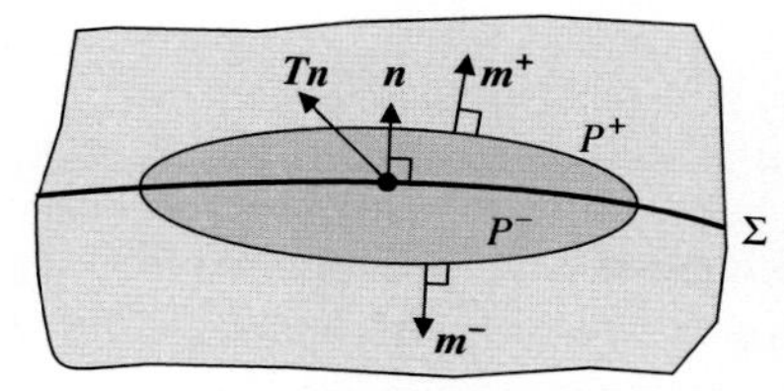

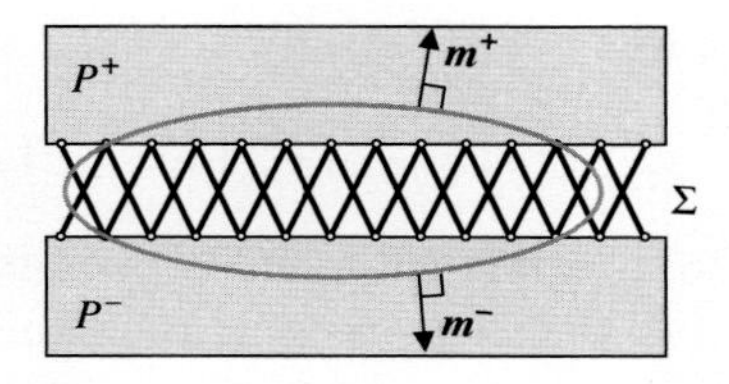

Figure 9.4. Elementary equilibrium considerations explain continuity of traction at a stationary discontinuity surface (*left*). This condition does not hold when the interface has finite thickness, as in the case of a structural interface (*right*).

Therefore, we can write the *local jump conditions in the material (left) and spatial (right) formats* as

$$
\begin{aligned}
&[\![\rho_0]\!] = 0 && [\![\rho\,\omega_n]\!] = 0, \\
&[\![\rho_0\, w^0_{n_0}\dot{\boldsymbol{x}} + \boldsymbol{S}\boldsymbol{n}_0]\!] = \boldsymbol{0} && [\![\rho\,\omega_n \boldsymbol{v} + \boldsymbol{T}\boldsymbol{n}]\!] = \boldsymbol{0}, \\
&[\![\rho_0\, w^0_{n_0}\dot{\boldsymbol{x}}^2/2 + \dot{\boldsymbol{x}}\cdot\boldsymbol{S}\boldsymbol{n}_0]\!] = 0 && [\![\rho\,\omega_n \boldsymbol{v}^2/2 + \boldsymbol{v}\cdot\boldsymbol{T}\boldsymbol{n}]\!] = 0,
\end{aligned}
\tag{9.36}
$$

where $\boldsymbol{T}$ and $\boldsymbol{S}$ are, respectively, the Cauchy and the first Piola-Kirchhoff stresses [Eqs. (3.98) and (3.106)], and $\boldsymbol{v}$, $\dot{\boldsymbol{x}}$, ρ and ρ_0 are the spatial and material descriptions of the velocity and mass density.

Note that balance of angular momentum is automatically satisfied when Eq. $(9.36)_2$ holds true [note also that internal energy and heat flux are not present because we are considering the isothermal case; see Chadwick (1976) for an account of temperature effects].[4]

For a stationary discontinuity, $w^0_{n_0} = 0$, and the conditions (9.36) reduce to the request that the stress vector (in the nominal and spatial versions) be continuous, that is

$$
[\![\boldsymbol{S}]\!]\boldsymbol{n}_0 = \boldsymbol{0}, \qquad [\![\boldsymbol{T}]\!]\boldsymbol{n} = \boldsymbol{0}, \tag{9.37}
$$

a condition that can be derived directly from equilibrium considerations but which does not hold when the interface has a finite thickness, as in the case considered by Bigoni and Movchan (2002) and Bertoldi et al. (2007a, 2007b, 2007c) (see Fig. 9.4).

We are interested now in the acceleration wave problem, where the mass density, the velocity and the stress are continuous. Taking the material-time derivative of

[4] The spatial versions of Eqs. (9.36) can be obtained directly from their material counterpart using Eqs. (9.20). For instance, multiplying Eq. $(9.36)_3$ by da_0, the elementary referential area of Σ and using Eqs. (3.105) and $(3.23)_1$ we obtain

$$
(\rho_0\, w^0_{n_0}\dot{\boldsymbol{x}} + \boldsymbol{S}\boldsymbol{n}_0)\, da_0 = \left(\frac{\rho_0\, w^0_{n_0}\dot{\boldsymbol{x}}}{J|\boldsymbol{F}^{-T}\boldsymbol{n}_0|} + \boldsymbol{T}\boldsymbol{n} \right) da,
$$

which, using the equation resulting from the scalar multiplication of Eq. (9.10) by $\boldsymbol{n}_0$ and noting that $\boldsymbol{n} = \boldsymbol{F}^{-T}\boldsymbol{n}_0/|\boldsymbol{F}^{-T}\boldsymbol{n}_0|$, gives Eq. $(9.36)_4$.

Eq. (9.36), we obtain[5]

$$[\dot{\boldsymbol{S}}]\boldsymbol{n}_0 = -\rho_0 w^0_{n_0}[\ddot{\boldsymbol{x}}], \qquad [\dot{\boldsymbol{T}}]\boldsymbol{n} = -\rho\,\omega_n[\boldsymbol{a}], \tag{9.38}$$

so using Eq. (9.38) together with the Maxwell and kinematic compatibility conditions [Eqs. (9.13) and (9.18)], we arrive at the conditions governing the *propagation of acceleration waves*

$$\begin{aligned} &[\dot{\boldsymbol{S}}]\boldsymbol{n}_0 = \rho_0 w^{02}_{n_0}\boldsymbol{g}_0 && [\dot{\boldsymbol{T}}]\boldsymbol{n} = \rho\,\omega_n^2\boldsymbol{g}, \\ &[\dot{\boldsymbol{F}}] = \boldsymbol{g}_0 \otimes \boldsymbol{n}_0 && [\boldsymbol{L}] = \boldsymbol{g} \otimes \boldsymbol{n}, \\ &[\ddot{\boldsymbol{x}}] = -w^0_{n_0}\boldsymbol{g}_0 && [\boldsymbol{a}] = -\omega_n\boldsymbol{g}, \end{aligned} \tag{9.39}$$

where

$$\boldsymbol{g}_0 = [\dot{\boldsymbol{F}}]\boldsymbol{n}_0, \qquad \boldsymbol{g} = [\boldsymbol{L}]\boldsymbol{n}. \tag{9.40}$$

As noted by Rice (1977), Eq. $(9.38)_1$ can be obtained directly as follows: Since the first Piola-Kirchhoff stress $\boldsymbol{S}$ is continuous, the Maxwell compatibility condition can be applied to $\boldsymbol{S}^T\boldsymbol{a}$, in which $\boldsymbol{a}$ is an arbitrary constant vector, to obtain

$$[\mathrm{Grad}(\boldsymbol{S}^T\boldsymbol{a})] = [\mathrm{Grad}(\boldsymbol{S}^T\boldsymbol{a})\boldsymbol{n}_0] \otimes \boldsymbol{n}_0, \tag{9.41}$$

where introducing the normal derivative, we may write

$$[\mathrm{Grad}(\boldsymbol{S}^T\boldsymbol{a})\boldsymbol{n}_0] = \frac{d\boldsymbol{S}^T\boldsymbol{a}}{d\boldsymbol{n}_0}. \tag{9.42}$$

Taking the trace of Eq. (9.41), the following condition on the divergence of the first Piola-Kirchhoff stress is obtained

$$[\mathrm{Div}\boldsymbol{S}] = [\frac{d\boldsymbol{S}}{d\boldsymbol{n}_0}]\boldsymbol{n}_0. \tag{9.43}$$

Applying now the kinematic compatibility conditions (9.18) to $\boldsymbol{S}^T\boldsymbol{a}$, where $\boldsymbol{a}$ is an arbitrary constant vector, we obtain

$$[\dot{\boldsymbol{S}}] = -w^0_{n_0}[\frac{d\boldsymbol{S}}{d\boldsymbol{n}_0}], \tag{9.44}$$

so we arrive at

$$w^0_{n_0}[\mathrm{Div}\boldsymbol{S}] = -[\dot{\boldsymbol{S}}]\boldsymbol{n}_0. \tag{9.45}$$

Taking the jump of the local balance of momentum (3.115), namely,

$$[\mathrm{Div}\boldsymbol{S}] = \rho_0[\ddot{\boldsymbol{x}}], \tag{9.46}$$

and applying this to Eq. (9.45), we obtain Eq. $(9.38)_1$.

Conditions similar to Eqs. (9.39) but of order n can be obtained easily under the assumption that the fields are continuous up to the order $n-1$ [Gurtin (1972); see also the nth-order rate problem in Petryk and Thermann (1985), Nguyen and

[5] Since $\boldsymbol{T}$ is assumed continuous, eqns. (3.66) and $(9.39)_4$ imply that

$$[\dot{\boldsymbol{n}}] = -(\boldsymbol{I} - \boldsymbol{n}\otimes\boldsymbol{n})[\boldsymbol{L}^T] = -(\boldsymbol{I} - \boldsymbol{n}\otimes\boldsymbol{n})(\boldsymbol{n}\otimes\boldsymbol{g}) = \boldsymbol{0}.$$

Triantafyllidis (1989), Cheng and Lu (1993), and Bigoni (1996)]. Restricted to the material representation, we obtain

$$
\begin{aligned}
&[\overset{(n)}{\rho_0}] = 0, \\
&[\rho_0 w^0_{n_0} \overset{(n+1)}{\boldsymbol{x}} + \overset{(n)}{\boldsymbol{S}} \boldsymbol{n}_0] = \boldsymbol{0}, \\
&[\frac{\rho_0 w^0_{n_0}}{2} \left(\overset{(n+1)}{\boldsymbol{x}}\right)^2 + \overset{(n+1)}{\boldsymbol{x}} \cdot \overset{(n)}{\boldsymbol{S}} \boldsymbol{n}_0] = 0,
\end{aligned}
\tag{9.47}
$$

where the symbol $\overset{(n)}{(\cdot)}$ denotes the nth-order derivative of $(\cdot)$.

A stationary second-order discontinuity is obtained in the particular case where $w^0_{n_0} = \omega_n = 0$,[6] so from Eqs. (9.39) we arrive at the conditions holding for *a stationary discontinuity surface*

$$
\begin{aligned}
&[\dot{\boldsymbol{S}}]\boldsymbol{n}_0 = \boldsymbol{0} \qquad && [\dot{\boldsymbol{T}}]\boldsymbol{n} = \boldsymbol{0}, \\
&[\dot{\boldsymbol{F}}] = \boldsymbol{g}_0 \otimes \boldsymbol{n}_0 \qquad && [\boldsymbol{L}] = \boldsymbol{g} \otimes \boldsymbol{n},
\end{aligned}
\tag{9.48}
$$

where $\boldsymbol{g}_0$ and $\boldsymbol{g}$ are defined by Eqs. (9.40).

Note that Eqs. $(9.48)_{1,2}$ imply the continuity of traction rates across a discontinuity surface, whereas Eqs. $(9.48)_{3,4}$ set compatibility of the velocity gradient across the same surface. These equations are the basis of the strain localisation analysis presented in Chapter 11.

9.2 Boundary value problems in finite, rate and incremental forms

Our objective here is to define the boundary value problems for finite deformation in non-linear elasticity and for the incremental and rate problems in elastoplasticity involving loading of a solid material. To begin, we note that

> *in problems of solid mechanics neither the trajectories nor the current configuration is known, so the Eulerian formulation of boundary value problems, which is based on the knowledge of these, turns out to be 'impractical', and therefore it will not be presented.*

In a loading program, displacements and nominal surface tractions (mixed boundary conditions) are prescribed; these are assumed to be sufficiently regular functions of place and time over specific portions ∂B_0^ξ and ∂B_0^σ of the boundary in the reference configuration ($\partial B_0 = \partial B_0^\xi \cup \partial B_0^\sigma$). For simplicity, we limit the presentation to *controlled nominal surface tractions* on ∂B_0^σ (in other words, deformation-sensitive loadings are not considered) so that the boundary conditions are

$$
\begin{aligned}
\boldsymbol{x} &= \boldsymbol{\xi}(\boldsymbol{x}_0, t), \qquad && \text{on } \partial B_0^\xi \times [0, \infty), \\
\boldsymbol{S}\boldsymbol{n}_0 &= \boldsymbol{\sigma}(\boldsymbol{x}_0, t), \qquad && \text{on } \partial B_0^\sigma \times [0, \infty),
\end{aligned}
\tag{9.49}
$$

where $\boldsymbol{x}$ and $\boldsymbol{x}_0$ are the places occupied by the material points in the current and reference configuration, respectively, $\boldsymbol{n}_0$ is the outward unit vector to ∂B_0^σ and $\boldsymbol{S} \in \mathsf{Lin}$ is the first Piola-Kirchhoff stress [Eq. (3.106)]. Note that the so-called dead loading is

[6] Vanishing of $w^0_{n_0}$ is equivalent from Eq. (9.20) to vanishing of $\omega_n = 0$.

a particular case of condition $(9.49)_2$ in which the surface nominal traction becomes independent of time t.

Pressure-loading boundary conditions are an example of deformation-sensitive boundary conditions. In this case, the Cauchy traction is prescribed to correspond to a given pressure p applied on the boundary of the current configuration ∂B of unit normal $\boldsymbol{n}$, that is,

$$\boldsymbol{Tn} = -p\boldsymbol{n}, \tag{9.50}$$

where the minus sign arises because a positive pressure generates a compressive traction. Employing Nanson's rule (3.21) and the definition of first Piola-Kirchhoff stress (3.106), Eq. (9.50) can be rewritten as

$$\boldsymbol{Sn}_0 = -pJ\boldsymbol{F}^{-T}\boldsymbol{n}_0, \tag{9.51}$$

so we can conclude that *differently* from Eq. $(9.49)_2$, the nominal traction $\boldsymbol{Sn}_0$ depends on the deformation gradient $\boldsymbol{F}$ and its determinant J.

Taking the material-time derivative of Eq. (9.51) and using Eq. (3.67), we arrive at the rate boundary condition for assigned *pressure loading*:

$$\dot{\boldsymbol{S}}\boldsymbol{n}_0 = -J\left[\dot{p}\boldsymbol{I} + p((\text{tr}\boldsymbol{L})\boldsymbol{I} - \boldsymbol{L}^T)\right]\boldsymbol{F}^{-T}\boldsymbol{n}_0. \tag{9.52}$$

The boundary conditions (9.49) have to be complemented with the initial conditions so that we simply may specify the initial (say, at time $t = 0$), position and velocity

$$\boldsymbol{x}(\boldsymbol{x}_0, 0) = \bar{\boldsymbol{x}}(\boldsymbol{x}_0), \qquad \dot{\boldsymbol{x}}(\boldsymbol{x}_0, 0) = \bar{\boldsymbol{v}}(\boldsymbol{x}_0), \tag{9.53}$$

so the functions $\bar{\boldsymbol{x}}$ and $\bar{\boldsymbol{v}}$ are prescribed functions in B_0.

The regular boundary value problem for a non-linear elastic material now can be set as

- Given a reference configuration B_0 with prescribed boundary conditions (9.49) and initial conditions (9.53),
- Find the *smooth* function $\boldsymbol{x}$ and the *continuous* function $\boldsymbol{S}$ of place and time defined in $B_0 \times [0, \infty)$ through

$$\boldsymbol{x} = \boldsymbol{g}(\boldsymbol{x}_0, t), \qquad \boldsymbol{S} = \boldsymbol{S}(\boldsymbol{x}_0, t) \qquad \text{(12 unknown field components)},$$

- Such that the following field equations hold in B_0:

$$\begin{aligned} &\text{Div}\boldsymbol{S} + \boldsymbol{b}_0 = \rho_0\ddot{\boldsymbol{x}} \qquad &&\text{(momentum balance)},\\ &\boldsymbol{S} = \hat{\boldsymbol{S}}(\text{Grad}\boldsymbol{x}) &&\text{(constitutive laws)}, \end{aligned} \tag{9.54}$$

  which correspond to 12 scalar equations.[7]

For quasi-static problems, the time can be interpreted as a scalar parameter governing the loading, so the loading initiates at $t = 0$ and reaches its final values at $t = t_f$. The *regular quasi-static boundary value problem of non-linear elasticity with*

[7] Note that the current density ρ does not appear in Eqs. (9.54), whereas the referential density ρ_0 is known; therefore, the balance of mass need not to be included.

controlled nominal surface tractions and displacements reduces to find, respectively, smooth and continuous functions $\boldsymbol{x}$ and $\boldsymbol{S}$ defined in $B_0 \times [0, t_f]$ such that

$$\begin{array}{lll} \mathrm{Div}\boldsymbol{S} + \boldsymbol{b}_0 = \boldsymbol{0} & \text{in } B_0 \times [0, t_f] & \text{(equilibrium equations)}, \\ \boldsymbol{S} = \hat{\boldsymbol{S}}(\mathrm{Grad}\boldsymbol{x}) & \text{in } B_0 \times [0, t_f] & \text{(constitutive laws)}, \\ \boldsymbol{x} = \boldsymbol{\xi}(\boldsymbol{x}_0, t) & \text{on } \partial B_0^{\xi} \times [0, t_f] & \text{(controlled boundary displacements)}, \\ \boldsymbol{S}\boldsymbol{n}_0 = \boldsymbol{\sigma}(\boldsymbol{x}_0, t) & \text{on } \partial B_0^{\sigma} \times [0, t_f] & \text{(controlled nominal tractions).} \end{array} \tag{9.55}$$

When there is no evolution of loading (in other words, the dependence on time is completely neglected), we arrive at the regular quasi-static boundary value problem of non-linear elasticity for mixed *dead loading* and *displacement boundary conditions*, summarized through the equations

$$\begin{array}{lll} \mathrm{Div}\boldsymbol{S} + \boldsymbol{b}_0 = \boldsymbol{0} & \text{in } B_0 & \text{(equilibrium equations)}, \\ \boldsymbol{S} = \hat{\boldsymbol{S}}(\mathrm{Grad}\boldsymbol{x}) & \text{in } B_0 & \text{(constitutive laws)}, \\ \boldsymbol{x} = \boldsymbol{\xi}(\boldsymbol{x}_0) & \text{on } \partial B_0^{\xi}, & \text{(displacement boundary conditions)}, \\ \boldsymbol{S}\boldsymbol{n}_0 = \boldsymbol{\sigma}(\boldsymbol{x}_0) & \text{on } \partial B_0^{\sigma}, & \text{(dead loading boundary conditions).} \end{array} \tag{9.56}$$

It is important to note that there are no reasons to believe that shocks or even weak discontinuities should be excluded from a static or a dynamic boundary value problem in non-linear elasticity. Such field irregularities cannot be treated under regularity assumptions; rather, discontinuity conditions must be supplemented, which for weak discontinuities can be found in Section 9.1. For our future purposes, Eqs. (9.49), (9.53) and (9.54) or (9.56) are sufficient. However, dealing with elastoplasticity, the situation changes. In fact, the reason why the initial boundary value problem [Eqs. (9.49), (9.53) and (9.54)] cannot be modified simply by replacing the constitutive equation $(9.54)_2$ with an incrementally non-linear rate constitutive equation such as that given by Eq. (8.102) is that *the regularity conditions, which are already a straitjacket in non-linear elasticity, become unacceptable for elastoplastic problems if some generality is retained.* This occurs for several reasons, one of which is that the plastic and elastic zones are often separated by a discontinuity surface. Therefore, to retain generality without introducing a great complication, we will limit the analysis to quasi-static evolutive boundary value problems of elastoplasticity.

9.2.1 Quasi-static first-order rate problems

We know from the example of Coulomb friction in Section 1.9 that there are constitutive models which can be formulated only in terms of rate equations. This is certainly true for elastoplasticity, where knowledge of the deformation does not determine the stress, except if the plastic part of deformation is known, but this can be determined in general only through an integration in time of rate equations (Fig. 9.5) along a deformation path. Therefore, for elastoplastic materials, we cannot avoid a rate formulation. Since in this case the treatment complicates greatly, we will refer only to the quasi-static situation (with the exception of some notion of wave propagation that will be given in Chapter 14) and to the so-called first-order rate problem or velocity

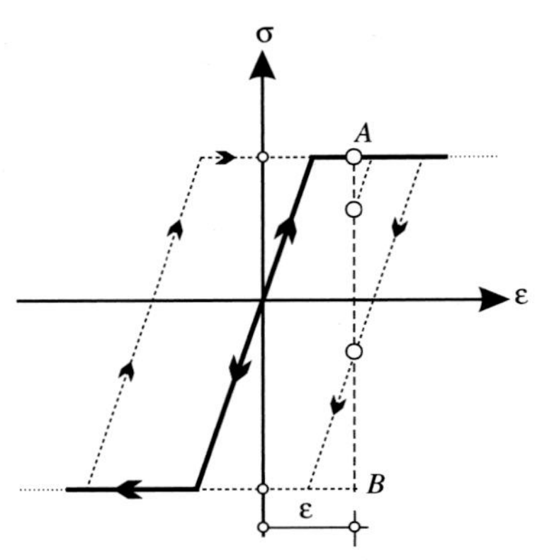

Figure 9.5. Knowledge of the strain ε is not sufficient to determine the stress state in elasto-plasticity. For instance, with reference to the uniaxial stress/strain curve reported in the figure, all the stress states represented by the points of segment AB correspond to the same deformation ε.

problem.[8] In this problem, a current configuration B (referred to as the 'reference configuration B_0') is assumed to be known, in which all fields $\boldsymbol{S}$ and $\boldsymbol{F}$ satisfying Eqs. (9.54), with $\ddot{\boldsymbol{x}} = \boldsymbol{0}$, are the solution of a quasi-static loading program continued until the configuration is reached. This configuration is taken as an initial state, and a *rate* solution of a quasi-static loading program of the type (9.49) is sought, where t does not play the role of 'physical time' but has to be understood as a scalar parameter governing the loading.

Neglecting for simplicity body forces, $\boldsymbol{b}_0 = \boldsymbol{0}$, the velocity boundary value problem is set as follows.

- *Given* a reference configuration B_0 and the following quasi-static rate boundary conditions (in which velocities and traction rates are prescribed on complementary regular subsets of the boundary):

$$\dot{\boldsymbol{x}} = \dot{\boldsymbol{\xi}}(\boldsymbol{x}_0, t), \quad \text{on} \quad \partial B_0^{\xi}; \qquad \dot{\boldsymbol{S}}\boldsymbol{n}_0 = \dot{\boldsymbol{\sigma}}(\boldsymbol{x}_0, t) \quad \text{on} \quad \partial B_0^{\sigma}, \tag{9.57}$$

- *Find* the following continuous, piece-wise smooth velocity field $\dot{\boldsymbol{x}}$ and piece-wise continuous stress rate field $\dot{\boldsymbol{S}}$

$$\dot{\boldsymbol{x}} = \dot{\boldsymbol{x}}(\boldsymbol{x}_0, t) \qquad \text{and} \qquad \dot{\boldsymbol{S}} = \dot{\boldsymbol{S}}(\boldsymbol{x}_0, t), \tag{9.58}$$

- *Satisfying*
  - Rate equilibrium conditions

$$\mathrm{Div}\,\dot{\boldsymbol{S}} = \boldsymbol{0}, \qquad \text{in } B_0 \setminus \Sigma_0 \tag{9.59}$$

  - Non-linear constitutive equations (8.102), namely,

$$\dot{\boldsymbol{S}} = \mathcal{C}(\mathrm{Grad}\,\dot{\boldsymbol{x}}) \tag{9.60}$$

[8] Higher-order problems also can be analysed; see Petryk and Thermann (1985), Nguyen and Triantafyllidis (1989), Cheng and Lu (1993), and Bigoni (1996).

- And appropriate discontinuity conditions across a possible discontinuity surface [Eq. (9.48_1) and continuity of velocity]

$$[\![\dot{\boldsymbol{S}}]\!]\boldsymbol{n}_0 = \boldsymbol{0} \quad \text{and} \quad [\![\dot{\boldsymbol{x}}]\!] = \boldsymbol{0}, \quad \text{on } \Sigma_0. \tag{9.61}$$

Note that except for the constitutive equation (9.60), the velocity problem is linear, and the preceding equations hold for the part of the body at yielding, that is, where $f_{\boldsymbol{S}}(\boldsymbol{S}, \mathcal{K}) = 0$, whereas in the elastic zone, that is, where $f_{\boldsymbol{S}}(\boldsymbol{S}, \mathcal{K}) < 0$, the preceding equations continue to hold, but with the constitutive equation (8.100), namely,

$$\dot{\boldsymbol{S}} = \mathbb{G}[\dot{\boldsymbol{F}}], \tag{9.62}$$

replacing Eq. (9.60).

In weak form, Eqs. (9.57), (9.59) and $(9.61)_1$ are equivalent to

$$\int_{B_0} \dot{\boldsymbol{S}} \cdot \nabla \boldsymbol{w} - \int_{\partial B_0^\sigma} \dot{\boldsymbol{\sigma}} \cdot \boldsymbol{w} = 0, \tag{9.63}$$

holding for every (continuous and piece-wise continuously twice differentiable) variation $\boldsymbol{w}$ of the velocity. In particular, $\nabla \boldsymbol{w} = \text{Grad}\, \boldsymbol{w}$ is the gradient (with respect to material points $\boldsymbol{x}_0$) of every field $\boldsymbol{w}$ defined in the reference configuration and vanishing on the portions of the boundary where displacements (and velocities) are prescribed.

9.2.2 Incremental non-linear elasticity

The incremental boundary value problem for *small elastic deformations superimposed on a given equilibrium state* can be set as following, referring to four configurations $B_\square$, B_0, B, and B' (see Fig. 9.6) (We are proceeding in a similar vein as when the relative Lagrangean description was introduced, i.e., Fig. 6.5.)

- The solution to the quasi-static boundary value problem (9.56) is assumed to be known, say, in terms of a configuration which is convenient to denote with B, referred to the two configurations B_0 and $B_\square$. Note that $B_\square$ has special properties for the constitutive description, for instance, it is the natural unloaded and isotropic (or orthotropic) configuration. These three configurations B, B_0 and $B_\square$ are all fixed in time; B and B_0 are arbitrary (not unloaded). Therefore, we have the deformations (6.118) and gradients (6.120), which we rewrite here

$$\begin{aligned} \boldsymbol{x}_0 &= \boldsymbol{g}_*(\boldsymbol{x}_\square), & \boldsymbol{F}_* &= \frac{\partial \boldsymbol{g}_*}{\partial \boldsymbol{x}_\square}, \\ \boldsymbol{x} &= \boldsymbol{g}(\boldsymbol{x}_0), & \boldsymbol{F} &= \frac{\partial \boldsymbol{g}}{\partial \boldsymbol{x}_0}, \\ \boldsymbol{x} &= \boldsymbol{g}_\square(\boldsymbol{x}_\square), & \boldsymbol{F}_\square &= \frac{\partial \boldsymbol{g}_\square}{\partial \boldsymbol{x}_\square}, \end{aligned} \tag{9.64}$$

so $\boldsymbol{F}_\square = \boldsymbol{F}\boldsymbol{F}_*$.

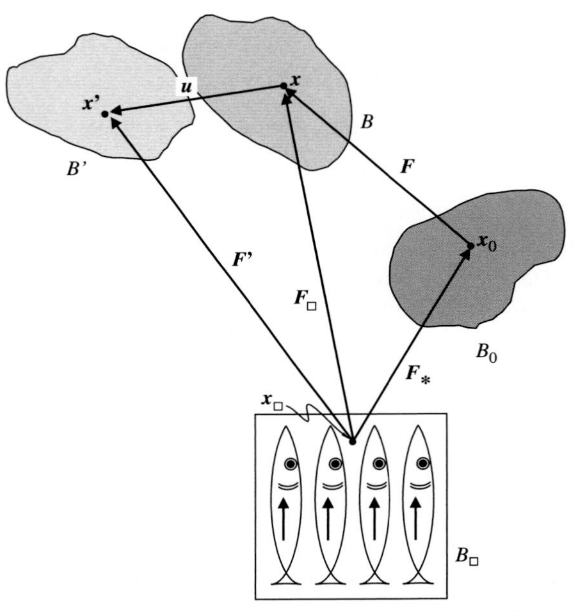

Figure 9.6. The route to obtain the equations of incremental non-linear elasticity: Four configurations are introduced, namely, the 'special' configuration for the constitutive description $B_\square$ (e.g., unloaded and orthotropic) and the other two reference configurations B_0 and B. The configuration B' moves in time and is displaced from the configuration B by superposition of the 'small' displacement $\boldsymbol{u}(\boldsymbol{x}_0,t)$, whereas all three configurations $B_\square$, B_0 and B are fixed. Eventually, B_0 is taken to coincide with B, so $\boldsymbol{F}=\boldsymbol{I}$, $J=1$, $\boldsymbol{F}_\square=\boldsymbol{F}_*$ and $J_\square=J_*$.

- A small,[9] time-dependent displacement $\boldsymbol{u}(\boldsymbol{x}_0,t)$ is superimposed to provide a dynamical perturbation to B, so

$$\boldsymbol{x}'=\boldsymbol{g}(\boldsymbol{x}_0)+\boldsymbol{u}(\boldsymbol{x}_0,t)=\boldsymbol{g}\left(\boldsymbol{g}_*(\boldsymbol{x}_\square)\right)+\boldsymbol{u}\left(\boldsymbol{g}_*(\boldsymbol{x}_\square),t\right), \tag{9.65}$$

  and therefore the perturbation in the deformation gradient of B owing to the superimposed displacement $\boldsymbol{u}$ is

$$\Delta\boldsymbol{F}=\boldsymbol{F}'-\boldsymbol{F}=\nabla\boldsymbol{u}, \qquad \text{where} \quad \nabla\boldsymbol{u}=\frac{\partial\boldsymbol{u}(\boldsymbol{x}_0,t)}{\partial\boldsymbol{x}_0}. \tag{9.66}$$

- In addition to the deformation gradient, the stress also is perturbed. Assuming the elastic law (6.121), the perturbation in the stress can be truncated at the first order and written as

$$\Delta\boldsymbol{T}_\square^{(m)}=\left(\boldsymbol{T}_\square^{(m)}\right)'-\boldsymbol{T}_\square^{(m)}=\mathbb{E}_\square[\Delta\boldsymbol{E}^{(m)}], \tag{9.67}$$

  where (truncating expressions at first-order in $\nabla\boldsymbol{u}$)

$$\mathbb{E}_\square=\frac{\partial\hat{\boldsymbol{T}}_\square^{(2)}}{\partial\boldsymbol{E}_\square^{(2)}}, \tag{9.68}$$

[9] With 'small' we intend that, all functions of $\boldsymbol{u}$ can be expanded into a Taylor series and in this expansion only the linear terms in $\nabla\boldsymbol{u}(\boldsymbol{x}_0,t)$ can be retained, while all higher-order terms neglected.

and

$$\Delta \boldsymbol{E}^{(m)} = \left(\boldsymbol{E}^{(m)}\right)' - \boldsymbol{E}^{(m)} = \boldsymbol{F}^T \nabla \boldsymbol{u} + \nabla \boldsymbol{u}^T \boldsymbol{F}. \tag{9.69}$$

The perturbation in the first Piola-Kirchhoff $\Delta \boldsymbol{S} = \boldsymbol{S}' - \boldsymbol{S}$ is given by Eq. (6.127) with the 'dot' replaced by Δ.

- Now B is taken to coincide with B_0, so $\boldsymbol{F} = \boldsymbol{I}$, $J = 1$, $\boldsymbol{F}_\square = \boldsymbol{F}_*$ and $J_\square = J_*$. The constitutive equations are governed by Eq. (6.130), the prestress is given by Eq. (6.129), which can be written together with the equations of motion as

$$\begin{aligned} &\Delta \boldsymbol{S} = \mathbb{G}[\nabla \boldsymbol{u}], \qquad \mathbb{G} = \frac{1}{J_\square}(\boldsymbol{F}_\square \boxtimes \boldsymbol{F}_\square)\mathbb{E}_\square (\boldsymbol{F}_\square \boxtimes \boldsymbol{F}_\square)^T + \boldsymbol{I} \boxtimes \boldsymbol{T}, \\ &\boldsymbol{T} = \frac{1}{J_\square} \boldsymbol{F}_\square \hat{\boldsymbol{T}}_\square^{(m)}(\boldsymbol{E}_\square^{(m)}, \boldsymbol{B}_\square) \boldsymbol{F}_\square^T, \\ &\mathsf{Div}\, \Delta \boldsymbol{S} = \rho_0 \ddot{\boldsymbol{u}}, \end{aligned} \tag{9.70}$$

where there is neither distinction between Div and div nor between ρ_0 and ρ, because B coincides with B_0.

In a quasi-static incremental problem, a time-like parameter always can be introduced to control the perturbation in the loading at the boundary, as done for the problem (9.55). In this case, the incremental displacement and its gradient become functions of this time-like parameter so that all quantities can be expanded into a Taylor series expansion around the value of t corresponding to the configuration B_0. In this way, all rate fields obtained in Chapter 3 can be employed as incremental fields. However, these rate fields also can be used as incremental fields for dynamic incremental loading. In these problems, time has the meaning of the actual physical entity, and the series expansion of functions has to be understood as an expansion in the modulus of incremental displacement. For instance, the Eulerian strain *rate* $\boldsymbol{D}$ can be understood as the *incremental* Eulerian strain for both quasi-static and dynamic incremental problems. In particular, we are interested in the regular incremental boundary value problem, which can be posed as follows:

The incremental versions of boundary conditions (9.49) and initial conditions (9.53) are

$$\boldsymbol{u} = \bar{\boldsymbol{u}}(\boldsymbol{x}_0, t), \quad \text{on } \partial B_0^\xi, \qquad \Delta \boldsymbol{S} \boldsymbol{n}_0 = \bar{\boldsymbol{\sigma}}(\boldsymbol{x}_0, t) \quad \text{on } \partial B_0^\sigma, \tag{9.71}$$

and

$$\boldsymbol{u}(\boldsymbol{x}_0, 0) = \bar{\boldsymbol{u}}(\boldsymbol{x}_0) \qquad \text{and} \qquad \dot{\boldsymbol{u}}(\boldsymbol{x}_0, 0) = \bar{\boldsymbol{w}}(\boldsymbol{x}_0), \tag{9.72}$$

where $\boldsymbol{u}$ and $\dot{\boldsymbol{u}}$ are the incremental displacement and velocity and $\Delta \boldsymbol{S}$ the increment of first Piola-Kirchhoff stress.

The regular initial boundary value problem of incremental finite elasticity in the absence of incremental body forces now can be set as follows.

- *Given* a reference configuration B_0 with prescribed boundary conditions (9.71) and initial conditions (9.72),
- *Find* the two smooth functions of place in B_0

$$\boldsymbol{u}(\boldsymbol{x}_0, t) \qquad \text{and} \qquad \Delta \boldsymbol{S} = \boldsymbol{S}(\boldsymbol{x}_0, t) \quad \text{(12 unknown components)},$$

- *Such that* the following field equations hold in B_0:

$$\begin{aligned} \mathrm{Div}\Delta \boldsymbol{S} &= \rho_0 \ddot{\boldsymbol{u}} && \text{(momentum balance)}, \\ \Delta \boldsymbol{S} &= \mathbb{G}[\mathrm{Grad}\boldsymbol{u}] && \text{(constitutive laws)}. \end{aligned} \tag{9.73}$$

For a quasi-static regular boundary value problem, $\ddot{\boldsymbol{u}} = \boldsymbol{0}$ in the preceding equations, t becomes a time-like parameter governing the loading, so $\Delta \boldsymbol{S}$ can be understood as $\dot{\boldsymbol{S}}$, and we obtain equations formally identical to the quasi-static first-order rate problem, where dots have to be understood as increments.

10 Global conditions of uniqueness and stability

Global uniqueness and stability conditions are introduced for elastic solids, and the rate form of these is derived for elastoplastic materials, which are governed by constitutive equations in rate form. Emphasis will be given to the non-associative flow rule, where the notion of Raniecki comparison solids needs to be introduced. In the particular case of the associative flow law, stability of equilibrium and stability of a deformation path are distinguished and clarified with a simple example: buckling of an inelastic column.

We begin with a premise on global uniqueness and stability for elastic solids subject to conservative loads, so we refer for simplicity to a regular, quasi-static boundary value problem of *finite elasticity* [Eqs. (9.56)], with prescribed displacement and dead loading at the boundary and dead body forces (in other words, forces independent of the configuration).[1]

Following a similar route to the Kirchhoff uniqueness theorem in the small strain theory, we postulate that there are two solutions to the finite elasticity problem, say,

$$\boldsymbol{g}_1, \quad \boldsymbol{u}_1, \quad \boldsymbol{F}_1, \quad \boldsymbol{S}_1, \qquad \text{and} \qquad \boldsymbol{g}_2, \quad \boldsymbol{u}_2, \quad \boldsymbol{F}_2, \quad \boldsymbol{S}_2. \tag{10.1}$$

Thus we may construct the difference fields

$$\Delta \boldsymbol{u} = \boldsymbol{u}_1 - \boldsymbol{u}_2, \qquad \Delta \boldsymbol{F} = \boldsymbol{F}_1 - \boldsymbol{F}_2, \qquad \Delta \boldsymbol{S} = \boldsymbol{S}_1 - \boldsymbol{S}_2, \tag{10.2}$$

which satisfy Eqs. (9.56) with homogeneous boundary conditions and null body forces. Therefore, application of the divergence theorem yields

$$\int_{B_0} \Delta \boldsymbol{S} \cdot \Delta \boldsymbol{F} = 0, \tag{10.3}$$

so a *sufficient* condition for uniqueness of solution $(\cdot)_1$ is that there is not kinematically admissible deformation corresponding to a deformation gradient $\boldsymbol{F}_2$ such that condition (10.3) is verified. Using the constitutive law (6.30), we arrive at

$$\int_{B^0} \left[\hat{\boldsymbol{S}}(\boldsymbol{F}_1) - \hat{\boldsymbol{S}}(\boldsymbol{F}_2) \right] \cdot [\boldsymbol{F}_1 - \boldsymbol{F}_2] \neq 0 \tag{10.4}$$

[1] The extension of the considerations developed here to the problem of controlled nominal tractions and displacement, problem (9.55), is straightforward.

for every deformation $\boldsymbol{g}_2$ (different from $\boldsymbol{g}_1$) and assuming the prescribed values $(9.56)_3$ on B_0^ξ.

Considering now a Green elastic material (4.4) with strain energy function $\hat{W}$ defined as

$$\boldsymbol{S} = \frac{\partial \hat{W}}{\partial \boldsymbol{F}}, \tag{10.5}$$

a local sufficient condition that implies condition (10.4) is

$$\left[\frac{\partial \hat{W}}{\partial \boldsymbol{F}}(\boldsymbol{F}_1) - \frac{\partial \hat{W}}{\partial \boldsymbol{F}}(\boldsymbol{F}_2)\right] \cdot [\boldsymbol{F}_1 - \boldsymbol{F}_2] > 0 \tag{10.6}$$

for all pairs of different $\boldsymbol{F}_1$ and $\boldsymbol{F}_2$ (which need not be the gradients of a deformation field).

A comparison with Eq. (2.159) shows that condition (10.6) represents the *strict convexity requirement* for $\hat{W}$, so for a strictly convex strain energy function of $\boldsymbol{F}$, the quasi-static, regular boundary value problem of elasticity with mixed boundary conditions of prescribed displacements and dead surface nominal tractions and body forces admits a unique solution (Hill, 1957).

Since

> *global uniqueness for problems involving large deformations is not in general to be expected,*

strict convexity of the strain energy function [Eq. (10.6)], ‘usually’ should be violated. This can be appreciated as follows: We know from Eq. (2.164) that strict convexity of a function is equivalent (with possible exceptions on a nowhere dense subset of the function domain) to positive definiteness of its Hessian, so strict convexity of the energy function is equivalent to positive definiteness of the fourth-order tensor

$$\frac{\partial \boldsymbol{S}}{\partial \boldsymbol{F}}. \tag{10.7}$$

Assuming $n = 2$ in Eq. (4.1), we can employ Eq. (4.5) to obtain for every tensor $\boldsymbol{A} \in \mathrm{Lin}$

$$\frac{\partial \boldsymbol{S}}{\partial \boldsymbol{F}}[\boldsymbol{A}] = \frac{\partial \boldsymbol{F}\boldsymbol{T}^{(2)}}{\partial \boldsymbol{F}}[\boldsymbol{A}] = \boldsymbol{A}\boldsymbol{T}^{(2)} + \boldsymbol{F}\left(\frac{\partial \boldsymbol{T}^{(2)}}{\partial \boldsymbol{F}}[\boldsymbol{A}]\right), \tag{10.8}$$

but taking the derivative and using the definition (6.101) of the elastic tensor $\mathbb{E}$ (which has the minor symmetries), we arrive from Eq. (10.8) at

$$\frac{\partial \boldsymbol{T}^{(2)}}{\partial \boldsymbol{F}}[\boldsymbol{A}] = \frac{\partial \boldsymbol{T}^{(2)}}{\partial \boldsymbol{E}^{(2)}}\left[\frac{\partial \boldsymbol{E}^{(2)}}{\partial \boldsymbol{F}}[\boldsymbol{A}]\right] = \mathbb{E}[\boldsymbol{F}^T\boldsymbol{A}], \tag{10.9}$$

so we conclude that the fourth-order tensor (10.7) coincides with the elastic tensor $\mathbb{G}$ [Eq. (6.111)]:

$$\mathbb{G} = \frac{\partial \boldsymbol{S}}{\partial \boldsymbol{F}} = (\boldsymbol{F} \boxtimes \boldsymbol{I})\mathbb{E}(\boldsymbol{F} \boxtimes \boldsymbol{I})^T + \boldsymbol{I} \boxtimes \boldsymbol{T}^{(2)}. \tag{10.10}$$

Finally, we arrive at the following statement about positive definiteness of $\mathbb{G}$, convexity of $\hat{W}$ and uniqueness:

$$\mathbb{G} \text{ positive definite } \underbrace{\overset{\Longrightarrow}{\Longleftarrow}}_{\text{'exceptions'}} \hat{W}(\boldsymbol{F}) \text{ strictly convex } \Longrightarrow \text{uniqueness}, \tag{10.11}$$

where 'exceptions' is a reminder for possible 'pathological situations' related to the fact that the converse implication may fail on a nowhere dense subset of Lin.

Positive definiteness of tensor $\mathbb{G}$ will be analysed in detail in Section 11.1 for elastoplasticity and denoted as 'PD'. The PD condition written for elastoplasticity is always more general than that for elasticity because elasticity always can be recovered, so the two conditions coincide when plastic strain and flow are absent. It will be shown that the PD condition is stronger than one may expect, but this can be understood immediately because PD implies uniqueness, and uniqueness, as said before, is certainly not the rule in a large strain formulation. To better elucidate this point, because positive definiteness has to hold for every second-order tensor, we apply tensor $\mathbb{G}$ to a tensor $\boldsymbol{F}^{-T}\boldsymbol{W}$ with $\boldsymbol{W} \in \text{Skw}$, thus obtaining (because $\mathbb{E}$ has the minor symmetries)

$$\mathbb{G}[\boldsymbol{F}^{-T}\boldsymbol{W}] = \boldsymbol{F}\mathbb{E}[\boldsymbol{W}] + \boldsymbol{F}^{-T}\boldsymbol{W}\boldsymbol{T}^{(2)} = \boldsymbol{F}^{-T}\boldsymbol{W}\boldsymbol{T}^{(2)}. \tag{10.12}$$

We consider an isotropically 'contracted' configuration, for which $\boldsymbol{F} = \lambda\boldsymbol{I}$ with $0 < \lambda < 1$, producing for an isotropic material an isotropic and negative (i.e., compressive) stress response, $\boldsymbol{T} = -\beta\boldsymbol{I}$ (with $\beta > 0$). For this deformation gradient and stress, we obtain from Eq. (10.12)

$$\mathbb{G}[\boldsymbol{F}^{-T}\boldsymbol{W}] = -\frac{\beta}{\lambda^2}\boldsymbol{F}^{-T}\boldsymbol{W}, \tag{10.13}$$

which shows that $-\beta/\lambda^2$ is a negative eigenvalue, so $\mathbb{G}$ is not positive definite for a deformation producing an isotropic compressive stress.

In order to introduce the concept of stability, we now employ the form (2.158) of convexity requirement, so we may rewrite condition (10.6) in the form

$$\hat{W}(\boldsymbol{F}_2) - \hat{W}(\boldsymbol{F}_1) - \boldsymbol{S}_1 \cdot (\boldsymbol{F}_2 - \boldsymbol{F}_1) > 0. \tag{10.14}$$

We identify fields $(\cdot)_1$ (the index 1 will be omitted) with a solution of the regular quasi-static problem (9.56), whereas fields $(\cdot)_2$ (the index $_*$ will be used instead of 2) represent merely admissible (i.e., sufficiently regular and not violating kinematic boundary conditions) kinematic fields. Integration of Eq. (10.14) over B_0 and application of the divergence theorem yields

$$\int_{B_0} [\hat{W}(\boldsymbol{F}_*) - \hat{W}(\boldsymbol{F})] > \int_{\partial B_0^{\sigma}} \boldsymbol{\sigma} \cdot (\boldsymbol{u}_* - \boldsymbol{u}) + \int_{B_0} \boldsymbol{b}_0 \cdot (\boldsymbol{u}_* - \boldsymbol{u}). \tag{10.15}$$

Equation (10.15) states that the increase in the strain energy for a departure from the solution of the quasi-static boundary value problem to the kinematics $(\cdot)_*$ is greater than the work done by the prescribed (dead) tractions and (dead) body forces. Therefore, the solution is stable.

In other words, introducing the *energy functional*

$$E(\boldsymbol{u}) = \int_{B_0} \hat{W}(\boldsymbol{F}) - \int_{\partial B_0^{\boldsymbol{\sigma}}} \boldsymbol{\sigma} \cdot \boldsymbol{u} - \int_{B_0} \boldsymbol{b}_0 \cdot \boldsymbol{u}, \tag{10.16}$$

Eq. (10.15) becomes

$$E(\boldsymbol{u}_*) > E(\boldsymbol{u}), \tag{10.17}$$

so *for a stable equilibrium configuration, the energy functional is minimised within the class of kinematically admissible deformations*. Note that the quantity

$$\Omega = -\int_{\partial B_0^{\boldsymbol{\sigma}}} \boldsymbol{\sigma} \cdot \boldsymbol{u} - \int_{B_0} \boldsymbol{b}_0 \cdot \boldsymbol{u} \tag{10.18}$$

is the potential energy associated with the 'loading device' applying nominal surface tractions $\boldsymbol{\sigma}$ and body forces $\boldsymbol{b}_0$ in a configurationally independent way. Therefore, an increment in E can be interpreted as the amount of energy supplied by 'external sources' to the deformed body and the loading device to produce a quasi-static deformation increment.

Problems of bifurcation and stability of solids subject to finite deformations include the case of an incompressible elastic cube subject to dead loading orthogonal to the faces (Marsden and Hughes, 1983, sec. 7.2) and the problem of kink band formation as analysed by Fu and Zhang (2006).

A problem that can be solved easily is the biaxial stretching of a Mooney-Rivlin membrane (Kearsley, 1986; Müller, 1996; Müller and Strehlow, 2004). With reference to the inset of Fig. 10.1, we consider a square membrane referred to a x_1–x_2-x_3 reference system (with the out-of-plane axis labelled x_3) and stretched in the plane by two equal forces. Assuming a Mooney-Rivlin constitutive law (4.27) and that the in-plane stretches are λ_1 and λ_2, the corresponding Cauchy stresses are

$$T_1 = -\pi + \mu_1\lambda_1^2 + \frac{\mu_2}{\lambda_1^2} \quad \text{and} \quad T_2 = -\pi + \mu_1\lambda_2^2 + \frac{\mu_2}{\lambda_2^2}, \tag{10.19}$$

whereas the out-of-plane stretch is $\lambda_3 = 1/(\lambda_1\lambda_2)$, and the corresponding stress T_3 has to be null, a condition which allows us to calculate π in the form

$$\pi = \frac{\mu_1}{\lambda_1^2\lambda_2^2} + \mu_2\lambda_1^2\lambda_2^2, \tag{10.20}$$

so the in-plane Cauchy stresses become

$$T_1 = \frac{\lambda_1^4\lambda_2^2 - 1}{\lambda_1^2\lambda_2^2}\left(\mu_1 - \mu_2\lambda_2^2\right) \quad \text{and} \quad T_2 = \frac{\lambda_1^2\lambda_2^4 - 1}{\lambda_1^2\lambda_2^2}\left(\mu_1 - \mu_2\lambda_1^2\right). \tag{10.21}$$

The components of the first Piola-Kirchhoff (transpose of the nominal) stress can be calculated from its definition [Eq. (3.106)], to be $S_1 = T_1/\lambda_1$ and $S_2 = T_2/\lambda_2$, so we obtain

$$\frac{S_1}{\mu_1} = \frac{\lambda_1^4\lambda_2^2 - 1}{\lambda_1^3\lambda_2^2}\left(1 - \frac{\mu_2}{\mu_1}\lambda_2^2\right) \quad \text{and} \quad \frac{S_2}{\mu_1} = \frac{\lambda_1^2\lambda_2^4 - 1}{\lambda_1^2\lambda_2^3}\left(1 - \frac{\mu_2}{\mu_1}\lambda_1^2\right). \tag{10.22}$$

Solutions corresponding to equal forces applied to the edges of the membrane can be obtained by imposing equality between the two nominal stresses, namely,

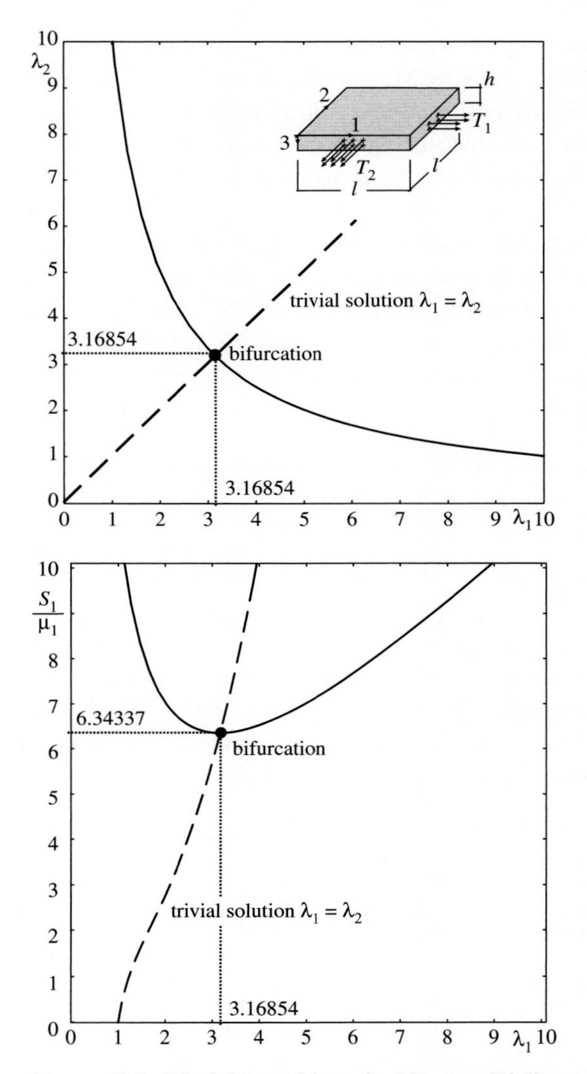

Figure 10.1. Biaxial stretching of a Mooney-Rivlin square membrane subject to equal forces (i.e., nominal stresses) along the edges (see the inset in the upper part). Top: solutions of Eq. (10.23) showing the trivial response $\lambda_1 = \lambda_2$ and the bifurcated solution. Bottom: The nominal stress response during biaxial stretching. Calculations refer to $\mu_2/\mu_1 = -0.1$.

$S_1 = S_2$, a condition that using Eq. (10.22) can be simplified to

$$(\lambda_1 - \lambda_2)\left[1 + \lambda_1^3\lambda_2^3 - \frac{\mu_2}{\mu_1}\left(\lambda_1^2 + \lambda_1\lambda_2 + \lambda_2^2 - \lambda_1^4\lambda_2^4\right)\right] = 0. \tag{10.23}$$

We observe from Eq. (10.23) that the trivial response $\lambda_1 = \lambda_2$ is always a solution. This is the unique solution in the special neo-Hookean case, $\mu_2 = 0$, but alternative solutions exist in other cases. For instance, assuming the values (4.36), $\mu_2/\mu_1 = -0.1$, the solutions of Eq. (10.23) are plotted in the upper part of Fig. 10.1, whereas the response in terms of nominal stress versus the stretch λ_1 is shown in the lower part.

The bifurcation occurs in tension at $S_1/\mu_1 = S_2/\mu_1 = 6.34337$ and $\lambda_1 = \lambda_2 = 3.16854$ and has been verified experimentally by Müller and Strehlow (2004).

From Fig. 10.1 we note that there are equal values of nominal stress corresponding to different stretches, so it becomes obviously possible to construct a $\Delta \boldsymbol{S} = \boldsymbol{0}$ and a corresponding $\Delta \boldsymbol{F} \neq \boldsymbol{0}$, and because these fields are uniform, they satisfy the bifurcation condition (10.3).

The direct application of the preceding notions to the analysis of bifurcation and stability of non-linear elastic solids is a difficult task, with several 'technical' difficulties [the interested reader is referred to Knops and Wilkes (1973), Ogden (1984) and Marsden and Hughes (1983)]. Difficulties become formidable for elastoplastic solids, where the *constitutive equations are in rate form*, so a *direct* analysis of *global* uniqueness and stability is not feasible.

We therefore will abandon concepts related to global bifurcation and stability and will focus instead on their rate or incremental counterparts.

10.1 Uniqueness of the rate problem

With reference to the Lagrangean description, we recall the first-order rate problem [Eqs. (9.57) through (9.61)], with reference to the velocity field $\dot{\boldsymbol{x}}$ when the constitutive equation (8.52) is employed in Eqs. $(9.57)_2$, $(9.59)_2$ and $(9.61)_2$. Therefore, the velocity problem can be stated as follows: Given a certain state of a body, find a continuous and piece-wise continuously twice differentiable (shortly, 'admissible') velocity field $\dot{\boldsymbol{\xi}}$ satisfying

$$
\begin{aligned}
&\dot{\boldsymbol{x}} = \dot{\boldsymbol{\xi}}(\boldsymbol{x}_0, t) && \text{on} \quad \partial B_0^{\xi},\\
&\boldsymbol{C}(\dot{\boldsymbol{F}})\boldsymbol{n}_0 = \dot{\boldsymbol{\sigma}}(\boldsymbol{x}_0, t) && \text{on} \quad \partial B_0^{\sigma},\\
&\mathrm{Div}(\boldsymbol{C}(\dot{\boldsymbol{F}})) = \boldsymbol{0} && \text{in} \quad B_0 \setminus \Sigma_0,\\
&[\![\boldsymbol{C}(\dot{\boldsymbol{F}})]\!]\boldsymbol{m}_0 = \boldsymbol{0} && \text{on} \quad \Sigma_0,
\end{aligned}
\tag{10.24}
$$

where $\dot{\boldsymbol{F}} = \nabla \dot{\boldsymbol{x}}$ in $B_0 \setminus \Sigma_0$ and $[\![\dot{\boldsymbol{x}}]\!] = \boldsymbol{0}$ across the singular surface Σ_0 of unit normal $\boldsymbol{m}_0$.

The preceding equations refer to the part of the body at yielding, that is, to the so-called plastic zone. In the elastic zone, the governing equations are the same with $\boldsymbol{C}$ replaced by $\mathbb{G}$, and singular surfaces are usually not present.

In order to obtain an exclusion condition for bifurcation in velocity, we follow the Hill (1950, 1958) argument,[2] which generalises to elastoplasticity the Kirchhoff theorem of linear elasticity. Suppose, therefore, that the velocity problem admits two solutions, say, $\dot{\boldsymbol{x}}_1$ and $\dot{\boldsymbol{x}}_2$, both satisfying Eqs. (10.24). Their difference $\Delta \dot{\boldsymbol{x}} = \dot{\boldsymbol{x}}_1 - \dot{\boldsymbol{x}}_2$ defines an admissible velocity field with gradient $\Delta \dot{\boldsymbol{F}}$. Also, the difference in stresses $\Delta \dot{\boldsymbol{S}}$ satisfies equilibrium Eqs. (9.59) and $(9.61)_1$. Moreover, the difference fields satisfy homogeneous conditions on the boundary. On application of the divergence

[2] Hill's (1950) proof was restricted to the infinitesimal theory and was based on a theorem owing to Melan (1938).

theorem, it follows that

$$\int_{B_0} \Delta\dot{\boldsymbol{S}} \cdot \Delta\dot{\boldsymbol{F}} = 0, \tag{10.25}$$

where $\Delta\dot{\boldsymbol{S}} = \mathcal{C}(\dot{\boldsymbol{F}}_1) - \mathcal{C}(\dot{\boldsymbol{F}}_2)$. It has to be stressed that *owing to the non-linearity of* $\mathcal{C}$, $\Delta\dot{\boldsymbol{S}}$ does not, in general, coincide with $\mathcal{C}(\Delta\dot{\boldsymbol{F}})$. Note also the analogy between conditions (10.25) and (10.3).

Therefore, *a sufficient condition to exclude bifurcations of the velocity problem* is

$$\boxed{\int_{B_0} \Delta\dot{\boldsymbol{S}} \cdot \Delta\dot{\boldsymbol{F}} > 0,} \tag{10.26}$$

for all pairs of distinct, admissible velocity fields assuming the given values $(10.24)_1$ on ∂B_0^{ξ}.

Note that the exclusion condition (10.26) would be true even replacing $>$ with $<$. We will see that this possibility can be excluded in terms of instability.

If the assumptions on the external loadings are relaxed so that we require neither the body forces nor the applied tractions to be independent of the deformation, Eq. (10.25) becomes

$$\int_{B_0} \Delta\dot{\boldsymbol{S}} \cdot \Delta\dot{\boldsymbol{F}} = \int_{B_0} \Delta\dot{\boldsymbol{b}}_0 \cdot \Delta\dot{\boldsymbol{x}} + \int_{\partial B_0} \Delta\dot{\boldsymbol{t}}_0 \cdot \Delta\dot{\boldsymbol{x}}, \tag{10.27}$$

where $\dot{\boldsymbol{b}}_0$ and $\dot{\boldsymbol{t}}_0$ are the rate of body forces and the rate of applied tractions on ∂B_0^{σ}, respectively. As a consequence, the exclusion condition (10.26) becomes

$$\int_{B_0} \Delta\dot{\boldsymbol{S}} \cdot \Delta\dot{\boldsymbol{F}} > \int_{B_0} \Delta\dot{\boldsymbol{b}}_0 \cdot \Delta\dot{\boldsymbol{x}} + \int_{\partial B_0} \Delta\dot{\boldsymbol{t}}_0 \cdot \Delta\dot{\boldsymbol{x}} \tag{10.28}$$

for every pair of admissible and distinct velocity fields. The interested reader is referred to Petryk (1993a) for further details.

10.1.1 Raniecki comparison solids

The difficulty in proceeding with the exclusion condition (10.26) is related to the non-linearity of the constitutive operator $\mathcal{C}$. To overcome this problem, Hill (1958) proposed introducing a *linear comparison solid*, which, when used to replace the original constitutive operator, provides a lower bound to (10.26). Results of Hill were restricted to the associative flow rule, where the comparison solid turns out to coincide with a linear solid defined by the constitutive tensor corresponding to the loading branch of the elastoplastic operator (8.53). The Hill comparison theorem later was generalised by Raniecki (1979) and Raniecki and Bruhns (1981) to cover non-associative flow rules. In particular, Raniecki introduced a family of linear comparison solids (briefly, 'Raniecki comparison solids') defined by the following constitutive tensor (for every $\psi \in \mathrm{Re}^+$):

$$\mathbb{R} = \mathbb{G} - \frac{1}{4\psi g}(\boldsymbol{M} + \psi\boldsymbol{N}) \otimes (\boldsymbol{M} + \psi\boldsymbol{N}), \tag{10.29}$$

such that the following *comparison theorem* holds true:

$$\boxed{\Delta\dot{\boldsymbol{S}} \cdot \Delta\dot{\boldsymbol{F}} \geq \Delta\dot{\boldsymbol{F}} \cdot \mathbb{R}[\Delta\dot{\boldsymbol{F}}]} \tag{10.30}$$

for every difference of tensors $\Delta\dot{\boldsymbol{F}} = \dot{\boldsymbol{F}}_1 - \dot{\boldsymbol{F}}_2$ and related difference $\Delta\dot{\boldsymbol{S}} = \boldsymbol{\mathcal{C}}(\dot{\boldsymbol{F}}_1) - \boldsymbol{\mathcal{C}}(\dot{\boldsymbol{F}}_2)$. Therefore, the exclusion condition (10.26) is necessarily satisfied when the stronger condition

$$\boxed{\int_{B_0} \dot{\boldsymbol{F}} \cdot \mathbb{C}^c[\dot{\boldsymbol{F}}] > 0} \tag{10.31}$$

holds true for all (not identically zero) continuous and piece-wise continuously twice differentiable velocity fields satisfying homogeneous conditions on ∂B_0^{ξ}. The comparison solid $\mathbb{C}^c$ in Eq. (10.31) is equal, by definition, to $\mathbb{R}$ in the current plastic zone and to $\mathbb{G}$ in the current elastic zone. Note that owing to the linearity of $\mathbb{C}^c$, the difference fields denoted by Δ in Eq. (10.26) do not appear in Eq. (10.31).

In the case of hyperelasticity and the associative flow rule, $\boldsymbol{N} = \boldsymbol{M}$, the comparison solid (10.29) reduces to the Hill comparison solid when $\psi = 1$.

The Raniecki's comparison theorem can be proved as follows: Three cases must be analysed:

1. $\boldsymbol{N} \cdot \dot{\boldsymbol{F}}_1 < 0$ and $\boldsymbol{N} \cdot \dot{\boldsymbol{F}}_2 < 0$ (unloading/unloading),
2. $\boldsymbol{N} \cdot \dot{\boldsymbol{F}}_1 > 0$ and $\boldsymbol{N} \cdot \dot{\boldsymbol{F}}_2 > 0$ (loading/loading),
3. $\boldsymbol{N} \cdot \dot{\boldsymbol{F}}_1 > 0$ and $\boldsymbol{N} \cdot \dot{\boldsymbol{F}}_2 < 0$ (loading/unloading).

In all cases, it suffices to prove that

$$\Delta\dot{\boldsymbol{S}} \cdot \Delta\dot{\boldsymbol{F}} - \Delta\dot{\boldsymbol{F}} \cdot \mathbb{R}[\Delta\dot{\boldsymbol{F}}] = \frac{1}{4\psi g}$$
$$\left(-4\psi < \boldsymbol{N} \cdot \dot{\boldsymbol{F}}_1 > \boldsymbol{M} \cdot \Delta\dot{\boldsymbol{F}} + 4\psi < \boldsymbol{N} \cdot \dot{\boldsymbol{F}}_2 > \boldsymbol{M} \cdot \Delta\dot{\boldsymbol{F}} + [(\boldsymbol{M} + \psi\boldsymbol{N}) \cdot \Delta\dot{\boldsymbol{F}}]^2\right) \geq 0, \tag{10.32}$$

which, taking into account that $\psi g > 0$ and analysing cases 1 through 3, follows directly. It is important to remark that the comparison theorem (10.30) holds true for every $\mathbb{G}$, in other words, regardless of the symmetries and the definiteness of the elastic tensor $\mathbb{G}$.

Finally, it may be worth noting that the exclusion condition (10.31) for bifurcations of the velocity problem may be shown to be sufficient to exclude second- and higher-order bifurcations (under specific regularity conditions; see Petryk and Thermann, 1985; Nguyen and Triantafyllidis, 1989; Cheng and Lu, 1993; Bigoni, 1996).[3]

10.1.2 Associative elastoplasticity

It may be interesting to keep contact with the case of the associative flow rule $\boldsymbol{Q} = \boldsymbol{P}$ and Green elasticity ($\mathbb{E}$, $\mathbb{B}$ and $\mathbb{G}$ have the major symmetry); see Petryk (1993a) for a detailed presentation. In this case, constitutive equations admit a velocity-gradient

[3] Moreover, the techniques introduced in this section to exclude bifurcation can be 'exported' to another, similar context. This is the problem of contact with friction of an elastic body with a constraint (Radi et al., 1999).

potential. With reference to Eq. (8.53), because

$$\frac{\partial <\boldsymbol{N}\cdot\dot{\boldsymbol{F}}>^2}{\partial\dot{\boldsymbol{F}}} =<\boldsymbol{N}\cdot\dot{\boldsymbol{F}}> \frac{\partial(\boldsymbol{N}\cdot\dot{\boldsymbol{F}}+|\boldsymbol{N}\cdot\dot{\boldsymbol{F}}|)}{\partial\dot{\boldsymbol{F}}} =<\boldsymbol{N}\cdot\dot{\boldsymbol{F}}> (1+\frac{\boldsymbol{N}\cdot\dot{\boldsymbol{F}}}{|\boldsymbol{N}\cdot\dot{\boldsymbol{F}}|})\boldsymbol{N} = 2<\boldsymbol{N}\cdot\dot{\boldsymbol{F}}>\boldsymbol{N}, \tag{10.33}$$

it follows that

$$\dot{\boldsymbol{S}} = \frac{\partial U}{\partial\dot{\boldsymbol{F}}}, \qquad U = \frac{\dot{\boldsymbol{F}}\cdot\mathbb{G}[\dot{\boldsymbol{F}}]}{2} - \frac{<\boldsymbol{N}\cdot\dot{\boldsymbol{F}}>^2}{2g}. \tag{10.34}$$

Let us now consider the following functional (Hill, 1958; 1959), defined for every admissible velocity field $\boldsymbol{v}$ satisfying Eq. $(10.24)_1$ on ∂B_0^{ξ}:

$$J(\boldsymbol{v}) = \int_{B_0} U(\boldsymbol{v}) - \int_{\partial B_0^{\sigma}} \dot{\boldsymbol{\sigma}}\cdot\boldsymbol{v}. \tag{10.35}$$

The vanishing of the first weak (so-called Gateaux[4]) variation of $J(\boldsymbol{v})$ with respect to every admissible variation $\boldsymbol{w}$ of $\boldsymbol{v}$ is equivalent to Eq. (9.63). Therefore, a velocity field is a solution of the rate problem if and only if it assigns to the functional (10.35) a stationary value. Moreover, when the uniqueness condition (10.26) holds true, the functional $J(\boldsymbol{v})$ can be proved to be strictly convex. It follows that in the range where Eq. (10.26) holds true, the unique solution assigns to the functional $J(\boldsymbol{v})$ a strict, absolute minimum (Hill, 1958, 1959).

For the linear comparison solid defined by $\psi = 1$, that is, by the loading branch of the constitutive operator, we have

$$\dot{\boldsymbol{S}} = \frac{\partial U^L}{\partial\dot{\boldsymbol{F}}}, \qquad U^L = \frac{\dot{\boldsymbol{F}}\cdot\mathbb{G}[\dot{\boldsymbol{F}}]}{2} - \frac{(\boldsymbol{N}\cdot\dot{\boldsymbol{F}})^2}{2g}, \tag{10.36}$$

so the exclusion condition (10.31) now corresponds to the positive definiteness of the quadratic functional (where U^L in the elastic zone has to be identified with the actual elastic potential), that is,

$$I^L(\boldsymbol{w}) = \int_{B_0} U^L(\boldsymbol{w}) > 0, \tag{10.37}$$

for every admissible field $\boldsymbol{w}$ vanishing on B_0^{ξ}.

The variational basis of the rate problem for the associative flow rule has important consequences on bifurcation. In fact, for a given deformation path, let us *assume* that a series of configurations continuously evolves in a parameter space satisfying $I^L > 0$ and that this is terminated by a configuration for which

$$\begin{cases} I^L(\boldsymbol{w}) \geq 0 & \text{for every } \boldsymbol{w}, \\ I^L(\boldsymbol{w}^*) = 0 & \text{for some } \boldsymbol{w}^* \neq \boldsymbol{0}, \end{cases} \tag{10.38}$$

[4] The first weak variation of a functional J in the direction $\boldsymbol{w}$ at $\boldsymbol{v}$ is

$$\delta J(\boldsymbol{v},\boldsymbol{w}) = \frac{d}{d\alpha} J(\boldsymbol{v}+\alpha\boldsymbol{w})|_{\alpha=0}.$$

where the admissible fields $\boldsymbol{w}$ vanish on B_0^{ξ}. As shown by Hill and Hutchinson (1975) and Young (1976), the first instant at which $I^L > 0$ fails to hold need not satisfy condition (10.38).[5] However, for situations in which Eq. (10.38) holds true, a *primary bifurcation* has been found, and I^L is stationary at the minimum point $\boldsymbol{w}^*$. This is necessary and sufficient for $\boldsymbol{w}^*$ to be a solution of the homogeneous problem. Therefore, a *critical point* has been detected. This may represent *either* a true *bifurcation point* for the comparison solid *or* a so-called limit point. In the former case, that is, when a bifurcation point in the comparison solid has been detected, the eigenmode $\boldsymbol{w}^*$ can be added to a solution $\boldsymbol{v}$ of the non-homogeneous problem to generate a bifurcated solution. In other words, if $\boldsymbol{v}$ is a solution, $\boldsymbol{v} + \gamma \boldsymbol{w}^*$, $\gamma \in \text{Re}$, is a family of possible solutions. In this family, that is, for certain values of γ, bifurcated solutions usually can be found that correspond to the plastic branch of the constitutive equations in the current plastic zone. Among these solutions, those which initiate a quasi-static post-bifurcation path represent genuine elastic-plastic bifurcations of the real elastoplastic solid and may occur under broad hypotheses (Hutchinson, 1973).

10.1.3 'In-loading comparison solid'

From the discussion relative to the associative flow rule, it should be clear that failure of the Hill/Raniecki exclusion condition (10.31) is in general not sufficient for bifurcation even in the case of associative elastoplasticity. This becomes indeed more evident for non-associative elastoplasticity, where owing to a lack of a variational structure of the governing incremental field equations, failure of condition (10.31) is far from implying bifurcation. We may note, however, that the comparison solid in the associative case plays a double role. On the one hand, it excludes bifurcation when used in Eq. (10.31), and on the other hand, it provides a bifurcated field for the comparison solid when Eq. (10.31) fails, which often can be 'adjusted' for the real elastoplastic solid. In the non-associative case, the Raniecki comparison solids are effective for excluding bifurcation but not for providing a bifurcated field that is useful for the real elastoplastic solid. Therefore, following Raniecki and Bruhns (1981), let us consider, even for non-associative flow rules, a fictitious incrementally linear solid with tangent constitutive tensor

$$\mathbb{C} = \mathbb{G} - \frac{1}{g} \boldsymbol{M} \otimes \boldsymbol{N} \tag{10.39}$$

corresponding to the loading branch of the constitutive operator (8.53) (briefly, 'in-loading comparison solid' or 'Hill comparison solid'). Let us consider a deformation path in which an elastoplastic solid is deformed in the plastic range, so the actual behaviour corresponds to the behaviour of the fictitious solid 'in-loading'. The first bifurcation for this comparison solid may correspond (in the sense already explained for the associative flow rule) to a possible bifurcation of the real elastoplastic solid. Therefore, for non-associative elastoplasticity,

[5] In Hill and Hutchinson (1975) and Young (1976), this possibility arises in connection with the achievement of a situation where surface bifurcation modes exist of arbitrarily short wavelength (see Chapter 12).

two bounds can be defined for bifurcation. The lower bound corresponds to the failure of the exclusion condition (10.31) for the optimal Raniecki comparison solid $\mathbb{R}$. *The upper bound corresponds to bifurcation of the 'in-loading comparison solid' defined by the incremental tensor* $\mathbb{C}$.

Whether or not some bifurcation actually occurs in the real elastoplastic solid within the two bounds is still an open question. No examples are in fact yet known.

10.2 Stability in the Hill sense

Let us consider a generic equilibrium configuration of a body at a fixed value of the loading parameter governing the deformation so that the prescribed displacements on the boundary are fixed and the prescribed nominal surface tractions correspond, momentarily, to dead loading. Roughly speaking, the configuration is called 'stable' when the effects of a small disturbance remain sufficiently small during the entire motion subsequent to application of the disturbance itself. Therefore, stability analysis involves considerations about the dynamics of the system. The definition of stability becomes a precise concept when the measures of the distances and class of perturbations are specified. However, path dependence of inelastic material makes a rigorous analysis awkward. Therefore, we content ourselves with presenting the simple analysis that was proposed by Hill (1958) in the context of the associative flow rule. In that context, Hill's analysis has a much more firm basis than in the case of a non-associative flow rule (Petryk, 1991, 1985b, 1993a, 1993b).

Let us confine our attention to *direct* paths of departure from the equilibrium configuration (thus excluding arbitrary circuitous paths). In this way, a perturbed motion is a priori assumed such that variations of the direction of the velocity field along the path are negligible. This is the so-called directional stability which was analysed by Hill in the framework of associative elastoplasticity. Along any admissible direct path starting from the equilibrium state under consideration, the work of deformation in the body can be written as

$$W = \int_{B_0} \boldsymbol{S} \cdot \mathrm{Grad}\, \Delta \boldsymbol{u} + \frac{1}{2} \int_{B_0} \Delta \boldsymbol{S} \cdot \mathrm{Grad}\, \Delta \boldsymbol{u} + o((\Delta t)^2), \tag{10.40}$$

where $\boldsymbol{S}$ is the first Piola-Kirchhoff stress at the equilibrium state, and $\Delta \boldsymbol{u}$ and $\Delta \boldsymbol{S}$ are the increments in displacement and stress reached along the path. Finally, Δt is the increment in the time-like parameter measuring the length of the path. In view of the fact that $\Delta \boldsymbol{u}$ vanishes on ∂B_0^{ξ}, from the principle of virtual power, the first integral in Eq. (10.40) is equal to the work done by external *dead* loads W^{load}. Therefore, the work difference can be written as

$$W - W^{\mathrm{load}} = \frac{(\Delta t)^2}{2} \int_{B_0} \dot{\boldsymbol{S}} \cdot \nabla \dot{\boldsymbol{u}} + o((\Delta t)^2). \tag{10.41}$$

It follows from Eq. (8.52) that if

$$\boxed{\int_{B_0} \nabla \boldsymbol{w} \cdot \boldsymbol{C}(\nabla \boldsymbol{w}) > 0, \quad \text{(stability in Hill's sense)}} \tag{10.42}$$

for every admissible (not identically zero) velocity field $\boldsymbol{w}$ taking zero values on ∂B_0^{ξ}, then any movement from the equilibrium configuration requires some additional energy to be supplied to the system from external sources. In this sense, Eq. (10.42) is a *sufficient condition for directional stability* of equilibrium. We will briefly refer to this condition as 'stability in the Hill sense', even if

- It is only a sufficient condition (for the assumed class of paths).
- Hill never proposed to use this condition in a broader context than associative elastoplasticity.

With reference to elasticity, with $>$ replaced with $\geq$, condition (10.42) was proposed by Hadamard (1903) and is usually called 'infinitesimal stability'. As noticed by Truesdell and Noll (1965, section 68) and Beatty (1987), the existence of a stored energy function is not essential in the preceding definition of stability. In the elastoplastic case, when the associative flow rule is assumed, condition (10.42) with $>$ replaced with $\geq$ also was proved to be necessary for stability under broad hypotheses (Petryk, 1993a, 1993b).

In the case of non-associative flow laws, two important points should be emphasised. First, instability as related to violation of condition (10.42) is not proven, so Eq. (10.42) may not be a necessary condition for stability. Second, condition (10.42) should not even be considered sufficient. There are in fact certain instability phenomena, such as flutter, which may occur even when condition (10.42) is satisfied (see Section 13.3).

It is worth noting that Eq. (10.31) implies Eq. (10.42), more explicitly

$$\boxed{\text{Exclusion condition for bifurcation} \Longrightarrow \text{stability in the Hill sense,}}$$

but the converse need not be true.

Note that the preceding statement is *not* equivalent to saying that uniqueness implies stability because the Raniecki condition is a mere *sufficient* condition for uniqueness.

10.2.1 Associative elastoplasticity

For the associative flow rule, the rate potential (10.34) can be used in Eq. (10.42), so the stability in the Hill sense corresponds to the positivity for the functional

$$I(\boldsymbol{w}) = \int_{B_0} U(\nabla \boldsymbol{w}) > 0, \quad \text{(stability in Hill's sense for } \boldsymbol{M} = \boldsymbol{N}) \tag{10.43}$$

for every admissible (not identically zero) velocity field $\boldsymbol{w}$ taking zero values on ∂B_0^{ξ}.

The critical instant when condition (10.43) fails to hold along a certain deformation path corresponds *usually* to positive semi-definiteness of I, so

$$\begin{cases} I(\boldsymbol{w}) \geq 0 & \text{for every } \boldsymbol{w} \\ I(\boldsymbol{w}^{\square}) = 0 & \text{for some } \boldsymbol{w}^{\square} \neq \boldsymbol{0} \end{cases} \tag{10.44}$$

(where the admissible fields $\boldsymbol{w}$ vanish on B_0^{ξ}), which means that $I(\boldsymbol{w})$ is stationary at the minimum $\boldsymbol{w} = \boldsymbol{w}^{\square}$. Under *both* the assumptions of dead loading on ∂B_0^{σ}, namely, $\dot{\boldsymbol{\sigma}} = \boldsymbol{0}$, *and* prescribed null velocities $\boldsymbol{v} = \boldsymbol{0}$ on ∂B_0^{ξ}, functional J [Eq. (10.35)], coincides with I. In these conditions, $\boldsymbol{w}^{\square}$ (modulo an arbitrary positive multiplier) is a non-trivial solution of the first-order rate problem with homogeneous boundary conditions. The configuration for which such $\boldsymbol{w}^{\square}$ exists is a so-called eigenstate, so condition (10.44) defines a *primary eigenstate*. From this perspective, the condition for the primary bifurcation (10.38) is equivalent to the condition for a primary eigenstate for the linear 'in-loading comparison solid'.

10.2.2 Stability of a quasi-static deformation process

Until now, we have presented stability of equilibrium at fixed load. The Shanley (1947) example (see the section 1.3) shows that for elastoplastic systems, there is the possibility of a continuous range of bifurcations without instability of equilibrium, at least in the fundamental path and near to this (Hutchinson, 1973, 1974). At a bifurcation point, more than one deformation path can be followed, so the question arises as to which of the available paths actually will be followed. It is clear that the notion of instability of equilibrium is not sufficient to answer this question. For materials provided with a strain rate potential (the associative flow law in our case), Petryk (1985a, 1991; see also 1993a) has solved this problem by introducing the concept of 'path stability' or 'stability in the Petryk sense', which is now discussed briefly.

While for elastic systems stability of a quasi-static deformation path is identified with the stability of the equilibrium states traversed by the system, the situation changes for path-dependent, incrementally non-linear materials (as in the case of elastoplastic solids). Roughly speaking, stability of equilibrium implies that vanishing small disturbing agents have a negligible effect at fixed loads, but the effect of these agents may become relevant when applied during a continued deformation path. In other words, vanishing small perturbing agents may bring the system to a position far from that corresponding, at the same load, to the fundamental path. Note that when applied at varying loads, a small disturbance can activate different branches of the constitutive response having a completely different stiffnesses. Following Petryk (1985a, 1991), a deformation process is identified with a quasi-static motion, and stability of motion against persistent disturbances is analysed, that is, the effects are considered on the motion of perpetually disturbing small influences. Therefore, using Petryk's words 'if application of infinitesimally perturbing forces in any finite interval of time can cause finite deviations from a theoretical, unperturbed deformation process, then such a process is regarded as being unstable and, consequently, practically unrealisable in a physical system.' To make this criterion more precise, we have to go back to the energy functional (10.16) and rewrite it as [controlled nominal tractions and body forces are assumed, so Ω, eqn. (10.18), remains unchanged]

$$E = \int_{B_0} \int_{\text{deform. path}} \boldsymbol{S} \cdot d\boldsymbol{F} + \Omega, \tag{10.45}$$

making clear that the work of deformation in the body is path-dependent or, in other words, has to be obtained through integration along the entire deformation history. The functional (10.45) is defined for every kinematically admissible deformation process at fixed or varying loads.

The first-order rate of Eq. (10.45) is

$$\dot{E}(\boldsymbol{v}) = \int_{B_0} \boldsymbol{S} \cdot \nabla \boldsymbol{v} + \dot{\Omega}, \tag{10.46}$$

where

$$\dot{\Omega}(\boldsymbol{v}) = -\int_{\partial B_0^{\sigma}} \boldsymbol{\sigma} \cdot \boldsymbol{v} - \int_{B_0} \boldsymbol{b}_0 \cdot \boldsymbol{v} - \int_{\partial B_0^{\sigma}} \dot{\boldsymbol{\sigma}} \cdot \boldsymbol{u} - \int_{B_0} \dot{\boldsymbol{b}}_0 \cdot \boldsymbol{u}. \tag{10.47}$$

Thus, since the virtual work equation can be written as

$$\int_{B_0} \boldsymbol{S} \cdot \nabla \boldsymbol{v} = \int_{\partial B_0^{\sigma}} \boldsymbol{\sigma} \cdot \boldsymbol{v} + \int_{B_0} \boldsymbol{b}_0 \cdot \boldsymbol{v}, \tag{10.48}$$

and the terms

$$-\int_{\partial B_0^{\sigma}} \dot{\boldsymbol{\sigma}} \cdot \boldsymbol{u} - \int_{B_0} \dot{\boldsymbol{b}}_0 \cdot \boldsymbol{u} \tag{10.49}$$

are independent of $\boldsymbol{v}$, equilibrium can be expressed as

$$\dot{E}(\boldsymbol{v}) = \text{const}. \tag{10.50}$$

for all admissible velocity fields $\boldsymbol{v}$.

The rate of $\dot{E}$ [Eq. (10.46)], defines $\ddot{E}(\boldsymbol{v})$. Under the assumption of the existence of a strain rate potential (for associative elastoplasticity in our context) and conservative incremental loading (as in the case of Ω), Petryk (1985a, 1991) has proven that along a deformation path (satisfying broad regularity properties), the value of $\ddot{E}(\boldsymbol{v})$ is minimised by the fundamental solution up to a certain critical state, beyond which the path is defined as unstable. In other words, a quasi-static deformation process is unstable in the energy sense when the respective velocity solution $\boldsymbol{v}^0$ does not minimise the value of $\ddot{E}(\boldsymbol{v})$ over all admissible fields $\boldsymbol{v}$, that is, when

$$\ddot{E}(\boldsymbol{v}) < \ddot{E}(\boldsymbol{v}^0) \quad \text{for some admissible } \boldsymbol{v}. \tag{10.51}$$

Although the definition of $\ddot{E}(\boldsymbol{v})$ is valid also for non-associative elastoplasticity, the minimum property is crucially connected to the existence of a strain rate potential (which does not exist in the absence of major symmetry in the tangent constitutive operator), and the concept of path stability looses its definition for non-associative flow laws.

10.2.3 An example: Elastoplastic column buckling

Instability and bifurcation can be illustrated with simple examples in the case of *associative elastoplasticity*. One of these is the two degrees of freedom Shanley (1947) model for inelastic column buckling (see Section 1.3). In that example, we observed that a continuous range of bifurcations can begin at the tangent modulus load in conditions of stable equilibrium. Equilibrium becomes unstable in the Hill sense at the reduced modulus load, and the continuous spectrum of bifurcations terminates at the load for purely elastic bifurcation. In the range of loads comprised between tangent and reduced-modulus loads, the equilibrium is stable in the Hill sense, but the fundamental deformation path becomes unstable in the sense made clear by Petryk (1985a), Eq. (10.51).

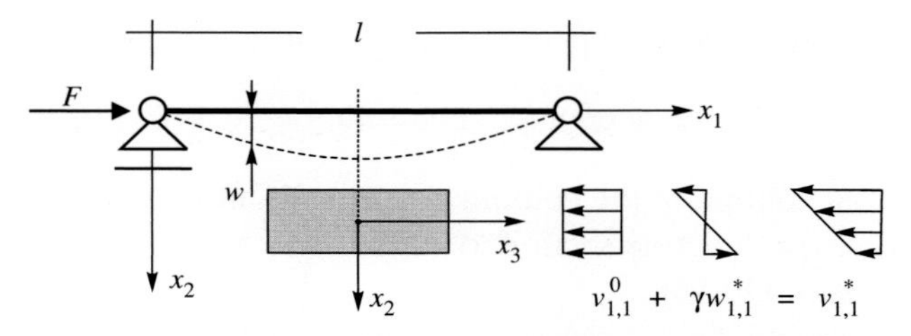

Figure 10.2. Column of length l subject to a compressive load F acting in the line of centroids along axis x_1. Axis x_3 is parallel to the weaker axis of the cross section (sketched as rectangular but generic in the mathematical treatment), so flexion owing to buckling occurs in the x_1–x_2 plane. Note the mode superposition at the primary bifurcation point, in which $v^0_{1,1}$ is the fundamental mode and $w^*_{1,1}$ is the eigenmode, multiplied by the amplitude γ. The buckling mode results as the sum of the fundamental mode plus the eigenmode 'calibrated' by the amplitude γ, namely, $v^*_{1,1} = v^0_{1,1} + \gamma w^*_{1,1}$.

Another simple example is the buckling of an elastoplastic column subject to compression. Here we follow Hill and Sewell (1960) and Petryk (1993) and consider a straight column of *current* length l constrained with two hinges at its ends and currently subject to a *uniform* compressive Cauchy stress $-\sigma$ corresponding to an axial force $F = \sigma A$, where A is the (uniform) area of the column cross section (Fig. 10.2). A Cartesian system x_1, x_2, x_3 is assumed with axis x_1 along the line of centroids of the cross sections and the axis x_3 parallel to the weaker principal axis of section at a centroid (note that the cross section is sketched as rectangular in Fig. 10.2 but remains arbitrary in the following calculations).

The column has been compressed of an amount sufficient to reach the current configuration and load F. We assume this configuration as reference, where only a Cauchy stress $-\sigma$ parallel to axis x_1 is present. Therefore, we assume a constitutive equation in the form of Eq. (8.58), so in the reference system x_1–x_2–x_3 we have

$$\dot{S}_{ij} = \overset{\circ}{\mathrm{K}}_{ij} - \sigma v_{i,1}\delta_{1j}, \tag{10.52}$$

and the exclusion condition (10.31), in the form of Eq. (10.37), becomes

$$I^L = \int_{\text{column}} \left[\overset{\circ}{\boldsymbol{K}} \cdot \boldsymbol{D} - \sigma \left(v_{1,1}^2 + v_{2,1}^2 + v_{3,1}^2 \right) \right] > 0, \tag{10.53}$$

so the critical stress σ for bifurcation is given by

$$\sigma_{cr} = \min \left\{ \frac{\int_{\text{column}} \overset{\circ}{\boldsymbol{K}} \cdot \boldsymbol{D}}{\int_{\text{column}} \left(v_{1,1}^2 + v_{2,1}^2 + v_{3,1}^2 \right)} \right\}. \tag{10.54}$$

Equation (10.54) coincides with eq. (7) given by Hill and Sewell (1960), but instead of employing their form of admissible velocity field, we postulate that the column is slender, so the shear distorsion at buckling is assumed to be negligible when compared with the flexural deformation, and the usual beam theory approximation is used:

$$v_1 = -x_2 w'(x_1), \qquad v_2(x_1) = w(x_1), \qquad v_3 = 0 \tag{10.55}$$

(where a prime denotes differentiation with respect to x_1), and consequently,

$$D_{11} = v_{1,1} = -x_2 w''(x_1), \qquad v_{2,1} = -v_{1,2} = w'(x_1), \qquad D_{22} = D_{12} = D_{3i} = 0. \quad (10.56)$$

Since the integral taken over the volume of the column can be split into an integral along the column axis and one over the cross section, condition (10.53) [equivalent to Eq. (10.54)] becomes, using Eqs. (10.56),

$$I^L(w(x_1)) = \int_l \left\{ (\mathbb{D}_{1111} - \sigma) I_3 [w''(x_1)]^2 - F[w'(x_1)]^2 \right\} dx_1 > 0, \qquad \forall w(x_1), \quad (10.57)$$

where

$$I_3 = \int_A x_2^2 \quad (10.58)$$

is the second moment of area about the weaker axis. Since

$$E_t = \dot{S}_{11}/v_{1,1} = \mathbb{D}_{1111} - \sigma \quad (10.59)$$

is the tangent modulus to the nominal stress versus logarithmic strain curve in uniaxial tension, from Eq. (10.57), uniqueness can be expressed as positivity of functional I^L

$$I^L(w(x_1)) = \int_l \left\{ I_3 E_t [w''(x_1)]^2 - F[w'(x_1)]^2 \right\} dx_1, \quad (10.60)$$

so the Euler-Lagrange equations (see, e.g., Sagan, 1961) are

$$w'''(x_1) + \frac{F}{I_3 E_t} w'(x_1) = 0, \quad (10.61)$$

from which the known solution in terms of critical load F_t and mode w^* is readily obtained

$$F_t = \frac{\pi^2 E_t I_3}{l^2}, \qquad w^*(x_1) = \sin \frac{\pi x_1}{l}, \quad (10.62)$$

satisfying the boundary conditions $w(0) = w(l) = 0$. The critical load (10.62) is the so-called Shanley or tangent modulus load, corresponding to a primary bifurcation point. At increasing axial force F, the eigenmode can be 'calibrated' by selecting some amplitude within an interval and superimposing it on the fundamental mode of uniform compression to generate a non-trivial solution describing incipient buckling without abrupt unloading at the instant of bifurcation (see Fig. 10.2).

Let us now examine the stability functional (10.44). In this case, we have to take care of the possibility of elastic unloading, which means we must consider the Macaulay bracket in Eq. (8.58), so the elastic modulus must replace the tangent modulus in the zone suffering elastic unloading. The critical load corresponding to a primary eigenstate corresponds to incipient buckling *at fixed load* and defines the so-called Engesser-Kármán (see Engesser, 1889; von Kármán, 1947) or reduced-modulus load F_r. Since the elastic modulus is (for associative elastoplasticity[6]) higher than the tangent modulus, it follows that

$$F_r \geq F_t. \quad (10.63)$$

[6] The incremental stiffness of the loading branch of non-associative elastoplasticity may be greater than the incremental stiffness at elastic unloading (Mróz, 1963, 1966).

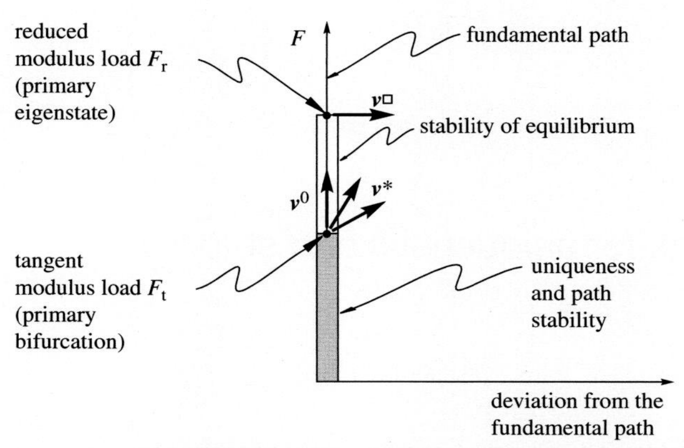

Figure 10.3. Elastoplastic column under axial load: Illustration of the tangent modulus (primary bifurcation) and the reduced-modulus (primary eigenstate) loads. Until $F < F_t$, the response is unique and stable, so both Hill's sufficient condition for uniqueness (10.37) [equivalent to Eq. (10.57)] and Hill's sufficient condition for stability (10.43) are satisfied. At $F = F_t$, Eq. (10.38) is satisfied, and bifurcations are possible at increasing load. When the load falls within the interval $[F_t, F_r[$, bifurcations are possible, but the equilibrium is still stable, so Hill's sufficient condition for stability (10.43) is satisfied. At $F = F_r$, stability of equilibrium is lost [Eq. (10.44)], and this instability can occur at fixed load. Note that when $F_t < F < F_r$, a continuous range of bifurcations at increasing loads is expected to become available. Moreover, the fundamental path becomes unstable in this range [Eq. (10.51)].

In conclusion, we remark that the difference between the tangent modulus and the reduced-modulus critical points is that *bifurcation* is possible at the former point *at increasing load*, but the equilibrium configuration remains stable, whereas *bifurcation and instability of equilibrium at fixed load* are possible at the latter; see Fig. 10.3.

It is important to highlight that the situation sketched in Fig. 10.3 *is a situation usual in associative elastoplasticity* and occurs much more generally than in the example of the column under axial load. However, for non-associative elastoplasticity, owing to the non-symmetry of the constitutive operator and the consequent lack of a variational basis, the situation is still not clear. In this case, summarizing, (1) the primary bifurcation point is replaced by *bounds* to an unknown value, (2) the Hill criterion of stability is not sufficient to exclude all forms of instability (e.g., flutter), and (3) the definition of path instability [Eq. (10.51)] cannot be applied owing to the lack of minimum properties of the solution in velocities.

11 Local conditions for uniqueness and stability

Local conditions sufficient and necessary for the uniqueness of the rate boundary value problem, namely, positive definiteness and non-singularity of the constitutive operator, are presented. For homogeneous problems with displacement prescribed on the whole boundary, these become strong ellipticity of the Raniecki comparison solid and ellipticity of the in-loading comparison solid. Strain localisation is also explained in terms of ellipticity loss. Conditions for the onset of flutter instability are finally investigated.

From the uniqueness and stability criteria considered in the preceding chapter, local conditions may be derived, which are treated herein. The importance of local conditions lies in the connection to material instabilities, namely, to instabilities which can develop from a point in a continuum and therefore result independent of the boundary conditions. For instance, we will see that loss of ellipticity corresponds to shear band formation. The following five local criteria will be analysed in this chapter:

- Positive definiteness of the constitutive operator (PD)
- Non-singularity of the constitutive operator (NS)
- Strong ellipticity (SE)
- Ellipticity (E)
- Flutter (F)

To begin providing an example of the preceding criteria in the simple case of the infinitesimal theory of isotropic elasticity, we recall the condition of positive definiteness (PD) [Eqs. (2.192)], which is the positiveness of the eigenvalues 2μ and $\lambda + 2\mu/3$ of the elastic fourth-order tensor. Non-singularity (NS) corresponds to the condition of non-vanishing of these eigenvalues, and in a similar vein, strong ellipticity (SE) corresponds to the positiveness of the eigenvalues μ and $\lambda + 2\mu$ of the acoustic tensor (2.176), whereas ellipticity (E) corresponds to the non-vanishing of the same eigenvalues. Flutter (F) occurs when the eigenvalues of the acoustic tensor become complex conjugate, which is simply impossible in the case of hyper-elasticity, where the acoustic tensor is symmetric, so flutter is excluded. The situation is sketched in Fig. 11.1 [note that nothing changes in the classification if the anisotropic elastic law given by Eq. (6.196) is considered instead of isotropic elasticity].

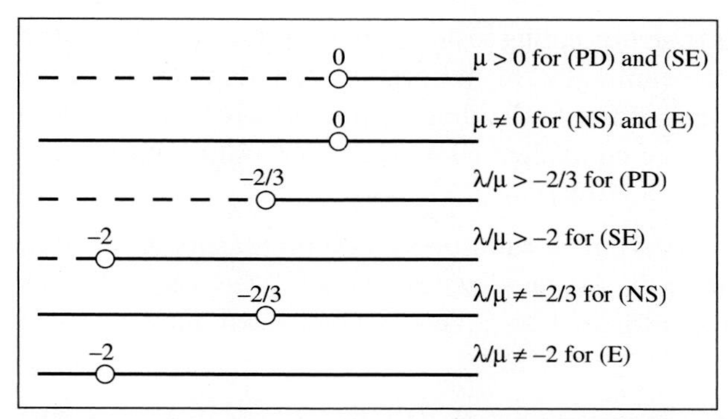

Figure 11.1. Conditions of positive definiteness (PD), non-singularity (NS), strong ellipticity (SE), and ellipticity (E) of the constitutive tensor of linear isotropic elasticity, with Lamé constants λ and μ (the solid lines correspond to the parameter range where the conditions hold). Note that (PD) and (NS) concern, respectively, positiveness and non-vanishing of the eigenvalues of the fourth-order elastic tensor, whereas (SE) and (E) concern, respectively, positiveness and non-vanishing of the eigenvalues of the second-order acoustic tensor.

In the following we will assume for simplicity that *the constitutive equations for a non-linear elastic or elastoplastic material evolve continuously with time (or with the loading parameter playing the role of time)*. Even under this hypothesis, the criteria just listed remain different, so the only connection we can draw in general is

$$\boxed{(\text{PD}) \Longrightarrow \begin{cases} (\text{SE}) \Longrightarrow (\text{E}) \\ (\text{NS}) \end{cases}}$$

For the *associative flow rule*, in a continuous deformation path initiating when (PD) [or (SE)] holds, failure of (PD) [or (SE)] will be shown to occur simultaneous to failure of (NS) [or (E)]. This will not be the case for non-associative flow rules, where (PD) and (NS), as well as (SE) and (E), are all different conditions, even in the case of a continuous loading path initiating when (PD) or (SE) holds.

We begin now with the positive definiteness (PD) condition, followed by the non-singularity (NS) condition. It will be shown that the strong ellipticity (SE) and ellipticity (E) conditions are analogous to the other two conditions but on the acoustic tensor instead of on the fourth-order tangent tensor. Finally, flutter instability (F) will be presented.

11.1 A local sufficient condition for uniqueness: Positive definiteness of the constitutive operator

It is easy to observe that when the Raniecki fourth-order tensor $\mathbb{R}$ [Eq. (10.29)], is positive definite at every point of the body, the uniqueness and stability criteria [Eqs. (10.31) and (10.42)], are both satisfied. Positive definiteness of tensor $\mathbb{R}$ implies positive definiteness of tensor $\mathbb{G}$, which is assumed in this chapter. Following Raniecki and Bruhns (1981), we will show that

$$\boxed{\text{Positive definiteness of } \mathbb{R}^{\text{opt}} \Longleftrightarrow \text{positive definiteness of } \mathbb{C} \text{ (PD condition)},}$$

where $\mathbb{R}^{opt}$ is the Raniecki solid corresponding to an optimal value of ψ to be defined later. In other words, positive definiteness of the best chosen Raniecki solid corresponds to positive definiteness of the 'in-loading comparison solid'. Obviously, positive definiteness of $\boldsymbol{\mathcal{C}}$ and $\mathbb{C}$ are equivalent (if $\boldsymbol{X}\cdot\boldsymbol{\mathcal{C}}(\boldsymbol{X}) = \boldsymbol{X}\cdot\mathbb{G}[\boldsymbol{X}]$, it suffices to consider $-\boldsymbol{X}$ to obtain $\boldsymbol{X}\cdot\boldsymbol{\mathcal{C}}(\boldsymbol{X}) = \boldsymbol{X}\cdot\mathbb{C}[\boldsymbol{X}]$).

Note that for the non-associative flow rule (only), certain tensors $\boldsymbol{X}$ make the *elastoplastic response, stiffer than the elastic response*, namely, $\boldsymbol{X}\cdot\mathbb{G}[\boldsymbol{X}] < \boldsymbol{X}\cdot\mathbb{C}[\boldsymbol{X}]$ (Mróz, 1963, 1966; Runesson and Mróz, 1989). This can be shown immediately by selecting tensor $\boldsymbol{X}$ such that

$$\boldsymbol{X}\cdot\boldsymbol{N} > 0 \quad \text{and} \quad \boldsymbol{X}\cdot\boldsymbol{M} < 0, \tag{11.1}$$

so the condition of plastic loading is satisfied, and

$$\boldsymbol{X}\cdot\boldsymbol{\mathcal{C}}(\boldsymbol{X}) = \boldsymbol{X}\cdot\mathbb{C}(\boldsymbol{X}) = \boldsymbol{X}\cdot\mathbb{G}[\boldsymbol{X}] + \frac{1}{g}(\boldsymbol{N}\cdot\boldsymbol{X})|\boldsymbol{M}\cdot\boldsymbol{X}| > \boldsymbol{X}\cdot\mathbb{G}[\boldsymbol{X}]. \tag{11.2}$$

This peculiarity, related to the difference between flow mode tensor $\boldsymbol{M}$ and yield function gradient $\boldsymbol{N}$, is also exhibited by the rate laws of friction Eqs. (1.39) and (1.40).

In order to show the equivalence between positive definiteness of $\mathbb{R}^{opt}$ and $\mathbb{C}$, under the assumption that $\mathbb{G}$ is positive definite, let us consider, for every tensor $\boldsymbol{X} \in \mathsf{Lin}$, the following inequality:

$$\boldsymbol{X}\cdot\mathbb{C}[\boldsymbol{X}] \geq \boldsymbol{X}\cdot\mathbb{R}[\boldsymbol{X}] = \boldsymbol{X}\cdot\tilde{\mathbb{G}}[\boldsymbol{X}] - \frac{(\boldsymbol{X}\cdot\boldsymbol{R})^2}{4\psi g}, \tag{11.3}$$

where $\boldsymbol{R} = \boldsymbol{M} + \psi\boldsymbol{N}$ and

$$\tilde{\mathbb{G}} = \frac{\mathbb{G} + \mathbb{G}^T}{2} \tag{11.4}$$

is the symmetric (with respect to the major symmetry) of part of $\mathbb{G}$; only this part plays a role (in fact, $\boldsymbol{X}\cdot\mathbb{G}[\boldsymbol{X}] = \boldsymbol{X}\cdot\tilde{\mathbb{G}}[\boldsymbol{X}]$). The Cauchy-Schwarz inequality in the metric $\tilde{\mathbb{G}}$ (see Section 2.12) can be used to yield

$$(\boldsymbol{X}\cdot\boldsymbol{R})^2 = (\boldsymbol{X}\cdot\tilde{\mathbb{G}}[\tilde{\mathbb{G}}^{-1}[\boldsymbol{R}]])^2 \leq (\boldsymbol{X}\cdot\tilde{\mathbb{G}}[\boldsymbol{X}])(\boldsymbol{R}\cdot\tilde{\mathbb{G}}^{-1}[\boldsymbol{R}]), \tag{11.5}$$

which, employed in Eq. (11.3), gives

$$\boldsymbol{X}\cdot\mathbb{R}[\boldsymbol{X}] \geq \frac{\boldsymbol{X}\cdot\tilde{\mathbb{G}}[\boldsymbol{X}]}{4\psi g}\left(4\psi g - \boldsymbol{R}\cdot\tilde{\mathbb{G}}^{-1}[\boldsymbol{R}]\right). \tag{11.6}$$

From Eq. (11.6), we note that $\boldsymbol{\mathcal{C}}$, $\mathbb{C}$ and $\mathbb{R}$ are positive definite when

$$g > \frac{(\boldsymbol{M} + \psi\boldsymbol{N})\cdot\tilde{\mathbb{G}}^{-1}[\boldsymbol{M} + \psi\boldsymbol{N}]}{4\psi}. \tag{11.7}$$

However, when Eq. (11.7) is violated, $\mathbb{R}$ is not positive definite. In other words, Eq. (11.7) is a necessary and sufficient condition for positive definiteness of $\mathbb{R}$. To show this, it suffices to note that

$$g \leq \frac{\boldsymbol{R}\cdot\tilde{\mathbb{G}}^{-1}[\boldsymbol{R}]}{4\psi}, \qquad \boldsymbol{X} \propto \tilde{\mathbb{G}}^{-1}[\boldsymbol{R}] \quad \Longrightarrow \quad \boldsymbol{X}\cdot\mathbb{R}[\boldsymbol{X}] \leq 0. \tag{11.8}$$

We now can consider the dependence on ψ and state that

$$g > \inf_{\psi>0} \left\{ \frac{(\boldsymbol{M}+\psi \boldsymbol{N}) \cdot \tilde{\mathbb{G}}^{-1}[\boldsymbol{M}+\psi \boldsymbol{N}]}{4\psi} \right\} \implies \boldsymbol{C} \text{ and } \mathbb{C} \text{ positive definite.} \tag{11.9}$$

The minimum problem is solved by

$$\psi = \sqrt{\frac{\boldsymbol{M} \cdot \tilde{\mathbb{G}}^{-1}[\boldsymbol{M}]}{\boldsymbol{N} \cdot \tilde{\mathbb{G}}^{-1}[\boldsymbol{N}]}}, \tag{11.10}$$

which defines $\mathbb{R}^{\text{opt}}$ and yields the following proposition:

$$\boxed{g > g_{cr}^{PD} = \frac{1}{2}\left(\sqrt{(\boldsymbol{M} \cdot \tilde{\mathbb{G}}^{-1}[\boldsymbol{M}])(\boldsymbol{N} \cdot \tilde{\mathbb{G}}^{-1}[\boldsymbol{N}])} + \boldsymbol{M} \cdot \tilde{\mathbb{G}}^{-1}[\boldsymbol{N}] \right) \geq 0 \iff \text{(PD).}} \tag{11.11}$$

Note that $g_{cr}^{PD} = 0 \iff \boldsymbol{M} \propto -\boldsymbol{N}$, a condition that never should be satisfied for realistic constitutive models.

In order to complete the proof of Eq. (11.11), it remains to show that when $g \leq g_{cr}^{PD}$, tensor $\mathbb{C}$ is not positive definite. To this purpose, it suffices to note that

$$g \leq g_{cr}^{PD}, \qquad \boldsymbol{X} \propto \sqrt{\boldsymbol{M} \cdot \tilde{\mathbb{G}}^{-1}[\boldsymbol{M}]}\, \tilde{\mathbb{G}}^{-1}[\boldsymbol{N}] + \sqrt{\boldsymbol{N} \cdot \tilde{\mathbb{G}}^{-1}[\boldsymbol{N}]}\, \tilde{\mathbb{G}}^{-1}[\boldsymbol{M}]$$
$$\implies \boldsymbol{X} \cdot \mathbb{C}[\boldsymbol{X}] \leq 0. \tag{11.12}$$

The condition (PD) has been expressed in terms of a critical value of the hardening modulus. In general, any condition expressed in this way is implicitly referred to a sufficiently regular deformation path in which the plastic modulus is a continuously varying function of the loading parameter.

A problem with the (PD) condition is that

the hypothesis that $\mathbb{G}$ be positive definite
is stronger than it may appear,

a statement already commented on in Chapter 10 [Eqs. (10.12) and (10.13)], but deserving further consideration. Therefore, let us express $\dot{\boldsymbol{S}} \cdot \dot{\boldsymbol{F}}$ going back to Eq. (6.107) and taking into account that $\dot{\boldsymbol{F}} = \boldsymbol{L}\boldsymbol{F}$ as

$$\dot{\boldsymbol{S}} \cdot \dot{\boldsymbol{F}} = \boldsymbol{F}^T \boldsymbol{L} \boldsymbol{F} \cdot \dot{\boldsymbol{T}}^{(2)} + \boldsymbol{L}\boldsymbol{F} \cdot \boldsymbol{L}\boldsymbol{K}\boldsymbol{F}^{-T} = \boldsymbol{F}^T \boldsymbol{D} \boldsymbol{F} \cdot \dot{\boldsymbol{T}}^{(2)} + \boldsymbol{L}\boldsymbol{K} \cdot \boldsymbol{L}, \tag{11.13}$$

where $\boldsymbol{K} = J\boldsymbol{T} \in \mathsf{Sym}$ is the Kirchhoff stress [Eq. (3.108)]. Equation (11.13), using Eq. (8.47) but restricted to the loading branch of the constitutive operator, may be written as

$$\mathbb{C}[\dot{\boldsymbol{F}}] \cdot \dot{\boldsymbol{F}} = \boldsymbol{D} \cdot \mathbb{H}[\boldsymbol{D}] + \underbrace{\frac{1}{2}\boldsymbol{D} \cdot \mathbb{W}[\boldsymbol{D}] - \boldsymbol{D} \cdot \mathbb{K}[\boldsymbol{W}] + \frac{1}{2}\boldsymbol{W} \cdot \mathbb{W}[\boldsymbol{W}]}_{\text{terms depending on spin}} - \frac{\hat{\boldsymbol{Q}} \cdot \mathbb{H}[\boldsymbol{D}]}{g} \boldsymbol{D} \cdot \mathbb{H}[\hat{\boldsymbol{P}}], \tag{11.14}$$

where $\boldsymbol{W} = \boldsymbol{L} - \boldsymbol{D}$ is the spin tensor, $\hat{\boldsymbol{Q}} = \boldsymbol{F}^{-T}\boldsymbol{Q}\boldsymbol{F}^{-1}$, $\hat{\boldsymbol{P}} = \boldsymbol{F}^{-T}\boldsymbol{P}\boldsymbol{F}^{-1}$ and

$$\mathbb{H} = (\boldsymbol{F} \boxtimes \boldsymbol{F})\mathbb{E}(\boldsymbol{F} \boxtimes \boldsymbol{F})^T, \qquad \mathbb{W} = \boldsymbol{K} \boxtimes \boldsymbol{I} + \boldsymbol{I} \boxtimes \boldsymbol{K}, \qquad \mathbb{K} = \boldsymbol{K} \boxtimes \boldsymbol{I} - \boldsymbol{I} \boxtimes \boldsymbol{K}, \tag{11.15}$$

where

- Tensor $\mathbb{H}$, which was already given [see Eq. (6.116)], always has the minor symmetries (because $\mathbb{E}$ has) and has the major symmetry in the case of a Green elastic material.
- Tensor $\mathbb{W}$ always has the major symmetry but not the minor symmetries and transforms symmetric tensors into symmetric tensors and skew tensors into skew tensors. Moreover, $\boldsymbol{W} \cdot \mathbb{W}[\boldsymbol{W}] = 2\boldsymbol{W} \cdot \boldsymbol{K}\boldsymbol{W}$.
- Tensor $\mathbb{K}$ always has the major symmetry but not the minor symmetries and transforms symmetric tensors into skew tensors and skew tensors into symmetric tensors. Moreover, $\boldsymbol{W} \cdot \mathbb{K}[\boldsymbol{W}] = \boldsymbol{D} \cdot \mathbb{K}[\boldsymbol{D}] = \boldsymbol{0}$ and $-\boldsymbol{D} \cdot \mathbb{K}[\boldsymbol{W}] = 2\boldsymbol{D}\boldsymbol{K} \cdot \boldsymbol{W}$. Finally, when $\boldsymbol{A}$ and $\boldsymbol{K}$ are coaxial, that is, $\boldsymbol{A}\boldsymbol{K} = \boldsymbol{K}\boldsymbol{A}$, $\mathbb{K}[\boldsymbol{A}] = \boldsymbol{0}$.

The spin $\boldsymbol{W}$ and the rate of deformation $\boldsymbol{D}$ are *independent* tensors. Therefore, a necessary condition for $\dot{\boldsymbol{S}} \cdot \dot{\boldsymbol{F}} > 0$, that is, for $\mathbb{C}$ to be positive definite, is that $\boldsymbol{W} \cdot \mathbb{W}[\boldsymbol{W}] > 0$ for every $\boldsymbol{W}$; in other words,

$$\boxed{(\text{PD}) \implies (\text{tr}\boldsymbol{T})\boldsymbol{I} - \boldsymbol{T} \ \text{pos. def.} \iff T_1 + T_2 > 0,\ T_1 + T_3 > 0,\ T_2 + T_3 > 0.} \tag{11.16}$$

Positive definiteness of $(\text{tr}\boldsymbol{T})\boldsymbol{I} - \boldsymbol{T}$ is equivalent to saying that the restriction of $\mathbb{W}$ to the space of skew tensors is positive definite. This condition was obtained by Hill (1967a), and it *holds true for every constitutive assumption*, so it is 'universal'. In other words, Eq. (11.16) reflects a purely 'geometric effect'.

Now let us assume positive definiteness of $(\text{tr}\boldsymbol{T})\boldsymbol{I} - \boldsymbol{T}$. Under this condition, the function (at fixed $\boldsymbol{D}$)

$$z(\boldsymbol{W}) = -\boldsymbol{D} \cdot \mathbb{K}[\boldsymbol{W}] + \frac{1}{2}\boldsymbol{W} \cdot \mathbb{W}[\boldsymbol{W}] \tag{11.17}$$

is strictly convex, and therefore, it admits a minimum point which corresponds to

$$\frac{\partial z(\boldsymbol{W})}{\partial \boldsymbol{W}} = \mathbb{W}[\boldsymbol{W}] - \mathbb{K}[\boldsymbol{D}] = \boldsymbol{0}. \tag{11.18}$$

Owing to the positive definiteness of $(\text{tr}\boldsymbol{T})\boldsymbol{I} - \boldsymbol{T}$, the restriction of $\mathbb{W}$ to the space of all skew tensors is invertible, namely,

$$\mathbb{W}\mathbb{W}^{-1} = \boldsymbol{I} \boxtimes \boldsymbol{I} - \boldsymbol{I} \underline{\overline{\otimes}} \boldsymbol{I}, \tag{11.19}$$

where

$$(\boldsymbol{I} \boxtimes \boldsymbol{I} - \boldsymbol{I} \underline{\overline{\otimes}} \boldsymbol{I})[\boldsymbol{A}] = (\boldsymbol{A} - \boldsymbol{A}^T)/2 \tag{11.20}$$

associates to every tensor its skew symmetric part, and therefore, condition (11.18) defines the minimum point of $z(\boldsymbol{W})$ in terms of $\hat{\boldsymbol{W}}$, that is,

$$\hat{\boldsymbol{W}} = \mathbb{W}^{-1}\mathbb{K}[\boldsymbol{D}], \tag{11.21}$$

as

$$z_{\min} = -\frac{1}{2}\boldsymbol{D} \cdot \left(\mathbb{K}\mathbb{W}^{-1}\mathbb{K}\right)[\boldsymbol{D}]. \tag{11.22}$$

The components of tensor $\mathbb{K}\mathbb{W}^{-1}\mathbb{K}$ can be calculated easily in the principal reference system of $\boldsymbol{K}$. This calculation reveals that $\mathbb{K}\mathbb{W}^{-1}\mathbb{K}$ has both the major and minor symmetries and has only the following non-null components:

$$(\mathbb{K}\mathbb{W}^{-1}\mathbb{K})_{ijij} = (\mathbb{K}\mathbb{W}^{-1}\mathbb{K})_{jiij} = \frac{(K_i - K_j)^2}{2(K_i + K_j)}, \qquad i \neq j;\ i,j \in [1,3]. \tag{11.23}$$

As a conclusion, the following inequality is obtained:

$$\mathbb{C}[\dot{\boldsymbol{F}}] \cdot \dot{\boldsymbol{F}} \geq \boldsymbol{D} \cdot \mathbb{L}[\boldsymbol{D}] - \frac{1}{g}(\boldsymbol{D} \cdot \mathbb{H}[\hat{\boldsymbol{P}}])(\boldsymbol{D} \cdot \mathbb{H}^T[\hat{\boldsymbol{Q}}]) \tag{11.24}$$

(with the equality holding true when $\boldsymbol{W} = \hat{\boldsymbol{W}}$), where

$$\mathbb{L} = \mathbb{H} + \frac{1}{2}\mathbb{W} - \frac{1}{2}\mathbb{K}\mathbb{W}^{-1}\mathbb{K}. \tag{11.25}$$

Therefore, under the condition that $(\text{tr}\boldsymbol{T})\boldsymbol{I} - \boldsymbol{T}$ be positive definite,

> *positive definiteness of the restriction of* $\mathbb{L}$ *to the space of all symmetric tensors is equivalent to positive definiteness of* $\mathbb{G}$ *on* Lin.

Therefore, assuming $\mathbb{L}$ positive definite, it is possible to show that

$$\boxed{\begin{array}{l} T_1 + T_2 > 0,\ T_1 + T_3 > 0,\ T_2 + T_3 > 0,\ g > g_{cr}^{PD} \geq 0, \\ g_{cr}^{PD} = \frac{1}{2}\left(\sqrt{(\hat{\boldsymbol{P}} \cdot \mathbb{H}^T\tilde{\mathbb{L}}^{-1}\mathbb{H}[\hat{\boldsymbol{P}}])(\hat{\boldsymbol{Q}} \cdot \mathbb{H}\tilde{\mathbb{L}}^{-1}\mathbb{H}^T[\hat{\boldsymbol{Q}}])} + \hat{\boldsymbol{Q}} \cdot \mathbb{H}\tilde{\mathbb{L}}^{-1}\mathbb{H}[\hat{\boldsymbol{P}}]\right) \iff \text{(PD)}, \end{array}} \tag{11.26}$$

where $\tilde{\mathbb{L}} = (\mathbb{L} + \mathbb{L}^T)/2$ and $\tilde{\mathbb{L}}^{-1}\tilde{\mathbb{L}} = \boldsymbol{I}\,\overline{\otimes}\,\boldsymbol{I}$. Condition (11.26) was obtained by Raniecki and Bruhns (1981) in the case of $\mathbb{L}$ having the major symmetry. Condition (11.26) is equivalent to Eq. (11.11) but more explicit in the sense that it singles out the effect of $(\text{tr}\boldsymbol{T})\boldsymbol{I} - \boldsymbol{T}$.

11.1.1 Uniaxial tension

In the case of uniaxial tension along axis 1, $T_1 = \sigma > 0$ and $T_2 = T_3 = 0$; thus $(\text{tr}\boldsymbol{T})\boldsymbol{I} - \boldsymbol{T}$ is positive semi-definite, and $\mathbb{G}$ is not invertible. Let us develop this point in detail. In this case, by direct calculation, we obtain

$$z(\boldsymbol{W}) = \sigma\left(W_{12}^2 + W_{13}^2 - 2D_{12}W_{12} - 2D_{13}W_{13}\right), \tag{11.27}$$

which is zero when $\boldsymbol{D} = \boldsymbol{0}$ and the only non-null component of $\boldsymbol{W}$ is W_{23}, representing a spin about axis 1. The minimum of $z(\boldsymbol{W})$ is

$$z_{\min} = -\sigma\left(D_{12}^2 + D_{13}^2\right). \tag{11.28}$$

It follows that

$$\dot{\boldsymbol{F}} \cdot \mathbb{C}[\dot{\boldsymbol{F}}] \geq \boldsymbol{D} \cdot \mathbb{H}[\boldsymbol{D}] - \frac{1}{g}(\boldsymbol{D} \cdot \mathbb{H}^T[\hat{\boldsymbol{Q}}])(\boldsymbol{D} \cdot \mathbb{H}[\hat{\boldsymbol{P}}]) + \sigma D_{11}^2, \tag{11.29}$$

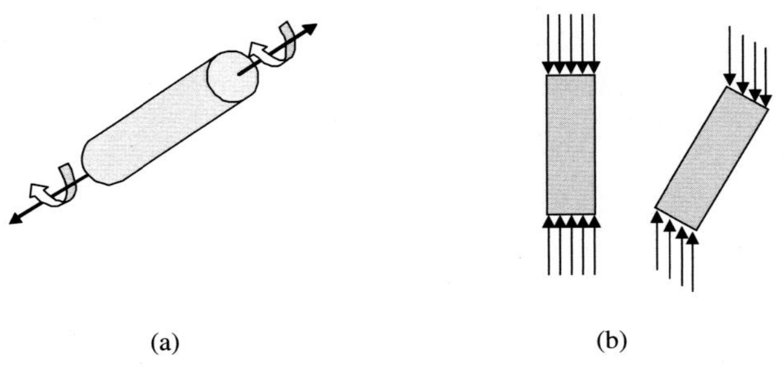

(a) (b)

Figure 11.2. (*a*) Neutral rotational equilibrium of a rod under tensile dead load. (*b*) Rotational instability of a rod under compressive dead load.

and therefore, when the restriction of

$$\mathbb{H} + \sigma \, \boldsymbol{e}_1 \otimes \boldsymbol{e}_1 \otimes \boldsymbol{e}_1 \otimes \boldsymbol{e}_1 - \frac{1}{g} \mathbb{H}[\hat{\boldsymbol{P}}] \otimes \mathbb{H}^T [\hat{\boldsymbol{Q}}] \tag{11.30}$$

to Sym is positive definite, condition (PD) is verified, except for a velocity gradient having W_{23} as the unique non-null component. However, integrals (10.31) and (10.42) can vanish only if $W_{23} \neq 0$ is the only non-null component of the velocity gradient everywhere in the body, and this corresponds to a loss of uniqueness consisting of arbitrary rigid rotations about the axis of tension (which is also an axis of neutral stability). This circumstance becomes particularly clear if we consider a bar pulled in tension and therefore subject to homogeneous stress and all-round dead loading (Fig. 11.2*a*). Analogously, $(\text{tr}\boldsymbol{T})\boldsymbol{I} - \boldsymbol{T}$ is negative semi-definite (and thus $\mathbb{G}$ is indefinite) in the important case of a bar subject to dead loading of uniaxial compression, even for a vanishing small value of axial force. This correctly corresponds to a well-known instability owing to rigid-body rotation (Fig. 11.2*b*).

Using Eq. (11.30), condition (PD)—except for rigid-body rotations about the axis of tension—can be written in explicit for uniaxial stress as

$$\boxed{\begin{aligned} & T_1 > 0, \; T_2 = T_3 = 0, \; g > g_{cr}^{PD} \geq 0, \; \mathbb{L} = \mathbb{H} + \sigma \, \boldsymbol{e}_1 \otimes \boldsymbol{e}_1 \otimes \boldsymbol{e}_1 \otimes \boldsymbol{e}_1, \\ & g_{cr}^{PD} = \tfrac{1}{2} \left(\sqrt{(\hat{\boldsymbol{P}} \cdot \mathbb{H}^T \tilde{\mathbb{L}}^{-1} \mathbb{H}[\hat{\boldsymbol{P}}])(\hat{\boldsymbol{Q}} \cdot \mathbb{H} \tilde{\mathbb{L}}^{-1} \mathbb{H}^T [\hat{\boldsymbol{Q}}])} + \hat{\boldsymbol{Q}} \cdot \mathbb{H} \tilde{\mathbb{L}}^{-1} \mathbb{H}[\hat{\boldsymbol{P}}] \right) \iff \text{(PD)} \end{aligned}} \tag{11.31}$$

where $\tilde{\mathbb{L}}$ is the symmetric part of $\mathbb{L}$, assumed positive definite.

11.1.2 The small strain theory

In the small strain theory, where we refer to constitutive Eqs. (8.66), it may be interesting to write g_{cr}^{PD} in terms of a critical value of the hardening modulus h_{cr}^{PD}. For a symmetric $\mathbb{E}$, this becomes

$$h_{cr}^{PD} = \frac{1}{2} \left(\sqrt{(\boldsymbol{P} \cdot \mathbb{E}[\boldsymbol{P}])(\boldsymbol{Q} \cdot \mathbb{E}[\boldsymbol{Q}])} - \boldsymbol{P} \cdot \mathbb{E}[\boldsymbol{Q}] \right) \geq 0, \tag{11.32}$$

which is never negative and becomes null for associative elastoplasticity. Therefore, in the infinitesimal theory (and assuming $\mathbb{E}$ symmetric), (PD) is always lost before softening. The critical modulus (11.32) was obtained by Mróz (1963), Maier and

Hueckel (1979) and Raniecki (1979). A derivation of the critical hardening modulus for plane strain and plane stress situations was given by Bigoni and Hueckel (1991a).

In closing this section, we note that under the hypotheses of homogeneity and all-round controlled nominal surface tractions, failure of (PD) implies failure of the Hill sufficient conditions for stability (10.42) and uniqueness (10.31). This situation is critical in the case of the associative flow rule, where instability and bifurcation may occur at loss of (PD) under broad hypotheses (Hill, 1967a; Miles, 1973). The situation is still not clear, however, for both stability and bifurcation in non-associative elastoplasticity. In this context, examples are not known in which any real bifurcation has been found in coincidence with failure of (PD).

11.2 Singularity of the constitutive operator

Loss of (PD) at a point of a body during a loading program is not in general a sufficient condition for bifurcation. There are, however, certain situations where loss of (PD) may become close to critical. These have been touched on in the closure of Section 11.1. Let us consider therefore a special class of problems where

1. controlled nominal tractions is prescribed on the entire boundary, and
2. material properties and deformation (and therefore stress) are homogeneous during a given loading path.

These situations have been analysed by Hill (1967a), Miles (1973), Raniecki and Bruhns (1981) and Ogden (1985). Under the preceding hypotheses, conditions (10.31) and (10.42) are equivalent to (PD). In other words, exclusion of bifurcation and stability in Hill's sense fails to hold when (PD) is lost, that is, when at least one $\boldsymbol{X}^* \neq \boldsymbol{0}$ exists such that

$$\boldsymbol{X}^* \cdot \mathbb{C}[\boldsymbol{X}^*] = 0. \tag{11.33}$$

Except for the associative case, this condition does not mean that $\mathbb{C}$ is singular; in other words, $\mathbb{C}[\boldsymbol{X}^*] \neq \boldsymbol{0}$ in general should be expected. Therefore, loss of (PD) is not directly connected to a bifurcation, even in the preceding special hypotheses. On the other hand, if we define the non-singularity condition

$$\boxed{\mathbb{C}[\boldsymbol{X}] \neq \boldsymbol{0}, \text{ for every (non-zero) } \boldsymbol{X} \in \mathsf{Lin} \quad \text{(NS condition)},} \tag{11.34}$$

we can understand that, in a sense, failure of this condition is critical for bifurcation of the homogeneous problem with controlled nominal surface tractions on the entire boundary. We note, in passing, that obviously (PD) implies (NS) and that when (PD) fails the first time in a continuous loading path (and for continuous dependence of constitutive equations on time), (NS) also fails for associative hyperelastic-plastic solids.

With respect to the constitutive operator (8.53), where $\mathbb{G}$ is assumed invertible, because singularity of $\mathbb{G}$ implies singularity of $\boldsymbol{C}$, we note that, assuming $\boldsymbol{N} \cdot \mathbb{G}^{-1}[\boldsymbol{M}] > 0$,

$$\dot{\boldsymbol{S}} = \boldsymbol{0} \iff \dot{\boldsymbol{F}} \propto \mathbb{G}^{-1}[\boldsymbol{M}],\ g = \boldsymbol{N} \cdot \mathbb{G}^{-1}[\boldsymbol{M}]. \tag{11.35}$$

Therefore,

$$\boxed{g \neq g_{cr}^{NS} = \boldsymbol{N} \cdot \mathbb{G}^{-1}[\boldsymbol{M}], \quad \iff \quad \mathbb{C} \text{ is not singular} \qquad \text{(NS condition)}.} \tag{11.36}$$

Note that $g_{cr}^{NS} < 0$ when $\boldsymbol{N} \cdot \mathbb{G}^{-1}[\boldsymbol{M}] < 0$. In this case, loss of (NS) does not occur (in the present constitutive framework limited to $g > 0$).

Loss of (NS) is critical in the sense explained by Raniecki and Bruhns (1981), namely, assuming that $\mathbb{C}[\boldsymbol{X}^*] = \boldsymbol{0}$, if $\dot{\boldsymbol{F}}$ is a solution of the problem in velocities (corresponding to plastic loading everywhere in the body), it satisfies

$$\boldsymbol{\mathcal{C}}(\dot{\boldsymbol{F}}) = \mathbb{C}[\dot{\boldsymbol{F}}] \quad \text{and} \quad \mathbb{C}[\dot{\boldsymbol{F}}]\boldsymbol{n}_0 = \dot{\boldsymbol{\sigma}} \text{ on } \partial B_0, \tag{11.37}$$

where $\boldsymbol{n}_0$ is the unit outward normal to ∂B_0. It follows that if $\boldsymbol{\mathcal{C}}(\dot{\boldsymbol{F}} + \gamma \boldsymbol{X}^*) = \mathbb{C}[\dot{\boldsymbol{F}}] + \gamma \mathbb{C}[\boldsymbol{X}^*]$ for at least some $\gamma \neq 0$,

$$\boldsymbol{\mathcal{C}}(\dot{\boldsymbol{F}} + \gamma \boldsymbol{X}^*)\boldsymbol{n}_0 = \mathbb{C}[\dot{\boldsymbol{F}}]\boldsymbol{n}_0 = \dot{\boldsymbol{\sigma}} \text{ on } \partial B_0, \tag{11.38}$$

and therefore, either a *bifurcation* or a *load maximum* has been reached in the loading program (Hill, 1967a).

Condition (11.36) may be further elaborated, assuming $T_1 + T_2 > 0$, $T_1 + T_3 > 0$ and $T_2 + T_3 > 0$, as follows. Let us observe that

$$\dot{\boldsymbol{S}}\boldsymbol{F}^T = \mathbb{H}[\boldsymbol{D}] + \boldsymbol{L}\boldsymbol{K} - \frac{1}{g}(\hat{\boldsymbol{Q}} \cdot \mathbb{H}[\boldsymbol{D}])\mathbb{H}[\hat{\boldsymbol{P}}], \tag{11.39}$$

where we may note that

$$\boldsymbol{L}\boldsymbol{K} = \frac{1}{2}(\mathbb{W} - \mathbb{K})[\boldsymbol{D}] + \frac{1}{2}(\mathbb{W} - \mathbb{K})[\boldsymbol{W}]. \tag{11.40}$$

For $\boldsymbol{W}$ in the form of Eq. (11.21), that is, $\boldsymbol{W} = \mathbb{W}^{-1}\mathbb{K}[\boldsymbol{D}]$, and $\mathbb{L}$ defined by Eq. (11.25), Eq. (11.39), using Eq. (11.40), becomes

$$\dot{\boldsymbol{S}}\boldsymbol{F}^T = \mathbb{L}[\boldsymbol{D}] - \frac{1}{g}(\hat{\boldsymbol{Q}} \cdot \mathbb{H}[\boldsymbol{D}])\mathbb{H}[\hat{\boldsymbol{P}}]. \tag{11.41}$$

Assuming $\mathbb{L}$ invertible, from Eq. (11.41) we conclude that under hypothesis (11.21), $\dot{\boldsymbol{S}} = \boldsymbol{0} \Longleftrightarrow g = \hat{\boldsymbol{Q}} \cdot (\mathbb{H}\mathbb{L}^{-1}\mathbb{H})[\hat{\boldsymbol{P}}]$ and $\boldsymbol{D} \propto \mathbb{L}^{-1}\mathbb{H}[\hat{\boldsymbol{P}}]$. Vice versa, assuming that $\dot{\boldsymbol{S}} = \boldsymbol{0}$ in Eq. (11.39), the minor symmetries of $\mathbb{H}$ imply that $\boldsymbol{W} = \mathbb{W}^{-1}\mathbb{K}[\boldsymbol{D}]$. We obtain, therefore, the following proposition:

$$\boxed{\begin{array}{l} T_1 + T_2 > 0,\ T_1 + T_3 > 0,\ T_2 + T_3 > 0: \\ g \neq g_{cr}^{NS} = \hat{\boldsymbol{Q}} \cdot (\mathbb{H}\mathbb{L}^{-1}\mathbb{H})[\hat{\boldsymbol{P}}], \quad \Longleftrightarrow \quad \mathbb{C} \text{ is not singular} \qquad \text{(NS condition).} \end{array}} \tag{11.42}$$

Obviously, conditions (11.36) and (11.42) are equivalent under the condition that $(\text{tr}\boldsymbol{T})\boldsymbol{I} - \boldsymbol{T}$ is positive definite.

11.2.1 Uniaxial tension

For uniaxial tension along axis 1, $T_1 = \sigma > 0$ and $T_2 = T_3 = 0$. Direct calculations on Eq. (11.39) yield that the critical condition is again given by Eq. (11.42) but with

$$\mathbb{L} = \mathbb{H} + \sigma\, \boldsymbol{e}_1 \otimes \boldsymbol{e}_1 \otimes \boldsymbol{e}_1 \otimes \boldsymbol{e}_1. \tag{11.43}$$

11.2.2 The small strain theory

In the case of the small strain theory [Eq. (8.66)] with $\mathbb{E}$ symmetric and positive definite, the condition (NS) becomes $h \neq 0$, and for $h = 0$ and $\boldsymbol{X}^* \propto \boldsymbol{P}$,

$$\mathbb{C}[\boldsymbol{X}^*] = \boldsymbol{0}.$$

As a conclusion and with the exception of associative plasticity, (PD) and (NS) clearly do not coincide in the infinitesimal theory, where (PD) is lost before softening and (NS) is always lost in the perfectly plastic case ($h = 0$).

11.3 Strong ellipticity

In Section 11.2 we analysed the special condition of a homogeneous body subject to all-round controlled nominal surface tractions. In that case, the (PD) and (NS) conditions play a special role. Now we analyse a dual case in which a homogeneous body is deformed in a homogeneous way under prescribed displacements on the entire boundary. As a result, we will show that strong ellipticity (SE) and ellipticity (E) play a role similar to those shown for (PD) and (NS) in Section 11.2.

Let us begin by considering the 'in-loading comparison solid', that is, the fictitious solid defined by the constitutive tensor $\mathbb{C}$. For this material, we will show the validity of the following *uniqueness theorem for the velocity problem, owing to van Hove (1947)*:

> For a homogeneous and homogeneously deformed body, characterised by an incrementally linear constitutive operator (here $\mathbb{C}$) and subject to prescribed velocity over the entire boundary, the strong ellipticity condition
>
> $$\boxed{\boldsymbol{g} \cdot \mathbb{C}[\boldsymbol{g} \otimes \boldsymbol{n}]\boldsymbol{n} > 0 \qquad \text{(SE condition)},} \tag{11.44}$$
>
> for every unit vector $\boldsymbol{n}$ and non-zero vector $\boldsymbol{g}$, implies that the velocity problem has at most one solution.

The condition of strong ellipticity may be expressed, in a different notation, as the positive definiteness of the *acoustic tensor* $\boldsymbol{A}(\boldsymbol{n})$ defined, for every unit vector $\boldsymbol{n}$ and $\boldsymbol{g} \in \mathcal{V}$, as

$$\boxed{\boldsymbol{A}(\boldsymbol{n})\boldsymbol{g} = \frac{1}{\rho_0}\mathbb{C}[\boldsymbol{g} \otimes \boldsymbol{n}]\boldsymbol{n},} \tag{11.45}$$

where ρ_0 is the mass density in the reference configuration (some time later and without loss of generality we will assume $\rho_0 = 1$ to simplify notation).

With reference to the 'in-loading' comparison solid (10.39), the acoustic tensor becomes

$$\boxed{\boldsymbol{A}(\boldsymbol{n}) = \boldsymbol{A}_E(\boldsymbol{n}) - \frac{1}{\rho_0 g}\boldsymbol{M}\boldsymbol{n} \otimes \boldsymbol{N}\boldsymbol{n},} \tag{11.46}$$

where $\boldsymbol{A}_E(\boldsymbol{n})$ is the *elastic acoustic tensor*

$$\boldsymbol{A}_E(\boldsymbol{n})\boldsymbol{g} = \frac{1}{\rho_0}\mathbb{G}[\boldsymbol{g} \otimes \boldsymbol{n}]\boldsymbol{n}, \tag{11.47}$$

assumed *positive definite* in the following.

For the proof of the van Hove theorem, we follow Hayes (1966). Let us consider the functional in Eq. (10.31) with $\mathbb{C}^c$ replaced by $\mathbb{C}$. When this functional is positive, for every admissible velocity field $\dot{\boldsymbol{x}}$ vanishing on ∂B_0, bifurcation is excluded (in the incrementally linear solid defined by $\mathbb{C}$). We can extend the definition of $\dot{\boldsymbol{x}}$ from $\bar{B}_0$ to all Euclidean space $\mathcal{E}$ simply by defining $\dot{\boldsymbol{x}} = \boldsymbol{0}$ on $\mathcal{E} \setminus \bar{B}_0$. The resulting field is admissible, and its gradient is discontinuous only on ∂B_0. Therefore, both $\dot{\boldsymbol{x}}$ and $\nabla\dot{\boldsymbol{x}}$ possess the three-dimensional Fourier transforms

$$\dot{\boldsymbol{x}}^*(\boldsymbol{x}_0) = \left(\frac{1}{2\pi}\right)^{\frac{3}{2}} \int_{\mathcal{E}} e^{i(\boldsymbol{x}_0 \cdot \boldsymbol{y})} \dot{\boldsymbol{x}}(\boldsymbol{y}) dv_{\boldsymbol{y}}, \tag{11.48}$$

$$\boldsymbol{X}(\boldsymbol{x}_0) = \left(\frac{1}{2\pi}\right)^{\frac{3}{2}} \int_{\mathcal{E}} e^{i(\boldsymbol{x}_0 \cdot \boldsymbol{y})} \nabla\dot{\boldsymbol{x}}(\boldsymbol{y}) dv_{\boldsymbol{y}}. \tag{11.49}$$

Before continuing, it should be noted that the fact that $\dot{\boldsymbol{x}} = \boldsymbol{0}$ on $\mathcal{E} \setminus \bar{B}_0$, Eqs. (11.48), (11.49) and the divergence theorem yield

$$\boldsymbol{X} = -i\dot{\boldsymbol{x}}^* \otimes \boldsymbol{x}_0, \tag{11.50}$$

where $i = \sqrt{-1}$ is the imaginary unit. Since homogeneity is satisfied in B_0, the generalised Parseval theorem gives

$$\int_{\mathcal{E}} \nabla\dot{\boldsymbol{x}} \cdot \mathbb{C}[\nabla\dot{\boldsymbol{x}}] = C_{ijhk} \int_{\mathcal{E}} \dot{x}_{i,j}\dot{x}_{h,k} = C_{ijhk} \int_{\mathcal{E}} X_{ij}\overline{X}_{hk} = \int_{\mathcal{E}} \boldsymbol{X} \cdot \mathbb{C}[\overline{\boldsymbol{X}}], \tag{11.51}$$

where $\overline{\boldsymbol{X}}$ is the complex conjugate of $\boldsymbol{X}$. Now we note that Eq. (11.50) implies

$$\boldsymbol{X}_R = \dot{\boldsymbol{x}}_I^* \otimes \boldsymbol{x}_0, \qquad \boldsymbol{X}_I = -\dot{\boldsymbol{x}}_R^* \otimes \boldsymbol{x}_0, \tag{11.52}$$

where the indices R and I stand for the real and imaginary parts, respectively. Using Eq. (11.52) in Eq. (11.51), it may be found that

$$\int_{B_0} \nabla\dot{\boldsymbol{x}} \cdot \mathbb{C}[\nabla\dot{\boldsymbol{x}}] = \int_{\mathcal{E}} (\dot{\boldsymbol{x}}_R^* \otimes \boldsymbol{x}_0) \cdot \mathbb{C}[\dot{\boldsymbol{x}}_R^* \otimes \boldsymbol{x}_0] + \int_{\mathcal{E}} (\dot{\boldsymbol{x}}_I^* \otimes \boldsymbol{x}_0) \cdot \mathbb{C}[\dot{\boldsymbol{x}}_I^* \otimes \boldsymbol{x}_0]. \tag{11.53}$$

As a conclusion, when (SE) holds, the preceding integral is greater than zero, and uniqueness follows.

Obviously, the van Hove theorem holds true for the Raniecki family of incrementally linear solids. Defining, therefore, the acoustic tensor relative to the generic Raniecki solid as

$$\boldsymbol{A}_R(\boldsymbol{n}, \psi)\boldsymbol{g} = \frac{1}{\rho_0}\mathbb{R}[\boldsymbol{g} \otimes \boldsymbol{n}]\boldsymbol{n}, \tag{11.54}$$

we may conclude that for a homogeneous elastoplastic body subject to prescribed velocities on the entire boundary, strong ellipticity of at least one of the Raniecki solids (SE$_\text{R}$) is sufficient for uniqueness. As a conclusion, we can state that

(SE$_\text{R}$) $\Longrightarrow$ uniqueness for an *elastoplastic solid* for van Hove's b.v.p.

(SE) $\Longrightarrow$ uniqueness for the 'in-loading comparison solid' for van Hove's b.v.p.

where by 'van Hove b.v.p.' we mean boundary value problems satisfying van Hove's conditions. An immediate consequence of the Raniecki comparison theorem is the following implication:

$$(SE_{\mathrm{R}}) \implies (SE). \tag{11.55}$$

Now we will determine the critical plastic moduli for loss of (SE) and loss of ($\mathrm{SE_R}$).

In order to write (SE) of $\mathbb{C}$ in terms of a critical value of the plastic modulus, let us consider, for every vector $\boldsymbol{x}$ and unit vector $\boldsymbol{n}$, the following inequality:

$$\boldsymbol{x}\cdot\boldsymbol{A}(\boldsymbol{n})\boldsymbol{x} \geq \boldsymbol{x}\cdot\boldsymbol{A}_R(\boldsymbol{n},\psi)\boldsymbol{x} = \boldsymbol{x}\cdot\tilde{\boldsymbol{A}}_E(\boldsymbol{n})\boldsymbol{x} - \frac{(\boldsymbol{x}\cdot\boldsymbol{R}\boldsymbol{n})^2}{4\psi g}, \tag{11.56}$$

where $\tilde{\boldsymbol{A}}_E(\boldsymbol{n}) = [\boldsymbol{A}_E(\boldsymbol{n}) + \boldsymbol{A}_E^T(\boldsymbol{n})]/2$ is the symmetric part of the elastic acoustic tensor.

We proceed analogously to the proof of Raniecki and Bruhns for (PD) using the Cauchy-Schwarz inequality in the metric $\tilde{\boldsymbol{A}}_E(\boldsymbol{n})$ (see Section 2.12; the dependence of $\tilde{\boldsymbol{A}}_E$ on $\boldsymbol{n}$ is omitted for simplicity in the following):

$$(\boldsymbol{x}\cdot\boldsymbol{R}\boldsymbol{n})^2 = (\boldsymbol{x}\cdot\tilde{\boldsymbol{A}}_E\tilde{\boldsymbol{A}}_E^{-1}\boldsymbol{R}\boldsymbol{n})^2 \leq (\boldsymbol{x}\cdot\tilde{\boldsymbol{A}}_E\boldsymbol{x})(\boldsymbol{R}\boldsymbol{n}\cdot\tilde{\boldsymbol{A}}_E^{-1}\boldsymbol{R}\boldsymbol{n}). \tag{11.57}$$

After calculations that parallel those reported in Section 11.1, we find that

$$g > \inf_{\psi>0}\left\{\frac{(\boldsymbol{M}\boldsymbol{n}+\psi\boldsymbol{N}\boldsymbol{n})\cdot\tilde{\boldsymbol{A}}_E^{-1}(\boldsymbol{M}\boldsymbol{n}+\psi\boldsymbol{N}\boldsymbol{n})}{4\psi}\right\} \implies \boldsymbol{A}(\boldsymbol{n}) \text{ pos. def.} \tag{11.58}$$

The minimum problem is solved by

$$\psi(\boldsymbol{n}) = \sqrt{\frac{\boldsymbol{M}\boldsymbol{n}\cdot\tilde{\boldsymbol{A}}_E^{-1}\boldsymbol{M}\boldsymbol{n}}{\boldsymbol{N}\boldsymbol{n}\cdot\tilde{\boldsymbol{A}}_E^{-1}\boldsymbol{N}\boldsymbol{n}}}, \tag{11.59}$$

which yields the following proposition:

$$\boxed{g > g_{cr}^{SE}(\boldsymbol{n}) \iff \boldsymbol{A}(\boldsymbol{n}) \text{ pos. def.} \quad \text{(at fixed } \boldsymbol{n}\text{)},} \tag{11.60}$$

where

$$\boxed{g_{cr}^{SE}(\boldsymbol{n}) = \frac{1}{2}\left(\sqrt{(\boldsymbol{M}\boldsymbol{n}\cdot\tilde{\boldsymbol{A}}_E^{-1}\boldsymbol{M}\boldsymbol{n})(\boldsymbol{N}\boldsymbol{n}\cdot\tilde{\boldsymbol{A}}_E^{-1}\boldsymbol{N}\boldsymbol{n})} + \boldsymbol{M}\boldsymbol{n}\cdot\tilde{\boldsymbol{A}}_E^{-1}\boldsymbol{N}\boldsymbol{n}\right) \geq 0.} \tag{11.61}$$

Note that $g_{cr}^{SE} = 0 \iff \boldsymbol{N} \propto -\boldsymbol{M}$, a condition which should never be satisfied for realistic constitutive models.

In order to complete the proof of Eq. (11.60), it may be checked that at $g = g_{cr}^{SE}(\boldsymbol{n})$, tensor $\boldsymbol{A}(\boldsymbol{n})$ loses positive definiteness for vectors $\boldsymbol{x}^*$ defined as

$$\boldsymbol{x}^* \propto \sqrt{\boldsymbol{M}\boldsymbol{n}\cdot\tilde{\boldsymbol{A}}_E^{-1}\boldsymbol{M}\boldsymbol{n}}\,\tilde{\boldsymbol{A}}_E^{-1}\boldsymbol{N}\boldsymbol{n} + \sqrt{\boldsymbol{N}\boldsymbol{n}\cdot\tilde{\boldsymbol{A}}_E^{-1}\boldsymbol{N}\boldsymbol{n}}\,\tilde{\boldsymbol{A}}_E^{-1}\boldsymbol{M}\boldsymbol{n} \tag{11.62}$$

and that for $g \leq g_{cr}^{SE}(\boldsymbol{n})$, $\boldsymbol{x}^*\cdot\boldsymbol{A}(\boldsymbol{n})\boldsymbol{x}^* \leq 0$.

It may be important to note from Eq. (11.61) that for every $\boldsymbol{n}$, failure of (SE) necessarily occurs for $g_{cr}^{SE}(\boldsymbol{n})$ positive.

All the preceding holds at $\boldsymbol{n}$ fixed and proves that for a given $\boldsymbol{n}$, loss of (PD) in the 'in-loading' comparison solid $\mathbb{C}$ and in the optimal Raniecki solid defined by Eq. (11.59) is equivalent. However, loss of (SE) in $\mathbb{C}$ actually will occur when g equals the maximum of $g_{cr}^{SE}(\boldsymbol{n})$ as a function of $\boldsymbol{n}$

$$\boxed{g > g_{cr}^{SE} = \max_{\boldsymbol{n},|\boldsymbol{n}|=1} g_{cr}^{SE}(\boldsymbol{n}) \Longleftrightarrow \text{(SE condition).}} \tag{11.63}$$

Therefore, differently from g_{cr}^{PD}, the critical value of the plastic modulus for loss of strong ellipticity g_{cr}^{SE} is not given in explicit terms but as the solution of a constrained maximisation problem.

Let us consider now strong ellipticity for Raniecki solids. From Eqs. (11.56) and (11.57), we obtain

$$\boxed{\boldsymbol{A}_R(\boldsymbol{n},\psi) \text{ pos. def.} \Longleftrightarrow g > g_{cr}^{SE_R}(\boldsymbol{n},\psi) = \frac{(\boldsymbol{M}\boldsymbol{n}+\psi\boldsymbol{N}\boldsymbol{n})\cdot\tilde{\boldsymbol{A}}_E^{-1}(\boldsymbol{M}\boldsymbol{n}+\psi\boldsymbol{N}\boldsymbol{n})}{4\psi},} \tag{11.64}$$

Taking into consideration the optimal ψ, we conclude that (SE$_R$) can be written as

$$\boxed{g > g_{cr}^{SE_R} = \inf_{\psi>0}\max_{\boldsymbol{n},|\boldsymbol{n}|=1} g_{cr}^{SE_R}(\boldsymbol{n},\psi), \Longleftrightarrow \text{(SE}_R\text{ condition).}} \tag{11.65}$$

The preceding critical plastic moduli were obtained by Bigoni and Zaccaria (1992a, 1992b). It may be worth noting that Eq. (11.55) is equivalent to

$$g_{cr}^{SE_R}(\boldsymbol{n},\psi) \geq g_{cr}^{SE}(\boldsymbol{n}), \tag{11.66}$$

holding for every ψ and $\boldsymbol{n}$. Moreover, we have seen that

$$g_{cr}^{SE}(\boldsymbol{n}) = \inf_{\psi>0} g_{cr}^{SE_R}(\boldsymbol{n},\psi). \tag{11.67}$$

Therefore, (SE) and (SE$_R$) *are equivalent criteria whenever*

$$\inf_{\psi>0}\max_{\boldsymbol{n},|\boldsymbol{n}|=1} g_{cr}^{SE_R}(\boldsymbol{n},\psi) = \max_{\boldsymbol{n},|\boldsymbol{n}|=1}\inf_{\psi>0} g_{cr}^{SE_R}(\boldsymbol{n},\psi), \tag{11.68}$$

a condition which has been proved to hold by Bigoni and Zaccaria only under very restrictive assumptions (small strain theory, isotropic elasticity and coaxiality of $\boldsymbol{P}$ and $\boldsymbol{Q}$; see Section 11.3.1), so a generalisation of their proof would be important.

Upto this point, we have given an interpretation of the (SE$_R$) and (SE) conditions as sufficient for uniqueness of a special class of problems (van Hove hypotheses).[1] In particular, the former condition is sufficient for uniqueness of the velocity problem for the elastoplastic solid and the latter for the fictitious solid corresponding to the loading branch of the constitutive tangent operator. However, as stressed by Bigoni and Zaccaria (1992b),

> *the strong ellipticity condition has a meaning even when a generic situation of inhomogeneous deformation and mixed boundary conditions is considered.*

[1] The van Hove theorem has been generalised in special cases by Ryzhak (1993, 1994).

In order to explain this point, let us introduce the semi-strong ellipticity condition (SSE), which is defined as in Eq. (11.44) except that $>$ is replaced with $\geq$. In other words, (SSE) is the condition of semi-positive definiteness of the acoustic tensor. In non-linear elasticity, the theorem of Cattaneo (1946; see also Truesdell and Noll, 1965, section 68bis) proves that (SSE) is a necessary condition for stability. In non-associative elastoplasticity, the theorem of Cattaneo was generalized by Ryzhak (1987).[2] Assuming positive definiteness of $\boldsymbol{A}_E(\boldsymbol{n})$, the Ryzhak theorem states that (SSE) is a necessary condition for semi-stability in Hill's sense, namely,

$$\boxed{\text{Semi-stability in the Hill sense} \Longrightarrow \mathbb{C} \text{ is (SSE)},}$$

where with the term 'semi-stability' we intend the sufficient stability condition [Eq. (10.42)] where $>$ is replaced with $\geq$. The (SSE) is a local criterion, and when it fails during a loading program of a generic (inhomogeneous) boundary value problem (with mixed boundary conditions), Hill's sufficient stability condition does not hold.

11.3.1 The small strain theory

In the special case of the small strain theory [Eq. (8.66)] and Green elasticity, the critical plastic modulus [Eq. (11.61)] simplifies to

$$g_{cr}^{SE}(\boldsymbol{n}) = \frac{1}{2}\left(\sqrt{(\mathbb{E}[\boldsymbol{P}]\boldsymbol{n}\cdot\boldsymbol{A}_E^{-1}\mathbb{E}[\boldsymbol{P}\boldsymbol{n}])(\mathbb{E}[\boldsymbol{Q}]\boldsymbol{n}\cdot\boldsymbol{A}_E^{-1}\mathbb{E}[\boldsymbol{Q}]\boldsymbol{n})} + \mathbb{E}[\boldsymbol{P}]\boldsymbol{n}\cdot\boldsymbol{A}_E^{-1}\mathbb{E}[\boldsymbol{Q}]\boldsymbol{n}\right) \geq 0, \tag{11.69}$$

where once again we note that $g_{cr}^{SE} = 0$ is equivalent to $\boldsymbol{P} \propto -\boldsymbol{Q}$, a condition which is excluded for realistic constitutive models. Assuming isotropic elasticity [Eq. (8.65)], the acoustic tensor can be written in the form of Eq. (2.176):

$$\boldsymbol{A}_E(\boldsymbol{n}) = (\lambda + \mu)\boldsymbol{n}\otimes\boldsymbol{n} + \mu\boldsymbol{I}, \tag{11.70}$$

with the inverse given by Eq. (2.177):

$$\boldsymbol{A}_E^{-1}(\boldsymbol{n}) = -\frac{\lambda+\mu}{\mu(\lambda+2\mu)}\boldsymbol{n}\otimes\boldsymbol{n} + \frac{1}{\mu}\boldsymbol{I}. \tag{11.71}$$

Under the preceding hypotheses, and assuming in addition coaxiality of $\boldsymbol{P}$ and $\boldsymbol{Q}$, Bigoni and Zaccaria (1992a, 1992b) have proved that Eq. (11.68) holds so that

> *in the case of the small strain theory, with isotropic elasticity and $\boldsymbol{P}$ and $\boldsymbol{Q}$ coaxial, (SE) and (SE_R) are equivalent criteria.*

Moreover, they obtained an explicit solution for the problem (11.63).

11.4 Ellipticity, strain localisation and shear bands

For associative hyperelastic-plastic behaviour, $\boldsymbol{N} = \boldsymbol{M}$, in a continuous deformation evolution initiating when (SE) holds, the acoustic tensor becomes singular as soon

[2] A simpler alternative proof can be inferred from Petryk (1985a, 1992) as an application of Graves' theorem. The discussion there was confined to symmetry of the constitutive operator, but in the (SSE) condition, only the symmetric part of the constitutive operator plays a role.

as (SE) fails. This is no longer true in the non-associative case, where (SE) is usually lost while the acoustic tensor is still non-singular. Therefore, we can introduce a condition analogous to (NS) but on the acoustic tensor (defined here relative to the 'in-loading' comparison solid). This is the condition of ellipticity

$$\boxed{\det \boldsymbol{A}(\boldsymbol{n}) \neq 0, \qquad \text{for all } \boldsymbol{n} \in \mathcal{V}, |\boldsymbol{n}| = 1 \qquad \text{(E condition).}} \tag{11.72}$$

Note that, without loss of generality,

we assume $\rho_0 = 1$ for the remainder of this section.

In a (sufficiently regular) deformation path, starting from a situation in which $\det \boldsymbol{A}(\boldsymbol{n}) > 0$ is satisfied, the acoustic tensor $\boldsymbol{A}(\boldsymbol{n})$ becomes singular when the plastic modulus reaches a critical value $g_{cr}^E(\boldsymbol{n})$. This critical hardening modulus was derived by Rice (1977) as follows:

The condition

$$\det \left(\boldsymbol{A}_E(\boldsymbol{n}) - \frac{1}{g} \boldsymbol{M}\boldsymbol{n} \otimes \boldsymbol{N}\boldsymbol{n} \right) > 0, \tag{11.73}$$

assuming $\boldsymbol{A}_E$ positive definite, can be written as

$$\det \boldsymbol{A}_E(\boldsymbol{n}) \det \left(\boldsymbol{I} - \frac{1}{g} \boldsymbol{A}_E^{-1}(\boldsymbol{n}) \boldsymbol{M}\boldsymbol{n} \otimes \boldsymbol{N}\boldsymbol{n} \right) > 0. \tag{11.74}$$

Making use of the identity (2.38), we obtain the critical hardening modulus for loss of (E) in the direction $\boldsymbol{n}$:

$$\boxed{g_{cr}^E(\boldsymbol{n}) = \boldsymbol{N}\boldsymbol{n} \cdot \boldsymbol{A}_E^{-1}(\boldsymbol{n}) \boldsymbol{M}\boldsymbol{n},} \tag{11.75}$$

which is greater than zero for associative elastoplasticity but exceptionally may result negative for a non-associative flow rule (a case not considered in the following).

Note that when $\det \boldsymbol{A}_E(\boldsymbol{n}) \neq 0$ and $g \neq \boldsymbol{N}\boldsymbol{n} \cdot \boldsymbol{A}_E^{-1} \boldsymbol{M}\boldsymbol{n}$, the elastoplastic acoustic tensor [Eq. (11.46)] can be inverted [as shown, for instance, by Benallal and Bigoni (2004)] to provide the expression

$$\boldsymbol{A}^{-1} = \boldsymbol{A}_E^{-1} + \frac{\boldsymbol{A}_E^{-1} \boldsymbol{M}\boldsymbol{n} \otimes \boldsymbol{A}_E^{-1} \boldsymbol{N}\boldsymbol{n}}{g - \boldsymbol{M}\boldsymbol{n} \cdot \boldsymbol{A}_E^{-1} \boldsymbol{N}\boldsymbol{n}}. \tag{11.76}$$

When the acoustic tensor is singular, that is, when $g = g_{cr}^E$, the eigenvector corresponding to the null eigenvalue is

$$\boldsymbol{g} \propto \boldsymbol{A}_E^{-1}(\boldsymbol{n}) \boldsymbol{M}\boldsymbol{n}. \tag{11.77}$$

As for the case of the (SE) condition, the critical plastic modulus for loss of (E) is the solution of the constrained maximisation problem:

$$\boxed{g > g_{cr}^E = \max_{\boldsymbol{n}, |\boldsymbol{n}|=1} g_{cr}^E(\boldsymbol{n}) \iff \det \boldsymbol{A}(\boldsymbol{n}) > 0, \quad \text{for all } \boldsymbol{n} \in \mathcal{V}, |\boldsymbol{n}| = 1 \implies \text{(E condition).}} \tag{11.78}$$

The condition of loss of ellipticity admits a particularly nice mechanical interpretation, namely,

shear band formation, or in other words localisation of deformation into a planar band, becomes possible at failure of (E).

This has been known since Hadamard (1903) but was investigated in the case of elastoplasticity by Nadai (1931, 1950), Hill (1952, 1962), Prager (1954), Thomas (1953, 1961), Mandel (1966a), Rudnicki and Rice (1975), Rice (1977) and Rice and Rudnicki (1980). Strain localisation may be linked directly to the initiation and growth of slip mechanisms and fractures in solids. It is observed experimentally in a wide range of materials (including metals, polymers, concretes and geomaterials) and is one of the most explored research fields in elastoplasticity since Rice's (1977) seminal paper.

It is important to understand how failure of ellipticity is connected to the emergence of localised deformations. To this purpose, let us consider an infinite body subjected to remote boundary conditions sufficient to impose continued quasi-static homogeneous deformation. At any instant of the deformation process, the uniform stress field trivially satisfies the equilibrium equations. At a certain point of the deformation process, let us assume that a non-trivial incremental solution becomes possible, consisting of a velocity gradient that is uniform except across a planar band, where it is discontinuous. Inside and outside the band, the incremental stress and strain fields remain uniform, so equilibrium and compatibility are satisfied. If the band has normal $\boldsymbol{n}_0$ in the material description, the nominal traction must remain continuous [Eq. (9.61)] across the band:

$$[\![\dot{\boldsymbol{S}}]\!]\boldsymbol{n}_0 = \boldsymbol{0}; \tag{11.79}$$

moreover, the jump in velocity gradient across the band must satisfy the Maxwell compatibility conditions [Eq. (9.13)].

If we express via constitutive Eqs. (8.53) the Piola-Kirchhoff stress rate in Eq. (11.79) in terms of jump of the gradient of velocity written using Eq. $(9.13)_1$, we arrive at

$$\left(\boldsymbol{\mathcal{C}}(\dot{\boldsymbol{F}} + \boldsymbol{g}\otimes\boldsymbol{n}_0) - \boldsymbol{\mathcal{C}}(\dot{\boldsymbol{F}})\right)\boldsymbol{n}_0 = \boldsymbol{0}. \tag{11.80}$$

This is a necessary condition for quasi-static shear banding or strain localisation into a planar band. Four cases need to be examined, corresponding to conditions of plastic loading (or elastic unloading) inside and outside the band and plastic loading inside (or outside) and elastic unloading outside (or inside) the band:

$$\begin{aligned}
&1.\ \left(\boldsymbol{\mathcal{C}}(\dot{\boldsymbol{F}} + \boldsymbol{g}\otimes\boldsymbol{n}_0) - \boldsymbol{\mathcal{C}}(\dot{\boldsymbol{F}})\right)\boldsymbol{n}_0 = \mathbb{C}[\boldsymbol{g}\otimes\boldsymbol{n}_0]\boldsymbol{n}_0 = \boldsymbol{0} && \text{(plastic/plastic)},\\
&2.\ \left(\boldsymbol{\mathcal{C}}(\dot{\boldsymbol{F}} + \boldsymbol{g}\otimes\boldsymbol{n}_0) - \boldsymbol{\mathcal{C}}(\dot{\boldsymbol{F}})\right)\boldsymbol{n}_0 = \mathbb{G}[\boldsymbol{g}\otimes\boldsymbol{n}_0]\boldsymbol{n}_0 = \boldsymbol{0} && \text{(elastic/elastic)},\\
&3.\ \left(\boldsymbol{\mathcal{C}}(\dot{\boldsymbol{F}} + \boldsymbol{g}\otimes\boldsymbol{n}_0) - \boldsymbol{\mathcal{C}}(\dot{\boldsymbol{F}})\right)\boldsymbol{n}_0 = \mathbb{C}[\boldsymbol{g}\otimes\boldsymbol{n}_0]\boldsymbol{n}_0 + (\mathbb{C}-\mathbb{G})[\dot{\boldsymbol{F}}]\boldsymbol{n}_0 = \boldsymbol{0} && \text{(plastic/elastic)},\\
&4.\ \left(\boldsymbol{\mathcal{C}}(\dot{\boldsymbol{F}} + \boldsymbol{g}\otimes\boldsymbol{n}_0) - \boldsymbol{\mathcal{C}}(\dot{\boldsymbol{F}})\right)\boldsymbol{n}_0 = \mathbb{G}[\boldsymbol{g}\otimes\boldsymbol{n}_0]\boldsymbol{n}_0 - (\mathbb{C}-\mathbb{G})[\dot{\boldsymbol{F}}]\boldsymbol{n}_0 = \boldsymbol{0} && \text{(elastic/plastic)}.
\end{aligned} \tag{11.81}$$

It is important to note that only case 1 corresponds to violation of (E) [Eq. (11.72), i.e., $\det \boldsymbol{A}(\boldsymbol{n}) = 0$]. Case 2 corresponds to a purely elastic loss of (E) [$\det \boldsymbol{A}_E(\boldsymbol{n}) = 0$]. However, we assume for simplicity that $\boldsymbol{A}_E(\boldsymbol{n})$ is positive definite, and therefore, case 2 is not considered. We now report the Rice and Rudnicki (1980) proof that

If $\det \boldsymbol{A}(\boldsymbol{n}) > 0$, so that case 1 is excluded, cases 3 and 4 are also excluded.

Therefore, we assume that (the index 0 of $\boldsymbol{n}_0$ is omitted for conciseness)

$$\det \boldsymbol{A}(\boldsymbol{n}) > 0 \iff 1 - \frac{\boldsymbol{N}\boldsymbol{n} \cdot \boldsymbol{A}_E^{-1}(\boldsymbol{n})\boldsymbol{M}\boldsymbol{n}}{g} > 0. \tag{11.82}$$

Let us analyse case 3. If Eq. (11.80) is satisfied under conditions 3, we have

$$\boldsymbol{N} \cdot \dot{\boldsymbol{F}} \leq 0, \qquad \boldsymbol{N} \cdot \dot{\boldsymbol{F}} + \boldsymbol{g} \cdot \boldsymbol{N}\boldsymbol{n} \geq 0, \qquad \boldsymbol{g} = \frac{\boldsymbol{N} \cdot \dot{\boldsymbol{F}} + \boldsymbol{g} \cdot \boldsymbol{N}\boldsymbol{n}}{g} \boldsymbol{A}_E^{-1}(\boldsymbol{n})\boldsymbol{M}\boldsymbol{n}, \tag{11.83}$$

from which it is immediate that $\boldsymbol{g} \cdot \boldsymbol{A}_E^{-1}(\boldsymbol{n})\boldsymbol{M}\boldsymbol{n} \geq 0$. But (excluding the trivial case $\boldsymbol{N} \cdot \dot{\boldsymbol{F}} + \boldsymbol{g} \cdot \boldsymbol{N}\boldsymbol{n} = 0$) the determinant

$$\det \left\{ \boldsymbol{I} - \frac{\boldsymbol{A}_E^{-1}(\boldsymbol{n})\boldsymbol{M}\boldsymbol{n}}{g} \otimes \left(\boldsymbol{N}\boldsymbol{n} + \frac{\boldsymbol{N} \cdot \dot{\boldsymbol{F}}}{g^2} \boldsymbol{g} \right) \right\}$$

$$= 1 - \frac{\boldsymbol{N}\boldsymbol{n} \cdot \boldsymbol{A}_E^{-1}(\boldsymbol{n})\boldsymbol{M}\boldsymbol{n}}{g} - \frac{\boldsymbol{N} \cdot \dot{\boldsymbol{F}}}{g g^2} \boldsymbol{g} \cdot \boldsymbol{A}_E^{-1}(\boldsymbol{n})\boldsymbol{M}\boldsymbol{n} \tag{11.84}$$

is positive, so Eq. (11.80) is not satisfied. Let us analyse case 4. If Eq. (11.80) is satisfied under conditions 4, we have

$$\boldsymbol{N} \cdot \dot{\boldsymbol{F}} \geq 0, \qquad \boldsymbol{N} \cdot \dot{\boldsymbol{F}} + \boldsymbol{g} \cdot \boldsymbol{N}\boldsymbol{n} \leq 0, \qquad \boldsymbol{g} = -\frac{\boldsymbol{N} \cdot \dot{\boldsymbol{F}}}{g} \boldsymbol{A}_E^{-1}(\boldsymbol{n})\boldsymbol{M}\boldsymbol{n}. \tag{11.85}$$

Using Eq. $(11.85)_3$ and Eq. $(11.85)_1$ in Eq. $(11.85)_2$ (excluding the trivial case $\boldsymbol{N} \cdot \dot{\boldsymbol{F}} = 0$), we obtain

$$1 - \frac{\boldsymbol{N}\boldsymbol{n} \cdot \boldsymbol{A}_E^{-1}(\boldsymbol{n})\boldsymbol{M}\boldsymbol{n}}{g} \leq 0, \tag{11.86}$$

which is a contradiction to condition (11.82). It follows that assuming $\boldsymbol{A}_E(\boldsymbol{n})$ positive definite and a continuous dependence of constitutive equations on time,

> *In a loading program controlled by a (regularly) varying parameter and starting from a situation of ellipticity with* $\det \boldsymbol{A}(\boldsymbol{n}) > 0$*, the first possibility of strain localisation always occurs at failure of ellipticty,* $\det \boldsymbol{A}(\boldsymbol{n}) = 0$*, in the comparison solid corresponding to the loading branch of the constitutive operator* $\mathbb{C}$.

11.4.1 The small strain theory

In the case of the small strain theory [Eq. (8.66)] and isotropic elasticity [Eq. (8.65)], Eq. (11.71) allows a simplification of Eqs. (11.75) and (11.77) which provide the critical plastic modulus for loss of ellipticity (shear band formation):

$$g_{cr}^E = \max_{\boldsymbol{n},\, |\boldsymbol{n}|=1} \left\{ -\frac{\lambda + \mu}{\mu(\lambda + 2\mu)} (\boldsymbol{n} \cdot \mathbb{E}[\boldsymbol{Q}]\boldsymbol{n})(\boldsymbol{n} \cdot \mathbb{E}[\boldsymbol{P}]\boldsymbol{n}) + \frac{1}{\mu} \mathbb{E}[\boldsymbol{Q}]\boldsymbol{n} \cdot \mathbb{E}[\boldsymbol{P}]\boldsymbol{n} \right\} \tag{11.87}$$

and

$$\boldsymbol{g} \propto -\frac{\lambda + \mu}{\mu(\lambda + 2\mu)} (\boldsymbol{n} \cdot \mathbb{E}[\boldsymbol{P}]\boldsymbol{n})\boldsymbol{n} + \frac{1}{\mu} \mathbb{E}[\boldsymbol{P}]\boldsymbol{n}, \tag{11.88}$$

where $\mathbb{E}$ has the isotropic form of Eq. (8.65). The constrained maximisation problem [Eq. (11.87)] was solved under various hypotheses on the form of $\boldsymbol{P}$ and $\boldsymbol{Q}$ by

Rudnicki and Rice (1975), Needleman and Ortiz (1991) and Ottosen and Runesson (1991). It was solved under the sole hypothesis of coaxiality of $\boldsymbol{P}$ and $\boldsymbol{Q}$ by Boehler and Willis (1991) and Bigoni and Hueckel (1990, 1991a, 1991b).

The solution of the constrained maximisation problem (11.87) has been obtained by Bigoni and Hueckel (1990, 1991a) in the following way: Problem (11.87) and the corresponding eigenvector (11.88) are re-formulated and simplified to

$$\frac{h^E_{cr}}{2\mu} = \max_{\boldsymbol{n},\,|\boldsymbol{n}|=1} \frac{h^E(\boldsymbol{n})}{2\mu}, \tag{11.89}$$

where the critical hardening modulus for localisation in a generic direction $\boldsymbol{n}$ is given by

$$\begin{aligned}\frac{h^E(\boldsymbol{n})}{2\mu} = {} & 2\boldsymbol{n}\cdot\boldsymbol{PQn} - (\boldsymbol{n}\cdot\boldsymbol{Pn})(\boldsymbol{n}\cdot\boldsymbol{Qn}) - \boldsymbol{P}\cdot\boldsymbol{Q} \\ & - \frac{\nu}{1-\nu}(\boldsymbol{n}\cdot\boldsymbol{Pn} - \mathrm{tr}\boldsymbol{P})(\boldsymbol{n}\cdot\boldsymbol{Qn} - \mathrm{tr}\boldsymbol{Q}),\end{aligned} \tag{11.90}$$

together with its associated eigenvector $\boldsymbol{g}(\boldsymbol{n})$, that is,

$$\boldsymbol{g}(\boldsymbol{n}) \propto 2\boldsymbol{Pn} - \frac{1}{1-\nu}(\boldsymbol{n}\cdot\boldsymbol{Pn} - \nu\mathrm{tr}\boldsymbol{P})\,\boldsymbol{n}, \tag{11.91}$$

where ν is Poisson's ratio of the isotropic elastic tensor $\mathbb{E}$. Now, the constrained maximisation problem (11.89) is reduced to the unconstrained maximisation of the functional

$$L(\boldsymbol{n},\beta) = \frac{h^E(\boldsymbol{n})}{2\mu} - \beta(\boldsymbol{n}\cdot\boldsymbol{n} - 1), \tag{11.92}$$

where β is a Lagrangean multiplier. The stationary points of Eq. (11.92) satisfy the conditions

$$\frac{\partial L(\boldsymbol{n},\beta)}{\partial \boldsymbol{n}} = \boldsymbol{0} \quad \text{and} \quad \frac{\partial L(\boldsymbol{n},\beta)}{\partial \beta} = 0, \tag{11.93}$$

which become

$$\begin{aligned}\boldsymbol{PQn} & + \boldsymbol{QPn} - \frac{1}{1-\nu}[(\boldsymbol{n}\cdot\boldsymbol{Qn})\boldsymbol{Pn} + (\boldsymbol{n}\cdot\boldsymbol{Pn})\boldsymbol{Qn}] \\ & + \frac{\nu}{1-\nu}[(\mathrm{tr}\boldsymbol{Q})\boldsymbol{Pn} + (\mathrm{tr}\boldsymbol{P})\boldsymbol{Qn}] - \frac{\beta}{2\mu}\boldsymbol{n} = \boldsymbol{0}\end{aligned} \tag{11.94}$$

and

$$\boldsymbol{n}\cdot\boldsymbol{n} = 1.$$

Assuming now coaxiality of $\boldsymbol{Q}$ and $\boldsymbol{P}$ so that $\boldsymbol{PQ} = \boldsymbol{QP}$, it is possible to represent Eqs. (11.94) in the principal reference system of $\boldsymbol{Q}$ and $\boldsymbol{P}$ and to prove that functional $L(\boldsymbol{n},\beta)$ admits a finite number of extrema. Therefore, the solution of the maximisation problem (11.87) can be pursued calculating the extrema and selecting those corresponding to the maximum values of $L(\boldsymbol{n},\beta)$. This calculation yields a number of findings, which are listed below.

For small strain, non-associative plasticity based on isotropic elasticity and coaxial flow-mode tensor $\boldsymbol{P}$ and yield function gradient $\boldsymbol{Q}$, the following features of strain localisation can be proven:

- For deviatoric associativity, the band normal always lies in a principal reference plane of $\boldsymbol{Q}$ and $\boldsymbol{P}$.
- Strain localisation cannot occur for positive hardening for associative plasticity, but it can occur during hardening, softening or perfectly plastic behaviour when the non-associative flow rule is adopted.
- The critical hardening modulus for strain localisation depends on the stress state and on the features of the assumed constitutive model.
- At least two shear bands always occur,[3] with inclination symmetric with respect to the principal axes of $\boldsymbol{Q}$ and $\boldsymbol{P}$.

The solution of the constrained maximisation problem (11.87) in the special cases of the associative flow rule, including associative flow with a specific anisotopic elastic law, and deviatoric associativity is reported below.

Solution for strain localisation for small strain, associative plasticity based on isotropic elastic law In the special case of the associative flow rule, namely, $\mathbf{P} = \mathbf{Q}$, the solution of problem (11.87) is expressed in the principal reference system of $\mathbf{Q}$, where its components are denoted with an index subscript Q_i. The solution is given for the critical hardening modulus, the squared components of the direction of propagation $\boldsymbol{n}$ and the mode of propagation $\mathbf{g}$. While there is a single critical hardening modulus, there are several, at least two for isotropic elasticity, associated shear band normals $\boldsymbol{n}$ and corresponding $\boldsymbol{g}$. The critical hardening modulus for strain localisation is the maximum among the three values:

$$h_E^{cr} = \max\{h_1, h_2, h_3\}. \tag{11.95}$$

For (i,j,k) circular permutations of (1,2,3), one has to perform the following calculations (repeated indices are not summed in the formulae below):
1. Set

$$x_k = \frac{Q_i + \nu Q_k}{Q_i - Q_j}. \tag{11.96}$$

2. Obtain h_k:

$$\frac{h_k}{\mu} = -2(1+\nu)Q_k^2 - \frac{2}{1-\nu}(Q_j + \nu Q_k)^2 \mathcal{H}(x_k - 1) - \frac{2}{1-\nu}(Q_i + \nu Q_k)^2 \mathcal{H}(-x_k); \tag{11.97}$$

3. Obtain the squares of the components of the shear band normal $\boldsymbol{n}$:

$$n_i^2 = \min\{< x_k >, 1\}, \qquad n_j^2 = 1 - n_i^2, \qquad n_k^2 = 0. \tag{11.98}$$

4. Obtain the components of the eigenvector $\mathbf{g}$:

$$g_i = (Q_i - Q_j)\,n_i, \qquad g_j = (Q_j - Q_i)\,n_j, \qquad g_k = 0. \tag{11.99}$$

In the preceding expressions, the symbol $< \cdot >$ denotes the Macaulay brackets [Eq. (2.103)] and $\mathcal{H}$ the Heaviside step function.
If $Q_i = Q_j$, a cone of solutions for the shear band normals exists in the sense that the eigenvectors i and j of $\boldsymbol{Q}$ are arbitrary. In this case, x_k can be taken indifferently

[3] For anisotropic elasticity, there can be only one shear band; see the discussion later and in Section 13.3.1.

equal to $+\infty$ or $-\infty$, and either $g_i = 1$, $g_j = 0$ or $g_i = 0$, $g_j = 1$ with, in both cases, $g_k = 0$.

Solution of strain localisation for small strain, associative plasticity based on a class of anisotropic elastic law Explicit solutions of the constrained maximisation problem (11.87) to incorporate the effects of geometrical terms were given by Rudnicki and Rice (1975) and Szabó (1994) and to incorporate anisotropic elasticity by Bigoni and Loret (1999) and Bigoni et al. (2000). In the case of a specific form of anisotropic elasticity (6.196), Bigoni and Loret (1999) have observed that the explicit solution found by Bigoni and Hueckel (1991a) for isotropic elasticity still can be used. The idea is the following:

First, we introduce the 'transformed' yield function gradient and plastic flow mode tensor

$$\tilde{\boldsymbol{Q}} = \boldsymbol{B}^{1/2}\boldsymbol{Q}\boldsymbol{B}^{1/2}, \qquad \tilde{\boldsymbol{P}} = \boldsymbol{B}^{1/2}\boldsymbol{P}\boldsymbol{B}^{1/2} \tag{11.100}$$

and the transformed shear band normal and eigenvector

$$\tilde{\boldsymbol{n}} = \frac{\boldsymbol{B}^{1/2}\boldsymbol{n}}{||\boldsymbol{B}^{1/2}\boldsymbol{n}||}, \qquad \tilde{\boldsymbol{g}} = \boldsymbol{B}^{1/2}\boldsymbol{g}. \tag{11.101}$$

Second, the 'in-loading' elastoplastic acoustic tensor (11.46) for the small strain theory may be written as

$$\boldsymbol{A}(\boldsymbol{n}) = (\boldsymbol{n} \cdot \boldsymbol{B}\boldsymbol{n})\boldsymbol{B}^{1/2}\boldsymbol{A}^{\text{iso}}(\tilde{\boldsymbol{n}})\boldsymbol{B}^{1/2}, \tag{11.102}$$

where

$$\boldsymbol{A}^{\text{iso}}(\tilde{\boldsymbol{n}}) = \boldsymbol{A}_E^{\text{iso}}(\tilde{\boldsymbol{n}}) - \frac{1}{g}\tilde{\boldsymbol{p}}(\tilde{\boldsymbol{n}}) \otimes \tilde{\boldsymbol{q}}(\tilde{\boldsymbol{n}}), \tag{11.103}$$

in which

$$\tilde{\boldsymbol{q}}(\tilde{\boldsymbol{n}}) = \mathbb{E}[\tilde{\boldsymbol{Q}}]\tilde{\boldsymbol{n}}, \quad \tilde{\boldsymbol{p}}(\tilde{\boldsymbol{n}}) = \mathbb{E}[\tilde{\boldsymbol{P}}]\tilde{\boldsymbol{n}}. \tag{11.104}$$

Third, we note that the condition of strain localisation $\boldsymbol{A}(\boldsymbol{n})\boldsymbol{g} = \boldsymbol{0}$ is equivalent to the condition of strain localisation $\boldsymbol{A}^{\text{iso}}(\tilde{\boldsymbol{n}})\tilde{\boldsymbol{g}} = \boldsymbol{0}$. However, now $\tilde{\boldsymbol{Q}}$ and $\tilde{\boldsymbol{P}}$ are not coaxial, even when $\boldsymbol{Q}$ and $\boldsymbol{P}$ are; fortunately, there are important cases in which these remain coaxial, for instance, for the associative flow rule $\boldsymbol{Q} = \boldsymbol{P}$ (or also in other situations exploited by Gajo et al., 2004).

The conclusion is (Bigoni and Loret, 1999) that for associative elastoplastic materials, with an anisotropic elastic law of the form (6.196), and with reference to the small strain theory, an explicit solution for strain localisation is obtained in the following steps: (1) Calculate the transformed quantities $\tilde{\boldsymbol{Q}}$ and $\tilde{\boldsymbol{P}}$ [Eqs. (11.100)], (2) solve the maximisation problem [Eq. (11.87)] in the transformed space, that is, using Eqs. (11.95) through (11.99) with $\tilde{\boldsymbol{Q}}$ replacing $\boldsymbol{Q}$, thus obtaining the critical hardening modulus h_{cr}^E and the transformed critical directions $\tilde{\boldsymbol{n}}$ and $\tilde{\boldsymbol{g}}$, and (3) retrieve $\boldsymbol{n}$ and $\boldsymbol{g}$ from Eqs. (11.101).

One interesting feature found by Bigoni et al. (2000) and related to the elastic anisotropy is the fact that

> *owing to elastic anisotropy*
> there are situations in which only one shear band forms (instead the 'usual two'),

as follows from the example reported in Section 13.3.1.

Solution of strain localisation at small strain for deviatoric associativity and isotropic elastic law Deviatoric associativity is represented by the laws (8.60) for $\boldsymbol{Q}$ and $\boldsymbol{P}$, where we note that $\chi_1 > 0$ and $\psi_1 > 0$. Considering the maximisation problem [Eqs. (11.89) and (11.90)], we can factorize $\chi_1\psi_1$ in Eq. (11.90) and maximise $h^E(\boldsymbol{n})/(2\mu\chi_1\psi_1)$ instead of $h^E(\boldsymbol{n})/(2\mu)$. With this expedient, we can work with the two tensors

$$\boldsymbol{P} = \hat{\boldsymbol{S}} + \frac{\chi_2}{3\chi_1}\boldsymbol{I} \qquad \text{and} \qquad \boldsymbol{Q} = \hat{\boldsymbol{S}} + \frac{\psi_2}{3\psi_1}\boldsymbol{I}, \tag{11.105}$$

(where $|\hat{\boldsymbol{S}}|$ is unprescribed), satisfying the following condition:

$$P_i - P_j = Q_i - Q_j, \qquad i,j \in [1,2,3], \quad i \neq j, \tag{11.106}$$

with reference to the principal components of $\boldsymbol{Q}$ and $\boldsymbol{P}$. It can be shown that the normal $\boldsymbol{n}$ to the shear band, solution of the constrained maximisation problem (11.87), always lies in a principal plane of $\boldsymbol{Q}$ and $\boldsymbol{P}$. We therefore may assume $n_k = 0$. The critical hardening modulus is obtained with the following algorithm:

1. For $i = 1$ to $i = 3$, with $i \neq j \neq k$, if $Q_i = Q_j$, then $h^{(k)} = -\infty$; else ($Q_i \neq Q_j$) compute:

$$\begin{aligned} &n_k = 0, \quad n_j^2 = 1 - n_i^2, \\ &n_i^2 = (1-\nu)\frac{P_iQ_i - P_jQ_j}{(Q_i - Q_j)^2} - \frac{Q_j + P_j}{2(Q_i - Q_j)} + \nu\frac{\text{tr}\boldsymbol{Q} + \text{tr}\boldsymbol{P}}{2(Q_i - Q_j)}. \end{aligned} \tag{11.107}$$

   If $n_i^2 \notin [0,1]$, then $h^{(k)} = -\infty$, else ($n_i^2 \in [0,1]$) compute:
   The hardening modulus corresponding to Eq. (11.107):

$$\frac{h^{(k)}}{2\mu} = \frac{n_i^4}{1-\nu}(Q_i - Q_j)^2 - P_iQ_i - P_kQ_k - \frac{\nu}{1-\nu}(P_i + P_k)(Q_i + Q_k). \tag{11.108}$$

   And the eigenvector:

$$\begin{aligned} g_i &= \left\{2P_i - \frac{1}{1-\nu}\left[n_i^2(Q_i - Q_j) + P_j\right] + \frac{\nu}{1-\nu}\text{tr}\boldsymbol{P}\right\} n_i \\ g_j &= \left\{2P_j - \frac{1}{1-\nu}\left[n_i^2(Q_i - Q_j) + P_j\right] + \frac{\nu}{1-\nu}\text{tr}\boldsymbol{P}\right\} n_j \\ g_k &= 0. \end{aligned} \tag{11.109}$$

2. For $i = 1$ to $i = 3$, with $i \neq j \neq k$, take $n_i = 1$ and $n_j = n_k = 0$, and compute
   The corresponding hardening modulus:

$$\frac{h^{(i)}}{2\mu} = -\frac{1}{1-\nu}\left[P_kQ_k + P_jQ_j + \nu(P_kQ_j + P_jQ_k)\right]. \tag{11.110}$$

And the eigenvector:

$$g_i = P_i + \frac{\nu}{1-\nu}(P_j + P_k), \quad g_j = g_k = 0. \tag{11.111}$$

3. Select the maximum of the hardening moduli $h^{(i)}$, which is the value corresponding to h_{cr}^E:

$$h_{cr}^E = \max\{h^{(1)}, h^{(2)}, \ldots, h^{(6)}\}. \tag{11.112}$$

4. If there are two or more identical values of $h^{(i)}$ simultaneously maximising Eq. (11.112), infinite shear bands become possible, with the normals describing either a cone or a sphere. These features can be detected from the structure of the corresponding eigenvectors g_i.

11.5 Flutter instability

In the case of non-associative elastoplasticity (or when the elastic tensor does not possess the major symmetry), the acoustic tensor is non-symmetric, and there is, in principle, the possibility of a particular type of instability. This is the so-called flutter instability (Rice, 1977) and corresponds to the occurrence of two complex conjugate eigenvalues of the acoustic tensor or, in other words, when (for at least one unit vector $\boldsymbol{n}$)

$\boldsymbol{A}(\boldsymbol{n})$ has complex conjugate eigenvalues $\Longleftrightarrow$ flutter instability (F condition).

This instability is typical of frictional problems (see the example in Section 1.13.5) and remains for different aspects not still completely understood. In particular, the following points will be addressed:

1. When flutter instability may occur for nonassociative elastoplasticity
2. What is known about its mechanical interpretation

Specific examples of elastoplastic materials displaying flutter instability are deferred to Section 13.3, whereas a dynamic interpretation of flutter instability, disclosing its mechanical meaning, is presented in Chapter 16.

11.5.1 Onset of flutter

There are two possible approaches to the problem. One is simply to consider flutter instability to be excluded when the eigenvalues of the acoustic tensor are real. Another is to consider the coalescence of two eigenvalues as a critical condition. In fact, when two eigenvalues are coincident, an appropriate, infinitesimally small disturbance may induce complex eigenvalues. The latter point of view was suggested by An and Schaeffer (1990). Proceeding with it, we should restrict the class of possible disturbances; otherwise, we will end up considering unstable a linear isotropic elastic material (in which the two eigenvalues μ of the acoustic tensor always coincide).

Different classes of disturbances have been considered: Loret (1992) analysed perturbations in the plastic flow direction, and Bigoni and Zaccaria (1994a) and Bigoni (1995) considered perturbations owing to effects of large strains. Bigoni and

Loret (1999) have analysed perturbations in terms of a small hyper-elastic anisotropy superimposed on the usual isotropic elastic-plastic models, an approach continued by Bigoni and Petryk (2002) and Piccolroaz et al. (2006). This kind of perturbation probably is the most convincing because it is *symmetric* and has a clear physical interpretation. The result is that flutter may be triggered by such a vanishing small perturbation (for non-associative flow rules) when two eigenvalues of the acoustic tensor coincide. The same perturbation has no effect when superimposed on a linear elastic law or on an associative elastoplastic model. These results *suggest* that

> *for non-associative elastoplasticity, the condition of coalescence of eigenvalues of the acoustic tensor can be considered critical for flutter even in cases in which complex eigenvalues are excluded in the absence of any perturbation.*

Following this criterion, the onset of flutter can be defined by finding the conditions of coalescence of eigenvalues. These can be obtained numerically for elastoplasticity at finite strain, although explicit formulae can be derived in two-dimensional cases, such as, for instance, in the case analysed in Section 13.3. For the moment, we begin with investigation of elastic-plastic solids with isotropic elasticity and subject to small strains.

11.5.2 Flutter instability for small strain elastoplasticity with isotropic elasticity

The results relative to small strain isotropic elasticity are mainly due to Loret et al. (1990); however, we follow Bigoni and Zaccaria (1994a) and consider the acoustic tensor corresponding to the plastic branch of the constitutive equation (8.66):

$$\boldsymbol{A}(\boldsymbol{n}) = (\lambda + \mu)\boldsymbol{n} \otimes \boldsymbol{n} + \mu \boldsymbol{I} - \frac{1}{g}\boldsymbol{p} \otimes \boldsymbol{q}, \tag{11.113}$$

where

$$\boldsymbol{q} = \lambda(\text{tr}\boldsymbol{Q})\boldsymbol{n} + 2\mu\boldsymbol{Q}\boldsymbol{n} \qquad \text{and} \qquad \boldsymbol{p} = \lambda(\text{tr}\boldsymbol{P})\boldsymbol{n} + 2\mu\boldsymbol{P}\boldsymbol{n} \tag{11.114}$$

are linear functions of $\boldsymbol{n}$. Assuming[4] that $\boldsymbol{n} \times \boldsymbol{q} \neq \boldsymbol{0}$, let us consider the following non-orthogonal dual bases of $\mathcal{V}$ so that $\boldsymbol{e}^i \cdot \boldsymbol{e}_j = \delta^i_j$ (see Section 2.6):

$$\begin{aligned} &\boldsymbol{e}_1 = \boldsymbol{n}, && \boldsymbol{e}_2 = \boldsymbol{q}, && \boldsymbol{e}_3 = \frac{\boldsymbol{n}\times\boldsymbol{q}}{|\boldsymbol{n}\times\boldsymbol{q}|}, \\ &\boldsymbol{e}^1 = \frac{(\boldsymbol{q}^2)\boldsymbol{n} - (\boldsymbol{q}\cdot\boldsymbol{n})\boldsymbol{q}}{\boldsymbol{q}^2 - (\boldsymbol{q}\cdot\boldsymbol{n})^2}, && \boldsymbol{e}^2 = \frac{\boldsymbol{q} - (\boldsymbol{q}\cdot\boldsymbol{n})\boldsymbol{n}}{\boldsymbol{q}^2 - (\boldsymbol{q}\cdot\boldsymbol{n})^2}, && \boldsymbol{e}^3 = \boldsymbol{e}_3. \end{aligned} \tag{11.115}$$

Projected onto the bases (11.115), the eigenvalue problem for Eq. (11.113) gives the characteristic equation

$$\det \begin{pmatrix} \lambda + 2\mu - \eta & -\frac{1}{g}\boldsymbol{p}\cdot\boldsymbol{n} & 0 \\ (\lambda + \mu)\boldsymbol{q}\cdot\boldsymbol{n} & \mu - \frac{1}{g}\boldsymbol{p}\cdot\boldsymbol{q} - \eta & 0 \\ 0 & -\frac{1}{g}\boldsymbol{p}\cdot\boldsymbol{e}_3 & \mu - \eta \end{pmatrix} = 0, \tag{11.116}$$

[4] In the special case $\boldsymbol{n} \times \boldsymbol{q} = \boldsymbol{0}$, the final results do not change. This particular case is straightforward and treated by Bigoni and Zaccaria (1994a) and Bigoni (1995).

where η is the generic eigenvalue of the acoustic tensor (11.113). The three solutions of the characteristic equation (11.116) are the eigenvalue μ and the two roots of the polynomial equation:

$$\eta^2 - \left(\lambda + 3\mu - \frac{1}{g}\boldsymbol{p}\cdot\boldsymbol{q}\right)\eta + (\lambda+2\mu)\left(\mu - \frac{1}{g}\boldsymbol{p}\cdot\boldsymbol{q}\right) + \frac{1}{g}(\lambda+\mu)(\boldsymbol{p}\cdot\boldsymbol{n})(\boldsymbol{q}\cdot\boldsymbol{n}) = 0. \tag{11.117}$$

Flutter corresponds to negative values of the discriminant Δ of the second-order polynomial in Eqn. (11.117), which can be written as

$$\Delta = \left(\lambda + 3\mu - \frac{1}{g}\boldsymbol{p}\cdot\boldsymbol{q}\right)^2 - 4\mu(\lambda+2\mu)\left(1 - \frac{g_{cr}^E(\boldsymbol{n})}{g}\right), \tag{11.118}$$

where $g_{cr}^E(\boldsymbol{n})$ is the critical plastic modulus for strain localisation (11.87) at fixed $\boldsymbol{n}$, that is,

$$g_{cr}^E(\boldsymbol{n}) = -\frac{\lambda+\mu}{\mu(\lambda+2\mu)}(\boldsymbol{p}\cdot\boldsymbol{n})(\boldsymbol{q}\cdot\boldsymbol{n}) + \frac{\boldsymbol{p}\cdot\boldsymbol{q}}{\mu}. \tag{11.119}$$

Therefore, we conclude that (Bigoni and Zaccaria, 1994a)

> *for a given direction **n**, flutter is always excluded for values of the plastic modulus less than or equal to the critical plastic modulus for localisation in a band orthogonal to that direction **n**.*[5]

Straightforward manipulation of the discriminant (11.118) allows us to obtain the necessary and sufficient conditions for flutter:

$$\boxed{(\boldsymbol{n}\cdot\boldsymbol{p})(\boldsymbol{n}\cdot\boldsymbol{q}) > 0, \qquad (\boldsymbol{n}\cdot\boldsymbol{p})(\boldsymbol{n}\cdot\boldsymbol{q}) - \boldsymbol{p}\cdot\boldsymbol{q} > 0, \qquad \text{and} \qquad g \in (g_1, g_2),} \tag{11.120}$$

where

$$\boxed{\left.\begin{array}{l} g_1 \\ g_2 \end{array}\right\} = \frac{1}{\lambda+\mu}\left(\sqrt{(\boldsymbol{n}\cdot\boldsymbol{p})(\boldsymbol{n}\cdot\boldsymbol{q})} \pm \sqrt{(\boldsymbol{n}\cdot\boldsymbol{p})(\boldsymbol{n}\cdot\boldsymbol{p}) - \boldsymbol{p}\cdot\boldsymbol{q}}\right)^2.} \tag{11.121}$$

If we assume deviatoric associativity (8.60), a simple calculation shows that Eq. $(11.120)_2$ is never satisfied; therefore, we reach the conclusion (Loret et al., 1990)

> *For elastic-plastic solids in the presence of isotropic elasticity and deviatoric associativity (8.60) with χ_1 and ψ_1 strictly positive, complex eigenvalues of the acoustic tensor are excluded.*

However, coincident eigenvalues are possible. These may be determined by requiring that the discriminant (11.118) be null.[6] This occurs when one of the following two conditions is satisfied

$$(\boldsymbol{n}\cdot\boldsymbol{p})(\boldsymbol{n}\cdot\boldsymbol{q}) = 0 \qquad \text{and} \qquad g = g_{cr}^C(\boldsymbol{n}) = -\frac{\boldsymbol{p}\cdot\boldsymbol{q}}{\lambda+\mu} \tag{11.122}$$

[5] This does not mean that flutter always occurs before strain localisation. Rather, it means that flutter at a certain $\boldsymbol{n}$ always occurs before localisation corresponding to that $\boldsymbol{n}$.

[6] We are thus interested in determining coincident roots of Eq. (11.117) and do not consider the situation when Eq. (11.117) has different roots but one coincides with μ.

or

$$(\boldsymbol{n}\cdot\boldsymbol{p})(\boldsymbol{n}\cdot\boldsymbol{q})=\boldsymbol{p}\cdot\boldsymbol{q} \quad \text{and} \quad g=g_{cr}^{C}(\boldsymbol{n})=\frac{\boldsymbol{p}\cdot\boldsymbol{q}}{\lambda+\mu}. \tag{11.123}$$

Assuming isotropic elasticity (8.65) and deviatoric associativity (8.60), it is easy to obtain

$$\boldsymbol{p}\cdot\boldsymbol{q}-(\boldsymbol{n}\cdot\boldsymbol{p})(\boldsymbol{n}\cdot\boldsymbol{q})=4\mu^{2}\chi_{1}\psi_{1}\left(\hat{\boldsymbol{S}}\boldsymbol{n}\cdot\hat{\boldsymbol{S}}\boldsymbol{n}-(\boldsymbol{n}\cdot\hat{\boldsymbol{S}}\boldsymbol{n})^{2}\right)\geq 0. \tag{11.124}$$

Therefore,

$$(\boldsymbol{n}\cdot\boldsymbol{p})(\boldsymbol{n}\cdot\boldsymbol{q})=0 \implies \boldsymbol{p}\cdot\boldsymbol{q}\geq 0 \implies g_{cr}^{C}(\boldsymbol{n})\leq 0, \tag{11.125}$$

so case (11.122) is not interesting. In case (11.123), Eq. (11.117) gives two coincident solutions equal to μ. Therefore,

the acoustic tensor (for certain **n***) has an eigenvalue equal to* μ*, with multiplicity 3, when condition (11.123) is satisfied.*

Now we determine the critical plastic modulus for such a coalescence. We begin by noting that the following condition holds true at coalescence:

$$\hat{\boldsymbol{S}}\boldsymbol{n}\cdot\hat{\boldsymbol{S}}\boldsymbol{n}=(\boldsymbol{n}\cdot\hat{\boldsymbol{S}}\boldsymbol{n})^{2} \tag{11.126}$$

and is verified if and only if $\boldsymbol{n}$ is an eigenvector of $\hat{\boldsymbol{S}}$, but in this case $\boldsymbol{n}$ is also an eigenvector of $\mathbb{E}[\boldsymbol{P}]$ and $\mathbb{E}[\boldsymbol{Q}]$.[7] Therefore, the critical plastic modulus for coalescence of eigenvalues is

$$\boxed{g_{cr}^{C}=\max_{i=1,2,3}\frac{(\mathbb{E}[\boldsymbol{P}])_{i}(\mathbb{E}[\boldsymbol{Q}])_{i}}{\lambda+\mu},} \tag{11.127}$$

where $\mathbb{E}$ has the isotropic form (8.65), and the index i, not summed, denotes principal components of $\mathbb{E}[\boldsymbol{P}]$ and $\mathbb{E}[\boldsymbol{Q}]$ (in the same reference system).

To summarise with the preceding specific example at hand, we may note that for deviatoric associativity, complex eigenvalues of the acoustic tensor are excluded, but coalescence of three eigenvalues may occur. When coalescence occurs and with reference to the preceding example, Bigoni and Loret (1999) have shown that a perturbation in terms of a small (appropriate) elastic anisotropy superimposed on the isotropic elastic law is sufficient to trigger flutter. Therefore, even if for this model complex eigenvalues are excluded, flutter as induced by physically motivated perturbations is possible, and the critical condition corresponds to coalescence of the three eigenvalues of the acoustic tensor.

When coalescence is considered, Bigoni and Loret (1999) have shown that this may occur without relation to the other criteria, namely, (PD), (NS), (SE) and (E). In conclusion,

coalescence of eigenvalues of the acoustic tensor and therefore flutter may occur even when (PD) is verified. Therefore, flutter instability is not excluded by the Hill sufficient condition of stability.

[7] Vice versa, all eigenvectors of $\mathbb{E}[\boldsymbol{P}]$ and $\mathbb{E}[\boldsymbol{Q}]$ are also eigenvectors of $\hat{\boldsymbol{S}}$ because χ_1 and ψ_1 have been assumed strictly positive.

A detailed analysis of flutter instability for a specific anisotropic elastic-plastic law is deferred to Section 13.3.1, where the fact that there is no hierarchical order between flutter and the other local criteria introduced in this section will be shown explicitly. The lack of this order should not surprise because flutter instability occurs in the structure presented in Section 1.13.5, although there is no bifurcation, and the stiffness matrix of the system remains positive definite.

11.5.3 Physical meaning and consequences of flutter

Flutter instability in a non-associative elastoplastic continuum should not be considered a strange phenomenon because non-associativity is the phenomenological counterpart of the frictional phenomena occurring at the micro-scale, and it is known from Bigoni and Noselli (2011; see the example in Section 1.13.5) that flutter can be induced by dry friction. However, the mechanical interpretation of flutter instability in a continuum is a topic not yet fully analysed which merits great attention. In particular, Bigoni and Willis (1994) have analysed a special wave propagation problem showing that flutter may correspond to an oscillating motion of material particles which blows up with time. However, the analysis is valid only for an incrementally *linear* material. As noted by Rice (1977) and clarified by Bigoni and Petryk (2002; see also Chapter 14), in the case of an elastoplastic solid, an oscillation may yield a crossing of the loading-branch constitutive cone and thus may invalidate an analysis based on a linear material. Only a partial answer has been given to this point. This is the numerical analysis of Simões (1997), in which a growth and a decay of an oscillation are found in a plane strain indentation of a Drucker-Prager ideally plastic material with a non-associative flow law. In this problem, flutter is excluded in the sense intended in this section, but it may occur as connected to the presence of a free boundary (Loret et al., 1995).

A fundamental step forward in the interpretation of flutter instability has been made by Piccolroaz et al. (2006), employing the perturbative approach proposed by Bigoni and Capuani (2002, 2005). They have shown that flutter in a continuum is related to a blowing up of planar waves organised along well-defined propagation directions. A detailed discussion of these results is deferred to Chapter 16.

11.6 Other types of local criteria and instabilities

The reader should be aware at this point that there may occur many types of instabilities in elastoplastic solids with different mechanical consequences. However, a number of these instabilities were not analysed for simplicity here. Some of these are mentioned briefly below.

Occurrence of a particular condition was noted by Ottosen and Runesson (1991), Brannon and Drugan (1993) and Bigoni (1995). It corresponds to the situation in which the acoustic tensor has two coincident eigenvalues with geometric multiplicity smaller than algebraic multiplicity; in other words, the matrix is defective; see, for instance matrix (2.63). This possibility is again excluded in the case of an isotropic elastic-plastic solid with deviatoric associativity. But it may occur in other, more general circumstances. Whether or not this occurrence may correspond to a material instability presently is not known.

Another interesting condition, where it is not known whether or not the condition may corresponds to an instability, is that the eigenvalues of the constitutive operator become complex conjugate. This possibility for small strain elastoplasticity or for a special large strain formulation has been investigated by Bigoni and Zaccaria (1994b); see Section 2.20.10.

A number of material instabilities may occur at a point of a boundary. These were investigated by Benallal et al. (1990) as related to the possibility of a surface instability, a condition found by Biot (1965) and presented in Chapter 12. The possibility of flutter instability in terms of complex conjugate velocities of propagation of Rayleigh waves at a free boundary of an elastoplastic solid was discovered by Loret et al. (1995).

Finally, cavitation has been thoroughly investigated for non-linear elastic solids [see Horgan and Polignone (1995) and references cited therein], but only scarcely analysed in the case of elastoplastic solids (Huang et al., 1991). For elastoplasticity, the effects of flow rule non-associativity should be dominant [see the related analysis by Bigoni and Laudiero (1989)].

11.7 A summary on local and global uniqueness and stability criteria

A summary of local and global uniqueness and stability criteria for elastoplastic solids with the non-associative flow rule is presented in Figs. 11.3 and 11.4, the latter figure being reserved to the case of the associative flow rule.

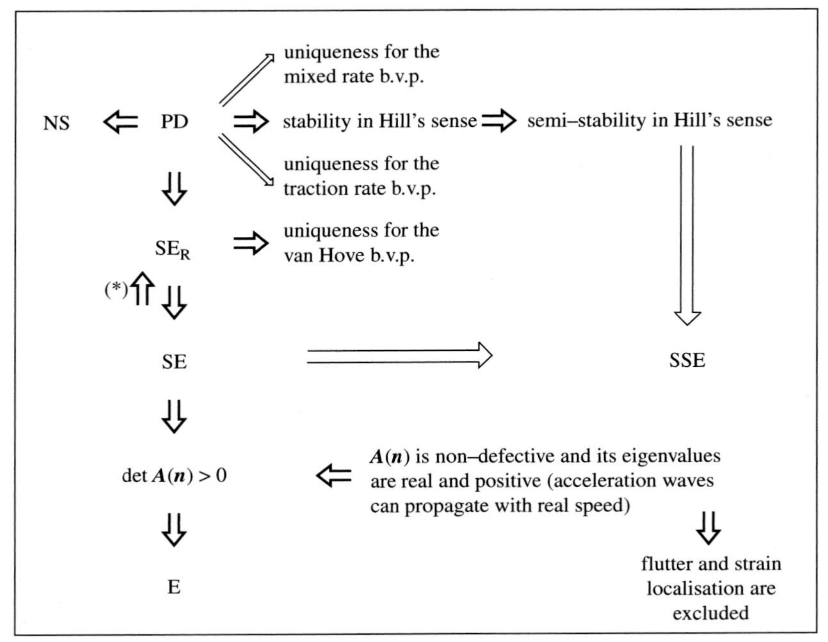

Figure 11.3. Hierarchy between uniqueness and stability criteria for non-associative elastoplasticity. The implication (*) holds whenever condition (11.68) is verified.

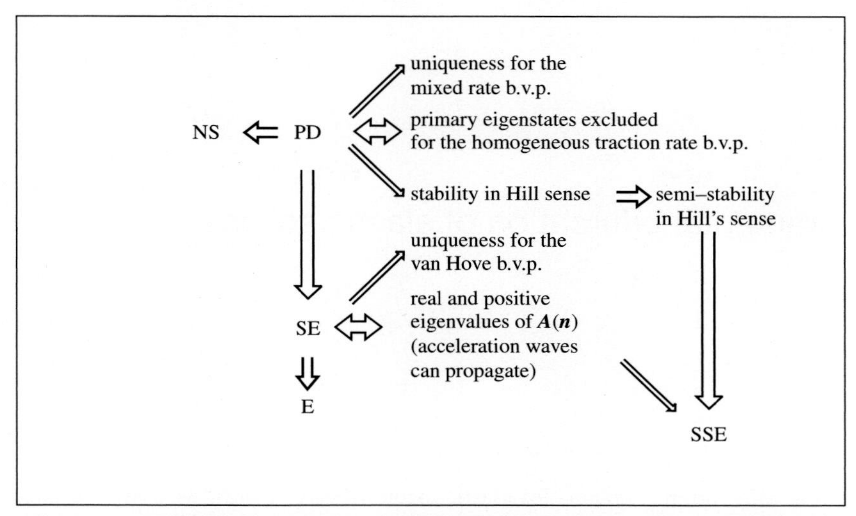

Figure 11.4. Hierarchy between uniqueness and stability criteria for associative elastoplasticity.

The hierarchy between the criteria is highlighted in the figures. We may conclude that although there are implications between the criteria, these implications are not strong enough to reduce the number of criteria. Each of these criteria has its own more or less explicit mechanical interpretation and therefore importance.

12 Incremental bifurcation of elastic solids

Several bifurcation problems for elastic incompressible solids, deformed in plane or axi-symmetric strain, are set and solved. In particular, an elastic block deformed under plane strain and a cylinder with circular cross section are considered, the former loaded biaxially and the latter uniaxially. Surface instabilities (at a free surface and at a surface dividing two elastic half spaces of different mechanical properties) and shear banding are also considered. The former is shown to represent an accumulation point for bifurcation stresses, and the latter is shown to correspond to an extreme form of instability. Finally, incremental bifurcations are analysed emerging from an inhomogeneous stress state, namely, finite plane strain bending of an incompressible elastic layer obeying neo-Hookean constitutive law.

In Chapters 10 and 11 we introduced sufficient conditions for uniqueness and stability of elastic and elastoplastic solids. In engineering applications the usual problem is to *find bifurcation loads and modes* during continued deformation of a solid body subjected to a prescribed loading program.

The purpose of this Chapter is to formulate and solve several bifurcation problems for incompressible materials deformed incrementally in plane strain or axi-symmetrically. In particular, the following five bifurcation problems are addressed: (1) a homogeneously stressed half space, loaded parallel to the free surface, (2) two homogeneously stressed elastic half spaces, loaded parallel to the surface separating them, (3) a homogeneously stressed block loaded parallel to two edges (a problem that will be generalized to include bifurcations of a layer on an elastic foundation, a layer on an elastic half space and a generic stack of layers), (4) a cylinder loaded under uniaxial compression parallel to its axis, and (5) an elastic layer subject to finite bending.

All these problems find important applications. Problem 1 corresponds to the well-known condition of surface instability and problem 2 to stationary Stoneley waves (Stoneley, 1924). Problem 3, generalised to include layered structures, finds a number of applications (sandwich panels in aircraft, submarine coatings, integrated circuits, MEMS and geological formations). Problems 1 through 3 have been formulated and analysed by Biot (1965) and subsequently by a number of authors, including Hill and Hutchinson (1975), Young (1976), Needleman (1979), Dorris and Nemat-Nasser (1980), Steif (1986a, 1986b, 1987, 1990), Papamichos et al. (1990), Dowaikh and Ogden (1991), Benallal et al. (1993), Triantafyllidis and

Lehner (1993), Triantafyllidis and Leroy (1994), Shield et al. (1994), Ogden and Sotiropoulos (1995) and Steigmann and Ogden (1996). In this chapter we will follow and slightly generalise Bigoni et al. (1997).

Problem 4, the bifurcation of a cylinder subject to uniaxial compression parallel to its axis, is a fundamental model in structural mechanics, leading for high aspect ratio to the Euler buckling model. It has been considered, under different perspectives, by a number of authors, including Wilkes (1955), Fosdick and Shield (1963), Cheng et al. (1971), Hutchinson and Miles (1974), Bassani et al. (1980), Haughton and Ogden (1979, 1980), Simpson and Spector (1984), Miles and Nuwayhid (1985), Davies (1991), Chau (1992, 1995), Pan and Beatty (1997a, 1997b) and Chau and Choi (1998). We will follow here Bigoni and Gei (2001) and Gei et al. (2004), and we will use the GBG constitutive model presented in Section 4.2.3. As a generalization to large strain of the Navier problem of bending, the finite flexure of an elastic layer considered by Rivlin (1949) provides an example of a problem where incremental bifurcations occur on an inhomogeneous state of stress. This bifurcation problem has been considered by a number of authors (Tryantafyllidis, 1980; Aron and Wang, 1995a, 1995b; Dryburgh and Ogden, 1999; Bruhns et al., 2002, 2003; Coman and Destrade, 2008) and only recently generalised to the case of multilayers by Roccabianca et al. (2010, 2011).

Although the models presented in this chapter refer to incompressible elasticity, several results may find applications in plastcity because they can be interpreted as bifurcation stresses calculated for the 'in loading comparison solid' of an incompressible elastoplastic solid with the associative flow law (where the two comparison solids coalesce into one) in the sense defined in Chapter 10.

12.1 The bifurcation problem

We are assuming now that a solid body is deformed quasi-statically under a prescribed continuous dead loading at a portion of its boundary, which is usually known as the 'fundamental (or trivial) deformation path'. We consider Eq. (10.24) governing quasi-static incremental equilibrium, where rates are to be understood as increments, and the non-linear constitutive operator $\boldsymbol{C}$ has to be replaced with the elastic tensor $\mathbb{G}$, valid for incremental elastic deformations.

In all the examples of this chapter, (1) we will consider absence of body forces and (2) we will look for regular solutions in a relative Lagrangean description, with the current state taken as reference (Section 9.2.2). Therefore, Eqs. (10.24) reduce to

$$\begin{aligned} &\boldsymbol{v} = \dot{\boldsymbol{\xi}}(\boldsymbol{x}_0), && \text{on} \quad \partial B_0^{\xi}, \\ &\mathbb{G}[\mathsf{grad}\boldsymbol{v}]\boldsymbol{n}_0 = \dot{\boldsymbol{\sigma}}(\boldsymbol{x}_0), && \text{on} \quad \partial B_0^{\sigma}, \\ &\mathsf{Div}(\mathbb{G}[\mathsf{Grad}\boldsymbol{v}]) = \boldsymbol{0}, && \text{in} \quad B_0, \end{aligned} \tag{12.1}$$

where the superimposed dot has to be regarded as an increment and $\boldsymbol{v}$ as the incremental displacement.

In the fundamental path, Eqs. (12.1) are trivially satisfied, but at each stage of this path, alternative solutions to Eqs. (12.1) will be sought, satisfying homogeneous boundary conditions. These solutions represent bifurcations of the incremental problem in the sense explained for comparison solids in associative elastoplasticity (Section 10.1.2), and these will make Hill's functional singular, so these will satisfy conditions (10.38). When a bifurcation is found, the deformed body therefore is called an 'eigen-configuration', and the bifurcation mode, in other words, the non-trivial solution of Eq. (12.1), is a so-called eigenmode.

When a bifurcation of the incremental first-order problem is found, this *normally* corresponds to a bifurcation of the non-linear problem admitting the incremental problem as a linearisation. However, strictly speaking, a bifurcation of the incremental first-order problem does not necessarily mean that the underlying non-linear bifurcation problem has a solution, and vice versa, when first-order bifurcations are excluded, this does not mean that the non-linear problem cannot admit a bifurcation. In the latter case, namely, when the first-order problem has a unique solution but the underlying non-linear problem admits a bifurcation, this bifurcation necessarily involves fields of order higher than the first, so a 'smooth bifurcation' is found (Bigoni, 1996).

12.2 Bifurcations of incompressible elastic solids deformed in plane strain

We refer here to the two-dimensional Biot (1965) theory of incremental incompressible elasticity, which was presented in Section 6.2.2; see Eqs. (6.155) through (6.177). This theory is of great importance because it provides a simple but rigorous approach to the behaviour of pre-stressed elastic solids.

We start from local uniqueness and stability criteria, applying concepts developed in Chapter 11, but now in a simpler setting, because the material is incrementally linear and the constitutive tensor has the major symmetry. Then we will treat four bifurcation problems for homogeneously stressed solids, namely, (1) bifurcation of an elastic half space (the so-called surface instability, or Rayleigh bifurcation, (2) bifurcation of two different half spaces in contact, or 'Stoneley bifurcation', (3) bifurcation of a uniaxially loaded elastic block, and finally, (4) we will set the problems of a layer on an elastic foundation, a layer on an elastic half space and a generic stack of layers, possibly connected through imperfect interfaces.

12.2.1 Local uniqueness and stability criteria for Biot plane strain and incompressible elasticity

Positive definiteness of $\mathbb{G}$

Positive definiteness of $\mathbb{G}$ is a local sufficient condition for uniqueness (see Section 11.1), which in our two-dimensional context, under the incompressibility constraint, can be written for all velocity gradients $(\nabla \boldsymbol{v})^T$ as

$$v_{j,i}\mathbb{G}_{ijkl}v_{l,k} > 0, \qquad v_{2,2} = -v_{1,1}, \tag{12.2}$$

which can be developed to yield the single condition

$$(\mathbb{G}_{1111} - 2\mathbb{G}_{1122} + \mathbb{G}_{2222})\, v_{1,1}^2 + \mathbb{G}_{2121}v_{1,2}^2 + 2\mathbb{G}_{1221}v_{1,2}v_{2,1} + \mathbb{G}_{1212}v_{2,1}^2 > 0, \tag{12.3}$$

an expression coincident with the analogous equation by Hill and Hutchinson [1975, their Eq. (2.9)[1]].

Since all components of the velocity gradient appearing in Eq. (12.3) (note that $v_{2,2}$ has been eliminated using incompressibility) are free parameters, the necessary and sufficient conditions for Eq. (12.3) to hold are

$$\mathbb{G}_{1111} - 2\mathbb{G}_{1122} + \mathbb{G}_{2222} > 0, \qquad \mathbb{G}_{1212} > 0, \qquad \mathbb{G}_{2121}\mathbb{G}_{1212} - \mathbb{G}_{1221}^2 > 0, \tag{12.4}$$

which, using the expressions (6.177) for the components of $\mathbb{G}_{ijkl}$, become

$$0 < T_1 + T_2 < 4\mu_*, \qquad \frac{T_1^2 + T_2^2}{T_1 + T_2} < 2\mu, \tag{12.5}$$

a condition given by Hill and Hutchinson [1975, their Eq. (3.9)] and in which we can recognise one of the conditions (11.16). Assuming $\mu > 0$, condition (12.5) can be written in terms of dimension-less constants (6.179) as

$$0 < \eta < 2\xi, \qquad \frac{k^2 + \eta^2}{2\eta} < 1, \tag{12.6}$$

which, for uniaxial plane strain tension/compression $\eta = k$, corresponds to the region between the vertical axis and the lines $\xi = k/2$ and $k = 1$ shown in Fig. 12.3.

Since in the particular case of Mooney-Rivlin (and neo-Hookean) material $0 < \eta = k < 1$ and $\xi = 1$, conditions (12.6) are satisfied:

Bifurcation is always excluded for plane strain uniaxial tension of a Mooney-Rivlin (and neo-Hookean) material.

For uniaxial plane strain tension parallel to the x_1 axis, the stress component T_1 is different from zero and $\eta = k$. For a J_2-deformation theory material, a substitution of Eqs. (6.188) into conditions (12.6) yields the exclusion condition for bifurcation

$$0 < \varepsilon_1 < N, \qquad \text{with } T_2 = 0, \tag{12.7}$$

where ε_1 is the logarithmic strain in the direction x_1 of the tensile load.

Eigenvalues and non-singularity of the constitutive operator (6.178)

Assuming $\mu \neq 0$, the incremental constitutive operator defined by Eqs. (6.178) can be written in matrix form as

$$\begin{bmatrix} \dot{t}_{11} \\ \dot{t}_{22} \\ \dot{t}_{12} \\ \dot{t}_{21} \end{bmatrix} = \begin{bmatrix} \mu(2\xi - k - \eta) & \mu & 0 & 0 \\ -\mu(2\xi + k - \eta) & \mu & 0 & 0 \\ 0 & 0 & \mu(1+k) & \mu(1-\eta) \\ 0 & 0 & \mu(1-\eta) & \mu(1-k) \end{bmatrix} \begin{bmatrix} v_{1,1} \\ \dot{p}/\mu \\ v_{2,1} \\ v_{1,2} \end{bmatrix}, \tag{12.8}$$

so the eigenvalues of the constitutive matrix are

- The two eigenvalues corresponding to the upper 2×2 block on the left of matrix (12.8)

$$\mu \frac{1 + 2\xi - k - \eta \pm \sqrt{(2\xi - k - \eta - 1)^2 - 4(2\xi + k - \eta)}}{2}, \tag{12.9}$$

[1] There is a misprint in that equation, their term $+1/2(\sigma_1 + \sigma_2)$ reads $+1/2(\sigma_1 - \sigma_2)$.

associated with the eigenvectors

$$v_{1,1} = 1 - (2\xi - k - \eta) \mp \sqrt{(2\xi - k - \eta - 1)^2 - 4(2\xi + k - \eta)},$$
$$\dot{p}/\mu = 2(2\xi + k - \eta), \qquad v_{2,1} = v_{1,2} = 0, \tag{12.10}$$

and

- The two eigenvalues corresponding to the lower 2×2 block on the right of matrix (12.8)

$$\mu\left(1 \pm \sqrt{k^2 + (1-\eta)^2}\right), \tag{12.11}$$

associated with the eigenvectors

$$v_{1,1} = \dot{p}/\mu = 0, \qquad v_{2,1} = k \pm \sqrt{k^2 + (1-\eta)^2}, \qquad v_{1,2} = 1 - \eta. \tag{12.12}$$

One of the two eigenvalues (12.9) and one of the eigenvalues (12.11) vanish, respectively, when

$$2\xi - \eta = 0, \qquad k^2 + \eta^2 = 2\eta, \tag{12.13}$$

so in the former case, the eigenvector

$$v_{1,1} = 2, \qquad \dot{p}/\mu = 2k, \qquad v_{2,1} = v_{1,2} = 0 \tag{12.14}$$

corresponds to a null increment of nominal normal stress increments, $\dot{t}_{11} = \dot{t}_{22} = 0$, for non-trivial $v_{1,1}$ and $\dot{p}$, whereas in the latter case, the eigenvector

$$v_{1,1} = \dot{p}/\mu = 0, \qquad v_{2,1} = k - 1, \qquad v_{1,2} = 1 - \eta \tag{12.15}$$

corresponds to a null increment of nominal shear stress increments, $\dot{t}_{12} = \dot{t}_{21} = 0$, for non-trivial $v_{2,1}$ and $v_{1,2}$. In these conditions, *the incremental solution of a uniformly deformed solid subjected to controlled nominal tractions over the whole boundary is not unique.*

Note that

> *starting from a situation for which the Hill exclusion condition (12.6) holds true, at the first time when this condition fails in a regular loading path, one of the two conditions (12.13) is verified, so a bifurcation (or a maximum load) occurs for a uniformly deformed solid subject to controlled nominal traction boundary conditions. In these circumstances, failure of the Hill exclusion condition is 'critical'.*

Strong ellipticity and ellipticity

Conditions of strong ellipticity and ellipticity have been presented in Sections 11.3 and 11.4, respectively. In the present two-dimensional setting under the incompressibility constraint, strong ellipticity condition (11.44) [or ellipticity, condition (11.72)] can be rewritten as

$$g_j n_i \mathbb{G}_{ijkl} n_k g_l > 0 \qquad [\text{or } \neq 0], \tag{12.16}$$

where the two unit vectors $\boldsymbol{n}$ and $\boldsymbol{g}$ are orthogonal to each other [so the incompressibility constraint is automatically satisfied; see the jump condition, Eq. (9.13), showing that for incompressible materials, $\boldsymbol{g}$ and $\boldsymbol{n}$ have to be normal to each other] and given in components by

$$\{\boldsymbol{n}\} = \{\cos\gamma, \sin\gamma\} \qquad \text{and} \qquad \{\boldsymbol{g}\} = \{-\sin\gamma, \cos\gamma\}, \tag{12.17}$$

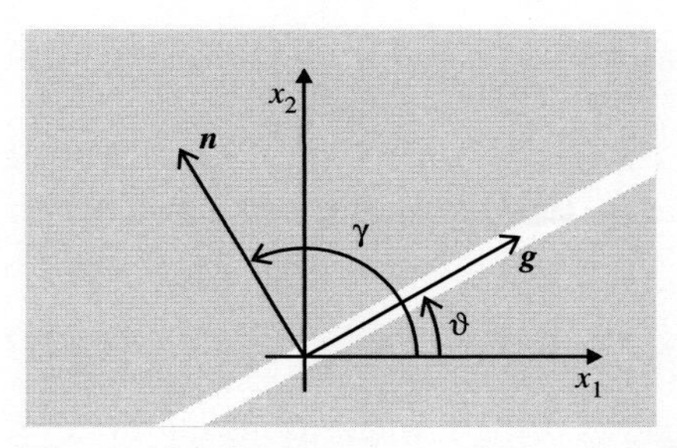

Figure 12.1. Unit vectors $\boldsymbol{n}$ (normal to the shear band) and $\boldsymbol{g}$ (parallel to the shear band) in the x_1–x_2 plane.

so γ is the angle between $\boldsymbol{n}$ and the x_1 axis; see Fig. 12.1. Representation (12.17) yields for every γ

$$\begin{aligned}&\mathbb{G}_{1212}\cos^4\gamma+\mathbb{G}_{2121}\sin^4\gamma+(\mathbb{G}_{1111}-2\mathbb{G}_{1122}-2\mathbb{G}_{1221}\\&\quad+\mathbb{G}_{2222})\cos^2\gamma\sin^2\gamma>0,\qquad[\text{or }\neq 0],\end{aligned}\tag{12.18}$$

which keeping into account the definition (6.177) of coefficients $\mathbb{G}_{ijkl}$, with the constants (6.179), becomes, for every γ,

$$\mu\sin^4\gamma\left[(1+k)\cot^4\gamma+2(2\xi-1)\cot^2\gamma+1-k\right]>0\qquad[\text{or }\neq 0],\tag{12.19}$$

equivalent[2] to the following three inequalities:

$$\mu>0\qquad[\text{or }\mu\neq 0],\qquad k^2<1,\qquad 2\xi>1-\sqrt{1-k^2}.\tag{12.21}$$

For the material model under consideration, we conclude that, *assuming* $\mu>0$, *ellipticity and strong ellipticity are equivalent criteria.*

Outside the elliptic boundary, localised solutions in terms of shear bands become possible. We postpone a discussion on these until after the full regime classification has been presented.

Thresholds for the loss of positive definiteness of $\mathbb{G}$ (PD) and ellipticity (E) have been calculated and reported for J_2-deformation theory and GBG materials in Figs. 5.6 and 5.7.

Regime classification

For a *uniformly pre-stressed solid* and in the absence of body forces, incremental equilibrium

$$\dot{t}_{ij,i}=0\tag{12.22}$$

yields the following two equations

$$\begin{aligned}\dot{p}_{,1}&=\mu[(1+k-2\xi)v_{1,11}-(1-k)v_{1,22}],\\\dot{p}_{,2}&=\mu[(1-k-2\xi)v_{2,22}-(1+k)v_{2,11}],\end{aligned}\tag{12.23}$$

[2] Note that the inequality (12.19) is equivalent to

$$\mu\cos^4\gamma\left[(1-k)\tan^4\gamma+2(2\xi-1)\tan^2\gamma+1+k\right]>0\qquad[\text{or }\neq 0],\qquad\forall\gamma.\tag{12.20}$$

which, together with the incompressibility constraint $v_{i,i} = 0$, provide a system of partial differential equations for v_1, v_2 and $\dot{p}$.

Since the material is incompressible, we can introduce a typical expedient of two-dimensional fluid mechanics (Ladyzhenskaya, 1963), namely, a *stream function* $\psi(x_1,x_2)$ with the property

$$v_1 = \psi_{,2}, \qquad v_2 = -\psi_{,1}, \tag{12.24}$$

so the incompressibility constraint is automatically satisfied, whereas the elimination of $\dot{p}$ in Eq. (12.23) gives the fourth-order partial differential equation

$$(1+k)\psi_{,1111} + 2(2\xi - 1)\psi_{,1122} + (1-k)\psi_{,2222} = 0, \tag{12.25}$$

derived by Biot [1965, p. 193, his Eq. (3.7); see also Hill and Hutchinson, 1975, their Eq. (3.3)].

Following Lekhnitskii (1981), Guz (1999), Cristescu et al. (2004) and more closely related to the notation used throughout this book, Radi et al. (2002) and Dal Corso et al. (2008a), a solution of Eq. (12.25) can be represented in terms of the analytical function F as

$$\psi(x_1,x_2) = F(x_1 + \Omega x_2), \tag{12.26}$$

where Ω is a complex constant satisfying the following bi-quadratic equation obtained inserting representation (12.26) in Eq. (12.25)

$$1 + k + 2(2\xi - 1)\Omega^2 + (1-k)\Omega^4 = 0. \tag{12.27}$$

The four roots Ω_j $(j = 1,\ldots,4)$ of Eq. (12.27) satisfy[3]

$$\Omega_j^2 = \frac{1 - 2\xi + (-1)^j\sqrt{4\xi^2 - 4\xi + k^2}}{1-k} \tag{12.28}$$

and are real or complex depending on the values of ξ and k. In compact form, we write

$$\Omega_j = \alpha_j + i\beta_j, \qquad j = 1,\ldots,4, \tag{12.29}$$

and define the four complex variables

$$z_j = x_1 + \Omega_j x_2 = x_1 + \alpha_j x_2 + i\beta_j x_2, \qquad j = 1,\ldots,4, \tag{12.30}$$

where $i = \sqrt{-1}$ is the imaginary unit, and $\alpha_j = \mathrm{Re}[\Omega_j]$ and $\beta_j = \mathrm{Im}[\Omega_j]$.

Through Eqs. (12.26) and (12.30), *the general solution of the differential Eq.* (12.25) can be written as

$$\psi(x_1,x_2) = \sum_{j=1}^{4} F_j(z_j). \tag{12.31}$$

The roots Ω_j, defined by Eq. (12.28) and changing their nature according to the values assumed by parameters ξ and k, can be classified as follows.

- In the elliptical complex regime (EC), defined as

$$k^2 < 1 \qquad \text{and} \qquad 1 - \sqrt{1-k^2} < 2\xi < 1 + \sqrt{1-k^2}, \tag{12.32}$$

[3] It may be instructive to compare Eq. (12.27) to Eq. (12.19) to obtain a relation between Ω_j and γ.

we have four complex conjugate roots

$$\Omega_1 = -\alpha + i\beta, \qquad \Omega_2 = \alpha + i\beta, \qquad \Omega_3 = \overline{\Omega}_1, \qquad \Omega_4 = \overline{\Omega}_2, \tag{12.33}$$

where

$$\left.\begin{array}{c}\beta\\ \alpha\end{array}\right\} = \sqrt{\frac{\sqrt{1-k^2} \pm (2\xi - 1)}{2(1-k)}} > 0. \tag{12.34}$$

- In the elliptical imaginary regime (EI), defined as

$$k^2 < 1 \qquad \text{and} \qquad 2\xi > 1 + \sqrt{1-k^2}, \tag{12.35}$$

we have four imaginary conjugate roots

$$\Omega_1 = i\beta_1, \qquad \Omega_2 = i\beta_2, \qquad \Omega_3 = \overline{\Omega}_1, \qquad \Omega_4 = \overline{\Omega}_2, \tag{12.36}$$

being $i = \sqrt{-1}$ and

$$\left.\begin{array}{c}\beta_1\\ \beta_2\end{array}\right\} = \sqrt{\frac{2\xi - 1 \pm \sqrt{4\xi^2 - 4\xi + k^2}}{1-k}} > 0. \tag{12.37}$$

- In the hyperbolic regime (H), defined as

$$k^2 < 1 \qquad \text{and} \qquad 2\xi < 1 - \sqrt{1-k^2}, \tag{12.38}$$

we have four real roots

$$\Omega_1 = \alpha_1, \qquad \Omega_2 = \alpha_2, \qquad \Omega_3 = -\Omega_1, \qquad \Omega_4 = -\Omega_2, \tag{12.39}$$

where

$$\left.\begin{array}{c}\alpha_1\\ \alpha_2\end{array}\right\} = \sqrt{\frac{1 - 2\xi \pm \sqrt{4\xi^2 - 4\xi + k^2}}{1-k}} > 0. \tag{12.40}$$

- In the parabolic regime (P), defined as

$$k^2 > 1, \tag{12.41}$$

we have two real and two imaginary roots

$$\Omega_1 = \alpha, \qquad \Omega_2 = i\beta, \qquad \Omega_3 = -\Omega_1, \qquad \Omega_4 = -\Omega_2, \tag{12.42}$$

where

$$\left.\begin{array}{c}\alpha\\ \beta\end{array}\right\} = \sqrt{\frac{\sqrt{4\xi^2 - 4\xi + k^2} \pm (1 - 2\xi)}{1-k}} > 0, \qquad \text{if } k < -1, \tag{12.43}$$

and

$$\left.\begin{array}{c}\alpha\\ \beta\end{array}\right\} = \sqrt{\frac{-\sqrt{4\xi^2 - 4\xi + k^2} \pm (1 - 2\xi)}{1-k}} > 0, \qquad \text{if } k > 1. \tag{12.44}$$

A sketch of the geometrical representation of the roots Ω_j in the complex plane is given in Fig. 12.2 with respect to the different regimes.

The regime classification in the k–ξ plane has been given by Radi et al. (2002, their fig. 2.3) and is now reported in Fig. 12.3.

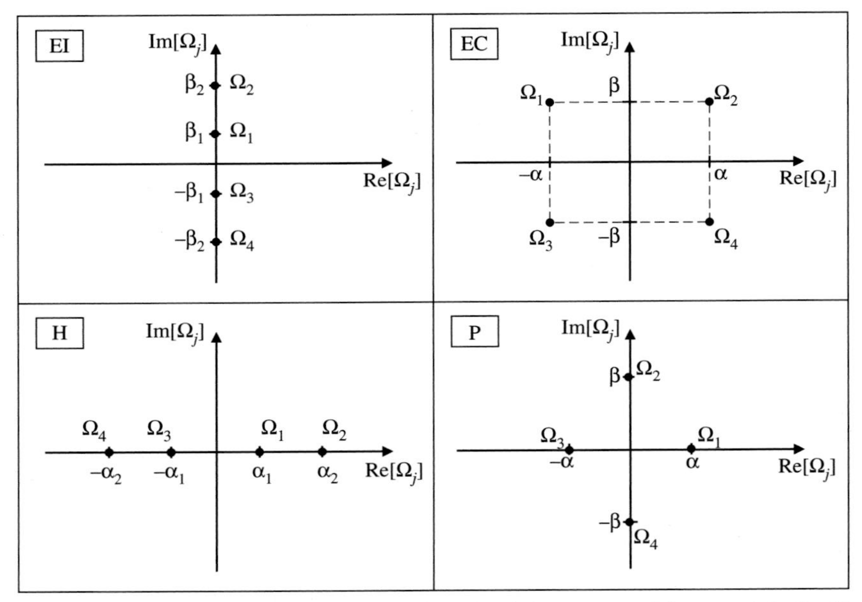

Figure 12.2. Plane strain deformation of an elastic incompressible material. Qualitative representation of the roots Ω_j in the complex plane within the different regimes: elliptical imaginary (EI), elliptical complex (EC), hyperbolic (H) and parabolic (P).

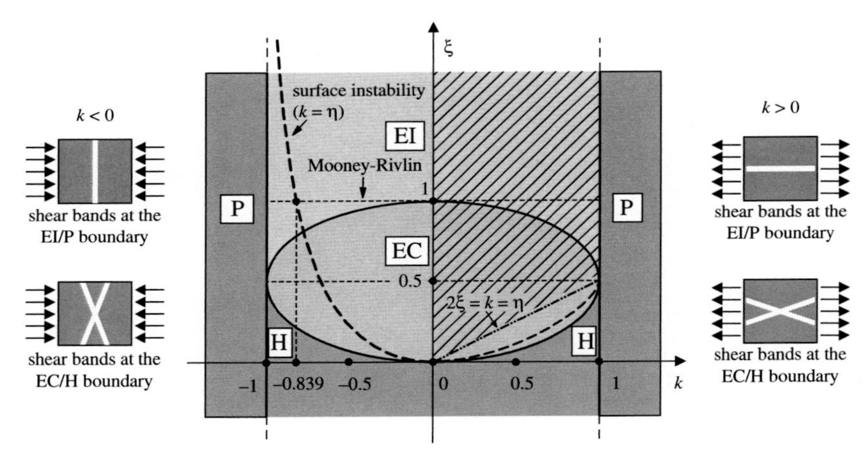

Figure 12.3. Regime classification in the $\xi = \mu_*/\mu$ versus $k = (T_1 - T_2)/(2\mu)$ parameter space for an elastic incompressible material deformed in plane strain. The surface bifurcation condition is also indicated (presented in Section 12.2.3), in the particular case where $k = \eta = T_1/2$, corresponding to a uniaxial principal stress aligned parallel to a free surface. In the same case, the exclusion condition (12.6) implies that every bifurcation is excluded in the region bounded by the ξ axis and the lines $\xi = k/2$ and $k = 1$.

The assumption of a specific material model determines the relation between ξ and k, which is graphically represented by a curve in the plane ξ versus k of Fig. 12.3. In particular, for every initially isotropic material, k depends on the in-plane stretch $\lambda > 0$ in the following way:

$$k = \frac{\lambda^4 - 1}{\lambda^4 + 1}. \tag{12.45}$$

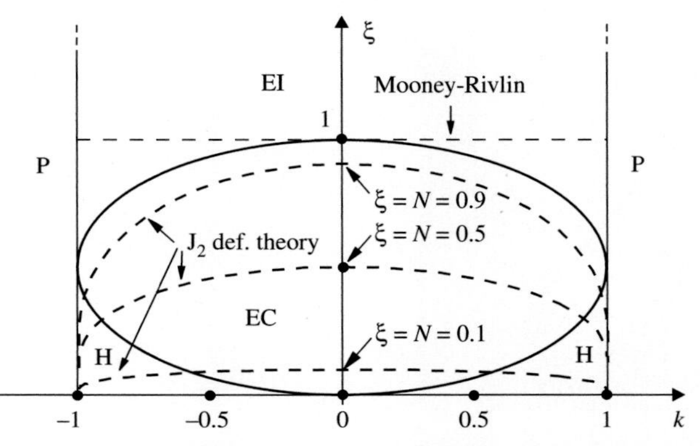

Figure 12.4. Relation between stress deviator k and anisotropy coefficient ξ for incompressible Mooney-Rivlin and J_2-deformation theory materials in plane strain. For a J_2-material, the hardening exponent N has been taken equal to 0.1, 0.5 and 0.9. Note that the EC/H boundary is crossed at finite stretch for every hardening exponent, whereas the EI/P boundary is approached at infinite stretch for a Mooney-Rivlin material.

Regarding ξ, this for a Mooney-Rivlin material (which coincides with a neo-Hookean material for plane isochoric deformations) can be determined from Eqs. (4.38) and $(6.183)_3$ to be constant and equal to one, namely,

$$\xi = 1, \qquad \text{Mooney-Rivlin, plane strain,} \tag{12.46}$$

whereas for a J_2-deformation material, Eqs. (6.186) and (4.39) yield

$$\xi = \frac{N(\lambda^4 - 1)}{2(\log\lambda)\,(\lambda^4 + 1)}, \qquad J_2\text{-deformation theory, plane strain.} \tag{12.47}$$

Note that $\xi = N$ when $\lambda \longrightarrow 1$. The curves in the plane ξ versus k described by Eqs. (12.45) through (12.47) are reported in Fig. 12.4 (for Mooney-Rivlin, a horizontal segment is obtained).

For a GBG material, we obtain (Dal Corso and Bigoni, 2010)

$$\xi = \frac{1}{2\sqrt{3}\epsilon_0}\left[\frac{\left(\frac{1+k}{1-k}\right)^{\frac{1}{2\sqrt{3}\epsilon_0}} + 1}{\left(\frac{1+k}{1-k}\right)^{\frac{1}{2\sqrt{3}\epsilon_0}} - 1}k - \frac{c+2}{c}|k|\right], \qquad \text{GBG material, plane strain,} \tag{12.48}$$

so the paths in the k–ξ plane are reported in Fig. 12.5 for the GBG materials with the parameters listed as (A), (B), and (C) in Table 5.17. Note that $\xi = -1/(\sqrt{3}c\epsilon_0) < 0$ when $\lambda \longrightarrow 1$.

Shear band inclination

The shear band inclinations, singled out by the angle γ^{SB} between the shear band normal $\boldsymbol{n}$ and the x_1 axis (Fig. 12.1), can be determined easily from Eq. (12.19), with the following results:

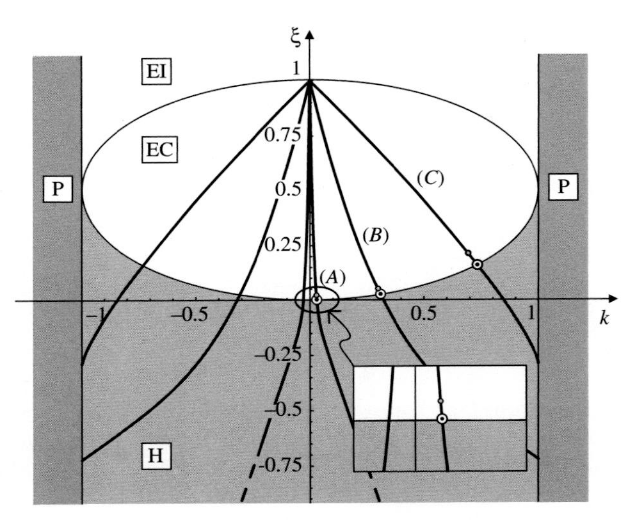

Figure 12.5. Relation between stress deviator k and anisotropy coefficient ξ during deformation of an elastic incompressible GBG material in plane strain. Material parameters correspond to the cases (A), (B) and (C) listed in Table 5.17. Note that small spots denote loss of positive definiteness of the constitutive operator, whereas the attainment of the EC/H boundary (indicated with large spots) is always approached for every material constant.

- Within the hyperbolic regime (H), ellipticity is always lost at (assuming for the moment $\gamma^{SB} \neq 0$)

$$\cot^2 \gamma^{SB} = \frac{1 - 2\xi \pm \sqrt{4\xi^2 - 4\xi + k^2}}{1 + k}, \tag{12.49}$$

or equivalently,

$$\tan^2 \gamma^{SB} = \frac{1 - 2\xi \pm \sqrt{4\xi^2 - 4\xi + k^2}}{1 - k}. \tag{12.50}$$

Therefore, within the hyperbolic regime, there are *four* shear bands.

- At the elliptical complex/hyperbolic boundary (EC/H), the following relation holds true

$$k = \mathsf{sign}(k)\, 2\sqrt{\xi\,(1 - \xi)}, \tag{12.51}$$

so the shear band inclination formula [Eq. (12.49)], gives (Hill and Hutchinson, 1975)

$$\tan^2 \gamma^{SB} = \frac{1 + \mathsf{sign}(k)\, 2\sqrt{\xi\,(1 - \xi)}}{1 - 2\xi}. \tag{12.52}$$

Thus *two* shear bands become possible. For instance, in the special case of $\xi = 0.25$, Eqn. (12.52) gives an inclination of the band normal $\gamma^{SB} = \pm 62.632^\circ$ for $k > 0$ and $\gamma^{SB} = \pm 27.368^\circ$ for $k < 0$, with respect to the direction of the x_1 axis [the band is inclined at $\pm(\pi/2 - \gamma^{SB})$ with respect to the x_1 axis]. Note that since ξ ranges between 0 and 1/2 in EC, for $k > 0$ ($k < 0$), the shear band is always inclined at an angle ranging between 45° and 0° (45° and 90°) with respect to the x_1 axis; see Fig. 12.6.

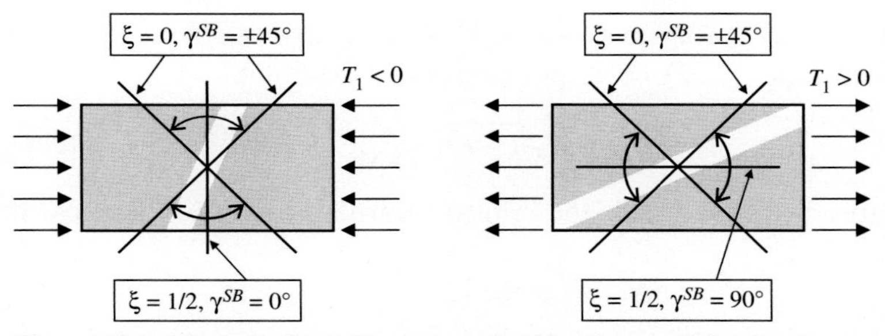

Figure 12.6. Plane strain deformation of an incompressible elastic material: possible shear band inclinations at the EC/H boundary, referred to a uniaxial compressive (tensile) state of stress $T_1 < 0$ ($T_1 > 0$) on the *left* (on the *right*).

- Within the parabolic regime (P), *two* shear bands become possible, with normals oriented at the angles γ^{SB} solution of

$$\cot^2\gamma^{SB} = \begin{cases} \dfrac{1-2\xi+\sqrt{4\xi^2-4\xi+k^2}}{1+k}, & \text{if} \quad k>1, \\ \dfrac{1-2\xi-\sqrt{4\xi^2-4\xi+k^2}}{1+k}, & \text{if} \quad k<-1, \end{cases} \tag{12.53}$$

or equivalently,

$$\tan^2\gamma^{SB} = \begin{cases} \dfrac{1-2\xi-\sqrt{4\xi^2-4\xi+k^2}}{1-k}, & \text{if} \quad k>1, \\ \dfrac{1-2\xi+\sqrt{4\xi^2-4\xi+k^2}}{1-k}, & \text{if} \quad k<-1. \end{cases} \tag{12.54}$$

- At the elliptical imaginary/parabolic boundary (EI/P), where

$$k^2 = 1 \quad \text{and} \quad 2\xi > 1, \tag{12.55}$$

we have only *one* shear band possible, aligned parallel to the x_1 axis, when $k=1$,

$$\gamma^{SB} = \frac{\pi}{2}, \tag{12.56}$$

or parallel to the x_2 axis when $k=-1$,

$$\gamma^{SB} = 0. \tag{12.57}$$

- At the hyperbolic/parabolic boundary (H/P), where

$$k^2 = 1 \quad \text{and} \quad 2\xi < 1, \tag{12.58}$$

three shear bands become possible. One is aligned parallel to the x_1 axis (x_2 axis), when $k=1$ ($k=-1$), whereas the other two are the solutions of

$$\cot^2\gamma^{SB} = 1-2\xi, \quad \text{for } k=1 \qquad \text{or} \qquad \tan^2\gamma^{SB} = 1-2\xi, \quad \text{for } k=-1. \tag{12.59}$$

For a Mooney–Rivlin material [Eqs. (12.45) and (12.46)], loss of ellipticity occurs in the limit of infinite stretch λ at the EI/P boundary; Eq. (12.45) gives $k=1$.

For a J_2-deformation theory material, loss of ellipticity always occurs at the EC/H boundary; see Eqs. (12.32), corresponding to the condition

$$2\xi = 1 - \sqrt{1-k^2}, \tag{12.60}$$

which, using relations (6.188), yields the critical logarithmic strain ε_1^E for loss of ellipticity

$$N = \varepsilon_1^E \tanh \varepsilon_1^E, \tag{12.61}$$

an equation equivalent to that given by Hutchinson and Tvergaard [1981, their Eq. (3.6)] and showing that the critical logarithmic deformation for ellipticity loss depends only on the hardening exponent N.

The inclination of the band normal with respect to the x_1 axis can be deduced by substituting condition (12.61) into Eq. $(6.188)_1$ and the resulting equation into Eq. (12.52) to obtain

$$\gamma^{SB} = \begin{cases} \arctan e^{\varepsilon_1^E}, & \text{for } \varepsilon_1^E > 0, \\ \dfrac{\pi}{2} - \arctan e^{\varepsilon_1^E}, & \text{for } \varepsilon_1^E < 0, \end{cases} \tag{12.62}$$

where it should be noted that since the inclination of the band is

$$\vartheta^{SB} = \frac{\pi}{2} - \gamma^{SB}, \tag{12.63}$$

Eq. (12.62) coincides with a result provided by Hutchinson and Tvergaard [1981, their Eq. (3.7)] and Radi et al. [2002, their Eq. (6.3)].

The situation when shear banding is an initial bifurcation mode

It has been already highlighted in Chapter 11 and it will be confirmed later in this chapter that shear banding is an extreme form of instability, usually occurring after infinite bifurcations are encountered in a loading path. One exception to this rule is when the problem corresponds to the van Hove conditions (see Section 11.3), generalised by Ryzhak (1993; 1994). In such cases, shear banding is an initial[4] instability mode.

Other remarkable exceptions occur in the special but important case when the Hill (1958, 1959) exclusion condition (12.6) fails simultaneously with ellipticity loss. For instance, considering Fig. 12.3, we note that for uniaxial tension, $\eta = k$, the exclusion condition (12.6) implies that every bifurcation is excluded for positive k in the elliptical regime and vales of ξ higher than the line $\xi = k/2$ (drawn dashed in Fig. 12.3). Therefore, as noted by Hill and Hutchinson (1975, their appendix AII) for $\xi \geq 0.5$ and $k = \eta > 0$, diffuse bifurcation modes are excluded in the elliptical regime for an elastic material (6.177) subject to uniaxial tension, so shear bands occur as an initial instability.

More in general, assuming $\eta = t\,k$, the Hill exclusion condition is satisfied within the region in the k–ξ plane bounded by the vertical ξ axis, by the inclined line $\xi = t\,k/2$ and by the vertical line at $k = 2t/(1+t^2)$. These two lines intersect at the point of

[4] We write 'an' initial and not 'the' initial because there can be simultaneity of shear banding with other instability modes.

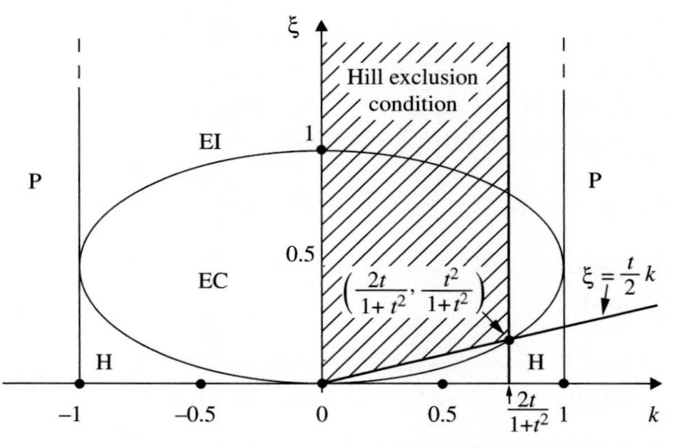

Figure 12.7. Plane strain deformation of an incompressible elastic material: The Hill exclusion criterion (12.6) represented in the k–ξ plane, assuming $\eta = t\,k$, with $t > 0$. For $t < 1$ (sketched in the figure), the exclusion region touches the elliptical complex/hyperbolic boundary (EC/H), whereas for $t > 1$, the exclusion region touches the elliptical complex/elliptical imaginary boundary (EC/EI). When $t \leq 1$, shear bands are an initial instability mode.

coordinates ($k = 2t/(1+t^2)$; $\xi = t^2/(1+t^2)$), which lies on the EC/H boundary for $t < 1$ or on the EC/EI boundary for $t > 1$ (see Fig. 12.7).

For a deformation path in the k–ξ plane terminating at the intersection point between the two lines and the EC/H curve reported in Fig. 12.7 and denoting failure of the Hill exclusion criterion, *shear bands are an initial bifurcation mode, so diffuse modes cannot occur previously*.

Situations where shear banding is an initial bifurcation mode have been analysed in elastoplasticity by Bigoni and Hueckel (1991a).

12.2.2 Bifurcations of layered structures: General solution

We know that a general solution of an incremental quasi-static problem in terms of stream function is given by Eq. (12.31), so as a particular case of this, we select the plane wave representation

$$\psi = -\frac{1}{ic_1}\left[b_1 e^{ic_1\Omega_1 x_2} + b_2 e^{ic_1\Omega_2 x_2} + b_3 e^{ic_1\Omega_3 x_2} + b_4 e^{ic_1\Omega_4 x_2}\right] e^{ic_1 x_1}, \tag{12.64}$$

where c_1 is a parameter, called (for reasons that will become transparent later) the 'wavenumber of the bifurcated mode'. With the purpose of keeping contact with Biot (1965), Hill and Hutchinson (1975) and Young (1976), we use instead of representation (12.64) the more general separate-variables representation

$$\psi = -\frac{1}{ic_1}\left[b_1 e^{ic_1\Omega_1 x_2} + b_2 e^{ic_1\Omega_2 x_2} + b_3 e^{ic_1\Omega_3 x_2} + b_4 e^{ic_1\Omega_4 x_2}\right] f(c_1 x_1), \tag{12.65}$$

but restricted to the three possibilities

$$f(c_1 x_1) = \begin{cases} \cos(c_1 x_1), \\ \sin(c_1 x_1), \\ e^{ic_1 x_1}, \end{cases} \tag{12.66}$$

all yielding through Eq. (12.25) the same characteristic equation (12.27), so all choices (12.66) represent 'equivalent' solutions.

From the definition of stream function (12.24) and the representation (12.65), we obtain the components of the incremental displacement as

$$v_1 = \tilde{v}_1(x_2) f(c_1 x_1) \quad \text{and} \quad v_2 = \tilde{v}_2(x_2) f'(c_1 x_1), \tag{12.67}$$

where a prime denotes differentiation with respect to the argument, thus, for $f(c_1 x_1) = \exp(i c_1 x_1)$, it is $f' = if$, and

$$\begin{aligned} \tilde{v}_1(x_2) &= -b_1 \Omega_1 e^{ic_1\Omega_1 x_2} - b_2 \Omega_2 e^{ic_1\Omega_2 x_2} - b_3 \Omega_3 e^{ic_1\Omega_3 x_2} - b_4 \Omega_4 e^{ic_1\Omega_4 x_2}, \\ \tilde{v}_2(x_2) &= -i\left[b_1 e^{ic_1\Omega_1 x_2} + b_2 e^{ic_1\Omega_2 x_2} + b_3 e^{ic_1\Omega_3 x_2} + b_4 e^{ic_1\Omega_4 x_2}\right]. \end{aligned} \tag{12.68}$$

An integration of Eqs. (12.23) yields the representation for the rate of in-plane mean stress

$$\dot{p} = f'(c_1 x_1) \sum_{j=1}^{4} q_j e^{ic_1\Omega_j x_2}, \tag{12.69}$$

with the coefficients q_j obtained from Eq. $(12.23)_1$ as

$$q_j/\mu = -c_1[(1+k-2\xi)\Omega_j - (1-k)\Omega_j^3] b_j, \tag{12.70}$$

or equivalently obtained from Eq. $(12.23)_2$ as

$$q_j/\mu = c_1[(1-k-2\xi)\Omega_j - (1+k)/\Omega_j] b_j. \tag{12.71}$$

Although not immediately evident, Eqs. (12.70) and (12.71) are identical. Therefore, considering Eq. (12.70), this can be transformed to

$$q_j/\mu = -c_1[k - (-1)^j \Lambda]\Omega_j b_j, \tag{12.72}$$

where

$$\Lambda = \sqrt{4\xi^2 - 4\xi + k^2}, \tag{12.73}$$

so that, eventually, we arrive at the representation for $\dot{p}$, that is,

$$\begin{aligned} \dot{p}/\mu = -c_1 \Big\{ &(k+\Lambda)(\Omega_1 b_1 e^{ic_1\Omega_1 x_2} + \Omega_3 b_3 e^{ic_1\Omega_3 x_2}) \\ &+(k-\Lambda)(\Omega_2 b_2 e^{ic_1\Omega_2 x_2} + \Omega_4 b_4 e^{ic_1\Omega_4 x_2}) \Big\} f'(c_1 x_1), \end{aligned} \tag{12.74}$$

which has been written neglecting an unessential constant term.

The incremental nominal stress now can be calculated using Eqs. (6.178) to yield

$$\begin{aligned} \dot{t}_{11} &= c_1 \mu (2\xi - k - \eta) \tilde{v}_1 f' + \dot{p}, \\ \dot{t}_{22} &= -c_1 \mu (2\xi + k - \eta) \tilde{v}_1 f' + \dot{p}, \\ \dot{t}_{12} &= \mu[-c_1(1+k)\tilde{v}_2 + (1-\eta)\tilde{v}_{1,2}] f(c_1 x_1), \\ \dot{t}_{21} &= \mu[-c_1(1-\eta)\tilde{v}_2 + (1-k)\tilde{v}_{1,2}] f(c_1 x_1), \end{aligned} \tag{12.75}$$

where

$$\tilde{v}_{1,2} = -ic_1 \sum_{j=1}^{4} b_j \Omega_j^2 e^{ic_1\Omega_j x_2}. \tag{12.76}$$

More explicit, using the identities

$$2\xi - \eta - \Lambda = 1 - \eta - (1-k)\Omega_4^2 \quad \text{and} \quad 2\xi - \eta + \Lambda = 1 - \eta - (1-k)\Omega_3^2 \quad (12.77)$$

and the fact that $\Omega_1^2 = \Omega_3^2$ and $\Omega_2^2 = \Omega_4^2$, Eqs. (12.75) become

$$\begin{aligned}
\dot{t}_{11}/(c_1\mu) &= -\big\{(2\xi - \eta + \Lambda)[b_1\Omega_1 e^{ic_1\Omega_1 x_2} + b_3\Omega_3 e^{ic_1\Omega_3 x_2}] \\
&\quad + (2\xi - \eta - \Lambda)[b_2\Omega_2 e^{ic_1\Omega_2 x_2} + b_4\Omega_4 e^{ic_1\Omega_4 x_2}]\big\} f'(c_1x_1), \\
\dot{t}_{22}/(c_1\mu) &= \big\{(2\xi - \eta - \Lambda)[b_1\Omega_1 e^{ic_1\Omega_1 x_2} + b_3\Omega_3 e^{ic_1\Omega_3 x_2}] \\
&\quad + (2\xi - \eta + \Lambda)[b_2\Omega_2 e^{ic_1\Omega_2 x_2} + b_4\Omega_4 e^{ic_1\Omega_4 x_2}]\big\} f'(c_1x_1), \\
\dot{t}_{12}/(ic_1\mu) &= \big\{[1 + k - (1-\eta)\Omega_1^2]\big(b_1 e^{ic_1\Omega_1 x_2} + b_3 e^{ic_1\Omega_3 x_2}\big) \\
&\quad + [1 + k - (1-\eta)\Omega_2^2]\big(b_2 e^{ic_1\Omega_2 x_2} + b_4 e^{ic_1\Omega_4 x_2}\big)]\big\} f(c_1x_1), \\
\dot{t}_{21}/(ic_1\mu) &= \big\{(2\xi - \eta + \Lambda)\big(b_1 e^{ic_1\Omega_1 x_2} + b_3 e^{ic_1\Omega_3 x_2}\big) \\
&\quad + (2\xi - \eta - \Lambda)\big(b_2 e^{ic_1\Omega_2 x_2} + b_4 e^{ic_1\Omega_4 x_2}\big)]\big\} f(c_1x_1).
\end{aligned} \quad (12.78)$$

To determine the conditions for bifurcation, we now have to consider specific boundary value problems and ensure that the solution represented through Eq. (12.67) and (12.78) satisfies the relevant boundary conditions. Imposing these, we always will find the trivial response $b_i = 0$ for $i = 1,\ldots,4$, but in addition to this, we will find alternative solutions occurring at a certain bifurcation pre-stress parameter k (or pre-stretch if k is represented as a function of this).

In all subsequent boundary value problems, we will consider rectangular (finite or semi-infinite) domains with a reference system aligned parallel to the edges. If one edge of this rectangular domain is identified with the x_1 axis, the choice $f(c_1x_1) = \sin(c_1x_1)$ makes explicit that we are considering sinusoidal bifurcation fields (12.67) and (12.75) satisfying null displacement v_1 and incremental shearing traction $\dot{t}_{12}$ at $c_1x_1 = n\pi$ for all integers n. Therefore, as mentioned earlier, c_1 plays the role of the wave number of the bifurcation field, whereas $2\pi/c_1$ represents the *wavelenght of the bifurcation field.*

12.2.3 Surface bifurcation

Let us consider an elastic half-space defined by the region $x_2 \leq 0$ and homogeneously pre-stressed of an arbitrary amount with a state of Cauchy stress having principal components aligned parallel and orthogonal to the surface $x_2 = 0$, where a dead loading is applied. Our problem is to find possible incremental bifurcations from this state, so two incremental boundary conditions have to be imposed:

- The nominal traction increment must be zero at the half space surface $x_2 = 0$, that is,

$$\dot{t}_{22}(x_1, x_2 = 0) = \dot{t}_{21}(x_1, x_2 = 0) = 0.$$

- All fields must decay exponentially when $x_2 \longrightarrow -\infty$.

First of all, we note that exponential decay is excluded in the hyperbolic regime owing to the real nature of Ω_j. Second, in the parabolic region, there is only one Ω_i corresponding to a decaying solution, either Ω_2 or Ω_4, and the solution may depend on one constant only, so it will be impossible in general to satisfy the two traction boundary conditions. We conclude that *surface instability may occur only in the elliptical regime.*

Note that the form of the velocity and incremental stress fields [Eqs. (12.67), (12.68) and (12.78)], used with the preceding boundary conditions implies that the incremental surface bifurcation problem corresponds to stationary (i.e., propagating at null speed) Rayleigh waves.

Restricting attention therefore to the elliptical regime, the decaying condition implies that $b_1 = b_2 = 0$, so only Ω_3 and Ω_4 have to be considered. Therefore, imposing the condition of null incremental nominal traction at the free surface $x_2 = 0$ of the half space yields, taking into account Eqs. (12.78), the following linear homogeneous system

$$\begin{bmatrix} (2\xi - \eta - \Lambda)\Omega_3 & (2\xi - \eta + \Lambda)\Omega_4 \\ 2\xi - \eta + \Lambda & 2\xi - \eta - \Lambda \end{bmatrix} \begin{bmatrix} b_3 \\ b_4 \end{bmatrix} = \boldsymbol{0}. \tag{12.79}$$

Thus, non-trivial solutions arise when

$$(2\xi - \eta - \Lambda)^2 \Omega_3 - (2\xi - \eta + \Lambda)^2 \Omega_4 = 0, \tag{12.80}$$

a condition which has been reduced in the elliptical regime by Needleman and Ortiz [1991, their Eq. (48)] to

$$4\xi - 2\eta = \frac{\eta^2 - 2\eta + k^2}{\sqrt{1 - k^2}}. \tag{12.81}$$

Assuming that the state of pre-stress is parallel to the free surface $x_1 = 0$, in other words, assuming that $\eta = k$, yields

$$\xi = \frac{k}{2}\left(1 - \sqrt{\frac{1-k}{1+k}}\right). \tag{12.82}$$

Note that

the bifurcation condition (12.80) is independent of the wavenumber c_1 [or wavelength $2\pi/c_1$] of the bifurcated mode, so at bifurcation, every wavelength corresponds to a possible bifurcation mode.

For this reason, the surface instability can be considered *a local instability criterion* because a bifurcation mode of infinitely small wavelength can 'fit' every curved boundary.

In the particular case of a Mooney-Rivlin (or neo-Hookean) material, $\xi = 1$, there is no bifurcation for uniaxial tensile stresses, whereas we obtain $T_1 \approx -1.678\mu$ for uniaxial compression.

Results pertaining to the J_2-deformation theory of plasticity can be obtained by inserting Eqs. (6.188) into Eq. (12.82), yielding the condition for surface bifurcation [Hutchinson and Tvergaard (1980), their Eq. (2.24)]:

$$\varepsilon_1^S \left(1 - e^{-2\varepsilon_1^S}\right) = N, \tag{12.83}$$

where ε_1^S is the critical logarithmic strain (parallel to the free surface) for surface instability and showing that this depends only on the hardening parameter N.

Equation (12.83) can be solved numerically, thus obtaining the results reported in Table 12.1, where also the critical logarithmic strain for ellipticity loss ε_1^E

Table 12.1. *Critical logarithmic strains for surface bifurcation ε_1^S and for ellipticity loss ε_1^E in a J_2-deformation theory material pre-stressed for plane strain uniaxial tension (positive sign) and compression (negative sign) parallel to the free surface of the elastic half space (orthogonal to the x_2 axis) as a function of the hardening parameter N*

		Uniaxial tension		Uniaxial compression	
N	ε_1^E	ε_1^S	ϑ^{SB}	ε_1^S	ϑ^{SB}
0.1	±0.322	0.252	±35.94°	−0.201	±54.06°
0.2	±0.463	0.377	±32.19°	−0.274	±57.81°
0.3	±0.577	0.484	±29.33°	−0.326	±60.67°
0.4	±0.678	0.582	±26.92°	−0.368	±63.08°
0.5	±0.772	0.675	±24.81°	−0.403	±65.19°
0.6	±0.861	0.766	±22.91°	−0.434	±67.09°
0.7	±0.948	0.855	±21.19°	−0.461	±68.81°
0.8	±1.032	0.943	±19.60°	−0.486	±70.40°
0.9	±1.116	1.031	±18.13°	−0.509	±71.87°

[Eq. (12.61)] and the shear band inclination ϑ^{SB} [in degrees, calculated from Eqs. (12.62) and (12.63) providing the inclination of the band normal, measured with respect to the x_1 axis] are reported (note that the absolute value of the critical logarithmic strain for loss of ellipticity remains the same for tension and compression, whereas the band inclination for compressive pre-strain is $\pi/2$ minus the value for positive pre-strain, ε_1^E).

Note that the critical logarithmic strain ε_1^{PD} for the Hill exclusion condition for bifurcation (12.6) to hold is given by Eq. (12.7), namely, $\varepsilon_1^{PD} = 0$ (so that bifurcation is never excluded in compression) and $\varepsilon_1^{PD} = N$ (so that bifurcation is excluded in tension at sufficiently small strain).

Table 12.1 confirms that for a uniaxial state of stress of a J_2-deformation theory material, loss of ellipticity always occurs after surface bifurcation.

More in general than for the J_2-deformation theory of plasticity, the condition for surface instability, Eq. (12.82) in the case $k = \eta$, is plotted dashed in Fig. 12.3, whereas the more general condition (12.81) is reported dashed in Fig. 12.8. From these figures, we can observe that

> *surface instability is always possible in a non-linear elastic half space pre-stressed parallel to the surface and, with the exception of the special cases treated in Section 12.2.1, before loss of ellipticity. In the special cases, surface instability occurs simultaneously with a shear band mode.*

12.2.4 Interfacial bifurcations

We now consider two elastic incompressible half spaces, one denoted by + and one by −, in contact at $x_2 = 0$ and look for incremental bifurcations in a form in which the mechanical fields correspond to stationary Stoneley waves (Stoneley, 1924). This is the so-called interfacial bifurcation problem. The half-spaces are

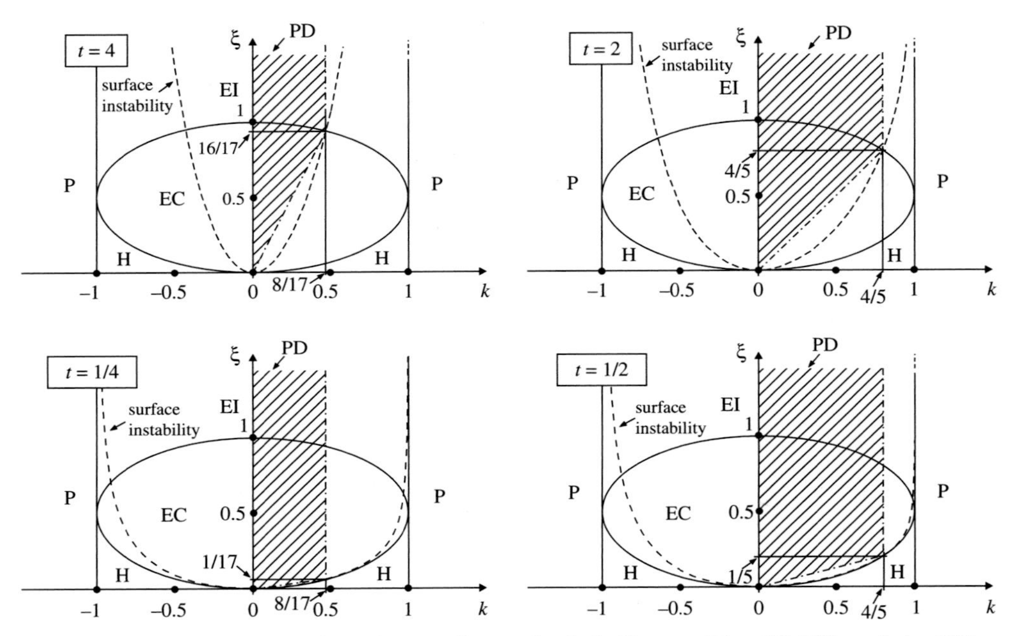

Figure 12.8. Surface bifurcation of a non-linear elastic half-space [Eq. (12.81)], and the Hill exclusion condition for bifurcation [Eq. (12.6)] in the k–ξ plane, together with the regime classification. The pre-stress parameter η has been linked to k as $\eta = tk$, and the cases $t = \{4, 2, 1/4, 1/2\}$ have been reported.

arbitrarily pre-stressed with a stress having principal components T_1 and T_2 in the x_1–x_2 reference system.

In this problem the incremental solution remains represented through Eq. (12.67) and (12.78) but must satisfy the following boundary conditions:

- Exponential decay of all fields when $x_2 \longrightarrow \pm\infty$ (where the + sign refers to the upper half-plane)
- Continuity of incremental displacement vector at the interface between the half spaces

$$v_1^+(x_1, x_2 = 0) = v_1^-(x_1, x_2 = 0), \qquad \text{and} \qquad v_2^+(x_1, x_2 = 0) = v_2^-(x_1, x_2 = 0);$$

- Continuity of incremental nominal traction at the interface between the half spaces

$$\dot{t}_{22}^+(x_1, x_2 = 0) = \dot{t}_{22}^-(x_1, x_2 = 0) \qquad \text{and} \qquad \dot{t}_{21}^+(x_1, x_2 = 0) = \dot{t}_{21}^-(x_1, x_2 = 0).$$

Note that, similar to the case of the elastic half space, the condition of exponential decay of the solution rules out the possibility of having interfacial bifurcations in the hyperbolic and parabolic regimes.

Imposition of the preceding conditions, using Eqs. (12.67) and (12.75), yields

$$[\boldsymbol{M}] \begin{bmatrix} b_1^+ \\ b_2^+ \\ b_3^- \\ b_4^- \end{bmatrix} = \boldsymbol{0}, \qquad (12.84)$$

where

$$[\boldsymbol{M}] = \begin{bmatrix} \Omega_1^+ & \Omega_2^+ & -\Omega_3^- & -\Omega_4^- \\ 1 & 1 & -1 & -1 \\ \frac{\mu^+}{\mu_-}(\Gamma - \Lambda)^+\Omega_1^+ & \frac{\mu^+}{\mu_-}(\Gamma + \Lambda)^+\Omega_2^+ & -(\Gamma - \Lambda)^-\Omega_3^- & -(\Gamma + \Lambda)^-\Omega_4^- \\ \frac{\mu^+}{\mu_-}(\Gamma + \Lambda)^+ & \frac{\mu^+}{\mu_-}(\Gamma - \Lambda)^+ & -(\Gamma + \Lambda)^- & -(\Gamma - \Lambda)^- \end{bmatrix} \tag{12.85}$$

in which

$$\Gamma = 2\xi - \eta. \tag{12.86}$$

Non-trivial solutions of system (12.84) are obtained when the determinant of the coefficient matrix $[\boldsymbol{M}]$ vanishes, that is,

$$\det[\boldsymbol{M}] = 0, \tag{12.87}$$

a condition that can be solved numerically for a given material, characterised by parameter ξ, for the unknown state of stress k and η.

Note that matrix $[\boldsymbol{M}]$ does not contain the wavenumber of the bifurcated mode c_1, so similar to surface instability, also interfacial ('Stoneley') bifurcation is independent of the wavelength of the bifurcation mode and can be interpreted as a local criterion of stability at the interface between two solids.

As examples of calculations, we report in Fig. 12.9 the critical logarithmic deformation ε_1^{ib} for interfacial bifurcation of two half spaces in contact, both made up of J_2-deformation theory material and subject to uniform compressive uniaxial stresses ($T_2 = 0$ in both half spaces) produced by a uniform negative pre-strain parallel to the x_1 axis, equal in the two half-spaces. The upper (lower) half space, which points correspond to $x_2 > 0$ (< 0), is denoted with $+$ ($-$), so the critical longitudinal logarithmic deformation for bifurcation is identical in the two half-spaces, that is, $\varepsilon_1^{ib} = \varepsilon_1^{ib-} = \varepsilon_1^{ib+}$, and equal with reversed sign to the transversal deformation, that is, $\varepsilon_1^{ib} = -\varepsilon_2^{ib} < 0$. Therefore, ε_1^{ib} is a function of the stiffness ratio K^-/K^+ and of the hardening exponents N^+ and N^- of the two media. Since the state of deformation is equal in the two half spaces, the state of stress is not, so $T_1^{ib-} \neq T_1^{ib+}$. However, the transversal stress is everywhere null $T_2^{ib-} = T_2^{ib+} = 0$.

Figure 12.9 is referred to $N^+ = 0.6$, whereas N^- has been taken equal to the values $\{0.2, 0.4, 0.6\}$. Since $K^-/K^+ > 1$ is investigated, the lower half-space ($-$) is taken to be stiffer than the upper half space ($+$); therefore, the compressive stress is higher (in absolute value) in the former half-space than in the latter.

It should be noted from Fig. 12.9 that all curves initiate at (are asymptotic to) the critical logarithmic strain for ellipticity loss (surface bifurcation) in the $-$ half-space (Table 12.1). Moreover, bifurcations are possible only at 'sufficient' stiffness contrast K^-/K^+ between the half spaces. Finally, we remark that interfacial bifurcations are also possible for tensile loads, but the results are not reported here.

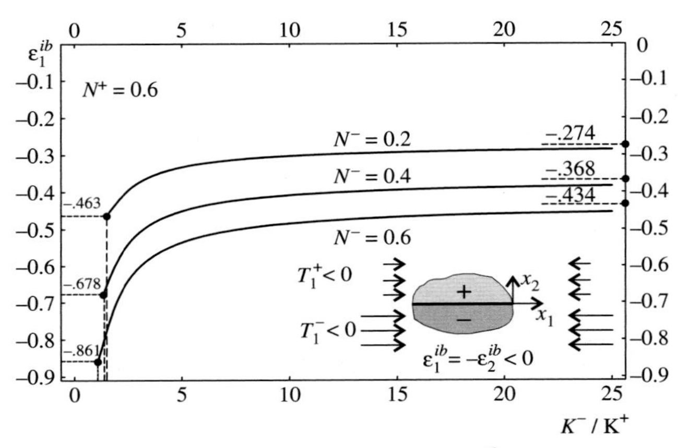

Figure 12.9. Critical logarithmic strain $\varepsilon_1^{ib} < 0$ for interfacial bifurcations of two J_2-deformation theory half-spaces uniformly deformed in compression, with perfect contact at $x_2 = 0$ and different stiffnesses $K^{\pm}$ and hardening exponents $N^{\pm}$. Bifurcation is possible within the elliptic regime and only when the stiffness contrast K^-/K^+ is sufficiently high.

12.2.5 Bifurcations of an elastic incompressible block

We now consider a rectangular elastic block placed between two rigid, parallel and smooth constraints, a problem which has been treated in great detail by Biot (1965), Hill and Hutchinson (1975) and Young (1976).

The elastic block is deformed in plane strain compression or tension through the relative movement the two rigid constraints (remaining parallel to each other) in a trivial mode for which the stress is uniform and uniaxial. At a certain stage of this loading program, the block has current length l and thickness h, as sketched in the inset of Fig. 12.10, and this configuration is taken as reference in a relative Lagrangean description. Assuming a reference system at the centre of the rectangle, the boundary conditions for the incremental bifurcation problem are

- Null incremental nominal normal tractions at $x_2 = \pm h/2$, namely,

$$\dot{t}_{22}(x_1, x_2 = \pm h/2) = 0 \quad \text{and} \quad \dot{t}_{21}(x_1, x_2 = \pm h/2) = 0. \qquad (12.88)$$

- Null incremental nominal shear traction and null horizontal incremental displacement at $x_1 = \pm l/2$, namely,

$$\dot{t}_{12}(x_1 = \pm l/2, x_2) = 0 \quad \text{and} \quad v_1(x_1 = \pm l/2, x_2) = 0. \qquad (12.89)$$

Selecting

$$f = \sin c_1 x_1 \quad \text{with} \quad c_1 = n\frac{2\pi}{l}, \qquad n = 1,2,3,...$$

or

$$f = \cos c_1 x_1 \quad \text{with} \quad c_1 = n\frac{\pi}{l}, \qquad n = 1,3,5,...$$

fields (12.67) and (12.78) automatically satisfy the boundary conditions (12.89), so the boundary conditions (12.88) can be imposed using Eq. $(12.78)_2$ and $(12.78)_4$, independent of the choice of function $f(c_1 x_1)$. This leads to the homogeneous algebraic

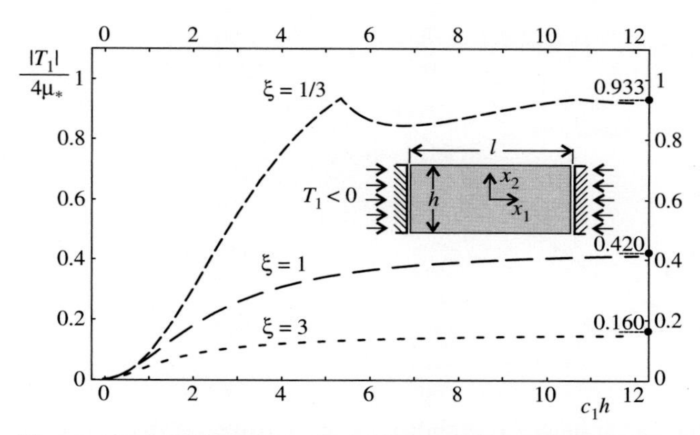

Figure 12.10. Bifurcation of an elastic block. Bifurcation stress $|T_1|/4\mu_*$ versus the wavenumber of the bifurcated field c_1 normalised through multiplication by the height of the block h for different values of the anisotropy ratio $\xi = \mu_*/\mu$.

problem

$$[\boldsymbol{M}]\begin{bmatrix} b_1 \\ b_2 \\ b_3 \\ b_4 \end{bmatrix} = \boldsymbol{0}, \tag{12.90}$$

where the matrix $[\boldsymbol{M}]$ is

$$\begin{bmatrix} (\Gamma-\Lambda)\Omega_1 e^{ic_1\Omega_1 h/2} & (\Gamma+\Lambda)\Omega_2 e^{ic_1\Omega_2 h/2} & (\Gamma-\Lambda)\Omega_3 e^{ic_1\Omega_3 h/2} & (\Gamma+\Lambda)\Omega_4 e^{ic_1\Omega_4 h/2} \\ (\Gamma-\Lambda)\Omega_1 e^{-ic_1\Omega_1 h/2} & (\Gamma+\Lambda)\Omega_2 e^{-ic_1\Omega_2 h/2} & (\Gamma-\Lambda)\Omega_3 e^{-ic_1\Omega_3 h/2} & (\Gamma+\Lambda)\Omega_4 e^{-ic_1\Omega_4 h/2} \\ (\Gamma+\Lambda) e^{ic_1\Omega_1 h/2} & (\Gamma-\Lambda) e^{ic_1\Omega_2 h/2} & (\Gamma+\Lambda) e^{ic_1\Omega_3 h/2} & (\Gamma-\Lambda) e^{ic_1\Omega_4 h/2} \\ (\Gamma+\Lambda) e^{-ic_1\Omega_1 h/2} & (\Gamma-\Lambda) e^{-ic_1\Omega_2 h/2} & (\Gamma+\Lambda) e^{-ic_1\Omega_3 h/2} & (\Gamma-\Lambda) e^{-ic_1\Omega_4 h/2} \end{bmatrix}, \tag{12.91}$$

which admits non-trivial solution when

$$\det[\boldsymbol{M}] = 0, \tag{12.92}$$

a condition which determines the critical value of k (or of the pre-stretch λ if k is taken to be a function of the stretch). Note that, differently from the cases involving semi-infinite media (the surface and interfacial bifurcations treated in Sections 12.2.3 and 12.2.4), the critical stress T_1 now depends on the wavenumber c_1 of the bifurcated field.

The absolute value of bifurcation stresses $|T_1|$ for an elastic block subjected to compression parallel to the x_1 axis, is reported in Fig. 12.10 versus the dimension-less parameter $c_1 h$ for different values of the anisotropy parameter ξ.

It can be observed that symmetric and anti-symmetric modes alternate for $\xi = 1/3$, and the transition from a mode to another produces a cusp in the graph. Only symmetric modes occur in the range explored in the figure, for $\xi = 1$ and 3.

Note that for low wavenumber (long wavelength), the elastic block behaves as an incompressible Kirchhoff plate deformed in plane strain, so the following asymptotic formula (a derivation is postponed to the next subsection) has been derived by Biot

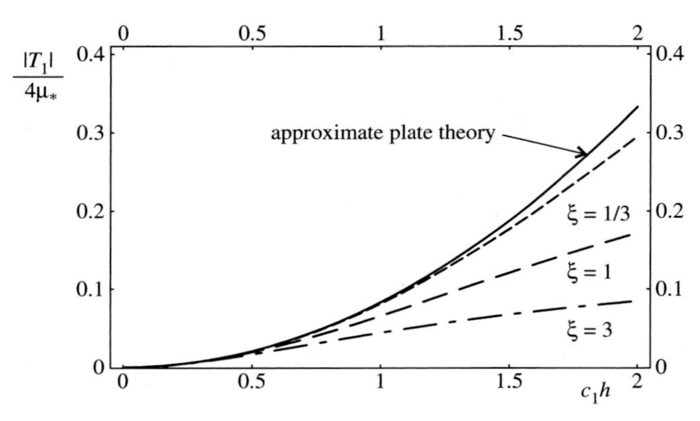

Figure 12.11. Asymptotic behaviour at small wavenumber c_1h of bifurcation stress $|T_1|/4\mu_*$ of an elastic incompressible block loaded in uniaxial plane strain compression. Different values of the anisotropy ratio $\xi = \mu_*/\mu$ are considered, together with the approximate Euler buckling formula (12.93).

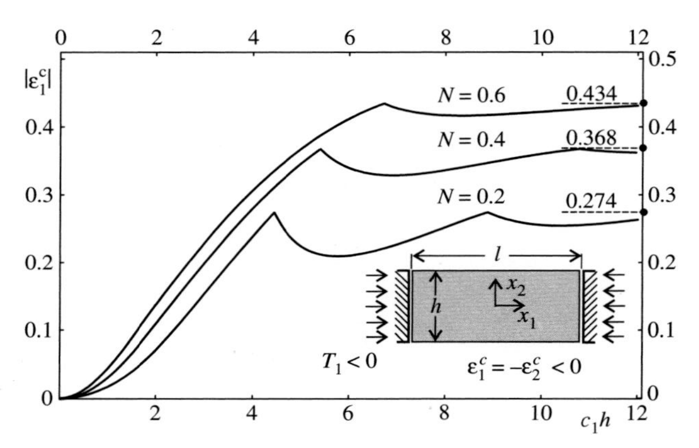

Figure 12.12. Bifurcation of an elastic incompressible block compressed parallel to the x_1 axis and obeying the J_2-deformation theory of plasticity. Absolute value of the bifurcation logarithmic strain ε_1^c versus the wavenumber of the bifurcated field c_1 normalised through multiplication by the current height of the block h for low $N = 0.2$, 'intermediate' $N = 0.4$ and high $N = 0.6$ strain hardening.

(1965) and by Bigoni et al. (1997; 2008) providing the Euler bucking load

$$T_1^{\text{buck}} = -\frac{\mu_*}{3}(c_1 h)^2, \tag{12.93}$$

which, as shown in Fig. 12.11, is a very accurate approximation when c_1h is smaller than 1/2 (the approximation gets better when ξ is decreased).

Results pertinent to a J_2-deformation theory block loaded in uniaxial plane strain are reported in Fig. 12.12, where the absolute value critical logarithmic strain for bifurcation ε_c is reported versus the dimension-less parameter c_1h at low ($N = 0.2$), high ($N = 0.6$) and 'intermediate' ($N = 0.4$) strain hardening.

12.2.6 Incompressible elastic block on a 'spring foundation'

A 'spring', or 'Winkler', foundation is a model of an interface elastically joining a solid with a rigid constraint in which the nominal traction on the solid is linearly related to the displacement against the rigid constraint. This model is common in the engineering literature and has been introduced in the present context by Suo et al. (1992) and Bigoni et al. (1997, 2008). Let us consider an elastic incompressible block subjected to uniform pre-stress k and η of current length l and thickness h with the upper surface subjected to dead loading and resting on a spring foundation (see the inset in Fig. 12.13).

Taking a reference system centred at the layer mid surface, the boundary conditions are

- Null incremental nominal traction vector at the layer's upper surface, that is,

$$\dot{t}_{2i}(x_1, x_2 = h/2) = 0, \qquad i = 1,2. \tag{12.94}$$

- Null incremental nominal shear traction at the layer's lower surface, that is,

$$\dot{t}_{21}(x_1, x_2 = -h/2) = 0. \tag{12.95}$$

- Linear relation between vertical components of incremental nominal tractions and displacement at the layer's lower surface, that is,

$$\dot{t}_{22}(x_1, x_2 = -h/2) = s v_2(x_1, x_2 = -h/2), \tag{12.96}$$

  where s is a stiffness coefficient, with the dimension of a stress divided by a length.

Note that condition (12.96) allows the possibility of an interpenetration between layer and substrate. This unphysical effect is the consequence of a simplification in the model, crucial to retain linearity in the incremental response. More complex models can be employed to avoid this effect and to introduce piece-wise linearity of the response, see for instance Radi et al., (1999).

In addition to the preceding boundary conditions [Eqs. (12.94) through (12.96)], we require as in the bifurcation problem of the block

- Null incremental nominal shear traction and null horizontal incremental displacement at $x_1 = \pm l/2$, namely,

$$\dot{t}_{12}(x_1 = \pm l/2, x_2) = 0 \qquad \text{and} \qquad v_1(x_1 = \pm l/2, x_2) = 0, \tag{12.97}$$

which is automatically satisfied taking $f(c_1 x_1)$ equal to $\sin c_1 x_1$ or $\cos c_1 x_1$, with, respectively, $c_1 = n2\pi/l$ or $c_1 = n\pi/l$. Therefore, only conditions (12.94) through (12.96) need to be imposed.

Results relative to the J_2-deformation theory of plasticity are reported in Fig. 12.13 for low $N = 0.1$ strain hardening. Note that results now depend on the dimension-less interfacial stiffness parameter sh/K, so when this stiffness is null, results relative to the elastic block are recovered (Fig. 12.12).

Assuming that the wavenumber of the bifurcated field c_1 is small (i.e., the wavelength is large) compared with the thickness of the block, namely, $c_1 h << 1$, we can

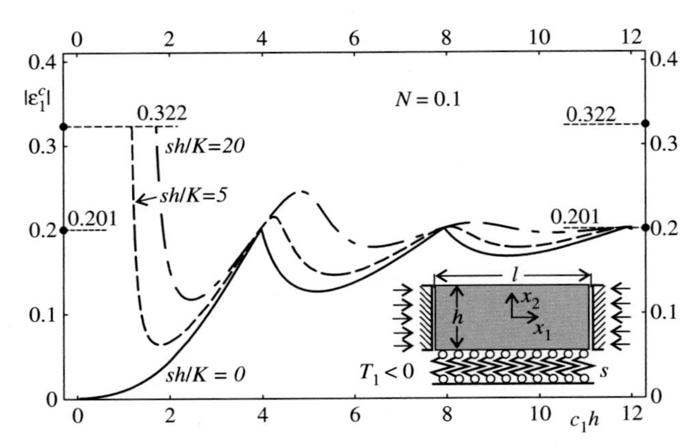

Figure 12.13. Bifurcation of a low-hardening ($N = 0.1$) J_2-deformation theory elastic incompressible block (of dimensions $l \times h$) on a spring (or Winkler) foundation of normal stiffness s. Bifurcation logarithmic strain ε_c versus the wavenumber of the bifurcated field c_1 normalised through multiplication by the width of the block l. Note that the solution depends on the dimension-less stiffness parameter of the foundation sh/K, so for $sh/K = 0$, the layer is disconnected from the foundation and behaves as an elastic block subject to uniaxial compression.

derive an asymptotic formula for the bifurcation stress of the block on spring foundation which reduces to Eq. (12.93) when the foundation stiffness vanishes, $s = 0$. Formulae of this kind have been derived by Biot (1937) and Gough et al. (1940) and generalised by Bigoni et al. (2008) and are often used in the engineering practice (Karam and Gibson, 1995; Kardomateas, 2005; Wilder et al., 2006). Here we follow the simple derivation given by Bigoni et al. (1997), employing the incremental kinematics of the Kirchhoff plate theory applied to an elastic block of infinite length, namely, an elastic layer.

Let us consider an elastic incompressible layer of current thickness h in the x_1–x_2 plane, where x_1 has the direction of the bond line and $x_2 \in [-h/2, h/2]$. Before bifurcation, the layer is subject to a state of compression parallel to x_1, so the only non-zero component of Cauchy stress is $T_1 < 0$. For a perturbation from this state, Biot (1965, section 2) has shown that incremental equilibrium requires

$$M_{,11} + m_{,1} + q + hT_1 \bar{v}_{2,11} = 0, \qquad (12.98)$$

where

$$M = \int_{-h/2}^{+h/2} x_2 \dot{t}_{11} dx_2, \qquad m = \dot{t}_{21} x_2 \Big|_{x_2 = h/2} - \dot{t}_{21} x_2 \Big|_{x_2 = -h/2},$$

$$q = \dot{t}_{22} \Big|_{x_2 = h/2} - \dot{t}_{22} \Big|_{x_2 = -h/2}, \qquad \bar{v}_2 = \frac{1}{h} \int_{-h/2}^{+h/2} v_2 dx_2. \qquad (12.99)$$

Because we have assumed that there is no tangential stiffness at the interface $m = 0$, and $q = -s\bar{v}_2$, therefore, Eq. (12.98) becomes

$$M_{,11} - s\bar{v}_2 + hT_1 \bar{v}_{2,11} = 0. \qquad (12.100)$$

The Kirchhoff plate theory deformation assumption is

$$\bar{v}_2 = w(x_1), \qquad v_1 = u_0(x_1) - x_2 w_{,1}, \tag{12.101}$$

so $\bar{v}_{2,1} + v_{1,2} = 0$. Also, each layer of the plate is assumed to be in a state of plane stress, thus

$$\dot{t}_{22} = T_2 = 0, \tag{12.102}$$

and therefore, from the incremental constitutive equations (6.178), $\dot{p} = 2\mu_* v_{1,1}$, so

$$\dot{t}_{11} = (4\mu_* - T_1) v_{1,1}. \tag{12.103}$$

Using Eq. (12.101) in Eq. (12.103), M can be calculated and substituted into Eq. (12.100) to obtain

$$D_* \frac{d^4 w}{dx_1^4} - T_1 h \frac{d^2 w}{dx_1^2} + sw = 0, \tag{12.104}$$

where

$$D_* = \frac{h^3}{12}(4\mu_* - T_1). \tag{12.105}$$

Note that Eq. (12.100) is exact (as remarked by Biot, 1965), whereas Eq. (12.104) is approximate because of both the assumptions of plane stress [Eq. (12.102)], and the plate theory deformation mode [Eq. (12.101)].

Seeking non-trivial solutions to Eq. (12.104) in the form

$$w(x_1) = A \sin(c_1 x_1), \tag{12.106}$$

where c_1 is the bifurcation mode wavenumber, we obtain

$$-T_1 h = D^* c_1^2 + \frac{s}{c_1^2}. \tag{12.107}$$

Finally, assuming that $4\mu_* >> |T_1|$ in the definition (12.105) of D_*, Eq. (12.107) can be simplified to

$$-\frac{T_1}{\mu_*} = \frac{1}{3}(c_1 h)^2 + \frac{hs}{\mu_*}\frac{1}{(c_1 h)^2}, \tag{12.108}$$

which reduces to Eq. (12.93) when $s = 0$. It has been shown by Bigoni et al. (1997, 2008) that Eq. (12.108) gives a good approximation to the layer on elastic foundation problem when h/l and hs/μ_* are sufficiently small.

12.2.7 Multi-layered elastic structures

For a multi-layered structure, made up of a stack of n elastic incompressible layers perfectly or imperfectly bonded to each other (Fig. 12.14). The 'imperfect bonding' is idealised with a spring-type interface defined by a constitutive relation similar to Eq. (12.96) and given below. The boundary conditions for the ith layer embedded in the multi-layer are the following (we consider for simplicity infinitely long layers of current thickness h_i, although it is not difficult to introduce a finite length l; moreover, we refer each layer to a reference system centred at the middle of the layer):

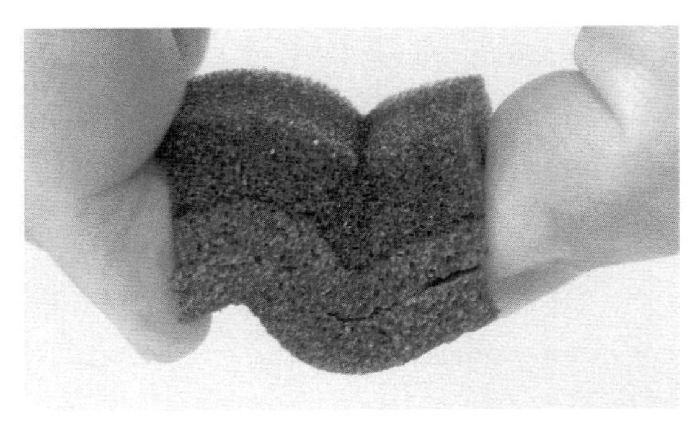

Figure 12.14. Buckling in a three-layer foam compressed by hand. Compare with Fig. 1.4.

- Continuity of incremental nominal traction at the interface between layers:

$$\begin{aligned} \dot{t}_{2j}^{(i-1)}(x_1, x_2 = -h^{(i-1)}/2) &= \dot{t}_{2j}^{(i)}(x_1, x_2 = h^{(i)}/2), \\ \dot{t}_{2j}^{(i+1)}(x_1, x_2 = h^{(i+1)}/2) &= \dot{t}_{2j}^{(i)}(x_1, x_2 = -h^{(i)}/2), \end{aligned} \tag{12.109}$$

  where $h^{(i-1)}$, $h^{(i)}$ and $h^{(i+1)}$ are the thicknesses of the layers $i-1$, i and $i+1$.
- For a perfectly bonded interface: continuity of incremental displacement occurs at the interface between layers:

$$\begin{aligned} v_j^{(i-1)}(x_1, x_2 = -h^{(i-1)}/2) &= v_j^{(i)}(x_1, x_2 = h^{(i)}/2), \\ v_j^{(i+1)}(x_1, x_2 = h^{(i+1)}/2) &= v_j^{(i)}(x_1, x_2 = -h^{(i)}/2). \end{aligned} \tag{12.110}$$

- For imperfectly bonded interface: at the interface between layers, the jump in incremental displacement is linearly related to the incremental nominal traction through a stiffness tensor s_{ji}:

$$\begin{aligned} \dot{t}_{22}^{(i)} &= -S_{2j}\left[v_j^{(i)}(x_1, x_2 = h^{(i)}/2) - v_j^{(i-1)}(x_1, x_2 = -h^{(i-1)}/2)\right], \\ \dot{t}_{21}^{(i)} &= -S_{1j}\left[v_j^{(i)}(x_1, x_2 = h^{(i)}/2) - v_j^{(i-1)}(x_1, x_2 = -h^{(i-1)}/2)\right], \end{aligned} \tag{12.111}$$

  where S_{kj} represents the interfacial stiffness.

In all cases, at each interface, four equations can be witten for the four unknowns constants $b_k^{(i)}$ $(k = 1, \ldots, 4)$ defining the amplitude of the bifurcation mode in each layer, for which the fields are described by Eqs. (12.68) and (12.78). Collecting all the equations for all layers in a matrix (and adding the equations holding at the external surfaces of the stack of layers), we end up with homogeneous system, so imposing vanishing of the determinant matrix yields the critical stress T_1 (or k) or the critical stretch for bifurcation.

As a simple example of calculation, we report in Fig. 12.15 the bifurcation of an elastic layer (of current thickness h) perfectly bonded to an elastic half space, both obeying a Mooney-Rivlin material model. The system is subjected to equal compressive strain in both the layer and the half space, corresponding to uniaxial compressive T_1-Cauchy stress parallel to the free surface $x_2 = h/2$. Since bifurcation is possible only for compressive loads, we report the critical values of $|T_1|$ normalized through division by the current shear stiffness μ as a function of the dimension-less wavenumber $c_1 h$ of the bifurcation mode. Note that since the two Mooney-Rivlin

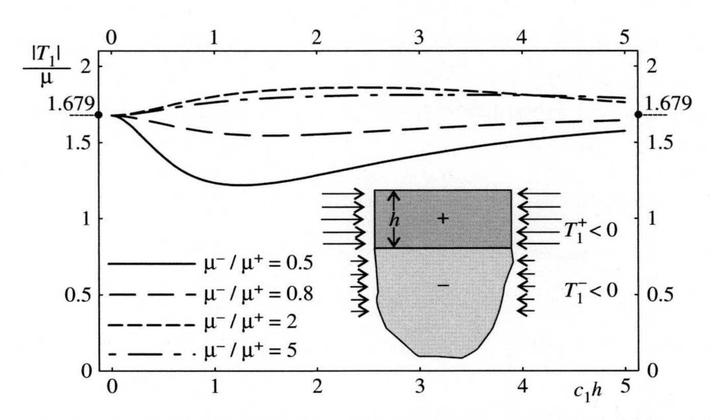

Figure 12.15. Bifurcation of an infinite layer on an elastic half space. Bifurcation stress $-T_2/\mu$ versus the wavenumber of the bifurcated field c_1 normalised through multiplication by the thickness of the layer h for Mooney-Rivlin materials with different stiffness ratios μ^-/μ^+.

materials are deformed at the same stretch, the values $|T_1^+|/\mu^+$ and $|T_1^-|/\mu^-$ are identical (so we simply write $|T_1|/\mu$). It may be interesting to note that in the limit $c_1h \to 0$ ($c_1h \to \infty$), the curves in the graph initiate at (are asymptotic to) the value of $|T_1|/\mu = 1.679$, corresponding to surface instability.

12.3 Bifurcations of an incompressible elastic cylinder

Bifurcations of an incompressible elastic cylinder are sought assuming the GBG material model [Eqs. (4.49) through (4.55)], formulated in the framework of J_2-deformation theory of plasticity, but with the aim of describing a smooth transition from hardening to softening behaviour in uniaxial compression and tension, see the behaviour of the model in uniaxial tension/compression (Section 5.3).

Let us consider an incompressible cylinder of radius $\bar{r}_0$ and height $\bar{h}_0$ in the undeformed, natural configuration B_0 whose points are labelled by $\boldsymbol{x}_0$ subject to an homogeneous, axi-symmetric deformation $\boldsymbol{g}$.

The current configuration $B = \boldsymbol{g}(B_0)$, whose points are denoted by $\boldsymbol{x}$, is described employing a cylindrical coordinates system (r,θ,z), with the z axis coincident with the axis of the cylinder and origin at its lower base. Let $\{\boldsymbol{e}_r, \boldsymbol{e}_\theta, \boldsymbol{e}_z\}$ be the orthonormal basis associated with the system just introduced.

Adopting the axial logarithmic strain $\boldsymbol{\varepsilon}$, employing the conditions of axi-symmetry and the incompressibility constraint, $J = 1$, allows us to express the Eulerian logarithmic strain $\boldsymbol{G}^{(0)}$ in the form given by Eq. (4.51), so the current radius $\bar{r}$ and height $\bar{h}$ of the elastic cylinder are given by

$$\bar{r} = \exp(-\varepsilon/2)\bar{r}_0 \qquad \text{and} \qquad \bar{h} = \exp(\varepsilon)\bar{h}_0, \tag{12.112}$$

respectively, where $\varepsilon < 0$ for axial compression. The lateral surface of the cylinder is traction-free and a uniaxial stress is assumed directed along its axis [see Eqs. (4.51) through (4.55)].

In an updated Lagrangian formulation and in the absence of body forces, let us consider an incremental displacement field $\boldsymbol{v}(\boldsymbol{x})$ superimposed on the current deformation. The incremental equilibrium equations require that the increment in the first Piola-Kirchhoff stress tensor $\dot{\boldsymbol{S}}$ satisfies the incremental equilibrium equations (9.59) (see the discussion in Section 9.2.2).

The incremental boundary value problem to be solved is finally completed by the following boundary conditions:

- Null incremental Piola tractions at the lateral surface of the cylinder,

$$\dot{S}_{rr} = \dot{S}_{\theta r} = \dot{S}_{zr} = 0, \qquad \text{at } r = \bar{r}. \tag{12.113}$$

- Perfectly smooth contact with a rigid, flat constraint on the faces $z = 0$ and $z = \bar{h}$ of the cylinder,

$$\dot{S}_{\theta z} = \dot{S}_{rz} = v_z = 0, \qquad \text{at } z = 0 \text{ and } z = \bar{h}. \tag{12.114}$$

The constitutive equations (4.49) yield a linear relationship between the incremental first Piola-Kirchhoff stress $\dot{\boldsymbol{S}}$ and the gradient of incremental displacement $\boldsymbol{L} = \operatorname{grad} \boldsymbol{v}$ and are expressed, in the specific axi-symmetric geometry, in terms of three incremental moduli, μ_i $(i = 1,2,3)$ (see also Bigoni and Gei, 2001), which in cylindrical components are

$$\begin{aligned}
&\dot{S}_{rr} = \dot{p} + 2\mu_2 L_{rr} + 2(\mu_1 - \mu_2) L_{\theta\theta},\\
&\dot{S}_{\theta\theta} = \dot{p} + 2\mu_2 L_{\theta\theta} + 2(\mu_1 - \mu_2) L_{rr},\\
&\dot{S}_{zz} = \dot{p} + (2\mu_1 - T_z) L_{zz},\\
&\dot{S}_{r\theta} = \dot{S}_{\theta r} = (2\mu_2 - \mu_1)(L_{r\theta} + L_{\theta r}),\\
&\dot{S}_{rz} = \left(\mu_3 + \frac{T_z}{2}\right) L_{rz} + \left(\mu_3 - \frac{T_z}{2}\right) L_{zr}, \quad \dot{S}_{zr} = \left(\mu_3 - \frac{T_z}{2}\right) L_{rz} + \left(\mu_3 - \frac{T_z}{2}\right) L_{zr},\\
&\dot{S}_{\theta z} = \left(\mu_3 + \frac{T_z}{2}\right) L_{\theta z} + \left(\mu_3 - \frac{T_z}{2}\right) L_{z\theta}, \quad \dot{S}_{z\theta} = \left(\mu_3 - \frac{T_z}{2}\right) L_{\theta z} + \left(\mu_3 - \frac{T_z}{2}\right) L_{z\theta},
\end{aligned} \tag{12.115}$$

where $\dot{p}$ is the Lagrange multiplier associated with the incompressibility constraint.

The incremental moduli are functions of the pre-stress (or equivalently pre-strain), which deeply influences the incremental response of the solid. They now can be expressed in terms of elastic strain energy $W(\varepsilon_e)$, function of the effective strain ε_e [Eq. $(4.40)_2$] and of the axial logarithmic strain, ε, as

$$\mu_1 = \frac{1}{3} W'', \qquad \mu_2 = \frac{1}{6}\left(W'' + \frac{W'}{\varepsilon_e}\right), \qquad \mu_3 = \frac{1}{2}\frac{W'}{\varepsilon_e}\, \varepsilon \coth\left(\frac{3}{2}\varepsilon\right), \tag{12.116}$$

where a prime denotes differentiation with respect to the variable ε_e. Therefore, employing the GBG strain energy [Eq. (4.49)], we can calculate the derivatives of W as

$$\begin{aligned}
W' &= K\left(1 - e^{-\varepsilon_e/\varepsilon_0}\right) e^{-\varepsilon_e/(c\varepsilon_0)},\\
W'' &= \frac{K}{c\varepsilon_0}\left[(1+c)e^{-\varepsilon_e/\varepsilon_0} - 1\right] e^{-\varepsilon_e/(c\varepsilon_0)}.
\end{aligned} \tag{12.117}$$

We may note from Eqs. (12.117) that since $\varepsilon_e \geq 0$ is a non-negative function of ε, $K > 0$ and therefore, $W' > 0$, we have

$$\mu_3(\varepsilon) > 0 \tag{12.118}$$

and

$$2\mu_2 - \mu_1 = \frac{W'(\varepsilon_e)}{3\varepsilon_e} > 0. \tag{12.119}$$

Moreover, for uniform axial loading occurring prior to bifurcation, $\varepsilon_e = |\varepsilon|$ and K/ε_0 represents an initial elastic modulus [see Eq. (4.56)], so

$$\mu_3(0) = \frac{K}{3\varepsilon_0} > 0 \tag{12.120}$$

plays the role of an initial shear modulus.

Before embarking in the bifurcation problem, it may be worth considering the issue of positive definiteness, strong ellipticity and ellipticity for a GBG elastic material subject to uniaxial stress.

Positive definiteness of the constitutive operator occurs when

$$\dot{\boldsymbol{S}} \cdot \boldsymbol{L} > 0 \tag{12.121}$$

for all gradients of incremental displacements $\boldsymbol{L} \in \mathsf{Lin} - \{0\}$, satisfying the incompressibility constraint

$$\mathrm{tr}\boldsymbol{L} = 0. \tag{12.122}$$

Considering a uniaxial state of stress and therefore employing the constitutive Eqs. (12.115) with Eq. (12.122) used to eliminate $L_{\theta\theta}$, condition (12.121) can be expanded to

$$\begin{aligned}&[L_{rr}\ \ L_{zz}][\boldsymbol{A}][L_{rr}\ \ L_{zz}]^T + 2(2\mu_2 - \mu_1)(L_{r\theta} + L_{\theta r})^2 + \\ &[L_{rz}\ \ L_{zr}][\boldsymbol{B}][L_{rz}\ \ L_{zr}]^T + [L_{\theta z}\ \ L_{z\theta}][\boldsymbol{B}][L_{\theta z}\ \ L_{z\theta}]^T > 0,\end{aligned} \tag{12.123}$$

where

$$[\boldsymbol{A}] = \begin{bmatrix} 8(2\mu_2 - \mu_1) & 4(2\mu_2 - \mu_1) \\ 4(2\mu_2 - \mu_1) & 2(2\mu_1 + 2\mu_2 - T_z) \end{bmatrix}, \qquad [\boldsymbol{B}] = \begin{bmatrix} 2\mu_3 + T_z & 2\mu_3 - T_z \\ 2\mu_3 - T_z & 2\mu_3 - T_z \end{bmatrix}. \tag{12.124}$$

First of all, we note that condition (12.123) is violated when $L_{r\theta} = -L_{\theta r}$, corresponding to an incremental rigid-body rotation Θ about the loading axis z,

$$v_\theta = \Theta r, \qquad v_r = v_z = 0 \quad \Rightarrow \quad \nabla \boldsymbol{v} = \Theta \boldsymbol{e}_\theta \otimes \boldsymbol{e}_r - \Theta \boldsymbol{e}_r \otimes \boldsymbol{e}_\theta, \tag{12.125}$$

a circumstance already evidenced in Section 11.1.1. In particular, the fact that the constitutive operator associates a null incremental nominal stress with arbitrary incremental rotations about the z axis is correct (see Fig. 11.2) but can be disregarded as inconsequential. Therefore, excluding the case $L_{r\theta} = -L_{\theta r}$ Eq. (12.123) is equivalent to the four conditions

$$T_z > 0, \qquad 3\mu_1 - T_z > 0, \qquad 2\mu_3 - T_z > 0 \quad \text{and} \quad 2\mu_2 - \mu_1 > 0, \tag{12.126}$$

so the constitutive operator cannot be positive definite when uniaxial compression is considered. This becomes evident by considering two rigid-body incremental rotations ϕ_{x_1} and ϕ_{x_2}, respectively, about two axes x_1 and x_2 orthogonal to the z axis

$$\boldsymbol{v} = \phi_{x_2} z \boldsymbol{e}_1 - \phi_{x_1} z \boldsymbol{e}_2 + \left(-\phi_{x_2} x_1 + \phi_{x_1} x_2\right) \boldsymbol{e}_z, \tag{12.127}$$

which in cylindrical coordinates becomes

$$\boldsymbol{v} = z\left(\phi_{x_2}\cos\theta - \phi_{x_1}\sin\theta\right)\boldsymbol{e}_r - z\left(\phi_{x_2}\sin\theta + \phi_{x_1}\cos\theta\right)\boldsymbol{e}_\theta - r\left(\phi_{x_2}\cos\theta - \phi_{x_1}\sin\theta\right)\boldsymbol{e}_z, \tag{12.128}$$

Thus, using Eq. (2.146), the gradient of incremental displacements writes as

$$\begin{aligned}\nabla\boldsymbol{u} = &\left(\phi_{x_2}\cos\theta - \phi_{x_1}\sin\theta\right)(\boldsymbol{e}_r\otimes\boldsymbol{e}_z - \boldsymbol{e}_z\otimes\boldsymbol{e}_r)\\ &+\left(\phi_{x_2}\sin\theta + \phi_{x_1}\cos\theta\right)(\boldsymbol{e}_z\otimes\boldsymbol{e}_\theta - \boldsymbol{e}_\theta\otimes\boldsymbol{e}_z),\end{aligned} \tag{12.129}$$

corresponding to $L_{rz} = -L_{zr}$ and $L_{\theta z} = -L_{z\theta}$, for which Eq. (12.123) provides

$$T_z\left(\phi_{x_1}^2 + \phi_{x_2}^2\right) > 0, \tag{12.130}$$

clearly violated for null or negative axial stress T_z.

Note that the assumed constitutive model satisfies condition (12.119), which is identical to condition (12.126)$_4$. Moreover, employing Eq. (4.54) we prove that Eq. (12.126)$_3$ is always verified for uniaxial tension, corresponding to positive ε, so that the only relevant conditions are Eqs. (12.126)$_1$ and (12.126)$_2$, the latter of which can be rewritten as

$$W'' > \frac{\varepsilon}{\varepsilon_e}W', \tag{12.131}$$

yielding

$$\varepsilon < \varepsilon_0 \log\left(1 + \frac{c}{1 + c\,\varepsilon_0}\right). \tag{12.132}$$

As a conclusion, a comparison between Eqs. (12.132) and (4.57) reveals that positive definiteness is lost during uniaxial tension at the peak of the nominal stress versus logarithmic strain curve, occurring before the peak of the Cauchy stress is attained [Eq. (4.58)]. Therefore, *bifurcations are excluded for uniaxial tensile stress during hardening in the nominal stress versus logarithmic strain curve.*

For the axi-symmetric state of stress assumed in Eq. (12.115), and owing to the incompressibility constraint, strong ellipticity and ellipticity write as in Eq. (12.16) with vector $\boldsymbol{n}$ and $\boldsymbol{g}$ given again by Eq. (12.17), where γ is now the angle between $\boldsymbol{n}$ and the unit vector $\boldsymbol{e}_r$. Therefore, in polar coordinates, Eq. (12.18) now is replaced by

$$\begin{aligned}&\mathbb{G}_{zrzr}\cos^4\gamma + \mathbb{G}_{rzrz}\sin^4\gamma + (\mathbb{G}_{zzzz} - 2\mathbb{G}_{zzrr} - 2\mathbb{G}_{zrrz}\\ &+\mathbb{G}_{rrrr})\cos^2\gamma\sin^2\gamma > 0 \qquad [\text{or } \neq 0],\end{aligned} \tag{12.133}$$

where the components of tensor $\mathbb{G}$ can be identified directly from Eqs. (12.115) to yield, instead of Eq. (12.19), the following condition

$$\mu_3\cos^4\gamma\left[\left(1+\frac{T_z}{2\mu_3}\right)\tan^4\gamma + 2\left(\frac{\mu_1}{\mu_3}+\frac{\mu_2}{\mu_3}-1\right)\tan^2\gamma + \left(1-\frac{T_z}{2\mu_3}\right)\right] > 0 \quad (\text{or } \neq 0), \tag{12.134}$$

holding for every γ and yielding the analogue of conditions (12.21), namely,

$$\mu_3 > 0 \quad (\text{or } \mu_3 \neq 0), \qquad \left(\frac{T_z}{2\mu_3}\right)^2 < 1, \qquad \frac{\mu_1}{\mu_3}+\frac{\mu_2}{\mu_3} > 1 - \sqrt{1-\left(\frac{T_z}{2\mu_3}\right)^2}. \tag{12.135}$$

As in plane strain, we see that in the present setting of an underlying axi-symmetric stress state, there is no practical difference between strong ellipticity and ellipticity.

Moreover, we can introduce two coefficients analogous to the coefficients k and ξ defined for plane strain, so the (strongly) elliptic region can be mapped with a result analogous to that reported in Fig. 12.3.

Loss of positive definiteness and of ellipticity of the tangent constitutive tensor are marked in the uniaxial stress versus logarithmic strain shown in Figs. 5.11 and 5.12.

Considering now to the bifurcation problem of the uniaxially compressed cylinder, we note that exploiting the condition of incompressibility of the incremental deformation [see Eq. (2.147)], namely,

$$\text{tr}\boldsymbol{L} = v_{r,r} + (u_r + v_{\theta,\theta})/r + v_{z,z} = 0, \tag{12.136}$$

the components of the incremental displacement $\boldsymbol{v}$ can be written in terms of two displacement potentials, $\Omega = \Omega(r,\theta,z)$ and $\Psi = \Psi(r,\theta,z)$, as

$$v_r = \Omega_{,rz} + \Psi_{,\theta}/r, \qquad v_\theta = \Omega_{,\theta z}/r - \Psi_{,r}, \qquad v_z = -\mathcal{M}(\Omega), \tag{12.137}$$

where

$$\mathcal{M}(\cdot) = (\cdot)_{,rr} + (\cdot)_{,r}/r + (\cdot)_{,\theta\theta}/r^2 \tag{12.138}$$

is the two-dimensional Laplacian operator in polar coordinates (see Section 2.17).

Bifurcations are sought in the separate variables form

$$\begin{cases} \Omega(r,\theta,z) = \omega(r)\cos n\theta \sin \eta z, \\ \Psi(r,\theta,z) = \psi(r)\sin n\theta \cos \eta z, \\ \dot{p}(r,\theta,z) = q(r)\cos n\theta \cos \eta z, \end{cases} \tag{12.139}$$

where[5] $\eta = k\pi/\bar{h}$ $(k = 1,2,\ldots)$ and n $(n = 0,1,2,\ldots)$ are, respectively, the longitudinal and circumferential wave numbers. The definition of η implies that the boundary conditions (12.114) are automatically satisfied.

Substitution of Eq. (12.139) into Eqs. (12.137) and (12.115) and the equilibrium equation (9.59) yields two ordinary differential equations for $\omega(r)$ and $\psi(r)$ and an expression for $q(r)$. The first two equations are

$$\begin{cases} (\mathcal{L}_n - \rho_3^2\eta^2)\,\psi = 0, \\ (\mathcal{L}_n + \rho_1^2\eta^2)(\mathcal{L}_n + \rho_2^2\eta^2)\,\omega = 0, \end{cases} \tag{12.140}$$

where

$$\mathcal{L}_n(\cdot) = d^2(\cdot)/dr^2 + d(\cdot)/dr/r - n^2(\cdot)/r^2 \tag{12.141}$$

is the Bessel operator, ρ_i^2 $(i = 1,2)$ are the solutions of the characteristic equation

$$(\mu_3 - T_z/2)\rho^4 + 2(\mu_1 + \mu_2 - \mu_3)\rho^2 + (\mu_3 + T_z/2) = 0 \tag{12.142}$$

and

$$\rho_3^2 = (\mu_3 + T_z/2)/(2\mu_2 - \mu_1). \tag{12.143}$$

[5] Note that the parameter η and the index k in Eq. (12.139) have nothing to do with the prestress parameters defined by Eqs. $(6.179)_{2,3}$.

The solutions for $\omega(r)$, $\psi(r)$ and $q(r)$ are

$$\begin{cases} \omega(r) = a_1 J_n(\rho_1 \eta r) + a_2 J_n(\rho_2 \eta r), \\ \psi(r) = b I_n(\rho_3 \eta r), \\ q(r) = (2\mu_1 - \mu_3 - T_z/2)\eta \mathcal{L}_n(\omega) - (\mu_3 - T_z/2)\mathcal{L}_n^2(\omega)/\eta, \end{cases} \tag{12.144}$$

where a_i $(i = 1,2)$ and b are arbitrary constants, and $J_n(x)$ and $I_n(x)$ are, respectively, the ordinary and the modified Bessel functions of order n.

It is worth noting that the nature of roots $\pm\rho_1$ and $\pm\rho_2$ of Eq. (12.142) defines the classification of regimes:

- Complex conjugate $\pm\rho_1$ and $\pm\rho_2$ in the elliptic complex regime (EC)
- Pure imaginary $\pm\rho_1$ and $\pm\rho_2$ in the elliptic imaginary regime (EI)
- Real $\pm\rho_1$ and $\pm\rho_2$ in the hyperbolic regime (H)
- Two real and two pure imaginary $\pm\rho_1$ and $\pm\rho_2$ in the parabolic regime (P)

We highlight that failure of ellipticity corresponds to localisation of deformation into shear bands. Therefore, because Eq. (12.119) holds and the investigation of bifurcation is restricted to the elliptic range, where $\mu_3 + T_z/2 > 0$ and the root ρ_3, Eq. (12.143) is always real.

Equations (12.137) through (12.144) fully specify the incremental displacement field and, through Eq. (12.115), the incremental stress state. Imposition of the incremental boundary conditions at the lateral surface [Eq. (12.113)], provides a homogeneous algebraic system for the constants a_i $(i = 1,2)$ and b. Non trivial solutions are obtained if the determinant of the associated matrix vanishes, which provides the bifurcation condition in terms of a critical value of axial logarithmic strain ε^{bif}. Once the current geometry and state is known, the bifurcation mode has to be selected in terms of the circumferential wave number n and the dimensionless parameter $\eta\bar{r}$.

12.3.1 Numerical results for bifurcations of an elastic cylinder subject to axial compression

Bifurcation axial stresses for samples with aspect ratios 1/2, 2/2, 4/2 and 5/2 have been computed and reported in Figs. 12.16 and 12.17, whereas the relative bifurcation modes are reported in Figs. 12.18 and sketched in a three-dimensional graphic in Fig. 12.19 (some results have been anticipated in Fig. 1.7). The bifurcation points are marked in Figs. 12.16 and 12.17 on the uniaxial stress versus logarithmic strain curves (obtained for the values K=1680 MPa, $c = 42$ and $\varepsilon_0 = 0.0045$ of strain energy parameters) with vertical segments, because they correspond to two different values of Cauchy (or true) and nominal stresses but to the same value of strain. Four different aspect ratios have been considered, for a sample of current height h and diameter d, namely, $h/d = 1/2, 2/2, 4/2$ and 5/2.

In the present bifurcation problem, shear banding (or localisation of deformation) occurs when the EC/H boundary is touched, that is, at $|\epsilon^{\text{E}}| = 0.0624$, as can be calculated from Eq. (12.142). The point corresponding to strain localisation is reported in Fig. 12.16, where it can be clearly appreciated that *shear banding always occurs in the strain softening regime.*

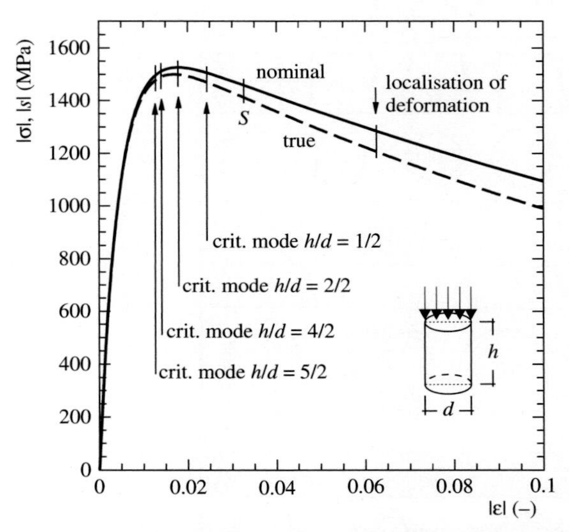

Figure 12.16. Bifurcation of an elastic incompressible cylinder subject to uniaxial compression. True $|\sigma|$ [Eq. (4.54)], and nominal $|s|$ [Eq. (4.55)], stress versus logarithmic strain curves (the former is dashed) with superimposed critical points for bifurcation. S denotes surface bifurcation that occurs at $|\varepsilon^{\mathrm{S}}| = 0.0325$, whereas localisation of deformation in shear bands (loss of ellipticity) occurs at $|\varepsilon^{\mathrm{E}}| = 0.0624$. The modes of bifurcation for the different aspect ratios h/d listed in the figure correspond (from left to right) with A, B, G and M in Table 12.2.

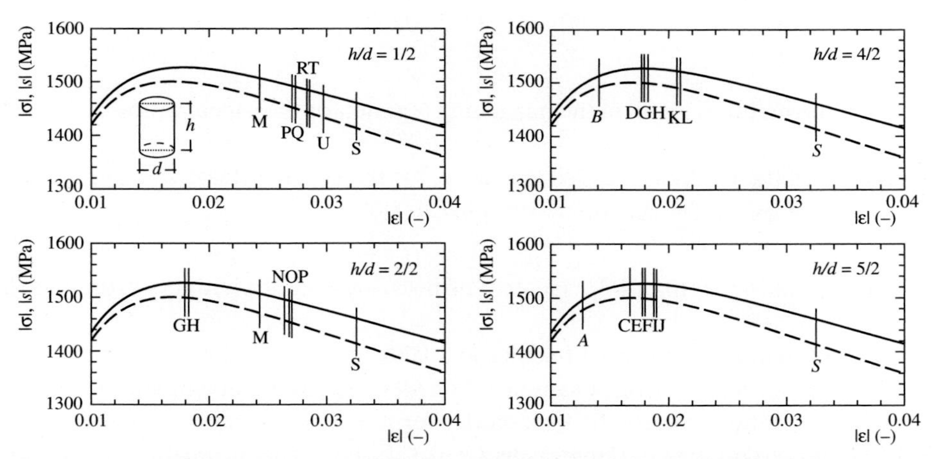

Figure 12.17. Bifurcation of an elastic incompressible cylinder subjected to uniaxial compression: Details of Fig. 12.16 for different aspect rations h/d. True $|\sigma|$ [Eq. (4.54)], and nominal $|s|$ [Eq. (4.55)], stress versus logarithmic strain curves (the former is dashed), with superimposed critical points for bifurcation. S denotes surface bifurcation that occurs at $|\varepsilon^{\mathrm{S}}| = 0.0325$, whereas localisation of deformation (loss of ellipticity) occurs at $|\varepsilon^{\mathrm{E}}| = 0.0624$ and therefore is not reported. Characteristics of modes A through U are reported in Table 12.2.

The critical or, in other words, occurring at the lowest strain bifurcation points for each of the four aspect ratios considered are reported in Fig. 12.16. All the four critical bifurcations correspond to an anti-symmetric mode, characterised by $n = 1$. Note that the critical bifurcation occurs

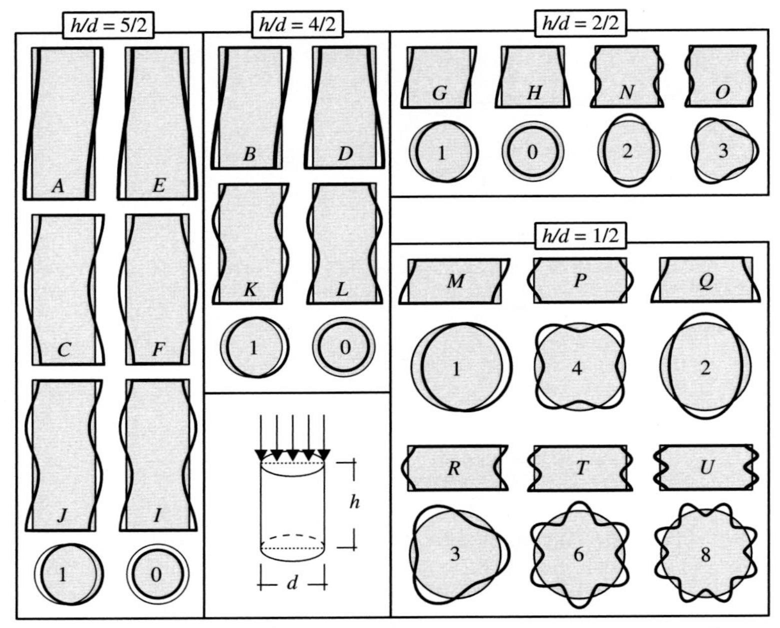

Figure 12.18. Sketch of bifurcation modes of an elastic incompressible cylinder subjected to uniaxial compression. The bifurcation modes are also reported in Fig. 12.19 and correspond to the loads indicated in Figs. 12.16 and 12.17 and Table 12.2, to which the letters are referred. Numbers refer to the circumferential wave number n.

- When the material is still in the hardening regime, for the aspect ratios $h/d = 4/2$ and 5/2
- At around the peak of stress/strain curve, for the aspect ratio $h/d = 2/2$
- During softening, for the aspect ratio $h/d = 1/2$

For aspect ratios higher than 4/2, a Euler-type buckling occurs, yielding a pronounced size effect.

Bifurcation modes with $n \neq 1$ become available at strains slightly higher than the critical strain, specially for thick samples. In order to present a complete picture of the bifurcation landscape, the first six modes for every aspect ratio are indicated in Fig. 12.17 and the relative parameters listed in Tab. 12.2. The bifurcation modes are sketched in Figs. 12.18 and 12.19, where capital letters refer to Fig. 12.17 and to the classification introduced in Table 12.2, whereas numbers denote the values of the circumferential wave number n.

For $h/d = 1/2$ (first plot in Fig. 12.17), the mode P, following the mode M, is a surface-type mode with double longitudinal wave number, corresponding to half wavelength, and $n = 4$. Moreover, the mode H (axi-symmetric) is almost coincident with mode G (anti-symmetric) for the aspect ratio 2/2 (second plot in Fig. 12.17).

After the sixth mode is attained, infinite bifurcation modes follow one on the other and become closer and closer towards the accumulation point S, representing the surface bifurcation (or surface instability) threshold ($|\varepsilon^S| = 0.0325$). Continuing along the uniaxial curve, strain localisation occurs as a final instability.

Table 12.2. *Bifurcation mode parameters of Fig. 12.17*

Mode	n	$\eta\bar{r}$	$\lvert\varepsilon^{\mathrm{bif}}\rvert$	Mode	n	$\eta\bar{r}$	$\lvert\varepsilon^{\mathrm{bif}}\rvert$
A	1	$\pi/5$	0.0127	K	1	$3\pi/4$	0.0207
B	1	$\pi/4$	0.0141	L	0	$3\pi/4$	0.0208
C	1	$2\pi/5$	0.0167	M	1	π	0.0243
D	0	$\pi/4$	0.0178	N	2	$3\pi/2$	0.0264
E	0	$\pi/5$	0.0178	O	3	$3\pi/2$	0.0268
F	0	$2\pi/5$	0.0179	P	4	2π	0.0270
G	1	$\pi/2$	0.0179	Q	2	π	0.0273
H	0	$\pi/2$	0.0182	R	3	2π	0.0283
I	0	$3\pi/5$	0.0187	T	6	3π	0.0284
J	1	$3\pi/5$	0.0190	U	8	4π	0.0297

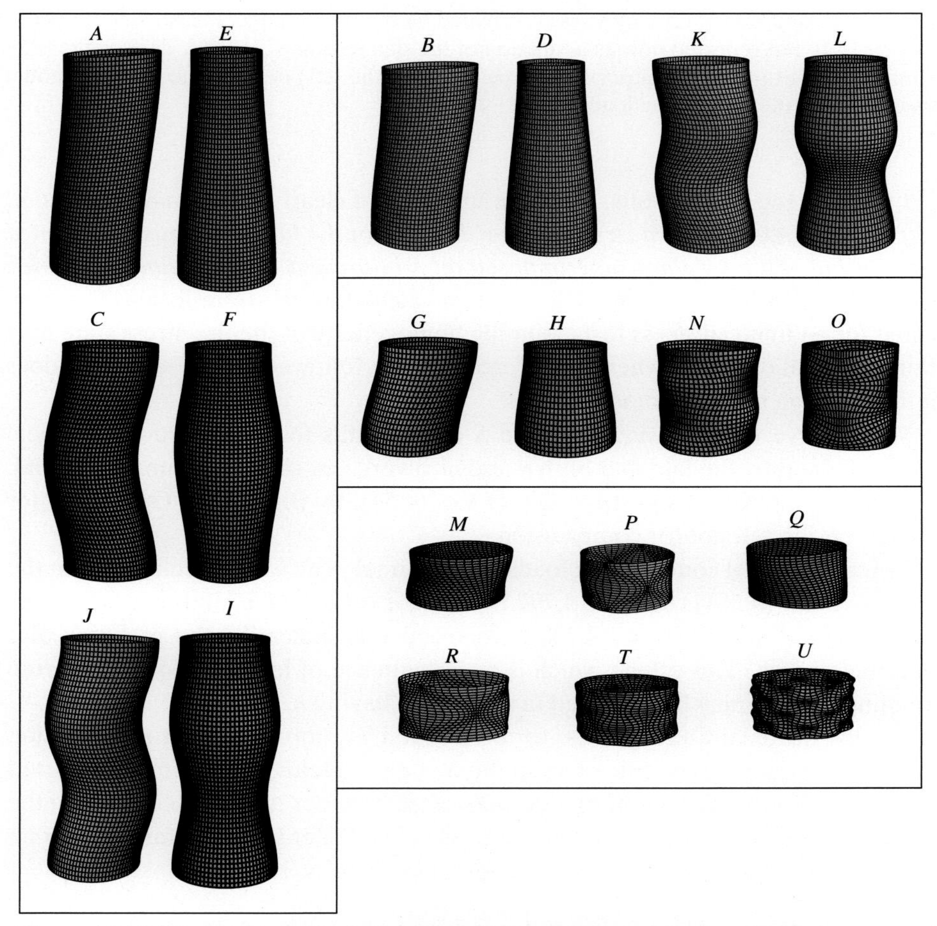

Figure 12.19. Three-dimensional sketches of bifurcation modes of an elastic incompressible cylinder subjected to uniaxial compression. The modes are also reported in Fig. 12.18 and correspond to the bifurcation loads indicated in Figs. 12.16 and 12.17 and Table 12.2, to which the letters are referred.

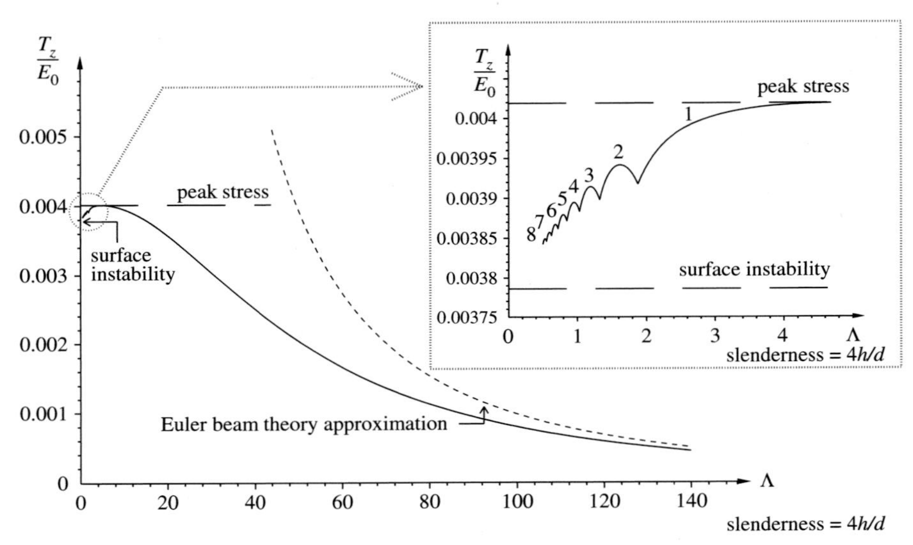

Figure 12.20. Bifurcation Cauchy stresses (divided by the initial tangent elastic modulus E_0) as in Fig. 12.16 but reported now as a function of the slenderness $\Lambda = 4h/d$. The number n in the inset (a detail of the initial portion of the curve on the left) denotes the circumferential wavenumber of the bifurcation load.

From the reported results, it can be understood clearly that *strain localisation in a cylinder under uniaxial stress never will occur in the homogeneously deformed specimen but will take place superimposed on a bifurcated deformation path which necessarily has to occur before*. Therefore, a calculation of strain localisation performed for an uniaxial stress test assuming homogeneity of the pre-stress state may retain some validity only when the bifurcated path followed by the specimen does not involve high strain inhomogeneities.

We observe that the surface mode S corresponds to a so-called orange-peel pattern, defined by the fact that both n and $\eta\bar{r}$ diverge. A similar feature was found, for a simpler uniaxial stress-strain law of Eq. (4.54), by Bigoni and Gei (2001) for uniaxial tension but not for compression.

In terms of total compression loads, the nominal peak load calculated from the constitutive law (4.55) is equal to 4,797 N. For $h/d = 4/2$ and 5/2, bifurcation occurs for a load of about 4,751 and 4,700 N, respectively, indicating the presence of a slenderness effect. This effect, which is a consequence of loss of uniqueness in the hardening branch, may be observed in specimens having $h/d > 1$.

Finally, the axial Cauchy stress for bifurcation T_z, normalised through division by $E_0 = K/\varepsilon_0$ [playing the role of an initial Young modulus, Eq. (4.56)] is reported in Fig. 12.20 as a function of the slenderness of the cylinder $\Lambda = 4h/d$ (a detail of the initiation of the curve is reported in the inset). The Euler buckling formula for an incompressible elastic beam of length h and diameter d is

$$\frac{T_z}{E_0} = \frac{\pi^2}{\Lambda^2}, \tag{12.145}$$

which also is included in Fig. 12.20 for comparison. We see that the Euler beam solution (12.145) provides a reasonable approximation only for high values of the slenderness and is not a safe solution (the beam model is stiffer than the continuous solution).

It is interesting to note from Fig. 12.20 that the curve providing bifurcation stresses is

- Tangent to the value of the peak load exhibited by the material stress/strain response, and
- Approaches a finite value corresponding to the surface bifurcation (marked 'surface instability') in the limit $\Lambda \to 0$.

The second feature, that the curve, differently from the Euler beam approximation (12.145), does not blow up to infinity when the slenderness approaches zero, is a consequence of the fact that a full three dimensional bifurcation analysis has been performed, so even an infinitely thick sample approaches its bifurcation threshold: the surface bifurcation. Note also the fact that during uniaxial compression, the end displacements of the cylinder are imposed (so the deformation is controlled before bifurcation); thus, for the very low slenderness corresponding to Λ on the left of the peak of the bifurcation curve (see the inset of Fig. 12.20), the stress raises above the surface instability threshold (occurring during softening; see Fig. 12.17) and then decreases until surface bifurcation is attained.

12.4 Bifurcation under plane strain bending

Bifurcations may occur during a loading path involving non-uniform stress/strain fields, as in the case of bending [see Kyriakides and Corona (2007) for an introduction to flexural instabilities in structural elements]. A simple example of this bifurcation is shown in Fig. 12.21, where a corrugated plastic tube (identical to that used for the experiment shown in Fig. 1.27) is subjected to a finite bending. We can see that, after an initial uniform bending (up to near an opening semi-angle of 20°), the deformation localises, forming 'snaps' in the corrugations. These subsequent snaps occur at the tensile side of the tube and can be interpreted as bifurcations originating from a bent configuration.

A behaviour similar to the corrugated tube but occurring at the compressive side of a rubber sample deformed under bending has been shown by Gent and Cho (1999; see also Gent, 2005) and corresponds to the formation of elongated creases such as those shown in Fig. 12.22. These creases can be interpreted as short-wavelength bifurcations occurring during bending.

We have repeated the experiments by Gent and Cho (including also coated layers, see Roccabianca et al., 2010) by gluing (using a commercial instant adhesive, Loctite) three natural rubber strips of dimensions $\{10, 15, 20\}$ mm $\times$ 4 mm $\times$ 100 mm (in which the larger dimension is that out of plane, taken sufficiently large to simulate the plane strain condition) along the 100 $\times$ 4 mm edges to two metallic platelets so that finite bending can be imposed using the screw-loading device shown in Fig. 12.23. With the device, the opening angle can be read on a scale marked on the lower (metallic) support. Creases are to be detected by direct visual inspection. In the test shown in Figs. 12.24 and 12.25 (the latter figure is a detail of the former, and both are referred to the 10-mm wide specimen), we observe these creases to appear at

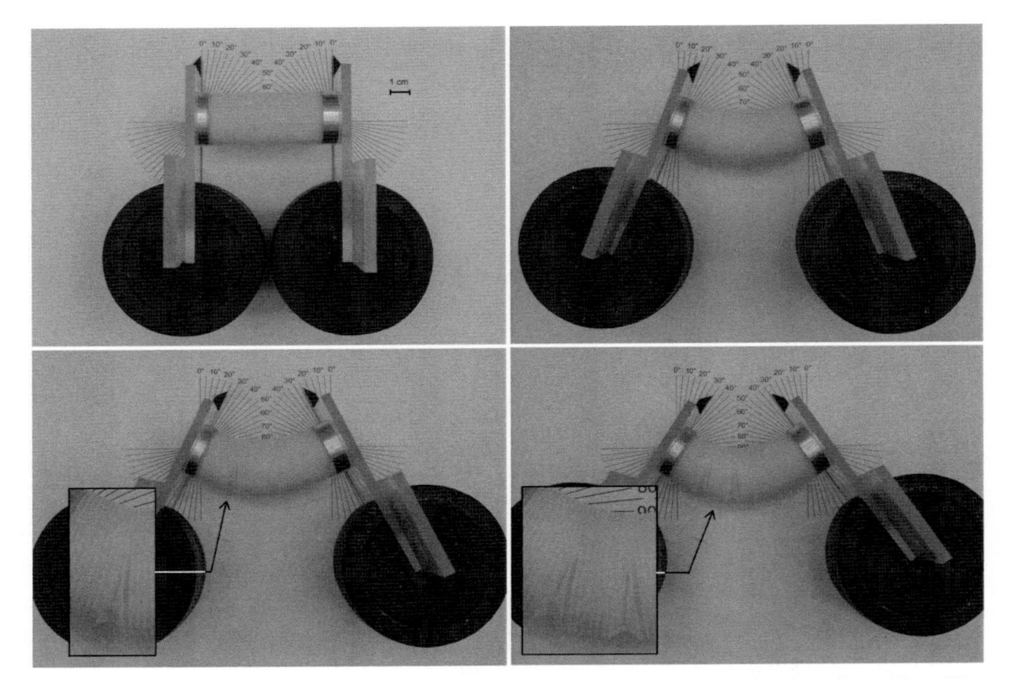

Figure 12.21. Behaviour of a corrugated plastic tube subjected to finite bending. *Top*: Undeformed configuration (*left*); bent configuration (before bifurcation) at an opening semi-angle of 20° (*right*). *Bottom*: First (on the left, at an opening semi-angle of 25°) and second (*right*, at an opening semi-angle of 35°) localisation of deformation, which may be interpreted as a bifurcation of the homogeneous bending solution. (See color plates section.)

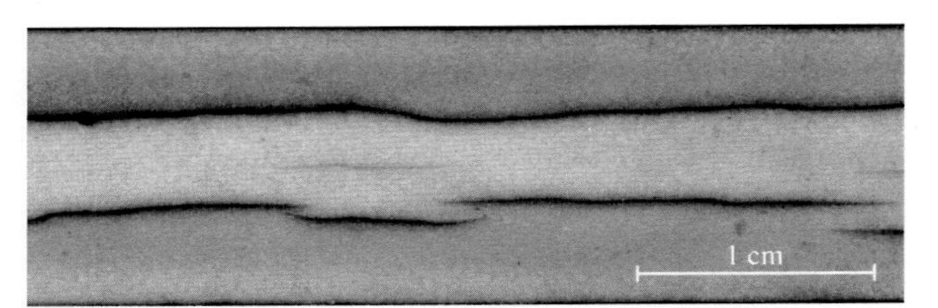

Figure 12.22. Creases at the compressive face of a natural rubber strip subject to finite bending. These may be interpreted as short-wavelength bifurcations emerging from a flexural deformation.

an opening semi-angle of 58°, whereas creases have appeared at 100° for the 15-mm-wide sample, and no creases have been observed in the 20-mm-wide rubber strip, which was bent up to the end, namely, to a semi-angle of 180°, corresponding to closure of the strip into a tube. These results demonstrate that *during bending, bifurcation may or may not occur depending on the geometry of the sample.* Thus, since the formation of creases set limits to the deformation capabilities of a system, a bifurcation analysis can become a useful design tool. Although the boundary conditions along the glued edges of the strips are different from those (corresponding to the setting in Section 5.5) that will be imposed for the bifurcation analysis, the three

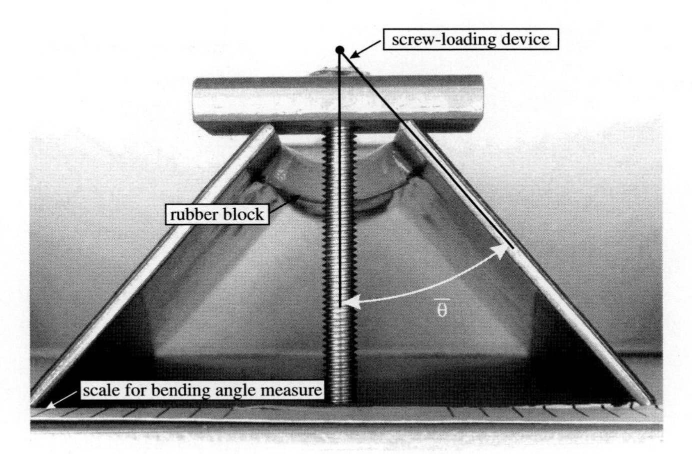

Figure 12.23. Screw-loading device used to impose finite bending (of a semi-angle $\bar{\theta}$, equal in the photo to 40°) to a 10 × 100 × 4-mm rubber block (see Fig. 5.20).

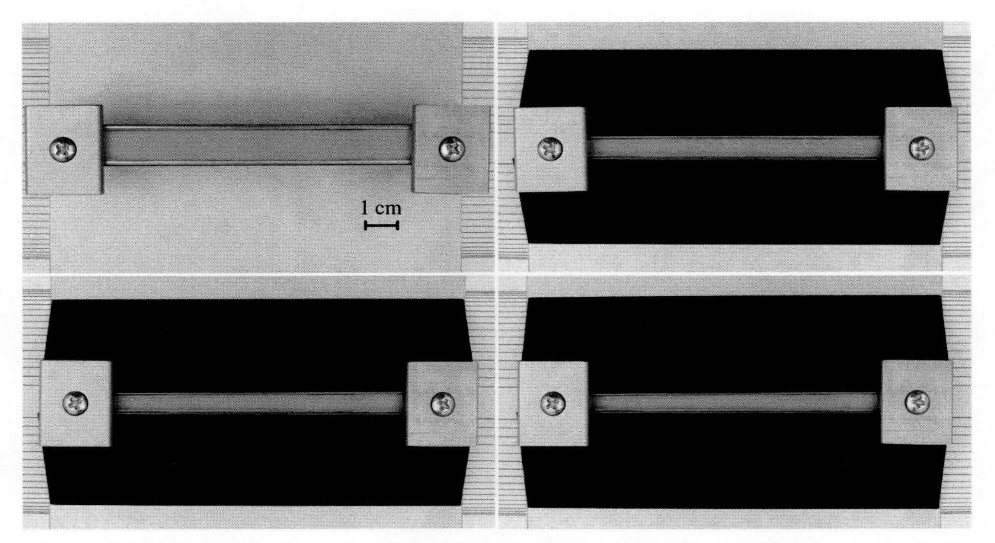

Figure 12.24. Finite bending of a 10 × 100 × 4-mm natural rubber strip, imposed with the device shown in Fig. 12.23. *From top to bottom and from left to right*: Specimen before loading; specimen bent at a semi-angle of 55° (creases are still not visible, see the detail reported in Fig. 12.25); specimen bent at a semi-angle of 60° (creases become visible, see the detail reported in Fig. 12.25); specimen bent at a semi-angle of 65° (creases invade the whole specimen; see the detail reported in Fig. 12.25).

above-mentioned experiments agree consistently[6] with the theory that will be developed below for bifurcation of an incompressible neo-Hookean block subjected to plane strain bending, as shown in Fig. 12.26. Differently from the case of multi-layers analysed by Roccabianca et al. (2010), for a rubber block, the stiffness parameter μ_0 of the the neo-Hookean constitutive model (Eq. (4.37)], does not affect bifurcation

[6] Experimental critical angles have been found by us to be always inferior to theoretical predictions, a trend also observed by Gent and Cho (1999).

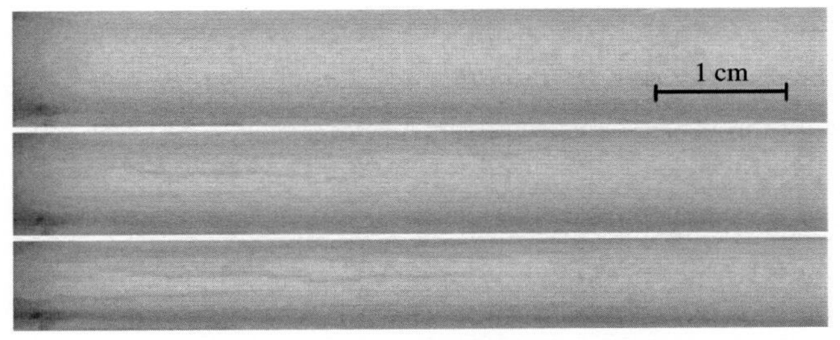

Figure 12.25. Details of Fig. 12.24. Creases become visible in the photo taken at an opening semi-angle of 60° (*centre*) and invade the whole sample at 65° (*bottom*), whereas they remain undetected at 55° (*top*).

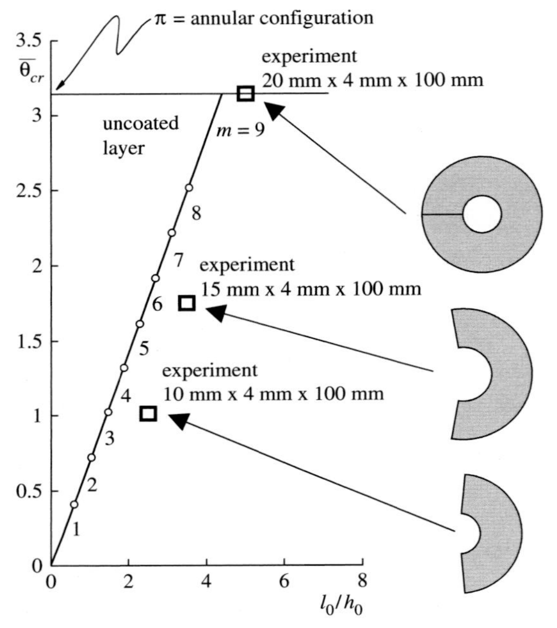

Figure 12.26. Experimental results versus theoretical predictions for bifurcation semi-angles $\bar{\theta}_{cr}$ of three natural rubber strips subjected to finite bending (the results of three experiments are indicated with squared spots). Configurations at bifurcation are reported on the right. $\bar{\theta}_{cr} = \pi$ means that the annular configuration can be reached without bifurcations. Parameter m defines the circumferential wavenumber of the bifurcation mode [Eq. (12.155)]. Note that the theoretical prediction of bifurcation is the envelope of different non-linear curves, but appears as a straight line in the plot.

angles, which depend only on the geometrical setting. In fact, the critical bifurcation semi-angle $\bar{\theta}_{cr}$ reported in Fig. 12.26 depends only on the geometrical parameter l_0/h_0, namely, the ratio between initial width and thickness of the rubber block. In the figure, the three experiments corresponding to $l_0/h_o = \{10/4, 15/4, 20/4\}$ are also reported. The critical configurations at bifurcation are shown on the right of the figure. Note from Fig. 12.26 that the curve representing theoretical values for the bifurcation bending semi-angles results from the envelope of different curves relative to different values of the parameter m, defining the circumferential wavenumber of the bifurcation mode [Eq. (12.155)]. Surprisingly, this envelope, which defines the bifurcation condition, can be very well approximated by the straight line of equation

$$\bar{\theta}_{cr} = 0.712\, l_0/h_0, \tag{12.146}$$

providing a simple tool for predicting bifurcation for a thick plate of incompressible neo-Hookean material subjected to bending.

The bifurcation analysis for a thick plate under finite bending and the route to arrive at results reported in Fig. 12.26 are explained in the following.

The problem of finding plane strain bifurcations during finite bending of an elastic layer has been addressed by Triantafyllidis (1980), Aron and Wang (1995a, 1995b), Dryburgh and Ogden (1999), Bruhns et al. (2002; 2003) and Coman and Destrade (2008). Three-dimensional incremental bifurcations have been addressed by Haughton (1999), whereas finite bending of layered structures and the related bifurcation problem has been solved by Roccabianca et al. (2010, 2011).

We consider plane strain incremental bifurcations of an incompressible neo-Hookean thick plate of initial (current) height l_0 (l) and width h_0 (h), subjected to finite bending, a solution given in Section 5.5. In this plane strain problem of continued deformation, the stress and strain are not uniform but known, and we seek incremental bifurcations from this state.

In polar coordinates, the gradient of incremental displacement $\boldsymbol{v}(\boldsymbol{x})$ is given by

$$\boldsymbol{L} = v_{r,r}\boldsymbol{e}_r \otimes \boldsymbol{e}_r + \frac{v_{r,\theta} - v_\theta}{r}\boldsymbol{e}_r \otimes \boldsymbol{e}_\theta + v_{\theta,r}\boldsymbol{e}_\theta \otimes \boldsymbol{e}_r + \frac{v_r + v_{\theta,\theta}}{r}\boldsymbol{e}_\theta \otimes \boldsymbol{e}_\theta, \tag{12.147}$$

and the incompressibility condition ($\mathrm{tr}\boldsymbol{L} = 0$) is

$$r v_{r,r} + v_r + v_{\theta,\theta} = 0. \tag{12.148}$$

For an incompressible isotropic elastic material, the components of the constitutive fourth-order tensor $\mathbb{G}$ can be written as functions of two incremental moduli, denoted μ and μ^* [Eqs. (6.177)]. In particular, the non-null components of $\mathbb{G}$ in Eqs. (6.177) may be expressed in polar coordinates as

$$\begin{aligned} &\mathbb{G}_{rrrr} = C_{\theta\theta\theta\theta} = 2\mu_*, && \mathbb{G}_{\theta r\theta r} = \mu - \Gamma, \\ &\mathbb{G}_{r\theta r\theta} = \mu + \Gamma, && \mathbb{G}_{r\theta\theta r} = \mathbb{G}_{\theta rr\theta} = \mu, \end{aligned} \tag{12.149}$$

where

$$\Gamma = (T_\theta - T_r)/2 \tag{12.150}$$

represents the state of pre-stress. For hyperelastic materials, μ and μ_* are given in terms of the strain-energy function $\hat{W}(\lambda)$ by Eq. (6.172), so the incremental constitutive equations in terms of first Piola-Kirchhoff tensor can be written as

$$\begin{aligned} \dot{S}_{rr} &= -\dot{p} + (p + 2\mu_*)v_{r,r}, \\ \dot{S}_{\theta\theta} &= -\dot{p} + (p + 2\mu_*)\frac{v_r + v_{\theta,\theta}}{r}, \\ \dot{S}_{r\theta} &= (\mu + \Gamma)\frac{v_{r,\theta} - v_\theta}{r} + (p + \mu)v_{\theta,r}, \\ \dot{S}_{\theta r} &= (p + \mu)\frac{v_{r,\theta} - v_\theta}{r} + (\mu - \Gamma)v_{\theta,r}. \end{aligned} \tag{12.151}$$

The incremental equations of equilibrium, in polar coordinates, are given by Eqs. (2.149), so a substitution of Eqs. (12.151) and an account of the fact that the state of pre-stress satisfies equilibrium, namely,

$$T_{r,r} = (T_\theta - T_r)/r, \tag{12.152}$$

yield

$$\begin{aligned}
\dot{p}_{,r} &= [(p+2\mu_*)_{,r} + \frac{2(p+2\mu_*)}{r}]v_{r,r} + (p+2\mu_*)v_{r,rr} \\
&\quad + (\mu+\Gamma)\frac{v_{r,\theta\theta} - v_{\theta,\theta}}{r^2} + (p+\mu)\frac{v_{\theta,r\theta}}{r}, \\
\dot{p}_{,\theta} &= [r(\mu-\Gamma)_{,r} + \mu - \Gamma](v_{\theta,r} + \frac{v_{r,\theta} - v_\theta}{r}) + r(\mu-\Gamma)v_{\theta,rr} + (\mu - 2\mu_*)v_{r,\theta r}.
\end{aligned} \tag{12.153}$$

Bifurcations are sought in the following separable-variable form

$$\begin{cases} v_r(r,\theta) = f(r)\cos n\theta \\ v_\theta(r,\theta) = g(r)\sin n\theta \\ \dot{p}(r,\theta) = k(r)\cos n\theta \end{cases} \tag{12.154}$$

where $f(r)$, $g(r)$ and $k(r)$ are real functions, and n is a real number to be determined by imposing boundary conditions. In particular, we assume

$$n = \frac{m\pi}{\bar{\theta}} \qquad (m \in \mathbb{N}), \tag{12.155}$$

so boundary conditions of vanishing shear stresses and normal incremental displacements along the boundaries $\theta = \pm\bar{\theta}$ are automatically satisfied.

A substitution of representations (12.154) into Eqs. (12.153) and use of the incompressibility condition, implying

$$g = -(f + rf')/r, \tag{12.156}$$

yield

$$\begin{aligned}
k' &= Df'' + \left(C_{,r} + D_{,r} + \frac{C+2D}{r}\right)f' + \frac{E(1-n^2)}{r^2}f, \\
k &= \frac{r^2 C}{n^2}f''' + \frac{F+3C}{n^2}rf'' + \left(\frac{F}{n^2} - D\right)f' - \frac{1-n^2}{n^2}\frac{F}{r}f,
\end{aligned} \tag{12.157}$$

where a prime denotes differentiation with respect to r. In terms of incremental moduli μ and μ_* and strain energy function $\hat{W}(\lambda)$, the parameters C, D, E and F take the form

$$\begin{aligned}
C &= \mu - \Gamma = \frac{\lambda}{\lambda^4 - 1}\hat{W}_{,\lambda}, \\
D &= 2\mu_* - \mu = \frac{\lambda}{2}\left(\lambda\hat{W}_{,\lambda\lambda} - \frac{2}{\lambda^4-1}\hat{W}_{,\lambda}\right), \\
E &= \mu + \Gamma = \frac{\lambda^5}{\lambda^4-1}\hat{W}_{,\lambda}, \\
F &= rC_{,r} + C.
\end{aligned} \tag{12.158}$$

By differentiating Eq. $(12.157)_2$ with respect to r and substituting into Eq. $(12.157)_1$, we obtain a single differential equation in terms of f

$$\begin{aligned} &Cr^4 f'''' + 2(F+2C)r^3 f''' + [(rF)_{,r} + 4F - 2n^2 D]r^2 f'' \\ &\quad + [(rF - 2rn^2 D)_{,r} - 2F]rf' + (1-n^2)(F - rF_{,r} - n^2 E)f = 0, \end{aligned} \tag{12.159}$$

governing our incremental bifurcation problem.

Since our goal is to solve Eq. (12.159) by employing the Runge-Kutta numerical method [in a way similar to that used by Dryburgh and Ogden (1999)], it becomes instrumental to rewrite this differential equation as a linear system of first-order ODEs. This can be achieved by introducing the vector

$$z(r) = [f(r) \;\; f'(r) \;\; f''(r) \;\; f'''(r)]^T. \tag{12.160}$$

Thus the solving differential system, equivalent to Eqs. (12.159), becomes

$$z' = \boldsymbol{A} z, \tag{12.161}$$

where the matrix $[\boldsymbol{A}]$ takes the form

$$[\boldsymbol{A}] = \begin{bmatrix} 0 & 1 & 0 & 0 \\ 0 & 0 & 1 & 0 \\ 0 & 0 & 0 & 1 \\ -A_{41} & -A_{42} & -A_{43} & -A_{44} \end{bmatrix}, \tag{12.162}$$

in which

$$\begin{aligned} A_{41}(r) &= (1-n^2)(F - rF_{,r} - n^2 E)/(Cr^4), \\ A_{42}(r) &= [(rF + 2rn^2 D)_{,r} - 2F]/(Cr^3), \\ A_{43}(r) &= [(rF)_{,r} + 4F - 2n^2 D]/(Cr^2), \\ A_{44}(r) &= 2(F+2C)/(Cr). \end{aligned} \tag{12.163}$$

For dead tractions on the external curved surfaces at $r = r_i$ and $r = r_e$, we have the boundary conditions

$$\begin{aligned} &Cr^3 f''' + (F+3C)r^2 f'' + [F - n^2(2C+D)]rf' - F(1-n^2)f = 0, \\ &r^2 f'' + rf' - (1-n^2)f = 0, \end{aligned} \tag{12.164}$$

which conveniently can be rewritten as

$$[\boldsymbol{B}][z] = [\boldsymbol{0}], \qquad \text{at } r = r_i, r_e, \tag{12.165}$$

where

$$[\boldsymbol{B}] = \begin{bmatrix} F(n^2-1) & r[F + n^2(2C+D)] & r^2(F+3C) & r^3 C \\ n^2 - 1 & r & r^2 & 0 \end{bmatrix}. \tag{12.166}$$

We are now in a position to set up a numerical strategy to solve system (12.161) with the boundary conditions (12.165). To this purpose, we note that by imposing the conditions at $r = r_i$, we find that the third and second derivatives of f can be expressed at $r = r_i$ as functions of f and f'

$$f''(r_i) = -\frac{r_i f' + (n^2 - 1)f}{r_i^2},$$
$$f'''(r_i) = \frac{[3C + n^2(2D + C)]r_i f' + 3C(n^2 - 1)f}{Cr_i^3}. \tag{12.167}$$

Therefore, we can proceed as follows.

- Employing a Runge-Kutta numerical integration, solve system (12.161) with the initial values

$$z^{(1,0)}(r_i) = [f(r_i) = 1, f'(r_i) = 0, \quad \underbrace{f''(r_i), f'''(r_i)}_{\text{from Eqs. (12.167)}} \quad]^T,$$

  and

$$z^{(0,1)}(r_i) = [f(r_i) = 0, f'(r_i) = 1, \quad \underbrace{f''(r_i), f'''(r_i)}_{\text{from Eqs. (12.167)}} \quad]^T.$$

  In this way, we find the two values of vector z at $r = r_e$, namely, $z^{(1,0)}(r_e)$ and $z^{(0,1)}(r_e)$.
- The solution at $r = r_e$ now can be represented as

$$z(r_e) = C_1 z^{(1,0)}(r_e) + C_2 z^{(0,1)}(r_e), \tag{12.168}$$

  where C_1 and C_2 are the two unknown constants corresponding to the amplitude of the bifurcated mode.
- The boundary conditions at $r = r_e$ [Eqs. (12.165)] now can be imposed using representation (12.168), thus finding

$$[\boldsymbol{M}][\; C_1 \quad C_2 \;]^T = \boldsymbol{0}, \tag{12.169}$$

  where

$$[\boldsymbol{M}] = \begin{bmatrix} B_{1k} z_k^{(1,0)} & B_{1k} z_k^{(0,1)} \\ B_{2k} z_k^{(1,0)} & B_{2k} z_k^{(0,1)} \end{bmatrix}_{r=r_e}, \tag{12.170}$$

  so the bifurcation is simply set by the non-trivial solution condition, which is determined by

$$\det[\boldsymbol{M}] = 0. \tag{12.171}$$

Equation (12.171) can be solved in terms of a critical semi-angle $\bar{\theta}_{cr}$ or a critical stretch λ_{cr} at $r = r_i$ for bifurcation. Examples of this calculation can be found in Fig. 12.27, whereas unit amplitude bifurcated modes (superimposed on the deformed configuration) are reported in Fig. 12.28, corresponding to the aspect ratios $l_0/h_0 = 0.42$ and 0.80, labelled a and b in Fig. 12.27. The numbers 1, 2 and 3 reported near the curves in Fig. 12.27 denote the circumferential wavenumber n, so we can conclude that $n = 1$ is critical for low values of aspect ratio l_0/h_0, whereas higher values of n become critical when l_0/h_0 is increased.

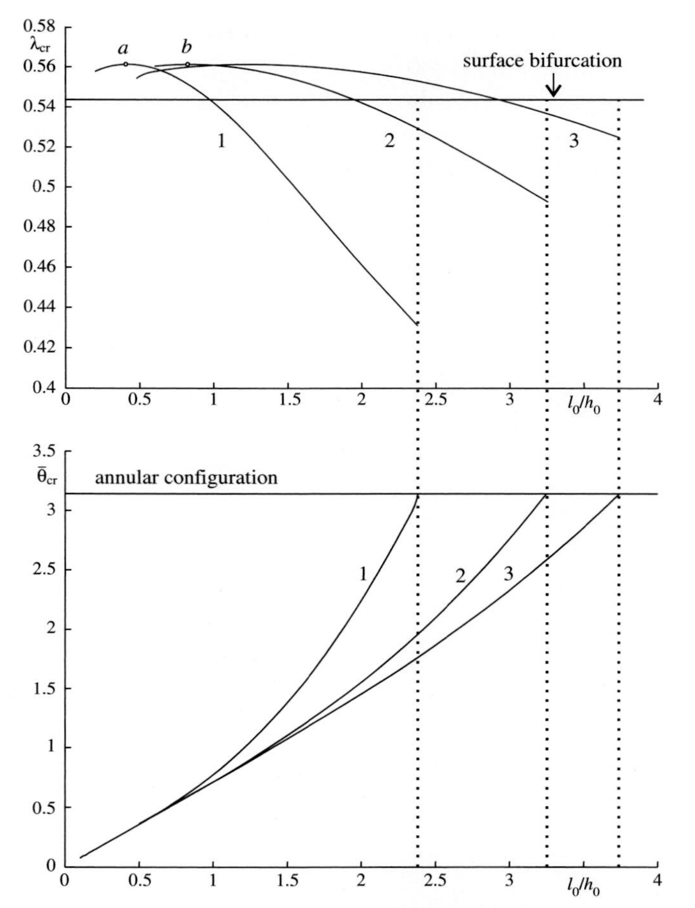

Figure 12.27. Bifurcation critical stretch (at the compressed side, *top*, λ_{cr}) and semi-angle (*bottom*, $\bar{\theta}_{cr}$) during plane strain finite bending of an elastic block as functions of the undeformed aspect ratio l_0/h_0. Points labelled *a* and *b* denote the two aspect ratios used to show critical modes in Fig. 12.28, whereas numbers 1, 2 and 3 denote circumferential wavenumbers.

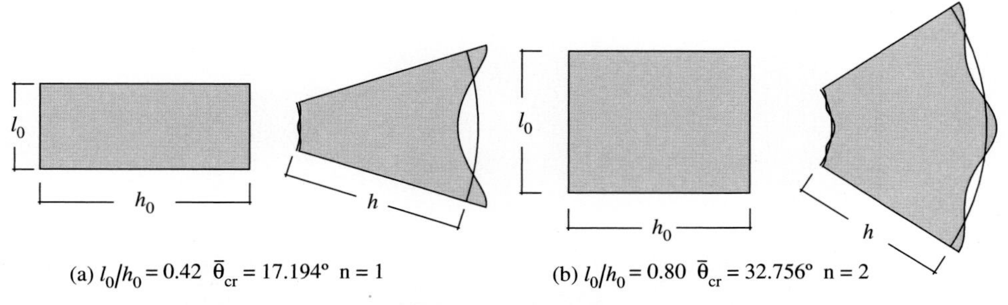

Figure 12.28. Bifurcation modes during plane strain finite bending of an elastic block, for $l_0/h_0 = 0.42$ (a) and 0.80 (b) corresponding to points labelled *a* and *b* in Fig. 12.27.

It should be noted that using a different constitutive model, namely, a form of J_2-deformation theory material, Triantafyllidis (1980) performed an asymptotic analysis showing that the first bifurcation encountered during finite bending corresponds to $n \to \infty$, in other words, to a surface bifurcation. Although we do not confirm this result,[7] we find that the condition for surface bifurcation is very close to the critical condition, occurring for a 'global' bifurcation mode. For instance, for $l_0/h_0 = 0.42$, the critical angle and stretch for 'global' bifurcation are $\bar{\theta}_{cr} = 17.19°$ and $\lambda_{cr} = 0.5601$, whereas the critical values for surface bifurcation are $\bar{\theta}^S = 18.57°$ and $\lambda^S = 0.5437$, respectively.

[7] The slight disagreement with Triantafyllidis (1980) also has been noticed by Coman and Destrade (2008).

13 Applications of local and global uniqueness and stability criteria to non-associative elastoplasticity

Applications of the local and global criteria for uniqueness and stability are presented for elastoplastic solids with non-associative flow law. We begin with the simple case of the small strain theory, and subsequently, we treat the problem of uniaxial tension and compression of a non-associative elastoplastic cylinder subjected to finite strains. We fully develop the comparison theorem analysis, including also local criteria. Finally, an example of flutter instability occurring in an elastoplastic material with non-associative flow rule is presented.

Local and global uniqueness and stability criteria were introduced in Chapters 10 and 11, with reference to non-associative elastoplasticity (Chapter 8). The incremental non-linearity of the rate-constitutive equations of plasticity and the lack of symmetry connected to the flow-rule non-associativity strongly complicate the bifurcation and instability analyses with respect to the case of incremental elasticity. Therefore, despite interest in the applications to bifurcation problems for geological and quasi-brittle materials, there have been only a few attempts to apply the comparison solids methodology to bifurcation problems (Bruhns and Raniecki, 1982; Kleiber, 1984; 1986; Tomita et al., 1988; Bigoni, 2000), so our interest in this chapter is to provide examples of the methodologies explained in Chapters 10 and 11. We will use the simplest constitutive setting, which is that of small strain Drucker-Prager elastoplasticity with deviatoric associativity, a context in which we will limit examples to local stability criteria, whereas the use of Raniecki comparison solids will be presented for a simple elastoplastic non-associative model at large strain. Finally, examples of flutter instability are given for a non-associative model in which flutter comes from a combination of the effects of flow-rule non-associativity and anisotropy of the elasticity (in the way shown by Bigoni and Loret, 1999). This final example will clarify that flutter instability has 'no connection' with the other stability criteria; thus, for instance, it may occur before loss of positive definiteness (PD) or after loss of ellipticity (E).

13.1 Local uniqueness and stability criteria for non-associative elastoplasticity at small strain

We begin with some words of caution as to the use of the small strain assumption in any bifurcation and stability analysis. Usually, 'geometrical effects' are crucially important in these problems. To convince oneself of this fact, it suffices to recall Euler buckling of rods. Roughly speaking, with respect to the small strain approximation, the various local stability thresholds presented in Chapter 11 contain 'geometrical terms' on the order of stress over elastic shear modulus. These terms become important when the critical hardening moduli become comparable with a representative stress level, a situation seen relatively frequent in elastoplasticity (Hill, 1958; 1978; Rudnicki and Rice, 1975).

Proceeding now with the case in which all 'geometrical terms' are neglected in the equations, we assume, for simplicity, isotropic elasticity (8.65) and deviatoric associativity in the form of Eq. (8.60), with the choice of the dilatancy and pressure-sensitivity parameters given by Eqs. (8.62). In other words, the yield function gradient $\boldsymbol{Q}$ and the plastic flow-mode tensor $\boldsymbol{P}$ can be 'visualised' as the normals to yield and plastic potential surfaces, respectively, both of the Drucker-Prager type (Section 7.3.1). The model reduces therefore to von Mises plasticity when $\mathrm{tr}\boldsymbol{P} = \mathrm{tr}\boldsymbol{Q} = 0$. The threshold for loss of positive definiteness of the tangent operator (PD) is given by Eq. (11.32), and the threshold for loss of non-singularity (NS) is simply[1] $h_{cr}^{NS} = 0$. The critical hardening modulus for coalescence of the three eigenvalues of the acoustic tensor, condition (C), can be calculated easily using Eqn. (11.127). Evaluation of thresholds for loss of strong ellipticity (SE) and ellipticity (E) requires the solution of a constrained maximisation problem for the critical modulus given by Eqs. (11.69) and (11.87). As already mentioned, both these maximisation problems admit an analytical solution, which for the loss of ellipticity can be found in Section 11.4.1, whereas for the loss of (SE) it can be found in Bigoni and Zaccaria (1992a, 1992b). A few values of the critical hardening moduli (divided by the elastic shear modulus μ) are reported in Table 13.1 relative to uniaxial tension. A material with Poisson's ratio $\nu = 0$ and 0.3 is considered for several values of pressure sensitivity $\mathrm{tr}\boldsymbol{Q}$ and dilatancy $\mathrm{tr}\boldsymbol{P}$.

Note that in the column (E), corresponding to loss of ellipticity, the angle (in degrees) has been reported in parentheses between the normal to the band and the direction orthogonal to the axis of tension (Fig. 13.1). Obviously, in the uniaxial stress problem analysed, infinite bands become possible at strain localisation, with normals describing a cone about the tension axis. Additional numerical results can be found in Rudnicki and Rice (1975),[2] Bigoni and Zaccaria (1992a, 1992b) and Bigoni and Loret (1999) and need not be reported here.

The results presented in the table are sufficient to draw the following conclusions:

- As expected, in the case of an associative flow rule, it is $h_{cr}^{PD} = h_{cr}^{NS} = 0$. Moreover, loss of (E), coincident with loss of (SE), is always excluded for positive hardening.
- For non-associative flow laws, the critical hardening modulus for (PD) is always positive. The critical hardening moduli for loss of (SE) and (E) are different and may take any sign, even positive. Strain localisation therefore may occur at positive hardening, before loss of (NS), occurring at $h = 0$.

[1] A complete discussion on the eigenvalues of the elastoplastic tangent operator for isotropic elasticity may be found in Bigoni and Zaccaria (1994 b) and has been also reported in Section 2.20.10.

[2] Rudnicki and Rice (1975) neglected a term in the analytical solution for h_{cr}^{E}, thus obtaining imprecise numerical results. The complete solution can be found in (Bigoni and Hueckel, 1990).

Table 13.1. Uniaxial tension [*Values of* h_{cr}/μ *defining loss of positive definiteness (PD), strong ellipticity (SE) and ellipticity (E) of the elastic-plastic constitutive operator and coalescence of three eigenvalues of the acoustic tensor (C), for Poisson's ratio* $\nu = 0$ *and* 0.3, *and several values of pressure sensitivity* tr $\boldsymbol{Q}$ *and plastic dilatancy* tr $\boldsymbol{P}$. θ *in degrees is given in parentheses.*]

tr $\boldsymbol{Q}$	tr $\boldsymbol{P}$	ν	(PD)	(SE)	(E) (θ)	(C)
0	0	0	0	−0.167	−0.167 (54.7°)	0.333
		0.3	0	−0.217	−0.217 (48.8°)	−0.467
0.3	0	0	0.015	−0.101	−0.104 (58.3°)	0.564
		0.3	0.047	−0.106	−0.130 (53.2°)	−0.166
0.3	0.3	0	0	−0.071	−0.071 (62.2°)	0.775
		0.3	0	−0.093	−0.093 (57.8°)	0.108
0.6	0	0	0.057	−0.021	−0.031 (62.2°)	0.795
		0.3	0.167	0.054	−0.018 (57.8°)	0.133
0.6	0.3	0	0.013	−0.027	−0.028 (66.4°)	0.986
		0.3	0.034	−0.020	−0.031 (62.9°)	0.382
0.6	0.6	0	0	−0.016	−0.016 (71.3°)	1.177
		0.3	0	−0.020	−0.020 (68.6°)	0.630
0.9	0	0	0.120	0.070	0.052 (66.4°)	1.026
		0.3	0.330	0.239	0.117 (62.9°)	0.434
0.9	0.3	0	0.049	0.029	0.024 (71.3°)	1.197
		0.3	0.115	0.082	0.054 (68.6°)	0.656
0.9	0.6	0	0.011	0.007	0.007 (77.8°)	1.368
		0.3	0.022	0.017	0.015 (76.1°)	0.878
0.9	0.9	0	0	−0.0005	−0.0005 (90.0°)	1.539
		0.3	0	−0.001	−0.001 (90.0°)	1.101

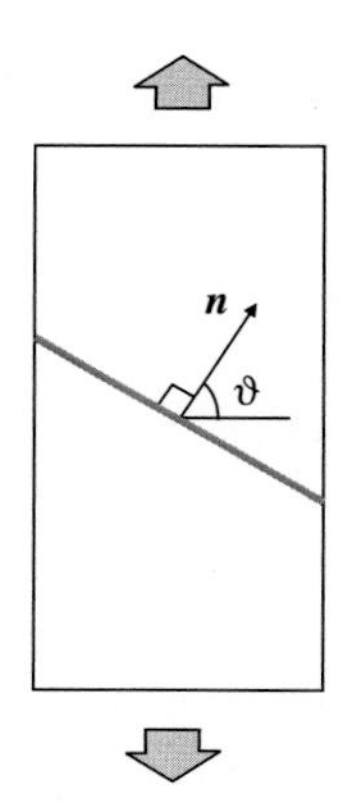

Figure 13.1. Convention for band inclination employed in Table 13.1.

- Loss of (PD), (NS) and (SE) always occurs for a positive plastic modulus, that is, before the snap-back limit $g = 0$ (null and negative values of g are not considered in our constitutive framework). On the other hand, loss of (E) *could* be excluded for strictly positive values of the plastic modulus g (even if this does not occur in the case treated by the table).

- Coalescence of eigenvalues of the acoustic tensor has no relation to the other stability criteria. It may occur at positive or negative hardening depending strongly on the value of Poisson's ratio. It may occur before (PD) or after (E) or between these thresholds.
- All stability thresholds are strongly influenced by
  - State of stress
  - Constitutive parameters (dilatancy, pressure sensitivity, Poisson's ratio)
  - Constitutive features not investigated in the example [yield surface curvature or vertex, non-coaxiality of $\boldsymbol{P}$ and $\boldsymbol{Q}$, elastic or plastic anisotropy (Bigoni and Loret, 1999; Bigoni et al., 2000)]

13.2 Axi-symmetric bifurcations of an elastoplastic cylinder under uniaxial stress

Bifurcations occurring during simple deformations such as the compression or tension of blocks deformed in plane strain or a cylinder subject to uniaxial compression were analysed for hyperelastic incompressible materials in Chapter 12. These analyses are also relevant to *associative* elastoplasticity, where they represent bifurcations of the comparison solid corresponding to the loading branch of the constitutive operator (e.g., the case of the J_2-deformation theory of plasticity corresponds to von Mises plasticity with hardening). As explained in Chapter 10, bifurcations detected in such a comparison solid also may represent 'genuine' elastoplastic bifurcations under broad hypotheses (Hutchinson, 1973).

The situation of non-associative elastoplasticity is much more complicated. In particular, we have seen that the search for bifurcations in an elastoplastic solid with a non-associative flow rule is replaced by the search for bifurcations in two linear comparison solids. One can be any member of the family of Raniecki comparison solids, and the other is simply the linear solid defined by the loading branch of the constitutive operator. Bifurcation in the former solid determines a lower bound to bifurcation stresses, and bifurcation in the latter provides an upper bound. Obviously, it will be convenient to determine the optimal within the family of Raniecki solids as a function of parameter ψ and with respect to the bifurcation problem under consideration.

Needleman (1979), Vardoulakis (1981, 1983), Chau and Rudnicki (1990) and Chau (1992, 1995) have investigated bifurcations in the 'in-loading comparison solid' of an elastoplastic material with non-associative flow rule, similar, in essence, to the Rudnicki and Rice (1975) model. The sole analytical analysis of the Raniecki bounds is that given by Bruhns and Raniecki (1982), whereas numerical results were presented by Tvergaard (1982), Kleiber (1984, 1986) and Tomita et al. (1988).

We develop below a simple example involving uniaxial tension and compression of a cylindrical bar. For this example, we give the (PD), (NS), (SE) and (E) thresholds and investigate the Raniecki bounds for assigned, axi-symmetric mode of bifurcation. It is important to realise that *priority is given to mathematical simplicity*; therefore, the example should be considered as a prototype, not intended to properly model any real material. Nevertheless, the main constitutive features such as pressure sensitivity and plastic dilatancy are taken into account. We base the analysis on the simple constitutive model suggested by Hill (1962) for an associative flow rule (Hutchinson, 1973; Neale, 1981). This model was presented in Section 8.2.3 and has the structure of Eq. (8.50) in the relative Lagrangean description (i.e., using the current configuration as reference, $\boldsymbol{F} = \boldsymbol{I}$) shown in Eq. (8.83), where $\mathbb{E}$ has the isotropic form of Eq. (8.65). When λ and μ are not considered *constants* and have a special dependence on deformation, Christoffersen (1991) has proved hyperelastic behaviour for the elastic part of the constitutive equation [Eq. (8.83)]. However, we will consider λ and μ constant in the following. Moreover, tensors $\boldsymbol{P}$ and $\boldsymbol{Q}$ are selected in the form of Eq. (8.60), proposed by Rudnicki and Rice (1975) [Eq. (8.62)], with *isochoric plastic deformation*

$$\boldsymbol{Q} = \frac{\operatorname{dev}\boldsymbol{T}}{2\sqrt{J_2}} + \frac{\alpha}{3}\boldsymbol{I}, \qquad \boldsymbol{P} = \frac{\operatorname{dev}\boldsymbol{T}}{2\sqrt{J_2}}. \tag{13.1}$$

The constitutive model given by Eq. (8.83) differs from that analysed in the case of the small strain theory to produce the results listed in Table 13.1 only by the presence of the 'geometrical term' $\boldsymbol{LT}$. The elastic acoustic tensor $\boldsymbol{A}_E(\boldsymbol{n})$ corresponding to Eq. (8.83) is (assuming a unit mass density)

$$\boldsymbol{A}_E(\boldsymbol{n}) = (\lambda + \mu)\boldsymbol{n} \otimes \boldsymbol{n} + \bar{\mu}\boldsymbol{I}, \tag{13.2}$$

where $\bar{\mu} = \mu + \boldsymbol{n} \cdot \boldsymbol{T}\boldsymbol{n}$, and it has the inverse

$$\boldsymbol{A}_E^{-1}(\boldsymbol{n}) = -\frac{\lambda + \mu}{\bar{\mu}(\lambda + \mu + \bar{\mu})}\boldsymbol{n} \otimes \boldsymbol{n} + \frac{1}{\bar{\mu}}\boldsymbol{I}. \tag{13.3}$$

Flutter instability for the model (8.83) was analysed by Bigoni and Zaccaria (1994a). They found that complex eigenvalues are excluded for deviatoric associativity [Eqs. (8.60) through (8.62)], but coalescence of eigenvalues still may occur. This will not be examined below.

Let us consider a circular cylindrical specimen of radius R and height L subjected to axi-symmetric deformation with uniaxial stress $\boldsymbol{T} = \sigma \boldsymbol{e}_z \otimes \boldsymbol{e}_z$ aligned along the cylinder axis $\boldsymbol{e}_z$ and traction-free lateral surface. The relative Lagrangean description is assumed, in which the current configuration is taken as reference ($\boldsymbol{F} = \boldsymbol{I}$). A cylindrical coordinate system (r, θ, z) is adopted with the z axis coincident with the axis of the cylinder. The ends ($z = 0, L$) are subject to flat, frictionless, rigid constraints, keeping null the nominal tangential traction: that is, $S_{rz} = 0$. The material constitutive model is assumed to correspond to the loading branch of constitutive equation (8.83) and therefore is given by

$$\dot{\boldsymbol{S}} = \mathbb{D}[\boldsymbol{D}] + \boldsymbol{LT}, \qquad \mathbb{D} = \mathbb{E} - \frac{1}{g}\mathbb{E}[\boldsymbol{Q}] \otimes \mathbb{E}[\boldsymbol{P}]. \tag{13.4}$$

We also will analyse bifurcations in the Raniecki solid, which can be obtained simply by replacing $\mathbb{D}$ in Eq. (13.4) with

$$\mathbb{D}^R = \mathbb{E} - \frac{1}{4\psi g}(\mathbb{E}[\boldsymbol{P}] + \psi\mathbb{E}[\boldsymbol{Q}]) \otimes (\mathbb{E}[\boldsymbol{P}] + \psi\mathbb{E}[\boldsymbol{Q}]). \tag{13.5}$$

Incremental equilibrium equations *for axi-symmetric deformations* (in cylindrical co-ordinates) are

$$\dot{S}_{rr,r} + \dot{S}_{rz,z} + \frac{1}{r}(\dot{S}_{rr} - \dot{S}_{\theta\theta}) = 0,$$

$$\dot{S}_{zr,r} + \dot{S}_{zz,z} + \frac{1}{r}\dot{S}_{zr} = 0. \tag{13.6}$$

Bifurcations are sought in terms of incremental fields satisfying

$$v_z = 0, \qquad \dot{S}_{rz} = 0 \qquad \text{at } z = 0 \text{ and } L,$$

$$\dot{S}_{rr} = 0, \qquad \dot{S}_{zr} = 0 \qquad \text{at } r = R. \tag{13.7}$$

The method of solution parallels that explained in Section 12.3 (see also Chau, 1995) and is based on introduction of the velocity potential $\Phi(r,z)$ so that

$$v_r = \Phi_{,rz}, \qquad v_z = -\frac{1}{A}[B\mathcal{N}(\Phi) + C\Phi_{,zz}], \tag{13.8}$$

where $\mathcal{N}(\Phi) = (r\Phi_{,r})_{,r}/r$ and

$$A = \mathbb{D}_{rrzz} + \mathbb{D}_{rzrz}, \qquad B = \mathbb{D}_{rrrr}, \qquad C = \mathbb{D}_{rzrz} + \sigma. \tag{13.9}$$

For subsequent reference, we introduce three additional quantities:

$$D = \mathbb{D}_{zzzz} + \sigma, \qquad E = \mathbb{D}_{rzrz}, \qquad F = \mathbb{D}_{zzrr} + \mathbb{D}_{rzrz}. \tag{13.10}$$

We note, in passing, that different but equivalent potentials can be introduced in the bifurcation analysis, as discussed by Miles and Nuwahyid (1985). In the axi-symmetric problem under consideration, the non-vanishing components of velocity gradient $\boldsymbol{L}$ are

$$L_{rr} = v_{r,r}, \qquad L_{\theta\theta} = \frac{v_r}{r}, \qquad L_{zz} = v_{z,z}, \qquad L_{rz} = v_{r,z}, \qquad L_{zr} = v_{z,r}. \tag{13.11}$$

Substitution of the constitutive law (13.4) into the equilibrium equations (13.6) reveals that Eq. $(13.6)_1$ is identically satisfied, whereas Eq. $(13.6)_2$ gives

$$\left[\mathcal{N}(\cdot) - \rho_1^2(\cdot)_{,zz}\right]\left[\mathcal{N}(\cdot) - \rho_2^2(\cdot)_{,zz}\right]\Phi = 0, \tag{13.12}$$

where ρ_1 and ρ_2 satisfy the condition

$$EB\rho_i^4 + (EC + DB - AF)\rho_i^2 + DC = 0, \qquad (i = 1,2). \tag{13.13}$$

The nature of the roots of Eq. (13.13) defines the regime classification as follows:

- Two ρ_i complex conjugate pairs in the elliptical complex regime
- Four pure imaginary ρ_i in the elliptical imaginary regime
- Four real ρ_i in the hyperbolic regime
- Two real and two pure imaginary ρ_i in the parabolic regime

The interest in this classification lies in the fact that when the boundary of the elliptical regime is touched, strain localisation may occur. In other words, loss of ellipticity, which is completely equivalent to failure of condition (11.72), occurs when at least two roots ρ_i become real. Diffuse bifurcated solutions are sought now *in the elliptical regime* of the form

$$\Phi(r,z) = \phi(r)\sin\eta z, \tag{13.14}$$

where $\eta = k\pi/L$ and $k = 1,2,\ldots n$.[3] The field described by Eq. (13.14) satisfies the boundary conditions [Eq. (13.7)] at $z = 0$ and $z = L$. The equilibrium equation (13.12) becomes

$$\left[\overline{\mathcal{N}}(\cdot) + \rho_1^2\eta^2\right]\left[\overline{\mathcal{N}}(\cdot) + \rho_2^2\eta^2\right]\phi = 0, \tag{13.15}$$

where

$$\overline{\mathcal{N}}(\phi) = \frac{1}{r}\frac{d}{dr}\left(r\frac{d\phi}{dr}\right). \tag{13.16}$$

The general solution of Eq. (13.15) for the cylinder can be expressed in terms of the Bessel function $J_0(x)$ of order 0:

$$\phi(r) = c_1 J_0(\eta r\rho_1) + c_2 J_0(\eta r\rho_2), \tag{13.17}$$

where constants c_1 and c_2 are, in general, complex.

Imposing boundary conditions at $r = R$ yields a 2×2 linear, homogeneous system. Bifurcation occurs when non-trivial solutions of this system are possible. This condition simply means that bifurcation is possible when $\det M_{ij} = 0$, where M_{ij} is defined as

$$M_{1j} = (\mathbb{D}_{rrrr} - \mathbb{D}_{rr\theta\theta})\rho_j J_1(\eta R\rho_j) + \left(\mathbb{D}_{rrzz}\left(\frac{B}{A}\rho_j^2 + \frac{C}{A}\right) - \mathbb{D}_{rrrr}\rho_j^2\right)\eta R J_0(\eta R\rho_j),$$

$$M_{2j} = \rho_j\left(1 - \frac{C}{A} - \frac{B}{A}\rho_j^2\right)J_1(\eta R\rho_j), \qquad (j = 1,2). \tag{13.18}$$

13.2.1 Results for the axi-symmetric bifurcations of a cylinder

In the interest of simplicity, we assume $\lambda = 0$ (corresponding to a null Poisson's ratio). Therefore, the model (8.83) reduces to

$$\frac{\dot{\boldsymbol{S}}}{2\mu} = \boldsymbol{D} + \frac{\boldsymbol{LT}}{2\mu} - \frac{2\mu}{g} < \boldsymbol{D}\cdot\boldsymbol{Q} > \boldsymbol{P}, \tag{13.19}$$

where $\boldsymbol{P}$ and $\boldsymbol{Q}$ take the form of Eq. (13.1), in which the stress is uniaxial:

$$\boldsymbol{P} = \frac{\operatorname{sign}\sigma}{2\sqrt{3}}(2\boldsymbol{e}_z\otimes\boldsymbol{e}_z - \boldsymbol{e}_r\otimes\boldsymbol{e}_r - \boldsymbol{e}_\theta\otimes\boldsymbol{e}_\theta), \qquad \boldsymbol{Q} = \boldsymbol{P} + \frac{\alpha}{3}\boldsymbol{I}. \tag{13.20}$$

The Raniecki comparison solids are defined by

$$\frac{\dot{\boldsymbol{S}}}{2\mu} = \boldsymbol{D} + \frac{\boldsymbol{LT}}{2\mu} - \frac{\mu}{2\psi g}(\boldsymbol{D}\cdot\boldsymbol{P} + \psi\boldsymbol{D}\cdot\boldsymbol{Q})(\boldsymbol{P} + \psi\boldsymbol{Q}); \tag{13.21}$$

moreover, we select for simplicity the value of ψ suggested by Bruhns and Raniecki (1982):

$$\psi = \sqrt{\frac{\boldsymbol{P}\cdot\boldsymbol{P}}{\boldsymbol{Q}\cdot\boldsymbol{Q}}} = \left(1 + \frac{2}{3}\alpha^2\right)^{-1/2}, \tag{13.22}$$

[3] Note that the parameter η and the index k in Eq. (13.14) have nothing to do with the pre-stress parameters defined by Eqs. $(6.179)_{2,3}$.

which is independent of the current stress level, and it is therefore *not* optimal both for (PD) and (SE_R).

The critical plastic modulus for failure of (PD) can be obtained from Eq. (11.31) by noting that

$$\mathbb{L} = \boldsymbol{I}\,\underline{\overline{\otimes}}\,\boldsymbol{I} + \frac{\sigma}{2\mu}\boldsymbol{e}_z \otimes \boldsymbol{e}_z \otimes \boldsymbol{e}_z \otimes \boldsymbol{e}_z \tag{13.23}$$

has the inverse

$$\mathbb{L}^{-1} = \boldsymbol{I}\,\underline{\overline{\otimes}}\,\boldsymbol{I} - \frac{\sigma}{2\mu+\sigma}\boldsymbol{e}_z \otimes \boldsymbol{e}_z \otimes \boldsymbol{e}_z \otimes \boldsymbol{e}_z. \tag{13.24}$$

For uniaxial tension, σ takes positive values, and $\mathbb{L}$ is always positive definite. Note also that, owing to coaxiality of $\boldsymbol{P}$, $\boldsymbol{Q}$ and $\boldsymbol{T}$, failure of (NS) occurs for a mode with null spin and always corresponds to *a maximum load*.

Failure of (SE) and (E) can be obtained from Eqs. (11.60) and (11.75) on the basis of the elastic acoustic tensor (13.2) and its inverse (13.3). Note that, for $\lambda = 0$, the acoustic tensor has eigenvalues equal to $\bar{\mu}$ and $\mu + \bar{\mu}$, which are always positive for tension and vanish in compression for the first time when $\boldsymbol{n} = \boldsymbol{e}_z$ and $\sigma = -\mu$. When these values are reached, a purely elastic loss of ellipticity occurs. In order to find the (SE) and (E) thresholds, the constrained maximisation problems (11.63) and (11.78) have to be solved. Owing to axial symmetry, this is an easy task in the present situation, where the unit vector $\boldsymbol{n}$ can be chosen without loss of generality in the form $\boldsymbol{n} = n_z\boldsymbol{e}_z + n_r\boldsymbol{e}_r$. Failure of ($SE_R$) can be obtained from Eq. (11.65) by solving the inf-max problem for $\boldsymbol{n}$ and ψ. However, for the present case of uniaxial tension, we have numerically proven the coincidence of (SE) and (SE_R), which holds for a certain optimal parameter ψ, which is a function of the current stress. Therefore, only (SE) will be reported in the figures. It is important to remark that both (PD) and (SE)≡(SE_R) are not relevant (except at $\sigma = 0$) for the Raniecki solid defined by the fixed value of ψ given by Eq. (13.22). This solid has its own peculiar curves for loss of (PD) and (SE)—not reported in the figures—which, by the way, are close to the curves (PD) and (SE) relative to the solid 'in loading' (coincident with the optimal Raniecki solid).

Curves corresponding to failure of (PD), (NS), (SE) and (E) are reported in Figs. 13.2 through 13.4, together with the curves corresponding to axi-symmetric bifurcation for a given bifurcation mode ηR. These curves are referred to a plane defined by axes $\sigma/2\mu$ (the axial stress divided by 2μ) and μ/g (the inverse of the plastic modulus, multiplied by μ). Note that the curves corresponding to (PD) and (NS) are reported only for tension because tensor $\mathbb{G}$ is not positive definite in compression. Note also that the reported range of variation of $\sigma/(2\mu)$ is very large and probably not sensible with the assumed simplified model. However, we prefer to show a picture as complete as possible. The range of variation of μ/g goes from 0, representative of purely elastic behaviour, that is, $g \to \infty$, to softening, that is, beyond the perfectly plastic limit $h = 0$, corresponding to $\mu/g = 1$. Note that the relation between h and g in this simple model is just

$$\frac{g}{\mu} = 1 + \frac{h}{\mu}. \tag{13.25}$$

Axi-symmetric bifurcations of a cylinder: Associative flow rule

Let us begin the discussion with Fig. 13.2, relative to the associative case, $\alpha = 0$. In this case, the plasticity is referred to a von Mises yield surface. Here the (PD) and

(NS) thresholds, as well as the (SE) and (E) thresholds, coincide. The (PD)≡(NS) curve approaches the horizontal axis at $\mu/g = 1$, corresponding to $h = 0$, which is the value of loss of (PD)≡(NS) for the small strain theory. This is obtained when the 'geometrical term' $\sigma/(2\mu)$ is set equal to zero. The curve relative to loss of (E)≡(SE) crosses the horizontal axis at $\mu/g = 1.2$, corresponding to $h/\mu = -0.167$, again the value relative to the small strain theory (see Table 13.1). The curve continues indefinitely in tension but terminates at $\sigma/(2\mu) = -0.5$ in compression, where failure of ellipticity of the elastic tensor occurs. Two other curves are reported in the graph, relative to axi-symmetric bifurcations of the cylindrical bar with prescribed modes $\eta R = 1$ and $\eta R = 2$. These cross (continuously) the horizontal axis a few decimals on the right side of the value corresponding to loss of (PD)≡(NS). As sketched in the figure, the bifurcation modes are *necking* modes for tension and *barrelling* modes for compression. Points situated at positive values of $\sigma/(2\mu)$ and on the left side of curve (PD)≡(NS) are representative of states for which (PD) is satisfied and *bifurcation is excluded a priori*. Points situated at $\sigma/(2\mu) > -0.5$ and on the left side of curve (SE)≡(E) are representative of *elliptical* states, for which *strain localisation is excluded*.

Let us assume that g is some continuous function of the stress state, which is infinity at the beginning of plastic flow and continuously decreases when the stress increases (or decreases, in compression). Let us consider now a monotonic loading program of uniaxial tension of a cylindrical specimen starting from some unloaded state, represented by the origin in the diagram of Fig. 13.2.

At the beginning of the loading program, the material is in the elastic range, and the point representative of the state simply moves up along the axis $\sigma/(2\mu)$, at $\mu/g = 0$. When the yield surface is reached, plastic deformation starts, and the point moves to the right in the diagram. Before the point touches the (PD)≡(NS) curve, the response is unique. As soon as the point reaches the (PD)≡(NS) curve, a maximum load occurs, and the loading program can be continued only if axial

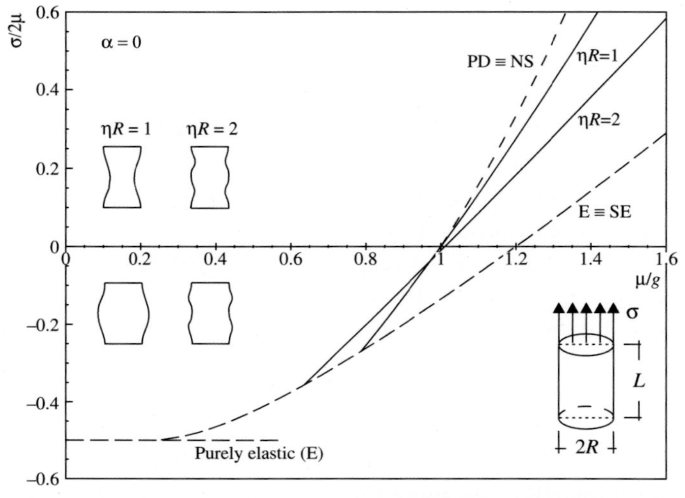

Figure 13.2. Axi-symmetric bifurcations and local stability criteria for an elastoplastic cylinder with associative flow rule, subject to uniaxial stress.

displacement is prescribed (thus, for a situation in which axial dead loading would be prescribed, instability would occur). After the (PD)≡(NS) curve is passed, the point representative of the state will touch the curve corresponding to necking bifurcation (with mode $\eta R = 1$). This bifurcation terminates the homogeneous deformation of the specimen. This might not be the first bifurcation encountered, because an earlier bifurcation may have occurred in another—not investigated—mode (represented by a curve lying between the two curves (PD)≡(NS) and corresponding to the mode $\eta R = 1$). The other bifurcation mode, $\eta R = 2$, occurs after that corresponding to $\eta R = 1$ in tension (but before in compression). It is clear from the graph that there is no way of touching the (SE)≡(E) curve in tension without encountering a bifurcation into a diffuse mode. Now let us analyse compression.

Here, (PD) is not defined, and in fact, bifurcation can occur immediately, even at vanishing small stress. This is the well-known case of Euler buckling of a very slender beam, which would be recovered here by analysing *anti-symmetric bifurcation modes*. However, if we concentrate on axi-symmetric bifurcation at fixed modes $\eta R = 1,2$ and consider a loading program at prescribed axial displacements, we see that there is only a range in which bifurcation with $\eta R = 2$ can be attained. Outside this range, loss of (SE)≡(E) will terminate the homogeneous response of the specimen. Once again, we stress that the picture of bifurcation is not complete because we expect that other axi-symmetric or anti-symmetric modes can be encountered before that corresponding to $\eta R = 2$.

Axi-symmetric bifurcations of a cylinder: Non-associative flow rule

We are now in a position to analyse the case of non-associativity. Here, the (PD), (NS), (SE) and (E) curves are separated, and (SE) and (E) do not cross the horizontal axis with continuity.[4] This reflects the fact that the assumed model has a Drucker-Prager yield surface. The normal to this surface in tension has a different inclination than in compression, and different values of thresholds (SE) and (E) result. For the same reason, the curves representative of diffuse bifurcation modes also do not cross the horizontal axis with continuity (even if the jump is so small that it cannot really be appreciated at the scale of the figure; see detail in Fig. 13.3).

Figure 13.3 is relative to $\alpha = 0.3$ and Fig. 13.4 to $\alpha = \sqrt{3}/2$. As for the case of an associative flow rule, the points where the (PD), (NS), (SE) and (E) curves approach the horizontal axis correspond to the values known from the small strain theory (and reported in Table 13.1 for $\alpha = 0.3$ and tension stress state). Points situated at positive values of $\sigma/(2\mu)$ and on the left side of the curve (PD) correspond to situations in which bifurcation is excluded a priori. The curve (NS) signals that a maximum load has been reached. Curve (E) is relative to strain localisation and is terminated in the negative part of the graph by the horizontal line corresponding to $\sigma/(2\mu) = -0.5$. The same line also terminates curve (SE). When curve (SE) is crossed, the sufficient stability condition of Hill is certainly lost, independent of the specific boundary conditions.

In Fig. 13.3, loss of (PD) occurs before (NS), which precedes (SE) and (E). This is completely different in Fig. 13.4, where failure of (PD) occurs before (SE) and (E), but (NS) is a line and follows (E) [the curve (NS) crosses the curve (E) where

[4] This also would occur for the associative case with pressure sensitivity.

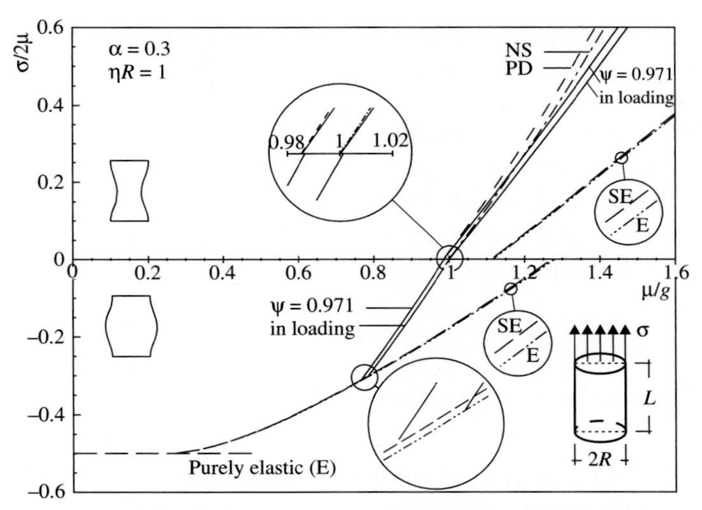

Figure 13.3. Bounds to axi-symmetric bifurcations and local stability criteria for an elasto-plastic cylinder subjected to uniaxial stress. Non-associative flow rule with $\alpha = 0.3$.

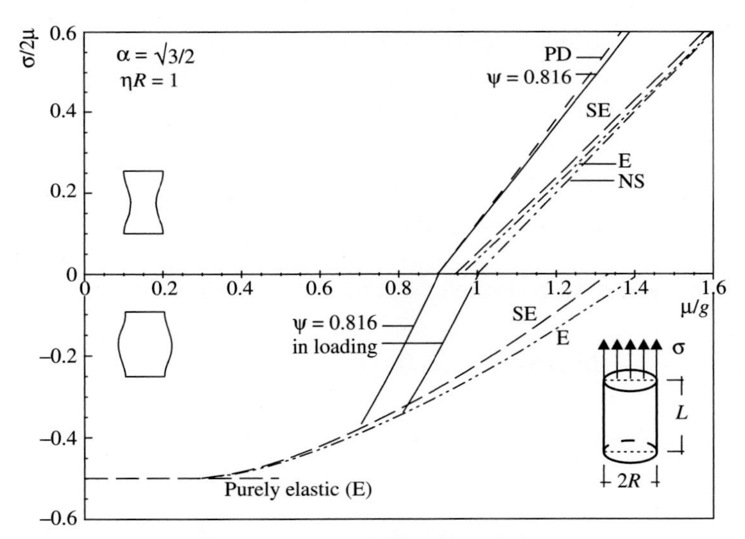

Figure 13.4. Bounds to axi-symmetric bifurcations and local stability criteria for an elasto-plastic cylinder subjected to uniaxial stress. Non-associative flow rule with $\alpha = \sqrt{3}/2$.

$\sigma/(2\mu) = 1$]. It also should be noted that the curves representative of the Raniecki lower bounds terminate in compression when loss of ellipticity occurs in the specific comparison solid (defined by $\psi = 0.971$ for Fig. 13.3 and $\psi = 0.816$ for Fig. 13.4), which does not coincide with the curve (SE) (relative to an optimal choice of ψ; see detail in Fig. 13.3).

Both the Raniecki bounds are reported in Figs. 13.3 and 13.4 for fixed mode $\eta R = 1$. They are separated and are observed both in tension and in compression.

Regarding Fig. 13.3, we note that in tension, the lower bound is represented by a curve starting between (PD) and (NS) and later crossing (NS). The upper bound is close to the right side of (NS). These bifurcations in tension occur well before the loss of (SE) and (E). In compression, just considering the fixed mode of bifurcation $\eta R = 1$, diffuse modes may go beyond the elliptic range, so there is a region in which (SE) and (E) are lost before barrelling bifurcation.

The curve corresponding to the lower bound $\psi = 0.816$ in Fig. 13.4 initiates in compression close (but does not touch) to the (SE) curve and continues (with a discontinuity crossing the horizontal axis) in tension between (PD) and (NS). The bifurcation curve relative to the upper bound, with $\eta R = 1$, is not found in tension in the elliptical regime.

13.3 Flutter instability for a finite-strain plasticity model with anisotropic elasticity

Following Piccolroaz et al. (2006), we now consider the model introduced in Section 8.2.3, which is a simple model based on anisotropic elasticity of the type (6.196), with Drucker-Prager yield criterion and non-associative flow rule, of the type in Eq. (8.60), now with the flow rule specified by eq. (8.61). The interesting feature is that this model displays flutter instability, which therefore results from the combination of non-associativity of the flow rule and anisotropy of the elastic law.

In particular, we refer to the model described by Eq. (8.83), but now with $\mathbb{E}$ given by the law (6.196). We will restrict the analysis to the loading branch of the constitutive operator, which is involved in the flutter analysis. Therefore, the acoustic tensor is

$$\boldsymbol{A}(\boldsymbol{n}) = \boldsymbol{A}_E(\boldsymbol{n}) - \frac{1}{\rho g}\left(\mathbb{E}[\boldsymbol{P}]\boldsymbol{n} \otimes \mathbb{E}[\boldsymbol{Q}]\boldsymbol{n}\right), \tag{13.26}$$

where ρ is the mass density and $\boldsymbol{A}_E(\boldsymbol{n})$ is the elastic acoustic tensor, defined as

$$\boldsymbol{A}_E(\boldsymbol{n}) = \frac{\lambda + \mu}{\rho}\boldsymbol{Bn} \otimes \boldsymbol{Bn} + \frac{\mu}{\rho}(\boldsymbol{n} \cdot \boldsymbol{Bn})\boldsymbol{B} + \frac{1}{\rho}(\boldsymbol{n} \cdot \boldsymbol{Tn})\boldsymbol{I}, \tag{13.27}$$

in which $\boldsymbol{B}$ assumes the transverse anisotropy form given by Eqs. (6.197) and (6.198).

13.3.1 Examples of flutter instability for plane problems

The *plane problem* is considered in which the elastic symmetry vector $\boldsymbol{b}$ and the propagation direction $\boldsymbol{n}$ lie in the plane spanned by $\boldsymbol{k}_1$ and $\boldsymbol{k}_2$, two unit eigenvectors of the Cauchy stress $\boldsymbol{T}$.

In the reference system $\{\boldsymbol{n},\boldsymbol{s},\boldsymbol{k}_3\}$, where $\boldsymbol{s} = \boldsymbol{k}_3 \times \boldsymbol{n}$, the acoustic tensor $\boldsymbol{A}(\boldsymbol{n})$ becomes

$$\begin{pmatrix} A^E_{nn} - \dfrac{1}{\rho g}(\boldsymbol{n}\cdot\boldsymbol{q})(\boldsymbol{n}\cdot\boldsymbol{p}) & A^E_{ns} - \dfrac{1}{\rho g}(\boldsymbol{n}\cdot\boldsymbol{q})(\boldsymbol{s}\cdot\boldsymbol{p}) & 0 \\ A^E_{ns} - \dfrac{1}{\rho g}(\boldsymbol{s}\cdot\boldsymbol{q})(\boldsymbol{n}\cdot\boldsymbol{p}) & A^E_{ss} - \dfrac{1}{\rho g}(\boldsymbol{s}\cdot\boldsymbol{q})(\boldsymbol{s}\cdot\boldsymbol{p}) & 0 \\ 0 & 0 & \dfrac{\mu b_2(\boldsymbol{n}\cdot\boldsymbol{Bn}) + \boldsymbol{n}\cdot\boldsymbol{Tn}}{\rho} \end{pmatrix}, \tag{13.28}$$

where

$$q \equiv \mathbb{E}[Q]n = \lambda(B \cdot Q)Bn + 2\mu BQBn,$$
$$p \equiv \mathbb{E}[P]n = \lambda(B \cdot P)Bn + 2\mu BPBn, \tag{13.29}$$

and A^E_{nn}, A^E_{ss}, A^E_{ns} are the in-plane components of the elastic acoustic tensor $\boldsymbol{A}^E(\boldsymbol{n})$, namely,

$$A^E_{nn} = \frac{\lambda+2\mu}{\rho}(\boldsymbol{n}\cdot\boldsymbol{Bn})^2 + \frac{1}{\rho}\boldsymbol{n}\cdot\boldsymbol{Tn},$$
$$A^E_{ss} = \frac{\lambda+\mu}{\rho}(\boldsymbol{s}\cdot\boldsymbol{Bn})^2 + \frac{\mu}{\rho}(\boldsymbol{n}\cdot\boldsymbol{Bn})(\boldsymbol{s}\cdot\boldsymbol{Bs}) + \frac{1}{\rho}\boldsymbol{n}\cdot\boldsymbol{Tn}, \tag{13.30}$$
$$A^E_{ns} = \frac{\lambda+2\mu}{\rho}(\boldsymbol{n}\cdot\boldsymbol{Bn})(\boldsymbol{s}\cdot\boldsymbol{Bn}).$$

Note that the out-of-plane eigenvalue A_{33} in Eq. (13.28) corresponds to a wave with out-of-plane amplitude ($\boldsymbol{g}$ proportional to $\boldsymbol{k}_3$) and is assumed to remain strictly positive.

From the matrix (13.28), we obtain the sum and the product of the two in-plane eigenvalues (squares of the acceleration waves' propagation velocities) c_1^2 and c_2^2 corresponding to waves with in-plane amplitude ($\boldsymbol{g}$ lying in the plane spanned by $\boldsymbol{k}_1$ and $\boldsymbol{k}_2$),

$$c_1^2 + c_2^2 = A^E_{nn} + A^E_{ss} - \frac{1}{\rho g}(f_1 - f_2),$$
$$c_1^2 c_2^2 = A^E_{nn}A^E_{ss} - (A^E_{ns})^2 + \frac{1}{\rho g}(A^E_{ns}f_3 - A^E_{ss}f_1 + A^E_{nn}f_2), \tag{13.31}$$

where

$$f_1 = (\boldsymbol{n}\cdot\boldsymbol{q})(\boldsymbol{n}\cdot\boldsymbol{p}), \qquad f_2 = -(\boldsymbol{s}\cdot\boldsymbol{q})(\boldsymbol{s}\cdot\boldsymbol{p}),$$
$$f_3 = (\boldsymbol{n}\cdot\boldsymbol{q})(\boldsymbol{s}\cdot\boldsymbol{p}) + (\boldsymbol{s}\cdot\boldsymbol{q})(\boldsymbol{n}\cdot\boldsymbol{p}). \tag{13.32}$$

A necessary and sufficient condition for the existence of complex conjugate eigenvalues a_1 and a_2 is represented by the simultaneous fulfilment of the following three conditions (Bigoni and Loret, 1999):

$$f_4 = (A^E_{nn} - A^E_{ss})^2\left[(f_1 + f_2 + 2ef_3)^2 - (1+4e^2)(f_1 - f_2)^2\right] > 0,$$
$$f_5 = (A^E_{nn} - A^E_{ss})(f_1 + f_2 + 2ef_3) > 0, \tag{13.33}$$
$$\frac{f_5 - \sqrt{f_4}}{(A^E_{nn} - A^E_{ss})^2 + 4(A^E_{ns})^2} < \rho^2 g < \frac{f_5 + \sqrt{f_4}}{(A^E_{nn} - A^E_{ss})^2 + 4(A^E_{ns})^2},$$

where

$$e = \frac{A^E_{ns}}{A^E_{nn} - A^E_{ss}}. \tag{13.34}$$

With reference to Fig. 13.5, let θ_σ and θ_n be the angles of inclination of the direction of elastic anisotropy $\boldsymbol{b}$ [Eq. (6.197)] and wave propagation normal $\boldsymbol{n}$ with respect to the stress principal axis $\boldsymbol{k}_1$.

Dividing all quantities having the dimension of a stress in Eqs. (13.28) through (13.33) by μ, the parameters on which the condition of flutter depends are the following:

- *Elastic parameters:* λ/μ, strength of anisotropy $\hat{b}$ [Eq. (6.198)] and orientation of the axis of elastic symmetry with respect to the principal stress axis $\boldsymbol{k}_1$, namely, θ_σ.

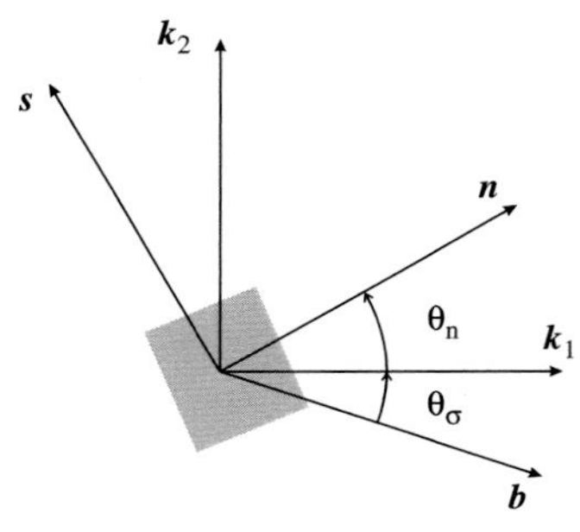

Figure 13.5. Principal stress axes $\boldsymbol{k}_1$ and $\boldsymbol{k}_2$, axis of elastic symmetry $\boldsymbol{b}$ and propagation direction $\boldsymbol{n}$, singled out by angles θ_σ and θ_n, respectively.

- *Plastic parameters:* plastic modulus g/μ, pressure sensitivity ψ and dilatancy χ parameters.
- *Principal normalised deviatoric stress values:* $\text{dev}T_1/|\text{dev}\boldsymbol{T}|$, $\text{dev}T_2/|\text{dev}\boldsymbol{T}|$, $\text{dev}T_3/|\text{dev}\boldsymbol{T}|$. However, these are not independent, so for the flow law and yield surface gradient $\boldsymbol{P}$ and $\boldsymbol{Q}$ given by Eqs. (8.60) and (8.61), flutter depends on the angle

$$\theta_L = \text{sgn}\left(\frac{\text{dev}T_1}{|\text{dev}\boldsymbol{T}|} + 2\frac{\text{dev}T_2}{|\text{dev}\boldsymbol{T}|}\right)\cos^{-1}\left(\sqrt{\frac{3}{2}}\frac{\text{dev}T_1}{|\text{dev}\boldsymbol{T}|}\right) \tag{13.35}$$

in the deviatoric plane, which is a 'modified Lode angle', defined for $\theta_L \in [-\pi, \pi]$ and in which $\text{sgn}(0) = 1$.

It is possible to analyse flutter for every propagation direction $\boldsymbol{n}$ while varying the plastic modulus g/μ and keeping fixed all remaining parameters in the preceding list, by use of inequalities (13.33). Therefore, the ranges in which flutter occurs can be plotted in the plane g/μ versus θ_n. Restricting the analysis to the small strain theory, where the Oldroyd flux $(3.156)_2$ is identified with $\dot{\boldsymbol{T}}$, analyses have been performed for simplicity with different values of the modified Lode parameter $\theta_L = \{60°, 30°, 0°, -30°, -60°\}$, as indicated in Fig. 13.6.

Results are reported in Figs. 13.7 and 13.8, the latter providing more details for four of the cases reported in the former figure. Different stress paths defined by the values of the modified Lode angle [Eq. (13.35)], reported in Fig. 13.6 are considered for different anisotropy inclinations θ_σ in Fig. 13.7 at given values of $\psi = 30°$ and $\chi = 0°$. In the graphs, the closed contours denote regions where flutter occurs in the plane defined by the normalised critical plastic modulus g/μ and the inclination of propagation direction θ_n.

Four details of Fig. 13.7 are reported in Fig. 13.8, where $\lambda/\mu = 1$, $\hat{b} = 80°$, $\psi = 30°$, and $\chi = 0°$, as in Fig. 13.7. The six regions in Fig. 13.8 correspond to the four cases $\theta_L = 0°$ and $\theta_\sigma = 15°$ (case 1), $\theta_L = \theta_\sigma = 30°$ (case 2), $\theta_L = 0$ and $\theta_\sigma = 45°$ (case 3), and $\theta_L = 0$ and $\theta_\sigma = 60°$ (case 4).

With reference to the cases 1 through 4, detailed in Fig. 13.8, we note that the critical values of plastic modulus for loss of positive definiteness of the constitutive operator g_{cr}^{PD} and for loss of ellipticity g_{cr}^{E} permitting shear bands with normal inclined

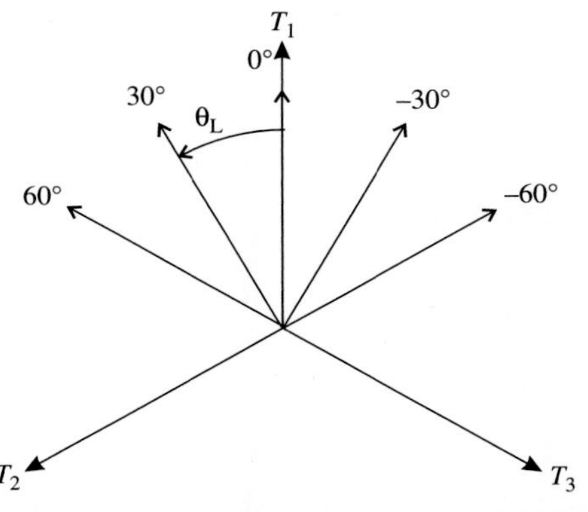

Figure 13.6. Stress directions in the deviatoric plane, defined by the modified Lode angle (13.35), considered for flutter analysis.

at θ_{nE} are

$$\begin{array}{llll} \text{Case 1:} & g_{cr}^{PD}/\mu = 0.42, & g_{cr}^{E}/\mu = 0.19, & \theta_{nE} = -28.0°, \\ \text{Case 2:} & g_{cr}^{PD}/\mu = 1.22, & g_{cr}^{E}/\mu = 0.18, & \theta_{nE} = -16.4°, \\ \text{Case 3:} & g_{cr}^{PD}/\mu = 1.03, & g_{cr}^{E}/\mu = 0.74, & \theta_{nE} = -32.0°, \\ \text{Case 4:} & g_{cr}^{PD}/\mu = 1.84, & g_{cr}^{E}/\mu = 1.57, & \theta_{nE} = -33.9°, \end{array} \tag{13.36}$$

so in all cases flutter may initiate when the constitutive operator is positive definite (therefore at an early stage of a deformation process) and may extend through a region possibly involving loss of ellipticity.

Note that

owing to anisotropy, only one shear band is found,

as first noticed by Bigoni et al. (2000).

Thresholds (13.36) have been represented graphically in Fig, 13.8, where light grey regions correspond to regions where flutter may occur with the constitutive operator still positive definite, whereas in the dark grey regions ellipticity is lost (horizontal lines marking ellipticity loss are denoted with ‘E (case i)’, where $i = 1, \ldots, 4$ stands for the number of the relevant case). In the same figure, three black spots and a white spot (referred to case 2) indicate the inclinations of shear bands at first loss of ellipticity. Note that the small flutter regions of cases 3 and 4 are beyond the positive definiteness threshold but still in the elliptical region. It may be important to remark that

the initial inclinations of propagation normals for flutter and shear bands are unrelated and remarkably different.

From the preceding analysis it can be deduced that the constitutive model allows one to approach flutter starting from a well-behaved state. Moreover, it may be interesting to note from Fig. 13.8 that there are overlapping regions corresponding to different stress states (cases 1 and 2). In these zones, the flutter may have identical characteristics even if the stress state is different.

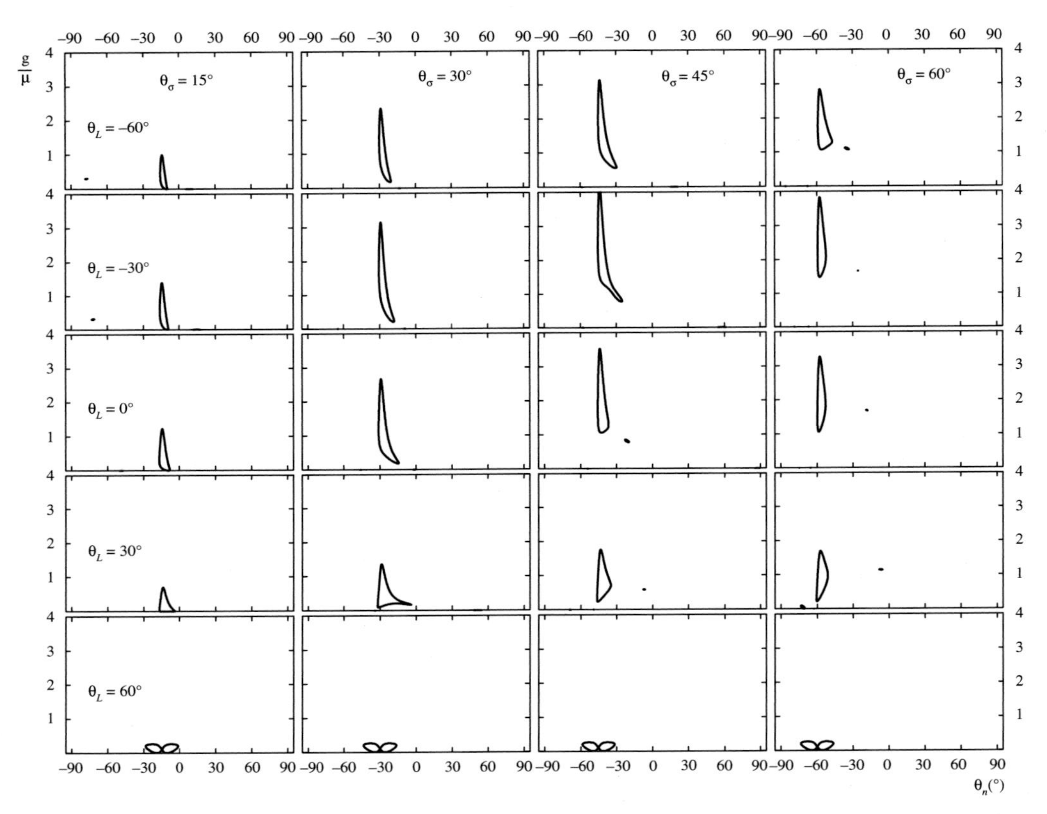

Figure 13.7. Regions of flutter instability (occurring for internal points) for a non-associative elastoplastic model in the g/μ versus θ_n plane, for the stress paths shown in Fig. 13.6 at various anisotropy inclinations θ_σ. The following values of material parameters have been considered: $\lambda/\mu = 1$, $\hat{b} = 80°$, $\psi = 30°$ and $\chi = 0°$.

13.3.2 Spectral analysis of the acoustic tensor

The spectral analysis of the acoustic tensor is instrumental to the development of the Green's function which will be presented in Chapter 16. The analysis is restricted to the in-plane components of the acoustic tensor $\boldsymbol{A}$:

$$\boldsymbol{A} = A_{11}(\boldsymbol{k}_1 \otimes \boldsymbol{k}_1) + A_{12}(\boldsymbol{k}_1 \otimes \boldsymbol{k}_2) + A_{21}(\boldsymbol{k}_2 \otimes \boldsymbol{k}_1) + A_{22}(\boldsymbol{k}_2 \otimes \boldsymbol{k}_2), \tag{13.37}$$

represented for later convenience in the principal stress basis $\boldsymbol{k}_1, \boldsymbol{k}_2$. The inverse of tensor (13.37) is given by the formula

$$\boldsymbol{A}^{-1} = \frac{1}{A_{11}A_{22} - A_{12}A_{21}} [A_{22}(\boldsymbol{k}_1 \otimes \boldsymbol{k}_1) - A_{12}(\boldsymbol{k}_1 \otimes \boldsymbol{k}_2) \\ - A_{21}(\boldsymbol{k}_2 \otimes \boldsymbol{k}_1) + A_{11}(\boldsymbol{k}_2 \otimes \boldsymbol{k}_2)]. \tag{13.38}$$

Now, the eigenvalues of the acoustic tensor (13.37) can be written in the form

$$\left.\begin{matrix} c_1^2 \\ c_2^2 \end{matrix}\right\} = \frac{A_{11} + A_{22} \pm \Delta}{2}, \qquad \Delta = \sqrt{(A_{11} - A_{22})^2 + 4A_{12}A_{21}}, \tag{13.39}$$

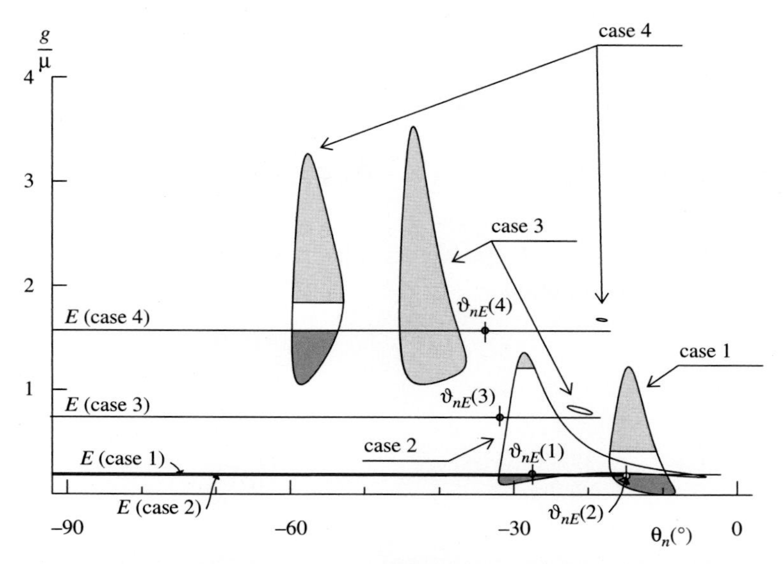

Figure 13.8. Regions of flutter instability (occurring for internal points) for a non-associative elastoplastic model in the g/μ versus θ_n plane, for $\lambda/\mu = 1$, $\hat{b} = 80°$, $\psi = 30°$, and $\chi = 0°$. Case 1: $\theta_L = 0°$ and $\theta_\sigma = 15°$. Case 2: $\theta_L = 30°$ and $\theta_\sigma = 30°$. Case 3: As in case 1, but $\theta_\sigma = 45°$. Case 4: As in case 1, but $\theta_\sigma = 60°$. The regions of positive definiteness of the constitutive operator are marked in light grey, whereas (E) denotes loss of ellipticity into shear bands (regions shaded in dark grey) inclined at $\theta_{nE}(i)$, where i=1, ... ,4 denotes the relevant case. Flutter may occur when the constitutive operator is positive definite.

so assuming non-defectiveness, the spectral representations of $\boldsymbol{A}$ and $\boldsymbol{A}^{-1}$ are

$$\boldsymbol{A} = c_1^2(\boldsymbol{v}_1 \otimes \boldsymbol{w}_1) + c_2^2(\boldsymbol{v}_2 \otimes \boldsymbol{w}_2), \tag{13.40}$$

and, assuming[5] $c_1^2 \neq 0$ and $c_2^2 \neq 0$,

$$\boldsymbol{A}^{-1} = \frac{1}{c_1^2}(\boldsymbol{v}_1 \otimes \boldsymbol{w}_1) + \frac{1}{c_2^2}(\boldsymbol{v}_2 \otimes \boldsymbol{w}_2), \tag{13.41}$$

where $\{\boldsymbol{v}_1, \boldsymbol{v}_2\}$ and $\{\boldsymbol{w}_1, \boldsymbol{w}_2\}$ are dual bases, thus satisfying $\boldsymbol{v}_i \cdot \boldsymbol{w}_j = \delta_{ij}$ $(i,j = 1,2)$, composed of right, $\boldsymbol{v}_i$, and left, $\boldsymbol{w}_i$, eigenvectors. This basis is given by

$$\boldsymbol{v}_1 = \boldsymbol{k}_1 + \frac{\Delta - (A_{11} - A_{22})}{2A_{12}}\boldsymbol{k}_2, \qquad \boldsymbol{v}_2 = \boldsymbol{k}_1 + \frac{-\Delta - (A_{11} - A_{22})}{2A_{12}}\boldsymbol{k}_2,$$
$$\boldsymbol{w}_1 = \frac{\Delta + (A_{11} - A_{22})}{2\Delta}\boldsymbol{k}_1 + \frac{A_{12}}{\Delta}\boldsymbol{k}_2, \qquad \boldsymbol{w}_2 = \frac{\Delta - (A_{11} - A_{22})}{2\Delta}\boldsymbol{k}_1 - \frac{A_{12}}{\Delta}\boldsymbol{k}_2, \tag{13.42}$$

[5] For $\Delta \to 0$ (coalescence of the eigenvalues), the tensor $\boldsymbol{A}$ becomes defective (except for the trivial case where $\boldsymbol{A}$ is isotropic), and each term in the spectral representation of $\boldsymbol{A}$, as well as of $\boldsymbol{A}^{-1}$, blows up, but $\boldsymbol{A}^{-1}$ continues to exist and to be defined correctly. Indeed, a substitution of Eqs. (13.39) and (13.42) or (13.43) into Eq. (13.41) leads to Eq. (13.38).

when $A_{12} \neq 0$, or by

$$v_1 = \frac{\Delta + (A_{11} - A_{22})}{2A_{21}} \boldsymbol{k}_1 + \boldsymbol{k}_2, \qquad v_2 = \frac{-\Delta + (A_{11} - A_{22})}{2A_{21}} \boldsymbol{k}_1 + \boldsymbol{k}_2,$$

$$w_1 = \frac{A_{21}}{\Delta} \boldsymbol{k}_1 + \frac{\Delta - (A_{11} - A_{22})}{2\Delta} \boldsymbol{k}_2, \qquad w_2 = -\frac{A_{21}}{\Delta} \boldsymbol{k}_1 + \frac{\Delta + (A_{11} - A_{22})}{2\Delta} \boldsymbol{k}_2, \tag{13.43}$$

when $A_{21} \neq 0$. Since the case $A_{21} = A_{12} = 0$ is trivial, the spectral analysis of the acoustic tensor (13.37) is now completed.

14 Wave propagation, stability and bifurcation

With reference to plane strain, incompressible elasticity, it is shown that bifurcation of elastic materials deformed incrementally can be interpreted as the occurrence of waves propagating at null speed. After the plane wave propagation is solved for an infinite medium as a perturbation superimposed on a finitely and homogeneously strained elastic material, a wave propagation analysis in elastoplasticity elucidates the meaning of divergence instability (occurrence of negative eigenvalues of the acoustic tensor) and the difficulties (related to the fact that the constitutive tangent operator is piece-wise linear) connected with the interpretation of flutter instability. Finally, the treatment of acceleration waves reveals that the condition of localisation of deformation in elastoplasticity can be understood as the condition of vanishing speed of acceleration waves.

Wave propagation in solids is a topic strictly connected with stability and bifurcation. It will be shown in this chapter that the condition for incremental bifurcation analysed in chapter 12 for elastic solids is equivalent to the condition of vanishing propagation speed for an incremental wave mode, whereas instability corresponds to a blow up of the wave mode amplitude during propagation.

The simple example of small-amplitude vibrations of a beam superimposed on a given axial stress ('pre-stress') is sufficient to clarify the above-mentioned issues. To this purpose, we reconsider the beam illustrated in Section 10.2.3, subjected to axial load F corresponding to a longitudinal Cauchy stress $-\sigma$ parallel to the beam axis x_1.

To derive approximate equations for transverse beam vibration in the x_1–x_2 plane (so that the inertia force reduces to $\rho_0 \ddot{v}_2$), we recall the equations of motion (3.115) which, assuming that all out-of-plane ($\dot{S}_{i3}$, $\dot{S}_{3j}$) as well as transverse ($\dot{S}_{22}$), stresses vanish, reduce to

$$\dot{S}_{11,1} + \dot{S}_{12,2} = 0, \qquad \dot{S}_{21,1} = \rho_0 \ddot{v}_2, \tag{14.1}$$

where ρ_0 is the referential description of the mass density. Equation (10.52) allows us to express $\dot{S}_{21}$ as a function of its transpose:

$$\dot{S}_{21} = \dot{S}_{12} - \sigma v_{2,1}. \tag{14.2}$$

Thus, multiplying Eq. (14.1)$_1$ by x_2, taking the derivative of this equation and of Eqs. (14.2) and (14.1)$_2$ with respect to x_1 and substituting, we obtain

$$\left(x_2\dot{S}_{11}\right)_{,11} + \left(x_2\dot{S}_{12}\right)_{,21} - \sigma v_{2,11} - \rho_0 \ddot{v}_2 = 0. \tag{14.3}$$

Let us now assume the simplified kinematics of the beam theory [Eqs. (10.55) and (10.56), where w is now a function also of time t] and the simple constitutive equations [Eqs. (10.52) and (10.59)] and we obtain

$$\dot{S}_{11} = -E_t x_2 w''(x_1,t), \qquad \dot{S}_{12} = 0, \qquad v_{2,11} = w''(x_1,t), \qquad \ddot{v}_2 = \ddot{w}(x_1,t). \tag{14.4}$$

A substitution of Eqs. (14.4) into Eq. (14.3) and an integration through the area A of the beam cross section yield the dynamic equations of motion of a beam subject to an axial compressive force F

$$w''''(x_1,t) + \frac{F}{E_t I_3} w''(x_1,t) + \frac{\rho_0 A}{E_t I_3}\ddot{w}(x_1,t) = 0, \tag{14.5}$$

where $w(x_1,t)$ is the lateral deflection of the beam, E_t its elastic modulus and I_3 is the moment of inertia around the x_3 axis [Eq. (10.58)].

Harmonic waves correspond to solutions of Eq. (14.5) in the form

$$w(x_1,t) = Be^{i(\kappa x_1 - \omega t)}, \tag{14.6}$$

where B is the amplitude, κ the wavenumber and ω the circular frequency, so a substitution into Eq. (14.5) yields

$$\bar{\kappa}^4 - f\bar{\kappa}^2 - \bar{\omega}^2 = 0, \tag{14.7}$$

where

$$\bar{\kappa}^4 = I_3\kappa^4, \qquad f = \frac{F}{E_t\sqrt{I_3}}, \qquad \bar{\omega}^2 = \omega^2\frac{\rho_0 A}{E_t}. \tag{14.8}$$

The characteristic Eq. (14.7) has the solutions

$$\kappa = \pm\frac{1}{\sqrt[4]{I_3}}\sqrt{\frac{f \pm \sqrt{f^2 + 4\bar{\omega}^2}}{2}}, \tag{14.9}$$

which correspond to two real $\pm\alpha$ and two purely imaginary $\pm i\beta$ solutions, so, representation (14.6) becomes

$$w(x_1,t) = (A_1 \sin\alpha x_1 + A_2\cos\alpha x_1 + A_3\sinh\beta x_1 + A_4\cosh\beta x_1)\,e^{-i\omega t}, \tag{14.10}$$

where coefficients A_j (j=1,...,4) are arbitrary amplitudes.

Since the beam is simply supported, the four null displacement and moment boundary conditions $w(0,t) = w(l,t) = 0$ and $w''(0,t) = w''(l,t) = 0$ plus the request for obtaining non-trivial solutions yield

$$A_2 = A_3 = A_4 = 0, \tag{14.11}$$

and the frequency equation

$$\sin\alpha l = 0, \qquad \alpha l = n\pi, \qquad n = 1,2,3,... \tag{14.12}$$

corresponding to

$$\bar{\omega}^2 = \frac{n^2\pi^2}{l^2 E_t}\left(\frac{n^2\pi^2 E_t I_3}{l^2} - F\right), \qquad n = 1,2,3,\ldots. \tag{14.13}$$

F is negative for tensile forces, and Eq. (14.13) always provides real solutions for $\bar{\omega}$. For positive values of the axial force corresponding to compressive loads, we find $\bar{\omega} = 0$ at critical load for bifurcation [Eq. $(10.62)_1$], so

bifurcation corresponds to a wave of vanishing speed

of mode given by Eq. $(10.62)_2$. For compressive loads higher than the critical load, $\bar{\omega}$ becomes purely imaginary, so the exponential in Eq. (14.10), instead generating a sinusoidal wave, produces one wave blowing up (and one decaying) in time, in other words, a divergence instability.

The situation for elastoplastic solids with non-associative flow rule is obviously much more complicated, as will be shown later with reference to shear band formation (occurring when an eigenvalue of the acoustic tensor vanishes), divergence instability (occurring when an eigenvalue of the acoustic tensor becomes negative) and flutter instability (occurring when two eigenvalues of the acoustic tensor become complex conjugate). It will be shown that the acoustic tensor introduced in Chapter 11 governs for elastoplastic solids the propagation of acceleration waves, so the condition of shear band formation, or ellipticity loss, is exactly the condition of vanishing speed of an acceleration wave. Moreover, divergence instability will be shown to represent an instability condition for elastoplastic solids, whereas the achievement of the same conclusion for flutter becomes less clear, owing to the incremental non-linearity of the constitutive equations (note that the non-linear analysis provided in Section 1.13.5 for the two degree of freedom structure in flutter condition fully confirms conclusions about instability that can be drawn from the linearised analysis, see Bigoni and Noselli, 2011).

14.1 Incremental waves and bifurcation

The aim of this section is to establish a contact between wave propagation and bifurcation analysis. It will be shown, with reference to plane strain problems of incompressible elasticity, that the conditions for incremental bifurcation found in Chapter 12 are equivalent to conditions of vanishing speed of propagation of incremental waves (a conclusion which remains valid in the compressible case). For instance, vanishing of the propagation speed of Rayleigh and Stoneley waves corresponds to surface and interface bifurcation, respectively (Chapter 12).

Let us consider an incremental time-harmonic motion with circular frequency ω such that the incremental displacement $\boldsymbol{u}(\boldsymbol{x}_0,t)$, as well as all other incremental fields, can be expressed as

$$\boldsymbol{u}(\boldsymbol{x}_0,t) \propto \hat{\boldsymbol{u}}(\boldsymbol{x}_0)\, e^{-i\omega t}, \tag{14.14}$$

and the equations of motion for incremental elasticity [Eqs. (9.73)], become, in the absence of body forces,

$$\begin{aligned} &\hat{\boldsymbol{u}} = \hat{\boldsymbol{\xi}}(\boldsymbol{x}_0) && \text{on} \quad \partial B_0^{\xi}, \\ &\mathbb{G}[\text{grad}\hat{\boldsymbol{u}}]\boldsymbol{n}_0 = \hat{\boldsymbol{\sigma}}(\boldsymbol{x}_0) && \text{on} \quad \partial B_0^{\sigma}, \\ &\text{Div}(\mathbb{G}[\text{Grad}\hat{\boldsymbol{u}}]) + \rho_0\omega^2\hat{\boldsymbol{u}} = \boldsymbol{0} && \text{in} \quad B_0. \end{aligned} \tag{14.15}$$

Therefore, for two-dimensional incompressible elasticity, with introduction of the stream function (12.24) (in which we have added the superscript $\hat{}$), Eq. (14.15)$_3$ yields the equivalent of Eq. (12.25), but now for time-harmonic dynamics

$$(1+k)\hat{\psi}_{,1111} + 2(2\xi-1)\hat{\psi}_{,1122} + (1-k)\hat{\psi}_{,2222} = -\frac{\rho_0\omega^2}{\mu}\left(\hat{\psi}_{,11} + \hat{\psi}_{,22}\right), \tag{14.16}$$

from which we note that (see, e.g., Renardy and Rogers, 1993)

> *since only the principal part of a differential operator plays a role in determining the character of a differential equation, the regime classification for the dynamic case remains the same as for the quasi-static case, see Section 12.2.1.*

Introducing function $\widehat{F}$ [Eq. (12.26)], into Eq. (14.16), we obtain

$$\left[1+k+2(2\xi-1)\Omega^2+(1-k)\Omega^4\right]\widehat{F}'''' + \left(1+\Omega^2\right)\frac{\rho_0\,\omega^2}{\mu}\widehat{F}'' = 0. \tag{14.17}$$

Equation (14.17) provides an explanation for the fact that only the principal part of the differential operator plays a role in determining the regime classification. In fact, since the term multiplying $\widehat{F}''$ is always positive (and null in the special quasi-static case), the character of the differential equation is determined by the sign of the term multiplying $\widehat{F}''''$, and the vanishing of this coefficient represents the extreme condition of loss of ellipticity.

We now select for $\hat{\psi}$ again a representation of the type (12.64), which means that function $\widehat{F}$ takes the form

$$\widehat{F} \propto e^{ic_1(x_1+\Omega x_2)}, \tag{14.18}$$

which substituted into Eq. (14.17) yields

$$1+k-2\hat{\omega}^2+2\left(2\xi-1-\hat{\omega}^2\right)\Omega^2+(1-k)\,\Omega^4 = 0, \tag{14.19}$$

an equation corresponding to Eq. (5.11) of Dowaik and Ogden (1990), where

$$\hat{\omega} = c\sqrt{\frac{\rho_0}{2\mu}}, \qquad c = \frac{\omega}{c_1}, \tag{14.20}$$

in which c is the *wave propagation speed*. The solution of Eq. (14.19) is

$$\Omega_j^2 = \frac{1-2\xi+\hat{\omega}^2+(-1)^j\sqrt{(1-2\xi+\hat{\omega}^2)^2-(1-k)(1+k-2\hat{\omega}^2)}}{1-k}, \tag{14.21}$$

which for $\hat{\omega}=0$, that is, in the quasi-static case, reduces to Eq. (12.28).

Therefore, the general solution to Eqs. (14.16) is represented in a form similar to Eq. (12.64), namely,

$$\hat{\psi} = \left(b_1 e^{ic_1\Omega_1 x_2} + b_2 e^{ic_1\Omega_2 x_2} + b_3 e^{ic_1\Omega_3 x_2} + b_4 e^{ic_1\Omega_4 x_2}\right) e^{ic_1 x_1}, \tag{14.22}$$

where the Ω_i's are given by the solutions (14.21), and the constants b_i can be determined by imposing boundary conditions, for instance, for surface (Rayleigh) waves decaying of the solution, when $x_1 \longrightarrow -\infty$ and null incremental nominal tractions at $x_2 = 0$ have to be imposed. In essence, the same boundary conditions employed for the bifurcation problems solved in Section 12.2 can be imposed to analyse surface waves, interfacial waves and waves propagating into a single: and multi-layer structures.

Since in the limit of null propagation speed $c = 0$, the roots (14.21) reduce to the roots (12.28), and function $\hat{\psi}$ [Eq. (14.22)] reduces to function ψ [Eq. (12.64)], it is easy to conclude that:

The bifurcation problems analysed in Section 12.2 for different boundary conditions correspond to incremental wave propagation problems in the limit of vanishing speed.

We will in the following analyse incremental plane waves propagating in an infinite pre-stressed elastic material, and we will show that the condition of vanishing speed corresponds to loss of ellipticity. We will later show that this analysis becomes much less straightforward when elastoplastic behaviour is involved, but in this case, acceleration wave of vanishing speed are shown again to correspond to loss of ellipticity.

14.2 Incremental plane waves

14.2.1 Non-linear elastic materials

We follow here Hayes and Rivlin (1961a; 1961b), Truesdell (1961) and Truesdell and Noll (1965), and we consider an infinite, homogeneously deformed and stressed elastic material, so equilibrium and compatibility are trivially satisfied. Any incremental dynamic solution must satisfy the equations of motion (9.73), considered in the absence of body forces. We assume this configuration as a reference configuration in a relative Lagrangean description and look for the possibility of the existence of incremental solutions in the following form

$$\boldsymbol{w} = \mathrm{Re}\{\boldsymbol{a} e^{ik(\boldsymbol{n}\cdot\boldsymbol{x} \pm ct)}\}, \tag{14.23}$$

where for $i = \sqrt{-1}$, $\boldsymbol{w}$ denotes the incremental displacement, $\boldsymbol{n}$ is the unit vector of propagation, $\boldsymbol{a}$ is the (possibly complex) wave amplitude vector, k is the (positive) wave number and c is the (possibly complex) wave speed. Note that B has been taken to coincide with B_0 in problem (9.73), so to simplify notation, we use $\boldsymbol{n}$ instead of $\boldsymbol{n}_0$ and $\boldsymbol{x}$ instead of $\boldsymbol{x}_0$.

Adopting the complex notation, the gradient and the time derivative of Eq. (14.23) are

$$\nabla \boldsymbol{w} = ik\,\boldsymbol{n} \otimes \boldsymbol{w} \qquad \text{and} \qquad \dot{\boldsymbol{w}} = \pm ick\,\boldsymbol{w}, \tag{14.24}$$

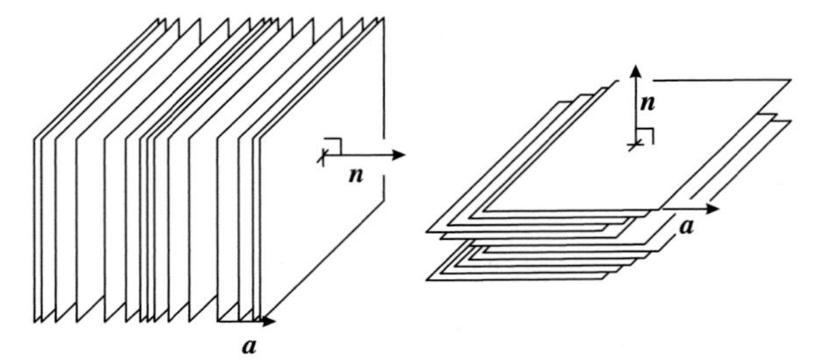

Figure 14.1. Sketch of propagation of plane waves in an elastic prestressed material. *Left*: longitudinal, or 'p', wave in which the direction of propagation $\boldsymbol{n}$ is parallel to the amplitude $\boldsymbol{a}$. *Right*: transverse, or 's', wave in which the direction of propagation $\boldsymbol{n}$ is orthogonal to the amplitude $\boldsymbol{a}$.

whereas the second gradient and second time derivative are

$$\nabla(\nabla \boldsymbol{w}) = -k^2 \boldsymbol{n} \otimes \boldsymbol{n} \otimes \boldsymbol{w} \qquad \text{and} \qquad \ddot{\boldsymbol{w}} = \mp c^2 k^2 \boldsymbol{w}. \tag{14.25}$$

Inserting Eq. (14.25) into the momentum balance (9.73) leads to the propagation condition

$$(\boldsymbol{A}_E(\boldsymbol{n}) - c^2 \boldsymbol{I})\boldsymbol{a} = \boldsymbol{0}, \tag{14.26}$$

where $\boldsymbol{A}_E(\boldsymbol{n})$ is the elastic acoustic tensor defined by eqn. (11.47), so the squared propagation velocities c^2 are the eigenvalues of the acoustic tensor.

Note that a wave is called 'longitudinal' ('p') or 'transverse' ('s') if the propagation direction $\boldsymbol{n}$ is parallel or orthogonal to the wave amplitude vector $\boldsymbol{a}$ (Fig. 14.1).

Therefore, we can draw the following conclusions.

- In the case of hyperelasticity, $\boldsymbol{A}_E(\boldsymbol{n})$ is symmetric, so the eigenvalues c^2 are always real and the eigenspaces orthogonal.
  - When the eigenvalues of the acoustic tensor are positive, the propagation occurs with positive speed, and the condition of strong ellipticity [Eq. (11.44), where $\mathbb{C}$ has to be replaced by $\mathbb{G}$] is verified.
  - If the acoustic tensor has a null eigenvalue, the condition of strain localisation or loss of ellipticity [Eq. (11.72), where $\boldsymbol{A}(\boldsymbol{n})$ has to be replaced by $\boldsymbol{A}_E(\boldsymbol{n})$] is met, and the propagation corresponds to a null speed.
  - If the acoustic tensor has a negative eigenvalue, as remarked by Hayes and Rivlin (1961a), a suitable combination of solutions of the type (14.23), similar to that considered in the next paragraph, yields a blowing up disturbance, a so-called divergence instability. Therefore, strong ellipticity [Eq. (11.44), where $\mathbb{C}$ has to be replaced by $\mathbb{G}$] can be viewed as a stability criterion.
- In the case of hypoelasticity, the acoustic tensor may be un-symmetric, in which case there is the possibility that two eigenvalues of the acoustic tensor (squared propagation velocities) are complex conjugated. Also in this case, a solution blowing up with time can be constructed, leading to so-called flutter instability. For un-symmetric acoustic tensor, however, strong ellipticity becomes only *sufficient but not necessary* to exclude *divergence* instability because violation

of strong ellipticity does not mean that an eigenvalue becomes negative (see remark 1 in Section 2.5). Moreover, strong ellipticity does not exclude the possibility either of flutter instability or of a defective acoustic tensor (see remark 3 in Section 2.5). For a hypo-elastic material, therefore, *a local stability condition related to incremental plane wave propagation is that all the eigenvalues of the acoustic tensor are real and positive, and if coincident, the corresponding acoustic tensor must be non-defective*. With the exception of the possibility of defectiveness of the acoustic tensor, this stability criterion has been proposed by Mandel (1966a), in the context of non-associative elastoplasticity (where the acoustic tensor is not symmetric).

14.3 Waves and material instabilities in elastoplasticity

From the preceding section we may conclude that for a homogeneous material characterised by a *linear* constitutive relationship between the increments in stress and strain, the propagation of a disturbance represented by a planar wave depends critically on the eigenvalues of the acoustic tensor. The waves propagate with a real speed when all eigenvalues of the acoustic tensor for any wavefront normal are real and positive and when the acoustic tensor is not defective (a particular case that will be for the moment ignored). Occurrence of negative or complex eigenvalues is related to 'divergence' or 'flutter' growth of disturbances, respectively, whereas the critical case when one eigenvalue is zero is associated with the failure of ellipticity and the onset of shear banding.

Following Bigoni and Petryk (2002), our interest now is to analyse the situation when elastoplasticity is involved, in which the rate response is governed by the incrementally non-linear operator $\mathcal{C}$ [Eq. (8.102)]. We will concentrate on the two instabilities occurring in an homogeneously deformed infinite medium (so that we are operating in the so-called van Hove conditions; see Section 11.3), namely,

- *Divergence instability*, occurring after shear banding, when the acoustic tensor has a negative eigenvalue
- *Flutter instability*, which may occur independent of shear banding (and is much less understood than divergence; see the discussion in Section 11.5.3)

In the case of elastoplasticity, interpretation of the above-mentioned instabilities in terms of the plane wave solution presented in the preceding section becomes less clear in view of the possibility of activating different constitutive cones at different material points, which would invalidate the wave solution itself. As a consequence, the physical interpretation of the occurrence of negative or complex eigenvalues of the acoustic tensor for an elastoplastic material becomes uncertain. A *reasonable conjecture* is that an arbitrary small disturbance in a uniformly strained material can grow and cross the domain of validity of the tangent stiffness moduli (e.g., by producing elastic unloading 'somewhere'). But what happens *after* this is presently unknown. In fact, two opposite situations may be imagined, in which the perturbing oscillation may further amplify or, contrarily, damp. Only the former possibility would correspond to, say, a 'genuine' instability.[1]

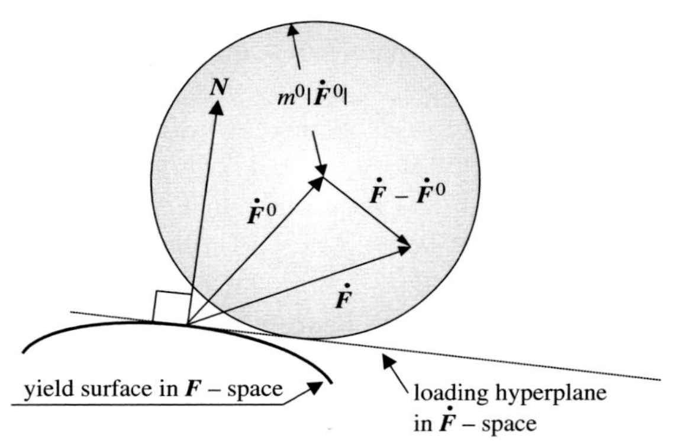

Figure 14.2. $\boldsymbol{F}$–space geometrical interpretation of the inequality in Eq. (14.27) for the case of of non-associative elastoplasticity [Eqs. (8.50)].

Let us consider a homogeneous, arbitrarily anisotropic, plastically deforming material. The configuration of a material element at time $t = 0$ is taken as a fixed reference configuration for the Lagrangean description adopted below. In a given state, the incremental response of the material is characterised by the time-independent constitutive rate equation (8.102) between the material derivatives of the first Piola-Kirchhoff stress and of the deformation gradient.

Quantities that appear in the fundamental motion, whose stability is to be examined, are distinguished by a superscript $\circ$. We shall assume that the fundamental (Lagrangean) velocity gradient $\dot{\mathbf{F}}^\circ$ corresponds to non-zero strain rate and lies inside a certain constitutive cone in $\dot{\mathbf{F}}$-space such that the operator $\boldsymbol{\mathcal{C}}(\cdot)$ restricted to that cone becomes a *linear* operator, represented by a fourth-order state-dependent tensor $\mathbb{C}$ of tangent moduli independent of $\dot{\mathbf{F}}$. It is convenient now to use an assumption weaker than full incremental non-liearity, specified as follows:

$$\boldsymbol{\mathcal{C}}(\dot{\mathbf{F}}) = \mathbb{C}[\dot{\mathbf{F}}] \qquad \text{if } |\dot{\mathbf{F}} - \dot{\mathbf{F}}^\circ| < m^\circ |\dot{\mathbf{F}}^\circ|, \tag{14.27}$$

where $m^\circ \leq 1$ is a positive constant such that the inequality defines a neighbourhood of $\dot{\mathbf{F}}^\circ$ contained in the mentioned constitutive cone. Fulfilment of the inequality in Eq. (14.27) with m° sufficiently small ensures that the strain-rate direction associated with $\dot{\mathbf{F}}$ is sufficiently close to the loading direction defined by $\dot{\mathbf{F}}^\circ$.

The simplest illustration of Eq. (14.27) is provided by the constitutive law of non-associative elastoplasticity with smooth yield and plastic potential surfaces (8.101), where $\boldsymbol{\mathcal{C}}$ at the yield point has two linear constitutive branches, singled out, with reference to Fig. 14.2, by the sign of the scalar product between $\dot{\mathbf{F}}$ and the normal $\mathbf{N}$ to the yield surface in $\mathbf{F}$-space.

For stress points lying on the yield surface $f = 0$, in the case of non-associative elastoplasticity at small strain (8.66) and under the loading assumption $\boldsymbol{D}^\circ \cdot \mathbb{E}[\boldsymbol{Q}] > 0$

[1] In Section 1.13.5 (see also Bigoni and Noselli, 2011) we analysed a two degree of freedom structure in flutter conditions, and we have seen that a non-linear analysis fully substantiates the conclusions about instability that can be drawn from the linearised analysis. Moreover, we have seen that our experiments provide full evidence to the instability. Therefore, the reading of the present section should stimulate consideration of phenomena related to non-linearity, not 'discourage' concluding instability when flutter is predicted from the linearised equations.

in the fundamental motion, Eq. (8.66) admits the linearised form corresponding to Eq. (14.27), namely,

$$\dot{\boldsymbol{T}} = \mathbb{E}[\boldsymbol{D}] - \frac{1}{g}(\boldsymbol{D} \cdot \mathbb{E}[\boldsymbol{Q}])\mathbb{E}[\boldsymbol{P}] \qquad \text{if } |\boldsymbol{D} - \boldsymbol{D}^\circ| < m_1^\circ |\boldsymbol{D}^\circ|, \tag{14.28}$$

where we can take

$$m_1^\circ = \frac{\mathbf{D}^\circ \cdot \mathbb{E}[\mathbf{Q}]}{|\mathbb{E}[\mathbf{Q}]|\,|\mathbf{D}^\circ|} \leq 1. \tag{14.29}$$

The fulfilment of the inequality in Eq. (14.28) with the value (14.29) of m_1° ensures that $\boldsymbol{Q} \cdot \mathbb{E}[\boldsymbol{D}] > 0$, as can be deduced from

$$-\boldsymbol{D} \cdot \mathbb{E}[\boldsymbol{Q}] = (\boldsymbol{D}^\circ - \boldsymbol{D}) \cdot \mathbb{E}[\boldsymbol{Q}] - \boldsymbol{D}^\circ \cdot \mathbb{E}[\boldsymbol{Q}] < m_1^\circ |\mathbb{E}[\boldsymbol{Q}]|\,|\boldsymbol{D}^\circ| - \boldsymbol{D}^\circ \cdot \mathbb{E}[\boldsymbol{Q}] = 0. \tag{14.30}$$

Generally, the variations in $\mathbb{C}$ along a deformation path may be neglected only if the path length in the deformation-gradient space is sufficiently small, less than some positive constant l°. Accordingly, we introduce the assumption that

$$\mathbb{C} = \text{const} \qquad \text{if } \int_0^t |\dot{\mathbf{F}}(\boldsymbol{x}_0, \tau)|\, d\tau \leq l^\circ. \tag{14.31}$$

We shall consider a plastically deforming, infinite, homogeneous medium, uniformly stressed at $t = 0$ in the assumed absence of body forces (as in the problem considered in Section 3.3). The equations of motion are Eqs. $(9.54)_1$, which *in rate form* become

$$\text{Div}\dot{\mathbf{S}} = \rho_0 \dddot{\mathbf{u}}. \tag{14.32}$$

The respective fundamental solution of the equations of motion (14.32), expressed in terms of displacements $\mathbf{u}$ from the position $\boldsymbol{x}_0$ of a material point in the *reference* configuration, is

$$\mathbf{u}^\circ(\boldsymbol{x}, t) = t\dot{\mathbf{F}}^\circ \boldsymbol{x}_0, \qquad \dot{\mathbf{F}}^\circ = \text{const}, \qquad t \geq 0. \tag{14.33}$$

Validity of the fundamental solution is restricted by the requirement $\det \mathbf{F}^\circ > 0$, with $\mathbf{F}^\circ = \mathbf{I} + t\dot{\mathbf{F}}^\circ$, which is ensured by Eq. (14.31) if l° is sufficiently small.

Let us examine now a perturbed velocity field, defined in the reference configuration as

$$\mathbf{v}(\boldsymbol{x}_0, t) = \mathbf{v}^\circ(\boldsymbol{x}_0) + \gamma \boldsymbol{w}(\boldsymbol{x}_0, t), \qquad \mathbf{v}^\circ(\boldsymbol{x}_0) = \dot{\mathbf{F}}^\circ \boldsymbol{x}_0, \tag{14.34}$$

where $\mathbf{v} = \dot{\mathbf{u}}$, γ is a positive constant parameter and $\boldsymbol{w}$ is assumed in the standard form of a harmonic wave vector [Eq. (14.23)]. The constant γ is taken sufficiently small to ensure validity of the constitutive equation (14.27), at least in some initial time interval $[0, t^*]$.

Under the additional assumption (14.31) that $\mathbb{C} = \text{const}$ in both space and time, the rate equation of motion (14.32) is satisfied by the velocity field (14.34) with the perturbation $\boldsymbol{w}$ given by Eq. (14.23), provided that the propagation condition (14.26) is satisfied, where the acoustic tensor (we write in the following $\boldsymbol{n}$ instead than $\boldsymbol{n}_0$) should be read as $\boldsymbol{A}^\circ(\boldsymbol{n})$, that is, the acoustic tensor related to the prescribed fundamental branch (14.27) of $\boldsymbol{C}$, defined by the identity

$$\mathbf{A}^\circ(\mathbf{n})\mathbf{b} \equiv \mathbb{C}[\mathbf{b} \otimes \mathbf{n}]\mathbf{n}, \tag{14.35}$$

to hold for every vector $\mathbf{b}$. Note that $\mathbb{C}[\boldsymbol{a}\otimes\boldsymbol{n}]$ need not be equal to $\boldsymbol{\mathcal{C}}(\boldsymbol{a}\otimes\boldsymbol{n})$ owing to non-linearity of the operator $\boldsymbol{\mathcal{C}}$.

The eigenvalues c^2 and eigenvectors $\mathbf{a}$ of the acoustic tensor $\mathbf{A}^\circ(\mathbf{n})$, which is non-symmetric in general (as in the case of non-associative elastoplasticity), may be real or complex conjugate. Let the velocity $c=\alpha+i\beta$ and the eigenvector $\mathbf{a}=\mathbf{p}+i\mathbf{q}$ satisfy Eq. (14.26), where α, β, $\mathbf{p}$ and $\mathbf{q}$ are real, and $\mathbf{a}$ is normalised in such a way that its modulus is equal to unity, $|\mathbf{a}|=1$. A perturbation (14.23) satisfying the rate equations of motion (14.32) is selected as a sum of two waves travelling in opposite directions

$$\begin{aligned} \boldsymbol{w}(\boldsymbol{x}_0,t) = e^{\beta kt}\{\mathbf{p}\cos[k(\mathbf{n}\cdot\boldsymbol{x}_0-\alpha t)]-\mathbf{q}\sin[k(\mathbf{n}\cdot\boldsymbol{x}_0-\alpha t)]\}\\ +e^{-\beta kt}\{\mathbf{p}\cos[k(\mathbf{n}\cdot\boldsymbol{x}_0+\alpha t)]-\mathbf{q}\sin[k(\mathbf{n}\cdot\boldsymbol{x}_0+\alpha t)]\}. \end{aligned} \tag{14.36}$$

It is immediate to verify that for this choice, the accelerations vanish at $t=0$, that is, $\dot{\boldsymbol{w}}(\boldsymbol{x}_0,0)=\mathbf{0}$, which is consistent with $\mathsf{Div}\mathbf{S}=\mathbf{0}$ at $t=0$. Note that this requirement may be overlooked if only the second equation in Eqs. (14.32) were taken into account. The sign of the imaginary part of the velocity β can be adjusted arbitrarily by replacing simply c by its conjugate. For a real wave speed, that is, for $\beta=0$, Eq. (14.36) describes a standing wave.

Suppose that the preceding perturbation in velocities is instantaneously superimposed at time $t=0$ on the fundamental solution, with the initial condition $\mathbf{u}(\boldsymbol{x}_0,0)=\mathbf{u}^\circ(\boldsymbol{x}_0,0)$ for displacements. Our aim is to examine the case when c^2 is not a non-negative real number, that is, when $\beta\neq 0$. If $\beta=0$, then either $c^2>0$, which is the regular case of propagation of a harmonic wave, or $c=0$, which corresponds to $\boldsymbol{w}$ independent of time in a dynamic solution and to bifurcation within a shear band of a quasi-static solution. Straightforward time integration of Eq. (14.36) yields the displacements in the subsequent motion free of further disturbances in the form

$$\begin{aligned} \mathbf{u}=\mathbf{u}^\circ+\frac{\gamma}{(\beta^2+\alpha^2)k}\Big(e^{\beta kt}\Big(\mathbf{r}\cos[k(\mathbf{n}\cdot\boldsymbol{x}_0-\alpha t)]-\mathbf{s}\sin[k(\mathbf{n}\cdot\boldsymbol{x}_0-\alpha t)]\Big)\\ -e^{-\beta kt}\Big(\mathbf{r}\cos[k(\mathbf{n}\cdot\boldsymbol{x}_0+\alpha t)]-\mathbf{s}\sin[k(\mathbf{n}\cdot\boldsymbol{x}_0+\alpha t)]\Big)\Big),\\ \mathbf{r}=\beta\mathbf{p}-\alpha\mathbf{q},\qquad \mathbf{s}=\alpha\mathbf{p}+\beta\mathbf{q},\qquad \beta\neq 0. \end{aligned} \tag{14.37}$$

It should be noted that the validity of the preceding perturbed solution is limited by the requirement $\det\mathbf{F}>0$ and by the assumptions (14.27) and (14.31), allowing us to perform the integration leading to Eq. (14.37) at fixed values of $\alpha,\beta,\mathbf{p}$ and $\mathbf{q}$. Clearly, the condition $\beta\neq 0$ is necessary and sufficient for the existence of perturbations that amplify exponentially in time for every non-zero value of k. There are two cases when $\beta\neq 0$:

- Some eigenvalue of $\mathbf{A}^\circ(\mathbf{n})$ is real and negative; then $\alpha=0$ (and $\mathbf{q}=\mathbf{0}$), which corresponds to monotonic growth of stationary waves, the divergence instability, or
- Two eigenvalues of $\mathbf{A}^\circ(\mathbf{n})$ are complex conjugate; then $\alpha\neq 0$ (and $\mathbf{a}$ is complex), which corresponds to growth of oscillations in both space and time, the flutter instability.

For a problem which is fully linear from the outset, the preceding is sufficient to conclude that instability will occur. However, this conclusion is now not immediate for the material constitutive law (8.102) with the *state-dependent non-linear* constitutive operator $\boldsymbol{\mathcal{C}}$. In fact, the following two important points remain to be analysed:

- It remains to be shown that infinitesimal disturbances lead to finite deviations from the fundamental path without violating the condition of the constitutive linearisation (14.27).
- The finite deviations should be attainable within a sufficiently small increment in $\mathbf{F}^{\circ}$ to satisfy the condition (14.31) of a fixed tangent moduli tensor $\mathbb{C}$.

These two questions are examined in the following.

14.3.1 Instability of uniform flow

Deformation paths of unrestricted length

In the stability analysis that follows, the velocity perturbation [Eq. (14.36)], superimposed on the fundamental motion will be considered for the imaginary part of the propagation velocity $\beta > 0$, because the sign of $\beta \neq 0$ is inessential. In view of the assumed sinusoidal form of all spatial perturbations in velocities, the amplitude factor $\gamma e^{\beta kt}$ plays the central role. When $\beta > 0$, a norm of the time-dependent spatial field $\boldsymbol{w}(\boldsymbol{x}_0, t)$ grows exponentially in time in the perturbed motion. This implies instability of the fundamental uniform flow (14.33) with respect to the velocity norm *if* the equations of motion (14.32) are treated as fully linear. In the case under examination, the inequality in Eq. (14.27), related to the piece-wise linearity of the constitutive rate equations, imposes a non-linear constraint on the velocity gradient. Our first aim is to show that the conclusion drawn for $\beta \neq 0$ about the instability of the uniform flow (14.33) remains valid if the inequality constraint (14.27) is imposed while the time domain is left unbounded, assuming $l^{\circ} = \infty$ in Eq. (14.31). Mathematically, this is an expected result because the considerations of stability may be a priori limited to a neighbourhood of the fundamental motion. However, the implications in the context of incrementally non-linear plasticity are less obvious, and therefore a detailed proof is provided below.

In the formal proof, it will be convenient to use the following semi-norm of a vector field $\boldsymbol{b}(\boldsymbol{x}_0)$ which is related to the constraint (14.27), namely,

$$\parallel \boldsymbol{b} \parallel = \sup_{\boldsymbol{x}_0} |\nabla \boldsymbol{b}|, \tag{14.38}$$

where ∇ is the gradient operator in the reference configuration. On introducing the equivalence between any two sinusoidal fields that differ merely by a rigid translation in space, Eq. (14.38) provides a norm of $\boldsymbol{w}$. With that identification and for $\mathbf{n}$ and k fixed in time, any other spatial norm of a time-dependent sinusoidal spatial field [Eqs. (14.36) or (14.37)] is equivalent to Eq. (14.38). Recall that two norms $\parallel \boldsymbol{w} \parallel$ and $\parallel \boldsymbol{w} \parallel^{*}$ in the linear space of vector fields $\boldsymbol{w}$ are equivalent if and only if

$$m_1 \parallel \boldsymbol{w} \parallel \leq \parallel \boldsymbol{w} \parallel^{*} \leq m_2 \parallel \boldsymbol{w} \parallel \tag{14.39}$$

for some positive constants m_1, m_2. In those circumstances, we may restrict ourselves to examining stability with respect to the norm (14.38). Moreover, we will use the triangle inequality and its consequence

$$| \| w_1 \| - \| w_2 \| | \leq \| w_1 + w_2 \| \leq \| w_1 \| + \| w_2 \|, \tag{14.40}$$

holding for every field w_1 and w_2. On introducing the notation

$$M = M(\mathbf{a},\mathbf{n}) = \sup_{\theta \in R} |\mathbf{p} \otimes \mathbf{n} \sin\theta + \mathbf{q} \otimes \mathbf{n} \cos\theta| \tag{14.41}$$

and noticing that

$$M = \sup_{x_0} |\mathbf{p} \otimes \mathbf{n} \sin[k(\mathbf{n} \cdot x_0 \pm \alpha t)] + \mathbf{q} \otimes \mathbf{n} \cos[k(\mathbf{n} \cdot x_0 \pm \alpha t)]|, \tag{14.42}$$

the semi-norm (14.38) of the field w defined by Eq. (14.36) is bounded through Eq. (14.40) by

$$2Mk \sinh(\beta kt) \leq \| w \|_t \leq 2Mk \cosh(\beta kt). \tag{14.43}$$

As a consequence, the distance in the sense of Eq. (14.38) between the fundamental and perturbed velocity solutions, [Eq. (14.33) and (14.34)], at time t is bounded by

$$2Mk\gamma \sinh(\beta kt) \leq \| \mathbf{v} - \mathbf{v}^\circ \|_t \leq 2Mk\gamma \cosh(\beta kt), \tag{14.44}$$

with $Mk\gamma > 0$ constant in time. From Eq. (14.36) it follows that the upper or lower bound is attained at instants t such that $\sin^2(k\alpha t) = 0$ or 1, respectively. For divergence instability, that is, $\alpha = 0$, the upper bound gives the exact value of the distance at every instant.

Regarding for simplicity all quantities as dimension-less, the following statement on the instability of the fundamental solution (14.33) with respect to the velocity-gradient distance is obtained.

> There exists a positive number ϵ such that for every positive number δ, however small, there is a velocity perturbation whose norm is arbitrarily small initially ($\| \mathbf{v} - \mathbf{v}^\circ \|_{t=0} < \delta$) and grows in a free dynamic motion to a finite value $\| \mathbf{v} - \mathbf{v}^\circ \|_{t=t^*} \geq \epsilon$ reached at a certain time $t^* > 0$. All this occurs without violating the constraint in Eq. (14.27) while t^* is not bounded from above.

This statement is formalised in the following proposition.

Proposition 14.3.1: If, for some $\mathbf{n}$, not all eigenvalues of $\mathbf{A}^\circ(\mathbf{n})$ are non-negative real numbers and $l^\circ = \infty$ in Eq. (14.31), then

$$\exists \epsilon > 0 \quad \forall \delta > 0 \, \exists \gamma, \, t^* > 0 : \| \mathbf{v} - \mathbf{v}^\circ \|_{t=0} < \delta \ \wedge \ \| \mathbf{v} - \mathbf{v}^\circ \|_{t=t^*} \geq \epsilon, \tag{14.45}$$

and the inequality constraint in Eq. (14.27), is satisfied for $t \leq t^*$.

Proof. Take a positive $\epsilon < m^\circ |\dot{\mathbf{F}}^\circ| / \sqrt{2}$ [see Eq. (14.27)][2] and for an arbitrary positive δ take a positive $\delta^* < \min(\delta, \epsilon)$. If some eigenvalue of $\mathbf{A}^\circ(\mathbf{n})$ is not a real

[2] The factor $1/\sqrt{2}$ is related to the ratio of the bound in Eq. (14.44), which is $\coth(\beta kt) < \sqrt{2}$ provided $\sinh(\beta kt) > 1$.

and non-negative number, then there exists $\beta > 0$ that defines the imaginary part of c. For an arbitrary $k > 0$, let

$$t^* = \frac{1}{k\beta} \sinh^{-1}\left(\frac{\epsilon}{\delta^*}\right), \qquad \gamma = \frac{\delta^*}{2Mk}. \tag{14.46}$$

By inspection and with the help of inequality (14.44), for these values of γ and t^* and for $\boldsymbol{w}$ defined by Eq. (14.36), we have

$$\| \mathbf{v} - \mathbf{v}^\circ \|_{t=0} < \delta \ \wedge \ \| \mathbf{v} - \mathbf{v}^\circ \|_{t=t^*} \geq \epsilon\,. \tag{14.47}$$

Moreover, from the upper bound in the inequality (14.44), it follows that

$$|\dot{\mathbf{F}} - \dot{\mathbf{F}}^\circ|_{\boldsymbol{x}_0,t} \leq \| \mathbf{v} - \mathbf{v}^\circ \|_t \leq 2Mk\gamma \cosh(\beta k t^*) \qquad \text{for } t \leq t^*\,. \tag{14.48}$$

Recalling now that for every x,

$$\sinh^{-1} x = \ln\left(x + \sqrt{x^2+1}\right),$$

and taking into account Eq. (14.46), we obtain

$$2Mk\gamma \cosh(\beta k t^*) = \epsilon + \frac{\delta^*}{\epsilon/\delta^* + \sqrt{(\epsilon/\delta^*)^2 + 1}} < \epsilon\sqrt{2} < m^\circ|\dot{\mathbf{F}}^\circ|. \tag{14.49}$$

Condition (14.49) implies the fulfilment of the inequality in Eq. (14.27) up to time t^* as required. This completes the proof of Proposition 14.3.1. □

In the preceding statement, the norm (14.38) can be replaced by any equivalent norm, which results straightforwardly by applying the inequalities (14.39).

It is clear that the conclusion about instability *at fixed k* requires an infinite time interval because $t^* \to \infty$ as $\delta \to 0$. Under the assumption of $\dot{\mathbf{F}}^\circ$ fixed, an unbounded time interval corresponds to an unbounded length of a deformation path, so the proof of Proposition 14.3.1 required the assumption $l^\circ = \infty$ in Eq. (14.31).

The question arises whether instability can be proven with respect to a distance in displacement gradient rather than in velocity gradient, which would correspond to a finite departure of the perturbed motion from the fundamental path. In a fully linear problem, an affirmative answer would be immediate. However, in view of the assumed constraint (14.27) on the velocity gradient, the initial disturbance in velocities cannot be replaced by another defined only in terms of displacements. This prompts us to retain the initial perturbation in velocities and to ask whether such arbitrarily small perturbation can induce a finite distance in displacement gradient from the fundamental solution.

Application of Eq. (14.40) to Eq. (14.37) provides the estimates of the distance between the fundamental and perturbed solutions in displacements, measured in the sense of Eq. (14.38), as follows:

$$2K\gamma \sinh(\beta k t) \leq \| \mathbf{u} - \mathbf{u}^\circ \|_t \leq 2K\gamma \cosh(\beta k t)\,, \tag{14.50}$$

with $K\gamma > 0$ constant in time, where

$$K = K(c,\mathbf{a},\mathbf{n}) = \frac{1}{\beta^2+\alpha^2} \sup_{\theta\in R} |\mathbf{r}\otimes\mathbf{n}\sin\theta + \mathbf{s}\otimes\mathbf{n}\cos\theta|, \tag{14.51}$$

in which **r** and **s** are defined by Eq. (14.37).

From the representation (14.37) it follows that the lower or upper bound in Eq. (14.50) is reached at instants t such that $\sin^2(k\alpha t) = 0$ or 1, respectively. In the case of divergence instability, that is, $\alpha = 0$, the lower bound gives an exact value of the distance at every instant.

The lower estimate (14.50) applied at the instant t^* defined by Eq. (14.46) in the proof of Proposition 14.3.1 yields

$$\| \mathbf{u} - \mathbf{u}^\circ \|_{t^*} \geq \epsilon_1 = \frac{K}{M}\frac{\epsilon}{k}. \tag{14.52}$$

From Eq. (14.52) and Proposition 14.3.1 we obtain the proof of the following statement.

Proposition 14.3.2: If, for some **n**, not all eigenvalues of $\mathbf{A}^\circ(\mathbf{n})$ are non-negative real numbers and $l^\circ = \infty$ in Eq. (14.31), then

$$\exists\epsilon_1 > 0 \quad \forall\delta > 0\,\exists\gamma,\, t^* > 0 : \| \mathbf{v} - \mathbf{v}^\circ \|_{t=0} < \delta \ \wedge \ \| \mathbf{u} - \mathbf{u}^\circ \|_{t=t^*} \geq \epsilon_1, \tag{14.53}$$

and the inequality constraint in Eq. (14.27) is satisfied for $t \leq t^*$.

This shows the instability (on an infinite time domain) of the uniform flow with respect to two distances: the velocity-gradient distance $\| \mathbf{v} - \mathbf{v}^\circ \|_{t=0}$, used to measure the strength of an initial disturbance, and the displacement-gradient distance $\| \mathbf{u} - \mathbf{u}^\circ \|_t$, equal to zero at $t = 0$, employed to define the current distance between the fundamental and perturbed motions.

As in Proposition 14.3.1, the norm (14.38) can be replaced above by any equivalent norm of the spatially periodic fields involved.

As mentioned earlier, the preceding results are not unexpected. However, a too facile analogy to a fully linear problem contains a pitfall which is avoided here by using the perturbation (14.36) imposed on the fundamental velocity solution.

Short deformation paths

Instability established for an infinite time interval may have no physical meaning if the fundamental solution cannot be extended indefinitely in time. This is indeed the case for the plastic flow with a constant velocity gradient $\dot{\mathbf{F}}^\circ \neq \mathbf{0}$ if a physical limit is imposed on the strain magnitude. Moreover, the tangent moduli tensor $\mathbb{C}$ will vary along a deformation path for any realistic material model. The variations may be neglected only if the path length in the deformation-gradient space is less than some value, denoted by l° in Eq. (14.31), depending on the desired accuracy of the constitutive description. Along the fundamental path (14.33), the current path length is $|\Delta\mathbf{F}^\circ| = t\,|\dot{\mathbf{F}}^\circ|$. In any perturbed motion that satisfies the inequality constraint in Eq. (14.27) up to an instant t^*, the path length is bounded by

$$\int_0^t |\dot{\mathbf{F}}(\boldsymbol{x}_0,\tau)|\,d\tau \leq t\,|\dot{\mathbf{F}}^\circ| + \int_0^t |\dot{\mathbf{F}}(\boldsymbol{x}_0,\tau) - \dot{\mathbf{F}}^\circ|\,d\tau < (1+m^\circ)|\dot{\mathbf{F}}^\circ|t^* \ \text{ for } t \leq t^*. \tag{14.54}$$

The inequality constraint in Eq. (14.31) thus will be satisfied in any time interval $[0,t^*]$ such that

$$t^* \leq t^\circ = \frac{l^\circ}{(1+m^\circ)|\dot{\mathbf{F}}^\circ|}. \tag{14.55}$$

The statement (14.45) can be extended to arbitrarily small t^* by appropriately adjusting the wave number k which was so far arbitrary. For any positive l° and for an associated positive value of t^* satisfying Eq. (14.55), Eq. $(14.46)_1$ now can be used to determine k instead of t^* as before. With this as the only change in the proof, from Proposition 14.3.1, we obtain the following corollary concerning instability of short deformation paths.

Proposition 14.3.3: If, for some $\mathbf{n}$, not all eigenvalues of $\mathbf{A}^\circ(\mathbf{n})$ are non-negative real numbers, then

$$\exists \epsilon > 0 \quad \forall \delta > 0 \forall t^* \leq t^\circ \exists k, \gamma > 0 : \| \mathbf{v} - \mathbf{v}^\circ \|_{t=0} < \delta \ \wedge \ \| \mathbf{v} - \mathbf{v}^\circ \|_{t=t^*} \geq \epsilon, \tag{14.56}$$

and the inequality constraint in (14.27) is satisfied for $t \leq t^*$.

This means that the final inequality in Eq. (14.56) can be reached for l° arbitrarily small for any given fundamental deformation rate $|\dot{\mathbf{F}}^\circ|$. It should be noted that $\delta \to 0$ implies $k \to \infty$ (the short-wavelength limit) if a finite velocity-gradient distance measured by Eq. (14.38) is to be reached at a finite t^*.

The statement (14.56) expresses nothing else than the lack of continuous dependence of the velocity solution on initial data with respect to the norm (14.38). This, in turn, is connected with the well-known concept of ill-posedness of the corresponding linear problem.

It may be interesting to note that it would be a flaw to formulate the statement (14.56) without any restriction on the velocity norm, just by analogy to the linear problem. Contrary to Proposition 14.3.1, the choice of a velocity norm in Proposition 14.3.3, where k can no longer be fixed in advance, is not fully arbitrary.[3] This can be seen from the inequality

$$\gamma \cosh(\beta k t) < \frac{1}{k} \frac{m^\circ}{2M} |\dot{\mathbf{F}}^\circ| \qquad \text{for } t \leq t^*, \tag{14.57}$$

obtained from Eq. (14.49). It implies that the amplitude of $(\mathbf{v} - \mathbf{v}^\circ)$ itself must tend to zero as $k \to \infty$, although the amplitude of the *gradient* of $(\mathbf{v} - \mathbf{v}^\circ)$ at a given t^* can remain finite in the limit. In particular, the difference in kinetic energy within any bounded spatial domain tends to zero in the limit, which raises some doubts about whether Proposition 14.3.3 may be interpreted as a proof of 'genuine' instability of plastic flow.

The next problem to be discussed is whether a property similar to Eq. (14.56) can be proven with respect to a distance in displacement gradient rather than in velocity gradient compared with Proposition 14.3.2. Perhaps unexpectedly, we will show that such extension is *not* possible under the assumptions introduced earlier.

[3] However, the norm (14.38) in Proposition 14.3.3 still can be replaced by any norm $\| \cdot \|^*$ of the *velocity-gradient* field such that Eq. (14.39) is satisfied for all k.

Observe first that under the inequality constraint in Eq. (14.27) satisfied up to time t^*, it is not possible to reach a given finite distance in the *displacement gradient* between the fundamental and perturbed motions within an arbitrarily small increment in $\mathbf{F}^\circ$. This can be seen from the following simple estimate

$$\| \mathbf{u} - \mathbf{u}^\circ \|_t \leq m^\circ |\Delta \mathbf{F}^\circ|_{t^*} \qquad \text{for } t \leq t^*, \tag{14.58}$$

obtained analogously to Eq. (14.54). The possibility of reaching a finite value of $\| \mathbf{u} - \mathbf{u}^\circ \|$ at arbitrarily small $|\Delta \mathbf{F}^\circ|$ would be related to the instability of equilibrium, which is different from the instability of plastic flow,[4] even when the plastic flow is regarded as quasi-static (see Section 10.2.2).

Moreover, it turns out that the proof of Proposition 14.3.2 [carried out for $l^\circ = \infty$ in Eq. (14.31)] cannot even be extended to any *bounded* increment in $\mathbf{F}^\circ$, that is, to any *finite* time interval $[0, t^*]$ at fixed $\dot{\mathbf{F}}^\circ$; this specifies limitations of the linear stability analysis in application to processes of plastic deformation. The same is shown for a finite l° if we take any positive t^* satisfying Eq. (14.55) to fulfil the inequality in Eq. (14.31). Namely, the following statement holds true:

Proposition 14.3.4: Irrespective of the type of eigenvalues of the acoustic tensor, if the inequality constraint in Eq. (14.27) is satisfied for $t \leq t^*$, then

$$\forall t^* \leq t^\circ \, \forall \epsilon_1 > 0 \quad \exists \delta > 0 \, \forall k, \gamma > 0 : \| \mathbf{v} - \mathbf{v}^\circ \|_{t=0} < \delta \;\Rightarrow\; \| \mathbf{u} - \mathbf{u}^\circ \|_{t \leq t^*} < \epsilon_1 . \tag{14.59}$$

Proof. The assumption $t^* < t^\circ$ in Eq. (14.59) implies, through Eq. (14.55), that Eq. (14.31) holds true, which jointly with the assumption (14.27) justifies the use of perturbation (14.36). Now, if $\beta = 0$, then the implication in eqn. (14.59) is trivial because in that case Eq. (14.36) describes a standing wave. It suffices thus to consider the case $\beta > 0$, which corresponds to a perturbed motion (14.37).

Suppose that Eq. (14.59) does not hold; thus

$$\exists t^* \leq t^\circ \, \exists \epsilon_1 > 0 \quad \forall \delta > 0 \, \exists k, \gamma > 0 \, \exists t \leq t^* : 2Mk\gamma < \delta \;\wedge\; \| \mathbf{u} - \mathbf{u}^\circ \|_t \geq \epsilon_1 , \tag{14.60}$$

whereas the inequality constraint in Eq. (14.27) is satisfied for $t \leq t^*$.

If $\delta \to 0$, then $\| \mathbf{u} - \mathbf{u}^\circ \|_{t \leq t^*}$ tends also to zero by the upper bound in Eq. (14.50), in contradiction to Eq. (14.60), unless $k \to \infty$ or $k \to 0$.

If $k \to \infty$ as $\delta \to 0$, then from the lower bound in Eq. (14.44) and the inequality in Eq. (14.27), we have

$$\gamma \sinh(\beta k t) \leq \frac{\| \mathbf{v} - \mathbf{v}^\circ \|_t}{2Mk} < \frac{m^\circ |\dot{\mathbf{F}}^\circ|}{2Mk}, \qquad \text{for } t \leq t^*. \tag{14.61}$$

Hence, at each positive $t \leq t^*$, the lower bound in Eq. (14.50) tends to zero as $k \to \infty$. Since the ratio of the bounds in Eq. (14.50) tends to 1 in the limit, the upper bound in Eq. (14.50) tends to 0 at each positive $t \leq t^*$ as $\delta \to 0$, which contradicts Eq. (14.60).

Finally, if $k \to 0$ as $\delta \to 0$, then $\| \mathbf{u} - \mathbf{u}^\circ \|_{t \leq t^*}$ tends to the lower bound in Eq. (14.50) by the continuity argument. But the lower bound tends to zero if $k \to 0$

[4] The situation is different in the special case of an incrementally *linear* material, where the inequality constraint in Eq. (14.27) is absent. Then the perturbation (14.23) can be superimposed on the degenerate fundamental motion of zero velocities, that is, on an equilibrium state. In this case, the preceding established instability concerns equilibrium as well.

as $\delta \to 0$, which leads again to the contradiction with Eq. (14.60). This completes the proof of Proposition 14.3.4. □

As a conclusion, a finite value of $\| \mathbf{u} - \mathbf{u}^\circ \|$ cannot be reached within a finite increment in $\mathbf{F}^\circ$ in the class of perturbed motions (14.37) for arbitrarily small initial perturbations in velocities if the inequality constraint in Eq. (14.27) is satisfied. The violation of the constraint in Eq. (14.27), implied by the instability of motion in the sense of Proposition 14.3.3, may be followed later by activation of constitutive branches other than the fundamental constitutive cone. No attempt is made here to analyse the subsequent non-linear behaviour of the material. In other words, Proposition 14.3.4 does not imply that the fundamental solution is stable for a fully non-linear constitutive law; rather, it shows that the boundary of the constitutive domain defined by inequalities in Eqs. (14.27) and (14.31) must be crossed before a finite deviation from the fundamental strain path is reached following an arbitrarily small disturbance.

14.3.2 A discussion on waves and instability in elastoplasticity

For a linear constitutive law, 'divergence' and 'flutter' instabilities (formally associated with the existence of negative or complex eigenvalues of the acoustic tensor) are related to monotonic or oscillatory growth of periodic initial disturbances in an infinite homogeneous medium.

The analysis performed through a superimposition of spatially sinusoidal perturbations on the fundamental velocity field (of constant gradient), with an amplitude sufficiently small to activate only the fundamental constitutive branch, allows us to reach the following conclusions:

- Assuming that the constitutive operator is independent of the current state but incrementally piece-wise linear (in a certain neighbourhood of the current velocity gradient), the instability of the fundamental motion can be proven on an infinite time interval (corresponding to deformation paths of unbounded length) (Propositions 14.3.1 and 14.3.2).
- The assumption that the constitutive operator is independent of the current state can be relaxed and applied only to sufficiently short paths in the deformation gradient space. In this context, a vanishingly small initial perturbation of the fundamental velocity field can be shown to grow to a finite perturbation of the velocity gradient (*not* of the velocity itself) in an arbitrarily short time interval if the superimposed wavelength is sufficiently short (Proposition 14.3.3). This rapid departure from the fundamental straining direction enables attaining a boundary of the domain of application of the fundamental tangent moduli within a small deformation increment.
- The linear stability analysis on a finite time interval is inconclusive if the strain distance is adopted for deviations from the fundamental solution. Instead of instability which might be expected by analogy to the fully linear problem, we have proven an opposite property, irrespective of the types of eigenvalues of the acoustic tensor (Proposition 14.3.4).

As a conclusion, the treatment developed in this section highlights the difficulties in analysing stability, as connected to the presence of both incremental non-linearity and dependence on the current state, typical of elastoplastic constitutive equations. These difficulties should provide a warning on the drawing of facile conclusions without a consideration of the fully non-linear behaviour, but should not discourage us from the use of linearised analyses, which remain a fundamental tool in the analysis of instabilities in elastoplasticity.

14.4 Acceleration waves

Propagation of acceleration waves is a fascinating topic which was initiated by Hadamard (1903) and is strictly connected to the results given in Section 9.1. We recall in fact that an acceleration wave is, by definition, 'a moving surface carrying second-order discontinuities and propagating in a deformed material'. Our interest here therefore is to specialise results provided in Section 9.1 to the case of elasticity and non-associative elastoplasticity.

Employing non-linear rate constitutive Eq. (8.102), we may relate a jump in the material time derivative of the first Piola-Kirchhoff stress to a jump in the material time derivative of the deformation gradient as

$$[\dot{\boldsymbol{S}}]=[\mathcal{C}(\dot{\boldsymbol{F}})], \tag{14.62}$$

which, keeping into account Eq. $(9.39)_3$, becomes

$$[\dot{\boldsymbol{S}}]=\mathcal{C}(\dot{\boldsymbol{F}}+\boldsymbol{g}_0\otimes\boldsymbol{n}_0)-\mathcal{C}(\dot{\boldsymbol{F}}). \tag{14.63}$$

Thus, employing Eq. $(9.39)_1$, we finally arrive at the *non-linear propagation condition*

$$\mathcal{C}(\dot{\boldsymbol{F}}+\boldsymbol{g}_0\otimes\boldsymbol{n}_0)\boldsymbol{n}_0-\mathcal{C}(\dot{\boldsymbol{F}})\boldsymbol{n}_0-\rho_0\, w_{n0}^{02}\boldsymbol{g}_0=\boldsymbol{0}. \tag{14.64}$$

Note that for elastoplastic material (8.50), the operator $\mathcal{C}$ becomes equal to $\mathbb{G}$ in the elastic zone, namely, where $f_{\boldsymbol{S}}(\boldsymbol{S},\mathcal{H})<0$.

14.4.1 Non-linear elastic material deformed incrementally

For a non-linear elastic material, the incremental elastic tensor $\mathbb{G}$ [Eq. (8.100)], replaces the non-linear operator $\mathcal{C}$ in Eq. (14.64), so *the propagation condition of acceleration waves* becomes

$$\left(\boldsymbol{A}_E(\boldsymbol{n}_0)-\rho_0\, w_{n0}^{02}\boldsymbol{I}\right)\boldsymbol{g}_0=\boldsymbol{0}, \tag{14.65}$$

which coincides with the propagation condition of plane incremental waves [Eq. (14.26)] except that ρ_0 is included there in $\boldsymbol{A}_E(n_0)$. Therefore, the loss of ellipticity (or shear band formation or strain localisation condition) coincides with the occurrence of vanishing speed of an acceleration wave.

14.4.2 Elastoplastic materials

For elastoplastic materials, fundamental works are those by Hill (1962), Mandel (1962a; 1962b) and Raniecki (1976). Owing to the incremental non-linearity of the elastoplastic constitutive tensor, the four following possibilities arise (we refer to Fig. 14.3 for the definition of zones + and −)

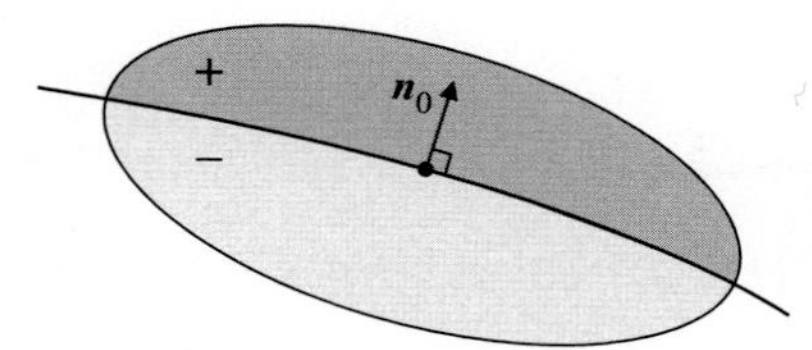

Figure 14.3. Notation for propagation of an acceleration wave.

- The state of the material is elastoplastic ahead (+) and behind (−) the wave, a condition defining a 'plastic wave'.
- The state of the material is elastic ahead (+) and behind (−) the wave, a condition defining an 'elastic wave'.
- The state of the material is elastoplastic ahead (+) and elastic behind (−) the wave, a condition defining an 'unloading wave'.
- The state of the material is elastic ahead (+) and elastoplastic behind (−) the wave, a condition defining a 'loading wave'.

The conditions of propagation, referred to the preceding four possibilities, are

- *Plastic wave*:
$$\left(\boldsymbol{A}(\boldsymbol{n}_0) - \rho_0 w_{n0}^{02} \boldsymbol{I}\right) \boldsymbol{g}_0 = \boldsymbol{0} \tag{14.66}$$
- *Elastic wave*:
$$\left(\boldsymbol{A}_E(\boldsymbol{n}_0) - \rho_0 w_{n0}^{02} \boldsymbol{I}\right) \boldsymbol{g}_0 = \boldsymbol{0} \tag{14.67}$$
- *Unloading wave*:
$$\boldsymbol{A}(\boldsymbol{n}_0)\boldsymbol{g}_0 + (\mathbb{C} - \mathbb{G})[\dot{\boldsymbol{F}}]\boldsymbol{n}_0 - \rho_0 w_{n0}^{02} \boldsymbol{g}_0 = \boldsymbol{0} \tag{14.68}$$
- *Loading wave*:
$$\boldsymbol{A}_E(\boldsymbol{n}_0)\boldsymbol{g}_0 - (\mathbb{C} - \mathbb{G})[\dot{\boldsymbol{F}}]\boldsymbol{n}_0 - \rho_0 w_{n0}^{02} \boldsymbol{g}_0 = \boldsymbol{0} \tag{14.69}$$

It is immediate to conclude that:

when the propagation speed is null, $w_{n0}^{0} = 0$, the propagation conditions (14.66) coincide with the four conditions for strain localisation presented in Section 11.4,

so *strain localisation can be viewed as the condition of vanishing propagation speed for an acceleration wave.*

Note that for simplicity of notation we will omit subscripts 0 in the following, so we will write $\boldsymbol{n}$ and $\boldsymbol{g}$ instead $\boldsymbol{n}_0$ and $\boldsymbol{g}_0$.

Plastic waves, first Mandel inequality The propagation condition (14.66) is equivalent to

$$\det\left(\boldsymbol{A}(\boldsymbol{n}) - \rho_0 w_{n0}^{02} \boldsymbol{I}\right) = 0, \tag{14.70}$$

so propagation is possible for every direction only when the elastoplastic acoustic tensor has positive eigenvalues.

Introducing now the tensor

$$\boldsymbol{G}(x) = \boldsymbol{A}_E(\boldsymbol{n}) - x\boldsymbol{I}, \qquad x = \rho_0 w_{n_0}^{02}, \tag{14.71}$$

and proceeding as for strain localisation [see Eq. (11.74)], condition (14.70) can be rewritten as

$$\det \boldsymbol{G}(x) \det\left(\boldsymbol{I} - \frac{1}{g}\boldsymbol{G}^{-1}(x)\boldsymbol{M}\boldsymbol{n} \otimes \boldsymbol{N}\boldsymbol{n}\right) = 0, \tag{14.72}$$

which, using property (2.38), can be cast in the form

$$\phi(x) = 0, \tag{14.73}$$

where

$$\phi(x) = \det \boldsymbol{G}(x)\left(1 - \frac{1}{g}\boldsymbol{N}\boldsymbol{n} \cdot \boldsymbol{G}^{-1}(x)\boldsymbol{M}\boldsymbol{n}\right). \tag{14.74}$$

Assuming hyperelasticity for the elastic part of the deformation, $\boldsymbol{A}_E(\boldsymbol{n})$ is symmetric, so function (14.74) can be written in the principal reference system of $\boldsymbol{G}$, which is also the principal reference system of $\boldsymbol{A}_E(\boldsymbol{n})$,

$$\begin{aligned}\phi(x) \; &= (A_{EI} - x)(A_{EII} - x)(A_{EIII} - x) \\ &\quad - \frac{1}{g}\left[(A_{EII} - x)(A_{EIII} - x)(\boldsymbol{N}\boldsymbol{n})_1(\boldsymbol{M}\boldsymbol{n})_1\right. \\ &\quad + (A_{EI} - x)(A_{EIII} - x)(\boldsymbol{N}\boldsymbol{n})_2(\boldsymbol{M}\boldsymbol{n})_2 \\ &\quad \left. + (A_{EI} - x)(A_{EII} - x)(\boldsymbol{N}\boldsymbol{n})_3(\boldsymbol{M}\boldsymbol{n})_3\right],\end{aligned} \tag{14.75}$$

where $(\boldsymbol{M}\boldsymbol{n})_i$ and $(\boldsymbol{N}\boldsymbol{n})_i$ are the components of $\boldsymbol{M}\boldsymbol{n}$ and $\boldsymbol{N}\boldsymbol{n}$ in the principal reference system of $\boldsymbol{A}_E(\boldsymbol{n})$, and the order of eigenvalues of $\boldsymbol{A}_E(\boldsymbol{n})$ has been chosen in such a way that

$$A_{EI} \geq A_{EII} \geq A_{EIII}. \tag{14.76}$$

We note that

$$g > 0, \qquad \phi(-\infty) > 0, \qquad \phi(\infty) < 0, \tag{14.77}$$

and that if strain localisation is excluded,

$$\phi(0) > 0. \tag{14.78}$$

Therefore, continuity of function $\phi(x)$ implies that *problem (14.72) has always at least one solution.*

Without loss of generality, the speeds of plastic waves can be ordered as

$$A_I \geq A_{II} \geq A_{III}. \tag{14.79}$$

Thus we can observe from Eqs. (14.77) that the following inequalities hold

$$\begin{aligned}(\boldsymbol{M}\boldsymbol{n})_3(\boldsymbol{N}\boldsymbol{n})_3 \geq 0 \;&\Longleftrightarrow\; \phi(A_{EIII}) \leq 0 \;\Longrightarrow\; A_{III} \leq A_{EIII}, \\ (\boldsymbol{M}\boldsymbol{n})_1(\boldsymbol{N}\boldsymbol{n})_1 \leq 0 \;&\Longleftrightarrow\; \phi(A_{EI}) \geq 0 \;\;\Longrightarrow\; A_I \geq A_{EI}.\end{aligned} \tag{14.80}$$

Moreover, considering again Eqs. (14.77), *if*

$$(\boldsymbol{M}\boldsymbol{n})_1(\boldsymbol{N}\boldsymbol{n})_1 \geq 0, \qquad (\boldsymbol{M}\boldsymbol{n})_2(\boldsymbol{N}\boldsymbol{n})_2 \geq 0, \qquad (\boldsymbol{M}\boldsymbol{n})_3(\boldsymbol{N}\boldsymbol{n})_3 \geq 0, \tag{14.81}$$

then

$$\phi(A_{EI}) \leq 0, \qquad \phi(A_{EII}) \geq 0, \qquad \phi(A_{EIII}) \leq 0. \tag{14.82}$$

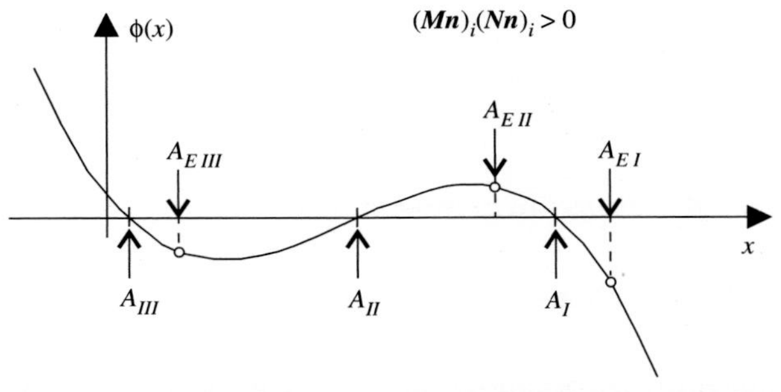

Figure 14.4. The behaviour of function $\phi(x)$ in the case when $(\boldsymbol{Mn})_i(\boldsymbol{Nn})_i > 0$, $i = 1,2,3$ determines the first Mandel inequality.

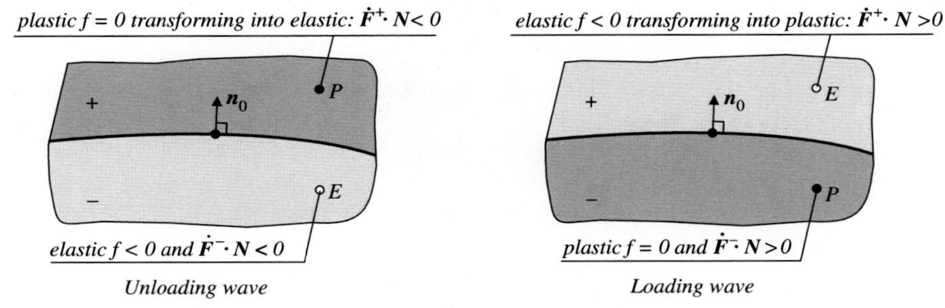

Figure 14.5. Unloading (*left*) and loading (*right*) acceleration waves in an elastoplastic material.

Thus, for a given propagation direction $\boldsymbol{n}$, *the plastic wave speeds are bounded by the respective elastic wave speeds according to*

$$A_{EI} \geq A_I \geq A_{EII} \geq A_{II} \geq A_{EIII} \geq A_{III}, \tag{14.83}$$

the so-called first Mandel inequality, which has been found by Mandel (1962a, 1962b, 1969) for *associative plasticity* where $(\boldsymbol{Mn})_i(\boldsymbol{Nn})_i = (\boldsymbol{Nn})_i^2 \geq 0$; thus Eq. (14.81) holds a priori (see Fig. 14.4).

For every propagation direction $\boldsymbol{n}$ for which $\boldsymbol{Nn} \neq \boldsymbol{0}$, it is always possible to consider $\boldsymbol{g}$ in the plane orthogonal to $\boldsymbol{Nn}$ so that $\boldsymbol{g} \cdot \boldsymbol{Nn} = 0$, which corresponds to a *neutral plastic wave*. Therefore,

a neutral plastic wave travels with the velocity of an elastic wave.

In the particular case of *isotropic linear elasticity*, $\boldsymbol{A}_E(\boldsymbol{n})$ is given by Eq. (2.176), so it is always possible to chose $\boldsymbol{g}$ parallel to $\boldsymbol{n} \times \boldsymbol{Nn}$, and therefore *a neutral plastic wave exists that propagates with the speed of a transversal elastic wave.*

Unloading waves, second Mandel inequality Considering now an unloading wave; the situation is that sketched in Fig. 14.5 (*left*).

In the elastic (plastic) zone, the material satisfies $f < 0$ ($f = 0$), and the propagation condition is given by Eq. (14.68). Since the wave yields unloading, the following conditions hold

$$\boldsymbol{N} \cdot \dot{\boldsymbol{F}} < 0, \qquad \boldsymbol{N} \cdot \dot{\boldsymbol{F}} + \boldsymbol{g} \cdot \boldsymbol{Nn} > 0, \tag{14.84}$$

so we can define the scalar r such that

$$r(\boldsymbol{g}) = 1 + \frac{\boldsymbol{N}\cdot\dot{\boldsymbol{F}}}{\boldsymbol{g}\cdot\boldsymbol{N}\boldsymbol{n}}, \qquad r \in [0,1]. \tag{14.85}$$

The propagation condition (14.68) therefore can be rewritten as

$$\left[\boldsymbol{A}_E(\boldsymbol{n}) - \frac{r(\boldsymbol{g})}{g}\boldsymbol{M}\boldsymbol{n}\otimes\boldsymbol{N}\boldsymbol{n} - \rho_0 w_{n_0}^{02}\boldsymbol{I}\right]\boldsymbol{g} = \boldsymbol{0}, \tag{14.86}$$

which is not a standard eigenvalue problem because r depends on $\boldsymbol{g}$. However, introducing tensor $\boldsymbol{G}(x)$ given by Eq. (14.71), a necessary condition for Eq. (14.86) to be satisfied, is obtained

$$\phi(x,r) = \det\boldsymbol{G}(x)\left(1 - \frac{r(\boldsymbol{g})}{g}\boldsymbol{N}\boldsymbol{n}\cdot\boldsymbol{G}^{-1}\boldsymbol{M}\boldsymbol{n}\right) = 0, \tag{14.87}$$

an equation that can be transformed similarly to Eq. (14.75), namely,

$$\begin{aligned}\phi(x,r) = {} & (A_{EI}-x)(A_{EII}-x)(A_{EIII}-x) \\ & - \frac{r(\boldsymbol{g})}{g}\left[(A_{EII}-x)(A_{EIII}-x)(\boldsymbol{N}\boldsymbol{n})_1(\boldsymbol{M}\boldsymbol{n})_1\right. \\ & + (A_{EI}-x)(A_{EIII}-x)(\boldsymbol{N}\boldsymbol{n})_2(\boldsymbol{M}\boldsymbol{n})_2 \\ & \left.+ (A_{EI}-x)(A_{EII}-x)(\boldsymbol{N}\boldsymbol{n})_3(\boldsymbol{M}\boldsymbol{n})_3\right],\end{aligned} \tag{14.88}$$

which reduces to Eq. (14.75) when $r = 1$.

Let us assume now that for a given $\boldsymbol{g} = \boldsymbol{g}^{\text{trial}}$, $r(\boldsymbol{g}^{\text{trial}})$ is determined and Eq. (14.87) is verified for a certain x such that the eigenvector, say, $\boldsymbol{g}$, solving Eq. (14.86) in general will be different from $\boldsymbol{g}^{trial}$. Fortunately, the modulus and the sign of $\boldsymbol{g}$ are arbitrary, so we can always define these to satisfy $\boldsymbol{g}\cdot\boldsymbol{N}\boldsymbol{n} = \boldsymbol{g}^{\text{trial}}\cdot\boldsymbol{N}\boldsymbol{n}$, and therefore, the propagation condition is finally satisfied with $r(\boldsymbol{g}) = r(\boldsymbol{g}^{\text{trial}})$.

We are now in a position to develop considerations similar to those leading to the first Mandel inequality, and because $r(\boldsymbol{g}) \in [0,1]$, ordering the speeds of unloading waves as

$$A_I^{\text{unload}} \geq A_{II}^{\text{unload}} \geq A_{III}^{\text{unload}} \tag{14.89}$$

yields

$$\begin{aligned}(\boldsymbol{M}\boldsymbol{n})_3(\boldsymbol{N}\boldsymbol{n})_3 \geq 0 \quad &\Longrightarrow \quad A_{III}^{\text{unload}} \leq A_{III}, \\ (\boldsymbol{M}\boldsymbol{n})_1(\boldsymbol{N}\boldsymbol{n})_1 \leq 0 \quad &\Longrightarrow \quad A_I^{\text{unload}} \geq A_{EI}.\end{aligned} \tag{14.90}$$

Finally, *if* condition (14.81) *holds* for a given propagation direction $\boldsymbol{n}$, *the unloading wave speeds are bounded by the respective plastic wave and elastic wave speeds according to*

$$A_{Ei} \geq A_i^{\text{unload}} \geq A_i, \quad i = I,\ II,\ III, \tag{14.91}$$

the so-called second Mandel inequality, which is *always verified for associative plasticity* where condition (14.81) holds a priori.

Loading waves, third Mandel inequality We now can proceed with loading waves (Fig. 14.5, *right*). The propagation condition now is given by Eq. (14.69). The fact that the wave yields a loading of the material implies that the following conditions hold true

$$\boldsymbol{N}\cdot\dot{\boldsymbol{F}} > 0, \qquad \boldsymbol{N}\cdot\dot{\boldsymbol{F}}+\boldsymbol{g}\cdot\boldsymbol{N}\boldsymbol{n} > 0. \tag{14.92}$$

Conditions (14.92) have been stated without any comment by Mandel (1962a, 1962b, 1966a, 1966b, 1969) and Raniecki (1976). However, the conditions (14.92) may not be intuitive at first glance because they both correspond to plastic loading. The key to understand these conditions is to remember that Eq. $(14.92)_2$ holds for states in the elastic range, that is, satisfying $f < 0$, and therefore, Eq. $(14.92)_2$ represents a loading condition for a material still in the elastic range, an observation implicitly reported only by Mandel (1969).

From the conditions (14.92), we can define a scalar r (again a function of $\boldsymbol{g}$) such that

$$r(\boldsymbol{g}) = \frac{\boldsymbol{N}\cdot\dot{\boldsymbol{F}}}{\boldsymbol{N}\cdot\dot{\boldsymbol{F}}+\boldsymbol{g}\cdot\boldsymbol{N}\boldsymbol{n}}, \qquad r(\boldsymbol{g}) \geq 0. \tag{14.93}$$

Thus $\boldsymbol{N}\cdot\dot{\boldsymbol{F}}$ can be obtained as a function of $\boldsymbol{g}$ and the propagation condition (14.69) can be rewritten as

$$\left[\boldsymbol{A}_E(\boldsymbol{n}) - \frac{1}{g[1-\frac{1}{r(\boldsymbol{g})}]}\boldsymbol{M}\boldsymbol{n}\otimes\boldsymbol{N}\boldsymbol{n} - \rho_0 w_{n0}^{02}\boldsymbol{I}\right]\boldsymbol{g} = \boldsymbol{0}, \tag{14.94}$$

which, as in the case of Eq. (14.86), is not a standard eigenvalue problem because r depends on $\boldsymbol{g}$. A new problem now is that the quantity $g[1-1/r(\boldsymbol{g})]$, playing the role of a 'modified plastic modulus', can be greater than $1/g$ (and also negative), so *the number of possible loading waves may be fewer than three even when three plastic waves exist for the same material.*

We now arrive at the equivalent of Eq. (14.88), which is

$$\begin{aligned} f(x,r) = {} & (A_{EI}-x)(A_{EII}-x)(A_{EIII}-x) \\ & - \frac{1}{g[1-1/r(\boldsymbol{g})]}[(A_{EII}-x)(A_{EIII}-x)(\boldsymbol{N}\boldsymbol{n})_1(\boldsymbol{M}\boldsymbol{n})_1 \\ & + (A_{EI}-x)(A_{EIII}-x)(\boldsymbol{N}\boldsymbol{n})_2(\boldsymbol{M}\boldsymbol{n})_2 \\ & + (A_{EI}-x)(A_{EII}-x)(\boldsymbol{N}\boldsymbol{n})_3(\boldsymbol{M}\boldsymbol{n})_3], \end{aligned} \tag{14.95}$$

again reducing to Eq. (14.75) when $r = 1$.

Ordering the elastic wave speeds as in Eq. (14.79) and *assuming for simplicity conditions* (14.81) (a priori valid for the associative flow rule), we arrive at the so-called third Mandel inequality:

$$\begin{aligned} & \text{For } \ 0 < r(\boldsymbol{g}) \leq 1, \\ & A_I^{\text{load}} \geq A_{EI} > A_I \geq A_{II}^{\text{load}} \geq A_{EII} \geq A_{II} \geq A_{III}^{\text{load}} \geq A_{EIII}, \\ & \text{For } \ r(\boldsymbol{g}) \geq 1, \\ & A_I \geq A_I^{\text{load}} \geq A_{EII} \geq A_{II} \geq A_{II}^{\text{load}} \geq A_{EIII} \geq A_{III} \geq A_{III}^{\text{load}}. \end{aligned} \tag{14.96}$$

As in the case of the unloading waves, if for a given $\boldsymbol{g} = \boldsymbol{g}^{\text{trial}}$, $r(\boldsymbol{g}^{\text{trial}})$ is determined and Eq. (14.95) vanishes for a certain x, the eigenvector, say, $\boldsymbol{g}$, solving Eq. (14.94) in general will be different from $\boldsymbol{g}^{\text{trial}}$. Fortunately, the modulus and the sign of $\boldsymbol{g}$ are arbitrary, so we can always define these to satisfy

$\boldsymbol{g} \cdot \boldsymbol{N}\boldsymbol{n} = \boldsymbol{g}^{\text{trial}} \cdot \boldsymbol{N}\boldsymbol{n}$, and therefore, the propagation condition is finally satisfied with $r(\boldsymbol{g}) = r(\boldsymbol{g}^{\text{trial}})$.

Acceleration waves are waves of second order. The case of nth-order waves in elastoplastic solids leads to conclusions similar to the conclusions for second-order waves. A treatment of these, restricted to the associative flow rule but including thermal effects, has been given by Raniecki (1975).

15 Post-critical behaviour and multiple shear band formation

Elastic one-dimensional models with spinodal stress/strain behaviour (involving softening and subsequent re-hardening) are employed to explain features of continued deformation after strain localisation. Global softening of the response is strongly influenced by localisation of deformation, and multiple localisation with stress oscillation is observed, as induced by the re-hardening subsequent to softening. Since one-dimensional models are not sufficient to describe the behaviour of real materials, which always deform at least in two dimensions, a simple technique (small strain, piece-wise uniform fields and fixed shear band width are assumed) is illustrated to compute the post-shear banding behaviour of a two-dimensional elastoplastic sample. Although the methodology probably is over-simplified, it is shown that it may capture important phenomena, in particular, softening and size effect, band saturation and post-saturation, leading to multiple band formation with possible stress fluctuations or delayed softening. All these phenomena are important in the understanding of the mechanical behaviour of many materials, in particular, granular materials.

Analysis of the behaviour of a material element *after* shear banding has occurred is crucial for an understanding of induced softening and size-effect phenomena, which may be relevant for different purposes, for instance, predicting possible catastrophic failure of structural elements (see the Introduction). Moreover, post-shear banding may involve multiple shear band formation, a phenomenon observed in different materials: foams (Moore et al., 2006), ductile metals (Hall, 1970), honeycombs (Papka and Kyriakides, 1999), sand (Finno et al., 1997), shape memory alloys (Shaw and Kyriakides, 1997) and the stacks of drinking straws shown in the Introduction (Section 1.6).

The development of a simple model, based on an incremental technique, for capturing post-critical shear band behaviour is postponed to a simple explanation provided by recurring to discrete and continuous one-dimensional models of softening elements through generalisation of the simple systems illustrated in Figs. 1.21 through 1.24 of the Introduction (Section 1.6). While these one-dimensional models can properly describe the behaviour of simple mechanical systems such as the stack of cans and the corrugated tube (Figs. 1.26 and 1.27), or can provide simple models for the understanding of certain basic phenomena (Braun and Kivshar, 1998), they cannot properly model the behaviour of a material element (which is at least two-dimensional). Therefore, a simplified model to attack the problem of post-localisation in two-dimensional samples eventually will be proposed in this chapter (following Gajo et al., 2004, see also Section 1.7).

15.1 One-dimensional elastic models with non-convex energy

Let us consider a 1D chain containing six (instead of three, as analysed in the Section 1.6) identical tri-linear elastic springs of the same type shown in the inset of Fig. 1.24 and described by Eq. (1.18) with $\beta = 2$, $\gamma = 1$ and $\theta = 15/2$ so that they have a non-convex strain energy and their force/displacement curve evidences a spinodal region. These six springs are connected in series and subjected to increasing displacements at the ends.[1] We present a simple treatment of this system, with the purpose of giving evidence to the limits of an analysis based on incremental bifurcation and of highlighting phenomena such as localisation and band accumulation owing to softening and subsequent re-hardening, whereas an exhaustive analysis of discrete systems with non-linear elements has been provided by Puglisi and Truskinowsky (2000; 2002) and Puglisi (2006).

Assuming absence of disturbances, jumps of the system are excluded initially, and the force can reach the peak of the force/displacement curve, following a stable path at the beginning, which later becomes meta-stable, without encountering any incremental bifurcation. The peak of the curve, corresponding to the end of phase I^6, is a bifurcation point, at which six different continuous bifurcation paths become possible (Fig. 15.1), of which the three re-entering paths cannot be followed if the end displacement is prescribed. Other configurations can be reached, allowing to the system the possibility of appropriate jumps in the force and displacement of internal nodes. A rule therefore is needed to select the post-bifurcation behaviour among the three continuous paths and other configurations involving finite jumps in the axial force and displacements of the internal nodes. We therefore can note that although the first bifurcation point can be detected through an incremental bifurcation analysis, the subsequent behaviour can be described correctly only allowing for the possibility of discontinuities in the behaviour.

Local stability of the six-spring system can be judged from the positive definiteness of the Hessian of the potential energy, which, for prescribed end displacements of a chain AB-BC-CD-DE-EF-FG of tri-linear springs, becomes the 5×5 tri-diagonal matrix [similar to that derived for the three-springs case, Eq. (1.20)]:

$$\begin{bmatrix} \frac{\partial^2 W_{BA}}{\partial \Delta_{BA}^2} + \frac{\partial^2 W_{CB}}{\partial \Delta_{CB}^2} & -\frac{\partial^2 W_{CB}}{\partial \Delta_{CB}^2} & 0 & 0 & 0 \\ -\frac{\partial^2 W_{CB}}{\partial \Delta_{CB}^2} & \frac{\partial^2 W_{CB}}{\partial \Delta_{CB}^2} + \frac{\partial^2 W_{DC}}{\partial \Delta_{DC}^2} & -\frac{\partial^2 W_{DC}}{\partial \Delta_{DC}^2} & 0 & 0 \\ 0 & -\frac{\partial^2 W_{DC}}{\partial \Delta_{DC}^2} & \frac{\partial^2 W_{DC}}{\partial \Delta_{DC}^2} + \frac{\partial^2 W_{ED}}{\partial \Delta_{ED}^2} & -\frac{\partial^2 W_{ED}}{\partial \Delta_{ED}^2} & 0 \\ 0 & 0 & -\frac{\partial^2 W_{ED}}{\partial \Delta_{ED}^2} & \frac{\partial^2 W_{ED}}{\partial \Delta_{ED}^2} + \frac{\partial^2 W_{FE}}{\partial \Delta_{FE}^2} & -\frac{\partial^2 W_{FE}}{\partial \Delta_{FE}^2} \\ 0 & 0 & 0 & -\frac{\partial^2 W_{FE}}{\partial \Delta_{FE}^2} & \frac{\partial^2 W_{FE}}{\partial \Delta_{FE}^2} + \frac{\partial^2 W_{GF}}{\partial \Delta_{GF}^2} \end{bmatrix},$$

[1] This part, including the numerical simulations of the continuous system, has been developed in cooperation with H. Petryk (Institute of Fundamental Technological Research, Polish Academy of Sciences, Warsaw, Poland) and A. Gajo (Univerity of Trento, Trento, Italy).

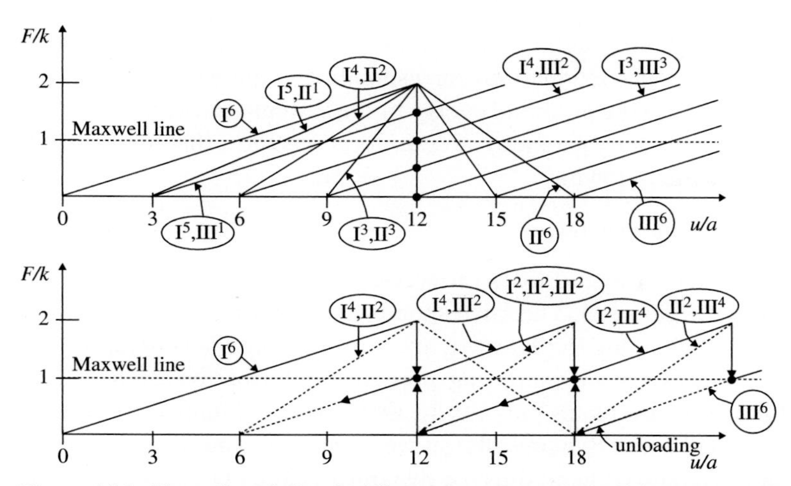

Figure 15.1. Six-spring chain of tri-linear (non-convex elastic) elements. *Top*: Different bifurcated paths possible at the peak. The configurations marked with a spot are all at least locally stable. *Bottom*: Global load/displacement behaviour assuming that the system follows the absolute minimum energy path after all the peaks in the force/displacement diagram are reached; note the hysteresis loop on loading-unloading.

which eigenvalues can be easily calculated, assuming the behaviour (1.18) and (1.19) where $\beta = 2$, $\gamma = 1$ and $\theta = 15/2$.

For the system under consideration, *all configurations corresponding to combinations of springs in phases I and III are at least locally stable, whereas all configurations containing at least one spring in phase II are unstable.*

Let us go back now to the three (non-re-entering) possibilities of continuous displacement emanating from the peak at the end of phase I^6 and note that all these contain at least one spring in phase II, and thus are all unstable. Therefore, excluding the possibility that the system can proceed along an unstable path or jump to an unstable configuration[2] the only possibility is that a jump occurs to one of the four configurations (marked with spots in Fig. 15.1) lying under the vertical of the peak and belonging to a stable path, which are:

$$\mathrm{I}^5\mathrm{III}^1, \quad \mathrm{I}^4\mathrm{III}^2, \quad \mathrm{I}^3\mathrm{III}^3, \quad \text{and} \quad \mathrm{I}^2\mathrm{III}^4.$$

After the system has switched to one of these stable paths, we can reasonably assume that this will be followed until the next peak in the force displacement curve is reached.

To decide which will be the path followed by the system among the four above-mentioned configurations is not possible on the basis of a quasi-static analysis, whereas a dynamical analysis involves the definition of masses and damping,[3] so in our position, only 'reasonable assumptions' can be introduced, that will always lead to situations that can be attained in reality.

[2] For instance, the configuration $\mathrm{I}^4\mathrm{II}^1\mathrm{III}^1$ is unstable and has not been reported in Fig. 15.1.
[3] In a dynamical motion, the response will be sensitive to the 'amount of damping' and to the type and magnitude of perturbations and/or imperfections, so it becomes more appropriate to speak of 'a possible path' rather than 'the path' of a system.

A sort of 'extremal choice', defining in a sense a limit behaviour, is to assume that the system will jump to a configuration corresponding to the absolute minimum of the energy. Thus, because the potential energy of the system is represented by the area under the paths, elementary geometrical considerations show that the preferred configuration becomes the following

$$\mathrm{I}^6 \longrightarrow \mathrm{I}^4\mathrm{III}^2 \qquad \text{minimum energy,}$$

touching the Maxwell line (which divides the force/displacement diagram into two equal-area zones; see Section 1.6). This corresponds to a deformation concentrated into *two* springs, both belonging to phase III, and representing *the volume fraction of the localisation band* (two springs suffering increasing deformation can be viewed as a band of width 2 or as two bands of width 1). Obviously, the jump of the system to the new situation will occur dynamically, so we have to assume that the energy jumps dissipate, for instance, as heat; thus the dynamics is quickly decayed, and the system again may follow, under the hypothesis of no external disturbances, the new meta-stable equilibrium branch until it becomes unstable.

Another choice to determine the behaviour of the system is the so-called maximum hysteresis or barrier free path noticed by Puglisi and Truskinowsky (2000). This corresponds to

$$\mathrm{I}^6 \longrightarrow \mathrm{I}^5\mathrm{III}^1 \qquad \text{maximum hysteresis,}$$

corresponding to the minimum band volume fraction. Comparing the two criteria of minimum energy and maximum hysteresis, we may conclude that the volume fraction of the localized bands depends on the assumed 'jump criterion'.

After the system has passed to the $\mathrm{I}^4\mathrm{III}^2$ configuration, the global elongation increases, as well as the force in the chain. In this way, assuming the absence of disturbances, the system reaches (along a meta-stable path) a second peak in the force/displacement curve, where again the continuous response becomes unstable. At this point, a 'multiple localisation of deformation' occurs, in the sense that following the path of minimum energy, the system will jump to the new path

$$\mathrm{I}^4\mathrm{III}^2 \longrightarrow \mathrm{I}^2\mathrm{III}^4 \qquad \text{minimum energy,}$$

in which two springs still remain in phase I, whereas four springs are subject to increasing elongation within phase III. In this situation, the volume fraction of the band is four, a phenomenon corresponding to a localisation band 'broadening' or 'accumulation'. Starting from the current configuration, the system displays a global elongation linked to a load increase until the third peak of the response curve is attained. At this point, the remaining two springs still in phase I move to phase III, so the system jumps to

$$\mathrm{I}^2\mathrm{III}^4 \longrightarrow \mathrm{III}^6 \qquad \text{minimum energy.}$$

In conclusion, under increasing elongation and assuming that a continuous response is possible only until it remains at least meta-stable and that configurations of absolute minimum of the energy are realised when jumps become necessary to avoid unstable configurations, the system traverses the following states

$$\mathrm{I}^6 \longrightarrow \mathrm{I}^4\mathrm{III}^2 \longrightarrow \mathrm{I}^2\mathrm{III}^4 \longrightarrow \mathrm{III}^6 \,.$$

If at the final state the displacement is reversed, the chain follows the 'extreme' hysteresis loop shown in Fig. 15.1 (*bottom*), touching the horizontal axis instead of the Maxwell line. Owing to the different choice of the 'jump criterion', the hysteresis that we have found differs from that suggested in Puglisi and Truskinowsky (2000) and Royer-Carfagni (2000). Although the latter is more close to experimental results, our hysteresis loop is determined by the 'limit' choice that the system follows the minimum energy path and cannot be excluded to occur in a real system.

The effects found with the chains of non-linear springs can be generalised for one-dimensional continuous systems and are explained in the next Section with reference to the so-called Ericksen's bar, where we will once more use for simplicity the criterion of absolute minimum of the potential energy.

Strain localisation in the Ericksen bar Ericksen (1975) has analysed the behaviour of a one-dimensional bar made up of an elastic material with non-convex strain energy (evidencing a spinodal region in the stress/strain behaviour with a peak, subsequent softening, followed by re-hardening), finding that the non-convexity can explain equilibrium of multiple phases in a way similar to that presented in the preceding example referred to a chain of non-linear springs. The Ericksen bar evidences strain localisation (so that the material outside a localisation band belongs to one phase, whereas the rest of material to another) and multiple band formation (corresponding to the co-existence of two different phases localised in different bands). As in the tri-linear chain model, we assume that displacements at the two ends of the bar are prescribed (the so-called loading in a hard device) in such a smooth way that the peak of the stress/strain curve can be reached without jumps to non-contiguous configurations.

Our interest here is to show with this one-dimensional model how bifurcations in the rate response can be detected and how the post-critical behaviour can be analysed *by introducing simple assumptions*, the most important of which is that the thickness of the localisation bands is fixed, an assumption avoiding pathological dependence of stability on the norm introduced to measure perturbations.

Therefore, we refer to the constitutive law assumed by Puglisi and Truskinowsky (2000), where the stress σ versus strain ε relation[4]

$$\sigma = (\varepsilon - c_4)\left[(\varepsilon - c_4)^2 - c_2^2\right]\frac{c_1}{c_2^4} + c_3 \tag{15.1}$$

follows from a 'double-well' energy of the type

$$W(\varepsilon) = \frac{c_1}{4}\left[\left(\frac{\varepsilon - c_4}{c_2}\right)^2 - 1\right]^2 + c_3(\varepsilon - c_4) + c_5, \tag{15.2}$$

where c_i, $i = 1,\ldots,5$ are positive constants, with c_1 and c_3 having the dimension of σ, whereas c_2 and c_4 are dimension-less. The stress/strain relation (15.1) reaches a maximum and a minimum when $\varepsilon = c_4 \pm c_2/\sqrt{3}$.

During a smooth (in the absence of disturbances) loading in a hard device, the first possibility for incremental bifurcation occurs when the peak of the stress/strain

[4] We have 'shifted' the constitutive equation proposed by Puglisi and Truskinowsky (2000), adding an additional dimension-less constant c_4.

curve is reached. At this point, deformation can localise in bands within which the strain continues to increase. The subsequent behaviour depends on the volume fraction of the bands, which is assumed by us to be concentrated in one band (so that the volume fraction now becomes equivalent to the band width). To analyse this situation without introducing non-local effects, we fix the band width of the transformed material at the instant when bifurcation in the incremental response occurs, and we plot stable or unstable deformation paths even when these are re-entering and therefore impossible to be followed for prescribed end displacements.

After the first band is formed, denoting by s the dimension-less ratio between the fixed band width and the initial length of the sample, we introduce the conventional strain ε_c defined as the total elongation of the bar divided by its initial length; thus

$$\varepsilon_c = (1-s)\varepsilon^e + s\varepsilon^i, \tag{15.3}$$

where ε^i and ε^e are the strains inside and outside the localised band, respectively. *Fixing a value of the volume fraction s of the material remaining subject to increasing deformation, the post-critical behaviour can be determined easily*, yielding a V-shaped curve in a σ vs. ε_c diagram. As an example, the global behaviour of a bar loaded in a hard device and made up of the material defined by Eq. (15.1) is shown in Fig. 15.2 for different values of the band width s. In particular, the dimensionless stress σ/c_1 is reported versus the conventional strain ε_c for

$$c_2 = 0.1, \qquad c_3/c_1 = 18.75, \qquad c_4 = 0.15.$$

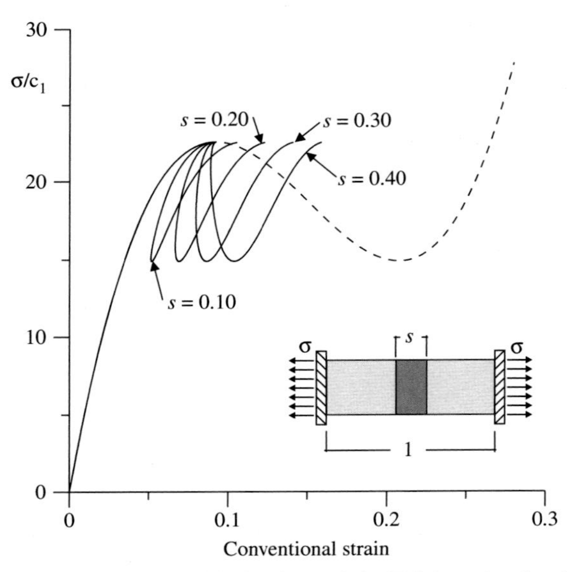

Figure 15.2. Global behaviour of the Ericksen bar loaded in a hard device: Effects of different volume fractions s of transforming material fixed after the point corresponding to the first rate bifurcation has been reached. This volume fraction, which in our model is for simplicity concentrated in one deformation band, strongly influences softening response. The homogeneous response is reported dashed. Compare with the example shown in Fig. 1.20.

It can be noted that *the post-localisation behaviour strongly depends on the volume fraction s of transforming material, yielding snap-back for small values of s.*

The post-localisation after formation of the first localisation of deformation terminates at a new peak in the loading curve, at which point the deformation inside the initial band is, say, 'saturated', and another incremental bifurcation becomes possible, thus determining formation of another band and therefore producing another V-shaped curve. This is different from the previous band, owing to the presence of the volume fraction of the previously transformed material, which remains in another phase. In a generic situation when the material in $n-1$ bands of width s_i $(i = 1,...,n-1)$ has been transformed already and the material in the nth band of width s_n is transforming, we may obtain the conventional strain ε_c as

$$\varepsilon_c = \underbrace{(1 - \sum_{j=1}^{n} s_j)\varepsilon^e}_{\text{unloading}} + \underbrace{s_n \varepsilon_n^i}_{\text{transforming}} + \underbrace{\varepsilon^i \sum_{j=1}^{n-1} s_j}_{\text{transformed}} . \tag{15.4}$$

When the V-shaped curve evidences snap-back (Fig. 15.2), the bar is assumed to 'jump' from the peak to the rising portion of the curve under the vertical below.

In the above-mentioned behaviour, there are infinite bifurcation paths emanating from the peak of the nominal-stress/conventional-strain diagram, each corresponding to different values of the band width parameter s; however, this parameter remains determined by a requirement on the energy (15.2), which in our case is assumed to reach an absolute minimum after the system jumps from the peak of the curve to a new equilibrium condition. Minimisation of the energy can be obtained numerically,[5] assuming increasing displacement at fixed s. An example of the global response obtained using the criterion of the absolute minimum of the energy is reported in Fig. 15.3 (*left*)—relative to the same material parameters used for Fig. 15.2—where σ/c_1 is plotted versus ε_c. In the same figure, the homogeneous response of the bar is also included for comparison. The global behaviour accounting for the jumps from the unstable configurations (reached at the peaks) to the stable (with minimum energy) configurations is reported in Fig. 15.3 (*right*). It is important to note in the latter figure that *all absolute minima lie on the Maxwell line*, an 'extreme' behaviour alternative to that obtained using the 'maximum hysteresis' criterion (Puglisi and Truskinovsky, 2000), predicting a response represented by a horizontal line joining the peak of the homogeneous response to the re-hardening phase. Note that both different behaviours are principle in possible.

Finally, we note that unloading is not reported in the figure. This would yield a behaviour similar to loading, but the curve would present relative minima and maxima lying below and on the Maxwell line, respectively.[6]

We note that our model is based on jumps to configurations of minimum potential energy occurring when a continuous stable or meta-stable path is not available, a

[5] Owing to the fact that the loading response is stiff, a high numerical accuracy is required, but this can be achieved in a standard way.

[6] Numerically, we have found that the ratio between the volume fractions of transforming material for two subsequent bands, that is, s_n/s_{n-1}, is constant; in particular, for our values of constants, we obtain $s_n/s_{n-1} = 0.866$.

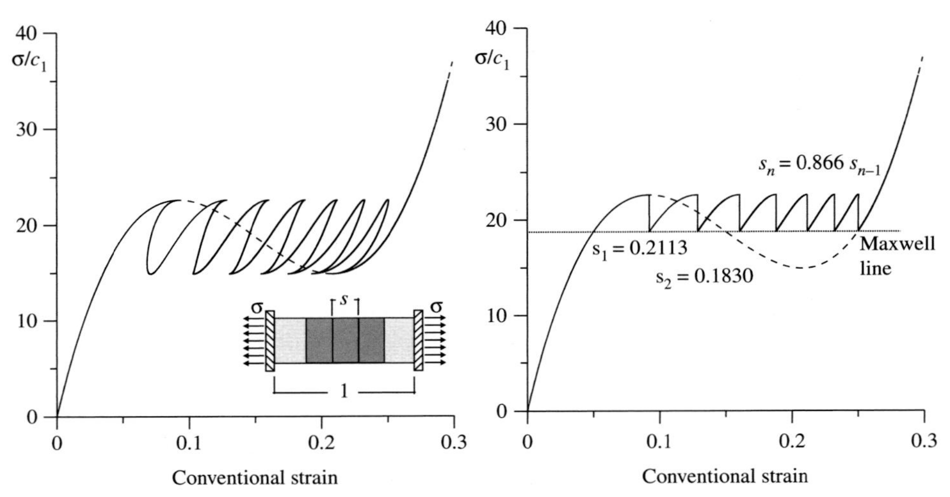

Figure 15.3. Global behaviour of the Ericksen bar loaded in a hard device in which the volume fraction of the transforming bands is concentrated in one band of fixed size, determined assuming the criterion of the absolute minimum of potential elastic energy. Bands of transformed material are assumed to accumulate contiguously. The post-critical behaviour is displayed on the left (the homogeneous behaviour is also shown dashed), whereas the post-critical behaviour is terminated at the Maxwell line on the right, which shows a possible behaviour of a real system. The global behaviour shares similarities with that found in our experiments on stacks of cans and corrugated tubes (Figs. 1.26 and 1.27).

behaviour that cannot be excluded to occur in a real system. The analysis of a real system requires the introduction of dynamic effects and viscosity, as in the cases analysed by Faciu and Suliciu (1994), Vainchtein and Rosakis (1999) and Vainchtein (2002). Our interest now is to consider simplified two-dimensional models for elastoplastic materials.

15.2 Two-dimensional elastoplastic modelling of post-shear banding

We have seen that under simplifying assumptions, both post-shear banding behaviour and multiple shear band formation can be treated for one-dimensional models. However, these models are first of all too simple to capture the behaviour of real material samples (which always should be treated at least as two-dimensional solids), and second, constitutive features of elastoplasticity which are believed to be of crucial importance in the modelling (such as, for instance, frictional behaviour, pressure sensitivity, and flow-rule non-associativity) cannot be incorporated properly.

The analysis of post-shear band formation for elastoplasticity has to be based on an incremental approach and is complicated by the following factors:

- Shear band is an ultimate mode of material instability involving loss of ellipticity of the governing equations; thus, except for special situations (e.g., van Hove conditions; see Section 11.3), a shear band is attained during a complex post-bifurcation behaviour (in which multiple global bifurcations are expected to occur).

- Shear banding usually involves plastic loading and/or elastic unloading inside and/or outside the band [as shown by Gajo et al. (2004), the condition of elastic unloading inside the band (Section 11.4), can occur in reality and plays an important role on the post-critical behaviour].
- At shear band formation, the volume fraction of shear bands can be determined through stability considerations [as shown previously with one-dimensional examples and more in general by Petryk and Thermann (2002)], but not their width (in the sense that the same volume fraction can correspond to a few 'thick' or to many 'thin' shear bands), which can be predicted only using micromechanical considerations or employing models incorporating an internal length scale (not considered here).
- Shear band propagation, interactions between shear bands and interactions with boundaries and/or inclusions certainly occur during post-critical behaviour, but these phenomena are almost unknown and cannot be incorporated in a simple way (this issue will be addressed in Chapter 16).
- Large strain effects, including, for instance, shear band rotation owing to deformation, can strongly affect the post-critical behaviour (they may explain difference between experiments performed under conditions of fixed or free top; see Fig. 1.11) but are very difficult to be incorporated.

An obvious possibility is to attack the post-critical behaviour with a finite-element technique (and introducing specific remedies to the ill-posedness related to the loss of ellipticity), but finite element techniques involve too gross approximations to correctly detail features such as those just listed. Therefore, Gajo et al. (2004) have proposed a simplifying approach based on the following hypotheses: (1) large strain effects are neglected, (2) all possible bifurcations occurring prior to strain localisation are disregarded (so that a sample is assumed to remain homogeneously deformed until localisation), (3) the width of the shear band is *imposed* on the basis of available experimental evidence, (4) Shear band propagation and/or interactions with boundaries are not considered, so the band is assumed to form instantaneously through the whole sample. This approach is motivated by the fact that experimental results (e.g., biaxial compression tests in sand) often show strain localisation to occur on samples in which the deformation fields still appear to be nearly uniform, so the effects of previous bifurcations cannot be decisive in setting the overall behaviour of the sample. Moreover, the approach is simple enough to permit a full consideration of all loading/unloading possibilities; thus it predicts a spectrum of phenomena (softening and size effects, formation of a single shear band or shear band saturation which may be persistent or not) and, applied to a specifically calibrated elastoplastic model, produces results in qualitative and *quantitative* agreement with experimental results in sand.

The simplified analysis of post-shear banding with treatment of the connected phenomena of band saturation and multiple shear band formation follows to a certain extent the analysis of post-shear banding in an infinite medium proposed by Hutchinson and Tvergaard (1981), Tvergaard (1982a) and Petryk and Thermann (2002) but is adapted to a finite-size sample. This analysis is detailed below.

15.2.1 Post-shear banding analysis

Referring for simplicity to small strain, the analysis of the post-localisation regime can be performed assuming the existence of two distinct homogeneous solutions, one outside the band (infinitesimal strain labelled $\boldsymbol{E}_o$) and the other inside the band (infinitesimal strain labelled $\boldsymbol{E}_i$). The increment of these strains must satisfy at every instant Maxwell kinematic compatibility conditions (9.13), so

$$\dot{\boldsymbol{E}}_i = \dot{\boldsymbol{E}}_o + \frac{1}{2}(\boldsymbol{g} \otimes \boldsymbol{n} + \boldsymbol{n} \otimes \boldsymbol{g}) \tag{15.5}$$

(note that in a small strain context the increment in the infinitesimal strain coincides with the Eulerian strain rate $\dot{\boldsymbol{E}} = \boldsymbol{D}$), where vector $\boldsymbol{n}$ is fixed and obtained initially imposing the singularity of the acoustic tensor [Eq. (11.46)], at ellipticity loss, whereas vector $\boldsymbol{g}$ is obtained imposing global kinematic compatibility for a specimen in a way that will be explained later. In addition to condition (15.5), stress compatibility requires continuity of traction at every instant [and thus also continuity of traction rates, Eq. (11.79)] across the interface

$$\boldsymbol{T}_o\boldsymbol{n} = \boldsymbol{T}_i\boldsymbol{n} \quad \rightsquigarrow \quad \dot{\boldsymbol{T}}_o\boldsymbol{n} = \dot{\boldsymbol{T}}_i\boldsymbol{n}. \tag{15.6}$$

We assume that the shear band is planar with constant inclination and thickness. While the band inclination is found from the solution of the condition of singularity of the acoustic tensor, *the thickness of the band is simply imposed*, assuming values deduced from experimental observations (for sand, see Roscoe, 1970; Muir Wood, 2002). In the post-localisation regime, because the material inside and outside the band can undergo either elastic unloading or plastic loading, the four different cases (11.81) may exist. Referring to the acoustic tensors of the material inside (denoted by index i) and outside (denoted by index o) the band, equipped with indices e and ep to specify the elastic and the tangent elastoplastic responses, and denoting with $\mathbb{E}$ the elastic fourth-order tensor of the small strain theory, conditions (11.81) can be rewritten as follows.

- *Plastic loading outside and inside the band*:

$$\boldsymbol{A}_i^{ep}(\boldsymbol{n})\boldsymbol{g} = \mathbb{C}_o[\dot{\boldsymbol{E}}_o]\boldsymbol{n} - \mathbb{C}_i[\dot{\boldsymbol{E}}_o]\boldsymbol{n} \tag{15.7}$$

- *Plastic loading outside and elastic unloading inside the band*:

$$\boldsymbol{A}_i^{e}(\boldsymbol{n})\boldsymbol{g} = \mathbb{C}_o[\dot{\boldsymbol{E}}_o]\boldsymbol{n} - \mathbb{E}_i[\dot{\boldsymbol{E}}_o]\boldsymbol{n} \tag{15.8}$$

- *Elastic unloading outside and plastic loading inside the band*:

$$\boldsymbol{A}_i^{ep}(\boldsymbol{n})\boldsymbol{g} = \mathbb{E}_o[\dot{\boldsymbol{E}}_o]\boldsymbol{n} - \mathbb{C}_i[\dot{\boldsymbol{E}}_o]\boldsymbol{n} \tag{15.9}$$

- *Elastic unloading outside and inside the band*:

$$\boldsymbol{A}_i^{e}(\boldsymbol{n})\boldsymbol{g} = \mathbb{E}_o[\dot{\boldsymbol{E}}_o]\boldsymbol{n} - \mathbb{E}_i[\dot{\boldsymbol{E}}_o]\boldsymbol{n}. \tag{15.10}$$

Although the second and fourth of these possibilities may seem unrealistic, the second will be found to occur and to be crucial in understanding the band saturation process, and the fourth also may occur in the post-saturation analysis as a consequence of the evolution of elastic properties (which may result, for instance from elastoplastic coupling; see Section 8.2.4).

It now would be possible to proceed following Hutchinson and Tvergaard (1982), Tvergaard (1982a) and Petryk and Thermann (2002), with reference to an infinite medium (in which case, different from the case of a finite sample, the thickness of the shear band becomes non-essential). However, our intention is to achieve a direct comparison with experimental results from biaxial plane strain tests, even at the cost of a sacrifice in the rigour of the analysis. Therefore, we introduce certain strong simplifying hypotheses, which we believe do not alter the qualitative results. In particular, we refer to a finite, rectangular specimen of initial height L_0 deformed in plane strain with a prescribed incremental displacement $\dot{u}_1$ at the top of the sample (the vertical direction is singled out by the unit vector $\boldsymbol{e}_1$), free of tractions at the two vertical lateral boundaries (parallel to $\boldsymbol{e}_1$) and containing a band of initial thickness s_0. Therefore, the plane strain condition (the out of plane direction is labelled 3) and the vanishing of the traction increment along the boundary parallel to $\boldsymbol{e}_1$ (having unit normal in the direction 2) are

$$\dot{E}_{33} = 0 \qquad \text{and} \qquad \dot{T}_{22} = 0. \tag{15.11}$$

The behaviour is assumed homogeneous until the instant of localisation, so the effects of all the diffuse bifurcations that in a finite specimen may occur before localisation are neglected. The available experimental results suggest in fact, that such bifurcations actually are likely to occur but do not much alter the overall behaviour, which remains dominated by the localisation phenomena. In these conditions, when a band of localised deformations is considered, a 'global' kinematic condition must be imposed for the deformed block, expressing the fact that the height variation of the block is the sum of the contributions inside and outside the band. This condition allows the determination of the modulus of vector $\boldsymbol{g}$ [Eq. (15.5)], through the relationship

$$L_0 \dot{E}^o_{11} + s_0 g_1 = \dot{u}_1, \tag{15.12}$$

where $\dot{E}^o_{11}$ is the component of $\dot{\boldsymbol{E}}_o$ parallel to the load axis.

A numerical integration of the constitutive rate equations before and after localisation now can be performed. Briefly [the interested reader is remanded for details to Gajo (2003)], for homogeneous response, the stress at the end of each integration step violates the yield condition, which therefore has to be enforced. This is obtained by introducing a generalisation of the cutting plane algorithm (Simo and Hughes, 1987) needed because the simulation of biaxial testing requires a mixed prescription of strain and stress. The same problem occurs after strain localisation outside the band, whereas inside the band different prescriptions arise from requirements that the tractions have to remain continuous across the band and the deformation rate has to take the form of Eq. (15.5).

During the process of post-localisation (as a result of the small strain hypothesis), the band inclination $\boldsymbol{n}$ remains fixed, whereas $\boldsymbol{g}$ is allowed to change both in direction and in modulus. At a generic step of numerical integration in the post-localisation regime, all Equations [(15.7) through (15.10)] hold true when the stress states inside and outside the band lie on the yield surface. In this case, all the mechanisms of deformation represented by Eqs. (15.7) through (15.10) are considered separately, according to the following procedure, referred for simplicity to the plastic/plastic mechanism but which can be extended easily to all the other particular cases.

- The vertical incremental displacement Δu_1 is assigned, and a tentative ΔE^o_{11} is assumed.
- ΔE^o_{22} is evaluated from the condition $\Delta T_{22} = 0$, using the tangent constitutive operator $\mathbb{C}_o$. Consequently, $\Delta \boldsymbol{E}_o$ is known.
- Imposing Eq. (15.7), $\boldsymbol{g}$ is calculated, so $\Delta \boldsymbol{E}_i$ is known.
- The conditions of plastic loading are checked inside and outside the band. If one of these is violated, the procedure is terminated, and another deformation mechanism is analysed. Otherwise, the current mechanism is considered possible, and the exact value of ΔE^o_{11} is obtained from Eq. (15.12). Finally, all quantities are rescaled to the correct values (this is an immediate operation because all equations here are linear).

It is possible that more than one mechanism of localised deformation is detected from the preceding screening. In such cases, a stability requirement could be employed to select one. A simple choice (used by Gajo et al., 2004) is the heuristic criterion by which the deformation mechanism corresponding to the steepest descent in the global force versus displacement curve for the sample is selected. The selection can be performed using the linearised equations. At the end of this screening, a tentative solution for the linearised problem is found. On the basis of this, a refined incremental solution is calculated enforcing the yield condition within a given tolerance, employing again the described extension of the cutting plane algorithm, which is needed to satisfy the condition that the stress state inside the band must satisfy continuity of tractions across the band. In this way, the boundary conditions at the points where the band emerges on the edges of the sample are violated, but this effect has been neglected from the beginning. Moreover, since the iterations with the cutting-plane algorithm slightly modify the values of the strain inside the band, at the end of the integration step, the Maxwell kinematic compatibility conditions [Eq. (15.5)], will not be exactly satisfied inside the band. If the error in terms of strain is large, then the strain increment necessary to satisfy the Maxwell kinematic compatibility conditions is applied inside the band, by using an implicit backward-Euler integration algorithm. Convergence can be ensured through use of the line search procedure, adjusting any Newton step ending ‘excessively far’ from the solution (Matthies and Strang, 1979; Johan et al., 1991).[7]

[7] A few iterations have been found to be sufficient for reaching a strict tolerance on residuals. However, this will be not the case when band saturation is involved. In fact, the elastoplastic acoustic tensor inside the band [Eq. (15.9)], may become singular when band saturation occurs. In this case, the final residual strain increment is evaluated by using the cutting plane algorithm (imposing strain increments only and consequently accepting a small error on the continuity of traction across the interface between inside and outside the band), instead of the above-described implicit backward-Euler integration algorithm. A maximum value of 1 Pa in the norm of the residual traction at band interface was tolerated in our computations.

15.2.2 Sharp shear banding versus saturation

The above-considered procedure allows computation of shear band initiation and post-critical behaviour of samples of finite dimensions, subject to loading under mixed boundary conditions. Within a sophisticate constitutive framework introduced to model granular material, Gajo et al. (2004)[8] have applied the above numerical procedure to find the global mechanical response of samples subject to plane strain biaxial tests and also to the so-called tri-axial tests (a radially confined cylindrical sample subject to vertical load). For dense granular materials, the formation of a single shear band is predicted, *initiating in the hardening regime and yielding a strong softening*, Fig. 1.19, and related size effects, with possible snap-back, Fig. 1.20.

For dense granular materials at low confining pressure or for loose granular materials, a special phenomenon has been observed, namely, the 'termination' of the shear band (which elastically unloads), with the immediate formation of a new one in a different position within the material outside the initial band, which is now in a condition of plastic loading. This occurrence was observed by Tvergaard (1982 a), employing a version of the Gurson model for porous plastic metals, who called it 'band saturation'. However, Tvergaard did not suggest any mechanical interpretation for this phenomenon[9], whilst Gajo et al. (2004) have interpreted it as a critical phenomenon discriminating between a 'sharp shear band' localization and an oscillatory behaviour either preluding again sharp shear banding or continuing up to large strain ('persistent shear banding', Fig. 1.28).

The interpretation of the saturation mechanism requires the definition of a tool for analyzing the post-saturation behaviour. This will be defined below, using a special numerical procedure, based on simplifying hypotheses.

15.2.3 Post-band saturation analysis

The analysis of the post shear band saturation regime can be performed in a similar way to post-localisation analysis. In particular, we assume at saturation the formation of a new band (outside the saturated band) and the consequent existence of three distinct homogeneous solutions: a first inside the band just formed ($\boldsymbol{E}_i$), a second inside the saturated band ($\boldsymbol{E}_s$), and a third outside both of these ($\boldsymbol{E}_o$).

The band just formed is taken parallel to the saturated, since it occurs in a material which is in the same condition for which the initial band occurred. Therefore,

[8] The framework includes density as an evolution variable and elastoplastic coupling to model stress-induced elastic anisotropy. These constitutive features are crucial for correctly determining (1) the conditions for the onset of strain localisation, (2) the evolution of the density within the shear band, and (3) the saturation and post-saturation mechanisms.

[9] Tvergaard (1982a) argued that band saturation could represent a spurious effect related to the fact that the non-associative plastic response may be stiffer than elastic. This 'strange' feature can be checked immediately to be possible with reference to the constitutive equation (8.101) by selecting and $\dot{\boldsymbol{F}}^*$ such that $\dot{\boldsymbol{F}}^* \cdot \boldsymbol{N} > 0$ and $\dot{\boldsymbol{F}}^* \cdot \boldsymbol{M} < 0$. It follows immediately that

$$\dot{\boldsymbol{F}}^* \cdot \mathbb{G}[\dot{\boldsymbol{F}}^*] < \dot{\boldsymbol{F}}^* \cdot \mathbb{G}[\dot{\boldsymbol{F}}^*] - \frac{1}{g}\left(\boldsymbol{N} \cdot \dot{\boldsymbol{F}}^*\right) \boldsymbol{M} \cdot \dot{\boldsymbol{F}}^*.$$

the deformation increments satisfy the following constraints

$$\dot{\boldsymbol{E}}_s = \dot{\boldsymbol{E}}_o + \frac{1}{2}[\boldsymbol{g}_s \otimes \boldsymbol{n} + \boldsymbol{n} \otimes \boldsymbol{g}_s], \qquad \dot{\boldsymbol{E}}_i = \dot{\boldsymbol{E}}_o + \frac{1}{2}[\boldsymbol{g}_i \otimes \boldsymbol{n} + \boldsymbol{n} \otimes \boldsymbol{g}_i], \tag{15.13}$$

where $\boldsymbol{g}_i$ and $\boldsymbol{g}_s$ are the deformation mode vectors, respectively for the newly formed and the saturated band. We denote here with the indices s, i and o the quantities defined inside the saturated band, inside the new band and outside the bands, respectively.

Given that the material inside each band (i.e. the saturated band and the band just formed) and outside the bands undergoes either elastic unloading or plastic loading, eight different cases may exist. As a result of the incremental piece-wise linearity of the constitutive relationship, the stress compatibility condition in the case, for example, of plastic loading inside the new band and elastic unloading elsewhere can be written as follows

$$\boldsymbol{A}_i^{\text{ep}}(\boldsymbol{n})\boldsymbol{g}_i = \mathbb{E}_o[\dot{\boldsymbol{E}}_o]\boldsymbol{n} - \mathbb{C}_i[\dot{\boldsymbol{E}}_o]\boldsymbol{n} \qquad \text{and} \qquad \boldsymbol{A}_s^{\text{e}}(\boldsymbol{n})\boldsymbol{g}_s = \mathbb{E}_o[\dot{\boldsymbol{E}}_o]\boldsymbol{n} - \mathbb{E}_s[\dot{\boldsymbol{E}}_o]\boldsymbol{n}, \tag{15.14}$$

and all the other possibilities of loading/unloading can be written down easily. Conditions (15.14) ensure continuity of the tractions at both the interfaces between the material outside the bands and the materials inside the new and the saturated bands. Consequently, in the special case in which the saturated and new band are adjacent, stress compatibility at the interface between them is fulfiled automatically.

If the mechanism of band saturation continues, thus forming more than one saturated band, and limiting the analysis to the initiation of the process, when only a few bands saturate, each new band behaves with a good approximation in the same way as previously analysed. Therefore, all the saturated bands are made up of the same material which suffered the same deformation history.

When there are n saturated bands and a new one has just formed, Eqs. (15.13) and (15.14) still hold true. Moreover, the global kinematic condition expressing the fact that the deformation of the sample may be written as the sum of the contributions inside all the bands plus that outside them is

$$L_0 \dot{E}_{11}^o + s_0(\boldsymbol{g}_i \cdot \boldsymbol{e}_1) + n s_0(\boldsymbol{g}_s \cdot \boldsymbol{e}_1) = \dot{u}_1. \tag{15.15}$$

In this way, the stress compatibility conditions [Eq. (15.14)], and the kinematic condition at the boundary [Eq. (15.15)], enable the evaluation of the deformation mode vectors $\boldsymbol{g}_i$ and $\boldsymbol{g}_s$.

During the process of band saturation, the continuously forming shear bands progressively transform the original material into a new one as the result of the plastic strains occurring inside each band: In loose granular material, the transformed material will be densified, whereas in dense granular material, it will be loosened. This process will be denoted as 'shear band broadening' (or 'shear band accumulation'), which does not necessarily mean that the subsequent bands form adjacent to each other.

During band broadening, the nominal stress describes a series of small oscillations resulting in the formation of horizontal 'ripples' in the stress/strain curve, a phenomenon similar to that evidenced in one-dimensional models (Section 15.1) and sharing similarities with 'stress serrations' observed in metal plasticity (Shaw and Kyriakides, 1997).

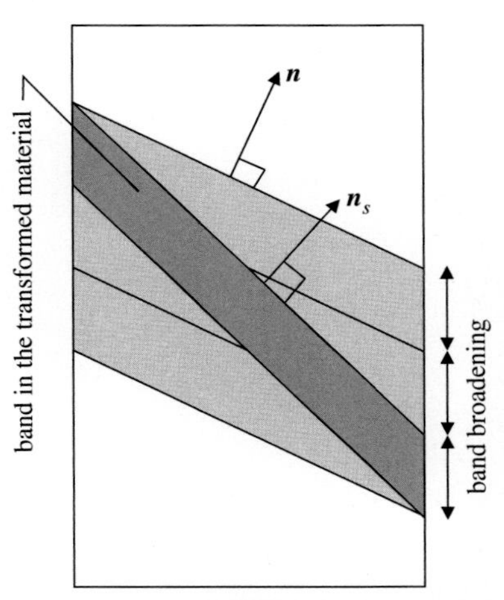

Figure 15.4. Sketch of formation of a differently inclined shear band within the material transformed by previous shear band accumulation.

After a finite-width zone of material is generated by the subsequent formation and saturation of adjacent bands, it is possible that a new shear band can initiate within this 'transformed' material. When this occurs in dense granular material, it usually terminates the mechanism of subsequent band saturation. In fact, when subjected to plastic loading again, the transformed material lies outside the elliptic domain (i.e., the hardening modulus is smaller than the critical hardening modulus), and a new shear band can initiate immediately.[10] In this case, with reference to the situation sketched in Fig. 15.4, the increments of strains satisfy

$$\dot{\boldsymbol{E}}_s = \dot{\boldsymbol{E}}_o + \frac{1}{2}[\boldsymbol{g}_s \otimes \boldsymbol{n} + \boldsymbol{n} \otimes \boldsymbol{g}_s] \qquad \text{and} \qquad \dot{\boldsymbol{E}}_i = \dot{\boldsymbol{E}}_s + \frac{1}{2}[\boldsymbol{g}_i \otimes \boldsymbol{n}_s + \boldsymbol{n}_s \otimes \boldsymbol{g}_i], \quad (15.16)$$

where $\boldsymbol{n}_s$ is the normal to the new band which initiates inside the transformed material, belonging to a zone bounded by planes of normal $\boldsymbol{n}$. Referring to the conditions of plastic loading inside the band in the transformed material and elastic unloading elsewhere, stress compatibility can be written as follows:

$$\boldsymbol{A}_i^{\text{ep}}(\boldsymbol{n}_s)\boldsymbol{g}_i = \mathbb{E}_s[\dot{\boldsymbol{E}}_s]\boldsymbol{n}_s - \mathbb{C}_i[\dot{\boldsymbol{E}}_s]\boldsymbol{n}_s \qquad \text{and} \qquad \boldsymbol{A}_s^{\text{e}}(\boldsymbol{n})\boldsymbol{g}_s = \mathbb{E}_o[\dot{\boldsymbol{E}}_o]\boldsymbol{n} - \mathbb{E}_s[\dot{\boldsymbol{E}}_o]\boldsymbol{n}. \quad (15.17)$$

The global kinematic constraint for the entire specimen is once more given by Eq. (15.15).

As a result, when a band saturates, two different deformation mechanisms may occur:

- A new band may initiate outside the material transformed by the previously saturated band.
- A new band may initiate within the material transformed by the previously saturated, adjacent bands with any inclination $\boldsymbol{n}_s$. This inclination has to be compatible with the thickness of the zone made up of saturated bands and when

[10] When the hardening modulus falls below the critical value for the onset of strain localisation, localisation with elastic unloading outside the band becomes possible for a finite fan of inclinations.

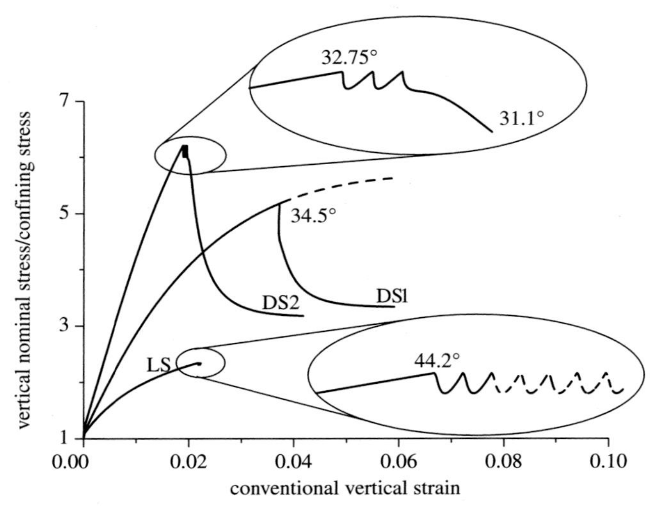

Figure 15.5. Simulated global nominal stress (made dimension-less through division by the confining stress) versus conventional strain behaviour for plane strain–confined compression tests (so-called biaxial test) on sand. Results are reported for dense sand at high (labelled 'DS1') and low (labelled 'DS2') confining pressure and loose sand (labelled 'LS'). DS1 is an example of sharp shear banding, whereas DS2 and LS are examples of multiple shear band formation with subsequent saturation. A sharp shear band forms within the saturated bands in DS2, whereas shear band saturation is persistent in LS.

the hardening modulus of the material is inferior to the critical, [Eq. (11.78); see Fig. 15.4].

The choice among the different possible deformation mechanisms again should be based on a stability criterion, and we have again employed the previously mentioned simple selection of the deformation mechanisms leading to the steepest descent in the conventional stress/strain curve.

The above-described procedure has been employed by Gajo et al. (2004; 2007) with a constitutive model capable of reproducing the behaviour of granular materials calibrated to simulate Hostun sand. Some of the results already have been mentioned in the Introduction (Figs. 1.19, 1.20 and 1.28) and are now summarised in Fig. 15.5, whereas corresponding modes of shear banding are reported in Fig. 15.6.

Three post-localization deformation mechanisms are found for sand: The dense sample tested at a high stress level (label 'DS1') shows bifurcated response with one shear band occurring before the peak of the homogeneous response[11] followed by a rapid drop of stress to a residual value. The dense sample tested at a lower stress level (labelled 'DS2') and the loose sample (labelled 'LS') show a response that we have described as band saturation. In particular, in the cases labelled 'DS2' and 'LS' the overall post-localisation global response changes from softening to hardening until a second shear band occurs in the material that, subsequent to the previous localisation, had been unloading elastically and is now again at yielding, realising

[11] Strain localisation at small strain cannot happen before the peak of the stress/strain curve for the associative flow rule (Section 11.4.1), so the fact that in this case strain localisation occurs before the peak is a consequence of the assumption of non-associative plastic flow.

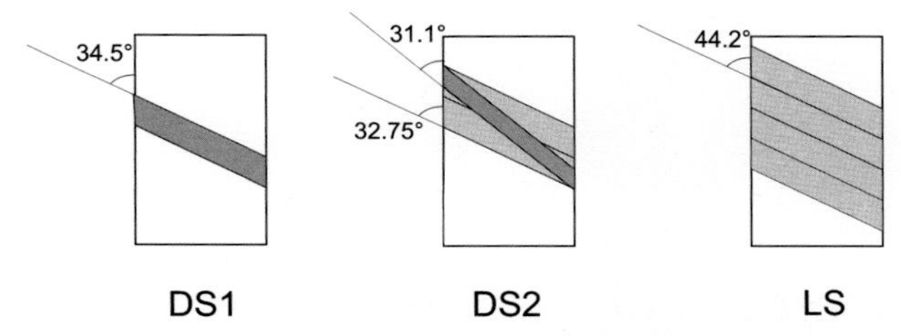

Figure 15.6. Shear bands obtained during the simulations of biaxial tests on sand (Gajo et al., 2004) reported in Fig. 15.5. A single shear band forms for dense sand at high confining stress, DS1, whereas band saturation and multiple shear bands form for dense sand at low confining stress, DS2, and for loose sand, LS. Multiple shear bands are not persistent in dense sand, where a differently inclined shear band forms in the material transformed by subsequent shear bands and terminates band saturation, whereas they are persistent in loose sand, where a 'gentle' stress oscillation is observed, similar to the stress fluctuation occurring during yielding of a ductile metal bar pulled in tension.

the condition of band saturation, yielding multiple shear banding. When a sufficient number of contiguous shear bands have accumulated, it may be possible for a new band to form at a different angle within this material, leading to global softening (as in the case labelled 'DS2'). In other cases, band saturation can continue (as in the case labelled 'LS'). These behaviours find a direct qualitative and quantitative experimental confirmation (Desrues and Hammad, 1989; see also Gajo et al., 2004).

16 A perturbative approach to material instability

A perturbative approach to material instabilities is introduced, in which a perturbing agent is superimposed on a homogeneously stressed and strained infinite medium. The perturbing agent may be a concentrated force, a dipole, but also a fracture, or a rigid inclusion, or a pre-existing shear band. It is shown that the technique is 'rich enough' to reveal phenomena which remain undetected with more conventional approaches. These involve effects of pre-stress on dislocation-induced distorsions; dynamics of shear bands; field solutions for materials in flutter conditions; interactions between shear bands, cracks and rigid inclusions; and features related to shear band growth. In particular, shear band growth is shown to have a definite tendency toward rectilinear propagation and to involve a stress field singularity akin to the singularity arising at a fracture tip. These results motivate the circumstance that shear bands are preferential near-failure deformation modes.

All analyses considered up to this point refer to situations of perfect systems loaded in perfect conditions. We are now in a position to judge the merits and the limits of the previous approach and to set up a new methodology that may capture aspects remaining undetected within the previously given framework. These aspects can be illustrated with reference to two examples: the local instability criteria of ellipticity loss and flutter instability. In particular, we know that both refer to *homogeneously deformed infinite bodies* and that the former is linked to the appearance of shear bands (or strain localisation), whereas the latter is linked to the generation of 'blowing-up incremental solutions'. The advantages of the usual approach based on the spectral analysis of the acoustic tensor are (1) simplicity and (2) that critical thresholds (in terms of plastic modulus) and critical modes (in terms of band inclination and flutter directions for shear band and flutter instability analyses, respectively) can be obtained easily. On the other hand, the drawbacks are the following:

- The shear band threshold is a sort of 'on-off' switch between stability and instability, so there is no direct evidence of any phenomena possibly occurring *before* the threshold is met.
- Dynamics effects are not given any evidence; for instance, there is no mechanical explanation for situations in which ellipticity is lost or flutter instability is met for problems of time-harmonic motion (where the classification of the character

of the differential equations remains the same as in the quasi-static limit, see Chapter 14).
- Since local criteria are defined at a point in a continuum, they cannot explain features related to shear band interaction (with other shear bands, or with inclusions, or with boundaries).

An example of the first point was given in the Introduction. In particular, for a Mooney-Rivlin material, the condition of ellipticity loss occurs from Eq. (12.45) when $k = 1$; thus, at a stretch λ, solution of

$$\frac{\lambda^4 - 1}{\lambda^4 + 1} = \pm 1, \tag{16.1}$$

which corresponds to either an infinite or a null value of stretch, a condition obviously not attainable in reality, so shear bands in the convetional sense are excluded. However, it is shown in Fig. 1.38 that the incremental response to a perturbation, in terms of a unit-force dipole superimposed on a uniformly pre-stretched Mooney-Rivlin material (at a stretch near 3), reveals zones of concentrated incremental deformation. Another example, paradigmatic of elastoplasticity, is illustrated in Fig. 16.1 (taken from Gajo et al., 2007, their fig. 9), where the $q = \sqrt{3J_2}$ versus axial strain behaviour is reported for a tri-axial test on dense sand, simulated within the constitutive modelling proposed by Gajo et al. (2004). In the figure, the critical hardening modulus for loss of ellipticity h^E_{cr} is also plotted against the hardening modulus h. The figure shows that the condition of loss of ellipticity is never attained, even if the hardening modulus h (divided by the elastic shear modulus μ in the figure) is very close to the critical value h^E_{cr}. In this situation, while the usual approach simply provides no information, a perturbative approach shows that localisation of deformation indeed can occur as the response of the system to a perturbation.

A perturbative approach to strain localisation has been proposed by Bigoni and Capuani (2002, 2005) in this way:

- The reponse to an incremental concentrated force applied quasi-statically or dynamically to a homogeneously pre-stressed infinite body (i.e., the incremental Green's function) is determined.
- A self-equilibrated distribution of Green's functions – for instance, a force dipole – is used as the perturbing agent of an infinite pre-stressed body.
- The incremental fields of the infinite body subject to the self-equilibrated perturbation are plotted until below the stability threshold under investigation, so peculiarities of the mechanical response up to the instability are revealed.

The perturbative approach has been applied by Bigoni and Capuani (2002) to the analysis of strain localisation and extended to time-harmonic oscillations by Bigoni and Capuani (2005). The approach reveals the effects related to dynamics (Fig. 1.40); thus, although the regime classification does not change between dynamics and time-harmonic motion (Chapter 14), we find interesting features in dynamics compared with the quasi-static situation, such as the fact that oscillations become focussed into well-defined patterns of planar bands.

The perturbative approach also has been employed by Piccolroaz et al. (2006) to reveal the mechanical meaning of flutter instability in granular materials. It is shown

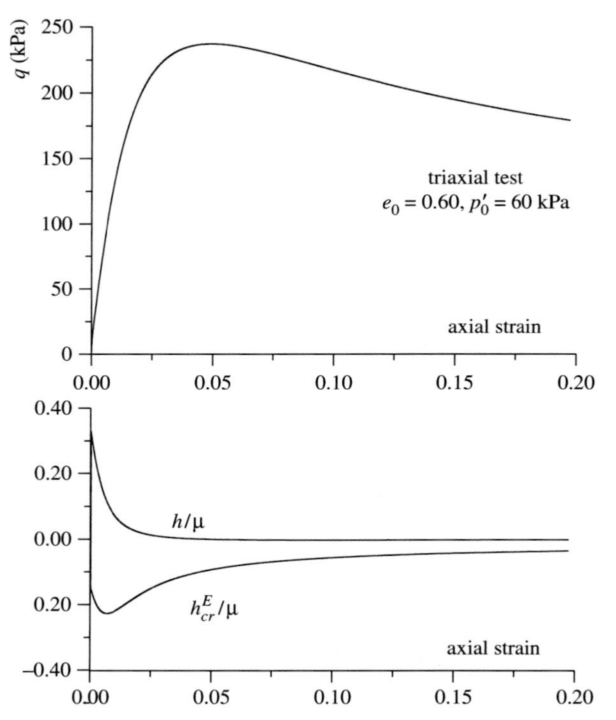

Figure 16.1. Simulation of a drained tri-axial compression test (in terms of $q = \sqrt{3J_2}$ versus axial strain) on a dense sand sample (relative density $e_0 = 0.60$) at low confining pressure ($p'_0 = 0.60\,\text{kPa}$), obtained with the non-associative elastoplastic model for granular material proposed by Gajo et al. (2004). The evolution is reported for the material hardening modulus h and the critical hardening modulus for strain localisation h^E_{cr} (both made dimension-less through division by the elastic shear modulus μ). Although the two values of h and h^E_{cr} are very close, shear bands are excluded according to the conventional approach because $h > h^E_{cr}$. Figure adapted from Gajo et al. (2007, their fig. 9).

in this case that flutter instability corresponds to blowing-up oscillations, but these oscillations self-organize along bands inclined with the inclination predicted by the flutter analysis (see Fig. 1.41 and the presentation that will be developed later in this Chapter, where we also will see that the perturbations can be given in a random distribution to simulate the effect of a dislocation cluster in a pre-stressed ductile material).

Obviously, perturbations different from concentrated forces can be envisaged. Radi et al. (2002) have investigated a mode I and mode II incremental perturbation on a crack parallel to the principal stress in a pre-stressed elastic material. Bigoni et al. (2007), Dal Corso et al. (2008) and Dal Corso and Bigoni (2009) have investigated mode I incremental loading of a rigid-line inclusion in a uniformly pre-stressed elastic material. These works evidence that when the pre-stress state is near the boundary of ellipticity loss (but still within the elliptic regime), localised deformations occur, which have inclinations akin to those which can be calculated at the loss of ellipticity in terms of acoustic tensor singularity.

Finally, these approaches have stimulated the idea of introducing a model of an already formed shear band in a pre-stressed material. This pre-existing shear band is viewed as a perturbing agent within an infinite, homogeneously stressed body and allows us to analyse the stress state near a shear band and its growing conditions (Bigoni and Dal Corso, 2008; Dal Corso and Bigoni, 2010). This investigation, which is presented in closure of this chapter, has provided justification to the fact that shear bands have a tendency towards rectilinear propagation and are preferred near-failure modes for ductile materials.

16.1 Infinite-body Green's function for a pre-stressed material

Green's functions for anisotropic materials have been found by Willis (1965, 1971, 1972, 1973) and Vogel and Rizzo (1973) and, when the anisotropy results from pre-stress, by Willis (1991) and Bigoni and Capuani (2002, 2005). We focus attention on two-dimensional Green's functions, following Bertoldi et al. (2005) for the quasi-static case and Piccoloraz et al. (2006) for the dynamic case. In both cases, the Green's function will be determined as the linearised response to a concentrated force incrementally loading an infinite (non-linear elastic or elastoplastic) material that is uniformly strained within the elliptical range. These Green's functions eventually will be used as perturbing agents to explore material instabilities.

16.1.1 Quasi-static Green's function

We will consider the *quasi-static* response of *two-dimensional infinite bodies homogeneously deformed within the elliptical regime*. The approach to Green's functions that will be followed will lead us to find results valid for incremental deformations of both homogeneously pre-stressed elastic and elastoplastic solids.

Let us start by considering the incremental equilibrium equations of a generic body characterised by a linear constitutive operator $\mathbb{G}$ as in Eq. (8.100) and assuming the current configuration as reference in a relative Lagrangean description so that

$$\text{div}\,(\mathbb{G}[\nabla \boldsymbol{v}]) + \boldsymbol{f} = \boldsymbol{0}, \tag{16.2}$$

where $\boldsymbol{f}$ denotes the body force and $\boldsymbol{v}$ is the incremental displacement (or velocity in a rate formulation). The fundamental solution or Green's function is the general solution of Eq. (16.2) to a concentrated force $\boldsymbol{f}^g = \boldsymbol{e}^g \delta(\boldsymbol{x})$ acting at the origin $\boldsymbol{x} = \boldsymbol{0}$ in the direction singled out by the unit vector $\boldsymbol{e}^g$, where $\delta(\boldsymbol{x})$ is the two-dimensional Dirac delta function. Note that on introducing an orthogonal reference system, the unit vector becomes the Kronecker delta $e_i^g = \delta_{ig}$.

The solution of the preceding problem is obtained by employing a plane wave expansion (Courant and Hilbert, 1962; Gel'fand and Shilov, 1964), which for a generic function h is written as

$$h(\boldsymbol{x}) = -\frac{1}{4\pi^2} \oint_{|\boldsymbol{n}|=1} \tilde{h}(\boldsymbol{n} \cdot \boldsymbol{x}) ds_{\boldsymbol{n}}, \tag{16.3}$$

where $\boldsymbol{n}$ is the radial unit vector centred at the origin of the position vector $\boldsymbol{x}$ (Fig. 16.2), and the superimposed tilde denotes the transformed function.

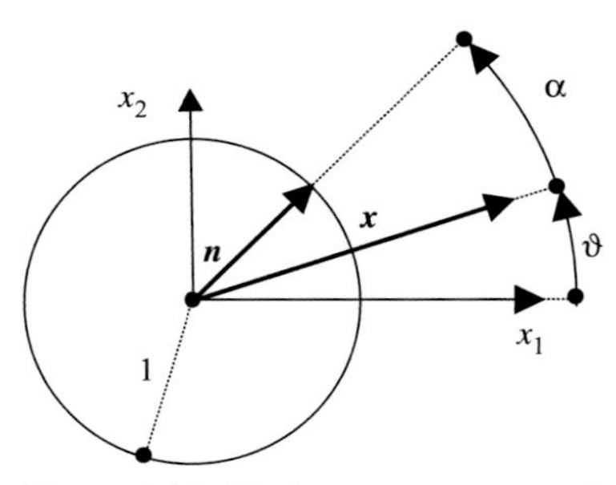

Figure 16.2. Reference system, unit vector $\boldsymbol{n}$, position vector $\boldsymbol{x}$ and angles α and θ.

Since in the transformed domain the following properties hold:

$$\tilde{\delta}(\boldsymbol{n}\cdot\boldsymbol{x}) = \frac{1}{(\boldsymbol{n}\cdot\boldsymbol{x})^2} \qquad \text{and} \qquad \nabla_{\boldsymbol{x}}\tilde{h}(\boldsymbol{n}\cdot\boldsymbol{x}) = \tilde{h}'(\boldsymbol{n}\cdot\boldsymbol{x})\,\boldsymbol{n}, \tag{16.4}$$

where the prime denotes differentiation with respect to the argument $\boldsymbol{n}\cdot\boldsymbol{x}$, the equilibrium Eq. (16.2) becomes

$$\boldsymbol{A}(\boldsymbol{n})(\tilde{\boldsymbol{v}}^g)'' + \boldsymbol{e}^g\tilde{\delta} = \boldsymbol{0}, \tag{16.5}$$

where the acoustic tensor $\boldsymbol{A}(\boldsymbol{n})$ [Eq. (11.45)] is now taken with unit mass density $\rho_0 = 1$.

When the acoustic tensor is not singular, $\det\boldsymbol{A}(\boldsymbol{n}) \neq 0$ for all unit vectors $\boldsymbol{n}$ – or, in equivalent words, *when the problem is elliptical*—eq. (16.5) can be integrated, yielding

$$\tilde{\boldsymbol{v}}^g(\boldsymbol{n}\cdot\boldsymbol{x}) = \boldsymbol{A}^{-1}(\boldsymbol{n})\boldsymbol{e}^g \log|\boldsymbol{n}\cdot\hat{\boldsymbol{x}}|, \tag{16.6}$$

where $\hat{\boldsymbol{x}}$ is a dimension-less measure of $\boldsymbol{x}$.

The Green's function finally can be obtained by inverse-transforming Eq. (16.6) to obtain

$$\boldsymbol{v}^g(\boldsymbol{x}) = -\frac{1}{4\pi^2}\oint_{|\boldsymbol{n}|=1} \boldsymbol{A}^{-1}(\boldsymbol{n})\,\boldsymbol{e}^g \log|\boldsymbol{n}\cdot\hat{\boldsymbol{x}}|ds_{\boldsymbol{n}}. \tag{16.7}$$

Introducing the radial (dimension-less) measure of the distance $\hat{r} = |\hat{\boldsymbol{x}}|$, the angle α between vectors $\boldsymbol{n}$ and $\boldsymbol{x}$ and the angle θ singling out $\boldsymbol{x}$ with respect to the x_1 axis (Fig. 16.2), the Green's function (16.7) can be expressed as

$$\boldsymbol{v}^g(\hat{r},\theta) = -\frac{\log\hat{r}}{4\pi^2}\int_0^{2\pi}\boldsymbol{A}^{-1}(\boldsymbol{n})\,\boldsymbol{e}^g d\alpha - \frac{1}{4\pi^2}\int_0^{2\pi}\boldsymbol{A}^{-1}(\boldsymbol{n})\,\boldsymbol{e}^g \log|\cos\alpha|d\alpha, \tag{16.8}$$

where it can be noted that the singularity is explicit and the spectral representation of the acoustic tensor is not needed.

It can be observed from Eq. (16.8) that

The Green's function (16.8) is symmetric if and only if $\boldsymbol{A}(\boldsymbol{n})$ is, which in (hyper-)elastoplasticity occurs for the associative flow rule, or, in other words:

$$u_i^g = u_g^i \iff \boldsymbol{A}(\boldsymbol{n}) = \boldsymbol{A}^T(\boldsymbol{n}) \iff \mathbb{G} = \mathbb{G}^T \Longleftarrow \text{associativity},$$
$$u_i^g \neq u_g^i \iff \boldsymbol{A}(\boldsymbol{n}) \neq \boldsymbol{A}^T(\boldsymbol{n}) \iff \mathbb{G} \neq \mathbb{G}^T \Longleftarrow \text{non-associativity},$$

where the transpose of a fourth-order tensor has been defined by Eq. (2.90) and involves only the major symmetry.

The Green's incremental displacement gradient $\nabla \boldsymbol{v}^g$ can be obtained either directly by differentiation of the Green's displacement $\boldsymbol{v}^g$ or by differentiation and subsequent transformation of $\tilde{\boldsymbol{v}}^g$, yielding in both cases

$$\nabla \boldsymbol{v}^g = -\frac{1}{4\pi^2} \oint_{|\boldsymbol{n}|=1} \boldsymbol{A}^{-1}(\boldsymbol{n})\, \boldsymbol{e}^g \otimes \frac{\boldsymbol{n}}{\boldsymbol{n}\cdot\boldsymbol{x}} ds_{\boldsymbol{n}}. \tag{16.9}$$

Finally, application of the constitutive tensor $\mathbb{G}$ to $\nabla \boldsymbol{v}^g$ yields the incremental first Piola-Kirchhoff Green's stress

$$\dot{\boldsymbol{S}}^g = -\frac{1}{4\pi^2} \oint_{|\boldsymbol{n}|=1} \mathbb{G}\left[\boldsymbol{A}^{-1}(\boldsymbol{n})\, \boldsymbol{e}^g \otimes \frac{\boldsymbol{n}}{\boldsymbol{n}\cdot\boldsymbol{x}}\right] ds_{\boldsymbol{n}}. \tag{16.10}$$

It is important to note that the Green's function [Eq. (16.7) or Eq. (16.8)] and all related quantities [Eqs. (16.9) and (16.10)] are defined for a generic acoustic tensor $\boldsymbol{A}(\boldsymbol{n})$ and thus also may be unsymmetric, as in the case for elastoplasticity with the non-associative flow rule, a situation that will be discussed in detail later.

It may be important at this stage to observe that

When the border of loss of ellipticity is approached, one eigenvalue of $\boldsymbol{A}(\boldsymbol{n})$ goes to zero, so the integrals in Eqs. (16.8) through (16.10) tend to become singular. This important circumstance is *linked to the appearance of shear bands*, and it has been thoroughly investigated by Bigoni and Capuani (2002; their appendix A).

The quasi-static Green's function for the loading branch of the elastic-plastic tangent operator

A generic inviscid elastoplastic constitutive equation relating the rates of first Piola-Kirchhoff stress $\dot{\boldsymbol{S}}$ and of deformation gradient $\dot{\boldsymbol{F}}$ can be written in the form of Eq. (8.101) so that the acoustic tensor corresponding to the loading branch of the constitutive equation (8.50) can be cast in the form of Eq. (11.46), taken now with unit mass density $\rho_0 = 1$.

Assuming ellipticity of both the elastic, $\det \boldsymbol{A}_E(\boldsymbol{n}) \neq 0$, and elastoplastic branch of Eq. (8.50), $g \neq \boldsymbol{N}\boldsymbol{n}\cdot\boldsymbol{A}_E^{-1}\boldsymbol{M}\boldsymbol{n}$, the elastoplastic acoustic tensor can be inverted [Eq. (11.76)]. Therefore, since Eq. (16.7) can be re-interpreted in terms of rates, a substitution of Eq. (11.76) into Eq. (16.8) yields for the plastic branch

$$\boldsymbol{v}^g_{ep} = -\frac{1}{4\pi^2} \oint_{|\boldsymbol{n}|=1} \left[\boldsymbol{A}_E^{-1}(\boldsymbol{n}) + \boldsymbol{B}_p(\boldsymbol{n})\right] \boldsymbol{e}^g \log|\boldsymbol{n}\cdot\hat{\boldsymbol{x}}|\, ds_{\boldsymbol{n}}, \tag{16.11}$$

where

$$\boldsymbol{B}_p = \frac{\boldsymbol{A}_E^{-1}(\boldsymbol{n})\boldsymbol{M}\boldsymbol{n} \otimes \boldsymbol{A}_E^{-1}(\boldsymbol{n})\boldsymbol{N}\boldsymbol{n}}{g - \boldsymbol{M}\boldsymbol{n}\cdot\boldsymbol{A}_E^{-1}(\boldsymbol{n})\boldsymbol{N}\boldsymbol{n}}. \tag{16.12}$$

It is therefore concluded that

The Green's function (16.11) can be written as

$$\boldsymbol{v}^g_{ep} = \boldsymbol{v}^g_e + \boldsymbol{v}^g_p, \tag{16.13}$$

where $\boldsymbol{v}_e$ is the Green's function relative to the elastic response and

$$\boldsymbol{v}_p^g(\hat{r},\theta) = -\frac{\log \hat{r}}{4\pi^2}\int_0^{2\pi} \boldsymbol{B}_p(\boldsymbol{n})\,\boldsymbol{e}^g\,d\alpha - \frac{1}{4\pi^2}\int_0^{2\pi} \boldsymbol{B}_p(\boldsymbol{n})\,\boldsymbol{e}^g \log|\cos\alpha|\,d\alpha, \tag{16.14}$$

(in which $\boldsymbol{n}$ is inclined at $\alpha+\theta$ with respect to x_1 axis) *can be interpreted as a sort of 'plastic correction'*, a circumstance having implications in the development of a boundary element approach to elastoplasticity (Bertoldi et al., 2005).

We finally remark that the Green's function (16.13) is defined for both associative $\boldsymbol{M}=\boldsymbol{N}$ and non-associative $\boldsymbol{M}\neq\boldsymbol{N}$ flow rules.

In the simple case of a linearly elastic isotropic material, the infinite-body Green's function for displacement, strains and stresses can be determined easily from Eq. (16.8). In fact, for a linear elastic isotropic material [Eq. (2.190)], the acoustic tensor and its inverse are given by Eqs. (2.176) and (2.177), respectively. Therefore, Eq. (16.8) provides the Green's function in the form

$$u_t^g(\hat{r},\theta) = -\frac{1}{4\pi^2\mu}\int_0^{2\pi}\left\{\delta_{tg} + \frac{\lambda+\mu}{\lambda+2\mu}\cos\left[\alpha+\theta+(1-g)\frac{\pi}{2}\right]\cos\left[\alpha+\theta+(1-t)\frac{\pi}{2}\right]\right\} \times \left(\log\hat{r} + \log|\cos\alpha|\right) d\alpha, \tag{16.15}$$

which, using the relations

$$\int_0^{\pi/2} \log(\cos\alpha)\left\{\begin{array}{c}\cos^2\alpha\\ \sin^2\alpha\end{array}\right\} d\alpha = \frac{\pi}{8}(\pm 1 - \log 4) \tag{16.16}$$

and neglecting constant terms, yields finally the Green's function for linear isotropic elasticity

$$u_t^g = -\frac{1}{8\pi\mu(1-\nu)}\left[\delta_{tg}(3-4\nu)\log\hat{r} - \cos\left(\theta+(1-g)\frac{\pi}{2}\right)\cos\left(\theta+(1-t)\frac{\pi}{2}\right)\right], \tag{16.17}$$

where ν denotes Poisson's ratio.

The Green's function for displacement gradient can be obtained through differentiation of Eq. (16.17) using the rule (2.136):

$$\begin{aligned}
u^1_{1,1} &= -\frac{[2(1-2\nu)+\cos 2\theta]\cos\theta}{8\pi(1-\nu)\mu r}, & u^2_{1,1} &= -\frac{\sin\theta\cos 2\theta}{8\pi(1-\nu)\mu r},\\
u^1_{1,2} &= -\frac{[4(1-\nu)+\cos 2\theta]\sin\theta}{8\pi(1-\nu)\mu r}, & u^2_{1,2} &= \frac{\cos\theta+\cos 3\theta}{16\pi(1-\nu)\mu r},\\
u^1_{2,1} &= u^2_{1,1}, & u^2_{2,1} &= \frac{[\cos 2\theta - 4(1-\nu)]\cos\theta}{8\pi(1-\nu)\mu r},\\
u^1_{2,2} &= u^2_{1,2}, & u^2_{2,2} &= \frac{[\cos 2\theta - 2(1-2\nu)]\sin\theta}{8\pi(1-\nu)\mu r},
\end{aligned} \tag{16.18}$$

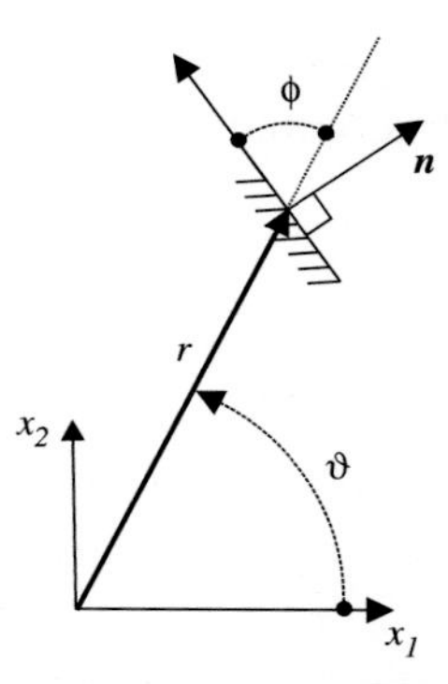

Figure 16.3. Definition of angles θ and ϕ; the latter is useful in the definition of the Green's applied tractions on an area element of unit normal $\boldsymbol{n}$.

so the Green's deformation field can be obtained through symmetrisation and the Green's stress field using the constitutive Eq. (2.190), thus yielding

$$\begin{aligned}
T_{11}^{1} &= -\frac{[2(1-\nu)+\cos 2\theta]\cos\theta}{4\pi(1-\nu)r}, & T_{11}^{2} &= -\frac{(2\nu+\cos 2\theta)\sin\theta}{4\pi(1-\nu)r},\\
T_{12}^{1} &= -\frac{[2(1-\nu)+\cos 2\theta]\sin\theta}{4\pi(1-\nu)r}, & T_{12}^{2} &= -\frac{[2(1-\nu)-\cos 2\theta]\cos\theta}{4\pi(1-\nu)r},\\
T_{22}^{1} &= \frac{(1-4\nu)\cos\theta+\cos 3\theta}{8\pi(1-\nu)r}, & T_{22}^{2} &= \frac{[\cos 2\theta-2(1-\nu)]\sin\theta}{4\pi(1-\nu)r}.
\end{aligned} \tag{16.19}$$

The components of the Green's functions for applied tractions on an area element of unit normal $\boldsymbol{n}$, that is, inclined at $\theta+\phi$, with the angle ϕ defined according to the convention explained in Fig. 16.3, can be calculated from Eqs. (16.19) in the form

$$\begin{aligned}
\tau_1^1 &= -\sin\phi\frac{1-2\nu+2\cos^2\theta}{4\pi(1-\nu)r}, & \tau_1^2 &= \frac{(1-2\nu)\cos\phi-2\sin\phi\cos\theta\sin\theta}{4\pi(1-\nu)r},\\
\tau_2^1 &= -\frac{(1-2\nu)\cos\phi+2\sin\phi\cos\theta\sin\theta}{4\pi(1-\nu)r}, & \tau_2^2 &= -\sin\phi\frac{1-2\nu+2\sin^2\theta}{4\pi(1-\nu)r}.
\end{aligned} \tag{16.20}$$

The case of an incompressible material can be obtained from Eqs. (16.20) simply by setting $\nu=1/2$. In view of applications to boundary elements, Bigoni et al. (2007) noticed that for $\phi=0$ and $\phi=\pi$, all components of the Green's function for tractions vanish in the incompressible case. Moreover, $\tau_2^1=\tau_1^2$ in the incompressible case or when $\phi=\pi/2$ and $\phi=3\pi/2$.

Quasi-static Green's function for incompressible incremental elasticity

The quasi-static Green's function for the incompressible Biot material deformed in plane strain [constitutive equations (6.178) and (6.179); see also Section 12.2] has been solved by Bigoni and Capuani (2002) and requires an ad hoc approach. Briefly, equilibrium equations expressed in terms of the stream function [Eqs. (12.25)], can be rewritten including a unit force $\delta(\boldsymbol{x})$ of components δ_{ig} as

$$(1+k)\psi^g_{,1111}+2(2\xi-1)\psi^g_{,1122}+(1-k)\psi^g_{,2222}+\frac{1}{\mu}\left[\delta_{1g}\delta_{,2}(\boldsymbol{x})-\delta_{2g}\delta_{,1}(\boldsymbol{x})\right]=0, \tag{16.21}$$

where ψ^g is the infinite-body Green's stream function. Enjoying properties (16.4), Eq. (16.21) becomes in the transformed domain the ordinary differential equation

$$L(\boldsymbol{n})\left(\tilde{\psi}^g\right)'''' = 2\frac{\delta_{1g}n_2 - \delta_{2g}n_1}{(\boldsymbol{n}\cdot\boldsymbol{x})^3}, \tag{16.22}$$

where a prime denotes differentiation, and

$$L(\boldsymbol{n}) = \mu n_2^4(1+k)\left(\frac{n_1^2}{n_2^2} - \gamma_1\right)\left(\frac{n_1^2}{n_2^2} - \gamma_2\right) > 0 \tag{16.23}$$

(a function strictly positive in the elliptic range), in which, recalling the definition (12.28) of Ω_i^2, we have

$$\gamma_1 = \frac{1-k}{1+k}\Omega_2^2, \qquad \gamma_2 = \frac{1-k}{1+k}\Omega_1^2. \tag{16.24}$$

A straightforward integration of Eq. (16.22) yields

$$\tilde{\psi}^g = \frac{\delta_{1g}n_2 - \delta_{2g}n_1}{L(\boldsymbol{n})}(\boldsymbol{n}\cdot\boldsymbol{x})\left(\log|\boldsymbol{n}\cdot\hat{\boldsymbol{x}}| - 1\right), \tag{16.25}$$

which, anti-transformed, leads to the Green's stream function

$$\begin{aligned}\psi^g = -\frac{r}{2\pi^2\mu(1+k)}\Bigg[&(\log\hat{r} - 1)\int_0^{\pi}\frac{\sin[\alpha+\theta+(1-g)\pi/2]\cos\alpha}{\Lambda(\alpha+\theta)}d\alpha\\ &+\int_0^{\pi/2}\frac{\sin[\alpha+\theta+(1-g)\pi/2]\cos\alpha\log(\cos\alpha)}{\Lambda(\alpha+\theta)}d\alpha\\ &-\int_0^{\pi/2}\frac{\cos[\alpha+\theta+(1-g)\pi/2]\sin\alpha\log(\sin\alpha)}{\Lambda(\alpha+\theta+\pi/2)}d\alpha\Bigg],\end{aligned} \tag{16.26}$$

where

$$\Lambda(\alpha) = \sin^4\alpha\left(\cot^2\alpha - \gamma_1\right)\left(\cot^2\alpha - \gamma_2\right) > 0. \tag{16.27}$$

Since from the definition of stream function (12.24) we have

$$v_1^g = \psi_{,2}^g, \quad v_2^g = -\psi_{,1}^g, \tag{16.28}$$

the Green's function for incremental displacements can be obtained by differentiating Eq. (16.26) and employing rule (2.136) to arrive at

$$\begin{aligned}v_j^g = \frac{1}{2\pi^2\mu(1+k)}\Bigg\{&\frac{\pi\,\delta_{jg}\log r}{[(2-j)\gamma_2+1-j]\sqrt{-\gamma_1}+[(2-j)\gamma_1+1-j]\sqrt{-\gamma_2}}\\ &-\int_0^{\frac{\pi}{2}}\left[K_j^g(\alpha+\theta)+(3-2j)(3-2g)K_j^g(\alpha-\theta)\right]\log(\cos\alpha)\,d\alpha\Bigg\},\end{aligned} \tag{16.29}$$

where

$$K_j^g(\alpha) = \frac{\sin[\alpha+(j-1)\frac{\pi}{2}]\sin[\alpha+(g-1)\frac{\pi}{2}]}{\Lambda(\alpha)}. \tag{16.30}$$

An application of the gradient rule (2.136) to Eq. (16.29) yields the Green's function for the gradient of incremental displacements:

$$v^j_{j,g} = \frac{1}{2\pi^2\mu(1+k)r}\left\{\frac{(3-2j)\pi\cos[\theta+(1-g)\frac{\pi}{2}]}{[(2-j)\gamma_2+1-j]\sqrt{-\gamma_1}+[(2-j)\gamma_1+1-j]\sqrt{-\gamma_2}}\right.$$
$$+\sin\left[\theta+(1-g)\frac{\pi}{2}\right]\int_0^{\frac{\pi}{2}}\Sigma\left(\alpha+\theta+(j-1)\frac{\pi}{2},\alpha+\theta\right)\log(\cos\alpha)\,d\alpha \tag{16.31}$$
$$\left.-\sin\left[\theta+(1-g)\frac{\pi}{2}\right]\int_0^{\frac{\pi}{2}}\Sigma\left(\alpha-\theta-(j-1)\frac{\pi}{2},\alpha+\theta\right)\log(\cos\alpha)\,d\alpha\right\},$$

where index j is not summed, and

$$\Sigma(\alpha,\beta) = \frac{\sin\alpha\,[2\cos\alpha\,\Lambda(\beta)-\sin\alpha\,\Lambda'(\beta)]}{\Lambda^2(\beta)} \quad \text{and} \quad \Lambda'(\beta) = \frac{\partial\Lambda(\beta)}{\partial\beta}. \tag{16.32}$$

The four components of the velocity gradient not included in Eq. (16.31) can be derived using the incompressibility constraint and the symmetry of Green's tensor (16.29); thus

$$v^1_{2,2} = v^2_{1,2} = -v^1_{1,1} \quad \text{and} \quad v^1_{2,1} = v^2_{1,1} = -v^2_{2,2}. \tag{16.33}$$

The solution for the Green's function set for an incompressible material is not yet complete because knowledge of the velocity gradient does not allow determination of the Green's function for in-plane incremental mean stress $\dot{p}$. To determine this Green's function, we consider again the equilibrium equations (12.23) written for the Green's function; thus the incremental concentrated forces have to be added:

$$\begin{aligned}\dot{p}^g_{,1} - \mu k v^g_{1,11} &= \mu[(1-2\xi)v^g_{1,11} - (1-k)v^g_{1,22}] - \delta_{1g}\delta(\boldsymbol{x}),\\ \dot{p}^g_{,2} - \mu k v^g_{1,12} &= \mu[(1-2\xi)v^g_{2,22} - (1+k)v^g_{2,11}] - \delta_{2g}\delta(\boldsymbol{x}).\end{aligned} \tag{16.34}$$

Introducing now the in-plane mean nominal stress

$$\dot{\pi} = \frac{\dot{t}_{11}+\dot{t}_{22}}{2} = \dot{p} - \frac{T_1-T_2}{2}v_{1,1}, \tag{16.35}$$

differentiating Eq. $(16.34)_1$ with respect to x_1, Eq. $(16.34)_2$ with respect to x_2 and summing the resulting equations yields

$$\begin{aligned}\frac{\dot{\pi}_{,11}+\dot{\pi}_{,22}}{\mu} &= 2(1-\xi)\left(v_{1,111}+v_{2,222}\right)\\ &\quad +k\left(v_{1,111}-v_{2,222}\right) - \frac{\delta_{1g}\delta_{,1}(\boldsymbol{x})+\delta_{2g}\delta_{,2}(\boldsymbol{x})}{\mu},\end{aligned} \tag{16.36}$$

which in the transformed domain becomes the ordinary differential equation

$$\begin{aligned}\frac{(\tilde{\pi}^g)''}{\mu} &= 2(1-\xi)\left[n_1^3\left(\tilde{v}^g_1\right)''' + n_2^3\left(\tilde{v}^g_2\right)'''\right]\\ &\quad +k\left[n_1^3\left(\tilde{v}^g_1\right)''' - n_2^3\left(\tilde{v}^g_2\right)'''\right] + 2\frac{n_1\delta_{1g}+n_2\delta_{2g}}{\mu(\boldsymbol{n}\cdot\boldsymbol{x})^3}.\end{aligned} \tag{16.37}$$

Since $\tilde{v}_i^g$ and its derivatives are known from differentiation of Eq. (16.25), Eq. (16.37) can be integrated and anti-transformed to give the Green's function for the in-plane mean nominal stress increment

$$\dot{\pi}^g = -\frac{1}{2\pi r}\left\{\cos\left[\theta - (g-1)\frac{\pi}{2}\right] + \frac{1}{\pi(1+k)}\int_0^{\pi} \frac{\tilde{K}_g(\alpha+\theta)}{\cos\alpha} d\alpha\right\}, \tag{16.38}$$

where

$$\tilde{K}_g(\alpha) = K_g^g(\alpha)\left[2\left(\frac{\mu_*}{\mu} - 1\right)\left(2\cos^2\alpha - 1\right) - k\right]\cos\left[\alpha + \delta_{2g}\frac{\pi}{2}\right], \tag{16.39}$$

completing the infinite-body Green's function set. In particular, application of the constitutive equations [Eqs. (6.178) and (6.179)] to the Green's velocity gradient and adding the Green's in-plane mean nominal stress (16.38) yields the Green's function for nominal stress:

$$\begin{aligned} \dot{t}_{11}^g &= (2\mu_* - p)\, v_{1,1}^g + \dot{\pi}^g, & \dot{t}_{12}^g &= (\mu - p)\, v_{1,2}^g + \left(\mu + \frac{\sigma}{2}\right) v_{2,1}^g, \\ \dot{t}_{21}^g &= (\mu - p)\, v_{2,1}^g + \left(\mu - \frac{\sigma}{2}\right) v_{1,2}^g, & \dot{t}_{22}^g &= (2\mu_* - p)\, v_{2,2}^g + \dot{\pi}^g. \end{aligned} \tag{16.40}$$

In the particular case of null pre-stress ($k = \eta = 0$) and isotropic constitutive equations ($\xi = 1$), the infinite-body Green's function set is

$$v_j^j = -\frac{\log\hat{r}}{4\pi\mu} + \frac{3-2j}{8\pi\mu}\cos 2\theta, \quad v_2^1 = v_1^2 = \frac{\sin\theta\cos\theta}{4\pi\mu}, \quad \dot{p}^1 = -\frac{\cos\theta}{2\pi r}, \quad \dot{p}^2 = -\frac{\sin\theta}{2\pi r}, \tag{16.41}$$

from which the Green's function for stress can be derived and results in the following form (expressed in Cartesian coordinates)

$$\begin{aligned} T_{11}^1 &= -\frac{x_1^3}{\pi r^4}, & T_{11}^2 &= -\frac{x_1^2 x_2}{\pi r^4}, \\ T_{12}^1 &= T_{11}^2, & T_{12}^2 &= -\frac{x_1 x_2^2}{\pi r^4}, \\ T_{22}^1 &= T_{12}^2, & T_{22}^2 &= -\frac{x_2^3}{\pi r^4}, \end{aligned} \tag{16.42}$$

which also can be obtained from Eqs. (16.19) taking $\nu = 1/2$.

Quasi-static dipoles in an infinite incompressible elastic medium Superimposing the effects, the Green's function (16.29) can be used to investigate the response of an incompressible non-linear elastic infinite medium that is homogeneously pre-stressed, and perturbed by a quasi-statically applied dipole (two equal and opposite concentrated forces placed at a distance, say, $2a$). In particular, the level sets of the modulus of incremental displacement vector induced by a dipole parallel to the horizontal axis x_1 are reported in Figs. 1.39 and 1.38, respectively, for an orthotropic ($\mu_*/\mu = 1/4$) and a Mooney-Rivlin ($\mu_*/\mu = 1$) material. The incremental response at null pre-stress is reported on the left, whereas the response for a pre-stress near the boundary of ellipticity loss ($k = 0.98$ and $k = 0.86$, respectively) is reported on the right.

It is evident from the figures (additional results can be found in Bigoni and Capuani, 2002) that

> *whilst at null pre-stress the incremental deformation field does not evidence any special structure, a shear band field emerges as a response to the dipole perturbation at a pre-stress level near (but still below) the boundary of ellipticity.*

Moreover,

> *shear bands also become visible for a Mooney-Rivlin material, where these remain simply excluded following the conventional approach based on the singularity of the acoustic tensor (see Section 11.4).*

A number of results concerning various dipole inclinations at different levels of pre-stress have been presented by Bigoni and Capuani (2002) and are not repeated here. It may be interesting instead to analyse the effect of a random distribution of randomly oriented dipoles. This random distribution may represent the presence of defects in materials, and in particular, these defects can be representative of a dislocation distribution.[1] In fact, there are analogies between the solutions of a concentrated force and of a dislocation in an elastic continuum and (as shown below)

> *when the pre-stress is absent and for isotropic incompressible elasticity, the far fields induced by a dislocation dipole on a single slip plane and a force dipole inclined at $\pi/4$ are identical.*

In an x–y reference system, the in-plane stress field produced by *a single straight edge dislocation* in a linearly elastic, isotropic medium is (Love, 1927; Hirth and Lothe, 1968)

$$\left\{\sigma^d_{xx}, \sigma^d_{yy}, \sigma^d_{xy}\right\} = \frac{b\mu}{2\pi(1-\nu)r^4}\left\{-y\left(3x^2+y^2\right),\, y\left(x^2-y^2\right),\, x\left(x^2-y^2\right)\right\}, \tag{16.43}$$

where b is the Burgers vector and μ and ν are the elastic shear modulus and Poisson's ratio. The corresponding strain field follows from Eq. (16.43) in the form

$$\left\{\epsilon^d_{xx}, \epsilon^d_{yy}, \epsilon^d_{xy}\right\} = \frac{1}{\bar{E}}\left\{\sigma^d_{xx} - \bar{\nu}\sigma^d_{yy},\; \sigma^d_{yy} - \bar{\nu}\sigma^d_{xx},\; (1+\bar{\nu})\sigma^d_{xy}\right\}, \tag{16.44}$$

where $\bar{E}$ and $\bar{\nu}$ are the modified Young modulus and Poisson's ratio, whereas the displacement field can be written as

$$u_x = \frac{b}{2\pi}\left[\arctan\frac{y}{x} + \frac{xy}{2(1-\nu)r^2}\right], \qquad u_y = -\frac{b}{8\pi(1-\nu)}\left[(1-2\nu)\log r^2 + \frac{x^2-y^2}{r^2}\right]. \tag{16.45}$$

A dislocation dipole on a single glide plane consists of two parallel edge dislocations lying in the same slip plane at a distance $2d$ and having opposite sign (see Fig. 16.4, *left*, for a sketch of the distortion induced in a crystal lattice, and Fig. 16.4, *right*, for its graphical conventional representation). In quasi-static conditions, such a simple dislocation structure is not stable, so the dislocations, unless pinned, 'attract'

[1] The analogy between dislocation and force dipoles has been suggested to me by Prof. A. B. Movchan (University of Liverpool, UK).

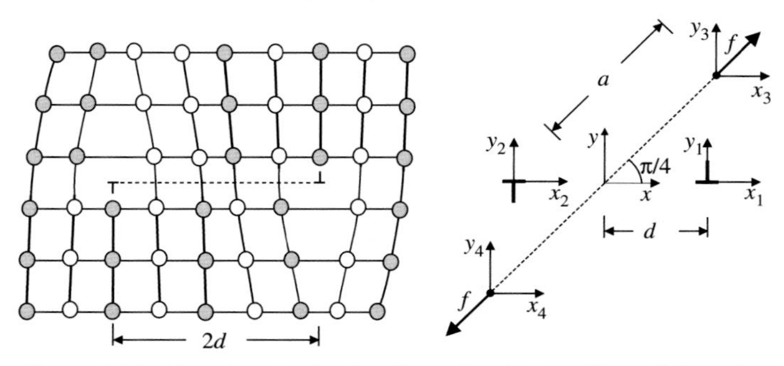

Figure 16.4. Sketch of the lattice distortion induced by a dislocation dipole lying on a single slip plane (*left*) and its conventional representation (*right*), together with a force dipole (inclined at $\pi/4$ with respect to x–y system). In linear isotropic *incompressible* elasticity, the far fields induced by the dislocation dipole and by the force dipole are identical.

each other to reduce their total elastic energy. In this way, they move towards each other until they combine and annihilate.

Leaving aside issues of stability, the stress field produced by the dislocation dipole simply can be obtained through superimposition of solution (16.43), which can be used with reference to the 'local' coordinates (Fig. 16.4 *right*):

$$x_i = x + (-1)^i d, \qquad y_i = y, \qquad r_i^2 = [x + (-1)^i d]^2 + y^2 = r^2 + (-1)^i 2xd + d^2, \quad (16.46)$$

where $i = 1,2$.

At a large distance from the dislocation dipole, parameter d/r can be considered small, and the stress fields can be expanded into a Taylor series to obtain the far field approximation

$$\left\{\sigma_{xx}^d, \sigma_{yy}^d, \sigma_{xy}^d\right\} \sim \frac{\mu d b}{\pi(1-\nu) r^6}\left\{2xy\left(r^2 - 4x^2\right), 2xy\left(r^2 - 4y^2\right), r^4 - 8x^2y^2\right\}, \quad (16.47)$$

which satisfies equilibrium equations.

The far-field stress field for a force dipole in linear elasticity can be obtained by superimposing two solutions in the form (16.19), each of which can be expressed in the local coordinate system shown in Fig. 16.4 (*right*); thus we may write

$$x_i = x + (-1)^i a\cos\alpha, \qquad y_i = y + (-1)^i a \sin\alpha,$$

$$r_i^2 = [x + (-1)^i a\cos\alpha]^2 + [y + (-1)^i a\sin\alpha]^2 = r^2 + (-1)^i 2a(x\cos\alpha + y\sin\alpha) + a^2, \quad (16.48)$$

where $i = 3,4$ and $\alpha = \pi/4$. At a large distance from the force dipole, the dimensionless parameter a/r becomes small, so a Taylor series expansion of the stress fields again gives representation (16.47) with the correspondence

$$fa = 2db\mu. \quad (16.49)$$

The effects of pre-stress have been explored by perturbing a uniform strain field with a random distribution of randomly oriented force dipoles (randomness has been obtained by using the 'pseudo-random' real number generation function available in Mathematica 5.2). In particular, a square 'window' of material of edge $20a$ ($200a$

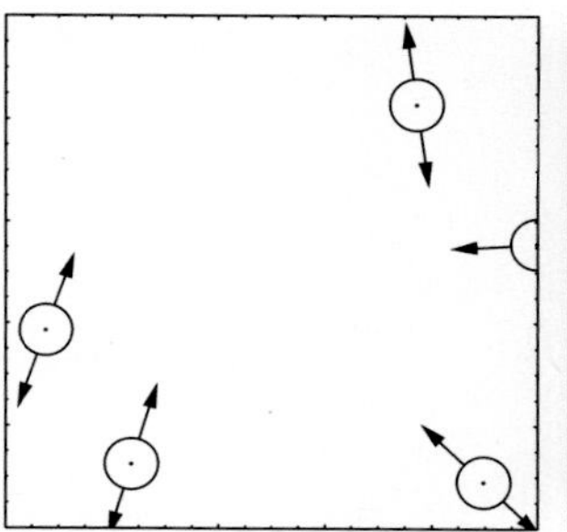
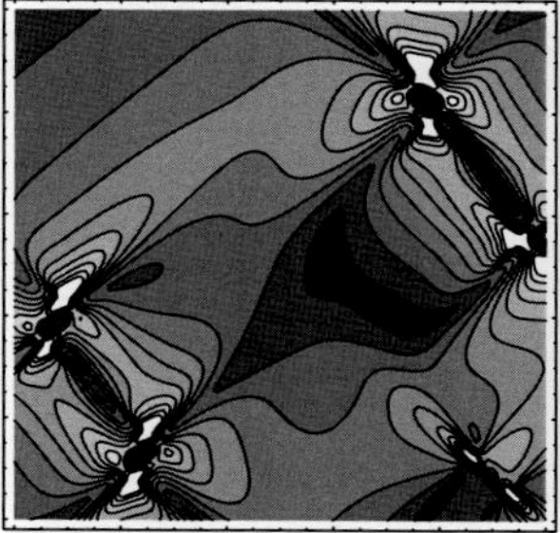

Figure 16.5. Effects of pre-stress on a material revealed by a perturbation consisting of five randomly placed and oriented dipoles. The five randomly selected dipoles are shown on the left, where the radius of the circles is the dipole distance and provides the bar scale of the representation, whereas the arrows evidence the force inclinations. The map of incremental displacements induced by the dipoles is shown in the centre at null pre-stress and on the left at a pre-stress near the border of ellipticity loss, $k = 0.86$, for an orthotropic material with $\xi = \mu_*/\mu = 1/4$.

for Fig. 16.6) has been considered in Fig. 16.5, where a is the half distance between the forces forming the dipole. Inside this window, one distribution of five dipoles have been randomly placed in Fig. 16.5, with random inclination of the forces (two distributions of 20 dipoles have been generated in Fig. 16.6). The dipole distributions and inclinations are shown in Figs. 16.5 and 16.6 on the left, whereas the maps of modulus of displacements are shown in the centre for null pre-stress and on the right for a pre-stress near the boundary of loss of ellipticity ($k = 0.86$). An orthotropic material with $\xi = \mu_*/\mu = 1/4$ has been considered. The figures reveal that the effect of pre-stress consists of *the emerging of a texture with privileged directions corresponding to the shear band inclinations that occur at the boundary of ellipticity loss.*

16.1.2 The dynamic time-harmonic Green's function for general non-symmetric constitutive equations

An initial static homogeneous deformation of an infinite body is considered in the absence of body forces, satisfying equilibrium in terms of first Piola-Kirchhoff stress [Eq. (3.115)], and taken as the reference state in an updated Lagrangian formulation. A dynamic perturbation is superimposed on this state, defined by an incremental displacement $\boldsymbol{v}$ satisfying the equations of incremental motion, written with reference to the constitutive equation (8.100) in which dotted symbols have to be interpreted now as incremental quantities rather than rates. Thus the incremental equations of motions are

$$\mathbb{G}_{ijkl}u_{k,lj} + f_i = \rho u_{i,tt}, \tag{16.50}$$

where u_i is the incremental displacement, the subscript t denotes the material time derivative and f_i and ρ are the incremental body forces and the mass density, respectively.

Equations (16.50) look like ordinary elastodynamics, except that

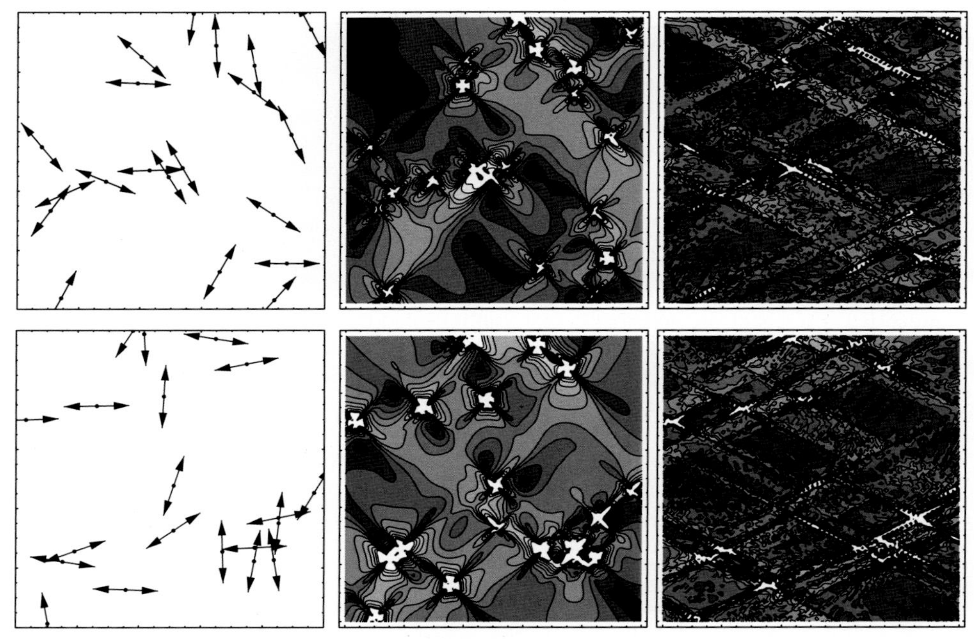

Figure 16.6. Effects of pre-stress on a material revealed by a perturbation consisting of 20 randomly placed and oriented dipoles. The 20 randomly selected dipoles are shown on the left, where the radius of the circles is the dipole distance and provides the bar scale of the representation, whereas the arrows evidence the force inclinations. The maps of incremental displacements induced by the dipoles are shown in the centre at null pre-stress and on the left at a pre-stress near the border of ellipticity loss, $k = 0.86$, for an orthotropic material with $\xi = \mu_*/\mu = 1/4$.

$\mathbb{G}_{ijkl}$ has neither the usual major $\mathbb{G}_{ijkl} \neq \mathbb{G}_{klij}$ nor the minor $\mathbb{G}_{ijlk} \neq \mathbb{G}_{ijkl} \neq \mathbb{G}_{jikl}$ symmetries.

Note that tensor $\mathbb{G}_{ijkl}$ can be identified with that provided by Eq. (8.100) or with the elastoplastic tangent operator restricted to the loading branch, but it also can be thought completely arbitrary in the following. The dynamic Green's function is sought under the time-harmonic assumption

$$u_j(\boldsymbol{x},t) = \hat{u}_j(\boldsymbol{x})e^{-i\omega t}, \qquad f_j(\boldsymbol{x},t) = \hat{f}_j(\boldsymbol{x})e^{-i\omega t}, \tag{16.51}$$

where ω is the circular frequency and t and $\boldsymbol{x}$ denote time and space variables, respectively, so that the time dependence can be removed from Eq. (16.50), which consequently becomes

$$\mathbb{G}_{ijkl}\hat{u}_{k,lj} + \rho\,\omega^2\hat{u}_i + \hat{f}_i = 0. \tag{16.52}$$

The Green's tensor $G_{ip}(\boldsymbol{x})$ is obtained by solving Eq. (16.52) under the hypothesis $\hat{f}_i = \delta_{ip}\delta(\boldsymbol{x})$, with $\delta(\boldsymbol{x})$ denoting the two-dimensional Dirac delta. We obtain

$$\mathbb{G}_{ijkl}G_{kq,lj}(\boldsymbol{x}) + \rho\,\omega^2 G_{iq}(\boldsymbol{x}) + \delta_{iq}\delta(\boldsymbol{x}) = 0. \tag{16.53}$$

Radon transform The Green's function is determined employing a Radon transform technique.[2] In particular, the Radon transform of a generic function $f(\boldsymbol{x})$, $\boldsymbol{x} \in \mathbf{R}^2$ is defined as

$$\mathcal{R}[f(\boldsymbol{x})] = \hat{f}(p,\boldsymbol{n}) = \int_{\mathbf{R}^2} f(\boldsymbol{x})\delta(p - \boldsymbol{n}\cdot\boldsymbol{x})\, d\boldsymbol{x}, \qquad p \in \mathbf{R},\ \boldsymbol{n} \in \mathbf{R}^2 \tag{16.54}$$

with the inverse

$$f(\boldsymbol{x}) = \frac{1}{4\pi^2} \int_{|\boldsymbol{n}|=1} \int_{-\infty}^{+\infty} \frac{\hat{f}'(p,\boldsymbol{n})}{(\boldsymbol{n}\cdot\boldsymbol{x} - p)}\, dp\, ds_{\boldsymbol{n}}, \tag{16.55}$$

where a prime denotes partial differentiation in the following way

$$\hat{f}'(p,\boldsymbol{n}) = \frac{\partial \hat{f}(p,\boldsymbol{n})}{\partial p}. \tag{16.56}$$

In addition to the linearity, we will make use of the following properties of the Radon transform:

- Derivative transforms

$$\mathcal{R}\left[f_{,j}(\boldsymbol{x})\right] = n_j \hat{f}'(p,\boldsymbol{n}), \qquad \mathcal{R}\left[f_{,lj}(\boldsymbol{x})\right] = n_l n_j \hat{f}''(p,\boldsymbol{n}) \tag{16.57}$$

- Transform of the two-dimensional Dirac delta function:

$$\mathcal{R}[\delta(\boldsymbol{x})] = \delta(p). \tag{16.58}$$

The Radon transform of Eq. (16.53) therefore is

$$\mathbb{G}_{ijkl} n_l n_j \hat{G}''_{kq}(p,\boldsymbol{n}) + \rho\omega^2 \hat{G}_{iq}(p,\boldsymbol{n}) + \delta_{iq}\delta(p) = 0, \tag{16.59}$$

where

$$\hat{G}''_{kq}(p,\boldsymbol{n}) = \frac{\partial^2}{\partial p^2} \hat{G}_{kq}(p,\boldsymbol{n}). \tag{16.60}$$

Equation (16.59) can be rewritten in tensorial form as

$$\boldsymbol{A}(\boldsymbol{n})\hat{\boldsymbol{G}}''(p,\boldsymbol{n}) + \omega^2 \hat{\boldsymbol{G}}(p,\boldsymbol{n}) + \frac{\delta(p)}{\rho}\boldsymbol{I} = \boldsymbol{0}, \tag{16.61}$$

where $\boldsymbol{A}(\boldsymbol{n})$ is the acoustic tensor corresponding to the fourth-order constitutive tensor $\mathbb{G}$.

Let us assume that $\boldsymbol{A}(\boldsymbol{n})$ has two non-null and distinct eigenvalues c_N^2 and corresponding left and right eigenvectors $\boldsymbol{w}_N$ and $\boldsymbol{v}_N$ ($N = 1,2$), which can be used as dual basis vectors (see Section 2.6), therefore satisfying $\boldsymbol{v}_N \cdot \boldsymbol{w}_M = \delta_{NM}$, $(N,M = 1,2)$. Employing the spectral representations of $\boldsymbol{A}(\boldsymbol{n})$ and $\boldsymbol{I}$,

$$\boldsymbol{A}(\boldsymbol{n}) = \sum_{N=1}^{2} c_N^2 \boldsymbol{v}_N \otimes \boldsymbol{w}_N, \qquad \boldsymbol{I} = \sum_{N=1}^{2} \boldsymbol{v}_N \otimes \boldsymbol{w}_N, \tag{16.62}$$

[2] The alternative approach based on plane wave expansion previously employed to obtain the quasi-static Green's function also can be used; see appendix A of Piccolroaz et al. (2006).

in Eqs. (16.61) and representing the transformed Green's function as

$$\hat{\boldsymbol{G}}(p,\boldsymbol{n}) = \sum_{N=1}^{2} \phi_N(p,\boldsymbol{n})\boldsymbol{v}_N \otimes \boldsymbol{w}_N, \tag{16.63}$$

where ϕ_N is a (for the moment unknown) function of p and $\boldsymbol{n}$, we obtain

$$\sum_{N=1}^{2} \left[c_N^2 \phi_N'' + \omega^2 \phi_N + \frac{\delta(p)}{\rho} \right] \boldsymbol{v}_N \otimes \boldsymbol{w}_N = \boldsymbol{0}, \tag{16.64}$$

which is equivalent to the following uncoupled system of two equations,

$$\phi_N'' + k_N^2 \phi_N + \frac{1}{\rho c_N^2} \delta(p) = 0, \qquad N = 1,2, \tag{16.65}$$

where the wavenumber $k_N = \omega / c_N$ has been introduced. Since we have chosen the harmonic time dependence to be of the form $e^{-i\omega t}$, the outgoing wave solution of Eq. (16.65) in the p coordinate is

$$\phi_N(p,\boldsymbol{n}) = -\frac{e^{ik_N|p|}}{2\rho\, i k_N c_N^2}. \tag{16.66}$$

Thus

$$\hat{\boldsymbol{G}}(p,\boldsymbol{n}) = -\sum_{N=1}^{2} \frac{e^{ik_N|p|}}{2\rho\, i k_N c_N^2} \boldsymbol{v}_N \otimes \boldsymbol{w}_N \tag{16.67}$$

and

$$\hat{\boldsymbol{G}}'(p,\boldsymbol{n}) = -\sum_{N=1}^{2} \frac{\operatorname{sgn}(p) e^{ik_N|p|}}{2\rho c_N^2} \boldsymbol{v}_N \otimes \boldsymbol{w}_N. \tag{16.68}$$

The anti-transform of Eq. (16.67) leads to

$$\boldsymbol{G}(\boldsymbol{x}) = -\frac{1}{4\pi^2} \sum_{N=1}^{2} \int_{|\boldsymbol{n}|=1} \int_{-\infty}^{+\infty} \frac{\operatorname{sgn}(p) e^{ik_N|p|}}{2\rho c_N^2 (\boldsymbol{n}\cdot\boldsymbol{x} - p)} \boldsymbol{v}_N \otimes \boldsymbol{w}_N \, dp\, ds_{\boldsymbol{n}}. \tag{16.69}$$

The integral in the variable p can be evaluated.[3] Thus, employing the cosine and sine integral functions

$$\operatorname{Ci}(z) = \int_{+\infty}^{z} \frac{\cos t}{t} dt, \qquad |\arg z| < \pi \qquad \text{and} \qquad \operatorname{Si}(z) = \int_{0}^{z} \frac{\sin t}{t} dt, \tag{16.73}$$

[3] To evaluate the integral in the variable p appearing in Eq. (16.69), the domain has to be split as

$$\int_{-\infty}^{+\infty} \frac{\operatorname{sgn}(p) e^{ik_N|p|}}{\xi - p} dp = -\int_{-\infty}^{0} \frac{e^{-ik_N p}}{\xi - p} dp + \int_{0}^{+\infty} \frac{e^{ik_N p}}{\xi - p} dp, \tag{16.70}$$

so the two integrals can be treated separately, that is,

$$-\int_{-\infty}^{0} \frac{e^{-ik_N p}}{\xi - p} dp = -e^{-ik_N \xi} \int_{k_N \xi}^{+\infty} \frac{e^{iq}}{q} dq, \tag{16.71}$$

where the substitution $q = k_N(\xi - p)$ has been made and

$$\int_{0}^{+\infty} \frac{e^{ik_N p}}{\xi - p} dp = -e^{ik_N \xi} \int_{-k_N \xi}^{+\infty} \frac{e^{iq}}{q} dq, \tag{16.72}$$

where the substitution $q = k_N(p - \xi)$ has been made. The two expressions [(16.71) and (16.72)] are used to obtain Eq. (16.74).

the Green's function finally can be written in the form

$$G(x) = -\frac{1}{8\pi^2}\sum_{N=1}^{2}\int_{|n|=1}[2\cos(k_N n\cdot x)\mathrm{Ci}(k_N|n\cdot x|) \tag{16.74}$$

$$+2\sin(k_N n\cdot x)\mathrm{Si}(k_N n\cdot x) - i\pi\cos(k_N n\cdot x)]\frac{v_N\otimes w_N}{\rho c_N^2}ds_n.$$

We introduce polar coordinates so that the position vector x has modulus $r = |x|$ and is inclined at angle θ to the x_1 axis. Taking the unit vector n inclined at $\alpha+\theta$ with respect to the x_1 axis (so that α is the angle between x and n) and noting that $\cos(\cdot)\,\mathrm{Ci}(\cdot)$ and $\sin(\cdot)\,\mathrm{Si}(\cdot)$ are even functions, we can rewrite Eq. (16.74) as

$$G(x) = -\frac{1}{8\pi^2}\sum_{N=1}^{2}\int_0^{2\pi}[2\cos(rk_N|\cos\alpha|)\mathrm{Ci}(rk_N|\cos\alpha|) \tag{16.75}$$

$$+2\sin(rk_N|\cos\alpha|)\mathrm{Si}(rk_N|\cos\alpha|) - i\pi\cos(rk_N|\cos\alpha|)]\frac{v_N\otimes w_N}{\rho c_N^2}d\alpha,$$

where k_N, v_N, w_N and c_N^2 depend on $\alpha+\theta$.

The acoustic tensor is a periodic function of α with period π because

$$\rho A_{ik}(n) = \mathbb{G}_{i1k1}n_1^2 + (\mathbb{G}_{i1k2}+\mathbb{G}_{i2k1})n_1n_2 + \mathbb{G}_{i2k2}n_2^2, \tag{16.76}$$

where $n_1 = \cos(\alpha+\theta)$ and $n_2 = \sin(\alpha+\theta)$, and also c_N, k_N, v_N and w_N are periodic functions of α with the same period. It follows that the integrand in Eq. (16.75) is π-periodic. Therefore,

> *the two-dimensional time-harmonic Green's function corresponding to a generic, completely non-symmetric constitutive fourth-order tensor, relating the increment of the first Piola-Kirchhoff stress to the deformation gradient increment [Eq. (8.100)],* can be written in the form

$$G(x) = -\frac{1}{4\pi^2}\sum_{N=1}^{2}\int_0^{\pi}[2\cos(rk_N|\cos\alpha|)\mathrm{Ci}(rk_N|\cos\alpha|) \tag{16.77}$$

$$+2\sin(rk_N|\cos\alpha|)\mathrm{Si}(rk_N|\cos\alpha|) - i\pi\cos(rk_N|\cos\alpha|)]\frac{v_N\otimes w_N}{\rho c_N^2}d\alpha,$$

where $k_N = \omega/c_N$ and c_N^2 are the eigenvalues of the acoustic tensor A [Eq. (11.45)] with corresponding left and right eigenvectors w_N and v_N, which are all quantities depending on n, in other words, on $\alpha+\theta$.

It can be noted that the integrand in Eq. (16.77) displays a logarithmic singularity at $r=0$ and $\alpha=\pi/2$ because (Lebedev, 1965)

$$\mathrm{Ci}(z) = \gamma + \log z - \int_0^z \frac{1-\cos t}{t}dt, \quad |\arg z| < \pi, \tag{16.78}$$

where γ is the Euler constant.

Time-harmonic Green's function for incompressible incremental elasticity
The dynamic time-harmonic Green's function for the incompressible Biot material deformed in plane strain [constitutive Eqs. (6.178) and (6.179); see also Section 12.2] has been solved by Bigoni and Capuani (2005) and requires a special approach to keep into account incompressibility. Equations of motion for superimposed small deformation expressed in terms of the stream function [Eqs. (12.25)], can be rewritten including circular frequency ω in Eq. (16.21) as

$$\begin{aligned}(1+k)\psi^g_{,1111}+2(2\xi-1)\psi^g_{,1122}+(1-k)\psi^g_{,2222}+\frac{1}{\mu}\left[\delta_{1g}\delta_{,2}(\boldsymbol{x})-\delta_{2g}\delta_{,1}(\boldsymbol{x})\right]\\ =-\frac{\rho\omega^2}{\mu}\left(\psi^g_{,11}+\psi^g_{,22}\right),\end{aligned} \tag{16.79}$$

where ψ^g is the time-harmonic infinite-body Green's stream function. Enjoying properties (16.4), Eq. (16.21) becomes in the transformed domain the ordinary differential equation

$$L(\boldsymbol{n})\left(\tilde{\psi}^g\right)''''+\rho\omega^2\left(\tilde{\psi}^g\right)''=2\frac{\delta_{1g}n_2-\delta_{2g}n_1}{(\boldsymbol{n}\cdot\boldsymbol{x})^3}, \tag{16.80}$$

where a prime denotes differentiation, and function $L(\boldsymbol{n})$, strictly positive in the elliptic range, is defined by Eq. (16.23).

On integration, Eq. (16.80) yields

$$\begin{aligned}\tilde{\psi}^g=\frac{\delta_{1g}n_2-\delta_{2g}n_1}{\omega\sqrt{\rho L(\boldsymbol{n})}}&\left[\mathrm{Ci}(\eta|\boldsymbol{n}\cdot\boldsymbol{x}|)\sin(\eta|\boldsymbol{n}\cdot\boldsymbol{x}|)-\mathrm{Si}(\eta|\boldsymbol{n}\cdot\boldsymbol{x}|)\cos(\eta|\boldsymbol{n}\cdot\boldsymbol{x}|)\right]\\ &+A_1\sin(\eta|\boldsymbol{n}\cdot\boldsymbol{x}|)+A_2\cos(\eta|\boldsymbol{n}\cdot\boldsymbol{x}|)+i\left[A_3\sin(\eta|\boldsymbol{n}\cdot\boldsymbol{x}|)+A_4\cos(\eta|\boldsymbol{n}\cdot\boldsymbol{x}|)\right],\end{aligned} \tag{16.81}$$

where A_j, $(j=1,\ldots,4)$ are arbitrary constants, and η is the wavenumber in the direction of $\boldsymbol{n}$, that is,

$$\eta=\omega\sqrt{\frac{\rho}{L(\boldsymbol{n})}}. \tag{16.82}$$

The arbitrary constants can be determined considering the far-field approximation for large $\boldsymbol{n}\cdot\boldsymbol{x}$ of function $\tilde{\psi}^g(\boldsymbol{n}\cdot\boldsymbol{x})$:

$$\tilde{\psi}^g=\frac{\delta_{1g}n_2-\delta_{2g}n_1}{\omega\sqrt{\rho L(\boldsymbol{n})}}\frac{\pi}{2}\left[\cos(\eta|\boldsymbol{n}\cdot\boldsymbol{x}|)+i\sin(\eta|\boldsymbol{n}\cdot\boldsymbol{x}|)\right]+O\left(\frac{1}{\boldsymbol{n}\cdot\boldsymbol{x}}\right) \tag{16.83}$$

(where an arbitrary harmonic solution has been neglected), which represents only outgoing waves. In order that Eq. (16.81) matches with Eq. (16.83), the arbitrary constants remain determined, so we arrive at

$$\begin{aligned}\tilde{\psi}^g=\frac{\delta_{1g}n_2-\delta_{2g}n_1}{\omega\sqrt{\rho L(\boldsymbol{n})}}&\left[\mathrm{Ci}(\eta|\boldsymbol{n}\cdot\boldsymbol{x}|)\sin(\eta|\boldsymbol{n}\cdot\boldsymbol{x}|)\right.\\ &\left.-\mathrm{Si}(\eta|\boldsymbol{n}\cdot\boldsymbol{x}|)\cos(\eta|\boldsymbol{n}\cdot\boldsymbol{x}|)-i\tfrac{\pi}{2}\sin(\eta|\boldsymbol{n}\cdot\boldsymbol{x}|)\right],\end{aligned} \tag{16.84}$$

which, anti-transformed, leads to the time-harmonic Green's stream function

$$\psi^g=-\frac{1}{2\pi^2\rho\omega c}\int_0^{\pi}\frac{\sin[\alpha+\theta+(1-g)\pi/2]}{\Lambda(\alpha+\theta)}\,\Xi\left(\frac{\omega r}{c}\frac{\cos\alpha}{\sqrt{\Lambda\alpha+\theta}}\right)d\alpha, \tag{16.85}$$

where the (strictly positive) function $\Lambda(\alpha)$ is defined by Eq. (16.27), whereas

$$\Xi(x) = \sin x \, \mathrm{Ci}(|x|) - \cos x \, \mathrm{Si}(|x|) - i\frac{\pi}{2}\sin x, \tag{16.86}$$

and c is the propagation speed of a transverse wave travelling parallel to the x_1 axis

$$c = \sqrt{\frac{\mu(1+k)}{\rho}} \tag{16.87}$$

such that $c\sqrt{\Lambda(\alpha)}$ is the velocity of propagation in the direction determined by the angle α.

From the definition of stream function (16.28), we may calculate the infinite-body Green's function for incremental displacements:

$$\begin{aligned} v_g^i(r,\theta) = &-\frac{(2\delta_{ig}-1)}{2\pi^2\mu(1+k)}\Bigg[\left(\log\hat{r}+\gamma\right)\int_0^{\pi} K_i^g(\alpha+\theta)\cos\left(\hat{r}\,\xi(\alpha,\alpha+\theta)\right)d\alpha \\ &+\int_0^{\frac{\pi}{2}} \log\left(\xi(\alpha,\alpha+\theta)\right) K_i^g(\alpha+\theta)\cos\left(\hat{r}\,\xi(\alpha,\alpha+\theta)\right)d\alpha \\ &+(2\delta_{ig}-1)\int_0^{\frac{\pi}{2}} \log\left(\xi(\alpha,\alpha-\theta)\right)K_i^g(\alpha-\theta)\cos\left(\hat{r}\,\xi(\alpha,\alpha-\theta)\right)d\alpha \\ &+\int_0^{\pi} K_i^g(\alpha+\theta)\,\mathrm{Im}\left(\hat{r}\,\xi(\alpha,\alpha+\theta)\right)d\alpha \\ &-i\frac{\pi}{2}\int_0^{\pi} K_i^g(\alpha+\theta)\cos\left(\hat{r}\,\xi(\alpha,\alpha+\theta)\right)d\alpha\Bigg], \end{aligned} \tag{16.88}$$

where function $K_i^g(\alpha)$ is given by Eq. (16.30), whereas

$$\xi(\alpha,\beta) = \frac{\cos\alpha}{\sqrt{\Lambda(\beta)}}, \tag{16.89}$$

and function

$$\mathrm{Im}(x) = \cos x \int_0^x \frac{\cos t - 1}{t}\,dt + \sin x \;\mathrm{Si(x)} \tag{16.90}$$

is non-singular.

An application of the gradient rule (2.136) to Eq. (16.88) yields the Green's function for the gradient of incremental displacements, which is not reported for conciseness (see Bigoni et al., 2007).

To complete the Green's function set, the Green's function for in-plane incremental nominal mean stress $\dot{\pi}$ [Eq. (16.35)] has to be determined. To this purpose, we note that Eq. (16.36) and its counterpart in the transformed domain, Eq. (16.37), still hold in the time-harmonic case, except that the Green's function $v^g_{j,jjj}$ is now the time-harmonic Green's function. Integrating and anti-transforming that equation yields

$$\dot{\pi}^g = (\dot{\pi}^g)_{\mathrm{static}} - \frac{\Omega}{2\pi^2(1+k)c}\int_0^{\pi} \frac{\tilde{K}_g(\alpha+\theta)}{\sqrt{\Lambda(\alpha+\theta)}}\;\Xi\left(\frac{\Omega r}{c}\frac{\cos\alpha}{\sqrt{\Lambda(\alpha+\theta)}}\right)d\alpha, \tag{16.91}$$

where the quasi-static Green's function, containing the singular terms, is given by Eq. (16.38) and function $\tilde{K}_g(\alpha)$ is given by Eq. (16.39).

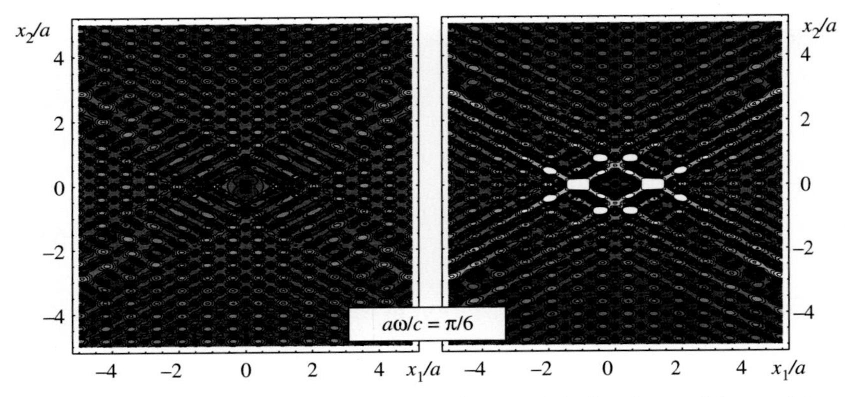

Figure 16.7. Real (*left*) and imaginary (*right*) parts of the level sets of the modulus of incremental displacement produced by a horizontal time-harmonic pulsating dipole (two pulsating unit forces placed at $x_1 = \pm a$). The maps are relative to an orthotropic material with $\xi = \mu_*/\mu = 1/4$ subjected to a pre-stress $k = 0.86$ very close to the boundary of ellipticity loss. A circular frequency ω (multiplied by the dipole half-distance a and divided by the velocity of wave propagation along the horizontal axis c) has been selected equal to $\pi/6$. Figure adapted from Bigoni and Capuani (2005, their fig. 7).

The Green's function (16.91) completes the determination of the time-harmonic Green's function set, whereas the Green's function for nominal stress can be obtained using Eqs. (16.40) or the procedure indicated by Bigoni et al. (2007).

The time-harmonic Green's function (16.88) has been employed to form a pulsating dipole, used as a perturbing agent on a uniformly pre-stressed elastic medium. An example is shown in Fig. 1.40, where the level sets of the real (*left*) and imaginary (*right*) parts of the incremental time-harmonic displacement are shown for an orthotropic material $\xi = \mu_*/\mu = 1/4$ near the border of ellipticity loss at $k = 0.86$. Now the solution becomes a function of the dimension-less circular frequency $a\omega/c$. A different frequency, namely, $a\omega/c = \pi/6$, is considered in Fig. 16.7, to be compared with Fig. 1.40. This comparison reveals that at higher frequency, a 'shadowing effect' occurs, so the waves are highly focussed and so narrow that almost nothing is visualised when the latter figure is plotted at the same scale of the former. Vice versa, when the frequency parameter $a\omega/c$ tends to zero, the solution (and therefore the maps of incremental velocity) tends to the quasi-static limit (Fig. 1.39, *right*).

We may conclude that

> *depending on the level of pre-stress and anisotropy, wave patterns are shown to emerge near the boundary of ellipticity loss, and these interact with shear band formation so that a focussing is observed of signals in the direction of shear bands. Varying the direction of the dynamic perturbation excites different wave patterns, which tend to degenerate to families of plane waves parallel to the shear bands when the elliptical boundary is approached, where the signal becomes localised in narrow 'channels'.*

16.1.3 Effects of flutter instability revealed by a pulsating perturbing dipole

The time-harmonic Green's function [Eq. (16.77)] can be used to analyse the effects of a perturbation superimposed on a given homogeneous deformation of an infinite body characterised by constitutive equations displaying flutter instability (Section 13.3). It becomes possible in this way to analyse the mechanical implications of flutter.

Following Piccolroaz et al. (2006), the analysis is limited to the loading branch of the elastoplastic constitutive operator, an assumptions introducing severe limits, as discussed in detail in Section 14.3.

We consider a dipole as the perturbing agent, acting along a line inclined at $\beta = 45°$ with respect to the x_1 axis. At this point, a preliminary normalisation of the Green's tensor and a study of the involved non-dimensional parameters becomes instrumental. In particular, introducing an arbitrary characteristic length a and, consequently, the dimension-less spatial variable $\bar{\boldsymbol{x}} = \boldsymbol{x}/a$, making use of the property

$$\delta(a\bar{\boldsymbol{x}}) = \frac{1}{a^2}\delta(\bar{\boldsymbol{x}}), \quad \bar{\boldsymbol{x}} \in \mathbf{R}^2, \tag{16.92}$$

Eq. (16.53) can be rewritten as

$$\bar{\mathbb{G}}_{ijkl}\frac{\partial^2 \bar{G}_{kq}(\bar{\boldsymbol{x}})}{\partial \bar{x}_j \partial \bar{x}_l} + \bar{\omega}^2 \bar{G}_{iq}(\bar{\boldsymbol{x}}) + \delta_{iq}\delta(\bar{\boldsymbol{x}}) = 0, \qquad \bar{\boldsymbol{x}} \in \mathbf{R}^2, \tag{16.93}$$

where

$$\bar{\mathbb{G}}_{ijkl} = \frac{\mathbb{G}_{ijkl}}{\mu}, \qquad \bar{\omega} = a\sqrt{\frac{\rho}{\mu}}\omega. \tag{16.94}$$

Thus a dimension-less version of the Green's tensor (16.74) reads

$$\begin{aligned}\bar{\boldsymbol{G}}(\bar{\boldsymbol{x}}) = -\frac{1}{8\pi^2}\sum_{N=1}^{2}\int_{|\boldsymbol{n}|=1}\Big[&2\cos(\bar{k}_N\boldsymbol{n}\cdot\bar{\boldsymbol{x}})\mathrm{Ci}(\bar{k}_N|\boldsymbol{n}\cdot\bar{\boldsymbol{x}}|)\\ &+2\sin(\bar{k}_N\boldsymbol{n}\cdot\bar{\boldsymbol{x}})\mathrm{Si}(\bar{k}_N\boldsymbol{n}\cdot\bar{\boldsymbol{x}}) - i\pi\cos(\bar{k}_N\boldsymbol{n}\cdot\bar{\boldsymbol{x}})\Big]\frac{\boldsymbol{v}_N \otimes \boldsymbol{w}_N}{\bar{c}_N^2}\,ds,\end{aligned} \tag{16.95}$$

where

$$\bar{k}_N = ak_N = \frac{\bar{\omega}}{\bar{c}_N}, \qquad \bar{c}_N = \sqrt{\frac{\rho}{\mu}}c_N \tag{16.96}$$

so that $\bar{c}_N^2$ are the eigenvalues of the dimension-less acoustic tensor $\bar{\boldsymbol{A}} = \rho\boldsymbol{A}/\mu$.

We begin with an analysis of the behaviour of the Green's function [Eq. (16.77)] outside and inside the flutter region.

As a reference, we consider case 3 shown in Fig. 13.8, in which the material is subject to the radial stress path corresponding to $\theta_L = 0$ in Fig. 13.6, and the direction of the axis of elastic symmetry is taken as inclined at $\theta_\sigma = 45°$ with respect to the principal stress direction $\boldsymbol{k}_1$. The employed material parameters are $\lambda/\mu = 1$, $\hat{b} = 80°$, $\psi = 30°$ and $\chi = 0°$. The dimension-less Green's tensor components have been computed for $\bar{\omega} = 1$ and for several values of the plastic modulus g/μ, including the values 3.53 and 1.5. These correspond, respectively, to situations near and inside the flutter region (see Fig. 13.8), but in both cases the tangent constitutive operator is positive definite (in other words, PD holds even when flutter is possible). The values of the components are plotted in Fig. 16.8 as functions of the distance from the singularity along a radial line inclined at −45° with respect to the x_1 axis, normalised through division by a. The real (imaginary) parts of the Green's function components

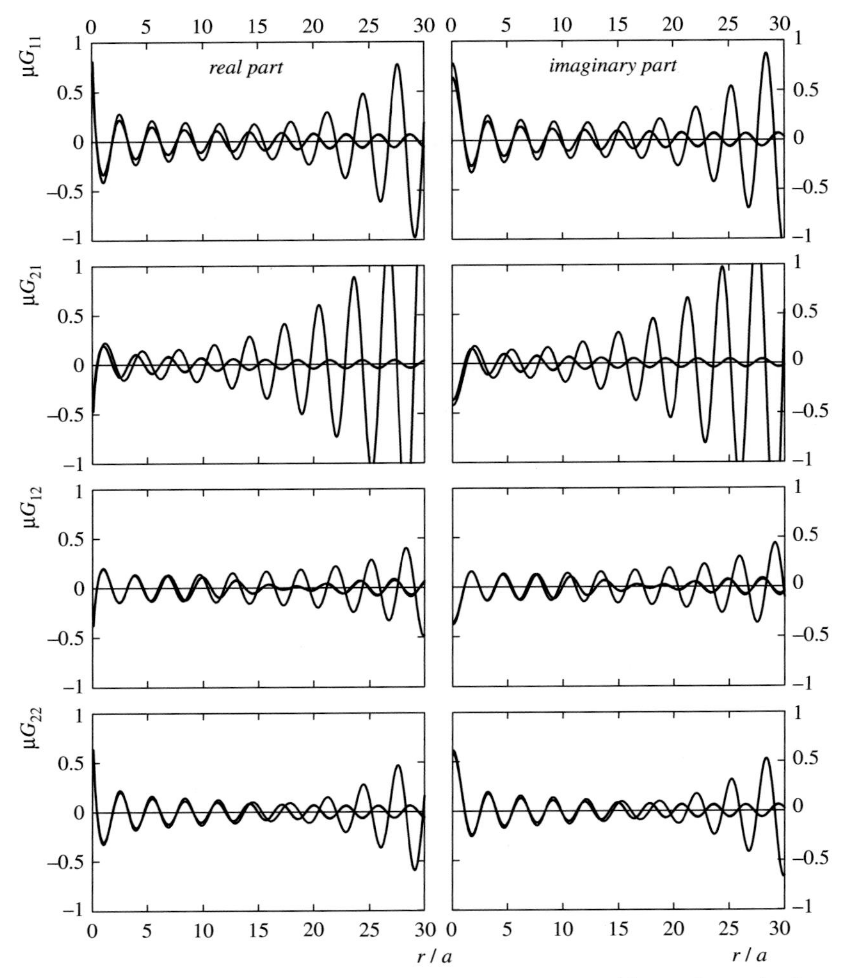

Figure 16.8. Dimension-less Green's tensor components (the real part is shown on the left and the imaginary part on the right in the figure) along a radial line inclined at $-45°$ with respect to x_1 axis for a non-associative elastoplastic material displaying flutter instability, case 3 of Fig. 13.8 and $\bar{\omega} = 1$. Two values of the plastic modulus $g/\mu = \{3.53, 1.5\}$ are considered, corresponding, respectively, to situations near and inside the flutter region. The blow-up of all components of the Green's tensor is evident within the flutter region, $g/\mu = 1.5$.

are plotted on the left (on the right) in the figure, the plots having been obtained starting from $x_1 = 1/10$ to exclude the singularity (in the real components of the Green's tensor).

Commenting on the results, first, we note from the figure that the Green's tensor is not symmetric (because the acoustic tensor is not), so $G_{12} \neq G_{21}$.

Second, results referring to values of plastic modulus g/μ higher than 3.53 and up to 7, not reported here for conciseness, produce curves practically coincident to those pertaining to $g/\mu = 3.53$; we therefore can conclude that blow-up is not evident when the material is outside the flutter region, but when inside, there is not much difference between the situations in which the material is far from and very near to the flutter boundary. This feature has been confirmed by us with several calculations (not reported here) and distinguishes flutter from shear banding, the latter becoming already visible when the condition of loss of ellipticity is approached from the interior of the elliptical range.

Third, *a blow-up of the solution with the space variable*, clearly visible in all components of the Green's tensor, is the characteristic feature of instability inside the flutter region, $g/\mu = 1.5$. This blow-up is similar to that evidenced by Bigoni and Willis (1994), but in a constitutive setting including viscosity, which is now absent.

It becomes evident that further exploration of flutter instability requires plotting of incremental displacement maps. These are obtained below employing a perturbation in the form of a pulsating dipole.

For this loading system, the level sets of the real part (Fig. 16.9, *left*) and the imaginary part (Fig. 16.9, *right*) of the components u_1 (first and third parts from the top of the figure) and u_2 (second and fourth parts from the top of the figure) of incremental displacements have been computed and plotted (note that a detail of Fig. 16.9 has been used to produce Fig. 1.41). The two upper parts of the figure refer to a situation far from flutter instability, whereas the two lower parts refer to a situation of flutter well inside the region of instability.

Moreover, the following parameters have been selected (results for different values of the parameters have been given by Piccolroaz et al., 2006):

$$\lambda/\mu = 1, \qquad \hat{b} = 80^\circ, \qquad \psi = 30^\circ, \qquad \chi = 0^\circ, \qquad \bar{\omega} = 1.$$

All components of incremental displacements have been plotted for the non-dimensional coordinates x_1/a and x_2/a ranging between -25 and 25.

Figure 16.9 refers to $g/\mu = 3$ (two upper parts), $g/\mu = 0.32$ (two lower parts) and to case 1 of Fig. 13.8, where $\theta_L = 0^\circ$ and $\theta_\sigma = 15^\circ$.

Note that the values of the plastic modulus selected for the example are all higher than the critical values for loss of positive definiteness of the constitutive operator (and thus of strong ellipticity and ellipticity) [see the values listed in Eq. (13.36)], so shear bands are excluded.

It can be observed from the upper parts of Fig. 16.9 (referring to a non-flutter situation) that the displacement maps are typical of an anisotropic material because 45° symmetry is not in evidence. Moreover, decay of the solution is appreciable when the distance from the dipole increases. Now, considering the lower parts of the figures, the effects of flutter instability become self-evident. In particular, we may observe a growth of the solution in space, which tends to degenerate into a system of blowing-up parallel plane waves. Results not reported here for brevity demonstrate that

> *the inclination of the blowing-up plane waves is almost independent of the dipole inclination, so it has to be considered a characteristic of the material, related to the particular stress state and constitutive features. We have observed that the inclination of the blowing-up waves corresponds to a value in the middle of the inclination fan of flutter (see Fig. 13.8).*

In particular, the inclinations of the plane waves at a sufficient distance from the dipole correspond to the mean value of flutter direction fan visible in Fig. 13.8 at the analysed g/μ values.

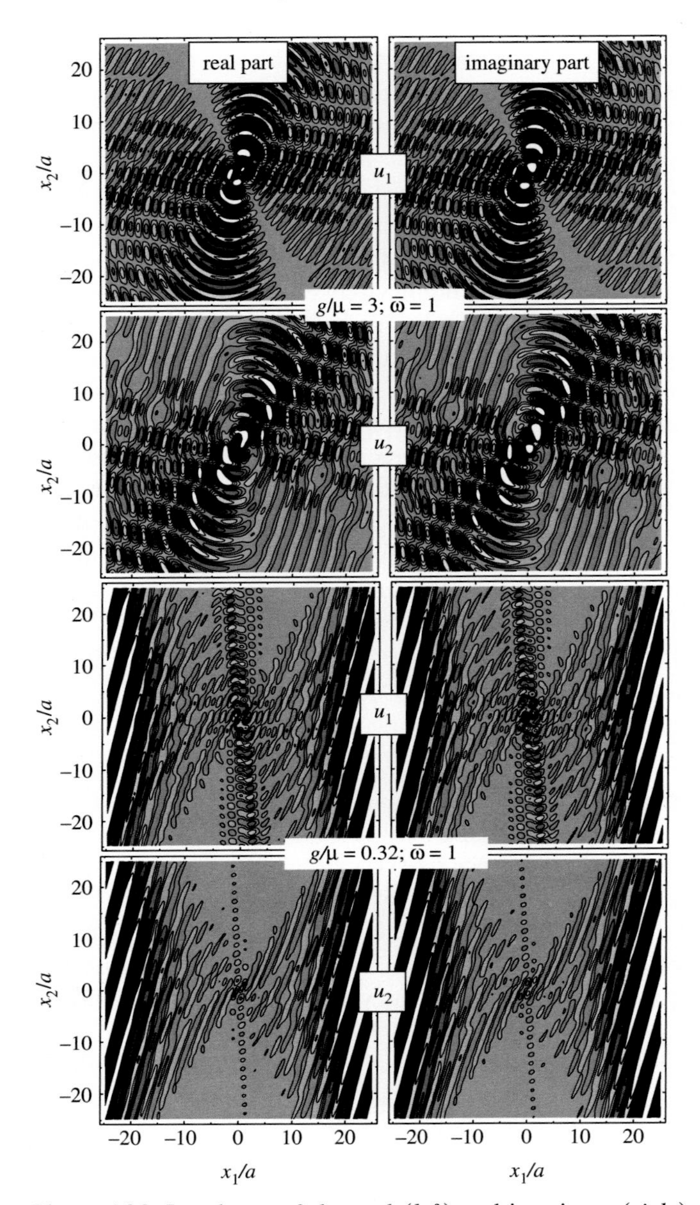

Figure 16.9. Level sets of the real (*left*) and imaginary (*right*) parts of the components of incremental displacements (u_1 first and third parts from the top, u_2 second and fourth parts) for a dipole inclined at $\beta = 45°$ for a non-associative elastoplastic material far from (upper two parts, $g/\mu = 3$) and inside (lower two parts, $g/\mu = 0.32$) the flutter region. Results pertain to case 1 of Fig. 13.8 for $\bar{\omega} = 1$. Note the system of blowing-up parallel waves revealing the effect of flutter.

As far as the effects of varying the non-dimensional frequency parameter $\bar{\omega}$ are concerned (not explored here but by Piccolroaz et al., 2006), we see that an increase in the frequency yields a narrowing of the distance between blowing-up plane waves. Moreover, on increase in frequency gives rise to the 'shadowing' effect already noted for shear bands (see also Bigoni and Capuani, 2005).

Compared with shear bands, we may observe that while these are already revealed when the boundary of the region of ellipticity is approached from the inside, flutter remains undetected. Beside this difference, there are, however, many similarities between the two phenomena: First of all, shear bands tend to blow up in space as the boundary of instability is approached and extend from a perturbation to infinity outside the elliptic range. Second, shear bands also tend to degenerate into families of plane waves parallel to a specific direction. Third, the signals tend to focus along well-defined patterns, both for shear bands and for flutter. Note, however, that flutter instability may occur much earlier than shear banding in a deformation process; moreover, waves near the loss of ellipticity threshold tend to blow up along the shear bands, but in contrast to flutter, they tend to decay in the parallel direction.

As a conclusion, we remark that flutter instability yields a self-organisation of dynamic disturbances along well-defined and blowing-up parallel waves, having inclinations corresponding to the mean value of the inclinations for which flutter is possible at the considered constitutive setting and stress state.

From the mechanical point of view, our results suggest that flutter yields a 'layering' of deformation patterns, with an inclination corresponding to the flutter direction, a spacing related to the frequency of the perturbing agency and possibly occurring early in a plastic deformation process.

It should be noted that the blow-up found in our approach to flutter will occur rapidly, and non-linearities neglected in our analysis (e.g., the possibility of elastic unloading and plastic reloading) soon may become important, possibly changing the overall mechanical response. Equally significant is the fact that the rate of growth increases with the frequency that is adopted. The governing equations of motion thus represent a problem that is *dynamically ill-posed* in the general transient case *unless* the tangent moduli in fact display a frequency dependence such that the flutter effect reduces as frequency increases [such a model was introduced by Bigoni and Willis (1994) in the context of a simple one-dimensional example].

16.2 Finite-length crack in a pre-stressed material

Until now we have considered perturbations in the form of concentrated forces. However, perturbations may be introduced in terms of fractures or inclusions in an infinite material that initially do not alter the uniformity of the state of pre-stress. The interest in the analysis of these perturbations lies in the possibility of analysing interactions between the singularities introduced, for instance, by a crack (or by a rigid line inclusion) and shear bands. We start with an analysis of a finite-length fracture, and then we move to the problem of a rigid-line inclusion and finally of a pre-existing shear band.

A homogenously pre-stressed and pre-strained, incompressible elastic infinite plane is considered, characterised by the constitutive equations [Eqs. (6.178) and (6.179)] of incremental, incompressible, orthotropic elasticity (see also Section 12.2), containing a crack of current length $2l$ taken parallel to the $\hat{x}_1$ axis in the $\hat{x}_1$–$\hat{x}_2$ reference system and loaded at infinity by a uniform nominal stress increment $\hat{t}_{2n}^{\infty}$, where $n = 1$ corresponds to mode II and $n = 2$ to mode I loading (Fig. 16.10). Obviously, the crack faces cannot be free of tractions because a dead loading is required to 'provide' the pre-stress state (with principal Cauchy components T_1 and T_2, assumed

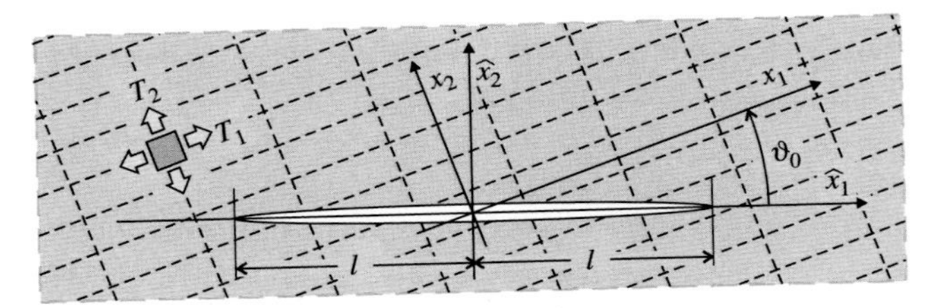

Figure 16.10. Crack of length $2l$ in a pre-stressed orthotropic (incompressible elastic) material inclined at an angle ϑ_0 (positive when anti-clockwise) with respect to the orthotropy axes x_1 and x_2. T_1 and T_2 denote the pre-stress state, expressed through the two in-plane principal Cauchy stresses aligned parallel to the x_1–x_2 reference system.

aligned parallel to the x_1–x_2 reference system, rotated at an angle ϑ_0 with respect to the $\hat{x}_1$–$\hat{x}_2$ system). An interesting exception to this rule occurs when the crack is aligned parallel to the x_1 axis, and the pre-stress is aligned parallel to the crack surfaces, namely, when the $\hat{x}_1$–$\hat{x}_2$ and x_1–x_2 systems coincide, that is, $\vartheta_0 = 0$, and $T_2 = 0$, corresponding to $\eta = k$. This situation has been considered by Guz (1999, and references quoted therein), Soós E (1996a, 1996b), Craciun and Soós (1998), Cristescu et al. (2004) and Radi et al. (2002). The case of a generic inclination ϑ_0, has been treated by Bigoni and Dal Corso (2008) for a pre-stressed material, and it is well known in linear, anisotropic elasticity (Savin, 1961; see also Sih and Liebowitz, 1968). Although the assumption of an inclined crack in a pre-stressed material may seem rather artificial, the treatment will provide the key for the modelling of a shear band formation and its propagation conditions, as shown at the end of this chapter.

Solution to the above-formulated crack problem is obtained by superimposing the trivial, unperturbed solution on the perturbation induced by the crack, the latter denoted with the apex $\circ$.

The unperturbed solutions are obtained by defining the uniform nominal stress field in the $\hat{x}_1$–$\hat{x}_2$ reference system

$$\hat{t}_{22} = \hat{t}_{22}^{\infty}, \qquad \hat{t}_{11} = 0, \qquad \hat{t}_{12} = \hat{t}_{21} = \hat{t}_{21}^{\infty} \tag{16.97}$$

so that $\hat{t}_{21}^{\infty} = 0$ ($\hat{t}_{22}^{\infty} = 0$) for mode I (mode II). The stress components in the $\hat{x}_1$–$\hat{x}_2$ reference system can be obtained through a rotation of the components in the pre-stress principal reference system x_1–x_2; thus, because the two systems are rotated at an angle ϑ_0 (taken positive when anticlockwise), we have

$$\hat{\boldsymbol{x}} = \boldsymbol{Q}^T \boldsymbol{x}, \qquad [\boldsymbol{Q}] = \begin{bmatrix} \cos\vartheta_0 & \sin\vartheta_0 \\ -\sin\vartheta_0 & \cos\vartheta_0 \end{bmatrix}, \tag{16.98}$$

so the nominal stress increment, incremental displacement and its gradient can be expressed in the $\hat{x}_1$–$\hat{x}_2$ reference system as

$$\hat{\boldsymbol{t}} = \boldsymbol{Q}^T \boldsymbol{t} \boldsymbol{Q}, \qquad \hat{\boldsymbol{v}} = \boldsymbol{Q}^T \boldsymbol{v}, \qquad \hat{\nabla}\hat{\boldsymbol{v}} = \boldsymbol{Q}^T \nabla \boldsymbol{v} \boldsymbol{Q}, \tag{16.99}$$

whereas the constitutive equations (6.175) transform to

$$\hat{\boldsymbol{t}} = \widehat{\mathbb{G}}[\hat{\nabla}\hat{\boldsymbol{v}}^T] + \dot{p}\boldsymbol{I}, \tag{16.100}$$

where the transformed fourth-order tensor $\widehat{\mathbb{G}}$ is given by

$$\widehat{\mathbb{G}} = (\boldsymbol{Q} \boxtimes \boldsymbol{Q})^T \mathbb{G} (\boldsymbol{Q} \boxtimes \boldsymbol{Q}), \qquad \text{or} \qquad \widehat{\mathbb{G}}_{ijhk} = Q_{li} Q_{mj} \mathbb{G}_{lmno} Q_{nh} Q_{ok}. \tag{16.101}$$

Note that the preceding definition (16.97) of mode I and II loadings is fully meaningful only when the constitutive equations [Eqs. (6.178) and (6.179)] are positively defined so that the Hill exclusion condition (12.6) holds true. For a non-positive definite constitutive equation, definition (16.97) would be better changed to one concerning the components of the incremental displacement gradient.

Assuming that condition (12.6) holds true, we can obtain directly from Eqs. (6.178) and (6.179) the components of the incremental displacement gradient and the incremental in-plane mean stress in the x_1–x_2 reference system:

$$\begin{aligned}
\dot{p}^{\infty} &= \frac{\hat{t}_{22}^{\infty}}{2} - \mu k v_{2,2}^{\infty}, \\
v_{2,2}^{\infty} &= -v_{1,1}^{\infty} = \frac{\hat{t}_{22}^{\infty} \cos 2\vartheta_0 - 2\hat{t}_{21}^{\infty} \sin 2\vartheta_0}{2\mu(2\xi - \eta)}, \\
v_{1,2}^{\infty} &= -\frac{(k+\eta)\left(\hat{t}_{22}^{\infty} \sin 2\vartheta_0 + 2\hat{t}_{21}^{\infty} \cos 2\vartheta_0\right)}{2\mu(k^2 - 2\eta + \eta^2)}, \\
v_{2,1}^{\infty} &= \frac{(k-\eta)\left(\hat{t}_{22}^{\infty} \sin 2\vartheta_0 + 2\hat{t}_{21}^{\infty} \cos 2\vartheta_0\right)}{2\mu(k^2 - 2\eta + \eta^2)}.
\end{aligned} \tag{16.102}$$

The components of the incremental displacement gradient in the $\hat{x}_1$–$\hat{x}_2$ reference system can be obtained through a rotation of Eqs. (16.102) by employing Eq. $(16.99)_3$.

It should be noticed from Eqs. (16.102) that in the absence of pre-stress, $k = \eta = 0$, Eqs. (16.102) fully determine the incremental displacement gradient. In this case, however, the incremental stress is related only to the symmetric part of the incremental displacement gradient, so an arbitrary incremental rotation can be added without altering the state of stress, a circumstance not possible when the pre-stress is different from zero. In other words, when pre-stress is present, loading (16.97) completely defines the incremental displacement gradient (and incremental mean stress) through Eqs. (16.102) so that incremental rigid body rotations remains determined.

16.2.1 Finite-length crack parallel to an orthotropy axis

Before proceeding with the solution for the inclined crack, it becomes instructive to begin with the simple case of null inclination, in which $\vartheta_0 = 0$ so that the orthotropy axes are aligned parallel and orthogonal to the x_1–x_2 axes, defining the pre-stress directions and coinciding with the $\hat{x}_1$–$\hat{x}_2$ axes.

The perturbed solution is derived separately for the two EI and EC regimes (beginning with EI); see Section 12.2.1.

From representation (12.26), the stream function ψ° [Eq. (12.24)], can be given in the form

$$\psi^{\circ}(z_1, z_2) = \text{Re}\left[\sum_{j=1}^{2} F_j(z_j)\right] \tag{16.103}$$

(note that the Ω_j's are in conjugated pairs in E) where $z_j = x_1 + \Omega_j x_2$, with Ω_j given by Eq. (12.28); thus the displacement field becomes

$$v_1^\circ(z_1,z_2) = \text{Re}\left[\sum_{j=1}^{2} \Omega_j F_j'(z_j)\right], \qquad v_2^\circ(z_1,z_2) = -\text{Re}\left[\sum_{j=1}^{2} F_j'(z_j)\right], \tag{16.104}$$

and its gradient can be written as

$$\begin{aligned} &v_{1,1}^\circ(z_1,z_2) = -v_{2,2}^\circ(z_1,z_2) = \text{Re}\left[\sum_{j=1}^{2} \Omega_j F_j''(z_j)\right], \\ &v_{1,2}^\circ(z_1,z_2) = \text{Re}\left[\sum_{j=1}^{2} \Omega_j^2 F_j''(z_j)\right], \qquad v_{2,1}^\circ(z_1,z_2) = -\text{Re}\left[\sum_{j=1}^{2} F_j''(z_j)\right]. \end{aligned} \tag{16.105}$$

The effects of the applied boundary conditions on the crack surfaces decay to zero at infinity, so from Eqs. (16.105) and the constitutive relation (6.178), we obtain

$$\lim_{|z_j|\to+\infty} F_j''(z_j) = 0, \qquad j = 1,2. \tag{16.106}$$

Mode I

To recover traction-free crack faces using superposition, the incremental nominal stress component $\dot{t}_{22}^\infty$ of reversed sign has to be prescribed at crack surfaces in the perturbed problem, namely, for mode I

$$\begin{aligned} &\dot{t}_{22}^\circ(x_1,0^\pm) = -\dot{t}_{22}^\infty, \qquad \forall |x_1| < l, \\ &\dot{t}_{21}^\circ(x_1,0^\pm) = 0, \qquad \forall x_1 \in \mathbb{R}. \end{aligned} \tag{16.107}$$

From Eqs. $(16.107)_2$ and (16.105), the following relation can be obtained, holding at every point x_1 of the real axis $\mathbb{R}$:

$$F_2''(x_1) = -\frac{2\xi - \eta + \Lambda}{2\xi - \eta - \Lambda} F_1''(x_1), \tag{16.108}$$

where Λ is defined by Eq. (12.73), whereas from Eq. $(16.107)_1$ the condition

$$\begin{aligned} \frac{\dot{t}_{22}^\infty}{\mu} = \text{Re}\Big\{&\Omega_1\left[4\xi - 1 - \eta + \Omega_1^2(1-k)\right]F_1''(x_1) \\ &+ \Omega_2\left[4\xi - 1 - \eta + \Omega_2^2(1-k)\right]F_2''(x_1)\Big\}, \end{aligned} \tag{16.109}$$

follows to hold true along the crack line $|x_1| < l$.

Within the EI regime [Eqs. (12.35)] and for mode I, the Riemann-Hilbert problem

$$-\frac{\beta_2\varepsilon_1^2 - \beta_1\varepsilon_2^2}{\varepsilon_2}\text{Re}\left[iF_1''(x_1)\right] = \frac{\dot{t}_{22}^\infty}{\mu}, \qquad \forall |x_1| < l, \tag{16.110}$$

where β_1 and β_2 are defined by Eq. (12.37) and

$$\varepsilon_n = 1 - \eta + (1-k)\beta_n^2, \quad n = 1,2, \tag{16.111}$$

has the following solution:

$$F_j''(z_j) = (-1)^k i \frac{\dot{t}_{22}^{\infty}}{\mu} \frac{\varepsilon_k}{\beta_2 \varepsilon_1^2 - \beta_1 \varepsilon_2^2} \left(1 - \frac{z_j}{\sqrt{z_j^2 - l^2}} \right), \qquad j,k = 1,2, \; j \neq k, \quad (16.112)$$

where the function $\sqrt{z_j^2 - l^2}$ is defined to have a unique branch cut along $x_2 = 0$, $-l < x_1 < l$, chosen[4] in such a way that $\sqrt{z_j^2 - l^2} = \sqrt{x_1^2 - l^2}$, for $x_1 > l$ and $x_2 = 0$.

The perturbed stream function becomes

$$\psi^{\circ} = -\frac{\dot{t}_{22}^{\infty}}{2\mu} \frac{\varepsilon_2}{\beta_2 \varepsilon_1^2 - \beta_1 \varepsilon_2^2} \sum_{j=1}^{2} \left(-\frac{\varepsilon_1}{\varepsilon_2} \right)^{j-1} \operatorname{Im} \left\{ z_j^2 - z_j \sqrt{z_j^2 - l^2} + l^2 \ln \left(z_j + \sqrt{z_j^2 - l^2} \right) \right\}, \quad (16.113)$$

and the incremental displacements assume the form

$$\begin{aligned} v_1^{\circ} &= -\frac{\dot{t}_{22}^{\infty}}{\mu} \frac{\varepsilon_2}{\beta_2 \varepsilon_1^2 - \beta_1 \varepsilon_2^2} \operatorname{Re} \left[\beta_1 \left(z_1 - \sqrt{z_1^2 - l^2} \right) - \frac{\beta_2 \varepsilon_1}{\varepsilon_2} \left(z_2 - \sqrt{z_2^2 - l^2} \right) \right], \\ v_2^{\circ} &= \frac{\dot{t}_{22}^{\infty}}{\mu} \frac{\varepsilon_2}{\beta_2 \varepsilon_1^2 - \beta_1 \varepsilon_2^2} \operatorname{Im} \left[\left(z_1 - \sqrt{z_1^2 - l^2} \right) - \frac{\varepsilon_1}{\varepsilon_2} \left(z_2 - \sqrt{z_2^2 - l^2} \right) \right]. \end{aligned} \quad (16.114)$$

Finally, for mode I in the EI regime, the incremental in-plane mean stress is given by

$$\dot{p}^{\circ} = \frac{\dot{t}_{22}^{\infty} \varepsilon_2}{\beta_2 \varepsilon_1^2 - \beta_1 \varepsilon_2^2} \left\{ \frac{\varepsilon_2 \beta_1 \delta_1 - \varepsilon_1 \beta_2 \delta_2}{\varepsilon_2} - \operatorname{Re} \left[\beta_1 \delta_1 \frac{z_1}{\sqrt{z_1^2 - l^2}} - \frac{\varepsilon_1 \beta_2 \delta_2}{\varepsilon_2} \frac{z_2}{\sqrt{z_2^2 - l^2}} \right] \right\}, \quad (16.115)$$

where

$$\delta_n = 2\xi - 1 - k - (1 - k) \beta_n^2, \quad n = 1,2, \quad (16.116)$$

[4] Note that function $\sqrt{z_j^2 - l^2}$ has been taken equal to $\sqrt{z_j - l}\sqrt{z_j + l}$, with the usual branch cut definition for the latter square root functions. Therefore, on the real axis,

$$x_2 = 0, \qquad \sqrt{z_j^2 - l^2} = \pm \sqrt{x_1^2 - l^2},$$

with the upper (lower) sign for $x_1 > l$ ($x_1 < -l$), whereas along the branch cut, when $\operatorname{Im}[\Omega_j] > 0$,

$$-l < x_1 < l, \qquad \sqrt{z_j^2 - l^2} = \pm i \sqrt{l^2 - x_1^2},$$

with the upper (lower) sign for $x_2 = 0^+$ ($x_2 = 0^-$).

whereas the incremental nominal stress components are

$$
\begin{aligned}
\dot{t}_{11}^{\circ} &= -\dot{t}_{22}^{\infty} \frac{\varepsilon_1 \varepsilon_2}{\beta_2 \varepsilon_1^2 - \beta_1 \varepsilon_2^2} \left\{ \beta_1 - \beta_2 - \mathrm{Re}\left[\beta_1 \frac{z_1}{\sqrt{z_1^2 - l^2}} - \beta_2 \frac{z_2}{\sqrt{z_2^2 - l^2}} \right] \right\}, \\
\dot{t}_{22}^{\circ} &= -\dot{t}_{22}^{\infty} \left\{ 1 + \frac{\varepsilon_2}{\beta_2 \varepsilon_1^2 - \beta_1 \varepsilon_2^2} \mathrm{Re}\left[\beta_1 \varepsilon_2 \frac{z_1}{\sqrt{z_1^2 - l^2}} - \frac{\varepsilon_1^2 \beta_2}{\varepsilon_2} \frac{z_2}{\sqrt{z_2^2 - l^2}} \right] \right\}, \\
\dot{t}_{12}^{\circ} &= -\dot{t}_{22}^{\infty} \frac{\varepsilon_2}{\beta_2 \varepsilon_1^2 - \beta_1 \varepsilon_2^2} \mathrm{Im}\left[\beta_1^2 \varepsilon_2 \frac{z_1}{\sqrt{z_1^2 - l^2}} - \frac{\varepsilon_1^2 \beta_2^2}{\varepsilon_2} \frac{z_2}{\sqrt{z_2^2 - l^2}} \right], \\
\dot{t}_{21}^{\circ} &= -\dot{t}_{22}^{\infty} \frac{\varepsilon_1 \varepsilon_2}{\beta_2 \varepsilon_1^2 - \beta_1 \varepsilon_2^2} \mathrm{Im}\left[\frac{z_1}{\sqrt{z_1^2 - l^2}} - \frac{z_2}{\sqrt{z_2^2 - l^2}} \right].
\end{aligned}
\tag{16.117}
$$

Within the EC regime [Eqs. (12.32)] and for mode I, the Riemann-Hilbert problem

$$
2\left[\alpha(\delta^2 - \chi^2) + 2\beta\delta\chi\right] \mathrm{Re}\left[\frac{F_1''(x_1)}{\chi - i\delta}\right] = \frac{\dot{t}_{22}^{\infty}}{\mu}, \qquad \forall |x_1| < l, \tag{16.118}
$$

where α and β are defined by Eq. (12.34) and

$$
\delta = 2(1-k)\alpha\beta, \qquad \chi = 2\xi - \eta, \tag{16.119}
$$

has the following solution:

$$
F_j''(z_j) = (-1)^k \frac{\dot{t}_{22}^{\infty}}{2\mu} \frac{\chi + (-1)^j i\delta}{\alpha(\delta^2 - \chi^2) + 2\beta\delta\chi} \left(1 - \frac{z_j}{\sqrt{z_j^2 - l^2}}\right), \qquad j,k = 1,2, \ j \neq k, \tag{16.120}
$$

where function $\sqrt{z_j^2 - l^2}$ has been defined as in Eq. (16.112). Note from Eqs. (12.37) and (12.34) that $z_j = x_1 + i\beta_j x_2$, with $\beta_j > 0$ in EI, and $z_j = x_1 + (-1)^j \alpha x_2 + i\beta x_2$, with $\beta > 0$ in EC, so for both regimes $\mathrm{Im}[z_j] = 0 \Longleftrightarrow x_2 = 0$.

The perturbed stream function becomes

$$
\begin{aligned}
\psi^{\circ} = &-\frac{\dot{t}_{22}^{\infty}}{4\mu[\alpha(\delta^2 - \chi^2) + 2\beta\delta\chi]} \sum_{j=1}^{2} \mathrm{Re}\left\{ \left[(-1)^j \chi + i\delta \right] \right. \\
&\left. \times \left[z_j^2 - z_j\sqrt{z_j^2 - l^2} + l^2 \ln\left(z_j + \sqrt{z_j^2 - l^2} \right) \right] \right\},
\end{aligned}
\tag{16.121}
$$

and the incremental displacements assume the form

$$
\begin{aligned}
v_1^\circ = & -\frac{\dot{t}_{22}^\infty}{2\mu} \frac{1}{\alpha(\delta^2-\chi^2)+2\beta\delta\chi} \left\{ (\alpha\chi - \beta\delta)\,\mathrm{Re}\left[\left(z_1 - \sqrt{z_1^2 - l^2}\right) + \left(z_2 - \sqrt{z_2^2 - l^2}\right)\right] \right. \\
& \left. + (\alpha\delta + \beta\chi)\,\mathrm{Im}\left[\left(z_1 - \sqrt{z_1^2 - l^2}\right) - \left(z_2 - \sqrt{z_2^2 - l^2}\right)\right]\right\}, \\
v_2^\circ = & -\frac{\dot{t}_{22}^\infty}{2\mu} \frac{1}{\alpha(\delta^2-\chi^2)+2\beta\delta\chi} \left\{ \chi\,\mathrm{Re}\left[\left(z_1 - \sqrt{z_1^2 - l^2}\right) - \left(z_2 - \sqrt{z_2^2 - l^2}\right)\right] \right. \\
& \left. + \delta\,\mathrm{Im}\left[\left(z_1 - \sqrt{z_1^2 - l^2}\right) + \left(z_2 - \sqrt{z_2^2 - l^2}\right)\right]\right\}.
\end{aligned}
\tag{16.122}
$$

Finally, for mode I in the EC regime, the incremental in-plane mean stress is given by

$$
\begin{aligned}
\dot{p}^\circ = & -\frac{\dot{t}_{22}^\infty}{2} \frac{1}{\alpha(\delta^2-\chi^2)+2\beta\delta\chi} \\
& \times \left\{ [(\beta\chi + \alpha\delta)\delta + (\alpha\chi - \beta\delta)k] \left(2 - \mathrm{Re}\left[\frac{z_1}{\sqrt{z_1^2 - l^2}} + \frac{z_2}{\sqrt{z_2^2 - l^2}}\right]\right) \right. \\
& \left. - [(\beta\chi + \alpha\delta)k - (\alpha\chi - \beta\delta)\delta]\,\mathrm{Im}\left[\frac{z_1}{\sqrt{z_1^2 - l^2}} - \frac{z_2}{\sqrt{z_2^2 - l^2}}\right]\right\},
\end{aligned}
\tag{16.123}
$$

whereas the incremental nominal stress components are

$$
\begin{aligned}
\dot{t}_{11}^\circ = & -\frac{\dot{t}_{22}^\infty}{2} \frac{\delta^2+\chi^2}{\alpha(\delta^2-\chi^2)+2\beta\delta\chi} \left\{ 2\alpha - \alpha\,\mathrm{Re}\left[\frac{z_1}{\sqrt{z_1^2 - l^2}} + \frac{z_2}{\sqrt{z_2^2 - l^2}}\right] \right. \\
& \left. - \beta\,\mathrm{Im}\left[\frac{z_1}{\sqrt{z_1^2 - l^2}} - \frac{z_2}{\sqrt{z_2^2 - l^2}}\right]\right\}, \\
\dot{t}_{22}^\circ = & -\frac{\dot{t}_{22}^\infty}{2} \left\{ 2 - \mathrm{Re}\left[\frac{z_1}{\sqrt{z_1^2 - l^2}} + \frac{z_2}{\sqrt{z_2^2 - l^2}}\right] \right. \\
& \left. - \frac{\beta(\delta^2-\chi^2) - 2\alpha\delta\chi}{\alpha(\delta^2-\chi^2)+2\beta\delta\chi}\,\mathrm{Im}\left[\frac{z_1}{\sqrt{z_1^2 - l^2}} - \frac{z_2}{\sqrt{z_2^2 - l^2}}\right]\right\},
\end{aligned}
\tag{16.124}
$$

$$\dot{t}_{12}^{\circ} = \frac{\dot{t}_{22}^{\infty}}{2} \left\{ \frac{(\alpha^2 - \beta^2)(\delta^2 - \chi^2) + 4\alpha\beta\delta\chi}{\alpha(\delta^2 - \chi^2) + 2\beta\delta\chi} \mathrm{Re}\left[\frac{z_1}{\sqrt{z_1^2 - l^2}} - \frac{z_2}{\sqrt{z_2^2 - l^2}} \right] \right.$$

$$\left. + \frac{2\alpha\beta(\delta^2 - \chi^2) - 2\delta\chi(\alpha^2 - \beta^2)}{\alpha(\delta^2 - \chi^2) + 2\beta\delta\chi} \mathrm{Im}\left[\frac{z_1}{\sqrt{z_1^2 - l^2}} + \frac{z_2}{\sqrt{z_2^2 - l^2}} \right] \right\},$$

$$\dot{t}_{21}^{\circ} = \frac{\dot{t}_{22}^{\infty}}{2} \frac{\delta^2 + \chi^2}{\alpha(\delta^2 - \chi^2) + 2\beta\delta\chi} \mathrm{Re}\left[\frac{z_1}{\sqrt{z_1^2 - l^2}} - \frac{z_2}{\sqrt{z_2^2 - l^2}} \right].$$

Mode II

The reverse of the incremental nominal stress component $\dot{t}_{21}^{\infty}$ has to be applied at the crack surfaces in the perturbed mode II solution, namely,

$$\dot{t}_{22}^{\circ}(x_1, 0^{\pm}) = 0, \qquad \forall x_1 \in \mathbb{R},$$
$$\dot{t}_{21}^{\circ}(x_1, 0^{\pm}) = -\dot{t}_{21}^{\infty}, \qquad \forall |x_1| < l. \tag{16.125}$$

Equations $(16.125)_1$, $(16.104)_4$, and $(16.125)_2$ provide the two conditions

$$F_2''(x_1) = -\frac{\Omega_1}{\Omega_2} \frac{2\xi - \eta - \Lambda}{2\xi - \eta + \Lambda} F_1''(x_1), \tag{16.126}$$

holding at every point x_1 of the real axis $\mathbb{R}$ and where Λ is defined by Eq. (12.73) and

$$\frac{\dot{t}_{21}^{\infty}}{\mu} = \mathrm{Re}\left\{ \left[1 - \eta - \Omega_1^2(1-k)\right] F_1''(x_1) + \left[1 - \eta - \Omega_2^2(1-k)\right] F_2''(x_1) \right\}, \tag{16.127}$$

holding for $|x_1| < l$.

Within the EI regime [Eqs. (12.35)] and for mode II, the Riemann-Hilbert problem

$$\frac{\beta_2\varepsilon_1^2 - \beta_1\varepsilon_2^2}{\beta_2\varepsilon_1} \mathrm{Re}\left[F_1''(x_1)\right] = \frac{\dot{t}_{21}^{\infty}}{\mu}, \qquad \forall |x_1| < l, \tag{16.128}$$

has the following solution:

$$F_j''(z_j) = (-1)^k \frac{\dot{t}_{21}^{\infty}}{\mu} \frac{\beta_k \varepsilon_j}{\beta_2\varepsilon_1^2 - \beta_1\varepsilon_2^2} \left(1 - \frac{z_j}{\sqrt{z_j^2 - l^2}} \right), \qquad j,k = 1,2, \; j \neq k, \tag{16.129}$$

so the perturbed stream function becomes

$$\psi^{\circ} = \frac{\dot{t}_{21}^{\infty}}{2\mu} \frac{\beta_2\varepsilon_1}{\beta_2\varepsilon_1^2 - \beta_1\varepsilon_2^2} \sum_{j=1}^{2} \left(-\frac{\varepsilon_2\beta_1}{\varepsilon_1\beta_2} \right)^{j-1} \mathrm{Re}\left[z_j^2 - z_j\sqrt{z_j^2 - l^2} + l^2 \ln\left(z_j + \sqrt{z_j^2 - l^2} \right) \right], \tag{16.130}$$

and the incremental displacements take the form

$$v_1^\circ = -\frac{\dot{t}_{21}^\infty}{\mu}\frac{\beta_1\beta_2\varepsilon_1}{\beta_2\varepsilon_1^2-\beta_1\varepsilon_2^2}\text{Im}\left[\left(z_1-\sqrt{z_1^2-l^2}\right)-\frac{\varepsilon_2}{\varepsilon_1}\left(z_2-\sqrt{z_2^2-l^2}\right)\right],$$
$$v_2^\circ = -\frac{\dot{t}_{21}^\infty}{\mu}\frac{\beta_2\varepsilon_1}{\beta_2\varepsilon_1^2-\beta_1\varepsilon_2^2}\text{Re}\left[\left(z_1-\sqrt{z_1^2-l^2}\right)-\frac{\beta_1\varepsilon_2}{\beta_2\varepsilon_1}\left(z_2-\sqrt{z_2^2-l^2}\right)\right]. \quad (16.131)$$

Finally, for mode II in the EI regime, the incremental in-plane mean stress is given by

$$\dot{p}^\circ = -\dot{t}_{21}^\infty\frac{\beta_1\beta_2\varepsilon_1}{\beta_2\varepsilon_1^2-\beta_1\varepsilon_2^2}\text{Im}\left[\delta_1\frac{z_1}{\sqrt{z_1^2-l^2}}-\frac{\varepsilon_2\delta_2}{\varepsilon_1}\frac{z_2}{\sqrt{z_2^2-l^2}}\right], \quad (16.132)$$

whereas the incremental nominal stress components are

$$\dot{t}_{11}^\circ = \dot{t}_{21}^\infty\frac{\beta_1\beta_2\varepsilon_1}{\beta_2\varepsilon_1^2-\beta_1\varepsilon_2^2}\text{Im}\left[\varepsilon_1\frac{z_1}{\sqrt{z_1^2-l^2}}-\frac{\varepsilon_2^2}{\varepsilon_1}\frac{z_2}{\sqrt{z_2^2-l^2}}\right],$$
$$\dot{t}_{22}^\circ = -\dot{t}_{21}^\infty\frac{\beta_1\beta_2\varepsilon_1\varepsilon_2}{\beta_2\varepsilon_1^2-\beta_1\varepsilon_2^2}\text{Im}\left[\frac{z_1}{\sqrt{z_1^2-l^2}}-\frac{z_2}{\sqrt{z_2^2-l^2}}\right],$$
$$\dot{t}_{12}^\circ = -\dot{t}_{21}^\infty\frac{\beta_1\beta_2\varepsilon_1\varepsilon_2}{\beta_2\varepsilon_1^2-\beta_1\varepsilon_2^2}\left\{\beta_1-\beta_2-\text{Re}\left[\beta_1\frac{z_1}{\sqrt{z_1^2-l^2}}-\beta_2\frac{z_2}{\sqrt{z_2^2-l^2}}\right]\right\},$$
$$\dot{t}_{21}^\circ = -\dot{t}_{21}^\infty\left\{1-\frac{\beta_2\varepsilon_1}{\beta_2\varepsilon_1^2-\beta_1\varepsilon_2^2}\text{Re}\left[\varepsilon_1\frac{z_1}{\sqrt{z_1^2-l^2}}-\frac{\beta_1\varepsilon_2^2}{\beta_2\varepsilon_1}\frac{z_2}{\sqrt{z_2^2-l^2}}\right]\right\}. \quad (16.133)$$

Within the EC regime [Eqs. (12.32)] and for mode II, the Riemann-Hilbert problem

$$-2\left[\alpha(\delta^2-\chi^2)+2\beta\delta\chi\right]\text{Re}\left[\frac{F_1''(x_1)}{(\alpha+i\beta)(\chi+i\delta)}\right]=\frac{\dot{t}_{21}^\infty}{\mu}, \qquad \forall |x_1|<l, \quad (16.134)$$

has the following solution:

$$F_j''(z_j) = -\frac{\dot{t}_{21}^\infty}{2\mu}\frac{[\alpha-(-1)^j i\beta][\chi-(-1)^j i\delta]}{\alpha(\delta^2-\chi^2)+2\beta\delta\chi}\left(1-\frac{z_j}{\sqrt{z_j^2-l^2}}\right), \qquad j=1,2, \quad (16.135)$$

so the perturbed stream function becomes

$$\psi^\circ = -\frac{\dot{t}_{21}^\infty}{4\mu}\frac{1}{\alpha(\delta^2-\chi^2)+2\beta\delta\chi}\sum_{j=1}^{2}\text{Re}\left\{[\alpha-(-1)^j i\beta][\chi-(-1)^j i\delta]\right.$$
$$\left.\times\left[z_j^2-z_j\sqrt{z_j^2-l^2}+l^2\ln\left(z_j+\sqrt{z_j^2-l^2}\right)\right]\right\}, \quad (16.136)$$

and the incremental displacements take the form

$$v_1^\circ = \frac{\dot{t}_{21}^\infty}{2\mu} \frac{\alpha^2+\beta^2}{\alpha(\delta^2-\chi^2)+2\beta\delta\chi} \left\{ \chi \,\mathrm{Re}\left[\left(z_1 - \sqrt{z_1^2 - l^2}\right) - \left(z_2 - \sqrt{z_2^2 - l^2}\right) \right] - \delta \,\mathrm{Im}\left[\left(z_1 - \sqrt{z_1^2 - l^2}\right) + \left(z_2 - \sqrt{z_2^2 - l^2}\right) \right] \right\},$$

$$v_2^\circ = \frac{\dot{t}_{21}^\infty}{2\mu} \frac{1}{\alpha(\delta^2-\chi^2)+2\beta\delta\chi} \left\{ (\alpha\chi - \beta\delta) \,\mathrm{Re}\left[\left(z_1 - \sqrt{z_1^2 - l^2}\right) + \left(z_2 - \sqrt{z_2^2 - l^2}\right) \right] - (\beta\chi + \alpha\delta) \,\mathrm{Im}\left[\left(z_1 - \sqrt{z_1^2 - l^2}\right) - \left(z_2 - \sqrt{z_2^2 - l^2}\right) \right] \right\}. \tag{16.137}$$

Finally, for mode II in the EC regime, the incremental in-plane mean stress is given by

$$\dot{p}^\circ = \frac{\dot{t}_{21}^\infty}{2} \frac{\alpha^2+\beta^2}{\alpha(\delta^2-\chi^2)+2\beta\delta\chi} \left\{ (\delta^2 - k\chi) \,\mathrm{Re}\left[\frac{z_1}{\sqrt{z_1^2 - l^2}} - \frac{z_2}{\sqrt{z_2^2 - l^2}} \right] + \delta(\chi + k) \,\mathrm{Im}\left[\frac{z_1}{\sqrt{z_1^2 - l^2}} + \frac{z_2}{\sqrt{z_2^2 - l^2}} \right] \right\}, \tag{16.138}$$

whereas the incremental nominal stress components are

$$\dot{t}_{11}^\circ = \frac{\dot{t}_{21}^\infty}{2} \frac{\alpha^2+\beta^2}{\alpha(\delta^2-\chi^2)+2\beta\delta\chi} \left\{ (\delta^2 - \chi^2) \,\mathrm{Re}\left[\frac{z_1}{\sqrt{z_1^2 - l^2}} - \frac{z_2}{\sqrt{z_2^2 - l^2}} \right] + 2\delta\chi \,\mathrm{Im}\left[\frac{z_1}{\sqrt{z_1^2 - l^2}} + \frac{z_2}{\sqrt{z_2^2 - l^2}} \right] \right\},$$

$$\dot{t}_{22}^\circ = \frac{\dot{t}_{21}^\infty}{2} \frac{(\alpha^2+\beta^2)(\delta^2+\chi^2)}{\alpha(\delta^2-\chi^2)+2\beta\delta\chi} \,\mathrm{Re}\left[\frac{z_1}{\sqrt{z_1^2 - l^2}} - \frac{z_2}{\sqrt{z_2^2 - l^2}} \right],$$

$$\dot{t}_{12}^\circ = -\frac{\dot{t}_{21}^\infty}{2} \frac{(\alpha^2+\beta^2)(\delta^2+\chi^2)}{\alpha(\delta^2-\chi^2)+2\beta\delta\chi} \left\{ 2\alpha - \alpha \,\mathrm{Re}\left[\frac{z_1}{\sqrt{z_1^2 - l^2}} + \frac{z_2}{\sqrt{z_2^2 - l^2}} \right] \right. \tag{16.139}$$

$$
- \beta \text{Im}\left[\frac{z_1}{\sqrt{z_1^2 - l^2}} - \frac{z_2}{\sqrt{z_2^2 - l^2}} \right] \Bigg\},
$$

$$
\dot{t}_{21}^{\circ} = \frac{\dot{t}_{21}^{\infty}}{2} \Bigg\{ -2 + \text{Re}\left[\frac{z_1}{\sqrt{z_1^2 - l^2}} + \frac{z_2}{\sqrt{z_2^2 - l^2}} \right]
$$

$$
- \frac{\beta(\delta^2 - \chi^2) - 2\alpha\delta\chi}{\alpha(\delta^2 - \chi^2) + 2\beta\delta\chi} \text{Im}\left[\frac{z_1}{\sqrt{z_1^2 - l^2}} - \frac{z_2}{\sqrt{z_2^2 - l^2}} \right] \Bigg\}.
$$

Incremental stress intensity factors

Similarly to the problem of fracture in small strain elasticity, Radi et al. (2002) have defined

$$
\dot{K}_I = \lim_{r \to 0} \sqrt{2\pi r}\, \dot{t}_{22}(r, \vartheta = 0) \qquad \text{and} \qquad \dot{K}_{II} = \lim_{r \to 0} \sqrt{2\pi r}\, \dot{t}_{21}(r, \vartheta = 0), \tag{16.140}
$$

for mode I and mode II incremental loading, respectively, in the polar coordinate system (r, ϑ) centred at the crack tip $(\hat{x}_1 = l, \hat{x}_2 = 0)$, so r denotes the radial distance from the crack tip and ϑ indicates values of the polar coordinate (anti-clockwise) angle singling out r from the $\hat{x}_1$ axis (so that $\vartheta = 0$ corresponds to points ahead of the crack tip, see Fig. 16.11 with $\vartheta_0 = 0$). From the preceding full–field solution (for uniform incremental loading at infinity), the results are

$$
\dot{K}_I = \dot{t}_{22}^{\infty} \sqrt{\pi l} \qquad \text{and} \qquad \dot{K}_{II} = \dot{t}_{21}^{\infty} \sqrt{\pi l} \tag{16.141}
$$

for mode I and mode II loading, respectively. Note that Eqs. (16.141) coincide with their counterpart in elasticity without pre-stress, except that now the nominal stress replaces the Cauchy stress.

The crack solution and the surface bifurcation condition

The previously obtained crack solution remains valid except when the surface bifurcation condition [Eq. (12.81)] is met. This condition corresponds to the fulfilment of

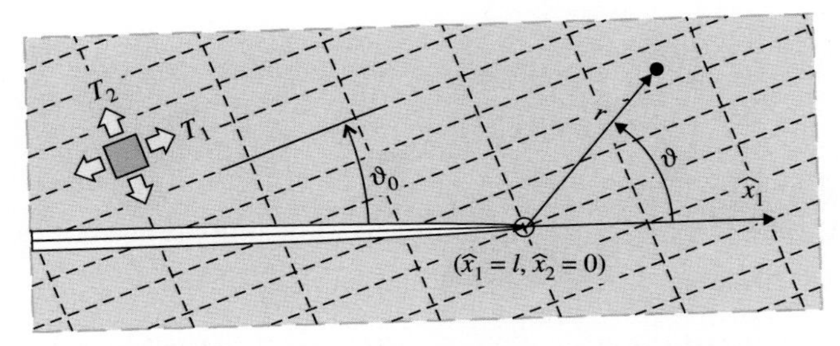

Figure 16.11. Polar coordinate system (r, ϑ) centred at the crack tip $(\hat{x}_1 = l, \hat{x}_2 = 0)$ for a fracture in an incompressible pre-stressed elastic material. The pre-stress axes x_1–x_2 are inclined at an angle ϑ_0 with respect the crack axis $\hat{x}_1$.

one of the two equations

$$\beta_2\varepsilon_1^2 - \beta_1\varepsilon_2^2 = 0 \qquad \text{or} \qquad \alpha(\delta^2 - \chi^2) + 2\beta\delta\chi = 0, \tag{16.142}$$

valid in EI and EC, respectively. When the surface bifurcation condition is approached, the fields solution of the crack problem tend to blow up, a peculiarity first noticed by Guz (1999, and references quoted therein).

For values of parameters ξ, k and η beyond the surface instability threshold, the solution obtained still works, from a purely mathematical point of view. However, the crack faces cannot be maintained straight after a surface bifurcation point has been passed, so the solution looses its physical meaning (the incremental energy release rate, which will be obtained in Section 16.2.4, becomes negative in this situation).

16.2.2 The inclined crack

We consider a crack inclined with respect to the x_1–x_2 axes defining the pre-stress directions and the orthotropy axes (Figs. 16.10 and 16.11). Therefore, the x_1–x_2 reference system has to be distinguished from the $\hat{x}_1$–$\hat{x}_2$ system, where the $\hat{x}_1$ axis is aligned parallel to the crack. The transformation between the two systems is expressed by Eq. (16.98), whereas the transformations between incremental displacement, its gradient, nominal stress and constitutive tensor are given by Eqs. (16.99) through (16.101).

The trick to solve the inclined crack problem can be deduced from Savin (1961; see also Sih and Liebowitz, 1968) and consists of the introduction of a function analogous to Eq. (16.113) [see also Eqs. (16.121), (16.130), and (16.136)], but now defined in the $\hat{x}_1$–$\hat{x}_2$ reference system, namely,

$$\widehat{\psi}^{\circ}_M(\hat{x}_1,\hat{x}_2) = \frac{\hat{t}^{\infty}_{2n}}{2\mu}\sum_{j=1}^{2}\mathrm{Re}\left\{A_j^M\left[\hat{z}_j^2 - \hat{z}_j\sqrt{\hat{z}_j^2 - l^2} + l^2\ln\left(\hat{z}_j + \sqrt{\hat{z}_j^2 - l^2}\right)\right]\right\} \tag{16.143}$$

(which automatically satisfies the decaying conditions of fields at infinity), where $n = 1$ and $M = II$ for mode II ($n = 2$ and $M = I$ for mode I), so $\hat{t}^{\infty}_{21}$ ($\hat{t}^{\infty}_{22}$) is the traction component parallel (orthogonal) to the crack line. Moreover, a new variable, replacing z_j [Eq. (12.30)], is introduced:

$$\hat{z}_j = \hat{x}_1 + W_j\hat{x}_2, \tag{16.144}$$

where

$$W_j = \frac{\sin\vartheta_0 + \Omega_j\cos\vartheta_0}{\cos\vartheta_0 - \Omega_j\sin\vartheta_0} \tag{16.145}$$

and the roots Ω_j are defined by Eq. (12.28).

Complex constants A_j^M in Eq. (16.143) can be obtained by imposing the boundary conditions on the crack faces, which are

$$\begin{aligned}
&\hat{t}^{\circ}_{21}(\hat{x}_1,0^{\pm}) = 0, && \hat{t}^{\circ}_{22}(\hat{x}_1,0^{\pm}) = \hat{t}^{\circ}_{22} = -\hat{t}^{\infty}_{22}, && \forall\,|\hat{x}_1| < l, && \text{for Mode I.}\\
&\hat{t}^{\circ}_{21}(\hat{x}_1,0^{\pm}) = \hat{t}^{\circ}_{21} = -\hat{t}^{\infty}_{21}, && \hat{t}^{\circ}_{22}(\hat{x}_1,0^{\pm}) = 0, && \forall\,|\hat{x}_1| < l, && \text{for Mode II.}
\end{aligned} \tag{16.146}$$

Imposing conditions (16.146) yields a linear algebraic system for the real and imaginary parts of constants A_j^M:

$$\begin{bmatrix} c_{11} & c_{21} & c_{12} & c_{22} \\ -c_{21} & c_{11} & -c_{22} & c_{12} \\ c_{31} & c_{41} & c_{32} & c_{42} \\ -c_{41} & c_{31} & -c_{42} & c_{32} \end{bmatrix} \begin{bmatrix} \mathrm{Re}[A_1^M] \\ \mathrm{Im}[A_1^M] \\ \mathrm{Re}[A_2^M] \\ \mathrm{Im}[A_2^M] \end{bmatrix} = \underbrace{\begin{bmatrix} -1 \\ 0 \\ 0 \\ 0 \end{bmatrix}}_{\text{for mode I}} \text{ or } \underbrace{\begin{bmatrix} 0 \\ 0 \\ -1 \\ 0 \end{bmatrix}}_{\text{for mode II}}, \tag{16.147}$$

where $M = I$ for mode I ($M = II$ for mode II) and coefficients c_{ij} are

$$\begin{aligned} 2\mu c_{1j} = {} & \widehat{\mathbb{G}}_{1112} - \widehat{\mathbb{G}}_{1222} - \mathrm{Re}[W_j]\big[\widehat{\mathbb{G}}_{1111} - 2\widehat{\mathbb{G}}_{1122} - \widehat{\mathbb{G}}_{1221} + \widehat{\mathbb{G}}_{2222} \\ & + \mathrm{Re}[W_j]\big(2\widehat{\mathbb{G}}_{1121} - 2\widehat{\mathbb{G}}_{2122} + \mathrm{Re}[W_j]\widehat{\mathbb{G}}_{2121}\big)\big] \\ & + \mathrm{Im}[W_j]^2\big(2\widehat{\mathbb{G}}_{1121} - 2\widehat{\mathbb{G}}_{2122} + 3\,\mathrm{Re}[W_j]\widehat{\mathbb{G}}_{2121}\big), \\ 2\mu c_{2j} = {} & \mathrm{Im}[W_j]\big[\widehat{\mathbb{G}}_{1111} - 2\widehat{\mathbb{G}}_{1122} - \widehat{\mathbb{G}}_{1221} + \widehat{\mathbb{G}}_{2222} \\ & + \mathrm{Re}[W_j]\big(4\widehat{\mathbb{G}}_{1121} - 4\widehat{\mathbb{G}}_{2122} + 3\,\mathrm{Re}[W_j]\widehat{\mathbb{G}}_{2121}\big) - \mathrm{Im}[W_j]^2\widehat{\mathbb{G}}_{2121}\big], \\ 2\mu c_{3j} = {} & -\widehat{\mathbb{G}}_{1221} + \mathrm{Re}[W_j]\big[\widehat{\mathbb{G}}_{1121} - \widehat{\mathbb{G}}_{2122} + \mathrm{Re}[W_j]\widehat{\mathbb{G}}_{2121}\big) - \mathrm{Im}[W_j]^2\widehat{\mathbb{G}}_{2121}, \\ 2\mu c_{4j} = {} & \mathrm{Im}[W_j]\big(-\widehat{\mathbb{G}}_{1121} + \widehat{\mathbb{G}}_{2122} - 2\,\mathrm{Re}[W_j]\widehat{\mathbb{G}}_{2121}\big), \qquad j = 1, \ldots, 4, \end{aligned} \tag{16.148}$$

and depend on the crack inclination ϑ_0 and the pre-stress and orthotropy parameters ξ, k and η.

The determinant of the coefficient matrix in Eq. (16.147) is null only when the surface instability condition [Eq. (12.81)] is met, so in all other cases, system (16.147) can be solved, and the solution of the inclined crack follows.

The perturbed incremental displacement along the crack faces can be obtained in the form

$$\begin{aligned} \hat{v}_1^{\circ M}(\hat{x}_1, \hat{x}_2 = 0^{\pm}) &= \frac{\hat{t}_{2n}^{\infty}}{2\mu}\mathrm{Re}\left[(W_1 A_1^M + W_2 A_2^M)\left(\hat{x}_1 \mp i\sqrt{l^2 - \hat{x}_1^2}\right)\right], \\ \hat{v}_2^{\circ M}(\hat{x}_1, \hat{x}_2 = 0^{\pm}) &= -\frac{\hat{t}_{2n}^{\infty}}{2\mu}\mathrm{Re}\left[(A_1^M + A_2^M)\left(\hat{x}_1 \mp i\sqrt{l^2 - \hat{x}_1^2}\right)\right], \end{aligned} \tag{16.149}$$

so the jump in incremental displacements across the crack surfaces ($|\hat{x}_1| < l$, $\hat{x}_2 = 0$) takes the form

$$\begin{aligned} [[\hat{v}_1^M]] &= \frac{\hat{t}_{2n}^{\infty}}{\mu}\mathrm{Im}[W_1 A_1^M + W_2 A_2^M]\sqrt{l^2 - \hat{x}_1^2}, \\ [[\hat{v}_2^M]] &= -\frac{\hat{t}_{2n}^{\infty}}{\mu}\mathrm{Im}[A_1^M + A_2^M]\sqrt{l^2 - \hat{x}_1^2}, \end{aligned} \tag{16.150}$$

where $n = 1$ and $M = II$ ($n = 2$ and $M = I$) for mode II (mode I).

It is worth noting that the following conditions, proven in the particular cases of null pre-stress ($k = \eta = 0$) or crack parallel to the orthotropy axes ($\vartheta_0 = 0$), have been verified numerically to hold:

$$\mathrm{Re}[A_1^I + A_2^I] = 0, \qquad \mathrm{Re}[W_1 A_1^{II} + W_2 A_2^{II}] = 0, \tag{16.151}$$

showing that the incremental perturbed displacement along the x_1 axis outside the crack is only longitudinal; that is, $\hat{v}_2^{\circ} = 0$ (transversal, i.e., $\hat{v}_1^{\circ} = 0$,) for mode I (for mode II), a circumstance noticed also by Broberg (1999, his section 4.14) for infinitesimal anisotropic elasticity.

In addition to Eqs. (16.151), the following conditions are obtained in the particular case of a crack parallel to the orthotropy x_1 axis, $\vartheta_0 = 0$,

$$\mathrm{Im}[W_1 A_1^I + W_2 A_2^I] = 0, \qquad \mathrm{Im}[A_1^{II} + A_2^{II}] = 0, \tag{16.152}$$

from which the solution obtained in Section 16.2.1 can be easily recovered.

Finally, *the incremental stress intensity factors for an inclined crack can be calculated and again result in the form* (16.141), *found for a crack parallel to the orthotropy axes.*

The inclined crack solution becomes particularly simple in the case where the pre-stress is null, $k = \eta = 0$. In particular, for mode I we have

$$A_j^I = -(-1)^j \frac{\cos 2\vartheta_0}{2\sqrt{1-\xi}} - i\frac{1-\xi-(-1)^j\sqrt{1-\xi}\sin 2\vartheta_0}{2(1-\xi)\sqrt{\xi}}, \qquad j = 1,2, \tag{16.153}$$

whereas for mode II we have

$$A_j^{II} = (-1)^j \left[\frac{\sin 2\vartheta_0}{2\sqrt{1-\xi}} + i\frac{\cos 2\vartheta_0}{2\sqrt{(1-\xi)\xi}}\right], \qquad j = 1,2. \tag{16.154}$$

The following properties also can be proven

$$W_1 A_1^I + W_2 A_2^I = 0 \quad \text{and} \quad A_1^{II} + A_2^{II} = 0. \tag{16.155}$$

An interesting feature that does not hold when the pre-stress is present and the crack is inclined can be deduced from Eqs. (16.150), $(16.155)_1$ and (16.154), namely, that a mode I (mode II) loading does not produce longitudinal v_1 (transversal v_2) incremental displacements along the crack line, so for $|\hat{x}_1| < l$ and $\hat{x}_2 = 0$, we have

$$[[\hat{v}]] = \left\{\frac{\hat{t}_{21}^{\infty}}{\mu\sqrt{\xi}}\sqrt{l^2 - \hat{x}_1^2},\ \frac{\hat{t}_{22}^{\infty}}{\mu\sqrt{\xi}}\sqrt{l^2 - \hat{x}_1^2}\right\}, \tag{16.156}$$

which is independent of the crack inclination ϑ_0.

As an example of the previous calculations, the deformed crack line and surfaces (incremental displacement components, reported on the vertical axis for $\hat{v}_2$ and on the horizontal axis for $\hat{x}_1 + \hat{v}_1$, are normalised through division by l) for mode I (*left*) and mode II (*right*) loading at infinity are illustrated in Fig. 16.12 for a Mooney-Rivlin material ($\xi = 1$) at null pre-stress $k = 0$ and at $k = 0.8$ for a crack parallel to the orthotropy x_1 axis, that is, $\vartheta_0 = 0$, and inclined at $\vartheta_0 = \pi/6$. Note that the mode II deformation at null pre-stress, $k = 0$, coincides with the horizontal axis and therefore is not visible.

Interesting features emerging from Fig. 16.12 are (1) the crack faces result displaced to the shape of an ellipse, (2) this ellipse degenerates into a segment for mode II and null pre-stress, and (3) the pre-stress introduces an incremental rigid-body rotation in the mode I and mode II solutions.

16.2.3 Shear bands interacting with a finite-length crack

In the spirit of the perturbative approach proposed by Bigoni and Capuani (2002), the role of shear banding in the incremental deformation fields around a crack of

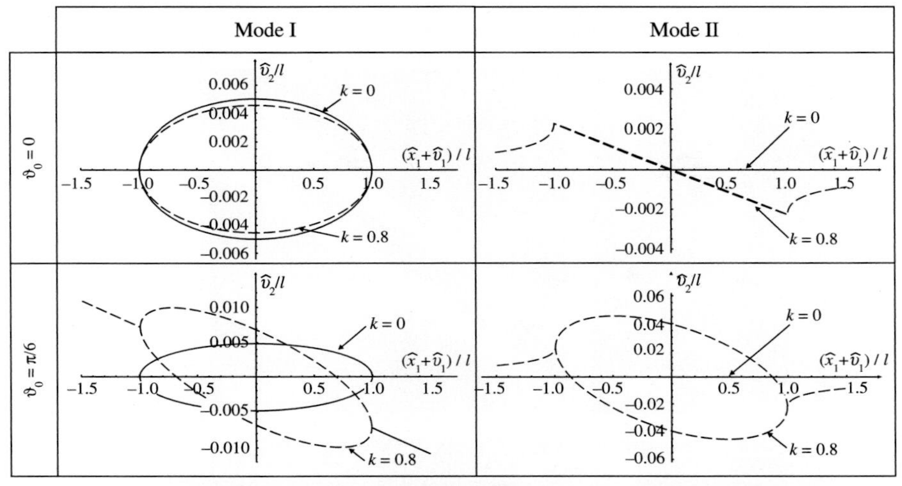

Figure 16.12. Deformed shape of a crack of length $2l$ subjected to mode I (*left*) and mode II (*right*) incremental loading ($\hat{t}_{22}^{\infty}/\mu = 0.01$ and $\hat{t}_{21}^{\infty}/\mu = 0.01$) in an infinite, elastic, and incompressible material. A Mooney-Rivlin material is considered with null pre-stress $k = 0$ (continuous curve) and a pre-stress defined by $k = 0.8$ (reported dashed). A crack is parallel to the x_1 orthotropy axis, $\vartheta_0 = 0$ (*top*), whereas a second crack is inclined at an angle $\vartheta_0 = \pi/6$ (*bottom*).

length $2l$ is investigated. This crack is assumed to be present in the material with a dead loading on its surfaces to maintain the state of pre-stress before and while the perturbation mode I and mode II loading is assigned.

The crack is considered within a J_2-deformation theory material, inclined at an angle corresponding to the shear band inclination ϑ^{SB} at the EC/H boundary [Eq. (12.62)]. In particular, for the two values of hardening exponent $N = 0.1$ and $N = 0.8$, the critical logarithmic strain for localisations (and the shear band inclination with respect to the x_1 axis) are $\varepsilon^{EL} \approx 0.322$ ($\vartheta^{SB} \approx 35.95°$) and $\varepsilon^{EL} \approx 1.032$ ($\vartheta^{SB} \approx 19.60°$), respectively (note that for a J_2-deformation theory material, the pre-strain, instead of the pre-stress, is used as the parameter controlling the current state).

The level sets of the modulus of incremental deviatoric strain have been mapped in Figs. 16.13 and 16.14 for low strain hardening $N = 0.1$ and high-strain hardening $N = 0.8$, respectively.

The investigation has been carried out with a choice of η, namely, $\eta/k = 0.311$ for $N = 0.1$ and $\eta/k = 0.775$ for $N = 0.8$, such that the Hill exclusion condition [Eq. (12.6)] is satisfied.

It can be concluded easily from Figs. 16.13 and 16.14 that

near the elliptical border, the deformation fields become highly focussed and aligned parallel to the shear band conjugate directions.

An analysis of the figures reveals that it becomes difficult to predict how the fracture will grow when loaded near the elliptic border. However, we have to keep in mind that the analysed crack has been taken aligned parallel to one shear band direction. It becomes instructive now to analyse the case of a horizontal crack (lying therefore in a symmetry axis with respect to the conjugate band directions), reported

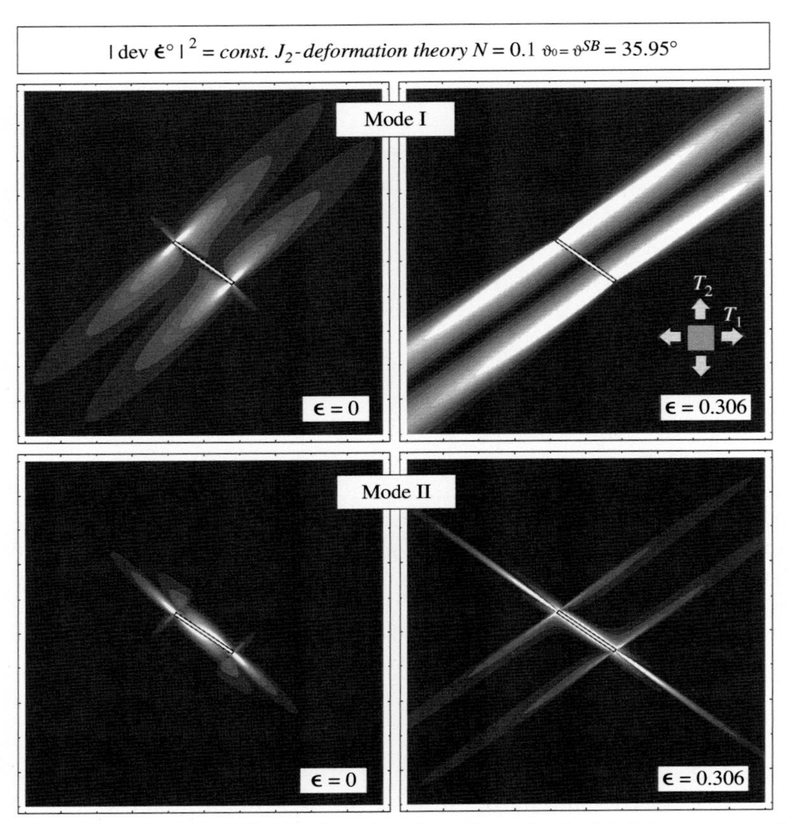

Figure 16.13. Interaction of shear bands and mechanical fields near a crack of length $2l$ (evidenced with a thin rectangle, providing the scale bar of the representation) in an infinite, elastic, and incompressible material. A J_2-deformation theory material has been considered at low strain hardening $N = 0.1$, at null pre-strain $\varepsilon = 0$ (*left*) and pre-strained near the elliptic border $\varepsilon = 0.306$ (*right*). The crack is aligned parallel to a shear band direction, $\vartheta_0 = \vartheta^{SB} = 35.95°$. Two parallel bands emerge for mode I incremental loading, whereas for mode II two conjugate band directions are visible.

in Fig. 16.15, for a J_2-deformation theory material at high-strain hardening $N = 0.8$, near the EC/H boundary, and loaded under incremental mode I. Results are qualitatively analogous for different values of strain hardening and for mode II loading, in particular, the mode II incremental deformation fields are dominated near the elliptic border by localised deformations aligned parallel with the two shear band conjugate directions.

We can observe that

two symmetric shear bands emerge near the crack tip,

and their interaction may lead to failure of the material under shear in front of the crack, a situation compatible with mode I growth, to be interpreted as a sort of 'alternating sliding off and cracking', as suggested by McClintock (1971), Kardomateas (1986) and Kardomateas and McClintock (1989). The situation is more complicated for mode II loading, but our results agree with the consideration made by Hallbäck

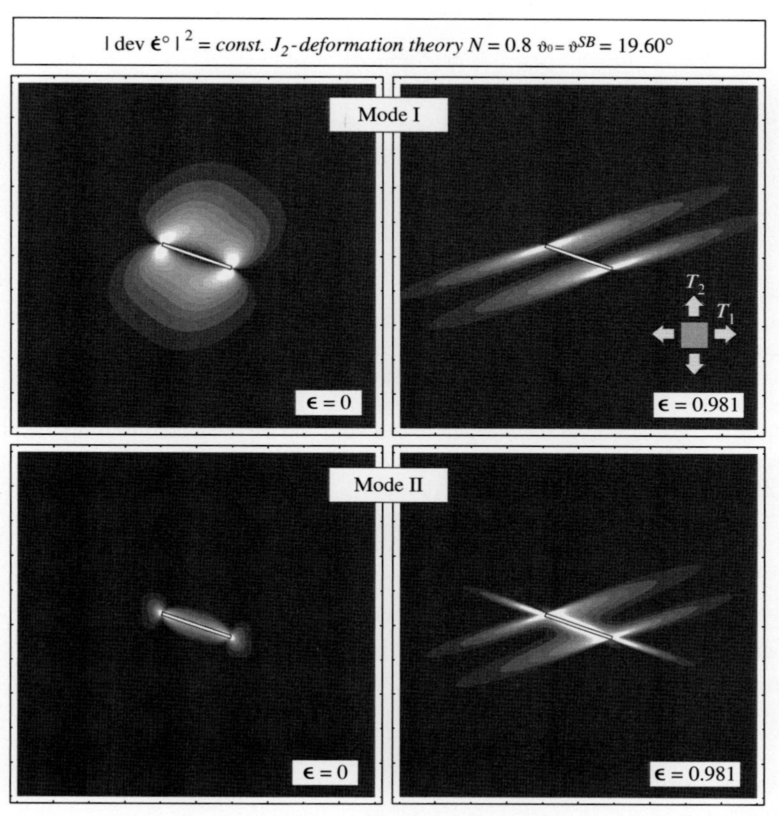

Figure 16.14. Interaction of shear bands and mechanical fields near a crack of length $2l$ (evidenced with a thin rectangle, providing the scale bar of the representation) in an infinite, elastic, and incompressible material. A J_2-deformation theory material has been considered at high-strain hardening $N = 0.8$, at null pre-strain $\varepsilon = 0$ (*left*) and pre-strained near the elliptic border $\varepsilon = 0.981$ (*right*). The crack is aligned parallel to a shear band direction, $\vartheta_0 = \vartheta^{SB} = 19.60°$. Two parallel bands emerge for mode I incremental loading, whereas for mode II two conjugate band directions are visible.

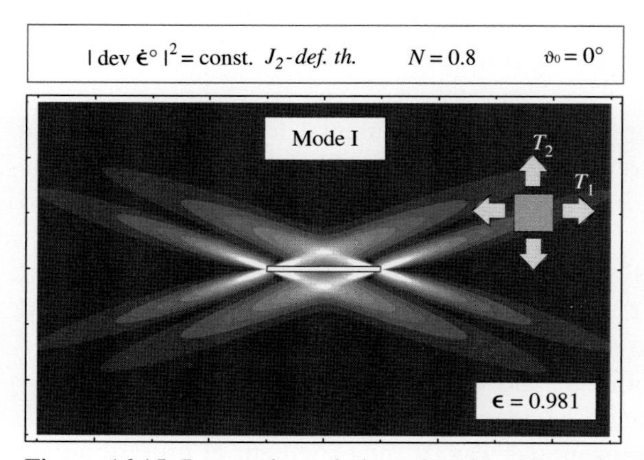

Figure 16.15. Interaction of shear bands and mechanical fields near a crack of length $2l$ under mode I incremental loading in an infinite, elastic, and incompressible material. A J_2-deformation theory material has been considered at high-strain hardening $N = 0.8$, pre-strained near the elliptic border $\varepsilon = 0.981$. The crack is horizontal, whereas the shear bands are inclined at $\pm 19.60°$. Note that four shear bands emerge.

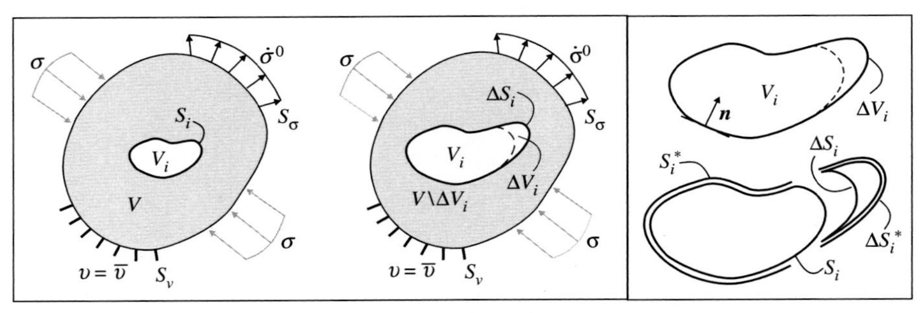

Figure 16.16. Two elastic, pre-stressed bodies are compared (*left*), having identical shape, boundary conditions, elastic properties, pre-stress and pre-strain but voids of different sizes. The detail of the void and its surface is reported on the right; note the unit normal vector, defined to point outward from the elastic body and toward the void. Incremental deformation of pre-stressed solids is considered, so the surface of the void can be subjected to finite dead loading, and surface ΔS_i^* must be subject to the nominal tractions present on the same surface embedded in the material in the configuration on the left.

and Nilsson (1994) that 'mode II failure results when the direction of the prospective shear band coincides with the crack surface direction, while mode I type failure occurs when the shear bands are inclined to the direction of crack surfaces.'

16.2.4 Incremental energy release rate for crack growth

We slightly generalise the energy release rate concept employed by Rice (1968) in the context of linear elasticity and start referring to Fig. 16.16 and comparing two incremental boundary value problems (for finite bodies subject to identical conditions on the external boundaries $S_\sigma \cup S_v$, namely, prescribed incremental nominal tractions $\dot{\boldsymbol{\sigma}}^0$ on S_σ and incremental displacements $\mathbf{v} = \bar{\mathbf{v}}$ on S_v) only differing in the sizes of the void they contain. Note that we are addressing an incremental problem, so the surface of the void can be loaded by dead loading. In particular, the void in the body on the right (of volume $V_i \cup \Delta V_i$, enclosed by surface $S_i^* \cup \Delta S_i^*$) has been obtained by increasing the size of the void in the body on the left (of volume V_i, enclosed by surface S_i).

Since we want to include pre-stress in an incremental formulation, nominal (finite) dead tractions identical to those existing within the material containing the void V_i must be applied on the surface ΔS_i^* of the material containing the void $V_i \cup \Delta V_i$.

We define the incremental displacement and nominal traction fields, solutions to the two problems, as $\boldsymbol{v}^0$ and $\dot{\boldsymbol{t}}^0$ for the problem on the left and $\boldsymbol{v} = \boldsymbol{v}^0 + \tilde{\boldsymbol{v}}$ and $\dot{\boldsymbol{t}} = \dot{\boldsymbol{t}}^0 + \tilde{\boldsymbol{t}}$ for the problem on the right. Since the void surfaces are subject to dead loading, $\dot{\boldsymbol{t}}^{0T}\boldsymbol{n} = \boldsymbol{0}$ and $\dot{\boldsymbol{t}}^T\boldsymbol{n} = \boldsymbol{0}$ within V_i and $V_i \cup \Delta V_i$, respectively.

The two bodies are assumed to be identically pre-stressed and pre-strained, although not necessarily in a homogeneous way. If the expedient of prescribing ad hoc dead tractions on ΔS_i^* is not considered and the void surface is free of tractions, in order to have identical pre-stress and pre-strain, the two current configurations shown in Fig. 16.16 must have special geometries and loadings, as is the case of a crack aligned parallel to a principal stress direction with the other principal stress to be null and, more important, of a shear band model (Section 16.4).

The incremental potential energy decrease for a void growth in an elastic (incompressible or compressible, generically anisotropic and pre-stressed) body assumes an expression analogous to that reported by Rice [1968, his Eq. (55), p. 207], namely,

$$-\Delta\dot{P} = \int_{\Delta V_i} \phi(\nabla \mathbf{v}^0) dV - \frac{1}{2} \int_{\Delta S_i^*} \boldsymbol{n} \cdot \boldsymbol{t}^0 \tilde{\boldsymbol{v}} dS, \tag{16.157}$$

a quantity which, when positive, implies void *growth*. Note that the scalar function ϕ is the incremental gradient potential defined as

$$\dot{t}_{ij} = \frac{\partial \phi(\nabla \boldsymbol{v})}{\partial v_{j,i}} + \dot{p}\,\delta_{ij}, \qquad \phi(\nabla \boldsymbol{v}) = \frac{1}{2} v_{j,i} \mathbb{G}_{ijhk} v_{k,h}. \tag{16.158}$$

Turning our attention now to a thin void inclusion, namely, a crack aligned parallel to the $\hat{x}_1$ axis (Fig. 16.10), the volume integral in Eq. (16.157) vanishes; thus, taking the limit of the length increase $\Delta l \to 0$ at fixed incremental stress intensity factor $\dot{K}$ [Eq. (16.157)] becomes

$$\dot{G} = -\frac{d\dot{P}}{dl} = \lim_{\Delta l \to 0} \frac{1}{2\Delta l} \int_0^{\Delta l} \hat{\dot{t}}_{2i}(r,0) [[\hat{\dot{v}}_i(\Delta l - r, \pi)]]\, dr, \tag{16.159}$$

where the symbol $\hat{\cdot}$ denotes that we are using the inclined crack solution, the repeated index is summed, and r denotes the radial distance from the crack tip, and the polar (anti-clockwise) coordinate θ singling out r from the $\hat{x}_1$ axis (so that $\theta = 0$ corresponds to points ahead of the crack tip) has been taken equal to 0 and π. Equation (16.159) defines

the incremental energy release rate for a mixed-mode growth of a crack in an elastic, incompressible or compressible body, generically anisotropic and pre-stressed.

The proof that the incremental energy release rate coincides with the path–independent incremental $\dot{J}$-integral

$$\dot{J} = \int_\Gamma \left(\hat{\phi}\, \hat{n}_1 - \hat{n}_j \hat{\dot{t}}_{ji} \frac{\partial \hat{\dot{v}}_i}{\partial \hat{x}_1} \right) d\Gamma \tag{16.160}$$

has not been explicitly obtained, but the validity of $\dot{G} = \dot{J}$ has been verified numerically.

The incremental energy release rate (16.159) can be developed making use of the asymptotic near-tip incremental nominal stress ahead of the crack

$$\hat{\dot{t}}_{22}(r,0) = \frac{\dot{K}_I}{\sqrt{2\pi r}}, \qquad \hat{\dot{t}}_{21}(r,0) = \frac{\dot{K}_{II}}{\sqrt{2\pi r}}, \tag{16.161}$$

and incremental displacement on the crack faces (where constants have been neglected)

$$\begin{aligned}
\hat{\dot{v}}_1(\Delta l - r, \pm\pi) &= \pm \frac{\sqrt{2l}\sqrt{\Delta l - r}}{2\mu} \mathrm{Im}\left[\hat{\dot{t}}_{22}^{\infty} (W_1 A_1^I + W_2 A_2^I) + \hat{\dot{t}}_{21}^{\infty} (W_1 A_1^{II} + W_2 A_2^{II}) \right], \\
\hat{\dot{v}}_2(\Delta l - r, \pm\pi) &= \mp \frac{\sqrt{2l}\sqrt{\Delta l - r}}{2\mu} \mathrm{Im}\left[\hat{\dot{t}}_{22}^{\infty} (A_1^I + A_2^I) + \hat{\dot{t}}_{21}^{\infty} (A_1^{II} + A_2^{II}) \right],
\end{aligned} \tag{16.162}$$

holding for 'small' Δl.

Employing the asymptotic near-tip representations [Eqs. (16.161) and (16.162)] in Eq. (16.159), we obtain

$$\dot{G} = -\dot{K}_I^2 \frac{\text{Im}\left[A_1^I + A_2^I\right]}{4\mu} + \dot{K}_{II}^2 \frac{\text{Im}\left[W_1 A_1^{II} + W_2 A_2^{II}\right]}{4\mu} + \dot{K}_I \dot{K}_{II} \frac{\text{Im}\left[W_1 A_1^I + W_2 A_2^I - A_1^{II} - A_2^{II}\right]}{4\mu}, \tag{16.163}$$

the incremental energy release rate for an inclined crack loaded in mixed mode in a pre-stressed, orthotropic and incompressible material.

From Eq. (16.163), *the incremental energy release rate for a mixed-mode loading of a crack parallel to the orthotropy axes* (i.e., $\vartheta_0 = 0$) can be made explicit:

$$\dot{G} = \frac{\Lambda}{\mu} \frac{\dot{K}_I^2 \sqrt{1-k} + \dot{K}_{II}^2 \sqrt{1+k}}{(2\xi - \eta + \Lambda)^2 \sqrt{2\xi - 1 - \Lambda} - (2\xi - \eta - \Lambda)^2 \sqrt{2\xi - 1 + \Lambda}}, \tag{16.164}$$

where there is no coupling between the two modes I and II.

Another interesting special case is that of null pre-stress (k=η=0), in which for an inclined crack ($\vartheta_0 \neq 0$) the following expression of the incremental energy release rate can be obtained:

$$\dot{G} = \frac{\dot{K}_I^2 + \dot{K}_{II}^2}{4\mu\sqrt{\xi}}, \tag{16.165}$$

which agrees with the known isotropic elasticity solution in the incompressible limit recovered for $\xi = 1$.

Note that both incremental energy release rates [Eq. (16.164) and (16.163)] generally (an exception to this rule will be shown in Fig. 16.17) blow up to infinity when the surface bifurcation [Eq. (12.81) or Eq. (16.142)] is approached, as in the case of the crack aligned parallel to one of the orthotropy axes. This feature is evident in the example reported below.

An example of calculation of incremental energy release rate for inclined (at $\vartheta_0 = \{0, \pi/4, \pi/2\}$) mode I and mode II cracks in an incrementally isotropic material ($\xi = 1$) as a function of the pre-stress parameter k is reported in Fig. 16.17, where $\dot{G}$ has been normalised through division by $\dot{K}_M^2$ and multiplication by 4μ. In order to explore the incremental energy release rate until close to the elliptical boundary (more precisely, to the EI/P boundary, see Section 12.2), we have taken $\eta = k > 0$, so the Hill condition (12.6) excludes all possible bifurcations within EI (Fig. 12.3). It may be interesting to observe from Fig. 16.17 that, with the exceptions of $\vartheta_0 = 0$ for mode I and $\vartheta_0 = \pi/2$ for mode II, the incremental energy release rate blows up to infinity when k approaches 1. These exceptions can be motivated by the circumstance that at the EI/P boundary only one shear band forms aligned parallel with the major principal (tensile in this case) stress component T_1. Therefore, for mode I (mode II), a crack parallel (orthogonal) to the shear band is not influenced by the progressive weakening in the shear band direction occurring when the elliptical boundary is approached.

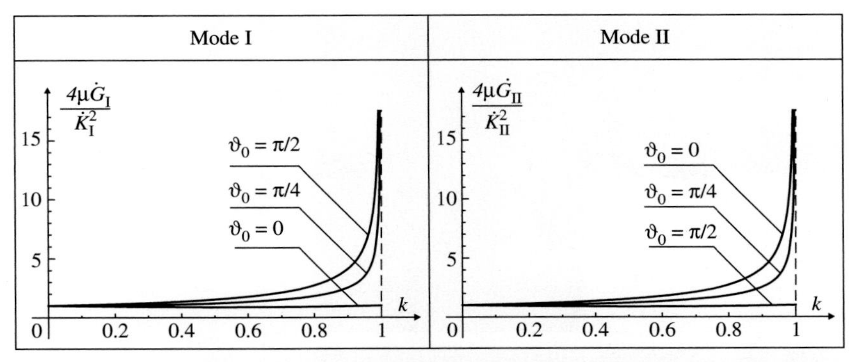

Figure 16.17. Incremental stress release rate for mode I and mode II cracks of legth $2l$ inclined at $\vartheta_0 = \{0, \pi/4, \pi/2\}$ with respect to the principal stress axes in an incrementally isotropic (incompressible and elastic) material ($\xi = 1$) as a function of the pre-stress parameter k, taken positive and equal to η, so that the Hill exclusion criterion (12.6) is satisfied.

Note that for a Mooney-Rivlin material, μ is a function of k, blowing up to infinity when the EI/P boundary ($k = 1$) is approached. As a consequence, for a Mooney-Rivlin material, the energy release rate remains finite when k tends to 1.

Note that for null pre-stress, $\eta = k = 0$, Eq. (16.165) shows that the incremental energy release rate blows up to infinity when ξ tends to zero, which corresponds to the EC/H boundary and to the appearance of the two shear bands inclined at $\pi/4$ with respect to the principal stress direction, typical of Mises plasticity.

Figure 16.17 reveals another interesting feature, namely, that the curves corresponding to $\vartheta_0 = \{0, \pi/4, \pi/2\}$ in mode I are identical to the curves corresponding, respectively, to $\vartheta_0 = \{\pi/2, \pi/4, 0\}$ in mode II. More in general, the following relation can be proven in the absence of pre-stress using Eqs. (16.153) and (16.154):

$$\frac{\dot{G}_I(\vartheta_0)}{\dot{K}_I^2} = \frac{\dot{G}_{II}(\pi/2 - \vartheta_0)}{\dot{K}_{II}^2}, \tag{16.166}$$

and this also has been found to hold numerically when the pre-stress is different from zero.

16.3 Mode I perturbation of a stiffener in an infinite non-linear elastic material subjected to finite simple shear deformation

Following Bigoni and Dal Corso (2009), we consider now a rigid line inclusion, a so-called stiffener, as a perturbing agent embedded in an elastic material, homogeneously pre-stressed within the elliptic regime by a simple shear parallel to the line inclusion. In fact, the inclusion remains neutral (in other words, it does not alter uniformity of stress; see Section 5.4) under the finite shear of a non-linear elastic material, providing the pre-stress to the material, but it reacts when subjected to an incremental uniform mode I perturbation.

The stiffener problem has been solved for linear, isotropic elasticity (without pre-stress; Muskhelishvili, 1953). Before embarking on the solution for pre-stressed

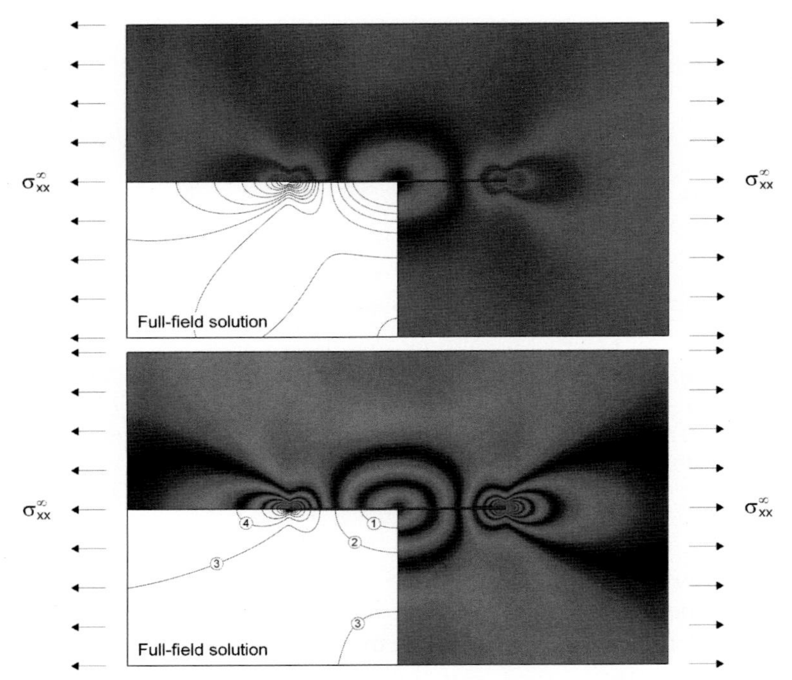

Figure 16.18. Photoelastic (isochromatic, *top*, and monochromatic, *bottom*) fringes revealing the stress field near a thin line inclusion (0.1-mm-thick steel platelet) embedded in an elastic matrix (a two-component 'soft' epoxy resin) and subjected to mode I uniform loading σ_{xx}^{∞}. A comparison is made with the elastic solution in plane strain, with Poisson's ratio equal to 0.45. (See color plates section.)

material, it may be important to mention that the stress field predicted by the linear elastic solution finds an excellent correspondence in photoelastic experiments provided by Noselli et al. (2010).

In particular, results of photoelastic experiments [see Coker and Filon (1957), Frocht (1965) and Dally and Riley (1965) for an introduction to photoelasticity] are reported in Figs. 16.18 and 16.19 and compared with the linear elastic solution.

While Fig. 16.18 demonstrates that the 'global' stress field is excellently captured by the elastic solution, a quantitative comparison with the *singular* full-field small strain solution and its asymptotic approximation is presented in Fig. 16.19. The latter figure shows the validity of even the asymptotic approximation used near the tip of the stiffener, which is quantitatively in agreement with experiments showing a near-tip stress concentration up to a factor of 7.

The rigid-line inclusion has length $2l$ (Fig. 16.20), and the elastic matrix has been subjected to a simple shear parallel to the inclusion and defined by the shear amount γ. The homogeneous state of pre-stress has the principal stress axes inclined at an angle ϑ_E [Eq. (5.43)] with respect to the inclusion line and is taken as the reference state[5] on which a perturbation corresponding to remote uniform mode I incremental deformation at infinity $\hat{v}_{2,2}^{\infty}$ is superimposed.

[5] The analysis can be carried out with respect to a generic, uniform state of pre-stress (with principal values inclined at ϑ_0 different from ϑ_E) not necessarily generated through a simple shear deformation.

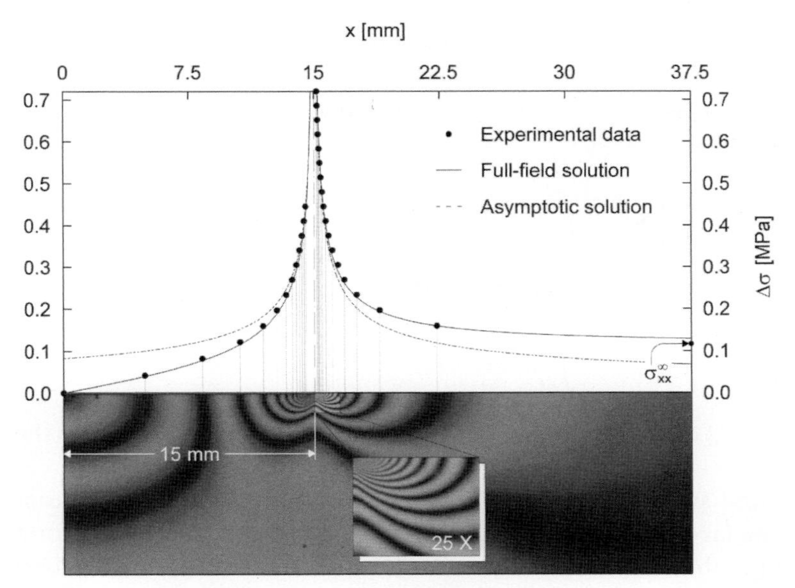

Figure 16.19. In-plane principal stress difference $\Delta\sigma = \sigma_1 - \sigma_2$ along the stiffener line: Comparison between the photoelastic experiment with monochromatic light and the full-field and asymptotic elastic solutions. A detail of the region near to the (*right*) stiffener tip has been captured using an optical microscope (photo reported in the insert), so a great number of fringes (20) have been detected, until a distance from the tip of the same order of magnitude as the stiffener thickness (0.1 mm). A stress concentration of 7 is visible. (See color plates section.)

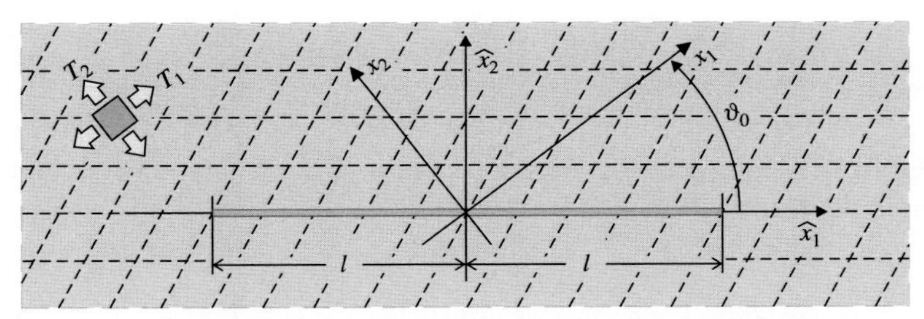

Figure 16.20. The rigid-line inclusion (of length $2l$) in an infinite non-linear elastic incompressible material deformed under simple shear parallel to the inclusion. The state of stress generated at a certain deformation γ has principal values inclined with respect to the $\hat{x}_1$–$\hat{x}_2$ reference system at an angle ϑ_0 taken equal to ϑ_E (although this identification is not necessary in the calculations).

An $\hat{x}_1$–$\hat{x}_2$ reference system located at the stiffener centre, taken with the inclusion line parallel to the $\hat{x}_1$ axis, is inclined at an angle ϑ_0 (assumed equal to ϑ_E) with respect to the x_1–x_2 system, defining the principal stress directions (Fig. 16.20).

The kinematic boundary conditions for a stiffener express the fact that this can only suffer an incremental rigid-body motion:

$$\begin{cases} \hat{v}_1(\hat{x}_1,0) = \hat{v}_1(0,0), \\ \hat{v}_2(\hat{x}_1,0) = \hat{v}_2(0,0) + \omega_S x_1, \end{cases} \qquad \forall\, |\hat{x}_1| < l, \tag{16.167}$$

so $\hat{v}_1(0,0)$, $\hat{v}_2(0,0)$ and ω_S represent unknown quantities to be determined as a part of the solution by imposing the quasi-static boundary conditions, ensuring equilibrium of the stiffener in terms of incremental 'global' axial and shearing forces and incremental moment, respectively:

$$\begin{cases} \dot{\mathcal{N}} = \int_{-l}^{l} [[\hat{t}_{21}(y,0)]]\, dy = 0, \\ \dot{\mathcal{T}} = \int_{-l}^{l} [[\hat{t}_{22}(y,0)]]\, dy = 0, \\ \dot{\mathcal{M}} = \int_{-l}^{l} [[\hat{t}_{22}(y,0)]]\, y\, dy = 0, \end{cases} \tag{16.168}$$

where the brackets $[[\cdot]]$ denotes the jump in the relevant argument, taken across the stiffener.

Owing to central symmetry considerations (with respect to the stiffener centre) involved in the far-field loading problem under analysis and the specific form of solution sought in the following, the boundary conditions (16.167) and (16.168) can be reduced to the following homogeneous incremental displacement gradient conditions:

$$\begin{cases} \hat{v}_{1,1}(\hat{x}_1,0) = 0, \\ \hat{v}_{2,1}(\hat{x}_1,0) = \omega_S, \end{cases} \qquad \forall\, |\hat{x}_1| < l, \tag{16.169}$$

plus the requirement that the normal stress increment $\hat{t}_{22}$ be continuous across the stiffener

$$[[\hat{t}_{22}(\hat{x}_1,0)]] = 0, \qquad \forall\, |\hat{x}_1| < l, \tag{16.170}$$

a condition allowing determination of ω_S.

Prescribing an incremental deformation $\hat{v}_{2,2}^{\infty}$ at infinity, analogous to the crack problem analysed in Section 16.2.2, the stream function of the perturbed problem $\widehat{\psi}^{\circ}$ can be sought in the form

$$\widehat{\psi}^{\circ}(\hat{x}_1,\hat{x}_2) = \frac{\hat{v}_{2,2}^{\infty}}{2} \sum_{j=1}^{2} \mathrm{Re}\left\{ D_j \left[\hat{z}_j^2 - \hat{z}_j\sqrt{\hat{z}_j^2 - l^2} + l^2 \ln\left(\hat{z}_j + \sqrt{\hat{z}_j^2 - l^2} \right) \right] \right\}, \tag{16.171}$$

satisfying automatically the decaying condition on the velocity, incremental strain and stress at infinity (in the elliptic regime) and providing a stress square-root singularity at the stiffener tips.

Imposing the stream function (16.171) to satisfy the boundary conditions along the stiffener line [Eqs. (16.169) and (16.170)] yields the following linear problem for the complex constants D_1 and D_2:

$$\begin{bmatrix} \mathrm{Re}[W_1] & -\mathrm{Im}[W_1] & \mathrm{Re}[W_2] & -\mathrm{Im}[W_2] \\ \mathrm{Im}[W_1] & \mathrm{Re}[W_1] & \mathrm{Im}[W_2] & \mathrm{Re}[W_2] \\ 0 & 1 & 0 & 1 \\ -c_{21} & c_{11} & -c_{22} & c_{12} \end{bmatrix} \begin{bmatrix} \mathrm{Re}[D_1] \\ \mathrm{Im}[D_1] \\ \mathrm{Re}[D_2] \\ \mathrm{Im}[D_2] \end{bmatrix} = \begin{bmatrix} 1 \\ 0 \\ 0 \\ 0 \end{bmatrix}, \tag{16.172}$$

where the real constants c_{1j} and c_{2j} ($j = 1,\ldots,4$) are defined by Eqs. (16.148) and depend on the stiffener inclination ϑ_0 and on the pre-stress and orthotropy parameters ξ, k and η. We introduce the normalised stiffener rotation Γ as

$$\Gamma = \frac{\omega_S}{\hat{v}_{2,2}^{\infty}} = -\mathrm{Re}[D_1 + D_2], \tag{16.173}$$

and since analytical proof looks awkward, we have checked numerically that (1) the solution of the present problem is independent of the in-plane stress parameter η, (2) the coefficients D_j $(j = 1,2)$ solving system (16.172) satisfy the following two equations:

$$\begin{aligned} &W_1^2 D_1 + W_2^2 D_2 = W_1 + W_2 + \Gamma W_1 W_2, \\ &W_1^3 D_1 + W_2^3 D_2 = W_1^2 + W_1 W_2 + W_2^2 + \Gamma W_1 W_2 (W_1 + W_2), \end{aligned} \tag{16.174}$$

and (3) the rotation parameter Γ (16.173) satisfies the conditions

$$\begin{aligned} &\Gamma\,(k=0,\vartheta_0) = \Gamma\,(k,\vartheta_0=0) = \Gamma\,(k,\vartheta_0=\pi/2) = 0, \\ &\Gamma = \Gamma(k,\vartheta_0) = -\Gamma\,(-k,\pi/2-\vartheta_0)\,. \end{aligned} \tag{16.175}$$

Defined in terms of incremental velocity gradient as in Dal Corso et al. (2008), the incremental stress intensity factor under mode I loading is

$$\dot{K}_{(\epsilon)I} = 2\mu \lim_{\hat{x}_1 \to l^+} \sqrt{2\pi\,(\hat{x}_1 - l)}\,\hat{v}_{2,2}(\hat{x}_1,\hat{x}_2 = 0) = 2\mu\,\hat{v}_{2,2}^{\infty}\sqrt{\pi l}, \tag{16.176}$$

resulting independent of the pre-stress parameters ξ, k and η (but μ may depend on the full set of current state variables) and of the angle ϑ_0 between the stiffener and the directions of principal stress T_1.

Calculation of the incremental energy release rate (not reported here, see Dal Corso and Bigoni, 2009) and the incremental axial force in the stiffener generalise results for a stiffener aligned parallel to the principal stress directions obtained by Bigoni et al. (2008) and Dal Corso et al. (2008). Those results now can be recovered by setting $\vartheta_0 = 0$ and thus obtaining

$$\begin{aligned} &D_1 = -D_2 = -\frac{1}{2\alpha}, \qquad \Gamma = 0, \qquad \text{in EC}, \\ &D_1 = -D_2 = -\frac{i}{\beta_1 - \beta_2}, \qquad \Gamma = 0, \qquad \text{in EI}, \end{aligned} \tag{16.177}$$

showing that in the case of a stiffener aligned parallel to pre-stress principal axes, there is no rigid rotation of the line owing to the symmetry of the problem. The rotation ω_S also is null in another case corresponding to $k = 0$ (and $\{\vartheta_0, \eta\} \neq 0$), that is,

$$D_1 = -D_2 = \frac{1}{2\sqrt{1-\xi}}(-\cos 2\vartheta_0 + i\sqrt{\xi}\sin 2\vartheta_0), \qquad \Gamma = 0. \tag{16.178}$$

The normalised stiffener rotation Γ (16.173) is reported in Fig. 16.21 for a J_2-deformation theory material, showing an anti-symmetric behaviour with respect to the shear parameter γ. Note that results reported in Fig. 16.21 are independent of the hardening parameter N, except for the fact that the curve terminates at failure of ellipticity $\gamma^{EL} = \gamma^{EL}(N)$.

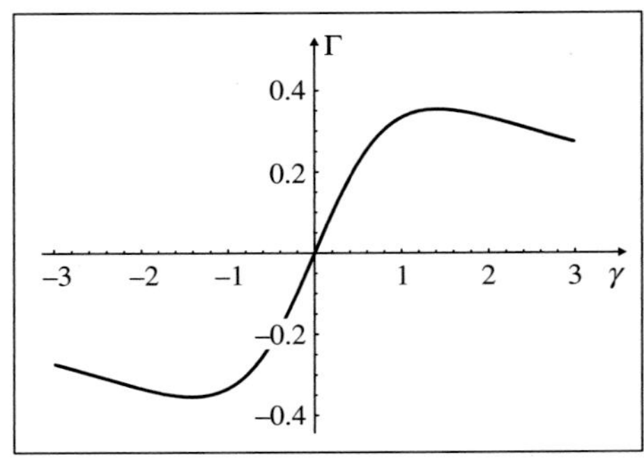

Figure 16.21. The normalised stiffener rotation $\Gamma = \omega_S / \hat{v}^{\infty}_{2,2}$ in a J_2-deformation theory material generated by an incremental mode I superimposed on a simple shear of finite amount γ.

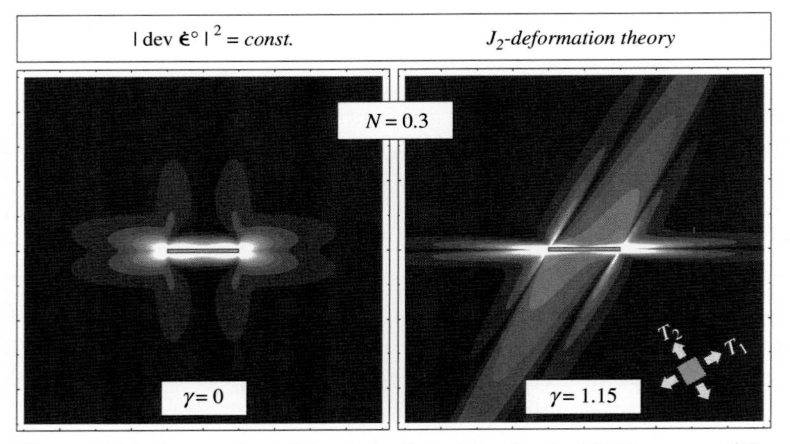

Figure 16.22. Interaction of shear bands and mechanical fields near a stiffener (evidenced with a thin rectangle, providing the scale bar of the representation) embedded in a J_2-deformation theory material subject to a finite simple shear of amount γ and a subsequent mode I incremental uniform remote load. Null shear is seen before the perturbation is considered on the left, whereas a shear equal to 0.95 times the amount at ellipticity loss γ^{EL} is considered on the right. Note that the pre-stress (of principal components T_1 and T_2) generated through the shear deformation is inclined with respect to the stiffener line (and sketched in the figure).

The previously obtained solution now can be employed to analyse the incremental strain field near the stiffener. In particular, level sets of the modulus of perturbed incremental deviatoric strain for J_2-deformation theory of plasticity material are reported in Fig. 16.22 for $N = 0.3$ strain hardening.

For a J_2-deformation material, the loss of ellipticity occurs at $\gamma^{EL} = 1.218$ for $N = 0.3$. Two values of amount of shear γ have been considered in Fig. 16.22, namely, $\gamma = 0$ corresponding to a material with null pre-stress[6] and $\gamma \simeq 0.95\gamma^{EL}$, namely, close to the boundary of ellipticity loss, before the mode I perturbation is applied.

[6] At null pre-stress, the J_2-deformation theory of plasticity material becomes incrementally rigid (because both μ and μ_* tend to infinity, but their ratio ξ tends to N). It is expedient, therefore, to plot results normalised through division by μ so that they tend to results pertaining to an incompressible

When the perturbation is applied at high pre-strain, the incremental deformation fields appear strongly focussed along the near-tip directions of the shear bands formally possible at ellipticity loss, as found by Bigoni et al. (2008) and Dal Corso and Bigoni (2009). Moreover, the results show that *the bands closest to the stiffener line result to be privileged, so a form of a 'thick' shear band parallel to the stiffener appears*, a finding in qualitative agreement with experimental results by Misra and Mandal (2007).

The stiffener solution in the parabolic and hyperbolic regimes Accepting non-decaying of the solution and infinite strain (and stress) increment along certain shear band lines, the previously obtained solution for a rigid-line inclusion in an infinite elastic material can be extended to the parabolic and hyperbolic regimes. Since beyond the elliptical range a problem is known to be ill-posed, analysis of the solution within the parabolic and hyperbolic regimes is instructive, revealing features related to ill-posedness.

To obtain a solution (which need not be unique) beyond the elliptical range, we have to go back to the representation of the stream function (16.171), where now the indices have to range from 1 to n so that

$$\widehat{\psi}^{\circ}(\hat{x}_1,\hat{x}_2) = \frac{\hat{v}^{\infty}_{2,2}}{2}\sum_{j=1}^{n}\text{Re}\left\{D_j\left[\hat{z}_j^2 - \hat{z}_j\sqrt{\hat{z}_j^2 - l^2} + l^2\ln\left(\hat{z}_j + \sqrt{\hat{z}_j^2 - l^2}\right)\right]\right\}, \quad (16.179)$$

where n indicates the number of non-conjugate roots [i.e., $n = 4$ ($n = 3$) in H (P)].

Outside the elliptical regime, the characteristic lines defined as

$$\hat{z}_j = \text{const} \quad \Rightarrow \quad \frac{d\hat{x}_2}{d\hat{x}_1} = -\frac{1}{W_j} \quad (16.180)$$

become real and correspond to four (two) different families in the hyperbolic (parabolic) regime, and in particular, their inclinations correspond to the following shear band inclinations with respect to the $\hat{x}_1$ axis

$$\widehat{\vartheta}_j^{SB} = -\arctan\left[\frac{1}{W_j}\right] = \vartheta_0 - \arctan\left[\frac{1}{\Omega_j}\right] = \vartheta_0 - \vartheta_j^{SB}, \quad (16.181)$$

where $j = 1,\ldots,4$.

The decaying of solution (16.179) is lost along the characteristic lines emanating from the stiffener tips, where, additionally, the increment of strain, and consequently stress, becomes infinite (whereas for the other characteristic lines cutting the stiffener, the solution always remains bounded). In contrast, incremental displacements remain continuous and finite everywhere, even along characteristics.

orthotropic material deformed in small strain. The axes of orthotropy are therefore inclined at 45° with respect to the stiffener line in Fig. 16.22.

The system of linear equations

$$\begin{cases} \sum_{j=1}^{n} \mathrm{Re}[W_j]\mathrm{Re}[D_j] - \mathrm{Im}[W_j]\mathrm{Im}[D_j] = 1 \\ \sum_{j=1}^{n} \mathrm{sign}[\mathrm{Im}[\Omega_j]] \left\{ \mathrm{Im}[W_j]\mathrm{Re}[D_j] + \mathrm{Re}[W_j]\mathrm{Im}[D_j] \right\} = 0 \\ \sum_{j=1}^{n} \mathrm{sign}[\mathrm{Im}[\Omega_j]]\mathrm{Im}[D_j] = 0 \\ \sum_{j=1}^{n} \mathrm{sign}[\mathrm{Im}[\Omega_j]] \left\{ -c_{2j}\mathrm{Re}[D_j] + c_{1j}\mathrm{Im}[D_j] \right\} = 0 \end{cases} \tag{16.182}$$

replaces system (16.172) and determines the n complex constants D_j, providing the solution. Note that the determination of these $2n$ real constants depends on four equations, so ∞^{2n-4} solutions are possible.

Focussing attention on the hyperbolic regime, where the roots W_j are real, the system (16.182) simplifies to

$$\begin{cases} \sum_{j=1}^{4} W_j \mathrm{Re}[D_j] = 1, \\ \sum_{j=1}^{4} W_j \mathrm{Im}[D_j] = 0, \\ \sum_{j=1}^{4} \mathrm{Im}[D_j] = 0, \\ \sum_{j=1}^{4} W_j^2 \left[2\left(\widehat{\mathbb{G}}_{2221} - \widehat{\mathbb{G}}_{2111} \right) - W_j \widehat{\mathbb{G}}_{2121} \right] \mathrm{Im}[D_j] = 0, \end{cases} \tag{16.183}$$

so the general solution of the linear system (16.183) can be written as

$$\begin{bmatrix} D_1 \\ D_2 \\ D_3 \\ D_4 \end{bmatrix} = \zeta_1 \begin{bmatrix} 1 \\ 0 \\ 0 \\ 0 \end{bmatrix} + \zeta_2 \begin{bmatrix} 0 \\ 1 \\ 0 \\ 0 \end{bmatrix} + \zeta_3 \begin{bmatrix} 0 \\ 0 \\ 1 \\ 0 \end{bmatrix} + \frac{1 - \zeta_1 W_1 - \zeta_2 W_2 - \zeta_3 W_3}{W_4} \begin{bmatrix} 0 \\ 0 \\ 0 \\ 1 \end{bmatrix}$$

$$+ i\varrho \begin{bmatrix} (c_{14} - c_{13}) W_2 + (c_{12} - c_{14}) W_3 + (c_{13} - c_{12}) W_4 \\ (c_{13} - c_{14}) W_1 + (c_{14} - c_{11}) W_3 + (c_{11} - c_{13}) W_4 \\ (c_{14} - c_{12}) W_1 + (c_{11} - c_{14}) W_2 + (c_{12} - c_{11}) W_4 \\ (c_{12} - c_{13}) W_1 + (c_{13} - c_{11}) W_2 + (c_{11} - c_{12}) W_3 \end{bmatrix}, \tag{16.184}$$

where ζ_1, ζ_2, ζ_3 and ϱ are arbitrary real constants.

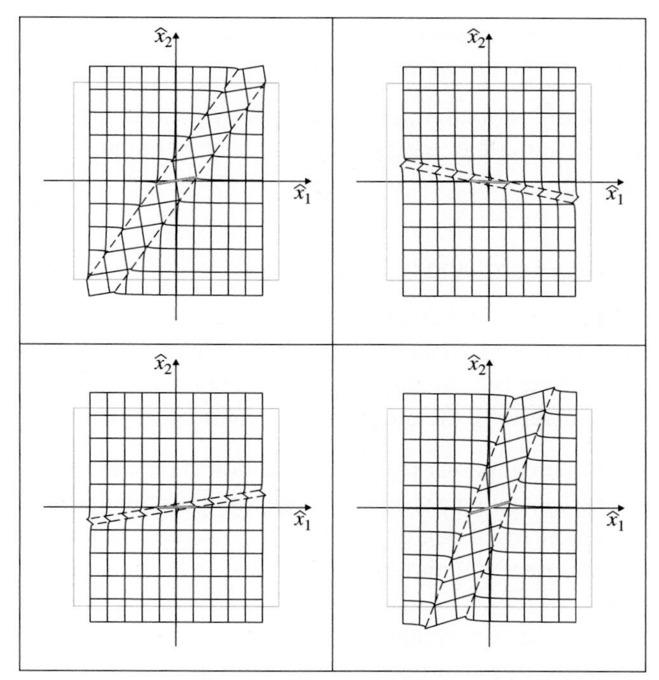

Figure 16.23. Deformed configurations (the grey squares represent the un-deformed, pre-stressed material) for a stiffener (evidenced with a thin rectangle, providing the bar scale of the representation) embedded in a J_2-deformation theory material with $N = 0.3$ (subject to a finite simple shear of amount $\gamma = 1.2\gamma^{EL}$) and a subsequent mode I incremental uniform remote load. Four solutions are reported among the ∞^4 possible *within the hyperbolic range*, where characteristics are inclined at $\{46.460°, -9.068°, 7.373°, 62.902°\}$ with respect to the $\hat{x}_1$ axis.

Since system (16.183) admits ∞^4 solutions, we have chosen to represent in Fig. 16.23 the deformed shape of an area near the stiffener for the four solutions corresponding (from left to right and from the upper part to the lower) to h ranging between 1 and 4 such that[7]

$$D_h = \frac{1}{W_h} \qquad \text{and} \qquad D_j = 0, \qquad j \neq h,\ j \in [1,4]. \tag{16.185}$$

We can note from Fig. 16.23 that for given uniform mode I remote loading of a stiffener embedded in a medium uniformly pre-strained beyond the elliptic range:

1. *An infinite number of solutions is possible.*
2. *These solutions do not decay at infinity.*

[7] The values in Eq. (16.185) are achievable for $\varrho = 0$ and

$$\zeta_h = \frac{1}{W_h}, \qquad \zeta_k = 0, \qquad k \neq h, \qquad \text{if } h = 1,2,3,$$

$$\zeta_1 = \zeta_2 = \zeta_3 = 0, \qquad \text{if } h = 4.$$

3. *They correspond to infinite incremental strain and stress along shear bands.*
4. *These shear bands emanate from the tips of the stiffener.*

The preceding conclusion, based on an analytical solution, fully explains the difficulties typically encountered in numerical analyses of ill-posed boundary value problems.

16.4 The stress state near a shear band and its propagation

We devote the last part of this chapter, to the analysis of a series of topics related to shear band growth in ductile materials, which can be addressed successfully using the perturbative approach, as indicated by Bigoni and Dal Corso (2008). In particular, we note the following problems concerning the mechanics of shear band growth:

- The analysis of the highly inhomogeneous stress/deformation state developing near a shear band tip (possibly involving a stress concentration) is crucial for an understanding of failure mechanisms of ductile material but is not possible using conventional approaches (numerical techniques can hardly have the appropriate resolution to detail this).
- An important feature that remains unexplained with conventional techniques but that can be investigated with the perturbative approach is the fact that shear bands grow quasi-statically and rectilinearly for remarkably long distances under mode II loading conditions, whereas the same feature is not observed in the akin problem of crack growth.
- Finally, and most important, the perturbative approach allows us to provide an explanation for why shear bands are preferential failure modes for quasi-statically deformed ductile materials.

We can provide answers to these problems employing the perturbative approach,[8] in which a pre-existing shear band is modelled as a slip surface for an incremental field in a pre-stressed material and perturbed with a mode II incremental loading. The problem is explained below; however, the major achievements are that a full-field solution is obtained for a finite-length shear band in an anisotropic, pre-stressed, non-linear elastic material incrementally loaded under mode II and revealing stress singularity, high inhomogeneity of the deformation and its focussing parallel and coaxially aligned with the shear band. Moreover, the incremental energy release rate is shown to blow up when the stress state approaches the condition for strain localisation (i.e., the elliptical boundary). These finding may explain the tendency towards rectilinear propagation under mode II and the fact that shear bands are preferred failure modes for ductile materials.

An infinite, incompressible elastic material obeying to the constitutive equations (6.178) and (6.179) is considered homogeneously and quasi-statically deformed in a given loading path directed towards the elliptical boundary. Inspired by the experimental observation that the sensibility of a material to shear banding is linked to pre-existing defects (Xue and Gray, 2006), we assume that there is an imperfection

[8] Before Bigoni and Dal Corso (2008), shear band growth had been considered only in a context pertaining to slope-stability problems in soil mechanics (Palmer and Rice, 1973; Rice, 1973; Puzrin and Germanovich, 2005).

present in the material, in the form of a thin zone of 'weak' material, which touches the EI/P or EC/H boundary (see Section 12.2) and is transformed into a shear band of length $2l$, whereas the surrounding material is still in the elliptical regime, although near the boundary of ellipticity loss. In this situation, we analyse the response to an incremental loading perturbation with the purpose of determining the stress state near a finite-length shear band and the shear-band growth conditions.

A 'shear band' is a thin layer of material across which certain components of the incremental nominal tractions vanish, namely, the incremental nominal shear component tangential to the shear band, for the material model [Eqs. (6.178) and (6.179)] considered here. It therefore becomes spontaneous to model a shear band in such a material as a 'slip discontinuity surface' across which the normal component of incremental displacement remain continuous, whereas the tangential incremental nominal stress component vanishes (Fig. 16.24). Such a discontinuity surface is not a crack because normal incremental traction can be transmitted across it, but it can behave equivalently to a crack under certain special symmetry conditions. This is the case when the shear band is aligned parallel to one of the principal orthotropy axes x_1 or x_2 corresponding either to a shear band formed at the EI/P boundary or to a shear band formed at the EC/H boundary at $\xi = 0$ (where two shear bands orthogonal to each other form simultaneously). In these cases, the slip surface model behaves as a crack when subject to a mode II loading increment and can be analysed directly with the solution developed in Section 16.2.1. In a more general case, a solution for a slip surface embedded in a pre-stressed material has to be developed and is proposed below.

We will analyse the symmetry case corresponding to the EI/P boundary, in which shear band and crack are equivalent models, and the generic situation, corresponding to the EC/H boundary.

In summary, for the proposed weak line model of a shear band of length $2l$, the incremental boundary conditions are the following:

- Null incremental nominal shearing tractions:

$$\hat{t}_{21}(\hat{x}_1, 0^{\pm}) = 0, \qquad \forall |\hat{x}_1| < l \tag{16.186}$$

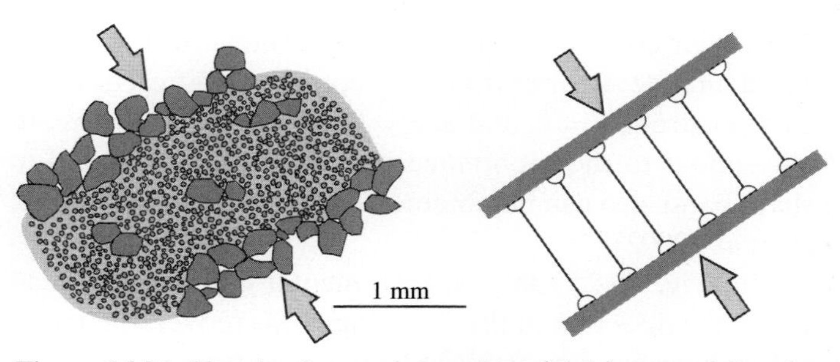

Figure 16.24. Sketch of a weak interface (*right*) to model a shear band (*left*, inspired by a deformation band observed in dry sandstone by Sulem and Ouffroukh, 2006). The hinged quadrilateral should be thought to have zero thickness, so materials in contact can freely slide incrementally along a weak surface, across which normal incremental displacement remains continuous.

- Continuity of the incremental nominal normal traction:

$$[[\hat{t}_{22}(\hat{x}_1,0)]] = 0, \qquad \forall |\hat{x}_1| < l \tag{16.187}$$

- Continuity of normal incremental displacement:

$$[[\hat{v}_2(\hat{x}_1,0)]] = 0, \qquad \forall |\hat{x}_1| < l \tag{16.188}$$

Before we proceed with the analysis, a digression becomes necessary. It is assumed in our model setting that a sliding surface abruptly forms when a weak thin zone of material touches the elliptical boundary. This model is obviously a strong idealisation because in reality the weak material approaches the elliptical boundary becoming incrementally less and less stiff in a continuous way. The abrupt formation of a sliding surface within an infinite solid may, depending on the stress conditions, generate a sudden 'spurious' interfacial instability so that in this condition the shear band model becomes over-simplified. Therefore, we have to limit the analysis to situations in which all instabilities are a priori excluded until the elliptical boundary is met, as is the case when the Hill exclusion condition [Eq. (12.6)], holds true. Fortunately, this condition is so general that all points of the EC/H and EI/P can be explored (by selecting appropriate values for η to enforce the validity of the Hill exclusion condition until the elliptical boundary is touched for the given loading path; see Section 12.2, Fig. 12.8).

Shear band at the EI/P boundary

All points of the elliptical imaginary/parabolic boundary can be approached while the Hill exclusion condition holds true when $\eta = k > 0$, corresponding to a uniaxial tensile stress state, $T_1 > 0$, $T_2 = 0$. In this situation, one shear band forms at the EI/P boundary, $k = 1$, parallel to the tensile loading direction (12.56), so the problem is symmetric, and the crack solution [Eqs. (16.132)], can be used. In fact, owing to symmetry, the normal displacement increment and all nominal incremental traction components are null (and therefore a fortiori continuous) at the shear band boundary under a mode II loading increment.

Equations (16.132) have been used to obtain the results shown in Fig. 16.25, where the level sets of incremental deviatoric strain are reported at different levels of pre-stress, namely, at null prestress, $k = 0$, and at $k = 0.95$, a value very close to the EI/P boundary. Results similar to those obtained in Fig. 16.25, but limited to fields near the tip of the shear band also can be obtained employing the asymptotic analysis presented by Radi et al. (2002).

It should be noticed from Fig. 16.25 that the incremental deformation field evidences a strong focussing in the direction of the shear band. Moreover, the incremental energy release rate for shear band growth can be deduced from the formula for crack advance under mode II, [Eq. (16.164)]. The energy released for an incremental advance of shear band has the typical behaviour shown in Fig. 16.17 (for mode II and $\vartheta_0 = 0$), evidencing an asymptote at the EI/P boundary (there are no qualitative changes when other values of the parameter $\xi \geq 0.5$ are considered, so the asymptote at $k = 1$ is always present).

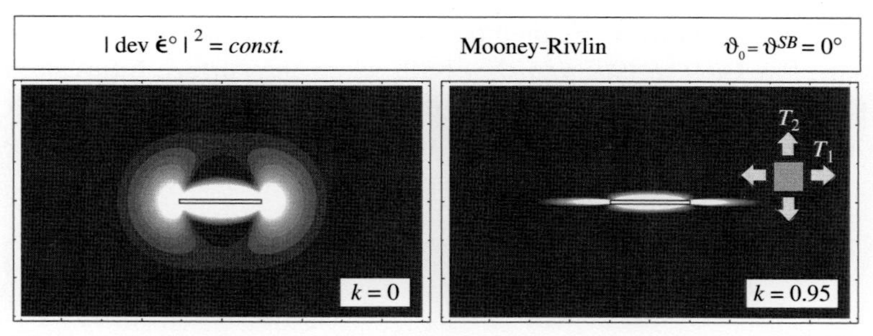

Figure 16.25. Level sets of the modulus of incremental deviatoric strain near a shear band of length $2l$ in an incrementally isotropic, $\xi = 1$, material without pre-stress, $k = 0$, and pre-stressed near, $k = \eta = 0.95$, the EI/P boundary. Mode II incremental loading is considered.

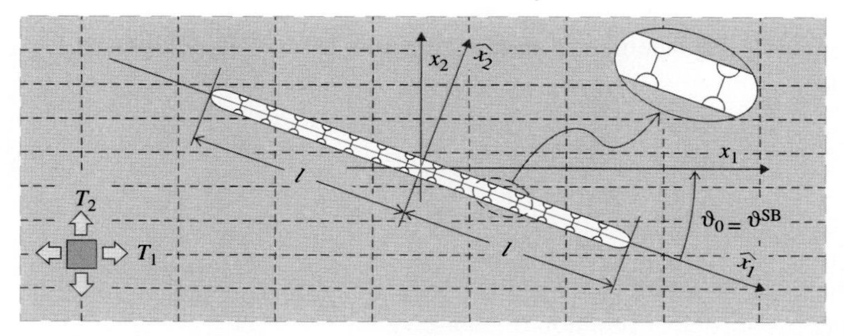

Figure 16.26. Shear band of length $2l$ in a prestressed, orthotropic material inclined at an angle ϑ_0 (positive when anti-clockwise) with respect to the orthotropy axes x_1 and x_2. T_1 and T_2 denote the prestress state, expressed through the two in-plane principal Cauchy stresses aligned parallel to the x_1–x_2 reference system.

Shear band at the EC/H boundary

According to Eq. (12.52), at the EC/H boundary, two shear bands form inclined with respect to the x_1–x_2 axes so that one of these is taken aligned parallel with the $\hat{x}_1$ axis inclined at ϑ^{SB} with respect the x_1 axis.[9] Using the weak line model, only mode II loading plays a role, whereas a mode I loading leaves the material unperturbed. Therefore, with reference to Fig 16.26 and employing a representation similar to Eq. (16.171), namely,

$$\widehat{\psi}^{\circ}(\hat{x}_1,\hat{x}_2) = \frac{\hat{t}_{21}^{\infty}}{2\mu} \sum_{j=1}^{2} \mathrm{Re}\left\{ B_j^{II} \left[\hat{z}_j^2 - \hat{z}_j \sqrt{\hat{z}_j^2 - l^2} + l^2 \ln\left(\hat{z}_j + \sqrt{\hat{z}_j^2 - l^2} \right) \right] \right\}, \quad (16.189)$$

and imposing the boundary conditions [Eqs. (16.186) through (16.188)] at a sliding surface yields the following algebraic system for the unknown constants B_j^{II}:

$$\begin{bmatrix} -c_{21} & c_{11} & -c_{22} & c_{12} \\ c_{31} & c_{41} & c_{32} & c_{42} \\ -c_{41} & c_{31} & -c_{42} & c_{32} \\ 0 & 1 & 0 & 1 \end{bmatrix} \begin{bmatrix} \mathrm{Re}[B_1^{II}] \\ \mathrm{Im}[B_1^{II}] \\ \mathrm{Re}[B_2^{II}] \\ \mathrm{Im}[B_2^{II}] \end{bmatrix} = \begin{bmatrix} 0 \\ -1 \\ 0 \\ 0 \end{bmatrix}, \quad (16.190)$$

[9] The analysis can be carried with respect to a generic uniform state of pre-stress with principal values inclined at ϑ_0 different from ϑ^{SB}.

where coefficients c_{ij} are again those defined by Eqs. (16.148). The determinant of the coefficient matrix in Eq. (16.190) vanishes both when the surface bifurcation condition [Eq. (12.81) or Eq. (16.142)] and when the EC/H boundary is met.

Similarly to the crack solution, the asymptotic fields near the shear band tip result for the incremental nominal stress to be given by

$$\hat{t}_{22}(r,0) = -\frac{\Upsilon \dot{K}_{II}}{\sqrt{2\pi r}}, \qquad \hat{t}_{21}(r,0) = \frac{\dot{K}_{II}}{\sqrt{2\pi r}}, \tag{16.191}$$

ahead of the tip, where

$$\Upsilon = \frac{\hat{t}^{\circ}_{22}}{\hat{t}^{\infty}_{21}} = c_{11}\mathrm{Re}[B_1^{II}] + c_{12}\mathrm{Im}[B_1^{II}] + c_{13}\mathrm{Re}[B_2^{II}] + c_{14}\mathrm{Im}[B_2^{II}], \tag{16.192}$$

and for the incremental displacements (where constants have been neglected),

$$\begin{aligned} \hat{v}_1(\Delta l - r, \pm\pi) &= \pm\frac{\hat{t}^{\infty}_{21}\sqrt{2l}\sqrt{\Delta l - r}}{2\mu}\mathrm{Im}\left[W_1 B_1^{II} + W_2 B_2^{II}\right], \\ \hat{v}_2(\Delta l - r, \pm\pi) &= \mp\frac{\hat{t}^{\infty}_{21}\sqrt{2l}\sqrt{\Delta l - r}}{2\mu}\mathrm{Im}\left[B_1^{II} + B_2^{II}\right], \end{aligned} \tag{16.193}$$

holding at the shear band surfaces for 'small' Δl.

The following properties of function Υ, namely,

$$\Upsilon(k=0,\vartheta_0) = \Upsilon(k,\vartheta_0=0) = \Upsilon(k,\vartheta_0=\pi/2) = 0, \tag{16.194}$$

have been proven, whereas the properties

$$\Upsilon = \Upsilon(k,\vartheta_0) = -\Upsilon(-k,\pi/2-\vartheta_0) \tag{16.195}$$

have been found numerically to hold, from which the identities

$$\Upsilon(k,\vartheta_0=\pi/4) = \frac{1-\sqrt{1-k^2}}{k} \quad \text{and} \quad \Upsilon(k,\vartheta_0=\pi/3) = \frac{\sqrt{3}(2+k-2\sqrt{1-k^2})}{4+5k} \tag{16.196}$$

follow with the help of a symbolic manipulator.

Employing the asymptotic near-tip representations (16.191) and (16.193) in Eq. (16.159), we obtain

$$\dot{G}^{SB} = \dot{K}_{II}^2\frac{\mathrm{Im}\left[W_1 B_1^{II} + W_2 B_2^{II}\right]}{4\mu}. \tag{16.197}$$

Note that the perturbed solution for the shear band model can be obtained alternatively, providing a mixed-mode loading to an inclined crack, in which the mode I loading component is 'calibrated' with respect to the mode II component in such a way as to eliminate the jump in normal incremental displacement along the crack faces generated by a pure mode II loading, in other words, to satisfy condition (16.188). All this procedure bears on the special feature found in the solution of the crack problem that a mode I loading uniform along the crack faces is sufficient to eliminate a mode II transversal mismatch in incremental displacements. In particular, Eq. (16.197) can be obtained from Eq. (16.163), considering a mixed mode

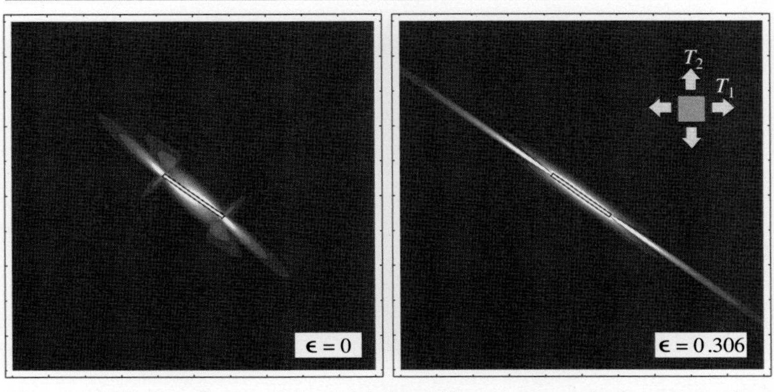

Figure 16.27. Level sets of the modulus of incremental deviatoric strain near a shear band of length $2l$ in a J_2-deformation theory material at low ($N = 0.1$) strain hardening, not pre-strained ($\varepsilon = 0$, *left*), and pre-strained until near the EC/H boundary ($\varepsilon = 0.306$, *right*).

defined by $\hat{t}_{22}^{\infty} = -\Upsilon \hat{t}_{21}^{\infty}$, so that the condition of continuity of transversal incremental displacement yields

$$\mathrm{Im}[A_1^{II} + A_2^{II}] - \Upsilon \mathrm{Im}[A_1^{I} + A_2^{I}] = 0, \tag{16.198}$$

and the constants defining the crack and shear band solutions are related through

$$B_j^{II} = A_j^{II} - \Upsilon A_j^{I}, \qquad j = 1,2. \tag{16.199}$$

Therefore, the difference between the crack and shear band problems lies in a uniform nominal normal stress increment applied at the crack surfaces.

Level sets of the modulus of incremental deviatoric strain for a J_2-deformation theory material (which is a particular case of the developed theory; see Sections 4.2.3 and 6.2.2) are reported in Fig. 16.27 for low ($N = 0.1$) and in Fig. 16.28 for high ($N = 0.8$) strain hardening.

In both cases, null pre-strain (and pre-stress) and a value of pre-strain near the EC/H boundary have been considered. Moreover, parameter η has been taken equal to $0.311k$ for $N = 0.1$ and equal to $0.775k$ for $N = 0.8$ to ensure the validity of the Hill exclusion condition (12.6) near the EC/H boundary. Note that for null pre-strain, $\varepsilon = 0$, the shear band model behaves as a fracture because the normal component of incremental displacement remains continuous for a crack in an orthotropic incompressible material at null pre-stress. Therefore, Figs. 16.27 and 16.28 (*left*) are identical to the analogue cases reported in Figs. 16.13 and 16.14 (*bottom, left*). The difference between the shear band model and the crack becomes evident comparing Figs. 16.13 and 16.14 (*bottom, right*) with Figs. 16.27 (right) and 16.28 (right), where in the former figures both conjugate directions of shear bands are activated under mode II loading, whereas only the direction aligned to the shear band is activated in the latter case.

We can conclude from Figs. 16.27 and 16.28 that

> *the incremental deformation field near a finite-length shear band is localised, elongated and evidences a strong focussing in the direction aligned parallel to the shear band.* This

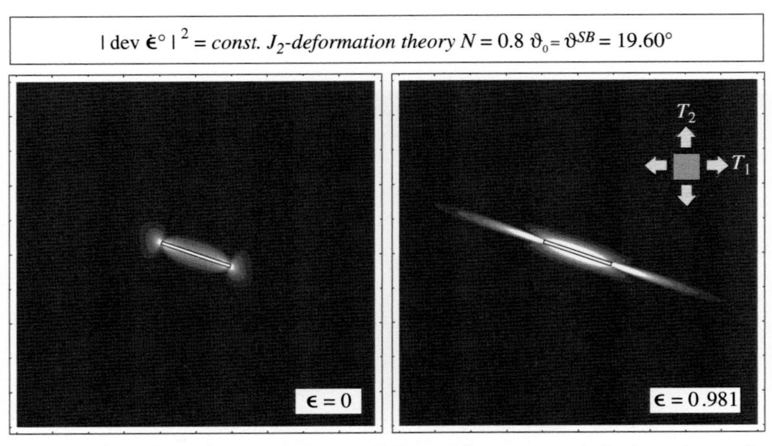

Figure 16.28. Level sets of the modulus of incremental deviatoric strain near a shear band of length $2l$ in a J_2-deformation theory material at high ($N = 0.8$) strain hardening, not pre-strained ($\varepsilon = 0$, *left*) and pre-strained until near the EC/H boundary ($\varepsilon = 0.981$, *right*).

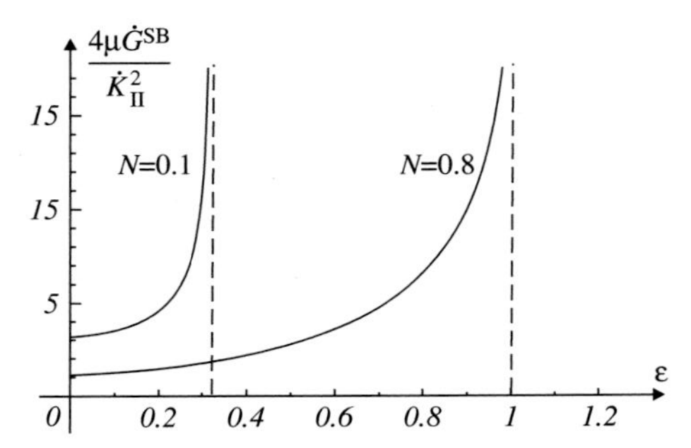

Figure 16.29. Incremental energy release rate for shear band growth in a J_2-deformation theory material at low ($N = 0.1$) and high ($N = 0.8$) strain hardening as a function of the pre-strain ε. The curve presents an asymptote at the EC/H boundary ($\varepsilon^{EL} \approx 0.322$, for $N = 0.1$, and $\varepsilon^{EL} \approx 1.032$, for $N = 0.8$) so that shear band growth becomes 'unrestrainable' when pre-stress approaches this point.

finding suggests that, *differently from mode II rectilinear crack propagation in a homogeneous material, that usually does not occur (because in first approximation, cracks deviate from rectilinearity following the maximum near-tip hoop stress inclination), shear band growth is very likely to occur aligned with the shear band itself. This observation explains the strong tendency that shear bands evidence towards rectilinear propagation for long (compared with their thickness) distances (see, e.g., Anand and Spitzig, 1980, 1982). Moreover, the focussing of incremental deformation and the stress singularity strongly promote shear band growth.*

To further analyse shear band growth, *the incremental energy release rate for an infinitesimal shear band advance* $\dot{G}^{SB}$ can be evaluated for an orthotropic pre-stressed material using Eq. (16.197), and it can be shown to blow up to infinity when the EC/H boundary is approached. In particular, calculations of the incremental energy release rate (made dimension-less by multiplication by $4\mu/\dot{K}_{II}^2$) for shear band growth in a

J_2-deformation theory material at low ($N = 0.1$) and high ($N = 0.8$) strain hardening are reported in Fig. 16.29, from which the following conclusion can be deduced:

> *It is assumed in fracture mechanics that a crack advances under small scale yielding when the energy release rate exceeds a critical threshold, believed to be a characteristic of the material. Whether this criterion can be generalised to the analysis of shear bands or not still can be a matter of discussion, but the important point is that the incremental energy release rate blows up to infinity when the Elliptic boundary is approached. In these conditions, a shear band can drive itself on and overcome possible barriers; in other words, it can grow 'unrestrainable', a finding which, together with the previous results on near-tip stress/deformation states, legitimises the common experimental observation that shear bands are preferred near-failure deformation modes.*

References

Adams, G. G. (1995). Self-excited oscillations of two elastic half-spaces sliding with a constant coefficient of friction. *J. Appl. Mech.* 62, 867–72.

Altenbach, H., and Tushtev, K. (2001). A new static failure criterion for isotropic polymers. *Mech. Compos. Mater.* 37, 475–82.

An, L., and Schaeffer, D. (1990). The flutter instability in granular flow. *J. Mech. Phys. Solids* 40, 683–98.

Anand, L., and Spitzig, W. A. (1980). Initiation of localized shear bands in plane strain. *J. Mech. Phys. Solids* 28, 113–28.

Anand, L., and Spitzig, W. A. (1982). Shear-band orientations in plane strain. *Acta Metall.* 30, 553–61.

Apostol, T. M. (1969). *Calculus*, Vols. I and II. Wiley, New York.

Argyris, J. H., Faust, G., Szimmat, J., Warnke, P., and Willam, K. (1974). Recent developments in the finite element analysis of prestressed concrete reactor vessels. *Nucl. Eng. Des.* 28, 42–75.

Aron, M., and Wang, Y. (1995a). Remarks concerning the flexure of a compressible nonlinearly elastic rectangular block. *J. Elasticity* 40, 99–106.

Aron, M., and Wang, Y. (1995b). On deformations with constant modified stretches describing the bending of rectangular blocks. *Q. J. Mech. Appl. Math.* 48, 375–87.

Auricchio, F., Brezzi, F., and Lovadina, C. (2004). Mixed finite element methods. In *Encyclopedia of Computational Mechanics*, edited by E. Stein, R. de Borst, and T. J. R. Hughes, Vol. 1: *Fundamentals*. Wiley, New York.

Bai, Y. L., and Dodd, B. (1992). *Adiabatic Shear Localisation: Occurrence, Theories and Applications*. Pergamon, Oxford, UK.

Baker, A. L. (1890). *Elliptic Function*. Wiley, New York.

Ball, J. M. (1977). Convexity conditions and existence theorems in nonlinear elasticity. *Arch. Rat. Mech. Anal.* 63, 337–403.

Bardet, J. P. (1990). Lode dependences for isotropic pressure-sensitive elastoplastic materials. *J. Appl. Mech.* 57, 498–06.

Bardi, F. C., Yun, H. D., and Kyriakides, S. (2003). On the axisymmetric progressing crushing of circular tubes under axial compression. *Int. J. Solids Structures* 40, 3137–55.

Bassani, J. L., Durban, D., and Hutchinson, J. W. (1980). Bifurcation at a spherical hole in an infinite elastoplastic medium. *Math. Proc. Camb. Phil. Soc.* 87, 339–56.

Batdorf, S. B., and Budianski, B. (1949). A mathematical theory of plasticity based on concept of slip. *NACA Tech Note* TN 1871.

Bazant, Z. P., and Mazars, J. (1990). France–U.S. workshop on strain localization and size effect due to cracking and damage. *J. Eng. Mech. ASCE* 116, 1412–24.

Beatty, M. F. (1987). Topics in finite elasticity: hyperelasticity of rubber, elastomers, and biological tissues-with examples. *Appl. Mech. Rev.* 40, 1699–1734.

Beatty, M. F. (2001). Hyperelastic Bell materials: Retrospection, experiment, theory. In *Nonlinear Elasicity: Theory and Applications*, edited by Y. B. Fu and R. W. Ogden. Cambridge University Press, Cambridge, UK.

Beck M. (1952). Die Knicklast des einseitig eingespannten, tangential gedrückten Stabes. *Z. Angew. Math. Phys.* 3, 225–8.

Bell, J. F. (1973). The experimental foundations of solid mechanics. In *Encyclopedia of Physics*, edited by C. Truesdell, Vol. VIa/1. Springer-Verlag, Berlin.

Beltrami, E. (1885). Sulle condizioni di resistenza dei corpi elastici. *Rend. Ist. Lomb.* XVIII, 704–14.

Benallal, A., and Bigoni, D. (2004). Effects of temperature and thermo-mechanical couplings on material instabilities and strain localization of inelastic materials. *J. Mech. Phys. Solids* 52, 725–53.

Benallal, A., Billardon, R., and Geymonat, G. (1988). Some mathematical aspects of the damage softening problem. In *Cracking and Damage*, edited by J. Mazar and Z. P. Bazant, Vol. 1, pp. 247–58.

Benallal, A., Billardon, R., and Geymonat, G. (1990). Phènoménes de localisation á la frontière d'un solide. *C. R. Acad. Sci., Paris* 310, 670–84.

Benallal, A., Billardon, R., and Geymonat, G. (1993). Bifurcation and localization in rate-independent materials: some general considerations. In CISM Lecture Notes No. 327, pp. 1–47. Springer, Berlin.

Bernstein, B. (1960). Hypo-elasticity and elasticity, *Arch. Rat. Mech. Anal.* 6, 89–104.

Bernstein, B., and Toupin, R.A. (1962). Some properties of the Hessian matrix of a strictly convex function. *J. für reine und angewndte Mathematik* 210, 65–72.

Bertoldi, K., Bigoni, D., and Drugan, W. J. (2007a). A discrete-fibers model for bridged cracks and reinforced elliptical voids. *J. Mech. Phys. Solids* 55, 1016–35.

Bertoldi, K., Bigoni, D., and Drugan, W. J. (2007b). Structural interfaces in linear elasticity. Part I. Nonlocality and gradient approximations. *J. Mech. Phys. Solids* 55, 1–34.

Bertoldi, K., Bigoni, D., and Drugan, W. J. (2007c). Structural interfaces in linear elasticity. Part II. Effective properties and neutrality. *J. Mech. Phys. Solids* 55, 35–63.

Bertoldi, K., Brun, M., and Bigoni, D. (2005). A new boundary element technique without domain integrals for elastoplastic solids. *Int. J. Numer. Meth. Eng.* 64, 877–906.

Bigoni, D. (1995). On flutter instability in elastoplastic constitutive models. *Int. J. Solids Structures* 32, 3167–89.

Bigoni, D. (1996). On smooth bifurcations in non-associative elastoplasticity. *J. Mech. Phys. Solids* 44, 1337–51.

Bigoni, D. (2000). Bifurcation and instability of non-associative elastoplastic solids. In *Material Instabilities in Elastic and Plastic Solids*, edited by H. Petryk, CISM Lecture Notes No. 414, pp. 1–52. Springer, Berlin.

Bigoni, D., and Capuani, D. (2002). Green's function for incremental nonlinear elasticity: Shear bands and boundary integral formulation. *J. Mech. Phys. Solids* 50, 471–500.

Bigoni, D., and Capuani, D. (2005). Time-harmonic Green's function and boundary integral formulation for incremental nonlinear elasticity: dynamics of wave patterns and shear bands. *J. Mech. Phys. Solids* 53, 1163–87.

Bigoni, D., Capuani, D., Bonetti, P., and Colli, S. (2007). A novel boundary element approach to time-harmonic dynamics of incremental non-linear elasticity: the role of pre-stress on structural vibrations and dynamic shear banding. *Comput. Meth. Appl. Mech. Engrg.* 196, 4222–49.

Bigoni, D., and Dal Corso, F. (2008). The unrestrainable growth of a shear band in a prestressed material. *Proc. R. Soc. Lond. A* 464, 2365–90.

Bigoni, D., Dal Corso, F., and Gei, M. (2008). The stress concentration near a rigid line inclusion in a prestressed, elastic material. Part II: Implications on shear band nucleation, growth and energy release rate. *J. Mech. Phys. Solids* 56, 839–57.

Bigoni, D., and Gei, M. (2001). Bifurcations of a coated, elastic cylinder. *Int. J. Solids Structures* 38, 5117–48.

Bigoni, D., Gei, M., and Movchan, A. B. (2008). Dynamics of a prestressed stiff layer on an elastic half space: filtering and band gap characteristics of periodic structural models derived from long-wave asymptotics. *J. Mech. Phys. Solids* 56, 2494–2520.

Bigoni, D., and Hueckel, T. (1990). A note on strain localization for a class of non-associative plasticity rules. *Ingenieur-Archiv* 60, 491–9.

Bigoni, D., and Hueckel, T. (1991a). Uniqueness and localization. I. Associative and non-associative elastoplasticity. *Int. J. Solids Structures* 28, 197–213.

Bigoni, D., and Hueckel, T. (1991b). Uniqueness and localization. II. Coupled elastoplasticity. *Int. J. Solids Structures* 28, 215–24.

Bigoni, D., and Laudiero, F. (1989). The quasi-static finite cavity expansion in a non-standard elastoplastic medium. *Int. J. Mech. Sci.* 31, 825–37.

Bigoni, D., and Loret, B. (1999). Effects of elastic anisotropy on strain localization and flutter instability in plastic solids. *J. Mech. Phys. Solids* 47, 1409–36.

Bigoni, D., Loret, B., and Radi, E. (2000). Localization of deformation in plane elastic-plastic solids with anisotropic elasticity. *J. Mech. Phys. Solids* 48, 1441–66.

Bigoni, D., Misseroni, D., Noselli, G., and Zaccaria, D. (2012). Effects of the constraint's curvature on structural instability: tensile buckling and multiple bifurcations. *Proc. Roy. Soc. A* doi: 10.1098/rspa.2011.0732.

Bigoni, D., and Movchan, A. B. (2002). Statics and dynamics of structural interfaces in elasticity. *Int. J. Solids Structures* 39, 4843–65.

Bigoni, D., and Noselli, G. (2010a). Localized stress percolation through dry masonry walls. Part I. Experiments. *Eur. J. Mechanics-A/Solids* 29, 291–8.

Bigoni, D., and Noselli, G. (2010b). Localized stress percolation through dry masonry walls. Part II. Modelling. *Eur. J. Mechanics-A/Solids* 29, 299–307.

Bigoni, D., and Noselli, G. (2011). The experimental evidence of flutter and divergence instabilities induced by dry friction. *J. Mech. Phys. Solids* 59, 2208–26.

Bigoni, D., Ortiz, M., and Needleman, A. (1997). Effect of interfacial compliance on bifurcation of a layer bonded to a substrate. *Int. J. Solids Structures* 34, 4305–26.

Bigoni, D., and Petryk, H. (2002). A note on divergence and flutter instabilities in elastic-plastic materials. *Int. J. Solids Structures* 39, 911–26.

Bigoni, D., and Piccolroaz, A. (2004). Yield criteria for quasi-brittle and frictional materials. *Int. J. Solids Structures* 41, 2855–78.

Bigoni, D., and Radi, E. (1993). Mode I crack propagation in elastic-plastic pressure-sensitive materials. *Int. J. Solids Structures* 30, 899–919.

Bigoni, D., Serkov, S. K., Movchan, A. B., and Valentini, M. (1998). Asymptotic models of dilute composites with imperfectly bonded inclusions. *Int. J. Solids Structures* 35, 3239–58.

Bigoni, D., and Willis, J. R. (1994). A dynamical interpretation of flutter instability. In *Localisation and Bifurcation of Rocks and Soils*, edited by R. Chambon, J. Desrues, and I. Vardoulakis, A.A. Balkema Scientific Publishers, Rotterdam.

Bigoni, D., and Zaccaria, D. (1992a). Strong ellipticity of comparison solids in elastoplasticity with volumetric non-associativity. *Int. J. Solids Structures* 29, 2123–36.

Bigoni, D., and Zaccaria, D. (1992b). Loss of strong ellipticity in non-associative elastoplasticity. *J. Mech. Phys. Solids* 40, 1313–31.

Bigoni, D., and Zaccaria, D. (1993). On strain localization analysis of elastoplastic materials at finite strain. *Int. J. Plasticity* 9, 21–33.

Bigoni, D., and Zaccaria, D. (1994a). On eigenvalues of the acoustic tensor in elastoplasticity. *Eur. J. Mechanics-A/Solids* 13, 621–38.

Bigoni, D., and Zaccaria, D. (1994b). Eigenvalues of the elastoplastic constitutive operator. *ZAMM* 74, 355–7.

Biot, M. A. (1937). Bending of an infinite beam on an elastic foundation. *J. Appl. Mech.* 59, A1–7.

Biot, M. A. (1965). *Mechanics of Incremental Deformations*. Wiley, New York.

Bishop, J. F. W., and Hill, R. (1951). A theory of the plastic distorsion of a polycrystalline aggregate under combined stresses, *Phil. Mag.* 42, 414–27.

Bodig, J., and Jayne, B. A. (1982). *Mechanics of Wood and Wood Composites*. Van Nostrand, New York.

Boehler, J. P. (1987). Introduction to the invariant formulation of anisotropic constitutive equations. In *Applications of Tensor Functions in Solid Mechanics*, CISM Course No. 292.

Boehler, J. P., and Willis J. R. (1991). An analysis of localization in highly pre-deformed sheet steel. Unpublished.

Bonet, J., and Burton, A. J. (1998). A simple orthotropic, transversely isotropic hyperelastic constitutive equation for large strain computations. *Comput. Meth. Appl. Mech. Engrg.* 162, 151–64.

Boyd, S., and Vandenberghe, L. (2004). *Convex Optimization*. Cambridge University Press, Cambridge, UK.

Bowen, R. M., and Wang, C. C. (1976). *Introduction to Vectors and Tensors*. Plenum Press, New York.

Brannon, R. M., and Drugan, W. J. (1993). Influence of non-classical elastic-plastic constitutive features on shock wave existence and spectral solutions. *J. Mech. Phys. Solids* 41, 297–330.

Braun, O. M., and Kivshar, Y. S. (1998). Nonlinear dynamics of the Frenkel-Kontorova model. *Physics Reports* 306, 1–108.

Broberg, K. B. (1999). *Cracks and Fracture*. Academic Press, San Diego, CA.

Bruhns, O., and Raniecki, B. (1982). Ein Schrankenverfahren bei Verzweigungsproblemen inelastischer Formänderungen. *ZAMM* 62, T111–13.

Bruhns, O., Xiao, H., and Meyers A. (2002). Finite bending of a rectangular block of an elastic Hencky material. *J. Elasticity* 66, 237–56.

Bruhns, O., Gupta, N. K., Meyers, A. T. M., and Xiao, H. (2003). Bending of an elastoplastic strip with isotropic and kinematic hardening. *Arch. Appl. Mech.* 72, 759–78.

Brun, M., Capuani, D., and Bigoni, D. (2003a). A boundary element technique for incremental, nonlinear elasticity. Part I. Formulation. *Comput. Meth. Appl. Mech. Engrg.* 192, 2461–79.

Brun, M., Bigoni, D., and Capuani, D. (2003 b). A boundary element technique for incremental, nonlinear elasticity. Part II. Bifurcation and shear bands. *Comput. Meth. Appl. Mech. Engrg.* 192, 2481–99.

Byrd, P. F., and Friedman, M. D. (1971). *Handbook of Elliptic Integrals for Engineers and Scientists*. Springer-Verlag, Berlin.

Capurso, M. (1972). On the extremal properties of the solution in dynamics of rigid-viscoplastic bodies allowing for large displacement effects. *Meccanica* 7, 236–47.

Capurso, M. (1979). Extremum theorems for the solution of the rate problem in elastic-plastic fracturing structures. *J. Struct. Mech.* 7, 411–34.

Cattaneo, C. (1946). Su un teorema fondamentale nella teoria delle onde di discontinuità. *Atti Acad. Naz. Lincei* (parts I and II). I, 67–72, 728–34.

Chadwick, P. (1976). *Continuum Mechanics.* Wiley, New York.

Chadwick, P., and Powdrill, B. (1965). Singular surfaces in linear thermoelasticity. *Int. J. Engng. Sci.* 3, 561–95.

Chau, K. T. (1992). Non-normality and bifurcation in a compressible pressure-sensitive circular cylinder under axisymmetric tension and compression. *Int. J. Solids Structures* 29, 801–24.

Chau, K. T. (1995). Buckling, barrelling, and surface instabilities of a finite, transversely isotropic circular cylinder. *Q. Appl. Math.* 53, 225–44.

Chau, K. T., and Choi, S. K. (1998). Bifurcations of thick-walled hollow cylinders of geomaterials under axisymmetric compression. *Int. J. Numer. Anal. Meth. Geomech.* 22, 903–19.

Chau, K. T., and Rudnicki, J. W. (1990). Bifurcations of compressible pressure-sensitive materials in plane strain tension and compression. *J. Mech. Phys. Solids* 38, 875–98.

Chen, P. (1973). Growth and decay of waves in solids. In *Encyclopedia of Physics*, edited by C. Truesdell, Vol. VIa/3. Springer-Verlag, Berlin.

Chen, W. F., and Saleeb, A. F. (1982). *Constitutive Equations for Engineering Materials: Elasticity and Modelling*. Wiley, New York.

Cheng, S. Y., Ariaratnam, S. T., and Dubey, R. N. (1971). Axisymmetric bifurcation in an elastic-plastic cylinder under axial load and lateral hydrostatic pressure. *Q. Appl. Math.* 29, 41–51.

Cheng, Y. S., and Lu, W. D. (1993). Uniqueness and bifurcation in elastic-plastic solids. *Int. J. Solids Structures* 30, 3073–84.

Christensen, R. M. (1997). Yield functions/failure criteria for isotropic materials. *Proc. R. Soc. Lond.* 453, 1473–91.

Christensen, R. M., Freeman, D. C., and DeTeresa, S. J. (2002). Failure criteria for isotropic materials, applications to low-density types. *Int. J. Solids Structures* 39, 973–82.

Christoffersen, J. (1991). Hyperelastic relations with isotropic rate forms appropriate for elastoplasticity. *Eur. J. Mechanics-A/Solids* 10, 91–9.

Christofferson, J., and Hutchinson, J. W. (1979). A class of phenomenological corner theories of plasticity. *J. Mech. Phys. Solids* 27, 465–87.

Ciarlet, P. G. (1988). *Mathematical Elasticity*. Vol. I *Three-Dimensional Elasticity*. North-Holland, Amsterdam.

Coffin, L. F., and Schenectady, N. Y. (1950). The flow and fracture of a brittle material. *J. Appl. Mech.* 17, 233–48.

Coker, E. G., and Filon, L. N. G. (1957). *A Treatise on Photo-elasticity*. Cambridge University Press, Cambridge, UK.

Coman, C. D., and Destrade, M. (2008). Asymptotic results for bifurcation in pure bending of rubber blocks *Q. J. Mech. Appl. Math.* 61, 395–414.

Courant, R., and Hilbert, D. (1962). *Methods of Mathematical Physics*, Vol. II. Wiley, New York.

Craciun, E. M., and Soós, E. (1998). Interaction of two unequal cracks in a prestressed fiber reinforced composite. *Int. J. Fracture* 94, 137–59.

Cristescu, N. D., Craciun, E. M., and Soós, E. (2004). *Mechanics of Elastic Composites*. Chapman & Hall/CRC, Boca Raton, FL.

Dafalias, Y. F. (1977). Il'iushin's postulate and resulting thermodynamic conditions on elastic-plastic coupling. *Int. J. Solids Structures* 13, 239–51.

Dal Corso, F., and Bigoni, D. (2009). The interactions between shear bands and rigid lamellar inclusions in a ductile metal matrix. *Proc. R. Soc. Lond. A* 465, 143–63.

Dal Corso, F., and Bigoni, D. (2010). Growth of slip surfaces and line inclusions along shear bands in a softening material. *Int. J. Fracture* 166, 225–37.

Dal Corso, F., Bigoni, D., and Gei, M. (2008). The stress concentration near a rigid line inclusion in a prestressed, elastic material. Part I. Full-field solution and asymptotics. *J. Mech. Phys. Solids* 56, 815–38.

Dally, J. W., and Riley, W. F. (1965). *Experimental Stress Analysis*. McGraw-Hill, New York.

Davies, P. J. (1991). Buckling and barrelling instabilities of nonlinearly elastic columns. *Q. Appl. Math.* 49, 407–26.

De Finetti, B. (1949). Sulle stratificazioni convesse. *Ann. Mat. Pura Appl.* 30, 173–83.

Den Hartog, J. P. (1952). *Advanced Strength of Materials*. McGraw-Hill, New York.

Deshpande, V. S., and Fleck, N. A. (2000). Isotropic constitutive models for metallic foams. *J. Mech. Phys. Solids* 48, 1253–83.

Desrues, J., Chambon, R., Mokni, M., and Mazerolle, F. (1996). Void ratio evolution inside shear bands in triaxial sand specimens studied by computed tomography. *Géotechnique* 46, 529–46.

Desrues, J., and Hammad, W. (1989). Shear band dependency on mean stress level. In *Numerical Methods for Localisation and Bifurcation of Granular Materials*, edited by E. Dembicki, G. Gudheus, and Z. Sikora, pp. 57–67. Gdansk, Poland.

Desrues, J., Lanier, J., and Stutz, P. (1985). Localisation of the deformation in tests on sand sample. *Eng. Fracture Mech.* 21(4), 909–21.

Dieter, G. E. (1961). *Mechanical Metallurgy*. McGraw-Hill, New York.

Dougill, J. W. (1976). On stable progressively fracturing solids. *Z. Angew. Math. Phys.* 27, 423–37.

Dorris, J. F., and Nemat-Nasser, S. (1980). Instability of a layer on a half space. *J. Appl. Mech.* 47, 304–12.

Dowaikh, M. A., and Ogden, R. W. (1990). On surface waves and deformations in a pre-stressed incompressible elastic solids. *IMA J. Appl. Math.* 44, 261–84.

Dowaikh, M. A., and Ogden, R. W. (1991). Interfacial waves and deformations in pre-stressed elastic media. *Proc. R. Soc. Lond. A* 443, 313–28.

Drescher, A., Vardoulakis, I., and Han, C. (1990). A biaxial apparatus for testing soils. *Geotech. Eng. J.* ASTM 13, 226–34.

Drucker, D. C. (1953). Limit analysis of two and three dimensional soil mechanics problems. *J. Mech. Phys. Solids* 1, 217–26.

Drucker, D. C. (1956). On uniqueness in the theory of plasticity. *Q. Appl. Math.* XIV, 35–42.

Drucker, D. C. (1964). On the postulate of stability of material in the mechanics of continua. *J. de Mécanique* 3, 235–49.

Drucker, D. C. (1973). Plasticity theory, strength differential (SD) phenomenon and volume expansion in metals and plastics. *Metall. Trans.* 4, 667–73.

Drucker, D. C., and Prager, W. (1952). Soil mechanics and plastic analysis for limit design. *Q. Appl. Math.* 10, 157–65.

Drugan, W. J. (2007). Elastic composite materials having a negative stiffness phase can be stable. *Phys. Rev. Lett.* 98(055502), 1–4.

Dryburgh, G., and Ogden, R. W. (1999). Bifurcation of an elastic surface-coated incompressible isotropic elastic block subject to bending. *Z. Angew. Math. Phys.* 50, 822–38.

Eekelen, H. A. M. (1980). Isotropic yield surface in three dimensions for use in soil mechanics. *Int. J. Numer. Anal. Meth. Geomech.* 4, 89–101.

Ehelers, W. (1995). A single-surface yield function for geomaterials. *Adv. Appl. Mech.* 65, 246–59.

Ekeland, I., and Temam, R. (1976). *Convex Analysis and Variational Problems*. North-Holland, Amsterdam.

Elishakoff, I. (2005). Controversy associated with the so-called "follower force": critical overview. *Appl. Mech. Rev.* 58, 117–42.

Engesser, F. (1889). Über die Knickfestigkeit gerader Stäbe. *Z. Architek. Ing.* 35, 455.

Ericksen, J. L. (1975). Equilibrium of bars. *J. Elasticity* 5, 191–201.

Everstine, G. C., and Pipkin, A. C. (1971). Stress channelling in transversely isotropic elastic composites. *ZAMP* 22, 825–34.

Faciu, C., and Suliciu, I. (1994). A Maxwellian model for pseudoelastic materials. *Scripta Metall. et Mater.* 31, 1399–404.

Feodosyev, V. I. (1977). *Selected Problems and Questions in Strength of Materials*. Mir Publisher, Moscow.

Finno, R. J., Harris, W. W., Mooney, M. A., and Viggiani, G. (1997). Shear bands in plane strain compression of loose sand. *Géotechnique* 47, 149–65.

Fosdick, R. L., and Shield, R. T. (1963). Small bending of a circular bar superposed on finite extension or compression. *Arch. Rat. Mech. Anal.* 12, 223–48.

Franceschini, G., Bigoni, D., Regitnig, P., and Holzapfel, G. A. (2006). Brain tissue deforms similarly to filled elastomers and follows consolidation theory. *J. Mech. Phys. Solids* 54, 2592–2620.

Franchi, A., Genna, F., and Paterlini, F. (1990). Research note on quasi-convexity of the yield function and its relation to Drucker postulate. *Int. J. Plasticity* 6, 369–75.

Frocht, M. M. (1965). *Photoelasticity*. Wiley, London.

Fu, Y. B., and Zhang, Y. T. (2006). Continuum-mechanical modelling of kink-band formation in fibre-reinforced composites. *Int. J. Solids Structures* 43, 3306–23.

Gajo, A. (2003). Instability phenomena in sand samples. Ph.D. thesis, University of Bristol, Bristol, UK.

Gajo, A., and Bigoni, D. (2008). A model for stress and plastic strain induced nonlinear hyperelastic anisotropy in soils. *Int. J. Numer. Anal. Meth. Geomech.* 32, 833–61.

Gajo, A., Bigoni, D., and Muir Wood, D. (2004). Multiple shear band development and related instabilities in granular materials. *J. Mech. Phys. Solids* 52, 2683–2724.

Gajo, A., Muir Wood, D., and Bigoni, D. (2007). On certain critical material and testing characteristics affecting shear band development in sand. *Géotechnique* 57, 449–61.

Gei, M., Bigoni, D., and Guicciardi, S. (2004). Failure of silicon nitride under uniaxial compression at high temperature. *Mech. Materials* 36, 335–45.

Gel'fand, I. M., and Shilov, G. E. (1964). *Generalized Functions*, Vol. 1: *Properties and Operations*. Academic Press, New York.

Gent, A. N. (2005). Elastic instabilities in rubber. *Int. J. Non-linear Mech.* 40, 165–75.

Gent, A. N., and Cho, I. S., (1999). Surface instabilities in compressed or bent rubber blocks *Rubber Chem. Technol.* 72, 253–62.

Gong, L., and Kyriakydes, S. (2005). Compressive response of open cell foams. Part II. Initiation and evolution of crushing. *Int. J. Solids Structures* 42, 1381–99.

Gough, G. S., Elam, C. F., and de Bruyne, N. A. (1940). The stabilisation of a thin sheet by a continuous supporting medium. *J. R. Aero. Soc.* 44, 12–13.

Goulbitsky, M., and Schaeffer, D. G. (1985). *Singularities and Groups in Bifurcation Theory*. Springer-Verlag, New York.

Green, A. E., and McInnis, B. C. (1967). Generalized hypoelasticity. *Proc. Roy. Soc. Edinburgh A Math. Phys. Sci.* 57, 220–30.

Green, A. E., and Zerna, W. (1968). *Theoretical Elasticity*, 2nd edn., Oxford University Press, Oxford, UK.

Gudehus, G. (1973). Elastoplastische Stoffgleichungen für trockenen Sand. *Ingenieur-Archiv* 42, 151–69.

Gudehus, G. (2004). A visco-hypoplastic constitutive relation for soft soils. *Soils Found.* 44, 11–25.

Gurtin, M. E. (1972). The linear theory of elasticity. In Flügge, S., ed., *Encyclopedia of Physics*, Vol. VIa/2, pp. 1–295. Berlin, Springer.

Gurtin, M. E. (1981). *An Introduction to Continuum Mechanics*. Academic Press, New York.

Gurson, A. L. (1977). Continuum theory of ductile rupture by void nucleation and growth. Part I. Yield criteria and flow rules for porous ductile media. *Int. J. Engng. Mat. Tech.* 99, 2–15.

Guz, A. N. (1999). *Fundamentals of the Three-Dimensional Theory of Stability of Deformable Bodies*. Springer-Verlag, Berlin.

Hadamard, J. (1903). *Lecons sur la Propagation des Ondes et les Équations de l' Hydrodynamique*. Hermann, Paris.

Haigh, B. P. (1920). Elastic limit of a ductile metal. *Engineering* 109, 158–60.

Hallbäck, N., and Nilsson, F. (1994). Mixed-mode I/II fracture behaviour of an aluminium alloy. *J. Mech. Phys. Solids* 42, 1345–74.

Hall, E. O. (1970). *Yield Point Phenomena in Metals and Alloys*. Macmillan, London.

Harren, S., Lowe, T. C., Asaro, R. J., and Needleman, A. (1989). Analysis of large-strain shear in ratedependent face-centred cubic polycrystals: correlation of micro- and macromechanics. *Phil. Trans. R. Soc. A* 328, 443–500.

Haughton, D. M. (1999). Flexure and compression of incompressible elastic plates. *Int. J. Engng. Sci.* 37, 1693–708.

Haughton, D. M., and Ogden, R. W. (1979). Bifurcation of inflated circular cylinders of elastic material under axial loading. II. Exact theory for thick-walled tubes. *J. Mech. Phys. Solids* 27, 489–512.

Haughton, D. M., and Ogden, R. W. (1980). Bifurcation of finitely deformed rotating elastic cylinders. *Q. J. Mech. Appl. Math.* 33, 251–65.

Hayes, M. (1966). On the displacement boundary-value problem in linear elastostatics. *Q. J. Mech. Appl. Math.* 19, 151–5.

Hayes, M., and Rivlin, R. S. (1961a). Propagation of a plane wave in an isotropic elastic material subject to pure homogeneous deformation. *Arch. Rat. Mech. Anal.* 8, 15–22.

Hayes, M., and Rivlin, R. S. (1961b). Surface waves in deformed elastic materials. *Arch. Rat. Mech. Anal.* 8, 358–80.

Haythornthwaite, R. M. (1985). A family of smooth yield surfaces. *Mec. Res. Commun.* 12, 87–91.

Hencky, H. (1924). Zur Theorie plastischer Deformationen und der hierdurch im Material hervorgerufenen Nachspannungen. *Z. Angew. Math. Mechanik* 4, 323–35.

Herrmann, G. (1971). Dynamics and stability of mechanical systems with follower forces. Techn. Rept. NASA CR-1782.

Herrmann, G., Nematnasser, S., and Prasad, S. N. (1966). Models demonstrating instability of nonconservative dynamical systems. Tech. Rept. No. 66-4, Str. Mech. Lab., Northwestern University.

Heyman, J. (1966). The stone skeleton. *Int. J. Solids Structures* 2, 249–79.

Hill, R. (1950a). *The Mathematical Theory of Plasticity*. Clarendon Press, Oxford.

Hill, R. (1950b). Inhomogeneous deformation of a plastic lamina in a compression test. *Phil. Mag.* 41, 733–44.

Hill, R. (1952). On discontinuous plastic states, with special reference to localized necking in thin sheets. *J. Mech. Phys. Solids* 1, 19–30.

Hill, R. (1958). A general theory of uniqueness and stability in elastic-plastic solids. *J. Mech. Phys. Solids* 6, 236–49.

Hill, R. (1959). Some basic principles in the mechanics of solids without a natural time. *J. Mech. Phys. Solids* 7, 209–25.

Hill, R. (1961). Discontinuity relations in mechanics of solids. In *Progress in Solid Mechanics* edited by I. N. Sneddon and R. Hill, Vol. II, pp. 247–76. Amsterdam, North-Holland.

Hill, R. (1962). Acceleration waves in solids. *J. Mech. Phys. Solids* 10, 1–16.

Hill, R. (1967a). Eigenmodal deformations in elastic/plastic continua. *J. Mech. Phys. Solids* 15, 371–86.

Hill, R. (1967b). On the classical constitutive laws for elastic/plastic solids. In *Recent Progress in Applied Mechanics, The Folke Odkvist Volume*, edited by B. Broberg, pp. 241–9. Almqvist & Wiksell, Stockholm.

Hill, R. (1967c). The essential structure of constitutive laws for metal composites and polycrystals. *J. Mech. Phys. Solids* 15, 79–95.

Hill, R. (1968). On constitutive inequalities for simple materials. Part I. *J. Mech. Phys. Solids* 16, 229–42.

Hill, R. (1978). Aspects of invariance in solid mechanics. In *Advances in Applied Mechanics*, edited by C.-S. Yih, Vol. 18, pp. 1–75. Academic Press, New York.

Hill, R., and Hutchinson, J. W. (1975). Bifurcation phenomena in the plane tension test. *J. Mech. Phys. Solids* 23, 239–64.

Hill, R., and Rice, J. R. (1973). Elastic potentials and the structure of inelastic constitutive laws. *SIAM J. Appl. Math.* 25, 448–61.

Hill, R., and Sewell, M. J. (1960). A general theory of inelastic column failure. Part I. *J. Mech. Phys. Solids* 8, 105–11.

Hirth, J. P., and Lothe, J. (1968). *Theory of Dislocations*. Wiley, New York.

Hirth, J. P., and Cohen, M. (1970). On the strength-differential phenomenon in hardened steel. *Metall. Materials Trans. B* 1, 3–8.

Hoger, A. (1987). The stress conjugate to logarithmic strain. *Int. J. Solids Structures* 23, 1645–56.

Holzapfel, G. A. (2000). *Nonlinear Solid Mechanics: A Continuum Approach for Engineering*, Wiley, Chichester, UK.

Hoek, E., and Brown, E. T. (1980). *Underground Excavations in Rock*. Institution of Mining and Metallurgy, London.

Hoque, E., and Tatsuoka F. (2004). Effects of stress ratio on small-strain stiffness during triaxial shearing. *Géotechnique* 54, 429–39.

Horgan, C. O., and Polignone, D. A. (1995). Cavitation in nonlinearly elastic solids: a review. *Appl. Mech. Rev.* 48, 471–85.

van Hove, L. (1947). Sur l'extension de la condition de Legendre du calcul des variations aux intégrales multiples à plusieurs fonctions inconnues. *Proc. Sect. Sci. K. Akad. van Wetenschappen* 50, 18–23.

Huang, K., Hutchinson, J. W., and Tvergaard, V. (1991). Cavitation instabilities in elastic-plastic solids. *J. Mech. Phys. Solids* 39, 223–41.

Hudson, J. A., Brown, E. T., and Fairhurst, C. (1971). Shape of the complete stress-strain curve for rock. In *Stability of Rock Slopes, Proceedings of the 13th Symposium on Rock Mechanics*, edited by E. J. Cording, pp. 773–95. University of Illinois Press, Urbana.

Hueckel, T., (1975). On plastic flow of granular and rock-like materials with variable elasticity moduli. *Bull. Pol. Acad. Sci.*, Ser. Techn. 23, 405–14.

Hueckel, T. (1976). Coupling of elastic and plastic deformation of bulk solids. *Meccanica* 11, 227–35.

Hueckel, T., and Maier, G. (1977). Incremental boundary value problems in the presence of coupling of elastic and plastic deformations: a rock mechanics oriented theory. *Int. J. Solids Structures* 13, 1–15.

Hunt, G. W., Peletier, M. A., and Wadee, M. A. (2000). The Maxwell stability criterion in pseudo-energy models of kink banding. *J. Struct. Geol.* 22, 669–81.

Hutchinson, J. W. (1970). Elastic–plastic behaviour of polycrystalline metals and composites. *Proc. R. Soc. Lond. A* 319, 247–72.

Hutchinson, J. W. (1973). Post-bifurcation behavior in the plastic range. *J. Mech. Phys. Solids* 21, 163–90.

Hutchinson, J. W. (1974). Plastic buckling. *Adv. Appl. Mech.*, 14, 67–144.

Hutchinson, J. W., and Koiter, W. T. (1970). Postbuckling theory. *Appl. Mech. Rew.* 23, 1353–66.

Hutchinson, J. W., and Miles, J. P. (1974). Bifurcation analysis of the onset of necking in an elastic/plastic cylinder under uniaxial tension. *J. Mech. Phys. Solids* 22, 61–71.

Hutchinson, J. W., and Neale, K. W. (1979). Finite strain J_2-deformation theory. In *Proceedings of the IUTAM Symposium on Finite Elasticity*, edited by D. E. Carlson and R. T. Shield, pp. 237–47. Martinus Nijhoff, The Hague.

Hutchinson, J. W., and Tvergaard, V. (1980). Surface instabilities on statically strained plastic solids. *Int. J. Mech. Sci.* 22, 339–54.

Hutchinson, J. W., and Tvergaard, V. (1981). Shear band formation in plane strain. *Int. J. Solids Structures* 17, 451–70.

Itskov, M., and Aksel, N. (2002). A closed-form representation for the derivative of non-symmetric tensor power series. *Int. J. Solids Structures* 39, 5963–78.

Johan, Z., Hughes, T. J. R., and Shakib, F. (1991). A globally convergent matrix-free algorithm for implicit time-marching schemes arising in finite element analysis in fluids. *Comput. Meth. Appl. Mech. Engrg.* 87, 281–304.

Jaumann, G. (1905). *Grundlagen der Bewegungslehre*. Springer, Berlin.

Kachanov, L. M. (1971). *Foundations of the Theory of Plasticity*. North-Holland, Amsterdam.

Kalish, D., and Rack, H. J. (1972). The strength differential in a maraging steel. *Metall. Materials Trans. B* 3, 2289–90.

Kardomateas, G. A. (1986). Fractographic observations in asymmetric and symmetric fully plastic crack growth. *Scripta Metall.* 20, 609–14.

Kardomateas, G. A. (2005). Wrinkling of wide sandwich panels/beams with orthotropic phases by an elasticity approach. *J. Appl. Mech.* 72, 818–25.

Kardomateas, G. A., and McClintock, F. A. (1989). Shear band characterization of mixed mode I and II fully plastic crack growth. *Int. J. Fracture* 40, 1–12.

Karam, G. N., and Gibson, L. J. (1995). Elastic buckling of cylindrical shells with elastic cores. Part 1. Analysis. *Int. J. Solids Structures* 32, 1259–83.

von Kármán, T. (1947). Discussion of 'Inelastic column theory'. *J. Areonaut. Sci.* 14, 267–8.

Kearsley, E. A. (1986). Asymmetric stretching of a symmetrically loaded elastic sheet. *Int. J. Solids Structures* 22, 111–19.

Key, S. W., and Krieg, R. D. (1982). On the numerical implementation of inelastic time dependent and time independent finite strain constitutive equations in structural mechanics. *Comput. Meth. Appl. Mech. Engrg.* 33, 439–52.

Kellog, O. D. (1953). *Foundations of potential theory*. Dover, New York.

Kleiber, M. (1984). Numerical study on necking-type bifurcations in void-containing elastic-plastic material. *Int. J. Solids Structures* 20, 191–210.

Kleiber, M. (1986). On plastic localization and failure in plane strain and round void containing tensile bars. *Int. J. Plasticity* 2, 205-21.

Knops, R. J., and Wilkes, E. W. (1973). Theory of elastic stability. In *Encyclopedia of Physics*, edited by S. Flügge, Vol. IVa/3. Springer-Verlag, Berlin.

Koiter, W. T. (1960). General theorems of elastic-plastic solids. *Prog. Solid Mech.* 165, 221.

Koiter, W. T. (1996). Unrealistic follower forces. *J. Sound Vibration* 194, 636–8.

Kuznetsov, V. V., and Levyakov, S. V. (2002). Complete solution of the stability problem for elastica of Euler's column. *Int. J. Non-Linear Mech.* 37, 1003–9.

Kyriakides, S., and Corona, E. (2007). *Mechanics of Offshore Pipelines.* Vol. 1: *Buckling and Collapse*. Elsevier, Amsterdam.

Lade, P. V. (1997). Modelling the strengths of engineering materials in three dimensions. *Mech. Cohes. Frict. Mat.* 2, 339–56.

Ladyzhenskaya, O. A. (1963). *The Mathematical Theory of Viscous Incompressible Flow.* Gordon & Breach, New York.

Lakes, R. S. (2001). Extreme damping in compliant composites with a negative-stiffness phase. *Phil. Mag. Lett.* 81, 95–100.

Lamb, H. (1928). *Statics*. Cambridge University Press, Cambridge, UK.

Laroussi, M., Sab, K., and Alaoui, A. (2002). Foam mechanics: nonlinear response of an elastic 3D-periodic microstructure. *Int. J. Solids Structures* 39, 3599–623.

Lebedev, N. N. (1965). *Special Functions and Their Applications*. Prentice-Hall, Englewood Cliffs, NJ.

Lee, E. H. (1969). Elastic-plastic deformation at finite strains. *J. Appl. Mech.* 36, 1–6.

Lekhnitskii, S. G. (1981). *Theory of Elasticity of an Anisotropic Body*. Mir Publisher, Moscow.

Lemaitre, J., and Chaboche, J. L. (1985). *Mechanics of Solid Materials.* Cambridge University Press, Cambridge, UK.

Levyakov, S. V., and Kuznetsov, V. V. (2010). Stability analysis of planar equilibrium configurations of elastic rods subjected to end loads. *Acta Mech.* 211, 73–87.

Lin, F.-B., and Bazant, Z. (1986). Convexity of smooth yield surface of frictional material. *J. Eng. Math.* 112, 1259–62.

Lode, W. (1926). Versuche über den Einfluß der mittleren Hauptspannung auf das Fließen der Metalle Eisen Kupfer und Nickel. *Z. Physik.* 36, 913–39.

Lorang, X., Foy-Margiocchi, F., Nguyen, Q. S., and Gautier, P. E. (2006). TGV disc brake squeal. *J. Sound Vibr.* 293, 735–46.

Loret, B. (1992). Does deviation from deviatoric associativity lead to the onset of flutter instability? *J. Mech. Phys. Solids* 40, 1363–75.

Loret, B., and Harireche, O. (1991). Acceleration waves, flutter instabilities and stationary discontinuities in inelastic porous media. *J. Mech. Phys. Solids* 39, 569–606.

Loret, B., Martins, J. A. C., and Simões, F. M. F. (1995). Surface boundary conditions trigger flutter instability in non-associative elastic-plastic solids. *Int. J. Solids Structures* 32, 2155–90.

Loret, B., Prevost, J. H., and Harireche, O. (1990). Loss of hyperbolicity in elastic-plastic solids with deviatoric associativity. *Eur. J. Mechanics-A/Solids* 9, 225–31.

Loret, B., Simões, F. M. F., and Martins, J. A. C. (1997). Growth and decay of acceleration waves in non-associative elastic-plastic fluid-saturated porous media. *Int. J. Solids Structures* 34, 1583–1608.

Loret, B., Simões, F. M. F., and Martins, J. A. C. (2000). Flutter instability and ill-posedness in solids and fluid-saturated porous media. In *Material Instabilities in Elastic and Plastic Solids*, CISM Lecture Notes No. 414, edited H. Petryk, pp. 109–207. Springer-Verlag, New York.

Love, A. E. H. (1927). *A Treatise on the Mathematical Theory of Elasticity*. Cambridge University Press, Cambridge, UK.

Lubliner, J. (1990). *Plasticity Theory*. Macmillan, London.

Lurie, A. I. (2005). *Theory of Elasticity*. Springer-Verlag, Amsterdam.

Macaulay, W. H. (1919). Note on the deflection of the beams. *Messenger of Math.* 48, 129–30.

Maddocks, J. H. (1984). Stability of nonlinear elastic rods. *Arch. Rat. Mech. Anal.* 85, 311–54.

Maier, G. (1967). On elastic-plastic structures with associated stress-strain relations allowing for work softening. *Meccanica* 2, 55–64.

Maier, G. (1970a). A minimum principle for incremental elastoplasticity with nonassociated flow-laws. *J. Mech. Phys. Solids* 18, 319–30.

Maier, G. (1970b). A matrix structural theory of piecewise linear elastoplasticity with interacting yield planes. *Meccanica* 5, 54–66.

Maier, G., and Hueckel, T. (1979). Non associated and coupled flow-rules of elastoplasticity for rock-like materials. *Int. J. Rock Mech. Min. Sci.* 16, 77–92.

Mandel, J. (1962a). Ondes plastiques dans un milieu indéfini à trois dimensions. *J. de Mécanique* 1, 3–30.

Mandel, J. (1962b). Propagation des surfaces de discontinuite dans un milieu elastoplastique. In *Stress Waves in Anelastic Solids*. edited by H. Kolsky and W. Prager, pp. 331–40. Springer, Berlin.

Mandel, J. (1966a). Conditions de stabilité et postulat de Drucker. In *Rheology and Soil Mechanics*. edited by J. Kravtchenko and P. M. Sirieys, pp. 58–68. Springer, Berlin.

Mandel, J. (1966b). Contribution theorique a l'etude de l'ecrouissage et des lois de l'ecoulement plastique, In *Proceedings of the 11th International Congress of Applied Mechanics (Munich 1964)*, pp. 502–509. Springer-Verlag, Berlin.

Mandel, J. (1969). Thermodynamique et ondes dans les milieux viscoplastiques. *J. Mech. Phys. Solids* 17, 125–40.

Marsden, J. E., and Hughes, T. J. R. (1983). *Mathematical Foundations of Elasticity*. Prentice-Hall, Englewood Cliffs, NJ.

Martins, J. A. C., Oden, J. T., and Simões, F. M. F. (1990). Recent advances in engineering science - a study of static and kinetic friction *Int. J. Engng. Sci.* 28, 29–92.

Matthies, H., and Strang, G. (1979). The solution of non-linear finite element equations. *Int. J. Numer. Meth. Eng.* 11, 1613–26.

McClintock, F. A. (1971). Plasticity aspects of fracture. In *Fracture. An Andvanced Treatise*, edited by H. Liebowitz, Vol. 3, pp. 47–225. Academic Press, New York.

Melan, E. (1938). Zur Plastizität des räumliche Kontinuums. *Ingenieur-Archiv* 9, 116–26.

Menétrey, Ph., and Willam, K. J. (1995). Triaxial failure criterion for concrete and its generalization. *ACI Struct. J.* 92, 311–18.

Michel, J. C., Lopez-Pamies, O., Castaneda P. P., and Triantafyllidis, N. (2007). Microscopic and macroscopic instabilities in finitely strained porous elastomers. *J. Mech. Phys. Solids* 55, 900–38.

Miles, J. P. (1973). Fluid-pressure eigenstates and bifurcation in tension specimens under lateral pressure. *J. Mech. Phys. Solids* 21, 145–62.

Miles, J. P., and Nuwayhid, U. A. (1985). Bifurcation in compressible elastic/plastic cylinders under uniaxial tension. *Appl. Sci. Res.* 42, 33–54.

von Mises, R. (1913). Mechanik der Festen Korper im plastisch deformablen Zustand. *Göttin. Nachr. Math. Phys.* 1, 582–92.

Misra, S., and Mandal, N. (2007). Localization of plastic zones in rocks around rigid inclusions: Insights from experimental and theoretical models. *J. Geophys. Res.-Solid Earth* 112, B09206.

Mooney, M. (1940). A theory of large elastic deformations. *J. Appl. Phys.* 11, 582–92.

Moore, B., Jaglinski, T., Stone, D. S., and Lakes, R. S. (2006). Negative incremental bulk modulus in foams. *Phil. Mag. Lett.* 86, 651–59.

Morrey, C. B., Jr., (1952). Quasi-convexity and the lower semicontinuity of multiple integrals. *Pacific J. Math.* 2, 25–53.

Mróz, Z. (1963). Non-associated flow laws in plasticity *J. de Mechanique* 2, 21–42.

Mróz, Z. (1966). On forms of constitutive laws for elastic-plastic solids. *Arch. Mech. Stosowanej* 18, 1–34.

Müller, I. (1996). Two instructive instabilities in non-linear elasticity: biaxially loaded membrane and rubber balloons. *Meccanica* 31, 387–95.

Müller, I., and Strehlow, P. (2004). *Rubber and Rubber Balloons*. Springer, Berlin.

Mullin, T., Deschanel, S., Bertoldi, K., and Boyce, M. C. (2007). Pattern transformation triggered by deformation. *Phys. Rev. Lett.* 99, 084301.

Muskhelishvili, N. I. (1953). *Some Basic Problems of the Mathematical Theory of Elasticity*. P. Nordhoff, Groningen.

Nadai, A. (1931). *Plasticity*. McGraw-Hill, New York.

Nadai, A. (1950). *Theory of Flow and Fracture of Solids*. McGraw-Hill, New York.

Neale, K. W. (1981). Phenomenological constitutive laws in finite plasticity *SM Archives* 6, 79–128.

Needleman, A. (1979). Non-normality and bifurcation in plane strain tension or compression. *J. Mech. Phys. Solids* 27, 231–54.

Needleman, A., and Ortiz, M. (1991). Effects of boundaries and interfaces on shear-band localization. *Int. J. Solids Structures* 28, 859–77.

Newman, K., and Newman, J. B. (1971). Failure theories and design criteria for plain concrete. In *Structure, Solid Mechanics and Engineering Design, Proceedings of the 1969 Southampton Civil Engineering Conference*, edited by M. Te'eni, pp. 963–995. Wiley Interscience, New York.

Nguyen, Q. S. (1995). *Stabilité des structures élastiques*. Springer-Verlag, Berlin.

Nguyen, Q. S. (2003). Instability and friction. *C. R. Acad. Sci., Paris* 331, 99–112.

Nguyen, Q. S., and Triantafyllidis, N. (1989). Plastic bifurcation and postbifurcation analysis for generalized standard continua. *J. Mech. Phys. Solids* 37, 545–66.

Noll, W. (1958). A mathematical theory of the mechanical behaviour of continuous media. *Arch. Rat. Mech. Anal.* 2, 197–226.

Noselli, G., Dal Corso, F., and Bigoni, D. (2011). The stress intensity near a stiffener disclosed by photoelasticity. *Int. J. Fracture* 166, 91–103.

Ogden, R. W. (1982). Elastic deformations of rubberlike solids. In *Mechanics of Solids, The Rodney Hill 60th Anniversary Volume*, edited by H. G. Hopkins and M. J. Sewell, pp. 499–537. Pergamon Press, New York.

Ogden, R. W. (1984). *Non-linear Elastic Deformations*. Ellis Horwood, Chichester, UK.

Ogden, R. W. (1985). Local and global bifurcation phenomena in plane-strain finite elasticity. *Int. J. Solids Structures* 21, 121–32.

Ogden, R.W. (2001). Elements of the theory of finite elasticity. In *Nonlinear Elasicity: Theory and Applications*, edited by Y. B. Fu and R. W. Ogden. Cambridge University Press, Cambridge, UK.

Ogden, R. W., and Sotiropulos, D. A. (1995). On interfacial waves in pre-stressed layered incompressible elastic solid. *Proc. R. Soc. Lond. A* 450, 319–41.

Oldroyd, J. C. (1950). On the formulation of rheological equations of state. *Proc. R. Soc. Lond. A* 200, 523–41.

Ol'khovik, O. (1983). Apparatus for testing of strength of polymers in a three-dimensional stressed state. *Mech. Compos. Mater.* 19, 270–75.

Ottosen, N. S. (1977). A failure criterion for concrete. *J. Eng. Mech. Div. ASCE* 103, 527–35.

Ottosen, N. S., and Runesson, K. (1991). Acceleration waves in elastoplasticity. *Int. J. Solids Structures* 28, 135–59.

Palmer, A. C., and Rice, J. R. (1973). The growth of slip surfaces in the progressive failure of overconsolidated clay. *Proc. R. Soc. Lond. A* 332, 527–48.

Pan, F., and Beatty, M. F. (1997a). Remarks on the instability of an incompressible and isotropic hyperelastic, thick-walled cylindrical tube. *J. Elasticity* 48, 217–39.

Pan, F., and Beatty, M. F. (1997 b). Instability of a Bell constrained cylindrical tube under end thrust. Part 1. Theoretical development. *Math. Mech. Solids* 2, 243–73.

Papamichos, E., Vardoulakis, I., and Mühlhaus, H.-B. (1990). Buckling of layered elastic media: a Cosserat-continuum approach and its validation. *Int. J. Numer. Anal. Meth. Geomech.* 14, 473–98.

Papka, S. D., and Kyriakides, S. (1999). Biaxial crushing of honeycomb. Part I. Experiments. *Int. J. Solids Structures* 36, 4367–96.

Paul, B. (1968). Macroscopic yield criteria for plastic flow and brittle fracture. In *Fracture an Advanced Treatise*, edited by H. Liebowitz, vol. I, pp. 313–496. Academic Press, New York.

Petryk, H. (1985a). On energy criteria of plastic instability. In *Plastic Instability, Proceedings of the Considère Memorial*, pp. 215–26. Ecole Nat. Ponts Chauss. Press, Paris.

Petryk, H. (1985b). On stability and symmetry conditions in time-independent plasticity. *Arch. Mech.* 37, 503–20.

Petryk, H. (1991). The energy criteria of instability in time-independent inelastic solids. *Arch. Mech.* 43, 519–45.

Petryk, H. (1992). Material instability and strain-rate discontinuities in incrementally nonlinear continua. *J. Mech. Phys. Solids* 40, 1227–50.

Petryk, H. (1993a). Theory of bifurcation and instability in time-independent plasticity. In *CISM Lecture Notes No. 327, Udine 1991*, edited by Q. S. Nguyen, pp. 95–152. Springer, Wien.

Petryk, H. (1993b). Stability and constitutive inequalities in plasticity. In *CISM Lecture Notes No. 336, Udine 1992*, edited by W. Muschik, pp. 255–329. Springer, Wien.

Petryk, H. (2000). General conditions for uniqueness in materials with multiple mechanisms of inelastic deformation. *J. Mech. Phys. Solids* 48, 367–96.

Petryk, H., and Thermann, K. (1985). Second-order bifurcation in elastic-plastic solids. *J. Mech. Phys. Solids* 33, 577–93.

Petryk, H., and Thermann, K. (1996). Post-critical plastic deformation of biaxially stretched sheets. *Int. J. Solids Structures* 33, 689–705.

Petryk, H., and Thermann, K. (2002). Post-critical plastic deformation in incrementally nonlinear materials. *J. Mech. Phys. Solids* 50, 925–54.

Pflüger, A. (1955). Zur Stabilität des tangential gedrückten Stabes. *Z. Angew. Math. Mechanik* 35, 191–91.

Phillips, A. (1974). Experimental plasticity. Some thoughts on its present status and possible future trends. In *Symposium on the Foundations of Plasticity (Warsaw, 1972)*, edited by A. A. Sawczuk, pp. 193–233. Nordhoff International Publishing, Leyden.

Piccolroaz, A., and Bigoni, D. (2009). Yield criteria for quasibrittle and frictional materials: a generalization to surfaces with corners. *Int. J. Solids Structures* 46, 3587–96.

Piccolroaz, A., Bigoni, D., and Gajo, A. (2006a). An elastoplastic framework for granular materials becoming cohesive through mechanical densification. Part I. Small strain formulation. *Eur. J. Mechanics-A/Solids* 25, 334–57.

Piccolroaz, A., Bigoni, D., and Gajo, A. (2006b). An elastoplastic framework for granular materials becoming cohesive through mechanical densification. Part II. The formulation of elastoplastic coupling at large strain. *Eur. J. Mechanics-A/Solids* 25, 358–69.

Piccolroaz, A., Bigoni, D., and Willis, J. R. (2006). A dynamical interpretation of flutter instability in a continuous medium. *J. Mech. Phys. Solids* 54, 2391–2417.

Pietruszczak, S., and Mróz, Z. (1981). Finite element analysis of deformation of strain-softening materials. *Int. J. Numer. Meth. Eng.* 17, 327–34.

Podio Guidugli, P. (2000). A primer in elasticity. *J. Elasticity* 58 1–104.

Podgórski, J. (1984). Limit state condition and the dissipation function for isotropic materials. *Arch. Mech. Soc.* 36, 323–42.

Podgórski, J. (1985). General failure criterion for isotropic media. *J. Eng. Mech. ASCE* 111, 188–99.

Poirier, C., Ammi, M., Bideau, D., and Troadec, J. P. (1992). Experimental study of the geometrical effects in the localization of deformation. *Phys. Rev. Lett.* 68, 216–19.

Poston, T., and Stewart, I. (1978). *Catastrophe Theory and its Applications.* Pitman, San Francisco.

Potier-Ferry, M. (1987). Foundations of elastic postbuckling theory. In *Buckling and Post-Buckling*, Lecture Notes in Physics 288, pp. 1–82. Springer-Verlag, Berlin.

Prager, W. (1949). Recent developments in the mathematical theory of plasticity. *J. Appl. Phys.* 20, 235–41.

Prager, W. (1954). Discontinuous fields of plastic stress and flow. In *2nd Nat. Congr. Appl. Mech.*, Ann Arbor, Michigan, pp. 21–32.

Price N. J., and Cosgrove, J. W. (1990). *Analysis of Geological Structures.* Cambridge University Press, Cambridge, UK.

Puglisi, G. (2006). Hysteresis in multi-stable lattices with non-local interactions. *J. Mech. Phys. Solids* 54, 2060–88.

Puglisi, G., and Truskinowsky, L. (2000). Mechanics of a discrete chain with bi-stable elements. *J. Mech. Phys. Solids* 48, 1–27.

Puglisi, G., and Truskinowsky, L. (2002). Rate independent hysteresis in a bi-stable chain. *J. Mech. Phys. Solids* 50, 165–87.

Puzrin, A. M., and Germanovich, L. N. (2005). The growth of shear bands in the catastrophic failure of soils. *Proc. R. Soc. Lond. A* 461, 1199–1228.

Radi, E., Bigoni, D., and Capuani, D. (2002). Effects of pre-stress on crack-tip fields in elastic, incompressible solids. *Int. J. Solids Structures* 39, 3971–96.

Radi, E., Bigoni, D., and Tralli, A. (1999). On uniqueness for frictional contact rate problems. *J. Mech. Phys. Solids* 47, 275–96.

Raniecki, B. (1976). Ordinary waves in inviscid plastic media. In *Mechanical Waves in Solids*, edited by J. Mandel and L. Brun, CISM Courses and Lectures No. 222. Springer-Verlag, Wien.

Raniecki, B. (1979). Uniqueness criteria in solids with non-associated plastic flow laws at finite deformations, *Bull. Acad. Polon. Sci. Ser. Sci. Tech.* 27, 391–99.

Raniecki, B., and Bruhns, O. T. (1981). Bounds to bifurcation stresses in solids with non-associated plastic flow law at finite strain. *J. Mech. Phys. Solids* 29, 153–71.

Raniecki, B., and Mróz, Z. (2008). Yield or martensitic phase transformation conditions and dissipation functions for isotropic, pressure-insensitive alloys exhibiting SD effect. *Acta Mech.* 195, 81–102.

Read, H. E., and Hegemier, G. A. (1984). Strain softening of rock, soil and concrete - A review article. *Mech. Materials* 3, 271–94.

Reiner, M. (1945). A mathematical theory of dilatancy, *Am. J. Math.* 67, 350–62.

Reiss, E. L. (1969). Column buckling: an elementary example of bifurcation. In *Bifurcation Theory and Nonlinear Eigenvalue Problems*. edited by J. B. Keller and S. Antman, pp. 1–16. W.A. Benjamin, New York.

Renardy, M., and Rogers, R. C. (1993). *An Introduction to Partial Differential Equations*. Springer-Verlag, New York.

Rice, J. R. (1968). Mathematical analysis in the mechanics of fracture. In *Fracture*, edited by H. Liebowitz, Vol. II, pp. 191–311. Academic Press, New York.

Rice, J. R. (1973). The initiation and growth of shear bands. In *Plasticity and Soil Mechanics* edited by A. C. Palmer, pp. 263–75. Cambridge University Engineering Department, Cambridge, UK.

Rice, J. R. (1977). The localization of plastic deformation. In *Theoretical and Applied Mechanics*, edited by W. T. Koiter, pp. 207–10. North-Holland, Amsterdam.

Rice, J. R., and Rudnicki, J. W. (1980). A note on some features of the theory of localization of deformation. *Int. J. Solids Structures* 16, 597–605.

Rittel, D. (1990). The influence of microstructure on the macroscopic patterns of surface instabilities in metals. *Scripta Metall. Mater.* 24, 1759–64.

Rittel, D., Aharonov, R., Feigin, G., and Roman, I. (1991). Experimental investigation of surface instabilities in cylindrical tensile metallic specimens. *Acta Metall. Mater.* 39, 719–24.

Rivlin, R. S. (1948a). Large elastic deformations of isotropic materials. I. Fundamental concepts. *Phil. Trans. R. Soc. Lond. A* 240, 459–90.

Rivlin, R. S. (1948b). The hydrodynamics of non-Newtonian fluids, *Proc. R. Soc. Lond.* 193, 260–81.

Rivlin, R. S. (1949). Large elastic deformations of isotropic materials. V. The problem of flexure. *Proc. R. Soc. Lond. A* 195, 463–73.

Rittel, D., and Roman, I. (1989). Tensile deformation of coarse-grained cast austenitic manganese steels. *Mat. Sci. Engng.* A110, 77–87.

Roberts, A. W., and Varberg, D. E. (1973). *Convex functions*. Academic Press, New York.

Roccabianca, S., Gei, M., and Bigoni, D. (2010). Plane strain bifurcations of elastic layered structures subject to finite bending: theory versus experiments. *IMA J. Appl. Math.* 75, 525–48.

Roccabianca, S., Bigoni, D., and Gei, M. (2011). Long wavelength bifurcations and multiple neutral axes of elastic layered structures subject to finite bending. *J. Mech. Materials Structures*, 6, 511–27.

Rogers, T. G. (1989). Squeezing flow of fibre-reinforced viscous fluid. *J. Eng. Math.* 23, 81–9.

Roscoe, K. H., and Burland, J. B. (1968). On the generalized stress-strain behaviour of 'wet' clay. In *Engineering Plasticity*, edited by J. Heyman and F. A. Leckie, Cambridge University Press, Cambridge, UK.

Roscoe, K. H., and Schofield, A. N. (1963). Mechanical behaviour of an idealised 'wet' clay. In *Proc. European Conf. on Soil Mechanics and Foundation Engineering, Wiesbaden*, Vol. 1, pp. 47–54. Deutsche Gesellshaft für Erd- und Grundbau e. V., Essen.

Rowe, P. W. (1962). The stress-dilatancy relation for static equilibrium of an assembly of particles in contact. *Proc. R. Soc. Lond., A* 269, 500–27.

Royer-Carfagni, G. F. (2000). Slip bands and stress oscillations in bars. *J. Elasticity* 59, 131–43.

Rudnicki, J. W., and Rice, J. R. (1975). Conditions for the localization of deformations in pressure-sensitive dilatant materials. *J. Mech. Phys. Solids* 23, 371–94.

Runesson, K., and Mróz, Z. (1989). A note on non-associated plastic flow rules. *Int. J. Plasticity* 5, 639–58.

Ryzhak, E. I. (1987). Necessity of Hadamard conditions for stability of elastic-plastic solids. *Izv. AN SSSR MTT* (Mechanics of Solids), 99–102.

Ryzhak, E. I. (1993). On stable deformation of "unstable" materials in a rigid triaxial testing machine. *J. Mech. Phys. Solids* 41, 1345–56.

Ryzhak, E. I. (1994). On stability of homogeneous elastic bodies under boundary conditions weaker than displacement conditions. *Q. J. Mech. Appl. Math.* 47, 663–72.

Sachkov, Y. L., and Levyakov, S. V. (2010). Stability of inflectional elasticae centered at vertices or inflection points. *Proc. Stelkov Inst. Math.* 271, 177–92.

Sagan, H. (1961). *Boundary and Eigenvalue Problems in Mathematical Physics.* Wiley, Chichester, UK.

Salencon, J. (1974). *Applications of the Theory of Plasticity in Soil Mechanics.* Wiley, Chichester, UK.

Sansour, C. (2001). On the dual variable of the logarithmic strain tensor, the dual variable of the Cauchy stress tensor, and related issues. *Int. J. Solids Structures* 38, 9221–32.

Savin, G. N. (1961). *Stress Concentration Around Holes.* Pergamon Press, London.

Scheidler, M. J. (1984). *Acceleration waves*, Ph.D. thesis, Rice University, Houston, TX.

Schofield, A. N., and Wroth, C. P. (1968). *Critical State Soil Mechanics.* McGraw-Hill, London.

Schneebeli, G. (1956). Une analogie mécanique pour les terres sans cohésion. *C. R. Acad. Sci., Paris* 243, 125–6.

Sfer, D., Carol, I., Gettu, R., and Etse, G. (2002). Study of the behaviour of concrete under triaxial compression. *J. Eng. Math.* 128, 156–63.

Shanley, F. R. (1947). Inelastic column theory. *J. Areonautic. Sci.* 14, 261–67.

Shaw, J. A., and Kyriakides, S. (1997). Initiation and propagation of localized deformation in elasto-plastic strips under uniaxial tension. *Int. J. Plasticity* 13, 837–71.

Shaw, M. C., and Sata, T. (1966). The plastic behaviour of cellular materials. *Int. J. Mech. Sci.* 8, 469–78.

Shearer, M., and Schaeffer, D. G. (1994). Unloading near a shear band in granular material. *Q. Appl. Math.* 52, 579–600.

Shield, R. T. (1955). On Coulomb's law of failure in soils. *J. Mech. Phys. Solids* 4, 10–16.

Shield, T. W., Kim, K. S., and Shield, R. T. (1994). The buckling of an elastic layer bonded to an elastic substrate in plane strain. *J. Appl. Mech.* 61, 231–5.

Sih, G. C., and Liebowitz, H. (1968). Mathematical theories of brittle fracture. In *Fracture. An Andvanced Treatise*, edited by H. Liebowitz, Vol. II, pp. 67–190. Academic Press, New York.

Simo, J. C. (1988). A framework for finite strain elastoplasticity based on maximum dissipation and the multiplicative decomposition. Part I. Continuum formulation. *Comput. Meth. Appl. Mech. Engrg.* 66, 199–219.

Simo, J. C., and Pister, K. S. (1984). Remarks on rate constitutive equations for finite deformation problems: computational implications. *Comput. Meth. Appl. Mech. Engrg.* 46, 201–15.

Simões, F. M. F. (1997). Instabilities in non-associated problems of solid mechanics. Ph.D. Thesis, Technical University of Lisbon, in Portuguese.

Simões, F. M. F., and Martins J. A. C. (1998). Instability and ill-posedness in some friction problems. *Int. J. Engng. Sci.* 36, 1265–93.

Simões, F. M. F., and Martins J. A. C. (2005). Flutter instability in a non-associative elastic-plastic layer: analytical versus finite element results *Int. J. Engng. Sci.* 43, 189–208.

Simões, F. M. F., Martins J. A. C., and Loret, B. (1999). Instabilities in elastic-plastic fluid-saturated porous media: harmonic wave versus acceleration wave analyses *Int. J. Solids Structures* 36, 1277–95.

Simo, J. C., and Hughes, T. J. R. (1987). General return mapping algorithms for rate-independent plasticity. In *Constitutive Laws for Engineering Materials: Theory and Applications*, edited by C. S. Desai et al., pp. 221–231. Elsevier Science, New York.

Simpson, H. C., and Spector, S. J. (1984). On barrelling instabilities in finite elasticity. *J. Elasticity* 14, 103–25.

Soós, E. (1996a). Resonance and stress concentration in a prestressed elastic solid containing a crack. An apparent paradox. *Int. J. Engng. Sci.* 34, 363–74.

Soós, E. (1996b). Stability, resonance and stress concentration in prestressed piezoelectric crystals containing a crack. *Int. J. Engng. Sci.* 34, 1647–73.

Spencer, A. J. M. (1984). Constitutive theory for strongly anisotropic solids. In *Continuum Theory of the Mechanics of Fibre-Reinforced Composites*, edited by A. J. M. Spencer, CISM Course No. 282. Springer-Verlag, New York.

Spencer, A. J. M. (1997). Fibre-streamline flows of fibre-reinforced viscous fluids. *Eur. J. Appl. Math.* 8, 209–15.

Spencer, A. J. M. (2004). Some results in the theory of non-Newtonian transversely isotropic fluids. *J. Non-Newtonian Fluid Mech.* 119, 83–90.

Spitzig, W. A., Sober, R. J., and Richmond, O. (1976). The effect of hydrostatic pressure on the deformation behavior of maraging and HY-80 steels and its implications for plasticity theory. *Metall. Trans.* 7A, 1703–10.

Sridhar, I., and Fleck, N. A. (2000). Yield behaviour of cold compacted composite powders. *Acta Mater.* 48, 3341–52.

Steif, P. S. (1986a). Bimaterial interface instabilities in plastic solids. *Int. J. Solids Structures* 22, 195–207.

Steif, P. S. (1986b). Periodic necking instabilities in layered plastic solids. *Int. J. Solids Structures* 22, 1571–78.

Steif, P. S. (1987). An exact two-dimensional approach to fiber micro-buckling. *Int. J. Solids Structures* 23, 1235–46.

Steif, P. F. (1990). Interfacial instabilities in an unbounded layered solid. *Int. J. Solids Structures* 26, 915–25.

Steigmann, D. J., and Ogden, R. W. (1997). Plane deformations of elastic solids with intrinsic boundary elasticity. *Proc. R. Soc. Lond. A* 453, 853–77.

Stoneley, R. (1924). Elastic waves at the surface of separation of two solids. *Proc. R. Soc. Lond. A* 106, 416–28.

Stronge, W. J., and Shim, V. P.-W. (1988). Microdynamics of crushing in cellular solids. *J. Eng. Mat. Tech.* 110, 185–90.

Stören, S., and Rice, J. R. (1975). Localized necking in thin sheets. *J. Mech. Phys. Solids* 23, 421–41.

Sugiyama, Y., Katayama, K., and Kinoi, S. (1995). Flutter of a cantilevered column under rocket thrust. *J. Aerospace Eng.* 8, 9–15.

Sugiyama, Y., Katayama, K., Kiriyama, K., and Ryu, B.-J. (2000). Experimental verification of dynamic stability of vertical cantilevered columns subjected to a sub-tangential force. *J. Sound Vibration* 236, 193–207.

Sulem J., and Ouffroukh, H. (2006). Hydromechanical behaviour of Fontainebleau sandstone. *Rock Mech. Rock Eng.* 39, 185–213.

Suo, Z., Ortiz, M., and Needleman, A. (1992). Stability of solids with interfaces. *J. Mech. Phys. Solids* 40, 613–40.

Szabó, L. (1994). Shear band formulation in finite elastoplasticity. *Int. J. Solids Structures* 31, 1291–1308.

Tarnai, T. (1980). Destabilizing effect of additional restraint on elastic bar structures. *Int. J. Mech. Sci.* 22, 379–90.

Tasuji, M. E., Slate, F. O., and Nilson, A. H. (1978). Stress-strain response and fracture of concrete in biaxial loading. *ACI J.* 75, 306–12.

Temme, N. M. (1996). *Special Functions.* Wiley, New York.

Thomas, T. Y. (1953). The effect of compressibility on the inclination of plastic slip bands in flat bars. *Proc. Nat. Acad. Sci. USA* 39, 266–73.

Thomas, T. Y. (1954). A discussion on load drop and related matters associated with the formation of a Lüders band. *Proc. Nat. Acad. Sci. USA* 40, 572–6.

Thomas, T. Y. (1961). *Plastic Flows and Fracture of Solids.* Academic Press, New York.

Timoshenko, S.P., and Gere, J.M. (1961). *Theory of Elastic Stability.* McGraw-Hill, New York.

Todhunter, I., and Pearson, K. (1960). *A history of the Theory of Elasticity and of the Strength of Materials from Galilei to Lord Kelvin*, Vol. I. Dover, New York.

Tomita, Y., Shindo, A., and Fatnassi, A. (1988). Bounding approach to bifurcation point of annular plates with nonassociated flow law subjected to uniform tension at their outer edges. *Int. J. Plasticity* 4, 251–63.

Torrenti, J. M., Desrues, J., Benaija, E. H. e Boulay, C. (1991). Stereophotogrammetry and localization in concrete under compression. *J. Eng. Mech. ASCE* 117, 1455–65.

Tresca, H. (1878). On further application of the flow of solids. *Proc. Inst. Mech. Eng.* 30, 301–45.

Triantafyllidis, N. (1980). Bifurcation phenomena in pure bending. *J. Mech. Phys. Solids* 28, 221–45.

Triantafyllidis, N., and Lehner, F. K. (1993). Interfacial instability of density-stratified 2-layer systems under initial stress. *J. Mech. Phys. Solids* 41, 117–42.

Triantafyllidis, N., and Leroy, Y. M. (1994). Stability of a frictional material layer resting on a viscous half-space. *J. Mech. Phys. Solids* 42 51–110.

Triantafyllidis, N., and Maker, B. N. (1985). On the comparison between microscopic and macroscopic instability mechanisms in a class of fiber-reinforced composites. *J. Appl. Mech.* 52, 794–800.

Truesdell, C. (1955). Hypo-elasticity. *J. Rat. Mech. Anal.* 4, 83–133 and 1019–20.

Truesdell, C. (1961). General and exact theory of waves in finite elastic strain. *Arch. Rat. Mech. Anal.* 8, 263–96.

Truesdell, C. (1966). *The Elements of Continuum Mechanics.* Springer-Verlag, Berlin.

Truesdell, C., and Noll, W. (1965). The non-linear field theories of mechanics. In *Encyclopedia of Physics*, edited by S. Flügge, Vol. III/3. Berlin, Springer-Verlag.

Truesdell, C., and Toupin, R. A. (1960). The classical Field theories. In Flügge, S., ed., *Encyclopedia of Physics*, III/1. Berlin, Springer-Verlag.

Tvergaard, V. (1981). Influence of voids on shear band instabilities under plane strain conditions. *Int. J. Fracture* 17, 389–407.

Tvergaard, V. (1982a). Influence of void nucleation on ductile shear fracture at a free surface. *J. Mech. Phys. Solids* 30, 399–425.

Tvergaard, V., and Needleman, A. (1980). On the localization of buckling patterns. *J. Appl. Mech.* 47, 613–19.

Vainchtein A. (2002). Dynamics of non-isothermal martensitic phase transition and hysteresis. *Int. J. Solids Structures* 39, 3387–408.

Vainchtein A., and Rosakis, P. (1999). Hysteresis and stick-slip motion of phase boundaries in dynamic models of phase transitions. *J. Non-linear Sci.* 9, 697–719.

Valanis, K. C. (1990). A theory of damage in brittle materials, *Eng. Fracture Mech.* 36, 403–16.

Vardoulakis, I. (1980). Shear band inclination and shear modulus of sand in biaxial tests. *Int. J. Numer. Anal. Meth. Geomech.* 4, 103–19

Vardoulakis, I. (1981). Bifurcation analysis of the plane rectilinear deformation on dry sand samples. *Int. J. Solids Structures* 11, 1085–1101.

Vardoulakis, I. (1983). Rigid granular plasticity model and bifurcation in the triaxial test. *Acta Mech.* 49, 57–79.

Vardoulakis, I., and Sulem, J. (1995). *Bifurcation Analysis in Geomechanics.* Blackie Academic & Professional, London.

Vogel, S. M., and Rizzo, F. J. (1973). An integral equation formulation of three dimensional anisotropic elastic boundary value problems. *J. Elasticity* 3, 203–16.

Wadee, M. A., Hunt, G. W., and Peletier, M. A. (2004). Kink band instability in layered structures. *J. Mech. Phys. Solids* 52, 1071–91.

Walley S. M. (2007). Shear localization: an historical overview. *Metall. Mater. Trans. A* 38A, 2629–53.

Wang, C. C. (1970). A new representation theorem for isotropic functions, Parts I and II. *Arch. Rat. Mech. Anal.* 36, 166–223.

Weissenberg, K. (1935). La mécanique des corps déformables. *Arch. Sci. phys. Nat. Genéve*, 5, 44–106, 130–71.

Weissenberg, K. (1949). Abnormal substances and abnormal phenomena of flow. In *Proc. Int. Congr. Rheology* (1948), Vol. I, pp. 29–46.

Westergaard, H. M. (1920). On the resistance of ductile materials to combined stresses. *J. Franklin Inst.* 189, 627–40.

Wilder, E. A., Guo, S., Lin-Gibson, S., Fasolka, M. J., and Stafford, C. M. (2006). Measuring the modulus of soft polymer networks via a buckling-based metrology. *Macromolecules* 39, 4138–43.

Wilkes, E. W. (1955). On the stability of a circular tube under end thrust. *Q. J. Mech. Appl. Math.* 8, 88–100.

Wilkinson, J. H. (1965). *The Algebraic Eigenvalue Problem*. Clarendon Press, Oxford, UK.

Willam, K. J., and Warnke, E. P. (1975). Constitutive model for the triaxial behaviour of concrete. Presented at the seminar on concrete structures subjected to triaxial stresses, ISMES, Bergamo, 1–30.

Willis, J. R. (1965). The elastic interaction energy of dislocation loops in anisotropic media. *Q. J. Mech. Appl. Math.* 18, 419–33.

Willis, J. R. (1969). Some constitutive equations applicable to problems of large dynamic plastic deformation. *J. Mech. Phys. Solids* 23, 405–19.

Willis, J. R. (1971). Interfacial stresses induced by arbitrary loading of dissimilar elastic half-spaces joined over a circular region. *J. Inst. Maths. Applics.* 7, 179–97.

Willis, J. R. (1972). The penny-shaped crack on an interface. *Q. J. Mech. Appl. Math.* 25, 367–85.

Willis, J. R. (1973). Self-similar problems in elastodynamics. *Philoso. Trans. R. Soc. Lond., A* 274, 435–91.

Willis, J. R. (1991). Inclusions and cracks in constrained anisotropic media. In *Modern Theory of Anisotropic Elasticity and Applications*, edited by J. J. Wu, T. C. T. Ting, and D. M. Barnett, pp. 87-102. SIAM, Philadelphia.

Wood, D. M. (1990). *Soil Behaviour and Critical State Soil Mechanics.* Cambridge University Press, Cambridge, UK.

Xue, Q., and Gray, G. T. (2006). Development of adiabatic shear bands in annealed 316L stainless steel: Part I. Correlation between evolving microstructure and mechanical behavior. *Metall. Mater. Trans. A* 37A, 2435–46.

Yang, W. H. (1980). A useful theorem for constructing convex yield functions. *J. Appl. Mech.* 47, 301–3.

Yasufuku, N., Murata, H., and Hyodo, M. (1991). Yield characteristics of anisotropically consolidated sand under low and high stress. *Soils and Foundations* 31, 95–109.

Young, N. J. B. (1976). Bifurcation phenomena in the plane compression test. *J. Mech. Phys. Solids* 24, 77–91.

Zaccaria, D., Bigoni, D., Misseroni, D., and Noselli, G. (2011). Structures buckling under tensile dead load. *Proc. R. Soc. Lond. A* 467, 1686–1700.

Zaremba, C. (1903). Remarques sur les travaux de M. Natanson relatifs á la thèorie de la viscositè. *Bull. Int. Acad. Sci. Cracovie*, 85–93.

Zhang, Z., and Clifton, R. J. (2007). Shear band propagation from a crack tip subjected to Mode II shear wave loading. *Int. J. Solids Structures* 44, 1900–26.

Zheng, Q. S. (1994). Theory of representations of tensor functions: a unified invariant approach to constitutive equations. *Appl. Mech. Rev.* 47, 545–87.

Ziegler, H. (1977). *Principles of Structural Stability*. Birkhäuser Verlag, Basel.

Zysset, P. K., and Curnier, A. (1995). An alternative model for anisotropic elasticity based on fabric tensors. *Mech. Materials* 21, 243–50.

Zytynski, M., Randolph, M. F., Nova, R., and Wroth, C. P. (1978). On modelling the unloading-reloading behaviour of soils.*Int. J. Numer. Anal. Meth. Geomech.* 2, 87–93.

Index

Printed in the United States
By Bookmasters